Chemistry

The Molecules of Life

Chemistry

The Molecules of Life

TRACE JORDAN

NEVILLE KALLENBACH

New York   Oxford
OXFORD UNIVERSITY PRESS

Oxford University Press is a department of the University of Oxford. It furthers the University's objective of excellence in research, scholarship, and education by publishing worldwide. Oxford is a registered trade mark of Oxford University Press in the UK and certain other countries.

Published in the United States of America by Oxford University Press
198 Madison Avenue, New York, NY 10016, United States of America.

© 2017 by Oxford University Press

For titles covered by Section 112 of the US Higher Education Opportunity Act, please visit www.oup.com/us/he for the latest information about pricing and alternate formats.

Library of Congress Cataloging-in-Publication Data

CIP data is on file at the Library of Congress.
9780199946174

9 8 7 6 5 4 3 2 1
Printed by LSC Comminations, Inc.
Printed in the United States

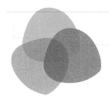

Brief Contents

Table of Contents

Chapter 4   Carbon: The Element of Life   93

Chapter 7   Monomers and Polymers   197

Chapter 8   The Unusual Nature of Water   229

Chapter 10   Measuring Concentration   291

Chapter 15   Drug Development   453

Preface

Why are many bacteria now resistant to antibiotic drugs that were effective in the past? We pose this question in the opening pages of *Chemistry: The Molecules of Life*. Providing an answer, which is the main focus of Chapter 1, requires an understanding of pharmaceuticals, biological molecules, and bacterial cells. This interdisciplinary case study serves as an introduction to our innovative approach to teaching chemistry in the 21st century. Instead of viewing chemistry as an isolated discipline, we emphasize its intersection with biology and medicine. By focusing on "the molecules of life," we demonstrate how chemistry is vital for understanding human health. *Chemistry: The Molecules of Life* introduces non-science majors to the fundamental chemical principles of biological molecules and everyday life.

We have written this textbook for students who have a variety of academic and career interests, not all of which necessarily align with the sciences or health professions. We believe that *all* undergraduate students—not just science majors—need to be educated about scientific knowledge and reasoning in order to make informed decisions about their personal well-being and important societal issues. Have you ever thought about how antibiotic medications work, or why it is important to take the entire dose that the doctor prescribed? Many of us also take a vitamin supplement, often in the form of a daily multivitamin. Do you know what roles these vitamins play in our body, or why the recommended daily dose for some vitamins is much lower than for others? Does the federal government regulate vitamin supplements with the same scrutiny that it applies to pharmaceuticals? Throughout this book, we examine relevant topics such as these and many more.

Scientific knowledge is constantly advancing. Almost daily, we hear or read stories in the media about a new scientific discovery, a new interpretation of what foods or activities are good or bad for us, or the societal impact of a new medical breakthrough. Given this rapid progress, we cannot learn science as a collection of facts to be memorized. After a short time, our knowledge will quickly become outdated. Instead, we need to understand *how* scientists investigate the natural world—the methods they use, the evidence they generate, and the conclusions they make based on that evidence. Throughout this textbook, we describe examples of scientific discovery. By analyzing these case studies, students will develop the critical thinking skills necessary to thoughtfully evaluate scientific information that is presented in the news media or obtained from other sources.

Approach

In *Chemistry: The Molecules of Life*, we use the chemistry of life to introduce, explain, and apply foundational chemical principles. Each chapter begins with a *framing question* to stimulate students' interest and motivate them to explore the chapter content that is necessary to answer the question. Many of these questions are pertinent to students' health, whereas others address foundational scientific topics. The scientific content in each chapter is introduced on a "need-to-know" basis that is related to the framing question. By the end of the chapter, students will be able to answer the question using the chemical concepts they have learned.

In addition to the core content, each chapter contains features that illustrate important themes. *Chemistry and Your Health* demonstrates the relevance of the chapter material by applying chemical concepts to students' health. Examples include omega-3 fatty acids, alcohol metabolism, and the potential hazards of indoor tanning beds. *Science in Action* develops students' understanding of scientific investigation—in other words, *how do we know what we know?* Some of these features focus on experimental techniques, such as using a mass spectrometer to measure the mass of atoms, or the application of X-ray diffraction to determine the structure of a molecule. Other examples describe an important historical experiment, such as Antoine Lavoisier's investigation regarding the role of atmospheric oxygen in chemical reactions. *Chemistry in Your Life*, which is included in selected chapters, uses chemical principles to explain students' life experiences. One such feature describes why chili peppers taste "hot," whereas another explains why hydrogen peroxide bubbles when you put it on a cut.

Overview of Chapters

Chapter 1 uses the example of antibiotic-resistant bacteria as an interdisciplinary overview of the molecules of life. This chapter also includes *Textbook Previews* that highlight the chapters in which scientific topics will be examined in greater depth.

The next four chapters of the textbook illustrate how atoms form the building blocks of molecules. Chapter 2 examines the chemical elements and the structure of atoms, including examples of elements that are beneficial or harmful to human health. Chapter 3 explains how and why atoms form chemical bonds. Chapter 4 investigates the unique chemistry of carbon and why life is "carbon based." Chapter 5 explores the remarkable diversity of molecular structures that can be made using only a handful of different atoms.

The textbook contains two chapters that describe the formation of molecules. Chapter 6 shows how chemical reactions produce changes in matter and energy, along with how these changes can be described in quantitative terms. Chapter 7 uses the example of sugars to illustrate the construction of very large molecules from multiple copies of smaller components.

The following four chapters describe the chemistry of water and solutions. Chapter 8 explains the unusual properties of water as deriving from interactions among H_2O molecules. Chapter 9 examines the behavior of molecules and ions in solution, with a focus on various types of vitamins. Chapter 10 provides a quantitative description of concentration, with applications to human health. Chapter 11 describes the chemistry of acids and bases, including the pH scale for measuring acidity.

The final four chapters of the textbook explore the molecular basis of life. Chapter 12 examines DNA from the perspective of the scientists who contributed to the discovery of its molecular structure. Chapter 13 explains how information stored within DNA is converted into various proteins, which perform most of the essential biological functions of cells. Chapter 14 uses the example of lactose intolerance to introduce the chemistry of enzymes and how they function as biological catalysts by accelerating chemical reactions. Chapter 15 returns to the theme of Chapter 1, antibiotics, by investigating of how these drugs work against various types of bacteria, with a special focus on the resurgent global problem of antibiotic-resistant tuberculosis.

Teaching and Learning Features

One difference between expert scientists and new learners is the ability to keep the "big picture" in mind. Therefore, each major section in a chapter is accompanied by a *Learning Objective* that focuses students' attention on the central concept. When reading the chapter, students can relate the scientific principles and skills to the overarching theme of the learning objective. In addition to the Learning Objectives, *The Key Idea* accompanies each chapter subheading to provide a one-sentence summary statement that highlights the main point of the section. Each subsection concludes with a set of *Core Concepts* that summarize the most important scientific principles of the section. In addition, a *Marginal Glossary* defines key scientific terms when they are introduced. Highlighting these glossary terms enables students to enrich and reinforce their scientific vocabulary.

The chapter art illustrates the amazing diversity of molecular structures, ranging from small molecules such as H_2O to large molecules such as DNA. With this art, we show how different representations of molecules are used to communicate various features of their structure. We also assist students in developing the skill of analyzing molecules in *three dimensions*, because the spatial geometry of a molecule is often essential for understanding its chemical and biological properties.

Each chapter also contains *Worked Examples* that pose a relevant question and then provide a step-by-step solution to help students develop their problem-solving skills. Each Worked Example is followed by a *Try It Yourself* exercise that enables students to practice what they have just learned. In additional, *Practice Exercises* provide students with further opportunities to answer both conceptual and numerical questions that relate directly to the chapter content.

Concept Questions within each chapter promote active learning by providing a basis for in-class exercises and discussions. These are not routine practice questions; instead, they stimulate students' thinking about the scientific principles and their applications. Some of these questions ask students to apply what they have learned to a new situation. Others prompt students to investigate a topic using the Internet and then report back to the class.

Each chapter concludes with a *Visual Summary* that presents key figures from the chapter accompanied by brief summaries of important scientific concepts. This novel feature is based on the insight that *visual learning* is an effective strategy for 21st-century students.

The *Learning Resources* section at the end of each chapter divides questions into three categories. *Reviewing Knowledge* questions ask students to recall and explain the core content within the chapter. *Developing Skills* questions provide multiple examples that enable students to developing their scientific reasoning through the application of chapter concepts. *Exploring Concepts* questions are more open-ended and analytical extensions of the chapter content. These questions often require students to investigate and critically evaluate a topic using the Internet.

SUPPORT PACKAGE

Oxford University Press has created a comprehensive set of ancillary resources to accompany *Chemistry: The Molecules of Life*. These resources are designed to help students master the concepts introduced in the text and to assist instructors in making chemistry accessible to those who are not science majors.

For Students

- **Laboratory Manual:** Written by the authors in conjunction with Bill Gunderson, Assistant Professor of Chemistry at Hendrix College, this laboratory manual includes more than 20 laboratory exercises that relate directly to topics explained in the text. Each laboratory exploration provides an overview of the experiment, a detailed description of the procedures, and data sheets that students use to record their observations and answer questions.

For Instructors

- **Sapling Learning Online Homework System:** Sapling Learning's online homework system includes more than 400 homework questions designed to test students' understanding of key concepts from the text. It also includes a set of interactive animations that allow students to simulate chemistry experiments online. Automatic homework grading, diagnostic feedback, and dedicated support from chemists provide instructors all the resources they need to assign homework that students will find useful for learning.

- **Lecture Notes:** Editable lecture notes in PowerPoint format make preparing lectures faster and easier than ever. Each chapter's presentation includes a succinct outline of key concepts and incorporates graphics from the chapter.

- **Digital Image Library:** The image library includes electronic files, again in PowerPoint format, of every illustration, graph, photo, figure caption, and table from the text, with both labeled and unlabeled versions. Images have been enhanced for clear projection in large lecture halls.

- **Test Bank:** The *Test Bank* includes more than 800 exam questions in multiple-choice and short-answer formats, provided in a series of editable Word files that can be easily customized.

Contact your local Oxford University Press sales representative or visit www.oup-arc.com/jordan to learn more and gain access to these resources.

Acknowledgments

Creating a textbook is a truly collaborative endeavor. We would like to thank the wonderful team at Oxford University Press USA for their encouragement, advice, and patience during the gestation of this project. First, we are grateful to Jason Noe, senior editor, for serving as our guide from the initial prospectus to the completion of the book. This project could not have reached the finish line without his unwavering support. We also benefited immensely from the insightful comments provided by our development editors, Anne Kemper, Naomi Friedman, and Maegan Sherlock. Andrew Heaton, associate editor, reviewed our text as it progressed through several drafts and then helped to prepare our final manuscript. Production editor Micheline Frederick, production team leader Theresa Stockton, and production manager Lisa Grzan worked together on a tight deadline to turn the manuscript into a finished book. Wesley Morrison provided eagle-eyes scrutiny of the pages during the proofreading stage. We are grateful to the design team led by Michele Laseau, art director, for the beautiful textbook design and page layout and Dragonfly Media Group's team of Craig Durant, Caitlin Duckwall, and Rob Duckwall, for the imaginative artwork they created. We would also like to thank Patrick Lynch, editorial director; John Challice, vice president and publisher; Frank Mortimer, director of marketing; David Jurman, marketing manager; Clare Castro, marketing manager; Ileana Paules-Brodet, marketing assistant; Meghan Daris, market development associate; and Bill Marting, national sales manager.

Other individuals also made important contributions to the project. Robert Weiss provided insightful editorial comments on many chapters. Anna Powers, Tania Lupoli, and Jennifer Sniegowski supplied ideas for end-of-chapter questions. Jennifer Lee invested many hours creating the art manuscript. Veronica Murphy gave valuable assistance by checking calculations, appendices, and page proofs. Our faculty colleagues at New York University, Bobby Arora, Daniela Buccella, and John Halpin, provided expert advice about specific chemistry topics.

We wish to express our deep appreciation to the many friends and colleagues who endured our fixation on this textbook for more years than we like to admit. We are grateful to George and the Chicken for their weekly welcome at our local diner. Finally, we thank Diana and Martha for supporting us throughout the entire journey.

Manuscript Reviewers

Throughout the development of this textbook, we have benefited from the insightful comments by many faculty members who provided expert reviews of the chapters. They not only informed us when a topic was not expressed clearly but also provided valuable suggestions for how to communicate it more effectively. The names of these individuals and their affiliations are listed below. We are enormously grateful to them for their guidance and advice.

Kate Aubrecht	Stony Brook University
Rita Bagwe	Great Basin College
Soumitra Basu	Kent State University
Robert Billmers	Rutgers University
Timothy Brewer	Eastern Michigan University
William Bryan	Saint Norbert College
Sarah Carberry	Ramapo College of New Jersey
Charles Carraher	Florida Atlantic University
Yuh-Cherng Chai	John Carroll University
Brent Chandler	Illinois College
Kaiguo Chang	New Mexico Highlands University
Li-Heng Chen	Aquinas College
Stephen Contakes	Westmont College
Paul Czech	Providence College
Cory DiCarlo	Grand Valley State University
Rodney Dixon	Towson University
Taela Donnelly	Wilkes University
Jason Dunham	Ball State University
Timothy Ehler	Buena Vista University
Andrew Frazer	University of Central Florida
Kenneth French	Blinn College
Richard Fronko	California State University, East Bay
Kimberly George	Marietta College
Marcia Gillette	Indiana University Kokomo
William Gunderson	Illinois College
Tamara Hamilton	Barry University
Alan Hazari	University of Tennessee
Thomas Holme	Iowa State University
Xiche Hu	University of Toledo
Amber Hupp	College of the Holy Cross
Kasem Kasem	Indiana University Kokomo
Daniel King	Eastern Mennonite University
Todd Knippenberg	High Point University
Punit Kohli	Southern Illinois University
Eric Lewis	Clarion University of Pennsylvania
Greg Love	East Tennessee State University
Helene Maire-Afeli	University of South Carolina Union

Lydia Martinez Rivera	The University of Texas at San Antonio
Kenneth Marx	University of Massachusetts Lowell
Forrest Gregg McIntosh	Winthrop University
Dorene Medlin	Albany State University
Zoltan Mester	Chapman University
Sheldon Miller	Chestnut Hill College
Ray Mohseni	East Tennessee State University
Robert Moran	Wentworth Institute of Technology
Basil Mugaga Naah	Wright State University
Ruth Nalliah	Huntington University
Daphne Norton	University of Georgia
Paul Okweye	Alabama A&M University
Sandra Olmsted	Augsburg College
Charlotte Otto	University of Michigan–Dearborn
Joyce Overly	Clarion University
Maria Pacheco	Buffalo State College
Felipe Pascal	Fairleigh Dickinson University
Karisa Pierce	Seattle Pacific University
Tamiko Porter	Indiana University–Purdue University Indianapolis
Mary Railing	Wheeling Jesuit University
Scott Reid	Marquette University
Arlie Rinaldi	The Claremont Colleges
Melinda Roberts	Texas Tech University
Katie Roles	Lake-Sumter State College
Sarah Rosenstein	Hamilton College
Diptirani Samantaray	Virginia State University
Joseph Scanlon	Ripon College
Allan Scruggs	Gonzaga University
Vasudha Sharma	Valencia College
Jennifer Sniegowski	Arizona State University Downtown Phoenix
Anne Marie Sokol	University at Buffalo
Craig Streu	St. Mary's College of Maryland
Mark Tapsak	Bloomsburg University of Pennsylvania
Nathan Tice	Butler University
Megan McLean Tichy	Santa Clara University
Marcia Tinone	University of Hartford
Petra van't Slot	Montclair State University
Alexandre Volkov	Oakwood University
Shelli Waetzig	College of the Holy Cross
Erin Whitteck	Butler University
Charles Wohlers	Bridgewater State University
Lou Wojcinski	Kansas State University
Mali Yin	Sarah Lawrence College
Kazushige Yokoyama	State University of New York at Geneseo

Chemistry

The Molecules of Life

Antibiotics and Resistance:

An Introduction to the Molecules of Life

CONCEPTS AND LEARNING OBJECTIVES

I n September 2015, *USA Today* published a news article entitled "MRSA Infections Remain a Back-to-School Risk." According to this article, an elite private school in Washington, D.C., postponed the beginning of the school year because two of its athletes acquired an MRSA infection. MRSA are bacteria that are resistant to many common antibiotics. How do bacteria become antibiotic-resistant, and why are MRSA infections becoming more common?

1.1   Why Are Some Infections Resistant to Antibiotics?

Learning Objective:
Show the importance of understanding scientific information and biological molecules in everyday life

Many of us exercise at the gym or participate in team sports. These activities can produce the occasional scrape or cut. You may have had this type of cut before, and it always heals quickly. But this time the injury isn't getting better—after several days the area around the wound is inflamed and your skin hurts when you touch it (Figure 1.1(a)). You go to the College Health Center, where the doctor says that you have a bacterial infection and prescribes a common antibiotic called cephalosporin. You take the antibiotic pills for several days and expect the infection to heal. But it doesn't . . . instead, it gets worse (Figure 1.1(b)).

Feeling worried, you hurry to the emergency room at the local hospital. You explain the history of your cut, the infection, and the ineffectiveness of the antibiotics you took. The ER doctor says that you probably have an MRSA infection. To check this diagnosis, she sends a swab from your infected wound to the hospital's lab for a diagnostic test. The doctor also recommends that you take a different antibiotic drug called vancomycin that has to be injected intravenously. After a few days and multiple doses of the drug, the swelling on your leg subsides. A couple of weeks later you return to playing sports. But now you are more careful to shower with soap and water after every game, and closely monitor yourself for any cuts and scrapes.

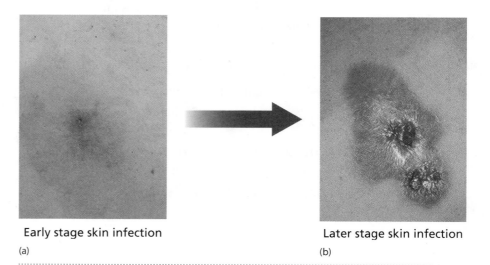

Early stage skin infection
(a)

Later stage skin infection
(b)

FIGURE 1.1   An MRSA infection. The appearance of the skin during (a) the early stage and (b) later stage of a bacterial infection with MRSA.

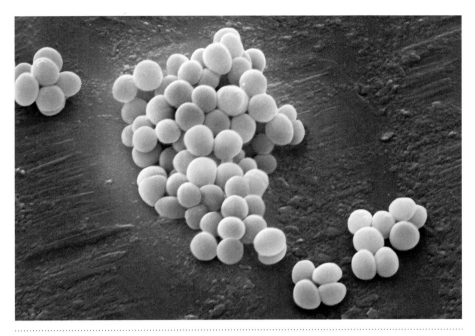

FIGURE 1.2 **A highly magnified microscope image of MRSA.** These bacteria are a drug-resistant form of *Staphylococcus aureus*. The name *Staphylococcus* is derived from the Greek words for "round grapes." The blue color has been added to enable us to see the bacteria, which are normally gold.

MRSA: An Antibiotic-Resistant Infection

The previous scenario is becoming more common in high schools, colleges, and professional sports. The infection is caused by **methicillin-resistant** ***Staphylococcus aureus*** **(MRSA)** (Figure 1.2). This is an antibiotic-resistant version of "*Staph*" bacteria that are commonly found on our skin. Bacteria are biological organisms composed of a single **cell**, which is the basic unit of life. MRSA bacteria are not killed by methicillin, an antibiotic that is closely related to penicillin.

The first outbreaks of MRSA occurred in hospitals, where they caused serious infection and were responsible for many patient deaths. More recently, MRSA has spread to local communities throughout the United States. It is particularly common in sporting and recreational environments with the potential for cuts, scrapes, and close person-to-person contact. However, everyday exposure to MRSA is not limited to these locations. During an investigation of shared bathrooms in 15 residence halls on a college campus, scientists obtained at least one sample of MRSA from all the tested sites, including the shower floors. In a 2013 report, the U.S. Centers for Disease Control and Prevention (CDC) estimated that over 80,000 serious MRSA infections occur in the United States each year, which result in more than 11,000 deaths.

This introduction raises several questions. Why is MRSA becoming more prevalent in our everyday environment? Why are many antibiotics no longer effective at treating some bacterial infections? How do bacteria acquire resistance to antibiotics? Answering these questions requires us to explore the main theme of this book—the molecules of life. What are the "molecules of life," and why are they so important for understanding our health and modern science?

THE KEY IDEA: Antibiotic-resistant infections such as MRSA are becoming increasingly common in our communities.

methicillin-resistant *Staphylococcus aureus* **(MRSA)** A type of infectious bacteria that is not harmed by the antibiotic methicillin and related drugs.

cell The basic unit of life.

Core Concepts

- **MRSA** stands for **methicillin-resistant Staphylococcus aureus**. These single-**cell** bacteria are resistant to treatment by methicillin, penicillin, and related antibiotics.
- MRSA infections are becoming increasingly common within our everyday environment.

FIGURE 1.3 Studying the molecules of life. (a) Scientific study of the molecules of life occurs at the intersection of chemistry, biology, and medicine. (b) DNA illustrates the chemical, biological, and medical importance of a biological molecule.

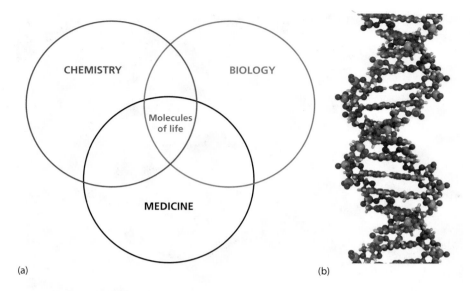

(a)   (b)

Why Study the Molecules of Life?

THE KEY IDEA: Scientific study of the molecules of life occurs at the intersection of chemistry, biology, and medicine.

Scientific research in the 21st century is providing unprecedented insights into the structure and function of biological molecules. These "molecules of life" are essential for every biological process that occurs in every living organism. They are involved in your eye's ability to detect light, in digesting your lunch, and in moving your muscles when you exercise. Scientific study of the "molecules of life" occurs at the intersection of chemistry, biology, and medicine (Figure 1.3(a)). To illustrate, consider the most famous of all biological molecules—DNA (Figure 1.3(b)). The double-helix structure of DNA is based on chemical principles of molecular structure. DNA has an essential biological role as the repository of genetic information. In addition, modern medicine is expanding its use of DNA tests to guide the choice of treatment based on a person's genetic characteristics.

This textbook has two main goals for what you will learn to do:

- Acquire scientific knowledge and skills concerning the chemistry of biological molecules and pharmaceuticals.

- Apply your scientific understanding to make informed decisions about your own health and the well-being of your family, your community, and society in general.

You may believe that only people who decide to pursue a career in science or medicine need to learn scientific thinking. But every one of us will have to make important decisions that require an understanding of scientific topics. These decisions include making choices about your own health, advising a family member, or casting a vote based on the policy proposals of a political candidate. Concept Question 1.1 contains a few examples for you to consider now, and you will encounter many more throughout the book.

CONCEPT QUESTION 1.1

Here are three scenarios that require decisions based on scientific information. In each case, read the question and ask yourself: What would you do?

- *You see an advertisement on TV for an "all-natural" supplement that will help you lose weight and boost your energy. You have been feeling tired recently, and this supplement sounds like a good solution. But then you start thinking more critically about the claims in the ad. What's the evidence that it works? Are the "before and after" pictures enough to convince you? Is something that is "natural" always good for you? What is the "active ingredient" in the supplement, how much does it contain, and how does this ingredient affect your body? What types of claims about the benefits of a supplement are legally allowed in TV ads? After considering all these questions, should you try this supplement?*

- *What advice would you give to your younger sister or cousin who asks you about going to a tanning salon so she can "look good" for her junior high prom? You have probably heard that ultraviolet light can cause skin cancer. What is cancer, and why does UV light trigger its development? Is there a connection between the use of tanning beds and the increasing incidence of skin cancer for young women in their 20s and 30s?*
- *Your grandmother tells you that she has recently been diagnosed with diabetes. From an Internet search, you learn that the disease is classified into two categories: type 1 and type 2. What's the difference, and which type does your grandmother have? You also read about the connection between diabetes and insulin. What is insulin, and what does it do? Does your grandmother need to start taking insulin now? If she has diabetes, does that mean you will get it too? Is there anything you can do now to reduce your risk of developing diabetes later in your life?*

Write your own example of a health-related issue or topic that you would like to learn more about. What questions do you have about this topic? What information would you like to know in order to make a well-informed decision when answering these questions?

Core Concepts

- Studying the molecules of life occurs at the intersection of chemistry, biology, and medicine.
- The scientific knowledge that you will gain about biological molecules and pharmaceuticals will enable you to make informed decisions about your own health and policy proposals that affect the well-being of society.

1.2 What is Science?

When you read the word "science," you probably think of a collection of facts in a textbook like this one. This is one meaning of science, which is based on the Latin word for "knowledge." But practicing scientists think of science as a *process*— a systematic method for investigating the natural world. The various stages of this process can be described by the **scientific method**, which is outlined in Figure 1.4. To illustrate the stages of the scientific method, we will return to a question posed at the introduction to this chapter—why are some bacteria resistant to antibiotics?

Learning Objective:
Describe the stages of the scientific method.

THE KEY IDEA: Science is a systematic method for studying the natural world.

scientific method A systematic process for investigating the natural world.

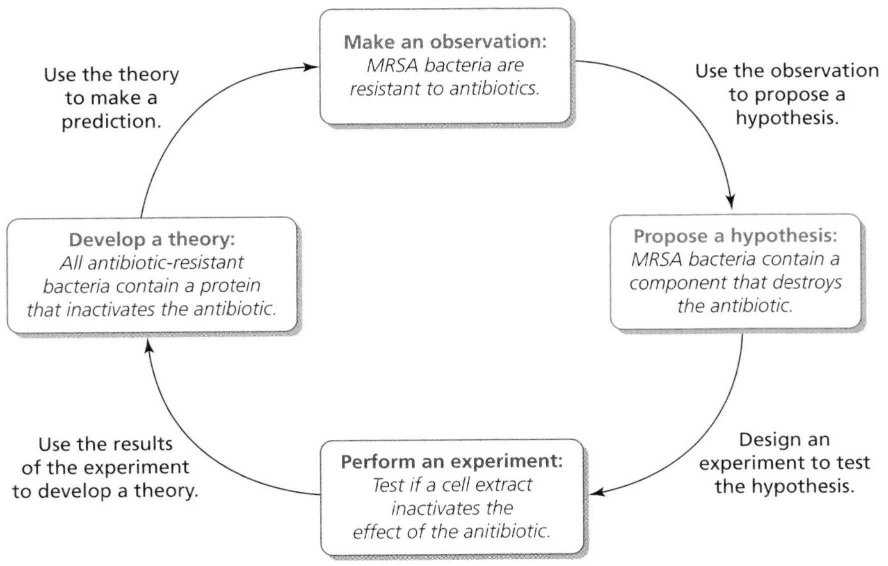

FIGURE 1.4 **The stages of the scientific method.** The application of the scientific method to investigating the cause of antibiotic resistance in bacteria. The circular diagram indicates that the results of one scientific investigation often serve as the starting point for subsequent research.

FIGURE 1.5 Performing an experiment. Science involves performing experiments to test hypotheses about the natural world.

observation The identification of a natural phenomenon that requires an explanation.

hypothesis An initial attempt to explain an observation.

experiment A procedure designed to test a hypothesis.

Scientific investigations often begin with an **observation** that requires an explanation. For example, why are MRSA bacteria resistant to some antibiotics? The next step is to propose a **hypothesis**, which is an informed first attempt to explain the observation. *Our hypothesis is that MRSA bacteria contain a biological molecule that destroys the antibiotic before it can do any harm to the cell.*

We can test this hypothesis by performing an **experiment**, as shown in Figure 1.5. First, we grow a large quantity of MRSA bacteria that have been taken from an infected patient. Next, we break apart the bacteria and make a *cell extract* that contains all the interior components of the cell (proteins, DNA, and so forth). There are no longer any living bacteria, but we have preserved all the molecules contained in their cells. We then add an antibiotic (methicillin) to the cell extract and let it incubate at 37°C (body temperature) for several hours. Finally, we take the total mixture (cell extract plus antibiotic) and test if it can kill bacterial cells that are still susceptible to antibiotics. If our hypothesis is correct, we predict that the bacterial cells will not be harmed.

After performing this experiment, we observe that *the mixture no longer kills bacterial cells,* even though it initially contained the antibiotic. This result confirms our original hypothesis. We must also perform a *control experiment* by following exactly the same procedure using the contents of *Staph* bacteria that are *not* drug resistant. In this case, we find that the mixture of cell extract and antibiotic *does* kill bacterial cells. The control experiment allows us to *compare results under different conditions.* In this case, the control tells us that the experimental procedure itself did not destroy the antibiotic.

The experiment confirms our hypothesis that MRSA bacteria contain some drug-destroying component that other *Staph* bacteria do not. However, a single experiment is not sufficient to adequately test a hypothesis. All experiments must be repeated multiple times to eliminate the possibility of error and ensure that the results can be faithfully reproduced.

After we have confirmed that the experiment is reproducible, we now need to investigate the type of biological molecule that produces the antibiotic resistance. For example, is the effect caused by the bacteria's DNA, by one of its proteins, or by some other biological molecule? Let's suppose further experiments show that antibiotic resistance in MRSA is caused by a protein that degrades the antibiotic before it can kill the bacteria. We investigate several different types of methicillin-resistant bacteria and isolate the same protein from all of them. When we place the isolated protein in a test tube and add an antibiotic, we find that the protein breaks down the drug into smaller pieces.

If a hypothesis has been confirmed by many experiments, it is sometimes elevated to a **theory**—a well-supported explanation of the natural world. After repeated experiments on different bacteria, we are ready to propose a theory: *All antibiotic-resistant bacteria contain a protein that inactivates the antibiotic.* This is a modest theory that describes a specific type of biological process. Some theories are much larger in scope, like the theory of evolution by natural selection. In everyday speech, it is common to dismiss a person's speculation by saying that it's "just a theory." But in the realm of science, a theory is held in the highest respect. *A theory is the best scientific explanation we currently have of the natural world.* A theory is valid as long as there is no evidence to contradict it. However, a theory can be disproven by new experimental results, which requires us to modify the original theory or replace it by a better one.

theory An explanation of the natural world that has been confirmed by multiple experiments.

A theory can be used to make predictions that are testable by experiments. For example, our theory predicts that all antibiotic-resistant bacteria contain a drug-destroying protein. However, when we test a new type of drug-resistant bacteria, the mixture of cell extract and antibiotic *does* kill other bacteria. *This experimental observation is not what we predicted.* In addition, we are not able to isolate the same protein from the new type of bacteria. These observations mean that we have to *reject our theory* since it does not work in all cases. Something different is happening in this new variety of drug-resistant bacteria, but what is it? We have to propose a new hypothesis, and the cycle of scientific investigation begins again.

Another term commonly used in science is **scientific law**. A law is a statement that summarizes a large number of observations but does not provide an explanation. As an example, all chemical reactions obey the *law of conservation of mass.* This law states that mass remains constant during a chemical reaction—that is, the mass of the products of the reaction is the same as the mass of the starting materials. However, this law does not tell us *why* mass is conserved. The explanation was provided later by the *atomic theory of matter*, which states that all matter is composed of tiny particles called atoms. A chemical reaction does not create or destroy atoms, so the number and type of atoms in the products are the same as in the starting materials. The atomic basis of matter (theory) explains why the mass remains constant in a chemical reaction (law).

scientific law A statement that summarizes a collection of scientific observations but does not provide an explanation for them.

The scientific method is a useful framework, but it doesn't convey everything about science, just as a food recipe does not capture the delicious taste of a dish made by a skilled chef. Scientific research is a complex and creative process that does not always follow a simple step-by-step procedure. Scientific discoveries sometimes happen by chance and not by testing a well-formulated hypothesis. In other cases, scientific knowledge advances through the development of new experimental techniques that expand our capabilities of observation and measurement. Many scientists use computer simulations in their research, which does not fit the standard model of hypothesis-driven experiments. Finally, modern science is becoming increasingly large-scale and collaborative. For example, the *Human Genome Project* was a massive international research project that determined the sequence of all the chemical building blocks that make up human DNA.

Core Concepts

- The **scientific method** describes the stages of scientific investigation.
- A **hypothesis** is an informed first attempt to explain an **observation**. A hypothesis must be testable by **experiments**.
- A **theory** is a well-supported explanation of the natural world. A theory remains valid if all the evidence supports it, but we reject the theory if it is contradicted by new experiments.
- A **scientific law** summarizes a large number of observations but does not provide an explanation. Laws can often be explained by theories.
- Scientific research is often more complex than the step-by-step procedure of the scientific method.

1.3 Antiseptics: Preventing Infections

The rest of this chapter examines two methods to reduce infections by bacteria. In this section, we see how **antiseptics** are used to *prevent* infection. The next section explores the use of **antibiotics** to *treat* bacterial infections that have already occurred. This presentation is intended as a preview of important chemical and biological concepts that will be explained in more depth within later chapters

Hand sanitizers are now a common sight in college lounges, residence halls, and cafeterias. You have probably used one of them yourself. Think about what the sanitizing gel feels like on your hands. Do you know what it contains? How does it work?

When you put a hand sanitizer on your fingers, you will notice a distinctive smell. This smell arises from *alcohol*, which functions as the antiseptic within the hand sanitizer. In general, *an antiseptic is a chemical substance that prevents infection* (*sepsis* is a medical term used for infection). The particular type of alcohol contained in hand sanitizers is called *ethanol* (the older name is *ethyl alcohol*). Ethanol is a **molecule** composed of **atoms** that are joined by **chemical bonds**. Its molecular structure can be represented by the drawing in Figure 1.6. In this structure, different types of atoms are indicated by various letters (C for carbon, O for oxygen, H for hydrogen), and chemical bonds between the atoms are shown as straight lines.

What do you notice about the molecular structure of ethanol? First, there are different numbers of each type of atom—six hydrogen atoms, two carbon atoms, and a single oxygen atom. The **chemical formula** for the molecule is C_2H_5OH, in which the subscript after the letter indicates how many times the atom occurs within the molecule. Second, each type of atom (H, C, and O) is joined to other atoms by a different number of chemical bonds (straight lines): Carbon is linked to its atomic neighbors by four bonds, oxygen by two bonds, and hydrogen by only one bond. The collection of atoms C—O—H is called a **functional group** because it is a characteristic of all molecules that are classified as alcohols.

How does ethanol kill bacterial cells? There are two mechanisms, both based on the chemical properties of ethanol and how it affects biological molecules. The first point of attack for antiseptics is the **cell membrane**, which forms a protective barrier between the inside of the cell and the outside environment. The membrane is composed of fat-like molecules called **lipids** that pack together to create its structure. Lipids do not dissolve easily in water, so the membrane remains

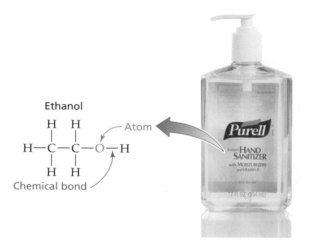

FIGURE 1.6 **Hand sanitizer contains ethanol as a disinfectant.** Ethanol is a molecule composed of atoms linked by chemical bonds.

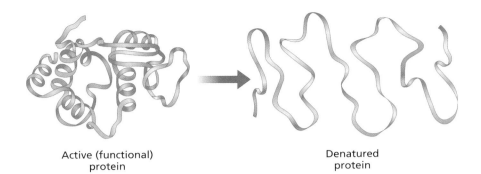

Active (functional)　　　　　　　　　　　　　Denatured
protein　　　　　　　　　　　　　　　　　　protein

FIGURE 1.7　An active and denatured protein. The active structure of a protein is folded in a specific way that is essential for the protein's biological function. Adding alcohol causes the protein to unfold, producing a denatured protein that is no longer functional.

intact. But ethanol molecules are able to dissolve lipids, which destroys the structure of the cell membrane. Without an intact membrane, the interior contents of the cell spill out, killing the bacteria.

If the cell membrane remains intact, ethanol can slip through the lipid molecules and enter the bacterial cell. Once within the cell, ethanol disrupts the protein molecules that bacteria require to survive and reproduce. A **protein** is a type of biological molecule that is constructed from a long chain of smaller molecules called **amino acids**. The protein chain must fold into a specific three-dimensional shape—the *native structure*—in order to perform its biological function. Most proteins reside within water, which supports formation of the protein's structure. However, ethanol causes proteins to unfold and lose their shape, producing a *denatured* protein that no longer functions (Figure 1.7). Without these working proteins, the bacterial cell cannot stay alive.

If ethanol is so deadly to bacterial cells, you may wonder why it is safe to use on our hands. The reason is that the outer layer of our skin is composed of dead cells that serve as a protective barrier for the sensitive cells beneath. Ethanol antiseptic cannot harm these surface cells because they are no longer alive. In addition, the ethanol in a hand sanitizer quickly evaporates after application, so our skin is not exposed for very long.

The ethanol in hand sanitizer is dissolved in water to make a **solution**. To be effective at killing bacteria, there has to be enough ethanol in the solution to disrupt the cell membrane and denature the bacterial proteins. If the amount of ethanol is too small, the bacteria will not be harmed. The specific amount of ethanol dissolved in a particular volume of the solution is measured by its **concentration**. In a hand sanitizer, ethanol is present in a high concentration of 63%. For comparison, the concentration of ethanol in vodka is approximately 40%.

In addition to hand sanitizers, you may have used so-called antibacterial soaps in an effort to prevent infection. Are these soaps more effective than regular soaps against bacteria? Are there any drawbacks to using them? You can explore these issues in Concept Question 1.2.

cell membrane A protective barrier surrounding a cell

lipids Fat-like molecules that do not dissolve easily in water.

protein A biological molecule constructed from a long chain of amino acids.

amino acid A group of molecules that are the building blocks of proteins.

solution A mixture that is formed when a substance dissolves in a liquid.

concentration A measure of the amount of substance dissolved in a specific volume of solution.

TEXTBOOK PREVIEW

Why does a protein stop working when it becomes denatured? In **Chapter 13,** we examine how proteins fold into specific three-dimensional shapes that are essential to their biological function.

TEXTBOOK PREVIEW

Why does ethanol dissolve in water, whereas other types of molecules (such as oils and fats) do not? We will study the properties of water in **Chapter 8** and the chemistry of solutions in **Chapter 9**. A discussion of solution concentration appears in **Chapter 10**.

CONCEPT QUESTION 1.2

All liquid hand soaps contain detergents, molecules that make greasy molecules more soluble in water and allow you to wash them off your hands. These detergents also destroy bacteria by dissolving the cell membrane.

Soaps that are called antibacterial contain at least one additional ingredient intended to boost the soap's ability to kill bacteria. One common additive is called triclosan, which is present in low concentrations (typically, between 0.1% and 0.45%). Check the label for the list of ingredients of the next soap bottle you purchase to see if it contains triclosan.

Are antibacterial soaps helpful or harmful? To provide an answer, we will examine three more specific questions:

(1)   Is antibacterial soap better than regular soap at eliminating bacteria from our hands?

(2)   Does the use of antibacterial soap promote the development of bacteria that are resistant to triclosan or other antibacterial drugs?

(3)   After we have finished using antibacterial soap, does triclosan have any harmful effects on the environment when it enters the water supply?

Investigate these questions online and discover the answers for yourself. Here is some guidance on how to proceed.

The explosion of websites on the Internet and the development of efficient search engines have given us immediate access to a vast amount of information. But simply finding information is not enough—you need to evaluate the reliability of what you read. That is a much harder task, especially since some of the websites may contain technical and scientific terms that you may not have previously encountered.

How can we find reliable information about health-related topics on the Internet? Good places to start include websites created by U.S. government agencies. Four of these websites are described next in alphabetical order. Use them to investigate the questions posed above, and write short answers to summarize the information you discover.

Centers for Disease Control and Prevention (CDC) www.cdc.gov

The CDC monitors and protects human health, with a particular focus on preventing the spread of infectious diseases.

Environmental Protection Agency (EPA) www.epa.gov

The EPA safeguards human health and the environment by enforcing regulations that are based on laws passed by Congress.

Food and Drug Administration (FDA) www.fda.gov

The FDA protects and promotes public health by regulating food products, pharmaceuticals, veterinary medicines, tobacco, and over-the-counter medications.

National Institutes of Health (NIH) www.nih.gov

The NIH supports research on health-related topics. The NIH campus contains over 20 different institutes devoted to advancing knowledge and prevention of a wide range of human diseases. The NIH also provides grant funding to researchers at colleges, universities, hospitals, and public health clinics.

As a follow-up exercise, use a search engine to see what information about this topic appears on a nongovernmental website—for example, the website of a hospital, a consumer group, or a manufacturer of antibacterial soaps. Who created the website and posted the information? Does the information presented agree or disagree with what you found on the websites for the CDC, EPA, FDA, and NIH?

Core Concepts

- An **antiseptic** is a chemical substance that prevents infection.
- Ethanol is the antiseptic ingredient in hand sanitizers. Ethanol is a **molecule** composed of **atoms** joined by **chemical bonds**.
- Ethanol kills bacterial cells in two ways. First, it disrupts the **cell membrane** of bacterial cells by dissolving the lipid molecules that produce the membrane's structure. Second, it disrupts the native structure of **proteins** and destroys their function.
- Hand sanitizers are a **solution** of ethanol dissolved in water. To be effective in killing bacteria, the solution contains a high **concentration** of ethanol.

1.4 Antibiotics: Treating Infections

Antiseptics such as ethanol are beneficial for disinfecting wounds and cuts on the skin's surface, which can prevent infections from happening. However, antiseptics do have one serious limitation—*they cannot be used inside the human body as an injection or a pill.* Their mode of action is nonspecific, so they kill human cells as well as bacteria. For this reason, antiseptics cannot be used to treat a patient who already has an internal infection.

We need a different type of weapon to combat bacterial infection, which is provided by antibiotics (a word that means "against life"). Antibiotics differ from antiseptics in two important ways. First, antibiotics kill bacteria at much lower amounts than antiseptics. Second, antibiotics can be used inside the body because they *selectively target bacterial cells.* However, we must recognize that all drugs, including antibiotics, have some unwanted side effects. For example, antibiotics that are taken to combat an infectious disease can also destroy the "good" bacteria in our digestive system.

We introduce antibiotics using **penicillin**, the most famous example of this class of drugs. The discovery, isolation, molecular structure, and antibacterial action of penicillin illustrate important general principles about antibiotics. Along the way we will introduce the contributions of several scientists as a way of illustrating "science in action"—the process by which scientific discoveries are made.

Learning Objective:

Characterize the discovery and function of penicillin.

penicillin An antibiotic that is produced by the *Penicillium* mold.

🔑 THE KEY IDEA: Penicillin was discovered through observation of its antibacterial activity.

Discovery and Isolation of Penicillin

In September 1928, Alexander Fleming (1881–1955; Figure 1.8) noticed something unusual when he returned to his London laboratory after a summer vacation. Fleming was studying bacteria in the hope of finding chemical substances that could treat bacterial infections. When serving in the Royal Army Medical Corps during World War I, he was horrified by watching many soldiers die from their infected wounds. When Fleming left the army at the end of the war, he was determined to find a treatment for the infections that had caused so much suffering and death.

What caught Fleming's eye on that September day? One of the Petri dishes (often called plates) on which he was growing bacteria contained a sample of blue-green mold. By chance, a spore of the mold had floated inside the laboratory and settled on the plate. When Fleming looked more closely, he saw something remarkable—there were no bacteria growing within a region surrounding the mold. Fleming became curious about this unexpected observation. Instead of throwing away the contaminated plate, he began to investigate why the area around the mold was devoid of bacteria.

Based on his observation, Fleming hypothesized that a substance from the mold had prevented bacteria from reproducing. To test this hypothesis, he grew a culture of the mold and found that the fluid around it killed a variety of bacteria. He identified the mold as belonging to the biological genus *Penicillium*, so he gave the name "penicillin" to the

FIGURE 1.8 Alexander Fleming. A photograph of Alexander Fleming inspecting a Petri dish with a culture of bacteria.

(a)

(b)

FIGURE 1.9 **Inhibition of bacterial growth by *Penicillium* mold.** (a) Acolony of *Penicillium* mold inhibits the growth of bacteria in a Petri dish. (b) Moldy bread that shows the characteristic blue-green color of *Penicillium* mold.

antibacterial substance produced by the mold. Figure 1.9(a) shows a modern version of Fleming's discovery, in which the growth of bacteria on the plate is inhibited in the region surrounding the *Penicillium* mold. You have probably seen *Penicillium* before as the greenish mold that grows on stale bread (Figure 1.9(b)).

Fleming reported the antibacterial activity of *Penicillium* mold in 1929, but his work received little attention at the time. Despite many attempts, Fleming could not isolate the chemical substance that produced the mold's antibacterial activity. Eventually, he decided that penicillin was a dead end and gave up his work on this subject.

The study of penicillin was picked up a decade later by a team of researchers at Oxford University, led by Howard Florey (1898–1968), trained as a physician, and Ernst Chain (1906–1979), an expert biochemist. They tackled the daunting challenge of obtaining enough penicillin to turn it from a biological curiosity into a potentially life-saving drug. Florey, Chain, and their colleagues cleverly designed new equipment to grow large quantities of *Penicillium* mold and isolate the active penicillin from a complex mixture. Because of the extensive bombing of Britain during World War II, Florey brought his mold samples to the United States, where the large-scale industrial production of penicillin became one of the highest priorities of the U.S. War Department. When the Allied forces landed on Normandy beaches in June 1944, many medics carried doses of penicillin to treat the wounded. In 1945, Fleming, Florey, and Chain received the Nobel Prize in Medicine for their discovery and development of penicillin.

Core Concepts

- Alexander Fleming discovered that *Penicillium* mold had the ability to prevent the growth of bacteria. He gave the name **"penicillin"** to the antibacterial substance produced by the mold, but he was unable to isolate it.
- Howard Florey, Ernst Chain, and their colleagues first isolated penicillin, which enabled it to be used as a drug. Mass production of penicillin began in the United States during World War II.

THE KEY IDEA: Penicillin has a complex molecular structure that is directly related to its effectiveness as an antibiotic.

What Is the Molecular Structure of Penicillin?

Although penicillin had been isolated and its antibacterial effects medically proven, one important question still remained: What was the structure of the penicillin molecule? There were two important reasons to know this information. First, the molecular structure could provide clues about how the drug kills bacteria. Second, the production of penicillin from mold remained difficult and time consuming. Knowing the molecular structure would open the possibility of faster production by synthesizing penicillin in a laboratory.

The challenge of determining the molecular structure of penicillin was taken up by Dorothy Hodgkin (1910–1994), who was also working at Oxford during the war years. Hodgkin and her collaborators deduced the structure of penicillin using a technique called **X-ray crystallography**, which involves bombarding crystals of the penicillin molecule with X-rays. For discovering the structure of penicillin and other biological molecules, Hodgkin was awarded the 1964 Nobel Prize in Chemistry and later honored with a special stamp issued by the British Post Office (Figure 1.10).

What does the penicillin molecule look like? Figure 1.11(a) illustrates the molecular structure of penicillin G, the most common form of this antibiotic. We can see that penicillin has a much more complicated molecular structure compared to ethanol. Figure 1.11(b) displays a *molecular model* of penicillin, in which the atoms are shown as balls and the chemical bonds as sticks. The molecular model depicts the accurate three-dimensional structure

FIGURE 1.10 Dorothy Hodgkin. A commemorative stamp showing Dorothy Hodgkin and a model of the penicillin molecule.

X-ray crystallography A technique in which scientists bombard crystals of a substance with X-rays to determine the structure of the substance.

Penicillin G

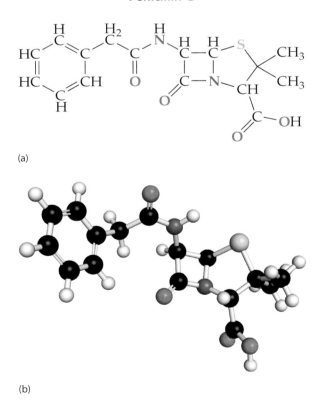

(a)

(b)

FIGURE 1.11 Molecular structure of penicillin G. This antibiotic is depicted as (a) a molecular structure drawing and (b) a molecular model. In these structures, each atom is color-coded according to its type—carbon atoms are shown as black, oxygen as red, nitrogen as blue, and a single sulfur atom as yellow (hydrogen atoms are shown as white in the molecular model).

FIGURE 1.12 **The effect of penicillin on a bacterial cell.** When treated with penicillin, bacterial cells burst open and spill their contents.

THE KEY IDEA: Penicillin disrupts the construction of the cell wall in certain bacteria.

cell wall A tough, mesh-like structure that surrounds many bacterial cells.

enzyme A biological molecule that causes a chemical reaction to occur faster.

catalyst A substance that increases the rate of a chemical reaction.

of the molecule, which is directly related to its effectiveness as an antibiotic. *The connection between the three-dimensional structure of molecules and their biological function is a key theme throughout this textbook.*

Core Concepts

- Dorothy Hodgkin determined the molecular structure of penicillin using the experimental technique of **X-ray crystallography**.
- Penicillin has a complex molecular structure that is directly related to its effectiveness as an antibiotic.

How Does Penicillin Work?

When penicillin is added to a collection of bacterial cells, the cells burst and their inner contents spill out (Figure 1.12). Why does this happen?

Penicillin disrupts the formation of the **cell wall**, a tough mesh-like structure that surrounds many bacterial cells. Each time a bacterial cell reproduces, a new cell wall needs to be made to surround the progeny cells. Building the cell wall requires the involvement of **enzymes**. *An enzyme is a biological molecule—usually a protein—that acts as a **catalyst** by making a chemical reaction happen faster.* A key enzyme in bacterial cells facilitates the chemical reaction that forms molecular linkages within the cell wall. Penicillin binds to this enzyme and prevents it from performing its biological task. As a result, the crosslinks are not added and the cell wall is weakened. The internal pressure within the cell causes the cell wall to rupture, and the interior contents of the cell explode through the hole, which kills the bacterium. Penicillin is most destructive to bacteria with a thick cell wall. By contrast, penicillin is much less effective against bacteria with a thin cell wall, and it has no effect on bacteria without any cell wall.

Core Concepts

- An **enzyme** is a biological **catalyst** that makes chemical reactions happen faster.
- Penicillin inhibits the function of a key enzyme that is required to make the bacterial **cell wall**. By weakening the cell wall, penicillin causes bacterial cells to rupture and die.

1.5   How Do Bacteria Become Resistant to Antibiotics?

Learning Objective:

Outline the origin and spread of antibiotic-resistant bacteria.

THE KEY IDEA: Antibiotic-resistant bacteria arise through genetic mutations and proliferate through evolution by natural selection.

When penicillin was first released to doctors and the public, it was viewed as a wonder drug—a "magic bullet" that could be used to treat a multitude of bacterial infections. Today, penicillin is used far less frequently because so many bacteria are now resistant to its effects. How did this happen?

The Origin and Spread of Antibiotic Resistance

Bacteria like *Staph* exist as many different varieties called **strains**. The difference between one strain of bacteria and another arises from variations in their DNA. Genetic information is stored within DNA as a sequence of chemical bases. When bacterial cells grow and multiply, the DNA is copied and passed from one generation to the next. However, this process is not perfect, and sometimes the DNA will acquire a "spelling mistake" called a **mutation**. For example, one type of base in the original DNA can be mutated into a different base in the copied DNA (Figure 1.13). This process is similar to what happens when you are typing a copy of a document and you make a typo by hitting the wrong key.

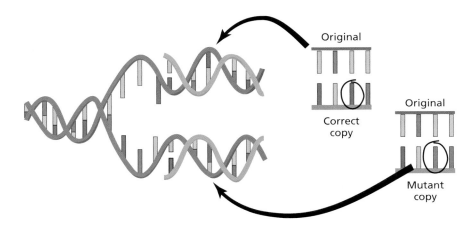

Original

Correct
copy

Original

Mutant
copy

FIGURE 1.13 DNA mutations produce genetic variation in bacteria. A DNA mutation can occur when DNA makes copies of itself. During the copying process, one of the chemical bases within the original DNA can be mutated into a different base.

We tend to think of mutations as harmful, but in some cases a mutation can be beneficial to an organism. Over time, a collection of DNA mutations can give bacteria the ability to disarm the effect of penicillin before it can cause any damage to the cell. We say that the bacterial strain has now become *antibiotic-resistant*.

Drug-resistant strains of bacteria were observed shortly after penicillin was first used. In fact, Alexander Fleming issued a warning about penicillin resistance in 1945, the year he won his Nobel Prize. In following decades, the frequency of drug resistance increased dramatically. Why did this happen?

One reason was the growing use of antibiotics in agriculture following World War II, a practice that continues today. Around 80% of the antibiotics sold in the United States are given to chickens, cows, and other animals that people eat. Antibiotics reduce the number of infections that animals acquire and also help them grow larger. But the widespread use of antibiotics in agriculture inevitably leads to some situations in which bacteria are exposed to a low level of antibiotics that is insufficient to destroy all possible strains. In this environment, certain strains of bacteria that can resist the harmful effects of antibiotics will survive and reproduce, while other strains are killed. This is an example of **evolution by natural selection**. The word "evolution" refers to change over time, while "natural selection" means that the presence of the antibiotic preferentially "selects" the drug-resistant strain for survival. Within a short period of time, the population of bacteria evolves to become increasingly drug-resistant.

Another biological mechanism in bacteria increases the spread of drug resistance. Unlike humans, who can only pass genes from parents to children, bacteria can easily swap genes by passing DNA from one cell to another. Sometimes the region of swapped DNA contains multiple genes that allow bacteria to withstand the effects of several different antibiotics. This effect gives rise to the phenomenon of *multidrug resistance*, known in the popular press as "superbugs."

We can now understand why the increased use of penicillin in the 1940s and 1950s generated more penicillin-resistant bacteria. In 1960, scientists tried to solve the problem by producing a new antibiotic called *methicillin*. But it took only two years until a new strain of *Staph* bacteria was discovered that was resistant to methicillin. This new strain was given the name MRSA, which we introduced at the beginning of the chapter. Today, MRSA generally refers to *Staph* bacteria that are resistant to penicillin, methicillin, and other similar antibiotics.

MRSA infections are only one of the antibiotic-resistant infections that are an increasing cause for concern. In a 2013 report, the CDC reviewed the health impact of the most worrisome drug-resistant pathogens (disease-causing agents), which consisted of 17 drug-resistant bacteria and one drug-resistant fungus. The CDC estimated that each year more than two million people in the United States are stricken with drug-resistant infections, and 23,000 die as a result.

TEXTBOOK PREVIEW

How do enzymes accelerate chemical reactions? How do drugs interfere with the function of enzymes? We examine the structure and function of enzymes in **Chapter 14,** and the action of drugs is described in **Chapter 15**.

strains Variations of an organism that arise from differences in their DNA.

mutation A change in the DNA of an organism.

evolution by natural selection A process by which certain individuals in a population preferentially survive and reproduce within a specific environment.

TEXTBOOK PREVIEW

What happens during DNA replication? How does a DNA mutation affect a biological organism? The structure of DNA and how it replicates are examined in **Chapter 12**. The mechanisms by which cells express genetic information are explained in **Chapter 13**.

In summary, the discovery and development of antibiotics to treat bacterial infections was one of the most important medical advances of the 20th century. Many millions of lives have been saved by antibiotics such as penicillin. But the widespread use of these powerful drugs triggered an increase of antibiotic-resistant bacteria. Ironically, we have now become the victims of our own success.

Core Concepts

- Bacteria exist as different **strains** that arise from variations in their genes. Genetic changes in bacteria are caused by **mutations** in their DNA. Some mutations occur randomly through DNA copying.
- Some DNA mutations enable bacteria to resist the effect of antibiotics. In an environment where antibiotics are present, these drug-resistant strains will preferentially survive and reproduce. This process is known as **evolution by natural selection**. Antibiotic-resistant genes can also be swapped between bacteria.

THE KEY IDEA: The development of new antibiotics is not keeping pace with the increase of antibiotic-resistant bacteria.

What Can Be Done About Antibiotic-Resistant Bacteria?

What can be done about antibiotic resistance today? This is a complex problem that involves chemistry, medicine, economics, agriculture, and human behavior. You may not realize it, but your own actions also contribute to making this situation better or worse.

One factor contributing to the spread of antibiotic resistance is the large quantities of antibiotics that are prescribed during routine trips to the doctor's office. In their 2013 report, the CDC estimated that roughly 50% of antibiotic prescriptions were unnecessary. If doctors and patients became more thoughtful about the need for antibiotics, their use could be restricted to the most important medical cases. Another problem is the continuing overuse of antibiotic drugs for farm animals. The FDA has proposed changes to the existing system of selling, administering, and tracking antibiotics in agricultural settings. Denmark and some other countries have implemented more restrictive policies to reduce the amount of antibiotics used in agriculture.

The problem of antibiotic resistance may sound easy to fix—the pharmaceutical industry should simply make new antibiotics. However, that option is not so easy. Figure 1.14 shows the number of new antibacterial drugs that have been approved by the FDA over the 30-year period from 1983–2012. Despite the growing problem of antibiotic-resistant bacteria, the number of approved antibiotics has *decreased*. Why has this happened?

Developing any new drug is complicated, lengthy, and expensive. It requires preparing sufficient amounts of the drug, testing it for safety in animals, and performing at least three stages of clinical trials with human patients. The FDA carefully scrutinizes each new drug to ensure that it is safe and effective. Only after the FDA has approved a drug can it be made available to doctors and patients. It typically takes more than 10 years and close to one billion dollars to bring a new drug to market. Even with all this investment of time and money, the pharmaceutical company has no guarantee of success.

Antibiotic drugs pose unique challenges. Bacteria that are already resistant to one class of drugs (e.g. penicillin and methicillin) are often also resistant to new variants of these drugs. Creating and testing a completely novel type of antibiotic requires a major investment of time and money. Regardless of how effective a new antibiotic may be, bacteria will eventually evolve to become resistant to the drug. Based on prior examples, antibiotic resistance

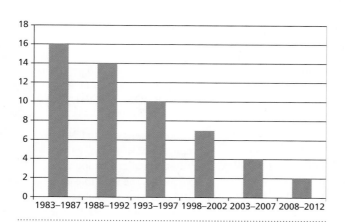

FIGURE 1.14 A decline in new antibiotics. A chart showing new antibiotic drugs approved by the Food and Drug Administration between 1983 and 2012. Each vertical bar represents the total number of new drugs approved within a 5-year period.

takes only a few years to develop following the introduction of a new drug. Even if a pharmaceutical company develops an effective new antibiotic, its availability may be strictly regulated to provide a "drug of last resort" to treat the most stubborn bacterial infections. In addition, antibiotics are prescribed only for a few days in order to treat an infection. By contrast, drugs that patients have to take for many years—such as cholesterol-lowering medications—provide a much more lucrative investment.

If we believe that developing new antibiotics is important to our society, we need to rethink the traditional approaches that are currently used for drug development. For example, Congress passed a new law in 2012 called Generating Antibiotic Incentives Now (GAIN). In 2014, President Barack Obama issued an Executive Order entitled "Combating Antibiotic Resistance," which calls for new coordination among federal agencies to address a challenge that "represents a serious threat to public health and the economy."

Since the problem of antibiotic resistance is so complex and multidimensional, the best way forward is for *everyone* to make a positive contribution to its solution. This approach is captured in the principle of *antibiotic stewardship* promoted by the CDC: "Stopping even some of the inappropriate and unnecessary use of antibiotics in people and animals would help greatly in slowing down the spread of resistant bacteria."

> **TEXTBOOK PREVIEW**
>
> How are new drugs developed? How are they tested for safety and effectiveness? In **Chapter 15,** we examine the process of drug discovery, development, testing, and approval.

Core Concepts

- Despite the rise of antibiotic-resistant bacteria, the number of new antibiotic drugs approved by the FDA has decreased in recent decades.
- The principle of antibiotic stewardship calls for a reduction of "inappropriate and unnecessary use of antibiotics in people and animals."

Antibiotic-Resistance Revisited

From reading this chapter, you can see that understanding antibiotic-resistant bacteria requires scientific insights from chemistry, biology, and medicine (Figure 1.15). Antiseptics and antibiotics are molecules that are described by the principles of chemistry. To comprehend these therapeutic molecules, we need to learn about atoms, chemical bonds, and molecular structure. The effect of antibiotics on bacteria requires an investigation of biological molecules and cells. In addition, the emergence of antibiotic-resistant bacteria is an example of evolution by natural selection, one of the unifying theories in biology. In the realm of medicine, we need to examine how new antibiotic drugs are developed, tested, and approved by the FDA. The threat of antibiotic-resistant bacteria is just one example of a complex, real-world scientific issue that you will need to confront in the 21st century. Our goal for this textbook is to provide you with the resources and skills to understand, evaluate, and take appropriate action on these issues.

THE KEY IDEA: The threat of antibiotic-resistant bacteria is a complex, real-world scientific issue that requires insights from chemistry, biology, and medicine.

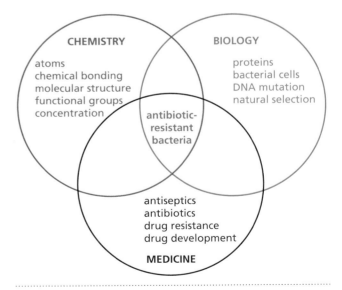

FIGURE 1.15 Understanding antibiotic-resistant bacteria. Analyzing this complex topic requires insights from chemistry, biology, and medicine.

Core Concepts

- The spread of antibiotic-resistant bacteria is a complex, real-world issue that requires scientific insights from chemistry, biology, and medicine.

CHAPTER 1
VISUAL SUMMARY

1.1 Why Are Some Infections Resistant to Antibiotics?

Learning Objective:

Show the importance of understanding scientific information and biological molecules in everyday life.

⊙ **MRSA** stands for **methicillin-resistant *Staphylococcus aureus***. These single-**cell** bacteria are resistant to treatment by methicillin, penicillin, and related antibiotics.

⊙ MRSA infections are becoming increasingly common within our local communities.

1.2 What Is Science?

Learning Objective:

Describe the stages of the scientific method.

⊙ The **scientific method** describes the stages of a scientific investigation.

⊙ A **hypothesis** is an informed first attempt to explain an **observation**. A hypothesis must be testable by an **experiment**.

⊙ A **theory** is a well-supported explanation of the natural world.

⊙ A **scientific law** summarizes a large number of observations but does not provide an explanation. Laws can often be explained by theories.

⊙ Scientific research is often more complex than the step-by-step procedure of the scientific method.

1.3 Antiseptics: Preventing Infections

Learning Objective:

Explain how the properties of ethanol enable it to function as an antiseptic.

⊙ An **antiseptic** is a chemical substance that prevents infection.

⊙ Ethanol is the antiseptic ingredient in hand sanitizers. Ethanol is a **molecule** composed of **atoms** joined by **chemical bonds**.

⊙ Ethanol kills bacterial cells in two ways. First, it disrupts the **cell membrane** of bacterial cells by dissolving the lipid molecules that produce the membrane's structure. Second, it unfolds the native structure of **proteins** and destroys their function.

1.4 Antibiotics: Treating Infections

Learning Objective:

Characterize the discovery and function of penicillin.

- ⊙ Alexander Fleming discovered that *Penicillium* mold had the ability to prevent the growth of bacteria. He gave the name **penicillin** to the antibacterial substance produced by the mold, but he was unable to isolate it.

- ⊙ Penicillin was isolated during World War II by Howard Florey and Ernst Chain at the University of Oxford. As a result, penicillin could now be used as an **antibiotic** drug.

- ⊙ Dorothy Hodgkin determined the molecular structure of penicillin using the experimental technique of **X-ray crystallography**.

- ⊙ Penicillin has a complex molecular structure that is directly related to its effectiveness as an antibiotic.

- ⊙ An **enzyme** is a biological **catalyst** that makes chemical reactions happen faster.

- ⊙ Penicillin inhibits the function of a key enzyme that is required to make the bacterial **cell wall**.

- ⊙ The effect of penicillin weakens the cell wall, which causes the cell to rupture and die.

1.5 How Do Bacteria Become Resistant to Antibiotics?

Learning Objective:

Outline the origin and spread of antibiotic-resistant bacteria.

- ⊙ Bacteria exists as different strains that arise from variations in their **DNA**. Genetic changes in bacteria are caused by DNA **mutations**. Mutations can occur randomly through DNA copying.

- ⊙ Some DNA mutations enable bacteria to resist the effect of antibiotics. In an environment where antibiotics are present, these drug-resistant strains will preferentially survive and reproduce. This process is called **evolution by natural selection**.

- ⊙ Despite the rise of antibiotic-resistant bacteria, the number of new antibiotic drugs approved by the FDA has decreased in recent decades.

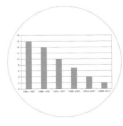

- ⊙ The principle of antibiotic stewardship calls for a reduction of "inappropriate and unnecessary use of antibiotics in people and animals."

LEARNING RESOURCES

Reviewing Knowledge

1.1: Introduction: Why Are Some Infections Resistant to Antibiotics?

1. What is MRSA? Why are MRSA infections potentially dangerous?

1.2: What Is Science?

2. Explain the differences between a *hypothesis*, a *theory*, and a *law* in science. Provide one example of each principle that is *not* discussed in this chapter.

1.3: Antiseptics: Preventing Infections

3. What molecule is used as an antiseptic in hand sanitizers? What are the two methods by which this molecule kills bacterial cells?

1.4: Antibiotics: Treating Infections

4. How did Alexander Fleming first discover the antibacterial effects of penicillin?
5. What did Howard Florey and Ernst Chain contribute to the development of penicillin as a medical therapy?
6. How did Dorothy Hodgkin determine the structure of penicillin?
7. When bacterial cells are treated with penicillin, they explode and their internal contents spill out. Explain why this happens.

1.5: How Do Bacteria Become Resistant to Antibiotics?

8. Describe how the increase of antibiotic-resistant bacteria can be explained by evolution by natural selection.
9. Write a short paragraph to summarize the challenges of developing new antibiotic drugs.

Developing Skills

10. When you receive an injection (like a flu shot) or give a blood sample, the doctor or nurse will disinfect the surface of your skin with an alcohol swab. The type of alcohol used in these swabs is called *isopropyl alcohol* ("rubbing alcohol'). The molecular structure of isopropyl alcohol is shown in the figure. Redraw this molecule and circle the alcohol functional group.

H
|
H—C—H
|
H—C—O—H
|
H—C—H
|
H

Isopropyl alcohol

11. The figure below shows an experiment to test the effectiveness of various antibiotics on a particular strain of bacteria. This experiment is based on the same principle as Fleming's discovery of the antibacterial effects of penicillin. Bacteria that have grown in the Petri dish are shown as a brown layer. Each white disc in the figure contains a different type of antibiotic. The clear area surrounding the disc shows how much the antibiotic has or has not prevented the growth of bacteria.

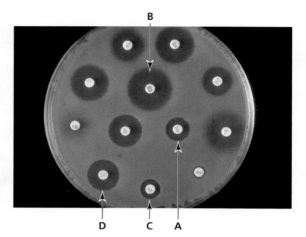

Use your observations of this experiment to rank the four labeled antibiotics (A, B, C, and D) in order of *increasing effectiveness* (lowest to highest). Write 1-2 sentences to explain your ranking.

Exploring Concepts

12. The following list provides the most commonly prescribed antibiotics in the United States (*brand name* / chemical name). Select one of these antibiotics, and use the Internet to discover its molecular structure and how it works to prevent the growth of bacteria. Write a one-paragraph summary of your findings. If possible, select an antibiotic that you have personally taken.

    Zithromax / azithromycin
    Amoxil / amoxicillin
    Cipro / ciprofloxacin
    Keflex / cephalexin

13. The CDC has released a comprehensive report entitled *Antibiotic Resistance Threats in the United States, 2013*. The report is available on the CDC website at www.cdc.gov. Read the Executive Summary of the report, followed by Sections 1 and 2. Based on what you have learned, find a creative way to inform your college community about antibiotic resistance. For example, you can write a letter to the school newspaper, create a YouTube video, or design a flyer for your campus.

14. The World Health Organization (WHO) is a division of the United Nations that is based in Geneva, Switzerland. Use the WHO website (http://www.who.int/en/) to investigate multiple drug-resistant tuberculosis (MDR-TB). What are the countries with the highest incidence of MDR-TB? What are the reasons for high infection in these countries? What is so-called XDR-TB? What steps are being taken to address the global problem of drug-resistant tuberculosis?

15. Use the Internet to investigate the federal legislation called GAIN (Generating Antibiotic Incentives Now). Write a one-paragraph summary of the main features of this law. Based on your research, evaluate whether the law has been effective at promoting the development of new antibiotics to address the growing threat of antibiotic-resistant bacteria. Write a second paragraph of analysis, and provide specific examples.

16. You have probably had numerous treatments with various antibiotics. Several of them likely happened when you were a young child and prone to infections. In addition to destroying the bacteria responsible for the infection, each of these antibiotic treatments kills some of the beneficial bacteria in your intestines. Propose a hypothesis for how multiple antibiotic treatments may affect the presence of drug-resistant bacterial strains within your body. Propose an experiment to test this prediction.

    Questions 17 and 18 ask you to apply the scientific method to current scientific research on antibiotics and resistance.

17. Exposing a culture of bacteria to an antibiotic prevents the growth of normally dividing cells. Penicillin, for example, kills bacteria by interfering with the construction of their cell wall during replication. However, bacterial cells have devised a trick to resist an antibiotic attack by allowing a small number of cells to become dormant. Bacterial cells that behave in this way are called *persister cells*—they cannot grow or divide, but they are still alive. If persister cells are not making new cell walls, penicillin cannot kill them. Once the antibiotic level decreases, the persister cells awake and resume their normal growth. However, persister cells are not truly drug resistant because most of the cells that grow after waking up can still be killed by the antibiotic.

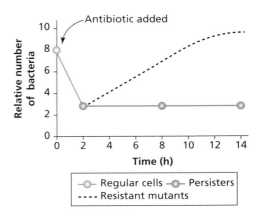

The graph above shows the response of different types of bacterial cells to treatment by an antibiotic. Regular cells (blue line) are killed by the antibiotic. Persisters (red line) are insensitive to the antibiotic but do not divide or grow. Resistant mutants (dashed line) grow in the presence of the drug.

(a) Propose a hypothesis about whether persister cells can also resist the effect of an ethanol antiseptic (as found in a hand sanitizer).

(b) Design an experiment to test your hypothesis. Explain how you would perform this experiment.

(c) Using the above figure as a guide, draw a graph of the results that you predict if your hypothesis is *correct*.

(d) Similarly, draw a graph of the results that you predict if your hypothesis is *incorrect*.

18. When bacteria are exposed to a surface, some cells stick on the surface and form a chemical coating that shields them from our immune defenses. This sticky layer of bacteria is called a *biofilm*; one familiar example is the dental plaque that forms on the surface of our teeth.

(a) Propose a hypothesis for why bacteria within a biofilm are especially difficult to kill with antibiotics.

(b) How could the biofilm be treated to make these bacteria more susceptible to antibiotics?

(c) Patients with cystic fibrosis suffer from the formation of mucus around lung tissues. The lungs are often infected by biofilms that are exceptionally difficult to treat with antibiotics. Use the Internet to discover what treatments are available for cystic fibrosis patients, and write a one-paragraph report.

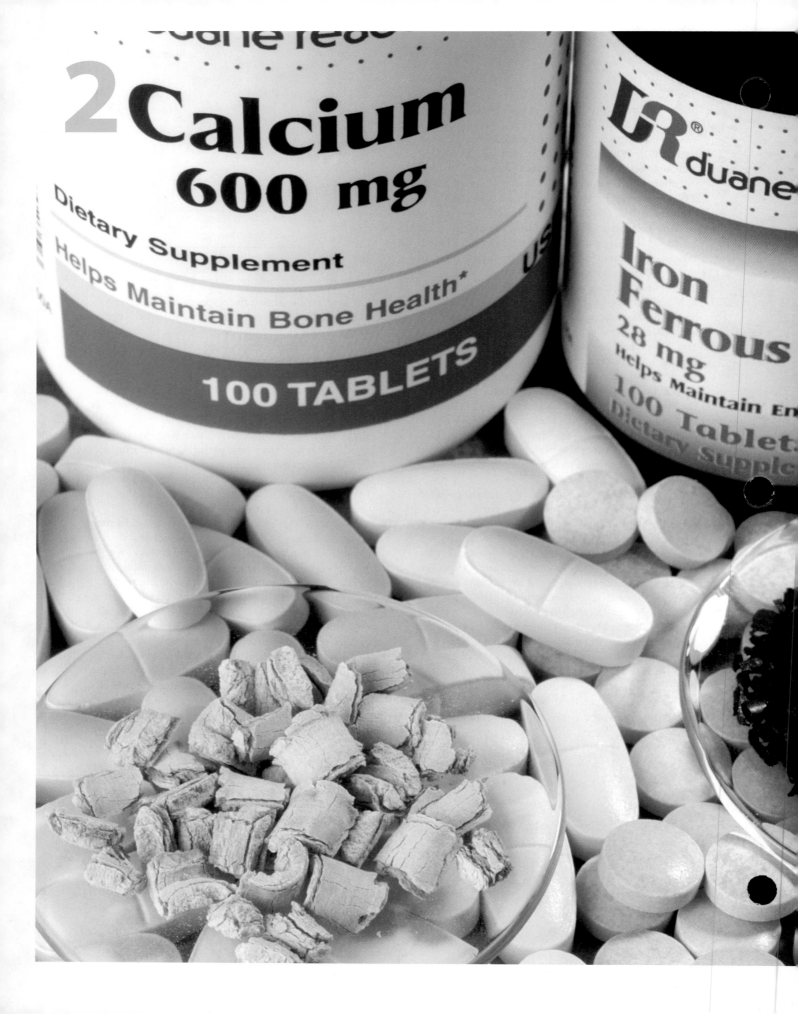

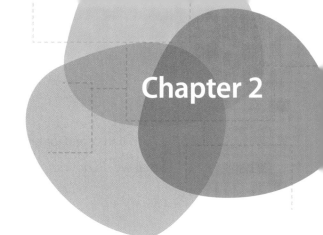

Chapter 2

Elements of Life and Death:

The Chemistry of Elements and Atoms

If you had cereal for breakfast, you might have eaten some iron in each spoonful. Iron is a metal that we do not normally associate with our diet. Why does breakfast cereal contain iron?

2.1 Why Is There Iron in My Cereal?

Learning Objective:

Illustrate the importance of iron for human health.

THE KEY IDEA: The human body needs chemical elements, including iron, for optimal health.

chemical symbol An abbreviation for an element that uses one or two letters.

hemoglobin A protein found in red blood cells that transports oxygen to the body's tissues.

The next time you eat cereal, look at the nutrition label on the side of the box; an example appears in Figure 2.1. For the cereal Total, the amount of iron in each serving (1 cup) provides 100% of the recommended dietary allowance (RDA), the daily amount of a nutrient that is required to maintain good health. The addition of iron to foods such as cereals and bread is called *fortification*. This process is carried out because iron is an important component of our diet. The added iron is usually in the form of iron atoms, which are denoted by the **chemical symbol** Fe (from *ferrum*, the Latin word for iron).

If you doubt the presence of iron in your cereal, you can extract the iron by following a few simple steps. Crush the cereal in water, and then use a strong magnet to pull the iron from the resulting slurry. Figure 2.2 illustrates what you will see after you have performed this procedure—a small collection of iron filings that were present in the cereal. You might have used similar iron filings in a physics experiment with magnets.

For the iron to be biologically useful, it must be absorbed by the body and then converted into an electrically charged form called an *ion*. An iron ion is formed when an iron atom loses two of its electrons; it is written as Fe^{2+} to indicate its positive charge.

Why does our body need iron? It is essential for the function of a protein called **hemoglobin** (Figure 2.3), which transports oxygen from the lungs to the body's tissues. Hemoglobin is abundant in red blood cells; in fact, it is responsible for their color. Hemoglobin is composed of four long molecular chains that are twisted into a complex three-dimensional shape. Each chain supports a ring-shaped molecule called a *heme*, which has an Fe^{2+} ion at its center. When oxygen transport occurs, one oxygen molecule binds to the Fe^{2+} ion within each of the four heme groups.

Some individuals have low levels of iron in their body, which leads to *iron deficiency anemia*. This condition inhibits hemoglobin's ability to transport

FIGURE 2.1 A nutrition label for Total cereal. The label indicates that the cereal contains iron. The amount in 1 cup provides 100% of the recommended dietary allowance (RDA).

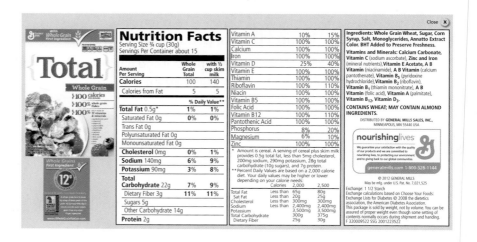

FIGURE 2.2 **Iron filings extracted from cereal.** A small amount of iron filings can be extracted from breakfast cereal, which demonstrates the presence of iron in this food.

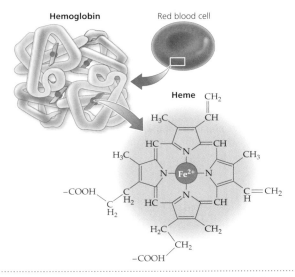

FIGURE 2.3 **Molecular structure of hemoglobin.** Hemoglobin is a protein that transports oxygen molecules and exists in high abundance within red blood cells. Each of the four heme groups within hemoglobin contains iron in the form of an Fe^{2+} ion. During oxygen transport, one oxygen molecule binds to each of the iron ions.

oxygen, and it also lowers the body's production of red blood cells. As a result, less oxygen is transported through the bloodstream, resulting in pale skin, fatigue, headaches, and shortness of breath. Most patients can remedy this condition by eating iron-rich foods or by taking an oral iron supplement.

This account of the role of iron in our body incorporates several important topics in chemistry. In this chapter, we examine the chemical elements—the building blocks of matter—and their relevance for human health. We explore how the elements are organized in the periodic table and how their chemical behaviors can be explained in terms of the atoms they contain. We also study how and why different elements form ions. These chemical principles will help us understand why some elements are beneficial and others are harmful.

Core Concepts

- Some cereals contain added iron because it is important for our health.
- Iron is an essential component of **hemoglobin**, a protein that transports oxygen within red blood cells. The protein consists of four molecular chains, each containing a heme group with an Fe^{2+} ion at its center.
- During oxygen transport, one oxygen molecule binds to the Fe^{2+} ion within each of the four heme groups.

2.2 Mixtures, Compounds, and Elements

During our daily lives, we encounter a multitude of substances—the air we breathe, the food we eat, the water we drink, and the clothes we wear. How can we begin to make sense of this great variety of stuff that makes up our world? This section describes how scientists classify the properties of matter and its different varieties.

Properties of Matter

Scientists define **matter** as *any substance that occupies space and has **mass**. The mass of an object measures the amount of matter that it contains.* The measurement units of mass are *grams* (g) for small quantities and *kilograms* (kg) for large quantities. The mass value reflects the number of atoms contained within an object. For example, an iron bar with a mass of 2 kg contains twice as many iron atoms as a bar with a mass of 1 kg. Because mass measures the amount of matter in an object,

Learning Objective:

Explain the organization of elements in the periodic table.

THE KEY IDEA: Properties of matter can be classified as physical or chemical.

matter Any substance that occupies space and has mass.

mass The amount of matter that an object contains.

FIGURE 2.4 **A comparison of the densities of foam, diamond, and iron.** For each material, the mass is different for the same volume (1 cm³).

weight The force exerted on an object by gravity.

physical property A characteristic that can be measured without changing the composition of the substance.

chemical property A characteristic of a substance that can be observed during a chemical reaction.

extensive property A property that depends on the amount of material; examples are length, mass, and volume.

intensive property A property that remains the same regardless of the amount of material (e.g., temperature).

volume The amount of space occupied by an amount of matter.

temperature A measurement of how hot or cold a substance is, relative to a standard scale.

density The amount of mass contained in a unit of volume.

it does not vary with the object's location. By contrast, the **weight** of an object can change with location. This change occurs because *weight is a measurement of the force exerted on an object by gravity.* Weight is measured in pounds (lb), which is familiar from our bathroom scales. On the Earth's surface, a mass of 1.0 kg has a weight of 2.2 lb.

Consider an astronaut with a mass of 75 kg. On Earth, the astronaut has a weight of 165 lb. When she is standing on the moon, her weight is only 27.5 lb because the gravitational attraction of the moon is only one-sixth as strong as that of the Earth (due to the moon's smaller size). When the astronaut is inside a rocket, traveling between the Earth and the moon, she has no weight because the planets are too far away to exert a gravitational force. You have likely seen images of astronauts floating weightlessly in space, which is a vivid reminder of the difference between mass and weight. The astronaut always has the same mass of 75 kg, which arises from the atoms that make up her body. However, she has different weights depending on her varying locations.

Matter has various **physical properties** and **chemical properties**. *A physical property is a characteristic of a substance that can be measured without changing the composition of the substance.* Physical properties can be divided into **extensive properties** and **intensive properties**. *An extensive property changes as the amount of material changes.* Examples of extensive properties are *length, mass,* and **volume**. It is possible to have a large or small volume of water, even though the substance remains the same. By contrast, *an intensive property remains the same regardless of how much material is present.* One example of an intensive property is **temperature**. For example, the temperature of a water sample does not depend on its volume.

The ratio of different properties can also be informative. For example, the **density** *of a substance is defined as the mass of a sample divided by its volume.* If we measure the mass in grams (g) and the volume in cubic centimeters (cm³), then we can calculate the density by using the following equation:

$$\text{density } (g/cm^3) = \frac{\text{mass } (g)}{\text{volume } (cm^3)}$$

For example, 1.00 cm³ of pure water has a mass of 1.00 g, so the density of water is 1.00 g/cm³ (grams per cubic centimeter). For comparison, a bar of gold has a much higher density of 19.3 g/cm³. Figure 2.4 illustrates the variability in density of three additional materials: foam, diamond, and iron. Mass and volume are both extensive properties, but their ratio—density—is an intensive property. The density of any substance remains constant regardless of the size of the sample.

WORKED EXAMPLE 2.1

Question: Pure gold has a density of 19.3 g/cm³. A U.S. $50 American Eagle gold coin has a mass of 33.9 g and a volume of 2.41 cm³. Is this coin made of pure gold?

Answer: We can calculate the density of the coin as the ratio of its mass and volume:

$$\text{density}(g/cm^3) = \frac{\text{mass } (g)}{\text{volume } (cm^3)} = \frac{33.3 \text{ g}}{2.41 \text{ cm}^3} = 14.1 \text{ g/cm}^3$$

The calculated density is *less* than the density of pure gold. Therefore, we conclude that the coin is not made from pure gold.

TRY IT YOURSELF 2.1

Question: Pure water has a density of 1.00 g/cm³. Solid materials with a larger density will sink in water, whereas solids with a smaller density will float on the surface. For each of the solid samples given below, predict whether it will sink or float.

   (a) Sample A has a mass of 36.0 g and a volume of 3.00 cm³.

   (b) Sample B has a mass of 125.0 g and a volume of 500.0 cm³.

   (c) Sample C has a mass of 55.0 g and a volume of 20.0 cm³.

PRACTICE EXERCISE 2.1

Consider the density of three substances.

   copper = 8.96 g/cm³   aluminum = 2.70 g/cm³   helium = 0.000179 g/cm³

Identify each of the following samples based on its density:

   (a) Sample A has a mass of 0.895 g and a volume of 5000 cm³.

   (b) Sample B has a mass of 448.0 g and a volume of 50.0 cm³.

   (c) Sample C has a mass of 675.0 g and a volume of 250.0 cm³.

CONCEPT QUESTION 2.1

We can perform an investigation of density using cans of soda (see the figure). If you place a can of regular Coca-Cola in water, the can will sink. However, if you place a can of Diet Coke in water, it will float below the surface. You can test this for yourself at home. Explain the reason for these different behaviors by considering the ingredients in each type of soda.

Unlike physical properties, which can be studied in isolation, *a chemical property of a substance can be observed only during a chemical reaction.* For example, when carbon exists in the form of diamond, we can measure its density (3.52 g/cm³) as a physical property. When diamond is burned with a hot flame, its carbon atoms (C) combine with oxygen gas (O_2) in the air to make carbon dioxide (CO_2). This behavior of diamond is a chemical property.

Other terms that describe the behavior of matter are **physical change** and **chemical change**. A physical change describes a change in physical properties without any modification in the composition of the substance. One example is a melting ice cube, when a solid form of H_2O (ice) turns into a liquid form of H_2O (water) without changing the H_2O molecules themselves. By contrast, a chemical change involves the transformation of one substance into another. This type of change occurs when electricity is passed through water, which converts the H_2O molecules into oxygen and hydrogen gases.

physical change A change in the physical properties of a substance that does not modify its composition.

chemical change The transformation of one substance to another via a chemical reaction.

Core Concepts

- ■ **Matter** is any substance that occupies space and has mass.
- ■ The **mass** of an object measures the amount of matter that it contains. The **weight** of an object measures the force exerted on the object by gravity. The mass remains the same regardless of the location of the object, but the weight can vary with location.
- ■ **Physical properties** can be measured without changing the composition of a substance. **Extensive properties** (e.g., length, mass, **volume**) depend on the amount of substance, whereas **intensive properties** (e.g., **temperature, density**) do not vary with the amount of substance.

■ The **chemical properties** of a substance are revealed during a chemical reaction.
■ A **physical change** describes a change in physical properties of the same substance, whereas a **chemical change** involves a transformation of one substance into another.

Classifying Matter

🔑 THE KEY IDEA: Matter can be classified into various categories—mixtures, pure substances, compounds, and elements.

mixture A physical combination of multiple substances that are not chemically bonded to one another.

homogeneous mixture A mixture in which the composition is the same throughout.

heterogeneous mixture A mixture in which the composition is not uniform, enabling the individual components to be identified.

pure substance A substance with a constant composition and characteristic chemical properties; it cannot be separated into components by physical methods.

compound A pure substance formed from two or more different elements that are chemically joined in a fixed ratio; for example water (H₂O).

Almost every type of matter we commonly experience is a type of **mixture**, which is *a physical combination of multiple substances that are not chemically bonded to one another*. For example, the air that we breathe is a mixture of different gases. A mixture can be either **homogeneous** or **heterogeneous**. The components of a homogeneous mixture are combined in such a way that the composition is the same throughout the mixture (*homo* means "same"). In these mixtures, the individual components are extremely difficult to identify. Examples of homogeneous mixtures are air, sea water, vinegar, and steel. By contrast, the components of a heterogeneous mixture are not uniform throughout the mixture, and it is possible to identify them (*hetero* means "different"). Examples are blood (a mixture of cells and fluid) and a bowl of cereal in milk.

The components of a mixture can be separated by physical methods such as boiling, freezing, and dissolving. Consider table salt, which has the chemical name *sodium chloride*. Table salt is obtained by evaporating sea water in large flat "pans" (Figure 2.5). The mixture of salts and water is spread in thin layers, and the liquid water evaporates to leave the solid salts. These dry salts are still a mixture because several different substances are present in sea water, including magnesium, calcium, and potassium. Some grocery stores sell "sea salt," which contains all of these ingredients (and several others). The mixed components of sea salt are separated from one another by further physical treatment, such as adding compounds that turn calcium salts into a solid and leave sodium salts in the liquid. At the end of these steps, we obtain pure sodium chloride.

Sodium chloride is one example of a **pure substance**. *A pure substance has constant composition and characteristic chemical properties; it cannot be separated into components by physical methods.* One type of pure substance is a **compound**. *A compound is a substance composed of two or more different elements that are chemically joined in a fixed ratio.* Some compounds are *molecules*, which are composed of atoms linked by chemical bonds to form a specific structure. Other compounds

FIGURE 2.5 Extracting salts from sea water by evaporation. This process is a large-scale industrial activity that yields important minerals economically because little energy is expended on the drying process.

are *salts*, which are made from ions. An ion is formed when an atom gains or loses one or more electrons, as we will see later in the chapter.

Water is a familiar molecular compound. A water molecule contains two types of atoms, hydrogen and oxygen. The composition of a compound can be specified by its *chemical formula*, a shorthand representation that uses the chemical symbols for the elements in the compound. For example, the chemical formula for water is H_2O. The letters in this formula tell us that water contains the elements hydrogen (H) and oxygen (O), whereas the subscript "2" indicates that there are two hydrogen atoms for each oxygen atom. We cannot write the chemical formula for water as HO_2 because that incorrectly specifies one atom of hydrogen and two atoms of oxygen. Table salt—sodium chloride—is an *ionic compound* with the formula NaCl. In this formula, the ratio of the two elements is 1:1, which means that the compound contains equal numbers of sodium ions and chloride ions.

Compounds can be decomposed by chemical methods that make use of reactions that break or form chemical bonds. For instance, H_2O molecules can be chemically decomposed by passing an electric current through liquid water to yield two gases—hydrogen and oxygen (Figure 2.6). These gases are elements. *An **element** contains only one type of atom and cannot be broken down into any simpler substance using chemical methods.* Elements are also examples of pure substances. Hydrogen and oxygen exist as molecules that each contain two atoms, as indicated by their chemical formulas: H_2 for hydrogen and O_2 for oxygen. Some elements, like helium and neon gases, occur naturally as individual atoms. Other elements, such as copper metal, consist of many trillions of atoms bonded in a repeating pattern to form a solid.

In summary, chemists describe any type of matter as either a mixture or a pure substance. In turn, a pure substance can be either a compound or an element. Physical methods are used to separate mixtures into pure substances, and chemical methods are used to transform compounds into elements. Figure 2.7 provides a flowchart for these terms, and Table 2.1 summarizes the definitions.

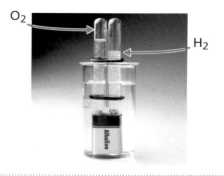

FIGURE 2.6 Decomposing H_2O into elements. Passing an electrical current from a battery through water causes the H_2O molecules to decompose into two gases—hydrogen (H_2) and oxygen (O_2). These gases contain only one type of atom and are classified as elements. Because the atoms in H_2O are present in a 2:1 ratio, the decomposition of water produces twice the volume of H_2 gas compared to the volume of O_2 gas.

element A substance composed of only one type of atom, which cannot be broken down into simpler substances using chemical methods.

TYPE OF MATTER	DEFINITION	EXAMPLES
Mixture	A physical blending of multiple substances that are not chemically bonded to one another. The components of a mixture can be separated by physical methods such as boiling, freezing, and dissolving.	An air sample; a cup of coffee containing milk and sugar.
Pure substance	A substance with constant composition and characteristic chemical properties; it cannot be separated into components by physical methods. This category includes both compounds and elements.	Distilled water; copper metal.
Compound	A pure substance composed of two or more elements that are chemically joined in a fixed ratio. Molecules and salts are both compounds. Compounds can be decomposed by chemical methods that make use of reactions that break or form chemical bonds.	Water (H_2O); sodium chloride (NaCl).
Element	A pure substance containing only one type of atom that cannot be broken down into any simpler substance using chemical methods.	Hydrogen gas (H_2); oxygen gas (O_2); copper metal (Cu).

TABLE 2.1   Definitions of *Mixture, Pure Substance, Compound,* and *Element*

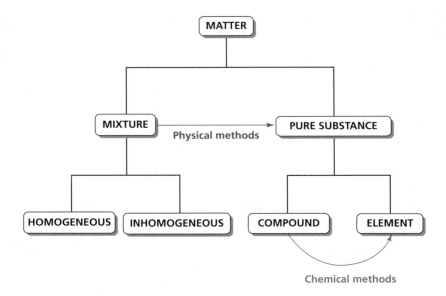

FIGURE 2.7 Classification of matter. Any type of matter can be categorized as a mixture, a pure substance, a compound, or an element. Mixtures can be separated into pure substances by physical methods. Compounds can be decomposed into elements by chemical methods.

WORKED EXAMPLE 2.2

Question: Classify the following types of matter as mixtures, pure substances, compounds, and/or elements:

(a) soil

(b) carbon dioxide

(c) sea water

(d) iron

(e) salad dressing

Answers: (a) mixture (b) pure substance, compound (c) mixture (d) pure substance, element (e) mixture.

TRY IT YOURSELF 2.2

Question: Classify the following mixtures as homogeneous or heterogeneous:

(a) pizza

(b) coffee

(c) air

(d) a bowl of mixed nuts

Core Concepts

- Matter can be classified as **mixtures**, **pure substances**, **compounds**, and **elements**.
- Mixtures can be separated into pure substances by physical methods. Compounds can be decomposed by chemical methods. Elements cannot be further simplified by chemical methods.

THE KEY IDEA: The periodic table organizes the elements into groups based on their chemical properties.

The Periodic Table

During the 18th and 19th centuries, chemists began to isolate the elements and study their chemical properties. Some observant scientists noticed similarities in the chemical behavior of the elements. The major breakthrough came from a

(a)

ОПЫТЪ СИСТЕМЫ ЭЛЕМЕНТОВЪ.

ОСНОВАННОЙ НА ИХЪ АТОМНОМЪ ВѢСѢ И ХИМИЧЕСКОМЪ СХОДСТВѢ.

		Ti = 50	Zr = 90	? = 180.
		V = 51	Nb = 94	Ta = 182.
		Cr = 52	Mo = 96	W = 186.
		Mn = 55	Rh = 104,4	Pt = 197,1.
		Fe = 56	Rn = 104,4	Ir = 198.
		Ni = Co = 59	Pl = 106,6	O = 199.
H = 1		Cu = 63,4	Ag = 108	Hg = 200.
	Be = 9,4 Mg = 24	Zn = 65,2	Cd = 112	
	B = 11 Al = 27,4	? = 68	Ur = 116	Au = 197?
	C = 12 Si = 28	? = 70	Sn = 118	
	N = 14 P = 31	As = 75	Sb = 122	Bi = 210?
	O = 16 S = 32	Se = 79,4	Te = 128?	
	F = 19 Cl = 35,6	Br = 80	I = 127	
Li = 7 Na = 23	K = 39	Rb = 85,4	Cs = 133	Tl = 204.
	Ca = 40	Sr = 87,6	Ba = 137	Pb = 207.
	? = 45	Ce = 92		
	?Er = 56	La = 94		
	?Yt = 60	Di = 95		
	?In = 75,6	Th = 118?		

Д. Менделѣевъ

(b)

FIGURE 2.8 **Mendeleev and the periodic table.** (a) A photograph of Dimitri Mendeleev, who first presented his periodic organization of the chemical elements in 1869. (b) Mendeleev's periodic table listed the elements in order of increasing atomic weight and organized them according to the periodic nature of their chemical properties. He also noted where elements appeared to be missing (indicated by ? in the table), and he predicted the chemical properties of the undiscovered elements.

Russian chemist named Dimitri Mendeleev (1834–1907; see Figure 2.8(a)). In 1869, Mendeleev made a card for each element and arranged the cards in order of the increasing weights of the atoms. When examining the compounds formed by these elements, he detected a striking trend—the elements appeared to exhibit a *periodicity* in their chemical properties. In other words, the chemical properties of the elements tended to repeat themselves in a regular pattern.

For example, Mendeleev put the card for sodium next to the one for lithium because both elements are soft metals that react violently with water to produce hydrogen gas. He emphasized not only the *types of reactions* shared by different elements, but also the numerical *ratios of the elements* that combine to form compounds. For example, sodium and lithium combine with oxygen to form compounds that have the ratio of two metal atoms to one oxygen atom: Na_2O and Li_2O. By contrast, magnesium and calcium form compounds that have the ratio of one metal atom to one oxygen atom: MgO and CaO. This observation suggests that Na and Li belong in one group of elements, whereas Mg and Ca belong in another. Although samples of carbon and silicon look totally different, Mendeleev put them into the same group because each element reacts with two oxygen atoms to form CO_2 and SiO_2, respectively.

Using this approach, Mendeleev created the first **periodic table**, which is the fundamental organizing principle in chemistry (Figure 2.8(b)). Although Mendeleev was not the first chemist to detect patterns among elements, he was the first to identify gaps in his table where no element was yet known and to predict the chemical properties of these missing elements. When the first of these missing elements—germanium (Ge)—was discovered in 1886, its observed behavior closely matched Mendeleev's prediction, thus convincing chemists that his table was scientifically valid.

periodic table An organized arrangement of the elements based on their physical and chemical properties.

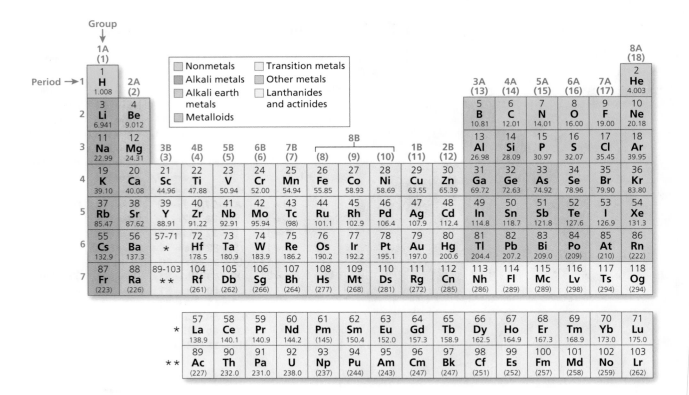

FIGURE 2.9 A modern periodic table of the elements. The vertical columns are groups of elements with similar chemical properties.

▪ **group** A column in the periodic table that contains elements with similar chemical properties.

▪ **period** A row in the periodic table.

▪ **noble gas** An unreactive (or mostly unreactive) element in Group 8A of the periodic table.

Figure 2.9 shows the modern periodic table, which contains 118 elements. Scientists have identified 94 naturally occurring elements, although some of them exist on Earth in only miniscule amounts. Using high-energy reactions, researchers have artificially created another 24 elements; four new elements were officially confirmed in January 2016. Instead of atomic weight, the modern table is based on a property called the *atomic number*, which we explain later in this chapter. The color-coding in the figure indicates that most elements are *metals*, which are substances that conduct electricity and can be hammered into different shapes. A much smaller number of elements, 17 in total, are classified as nonmetals. Significantly, the nonmetals contain the elements that are used most frequently to make biological molecules: hydrogen (H), carbon (C), oxygen (O), nitrogen (N), phosphorus (P), and sulfur (S).

The columns in the periodic table are called **groups**. *Elements in the group have similar chemical properties.* The *main-group elements* include the two groups on the left side (Groups 1A, 2A) plus the six groups on the right side (Groups 3A to 8A). The rows in the table are called **periods**. The first three periods contain only main-group elements. The fourth period begins to incorporate the so-called *transition elements*, which include familiar metals such as iron and copper.

Figure 2.10 compares elements from three groups in the periodic table. Note the similarities and differences among the elements within each group. For example, the elements in Group 4A are all solids although their chemical properties change as we proceed down the group. Their relative ability to conduct electricity illustrates this change: Carbon (C) conducts poorly, silicon (Si) is a semiconductor that is used in the electronic industry, and lead (Pb) is a fully conducting metal. In Group 6A, the first element—oxygen (O)— exists as a gas, whereas sulfur (S) appears in nature as a yellow solid. At first glance, these two elements seem to have little in common. However, their similarity is revealed by their chemical reactions because oxygen and sulfur both form compounds with two atoms of hydrogen (H_2O and H_2S). Finally, the elements in Group 8A are called the **noble gases**. Named after the noble class in European society, these gases either do not react with other atoms or do so only under extreme conditions.

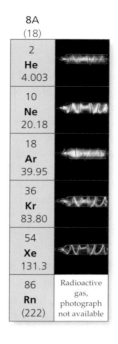

FIGURE 2.10 **Three groups in the periodic table.** A comparison of the elements in Groups 4A, 6A, and 8A of the periodic table.

Why do elements exhibit this periodicity in their chemical properties? The organization of the periodic table is based on the *atomic structure* of each element. The next section explores the relationship between atomic structure and chemical properties.

WORKED EXAMPLE 2.3

Question: The elements in the second period of the periodic table form chemical bonds with hydrogen atoms according to the following pattern:

$$CH_4 \quad NH_3 \quad H_2O \quad HF$$

What do you predict as the chemical formula of the compound produced when silicon (Si) forms a compound with hydrogen atoms?

Answer: The periodic table tells us that silicon belongs to Group 4A, as does carbon. Consequently, silicon and carbon have similar chemical properties. Therefore, we predict that the chemical formula for the silicon compound is SiH_4.

TRY IT YOURSELF 2.3

Question: What do you predict as the chemical formula when phosphorus (P) forms a compound with hydrogen atoms?

PRACTICE EXERCISE 2.2

You are presented with an unknown chemical element, which we call X. You react X with hydrogen and observe that it forms a compound with the chemical formula H_2X. To what group in the periodic table does X belong?

Core Concepts

- The **periodic table** organizes the chemical elements according to their physical and chemical properties.
- Elements in the same **group** of the periodic table exhibit similar chemical behavior.

2.3
Atomic Structure

Learning Objective:
Describe the structure of atoms in terms of their subatomic particles.

■ **atom** The fundamental component of matter.

■ **scanning tunneling microscope (STM)** A microscope that uses a moving metal probe to monitor the landscape of atoms arranged on a surface.

◎⊶ THE KEY IDEA: Scientific notation and scientific units are used to describe numerical quantities.

■ **scientific notation** The expression of a quantity as a number that is multiplied by 10 raised to a power.

■ **scientific unit** A unit of measurement that incorporates a power of 10.

Modern chemistry is based on the *atomic theory of matter*, which states that all matter is composed of discrete units called **atoms** (the term derives from the Greek word for "indivisible"). The chemical properties of an element arise from the characteristics of its constituent atoms. These atoms are so tiny that it is hard to imagine anything so small. The first scientists who studied atoms could not observe them directly, so their existence and structure had to be inferred from experiments and theoretical calculations. Today, it is possible to produce images of individual atoms using a **scanning tunneling microscope (STM)**. The STM uses a moving metal probe to construct the landscape of atoms on a surface. Figure 2.11 provides an STM image of a surface of silicon atoms, in which each "bump" corresponds to an individual atom. Given the tiny dimensions of atoms, we begin this section by explaining the methods that scientists use to describe the realm of the very small.

Scientific Notation and Scientific Units

Most of the biological and chemical entities that we discuss in this text—cells, viruses, molecules, and atoms—are too small to be seen with the naked eye. Figure 2.12 compares the sizes of several objects, ranging from a human body to a single atom. Because the range of sizes is so large, we describe them using **scientific notation** and **scientific units**. We encourage you to review these topics by studying Appendix A.

In general, scientific notation expresses any quantity as a number between 1.00 and 9.99 (called the *coefficient*) that is multiplied by 10 raised to a power (called the *exponent*). The exponent is always a whole number and written as a superscript.

$$(1.00 - 9.99) \times 10^{\text{power}}$$

Exponent

Coefficient

For example, suppose that we want to write the diameter of a hydrogen atom using the standard size unit of meters (m). Using decimals, we write the approximate size as:

$$0.00000000012 \text{ m}$$

This number takes up a lot of space, and we can easily lose track of some zeros. It is much more convenient to write the number in scientific notation:

$$1.2 \times 10^{-10} \text{ m}$$

The negative exponent means that the power of 10 is a reciprocal—in other words, $10^{-10} = 1/10^{10}$. Because we are dividing by a very large number (10^{10}), the resulting number is very small.

Obviously, the meter is too large a unit for measuring a hydrogen atom. Consequently, we can define

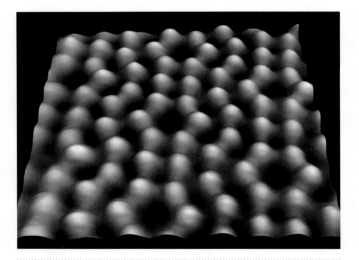

FIGURE 2.11 Visualizing atoms with the scanning tunneling microscope (STM). A surface of silicon atoms, visualized using an STM.

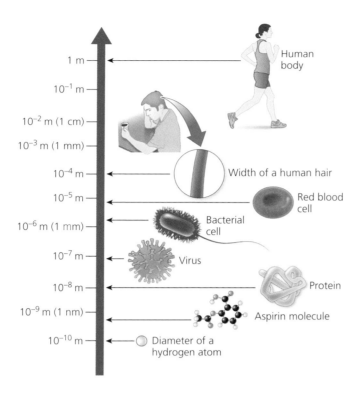

FIGURE 2.12 Size scale. A comparison of sizes of objects using a scale in powers of 10.

another unit that is more appropriate for this size scale. A *picometer* (pm) is one trillionth of a meter:

$$1 \text{ pm} = 10^{-12} \text{ m}$$

In effect, a *scientific unit incorporates a power of 10 into its definition*. We can convert between meters and picometers by using a **conversion factor**. In general, *a conversion factor is a ratio of two quantities that is equal to one*. Two conversion factors are possible from the definition of a picometer:

conversion factor A ratio of two different units that is equal to one.

$$\frac{1 \text{ pm}}{10^{-12} \text{ m}} = 1 \qquad \text{and} \qquad \frac{10^{-12} \text{ m}}{1 \text{ pm}} = 1$$

Which of these conversion factors should we use? In general, *select the conversion factor that cancels the given unit (m) and introduces the new unit (pm)*. In other words, the given units are on the *bottom* of the ratio (the denominator), where they cancel, and the units for the answer are on the *top* of the ratio (the numerator). The conversion calculation is provided below, with the two units highlighted in different colors:

new unit (pm)

$$\text{diameter of an H atom} = 1.2 \times 10^{-10} \text{ m} \times \frac{1 \text{ pm}}{10^{-12} \text{ m}} = 1.2 \times 10^{-10} \times 10^{12} \text{ pm} = 1.2 \times 10^{2} \text{ pm}$$

given unit (m)

The unit conversion tells us that the diameter of a hydrogen atom is 1.2×10^{2} pm, or 120 pm. This description is more convenient and informative than measuring the atoms in units of meters.

PRACTICE EXERCISE 2.3

Perform the following unit conversions:

(a) A red blood cell has an approximate diameter of 8×10^{-6} m. Use a conversion factor to convert this size into units of micrometers (μm), where 1 μm $= 10^{-6}$ m.

(b) A cold virus has an approximate diameter of 3×10^{-8} m. Use a conversion factor to convert this size into units of nanometers, where 1 nm $= 10^{-9}$ m.

Core Concepts

■ Atoms are so small that we must describe them using **scientific notation** and **scientific units**. The diameter of a hydrogen atom is 1.2×10^{-10} m, which can be written as 120 pm.

The Composition of Atoms

*Atoms are composed of three kinds of subatomic particles: **protons**, **neutrons**, and **electrons**.* Figure 2.13 illustrates the arrangement of these particles in an atom. Protons and neutrons cluster in the **nucleus**, a small core at the center of the atom that contains almost all the atom's mass. Electrons surround the nucleus and occupy a volume of space that defines the atom's size.

One fundamental property of particles is their *mass*. A proton's mass, written as m_p, is incredibly small. Expressed in the standard mass unit of grams (g):

$$m_p = 0.00000000000000000000000167 \text{ g}$$

We can rewrite this mass more compactly using scientific notation:

$$m_p = 1.67 \times 10^{-24} \text{ g}$$

The mass of a neutron is almost exactly the same as that of a proton. By contrast, the mass of an electron (m_e) is much smaller—note the difference in the power of 10:

$$m_e = 9.11 \times 10^{-28} \text{ g}$$

THE KEY IDEA: Atoms are composed of protons, neutrons, and electrons.

proton A positively charged subatomic particle located in the nucleus of an atom.

neutron An electrically neutral subatomic particle located in the nucleus of atom.

electron A negatively charged subatomic particle that surrounds the nucleus.

nucleus A small region at the center of the atom that contains almost all of the atom's mass.

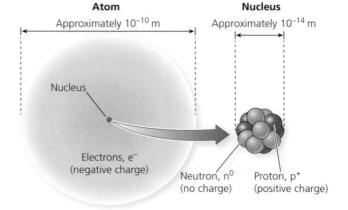

Atom
Approximately 10^{-10} m

Nucleus
Approximately 10^{-14} m

Nucleus

Electrons, e⁻
(negative charge)

Neutron, n⁰
(no charge)

Proton, p⁺
(positive charge)

FIGURE 2.13 Atomic structure. An atom is composed of protons (p⁺), neutrons (n⁰), and electrons (e⁻). The protons and neutrons make up the dense nucleus, which is surrounded by the electrons. This diagram of the atom is not to scale.

TABLE 2.2	Relative Mass and Charge for the Proton, Neutron, and Electron	
PARTICLE	MASS (AMU)	CHARGE
Proton (p$^+$)	1.0	+1
Neutron (n^0)	1.0	0
Electron (e$^-$)	0.00055	−1

The mass of an electron is slightly more than 0.05% of the mass of a proton. For this reason, the electron makes very little contribution to the mass of an atom. In general, an atom's mass arises from the combined mass of the protons and neutrons in its nucleus. The total mass of the electrons becomes perceptible only in the very largest atoms, such as uranium.

Because the masses of subatomic particles are so small, we measure them using a quantity called the *atomic mass unit (amu)*. Protons and neutrons have a mass of approximately 1.0 amu, whereas the mass of an electron is only 0.00055 amu. More than 1800 electrons are required to equal the mass of a single proton or neutron.

Particles can also have *electrical charges*, which are classified as positive or negative. Charges of the same type repel each other, and charges of the opposite type attract each other. We use *relative charge* to characterize the charges on the subatomic particles. A proton has a *positive* charge of +1, and an electron has a *negative* charge of −1. As its name suggests, a neutron is *electrically neutral*, which means that it has no charge.

Table 2.2 compares the relative properties of mass and charge for the proton (p$^+$), neutron (n^0), and electron (e$^-$). Because the nucleus includes protons and neutrons, it has a net positive charge. Atoms are electrically neutral, so the positive charge of the nucleus must be counterbalanced by the negative charge of the electrons. For any neutral atom, *the number of protons within the nucleus equals the number of electrons outside the nucleus*.

Core Concepts

- Atoms are composed of three subatomic particles: **protons, neutrons**, and **electrons**.
- Protons and neutrons constitute the dense **nucleus** of the atom, whereas electrons are located outside the nucleus.
- A proton has a mass of 1.0 amu and an electrical charge of +1. A neutron has a mass of 1.0 and no electrical charge. An electron has a much smaller mass (0.00055 amu) and an electrical charge of −1.

Characterizing Atoms

How can we distinguish an atom of carbon from an atom of oxygen or hydrogen? We identify and characterize atoms based on their subatomic particles. The **atomic number (Z)** is the number of protons in the nucleus. *Each element has a unique atomic number that defines its chemical identity.* Because atoms are electrically neutral, the atomic number also specifies the number of electrons. As we will learn in the next section, the electrons in an atom determines the atom's chemical properties—how it reacts (or doesn't react) with other atoms. The **mass number (A)** equals the sum of the protons and neutrons within the nucleus. The mass of the electrons is so small that it can be neglected in most cases. To calculate the number of neutrons in the nucleus, subtract the atomic number from the mass number: *number of neutrons = A − Z*.

Figure 2.14 illustrates the atoms of three different elements—hydrogen, carbon, and uranium. Hydrogen (Z = 1) is the simplest possible atom, with a

THE KEY IDEA: An atom is characterized by its atomic number and mass number.

atomic number (Z) The number of protons in the nucleus of an atom.

mass number (A) The sum of the protons and neutrons in the nucleus of an atom.

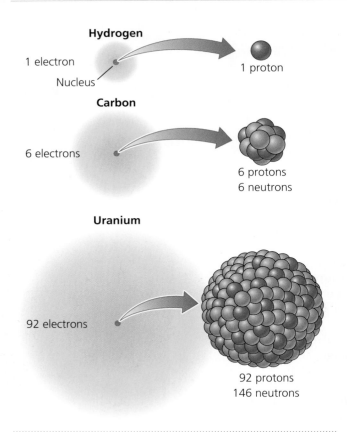

FIGURE 2.14 Three types of atom. Atomic structures for the chemical elements hydrogen, carbon, and uranium. These atoms differ in their number of protons, neutrons, and electrons.

single proton for its nucleus (it does not have a neutron) and a single electron. The carbon atom (Z = 6) is larger; it has 6 protons and 6 neutrons in its nucleus (A = 12) plus 6 electrons. The third element, uranium (Z = 92), is much larger, with a nucleus containing 92 protons and 146 neutrons (A = 238) plus 92 electrons.

The atomic number, mass number, and chemical identity of an atom can be summarized in a convenient shorthand called an **atomic symbol**. As shown in Figure 2.15(a), an atomic symbol has three components:

- The letter(s) denoting the chemical element (X)
- A subscript for the atomic number (Z)
- A superscript for the mass number (A)

Figure 2.15(b) illustrates the use of atomic symbols for the three elements in Figure 2.14.

The number of protons or electrons in an atom cannot change without changing the chemical identity of the atom. However, the number of neutrons in the nucleus has little influence on the atom's chemical properties. It is possible for atoms of the same element to have different mass numbers, corresponding to different numbers of neutrons. These variants of the same element are called **isotopes**—they are *atoms with the same atomic number but different mass numbers*. The name derives from *iso* (same) and *topos* (place), because isotopes occupy the same place in the periodic table.

For example, the nucleus of a carbon atom always has 6 protons (Z = 6), but it can have 6, 7, or 8 neutrons. This variability results in *three carbon isotopes* with mass numbers of 12, 13, and 14, respectively. Because the atomic number uniquely specifies a particular element, isotopes are often described by their mass numbers only—for example, carbon-12 or carbon-14.

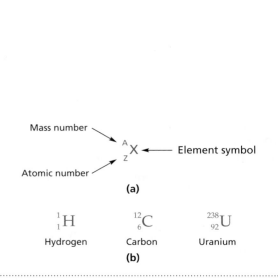

(a)

$${}^{1}_{1}\text{H} \qquad {}^{12}_{6}\text{C} \qquad {}^{238}_{92}\text{U}$$

Hydrogen   Carbon   Uranium

(b)

FIGURE 2.15 Atomic symbols. (a) The atomic symbol provides a convenient shorthand for writing the atomic number, mass number, and chemical identity of an atom. (b) Three atomic symbols that describe the atoms depicted in Figure 2.14.

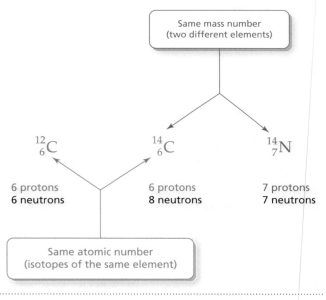

FIGURE 2.16 Isotopes. The two isotopes of carbon have the same atomic number (Z = 6) but different mass numbers (A= 12 or A = 14). Carbon-14 and nitrogen-14 have the same mass number, but they are different elements because their atomic numbers are not the same.

Figure 2.16 uses atoms of carbon and nitrogen to illustrate the relationship between atomic number, mass number, and chemical identity. The two isotopes of carbon have the same atomic number ($Z = 6$), even though their mass numbers are different ($A = 12$ and $A = 14$). The nitrogen atom has the same mass number ($A = 14$) as one of the carbon isotopes, but its atomic number is different ($Z = 7$). To summarize, the two carbon atoms have similar chemical properties but different masses. In contrast, the nitrogen atom and one of the carbon atoms have the same mass but different chemical properties.

atomic symbol A notation for an atom that includes the element, the atomic number, and the mass number.

isotope Variants of an atom that have the same atomic number but different mass numbers (i.e., different numbers of neutrons).

WORKED EXAMPLE 2.4

Question: The atomic symbols for two isotopes of helium are as shown. How many protons and neutrons are contained within each nucleus?

$$^3_2\text{He} \qquad ^4_2\text{He}$$

Answer: Because they are the same element, each isotope of helium has the same atomic number (Z), indicating the presence of 2 protons in the nucleus. The mass number (A) of the lighter isotope is 3, and so the number of neutrons is (A – Z) = (3 – 2) = 1. For the heavier isotope, with mass number 4, the number of neutrons is (4 – 2) = 2.

TRY IT YOURSELF 2.4

Question: How many protons and neutrons exist within the following isotopes: oxygen-16 and oxygen-18? Use the periodic table as a guide.

PRACTICE EXERCISE 2.4

Four atomic isotopes are provided in the accompanying table. For each isotope, complete the table by adding the atomic symbol, the number of protons, and the number of neutrons.

ISOTOPE	ATOMIC SYMBOL	NUMBER OF PROTONS	NUMBER OF NEUTRONS
uranium-235			
iron-56			
sulfur-32			
krypton-86			

Core Concepts

- An **atomic symbol** $\left(^A_Z\text{X} \right)$ contains three components: the element symbol, the **atomic number (Z)**, and the **mass number (A)**. The atomic number equals the number of protons, and the mass number equals the total number of protons plus neutrons.
- The identity of an element is defined by its atomic number.
- **Isotopes** are atomic variants of the same element. They have the same atomic number but different mass numbers (arising from varying numbers of neutrons).

relative atomic mass   A unitless number that compares the mass of one atom of the element to 1 amu.

atomic mass unit (amu)   The unit used to measure the mass of subatomic particles; 1 amu is defined as 1/12 the mass of a ^{12}C atom.

atomic mass   The weighted average of all the naturally occurring isotopes of an element, based on their mass and relative abundance.

Relative Atomic Mass

We can specify the mass of an element by using the **relative atomic mass**. This mass has no units because it compares the mass of one atom of the element to the reference value of one **atomic mass unit** (1 amu).

$$\text{relative atomic mass} = \frac{\text{mass of one atom of the element (amu)}}{1 \text{ amu}}$$

It makes conceptual sense to use the lightest atom—hydrogen—as the basis for the atomic mass unit. However, it is difficult to measure such a light mass. Consequently, modern chemistry bases its mass standard on the carbon-12 isotope (^{12}C), which can be measured with high precision. *The atomic mass unit (1 amu) is defined as 1/12 of the mass of a ^{12}C atom.* The approximate value of the atomic mass unit is:

$$1 \text{ amu} = 1.661 \times 10^{-24} \text{g}$$

The **atomic mass** of any atom is numerically equal to its relative atomic mass, but with the addition of amu as the unit of measurement. By this definition, the atomic mass of the carbon-12 isotope is exactly 12.00 amu. We should note the distinction between *mass number* (a whole number of nuclear particles with no units) and the *atomic mass* (a decimal number with units of amu). For example, the phosphorus-31 isotope (^{31}P) has a mass number of 31 (15 protons plus 16 neutrons) and has an atomic mass of 30.97 amu.

In the periodic table, each element is associated with two numbers. Figure 2.17 shows these numbers for carbon. The number above the element symbol is the atomic number. It is always a whole number because it counts the number of protons in the nucleus. The number below the element symbol is the relative atomic mass of the element, written as a decimal using four digits. For example, the relative atomic mass of carbon is 12.01, corresponding to an atomic mass of 12.01 amu. Why is this number slightly higher than the mass of the carbon-12 isotope discussed above?

Most natural samples of an element contain a *mixture of isotopes* in a ratio that is called the *natural abundance*. For example, the mass of a carbon sample contains contributions from two stable isotopes. Almost 99% of the atoms are ^{12}C, and slightly more than 1% are ^{13}C. The amount of ^{14}C is so tiny that it makes a negligible contribution to the mass and can be ignored. Worked Example 2.5 shows how to calculate a *weighted average* of carbon's atomic mass by multiplying the mass of each isotope by its relative abundance. The small influence of the heavier ^{13}C isotope causes the atomic mass of carbon to be 12.01 amu, which is slightly larger than the mass of 12.00 amu for the ^{12}C isotope alone.

The use of "average" atomic masses in the periodic table makes practical sense because we typically deal with gram-sized quantities of chemical samples in

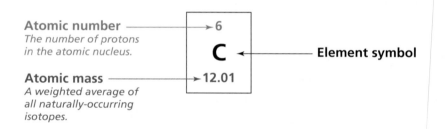

Atomic number
The number of protons in the atomic nucleus.

Atomic mass
A weighted average of all naturally-occurring isotopes.

6

C   Element symbol

12.01

FIGURE 2.17 Characterizing elements. The atomic number and relative atomic mass of carbon as presented in the periodic table.

the laboratory. These samples contain a mixture of different isotopes that reflect their natural abundance. Importantly, the value of 12.01 represents the *average* atomic mass for carbon and not the mass of any *particular* carbon atom. In a similar manner, the U.S. Census reported that the average family size per household is now roughly 2.5 persons, but we don't expect to find half a person in anyone's home.

WORKED EXAMPLE 2.5

Question: Calculate the relative atomic mass of carbon in the periodic table using the isotope data given here.

ISOTOPE	RELATIVE ATOMIC MASS	NATURAL ABUNDANCE
^{12}C	12.00 amu	98.89%
^{13}C	13.00 amu	1.110%

Note: *As stated above, we do not include the ^{14}C isotope because its natural abundance is too small to affect the mass calculation.*

Answer: We calculate the relative atomic mass as a weighted average by multiplying each isotope mass by its natural abundance. This calculation is easier if we use decimals rather than percentage values. For example, 98.89% = 98.89/100 = 0.9889.

$$\text{relative atomic mass} = (12.00 \times 0.9889) + (13.00 \times 0.1110)$$
$$= 11.87 + 0.1443$$
$$= 12.01$$

The calculation gives a relative atomic mass of 12.01, which is the value provided in the periodic table.

TRY IT YOURSELF 2.5

Question: Chlorine (Cl) has two stable isotopes, which are described in the following table:

ISOTOPE	RELATIVE ATOMIC MASS	NATURAL ABUNDANCE
^{35}Cl	34.97	75.78%
^{37}Cl	36.97	24.22%

(a) *Without performing a calculation*, deduce whether the relative atomic mass of naturally occurring chlorine will be closer to the mass of the ^{35}Cl isotope or to that of the ^{37}Cl isotope. Explain how you made your deduction. (b) Use the masses and natural abundances of the chlorine isotopes to calculate the relative atomic mass. Did your calculation match your deduction in part (a)?

PRACTICE EXERCISE 2.5

(a) Use the periodic table to find the relative atomic mass of boron (B). (b) Boron has two stable isotopes. One isotope (^{10}B) has a relative atomic mass of 11.01 and a natural abundance of 80.22%. The second isotope of boron has a natural abundance of 19.78%. Calculate the relative atomic mass of the second isotope.

SCIENCE IN ACTION

Measuring the Mass of Atoms

How do we know the mass of atoms? In the 19th and early 20th centuries, chemists had to determine these values indirectly by carefully measuring the mass of highly purified compounds and determining how they react with other compounds of known mass. This slow and laborious work was replaced by the invention of the *mass spectrometer* (Figure 1), which allows chemists to measure atomic masses directly.

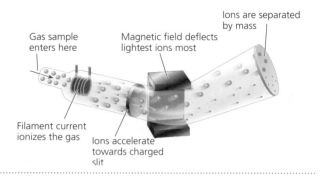

FIGURE 1 A schematic diagram of a simple mass spectrometer. The injected gas sample is ionized, and the ions are passed through a charged slit. The ions are deflected by a magnet, and the amount of deflection depends on the mass of the ion.

We will illustrate the operation of the mass spectrometer by using a sample of neon gas. After the gas sample is injected, the atoms are converted into ions by stripping off an electron; this process is called *ionization*. Because the mass of the electron is so small, ionization does not affect the measurement of the atomic mass.

$$\text{Ne} \xrightarrow{\text{Ionization}} \text{Ne}^+ + \text{e}^-$$

Atom   Ion

The ions are then accelerated toward a charged slit to reach a uniform velocity. After passing through the slit, the ions are subjected to the influence of a magnet, which causes them to move in a circle. If the velocity and charge remain the same, *then the radius of the circle depends on the mass of the ion*. This step separates the mixture of isotopes in the gas sample because the lighter ions are deflected more than the heavier ions. A detector senses the mass of each type of ion and its relative number.

The data output is called a *mass spectrum*, and the data for neon are shown in Figure 2. The locations of the peaks on the horizontal axis indicate the mass of each isotope in the gas sample,

and the height of each peak measures its relative abundance. This mass spectrum shows the two most abundant isotopes of neon: ^{20}Ne and ^{22}Ne.

Suppose that the mass spectrometer can measure the atomic mass and the natural abundance to one decimal place. The data from the mass spectrum are summarized in the following table:

ISOTOPE	RELATIVE ATOMIC MASS	NATURAL ABUNDANCE
^{20}Ne	20.0	91.0 %
^{22}Ne	22.0	9.0 %

Based on these data, we can calculate the atomic mass of neon as the weighted average of the two isotopes.

$$\text{atomic mass of Ne} = (0.910 \times 20.0\ \text{amu}) + (0.090 \times 22.0\ \text{amu}) = 20.2\ \text{amu}$$

Using a more sensitive mass spectrometer allows scientists to detect additional rare isotopes. In fact, neon has another rare isotope, ^{21}Ne, with a natural abundance of only 0.27%. With these improved data, scientists have measured the atomic mass of neon precisely as 20.1797 amu. Similar methods have been employed to determine the atomic masses of all of the elements in the periodic table. These masses closely correspond to the older values obtained by chemical measurements.

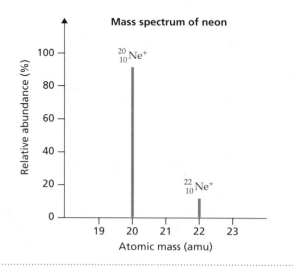

FIGURE 2 The mass spectrum of neon. The mass spectrum shows the two most abundant isotopes and their relative abundance.

....
CONCEPT QUESTION 2.2
....

Oxygen atoms exist as three stable isotopes: ^{16}O, ^{17}O, and ^{18}O. The periodic table gives the atomic mass of oxygen as 16.00, which appears to ignore the contributions from ^{17}O and ^{18}O. Propose a hypothesis to explain this observation.

....

Core Concepts

■ The **atomic mass** of an element is the weighted average of all the isotopes that occur within a natural sample of the element.

....

2.4 Electrons in Atoms

A central goal in chemistry is to explain the chemical behaviors of the elements in terms of their constituent atoms. For example, why are the noble gases chemically unreactive? *The properties of an element arise from the configuration of electrons in its atoms.* However, describing the behaviors of electrons is challenging because they obey a different set of scientific rules than do the everyday objects we see around us.

Scientists commonly use *models* to make sense of the natural world. In this section, we present two models for describing electrons in atoms. The first is a simplified model that specifies the location of electrons within different "shells" that surround the nucleus. Despite this simplicity, the *shell model* explains many properties of atoms. We then explain a more sophisticated model based on the theory of *quantum mechanics*, which provides a more accurate representation of electron behavior. The shell model does contain some aspects of quantum mechanics, but it does not apply this theory to its fullest extent.

Learning Objective:
Relate the electron configuration of an atom to its chemical properties.

The Electron Shell Model

In our first model, the electrons within an atom are organized into energy levels called **electron shells**. These electron shells are **quantized**—in other words, they can have only discrete values. Electrons are stable when they reside within an electron shell, but they are not stable at any location *in between* these shells. Because the electron shells are discrete, they can be identified by a *quantum number* *n* where *n* = 1, 2, 3 and so on. As an analogy, the quantization of electrons shells is similar to the rungs of a ladder—you can stand comfortably on the first rung, the second rung, and so on, *but you cannot stand in between the rungs*.

Each electron shell has a maximum number of electrons that it can contain; Table 2.3 provides the capacity of the first three electron shells. The rules for the first two shells are straightforward, but the rule for the third shell is an approximation. The noble gas argon forms stable atoms with 8 electrons in the third electron shell. The atom is stable because the 8 electrons occupy the lowest energy positions within the shell. However, it is possible to add 10 more electrons to higher-energy positions, giving a total of 18 electrons in the third shell.

THE KEY IDEA: The chemical elements can be described by the configuration of electrons in quantized electron shells.

electron shell A quantized energy level where electrons are located.

quantized A characteristic of electron shells in which they can only have integer values.

TABLE 2.3 **Maximum Electron Capacity of Electron Shells (*n* = 1, 2, and 3).**

ELECTRON SHELL (*n*)	MAXIMUM ELECTRON CAPACITY
1	2
2	8
3	8*

*The rule for the third shell is an approximation.

Hydrogen (H)   Helium (He)
Z = 1   Z = 2

FIGURE 2.18 Electron configurations of hydrogen (H) and helium (He) atoms. The first electron shell has a maximum capacity of 2 electrons.

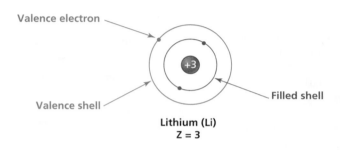

Valence electron

+3

Valence shell

Filled shell

Lithium (Li)
Z = 3

FIGURE 2.19 Electron configuration of a lithium atom. Lithium has a filled first electron shell plus a single valence electron in the second electron shell (the valence shell).

■ **electron configuration** The location of electrons within specific shells of an atom.

■ **valence shell** The outer electron shell of an atom.

■ **valence electrons** The electrons located in the outer shell; they determine the chemical properties of the atom.

The location of electrons within specific shells is called the **electron configuration** of an atom. We begin with the two simplest elements—hydrogen and helium (Figure 2.18). Hydrogen has one electron, located within the first electron shell. Helium has two electrons, both of which fit within the first electron shell. What do we know about the chemical properties of helium? It is a noble gas that does not react with other elements. We can explain this behavior of helium by its electron configuration: *Atoms with a filled electron shell are stable and unreactive.* By contrast, a hydrogen atom does *not* have a filled electron shell—the shell has the capacity to hold an additional electron, so it reacts with other atoms. As we will see in Chapter 3, hydrogen atoms can mimic the stable electron configuration of helium by sharing two electrons to form a chemical bond.

Let's move on to the next element: lithium (Z = 3). What is its electron configuration? The answer is shown in Figure 2.19. Two of lithium's electrons can occupy the first electron shell (n = 1). However, the third electron cannot fit there, so it must be placed in the second electron shell (n = 2). The second shell has a higher energy, and it is located farther from the nucleus. Because the first electron shell is filled, *the chemical properties of lithium are determined by the single electron in the second shell.* The outer electron shell of an atom is called the **valence shell**, and the electrons it contains are called **valence electrons**. The word "valence" means "power"; its usage here reflects the capacity of the outer electrons to participate in chemical bonding with other atoms. The valence electrons determine the chemical properties of an atom, as we shall see in Chapter 3.

We can "build up" the atoms for the second row of the periodic table by successively adding one more electron for each increase in atomic number. To visualize this process, imagine a room with three rows of seats: two seats in the first row, eights seats in the second row, and another eight seats in the third row. As people enter the room, they are instructed to fill the seats starting at the front of the room. After the initial two people fill the first row of seats, following individuals have to start sitting in the second row. After that row is filled with eight people, the next ones to enter the room have to sit in the third row. A similar process happens when successive electrons occupy electron shells. Figure 2.20 provides the electron configurations for the elements lithium to neon. When we reach Z = 10, we encounter another noble gas—neon. We can explain neon's lack of reactivity by noticing it has eight valence electrons in the second electron shell, which fills the shell to its maximum capacity.

Elements in the third period of the periodic table, beginning with sodium, start to fill the third electron shell. Sodium has one valence electron, the same as lithium, so these two elements have similar chemical properties. Adding electrons to the third electron shell produces a sequence of eight elements, ending with the noble gas argon. The filling of electron shells explains the periodicity of the chemical elements that Mendeleev cleverly deduced without any knowledge of atomic structure.

PRACTICE EXERCISE 2.6

We can extend the principle of electron configurations to the fourth period of the periodic table. (a) Using Figure 2.20 as a guide, draw the electron configuration for potassium (symbol K, Z = 19). (b) Based on your answer to part (a), explain why potassium has similar chemical properties to lithium and sodium.

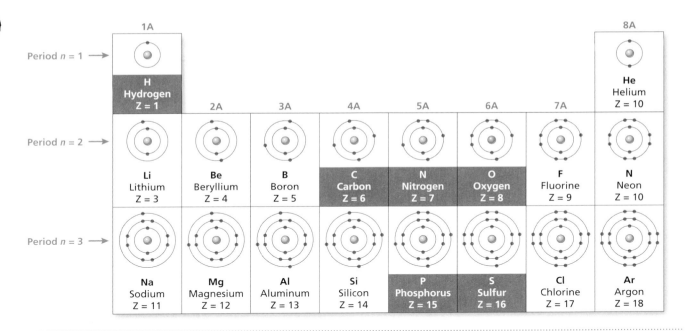

FIGURE 2.20 Electron configurations for atoms with atomic numbers Z = 1 through Z = 18. These atoms represent elements within the first three periods of the periodic table. Elements commonly found in biological molecules and pharmaceuticals are highlighted with red boxes.

Core Concepts

- In the electron shell model, the electrons in an atom are located within **quantized** electron shells. Each **electron shell** is identified by a *quantum number n = 1, 2, 3*, and so on. The maximum electron capacity of the first three shells is 2, 8, and 8, respectively.
- The **electron configuration** of an atom describes the location of its electrons within specific electron shells.
- Noble gases are chemically unreactive because they have filled electron shells. Helium has 2 electrons, which fill the first electron shell. Neon has a total of 10 electrons, with 2 electrons in the first shell and 8 electrons filling the second shell.
- The chemical properties of an element are determined by its **valence electrons**, which reside in the **valence shell** of the atom. Elements with the same number of valence electrons (e.g., lithium and sodium) have similar chemical properties and belong to the same group of the periodic table.

Forming Ions

Atoms are always electrically neutral—the number of positive charges from the protons in the nucleus is exactly balanced by the number of negative charges from the electrons. In nature, however, chemical elements often exist as *ions*. An **ion** is *formed when an atom or group of atoms gains or loses one or more electrons, thereby acquiring a net electrical charge.* What causes the formation of ions?

We can use the electron configuration of an atom to explain why it may form an ion. As we discussed in the previous section, atoms with filled outer electron shells are stable. However, most atoms lack a filled outer electron shell. One way in which atoms become more stable is to gain or lose electrons to achieve a filled valence shell. As an example, recall that a lithium atom (Z = 3) has one valence electron; its other two electrons are in a filled electron shell. If lithium can lose this outer electron, then it will have the stable electron configuration of helium. The result of this electron loss is the formation of a *lithium ion*, Li^+ (Figure 2.21). This ion has a net charge of +1 because it has three protons in its nucleus (+3)

THE KEY IDEA: Some atoms gain or lose one or more electrons to form ions.

ion An atom (or group of atoms) that has lost or gained one or more electrons and acquired an electrical charge.

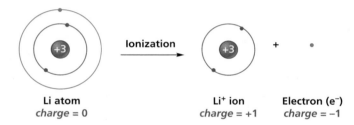

FIGURE 2.21 Ionization of lithium. A lithium atom (Li) loses an electron to form a lithium ion (Li⁺). The ion has the same electron configuration as helium.

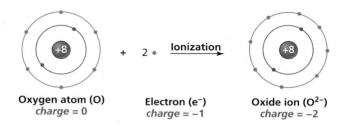

FIGURE 2.22 Ionization of oxygen. An oxygen atom (O) gains 2 electrons to form an oxide ion (O^{2-}). The ion has the same electron configuration as neon.

but only two electrons (–2). In addition, the Li⁺ ion is smaller than the Li atom because it has lost the electron in its outer shell. We can write this *ionization* process as:

$$Li \rightarrow Li^+ + e^-$$

Ions can also form by gaining electrons. For example, an oxygen atom has 6 electrons in its valence shell. By acquiring 2 electrons, it can achieve the stable electron configuration of the noble gas neon (Figure 2.22). The resulting ion, O^{2-}, is called an *oxide ion*. This ion has a net charge of –2 because it contains 8 protons (+8) and 10 electrons (–10). The ionization of oxygen is written as:

$$O + 2\,e^- \rightarrow O^{2-}$$

Main block elements on the left and right sides of the periodic table tend to form ions. For example, elements in Groups 1A and 2A lose electrons to form positively charged ions (e.g., Li⁺ and Ca^{2+}), whereas elements in Groups 6A and 7A gain electrons to form negatively charged ions (e.g., O^{2-} and Cl⁻). A carbon atom does not form ions easily because it has a half-filled valence shell of 4 electrons. Carbon has two options to form an ion with a filled valence shell: lose 4 electrons to become C^{4+} or gain 4 electrons to become C^{4-}. Both of these changes require large amounts of energy, so they occur only under extreme conditions.

The transition elements in the middle of the periodic table can also form ions. In the chapter introduction, we mentioned that the iron found within the heme groups of hemoglobin exists as an Fe^{2+} ion. However, the electron configuration of the transition elements is more complex, and their ions are not straightforward to predict.

WORKED EXAMPLE 2.6

Question: In the United States, fluoride is added to our water supply to build healthy teeth. Write the chemical equation for the ionization of fluorine atoms to form fluoride ions. What is the charge of the fluoride ion?

Answer: Fluorine (F) is in Group 7A of the periodic table, so a fluorine atom has seven valence electrons. The atom could obtain a stable electron configuration like neon (a noble gas) by gaining one additional electron. The equation for the ionization of fluorine is written as:

$$F + e^- \rightarrow F^-$$

The net charge of the fluoride ion is –1.

TRY IT YOURSELF 2.6

Question: Magnesium ions are found in cells, and they have several important biological functions. Write an equation for the ionization of a magnesium atom to form a magnesium ion. What is the charge of the magnesium ion?

PRACTICE EXERCISE 2.7

Four atoms are provided in the table below. For each atom, complete the table by adding the chemical formula for the ion that is formed by the atom. Also in the table, indicate the total number of protons and the total number of electrons that each ion contains.

ATOM	ION	NUMBER OF PROTONS	NUMBER OF ELECTRONS
Sulfur (S)			
Aluminum (Al)			
Iodine (I)			
Hydrogen (H)			

Core Concepts

■ Some atoms gain or lose one or more electrons to form a charged **ion** (e.g., Li^+, O^{2-}).
■ Ionization occurs because the ion has a more stable electron configuration that resembles a noble gas.

The Quantum Mechanical Model

The shell model of the atom enables us to explain many chemical properties of the elements. However, this model has several shortcomings. To accurately describe the properties of electrons in atoms, we need to employ a theory called **quantum mechanics**. This theory requires us to abandon our picture of electrons as tiny charged particles; instead, it describes electrons as having the properties of *waves*. Unlike a particle, which occupies a particular location, an electron wave spreads out over a region of space (Figure 2.23). It is no longer possible to accurately define the position of the electron. Instead we have to describe the *probability* of the electron being located within a particular region of space.

Applying the principles of quantum mechanics to electron waves provides a different perspective of electrons in atoms. We begin with the simplest example of a hydrogen atom with one electron. The spatial distribution of the electron is described mathematically by an **electron orbital**. When the hydrogen atom exists in its lowest energy state, the orbital has the shape of a sphere that indicates the *probability* of the electron being in a particular place. We can represent this probability as a "cloud" of **electron density**, as illustrated in Figure 2.24(a). Darker regions of the cloud indicate a higher electron density, which corresponds to a greater probability of finding the electron at these positions. When we begin considering atoms with more electrons, the description of the electron orbitals becomes more complex. Many orbitals are no longer spherical; instead, they are oriented along a particular direction in space, as shown in Figure 2.24(b).

Quantum mechanics provides the most complete and accurate model of the atom. Scientists can use this theory to calculate the location and energy of the electron shells that provide the foundation of our

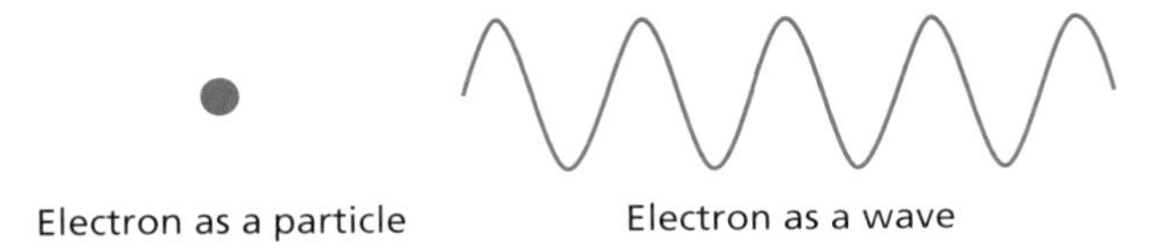

THE KEY IDEA: In the quantum mechanical model of the atom, electrons are described as waves instead of particles.

quantum mechanics A theory that describes electrons as waves.

electron orbital A quantum mechanical description of an electron's probable location.

electron density The probability of finding an electron at a specific location within an orbital.

Electron as a particle   Electron as a wave

FIGURE 2.23 Two descriptions of an electron. An electron can have the properties of a particle or a wave. Quantum mechanics describes the behavior of electrons in terms of their wave properties.

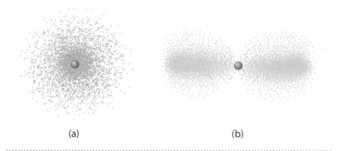

(a)   (b)

FIGURE 2.24 Electron orbitals. (a) The electron orbital for a hydrogen atom describes the probability of the single electron being located within a particular region of space. The distribution of electron density for a hydrogen atom has a spherical shape in three dimensions. (b) Some electron orbitals for other atoms are shaped like the number 8. In this case, the electron density is *not* spherical and has a specific orientation in space.

previous model. However, applying quantum mechanics to atoms and molecules is both mathematically and conceptually challenging. In most cases, it is sufficient to describe electrons as particles and to use the shell model of atoms, even though we must acknowledge the inaccuracy and limitations of such simplifications. In later chapters, we will encounter examples that require us to use quantum mechanics to explain certain features of molecular structure.

Core Concepts

- **Quantum mechanics** describes electrons as waves. Unlike a particle, a wave is spread out over space, and it does not have a well-defined location.
- The **electron orbital** describes the probability of an electron being located in a particular region of space. A higher **electron density** corresponds to a greater probability of finding the electron at a particular location.

2.5 Elements of Life and Death

Learning Objective:

Illustrate the beneficial and harmful roles of chemical elements in the human body.

🔑 THE KEY IDEA: Biological organisms require approximately 25 chemical elements in varying amounts.

Why are some chemical elements essential for life, whereas others are deadly? This section provides examples of elements that are beneficial to our health as well as some that are harmful.

The Chemical Elements of Life

The periodic table contains 118 chemical elements, but biological organisms (including humans) use only around 25 of them as the chemical basis of life. Figure 2.25 provides a "biological periodic table" that highlights these chemical elements. The six elements shown in orange are the nonmetal "building block" elements that are used to make biological molecules:

Hydrogen (H), carbon (C), nitrogen (N), oxygen (O), phosphorus (P), and sulfur (S)

As we will see in Chapter 3, these six elements readily form chemical bonds to create molecules.

Biological cells also contain elements that are classified as *minerals*, which is a general term for a nonliving substance found in nature. Examples of minerals include sodium, potassium, calcium, and chlorine. These elements readily lose their valence electrons to form ions. For example, flows of sodium and potassium ions are the basis for transmitting electrical signals in nerve cells, which affect bodily functions ranging from the movement of muscles to thought processes in the brain. Elements shown in blue are called *trace elements* because they occur

FIGURE 2.25 A periodic table showing the elements used in biological cells. Building block elements are shown in orange, minerals in green, and trace elements in blue.

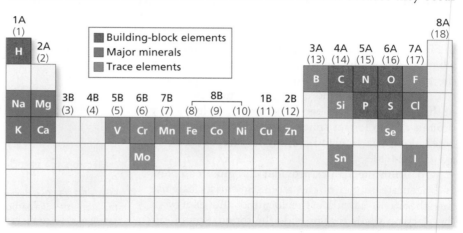

only in tiny amounts in the body, although they still have vital biological functions. For example, metals such as manganese, copper, and zinc play essential roles in the ability of enzymes to increase the rates of chemical reactions within the cell.

CONCEPT QUESTION 2.3

Iodine is one of the chemical elements that our body requires for good health. For this reason, iodine is added to some types of salt, which are labeled as "iodized."

Use the Internet to answer the following questions:
(a)   How and where is iodine used within the human body?
(b)   What are the health consequences of iodine deficiency?

Core Concepts

- Biological organisms, including humans, use approximately 25 elements as the chemical basis of life.
- Six elements form the atomic building blocks of most biological molecules: H, C, N, O, P, and S.

Why Do We Need Calcium?

Calcium is found in Group 2A of the periodic table, and its atoms have two valence electrons. A calcium atom loses these electrons to form a calcium ion, Ca^{2+}. The ionization reaction is written as:

$$Ca \rightarrow Ca^{2+} + 2\ e^-$$

Ca^{2+} ions play an essential role in forming the solid structures of bones and teeth, which account for most of the body's calcium. A smaller quantity of Ca^{2+} ions serves a variety of other biological purposes, including acting as a chemical signaling agent.

Bone consists of a network of fibers that include calcium, phosphorus, and a tough protein called *collagen*, which is also found in hair and nails. Bone is the hardest and most durable substance in the body, with a mechanical strength similar to that of steel. Unlike steel, however, bone structure is constantly built up and degraded throughout our lifetime. As we age, the bone structure dissolves away faster than it can be rebuilt, leading to a decrease in bone density. In serious cases, this process leads to a condition called *osteoporosis*, which means "porous bones." Figure 2.26 illustrates the structural difference between a normal hip bone and one affected by osteoporosis.

According to the U.S. National Institutes of Health, osteoporosis threatens 34 million Americans, primarily older women. This condition is characterized by reduced bone mass, which greatly increases the risk of fractures, sometimes from routine activities. Influences on osteoporosis include diet, exercise, genetic predisposition, and hormonal changes. One factor affecting risk for the disorder is the amount of calcium that is stored earlier in life. We can think of consuming calcium at a younger age as contributing to a "calcium retirement account" that we can draw upon later in life. The recommended daily intake of calcium is 1300 milligrams (1.3 g) between the ages of 9 and 18, and it should continue at 1000 milligrams (1.0 g) from ages 19 to 50.

THE KEY IDEA: Calcium is an essential component of human bones and teeth.

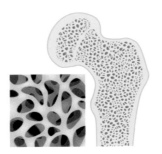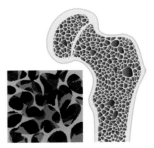

Healthy bone   Osteoporosis

FIGURE 2.26 Osteoporosis. As a person ages, the normal structure of a bone (left) can gradually lose its density and deteriorate into osteoporosis (right).

Core Concepts

- A calcium atom ionizes to form Ca^{2+} ions. These ions play an essential role in forming the solid structures of bones and teeth.
- Osteoporosis is a reduction of bone density that can occur when a person ages. The risk of developing osteoporosis can be reduced by consuming sufficient calcium at a young age.

Why Is Arsenic a Poison?

🔑 THE KEY IDEA: Arsenic is a poison because it replaces phosphorus, an essential element for life.

Arsenic is a potent poison. What makes it so toxic? We can answer this question by referring to the periodic table. Arsenic (As) belongs to the same chemical group as phosphorus (P), and the two elements have similar chemical properties. Phosphorus is an important element in cells because it combines with oxygen to make a *phosphate ion*, PO_4^{3-}. Phosphate is an example of a *polyatomic ion* because it contains more than one atom (*poly* means "many"). The phosphate ion is a component of many biological molecules. For example, three phosphate ions are contained in *adenosine triphosphate* (ATP), which provides an essential source of chemical energy for many cellular functions.

Like phosphorus, arsenic combines with oxygen to form an *arsenate ion*, AsO_4^{3-}. Figure 2.27 compares the two ions. Arsenate can replace phosphate in ATP, but the molecule then becomes unstable and breaks apart. As a result, the cells lose their primary source of energy, which causes them to malfunction or die. Arsenic also disrupts other cellular processes, and many of these effects arise from its chemical similarity to phosphorus.

Arsenic poisoning is a particularly serious problem in Bangladesh, where arsenic contaminates the water supply in shallow wells called tube wells (Figure 2.28(a)). The arsenic is a naturally occurring component of underground rock, which slowly leaches into the water supply. A person who ingests a large amount of arsenic from this contaminated water can suffer from arsenic poisoning, which produces painful sores on the hands and feet (Figure 2.28(b)). This problem is being addressed with simple kits that filter the water and remove the arsenic, although disposing of the trapped arsenic on a large scale remains a serious challenge.

Arsenic contamination is not just a concern in faraway countries. Potentially dangerous arsenic levels also occur in some sources of drinking water within the United States. Some arsenic is naturally present in rocks, and it can seep into groundwater. In other locations, contamination is caused by the use of arsenic-containing compounds to process metal ores. Another potential source of arsenic exposure is rice. In some southern states, rice is now grown in fields where cotton was formerly planted. Farmers treated these fields with arsenic to rid the cotton plants of insect pests. Because arsenic remains in soil for long periods of time, it can enter rice plants grown many years later.

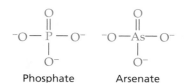

Phosphate   Arsenate

FIGURE 2.27 Phosphorus and arsenic have similar chemistry. Phosphate (PO_4^{3-}) and arsenate (AsO_4^{3-}) ions have similar formulas and structures. As discussed in Chapter 1, each straight line joining two atoms represents a chemical bond.

FIGURE 2.28 Arsenic can contaminate drinking water. (a) A child in Bangladesh drinks from a tube well. Tube wells are usually shallow, and arsenic accumulates in the wells after extended use. (b) One symptom of arsenic poisoning is development of sores on the hands and feet.

CHEMISTRY AND YOUR HEALTH

Why Is There Mercury in the Fish that We Eat?

Since ancient times, mercury has fascinated people because it is a liquid metal at room temperature (Figure 1). For centuries, scientists used mercury in thermometers (for measuring temperature) and barometers (for measuring air pressure). Today we recognize mercury as a dangerous environmental toxin. In particular, there is a growing concern about high levels of mercury in the fish we eat. Why is mercury poisonous, and how does it get into fish?

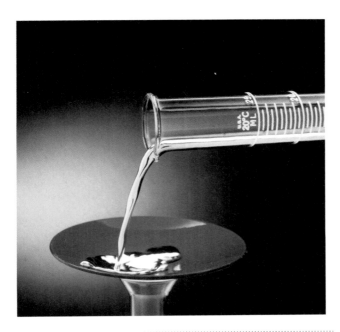

FIGURE 1 Liquid mercury. Mercury is an unusual metal because it is a liquid at room temperature. The original name for this element was "quicksilver."

Mercury is found in three forms: metal atoms (Hg) and two forms of ions in which the metal has lost either one electron (Hg^+) or two electrons (Hg^{2+}). The neutral form of mercury is relatively safe (it is used for dental fillings), but both mercury ions are poisonous. The Hg^+ ion can combine with a carbon-containing chemical group called a methyl group (discussed in Chapter 4) to form a very toxic compound called *methyl mercury*. This form of mercury poses the greatest threat to humans.

What makes methyl mercury so toxic? This compound has a strong chemical affinity for enzymes in cells that utilize the element *selenium* (Se), which is located in Group 6A of the periodic table below oxygen and sulfur. These enzymes play a vital role in

preventing and reversing chemical reactions that cause damage within the body's cells. Mercury inactivates the enzymes, which leads to an increase in these damaging reactions. The most pronounced effects of this damage occur in the brain, which is why mercury poisoning affects cognition and behavior. The effects are most severe in children, who can have developmental delays.

Some atmospheric mercury derives from sources such as forest fires that are not industrial in origin. Human emissions of mercury—called anthropogenic sources—are illustrated in Figure 2. The industrial source of much atmospheric mercury in the United States is burning coal in the power plants that supply approximately 40% of our electrical energy. Coal contains tiny amounts of mercury as an impurity. The quantity of mercury varies according to the geographical location of the coal deposit. Although the amounts are very small as a proportion of the coal, burning coal on a massive scale releases tons of mercury into the atmosphere. Another source of mercury arises from burning municipal waste that includes disposed materials such as batteries, fluorescent lights, and electrical switches.

Mercury released from burning coal or waste is dispersed as particles by winds, and it enters lakes and oceans remote from the site of combustion. Mercury in the ocean accumulates in fish, especially in large fish that feed on smaller ones and thus build up their mercury levels. Tuna has one of the highest levels of mercury content, and consumers are advised to limit the amount of tuna they eat.

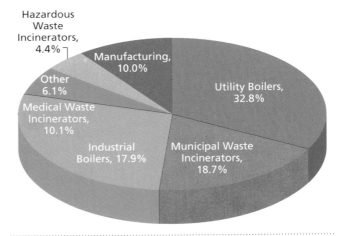

FIGURE 2 Anthropogenic sources of mercury in the United States. Coal-based power plants are the largest source of anthropogenic mercury emissions in the United States.

Core Concepts

- Arsenic has similar chemical properties as phosphorus, an essential element for life. Arsenic acts as a poison by replacing phosphorus in ATP, which causes the molecule to become unstable and deprives the cell of a vital source of energy.

CHAPTER 2
VISUAL SUMMARY

2.1 Why Is There Iron in My Cereal?

Learning Objective:

Illustrate the importance of iron for human health.

⊙ Iron is an essential component of **hemoglobin**, a protein that transports oxygen within red blood cells. Hemoglobin contains four ring-shaped heme groups with an Fe^{2+} ion at the center. Each Fe^{2+} ion binds an oxygen molecule for transport. A lack of dietary iron can lead to iron deficiency anemia.

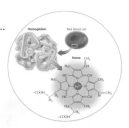

2.2 Mixtures, Compounds, and Elements

Learning Objective:

Explain the organization of elements in the periodic table.

⊙ **Physical properties** are either extensive or intensive. An **extensive property** (e.g., **mass**) depends on the amount of material, whereas an **intensive property** (e.g., **density**) is independent of the amount.

⊙ The **chemical properties** of a substance are revealed during a chemical reaction.

⊙ **Matter** can be classified as **mixtures**, **pure substances**, **compounds**, and **elements**.

⊙ Mixtures can be separated into pure substances by physical methods. Compounds can be decomposed by chemical methods. Elements cannot be further simplified by chemical methods.

⊙ The **periodic table** organizes the elements according to their physical and chemical properties.

⊙ Elements in the same **group** of the periodic table exhibit similar chemical behavior.

2.3 Atomic Structure

Learning Objective:

Describe the structure of atoms in terms of their subatomic particles.

⊙ **Atoms** are so small that we describe them using **scientific notation** and **scientific units**. The diameter of a hydrogen atom is 1.2×10^{-10} m, or 120 pm.

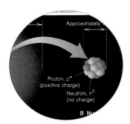

- Atoms are composed of **protons, neutrons**, and **electrons**.

- Protons and neutrons constitute the dense **nucleus** of the atom, whereas electrons are located outside the nucleus.

- An **atomic symbol** contains the element symbol, the **atomic number** (Z), and the **mass number** (A).

- The chemical identity of an element is defined by its atomic number.

- **Isotopes** of an element have the same atomic number but different mass numbers.

- The **atomic mass** of an element is the weighted average of all the isotopes that occur within a natural sample of the element.

2.4 Electrons in Atoms

Learning Objective:

Relate the electron configuration of an atom to its chemical properties.

- In the electron shell model, the electrons in an atom are located within **quantized electron shells**. The maximum electron capacity for the first three shells is 2, 8, and 8, respectively.

- Noble gases are chemically unreactive because they have filled electron shells.

- Elements with the same number of **valence electrons** have similar chemical properties.

- Some atoms gain or lose one or more electrons to form charged **ions** (e.g., Li^+, O^{2-}).

- Ionization occurs because the ion has a stable **electron configuration** that resembles a noble gas.

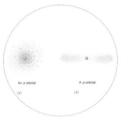

- **Quantum mechanics** describes electrons as waves.

- The **electron orbital** describes the probability of an electron being located in a particular region of space.

2.5   Elements of Life and Death

Learning Objective:

Illustrate the beneficial and harmful roles of chemical elements in the human body.

⊙ Biological organisms, including humans, use approximately 25 elements as the chemical basis of life.

⊙ Six elements form the atomic building blocks of most biological molecules: H, C, N, O, P, and S.

⊙ A calcium atom ionizes to form Ca^{2+} ions. These ions are used to form the solid structures of bones and teeth.

⊙ Osteoporosis is a reduction of bone density that occurs when a person ages.

⊙ Arsenic acts as a poison by replacing phosphorus in ATP, the energy currency of the cell, which causes the molecule to become unstable. Arsenic and phosphorus belong to the same group of the periodic table and have similar chemical properties.

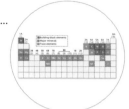

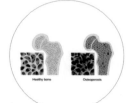

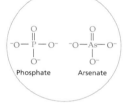

LEARNING RESOURCES

Reviewing Knowledge

1. What is the RDA for a nutrient in our diet?

2.1: Why Is There Iron in My Cereal?

2. What is the biological function of hemoglobin? What is the role of iron in this protein?
3. Describe the health consequences of iron deficiency. How can these be remedied?

2.2: Mixtures, Compounds, and Elements

4. What is the definition of *matter*?
5. A physical property can be intensive or extensive. Explain the difference between these two properties, and provide one example of each.
6. What condition is necessary to observe the chemical properties of a substance? How does this differ from the observation of a physical property?
7. What is the difference between a physical change and a chemical change?
8. Define the following terms used to describe matter: *mixture, pure substance, compound, element.*
9. Which methods are used to separate the components of a mixture? Provide one example.

10. Explain how the decomposition of water by an electrical current illustrates the difference between compounds and elements.
11. What principle is used to organize elements into groups in the modern periodic table?
12. In their elemental forms, oxygen exists as a colorless gas and sulfur exists as a yellow solid. Despite this difference in appearance, both elements are placed within the same group of the periodic table. Why?

2.3: Atomic Structure

13. Compare the properties of the proton, neutron, and electron in terms of (a) mass and (b) electrical charge.
14. Define the atomic number and mass number of an atom. Which number defines the identity of a chemical element?
15. What are isotopes? Provide some examples.
16. In the periodic table, the relative atomic mass is specified as a decimal (e.g., 12.01 for carbon). How is this mass determined?

2.4: Electrons in Atoms

17. What are the maximum electron capacities of the first, second, and third electron shells?

18. Define the terms *valence shell* and *valence electrons*.
19. Helium and neon are both unreactive noble gases. Explain this chemical property in terms of the electron configurations of these atoms.
20. Explain why some chemical elements tend to form ions. Provide one example.
21. How does the quantum mechanical model of the atom differ from the electron shell model?

2.5: Elements of Life and Death

22. What are the six chemical elements that are commonly used to build biological molecules?
23. What is the major role of calcium in the human body?
24. Why is arsenic such a potent poison?

Developing Skills

25. Humans may travel to Mars in future decades. Mars is smaller than Earth, and the force of gravity on Mars is only 38% of the Earth's gravity. Suppose that a 70-kg astronaut embarks on a mission to Mars. (a) What is the astronaut's weight on Earth (in pounds)? On the Earth's surface, a mass of 1.0 kg has a weight of 2.2 lb. (b) What will be the astronaut's mass (in kg) and weight (in pounds) when she stands on the surface of Mars?
26. The *diameter* of a hemoglobin protein is approximately 0.000000005 m. (a) Write this size in scientific notation. (b) Use a conversion factor to convert the diameter of hemoglobin to nanometers (nm), where $1 \text{ nm} = 10^{-9} \text{ m}$.
27. One type of cell in the human immune system is called a macrophage (the term means "big eater"). The size of a macrophage is 0.000021 m. (a) Write this size in scientific notation. (b) Use a conversion factor to convert the size of a macrophage to micrometers (μm), where $1 \text{ μm} = 10^{-6} \text{ m}$.
28. A solid sample of a material has a mass of 10.7 g and a volume of 5.4 cm^3. What is the density of this material? Is it more or less dense than water, which has a density of 1.00 g/cm^3?
29. The density of pure water is 1.00 g/cm^3, and the density of pure ethanol is 0.789 g/cm^3. (a) You are given a container of pure water that has a mass of 1 kilogram (1 kg = 1000 g). What is the volume of water in the container? Assume that the mass of the container has been subtracted from the measurement. (b) What volume of ethanol is required to obtain the same mass as the amount of water in part (a)?
30. Classify each of the following processes as a physical change or a chemical change:
    (a) Water boils in a beaker.
    (b) Raindrops freeze to form snowflakes.
    (c) Bubbles form in a bottle of soda after it is opened.
    (d) A balloon filled with hydrogen gas explodes after being lit with a match.
    (e) Mold forms on cheese.

31. State whether each of following chemical formulas represents a compound or an element:
    (a) H_2
    (b) N_2O
    (c) Li
    (d) CO
    (e) S_8
32. The element phosphorus (P) is found in some biological molecules. Which of the following elements does the chemistry of phosphorus most closely resemble: carbon, nitrogen, oxygen, or fluorine? Explain your choice.
33. Silicon is widely used in the semiconductor industry. Which of the following third period elements has chemical properties that are most similar to silicon: gallium (Ga), germanium (Ge), arsenic (As), or selenium (Se)? Explain your choice.
34. The atomic symbols of two unidentified elements (X and Y) are given below:

    $$ {}^{14}_{6}\text{X} \qquad {}^{14}_{7}\text{X} $$

    (a) Do these symbols represent the same element or different elements? Explain your answer.
    (b) Identify the element(s) represented by the atomic symbols.
    (c) How many protons and neutrons does each of these atoms contain?
35. One atom of the unstable new element with Z = 109 was reported, which was sufficient to claim that the element had been synthesized. The atom was detected in a mass spectrometer, and the mass found was 266. How many neutrons are present in this atom?
36. In 2006, a former Russian spy was fatally poisoned in London when a radioactive isotope called polonium-210 (^{210}Po) was put into his cup of tea. How many protons and neutrons does the nucleus of polonium-210 contain?
37. An atom has 15 protons and 16 neutrons. Identify the chemical element, and write the atomic symbol for this atom.
38. Scientists often abbreviate isotopes using only the element name and mass number, as shown in the left-hand column of the accompanying table. Write the complete atomic symbol for each of these atoms. For each example, also provide the number of protons and neutrons in the atomic nucleus.

ISOTOPE	ATOMIC SYMBOL	NUMBER OF PROTONS	NUMBER OF NEUTRONS
Carbon-13			
Chlorine-37			
Xenon-130			

39. Sulfur has four stable isotopes with mass numbers 32, 33, 34, and 36. Complete the following table by providing the atomic symbol for each isotope, together with the number of protons and neutrons that it contains.

MASS NUMBER	ATOMIC SYMBOL	NUMBER OF PROTONS	NUMBER OF NEUTRONS
32			
33			
34			
36			

40. A sample of naturally occurring silver contains two stable isotopes.

ISOTOPE	RELATIVE ATOMIC MASS	NATURAL ABUNDANCE
^{107}Ag	106.9	51.84%
^{109}Ag	108.9	48.16%

(a) *Without performing a calculation*, estimate the relative atomic mass of naturally occurring silver. Explain how you obtained your estimate. (b) Use the masses and natural abundances of the silver isotopes to calculate the relative atomic mass of naturally occurring silver. Is your calculation close to your estimate?

41. A sample of naturally occurring silicon contains three stable isotopes.

ISOTOPE	RELATIVE ATOMIC MASS	NATURAL ABUNDANCE
^{28}Si	27.98	92.23%
^{29}Si	28.98	4.68%
^{30}Si	29.97	3.09%

(a) *Without performing a calculation*, estimate the relative atomic mass of naturally occurring silicon. Explain how you obtained your estimate. (b) Use the masses and natural abundances of the silicon isotopes to calculate the relative atomic mass of naturally occurring silicon. Is your calculation close to your estimate?

42. Copper has two stable isotopes, ^{63}Cu and ^{65}Cu. The relative atomic mass of copper is 63.55. (a) *Without*

performing a calculation, deduce which isotope has a larger natural abundance. Explain your answer. (b) It is possible to calculate the natural abundance of each isotope. Use x for the natural abundance of ^{63}Cu, and $(1-x)$ for the natural abundance of ^{65}Cu. Write an equation that includes the mass of each isotope, its natural abundance, and the relative atomic mass. Use this equation to calculate the natural abundances of ^{63}Cu and ^{65}Cu. Does your calculation confirm your deduction in part (a)?

43. Bromine (Br) belongs to a group of the periodic table called the halogens, which also includes fluorine (F) and chlorine (Cl). Predict how many valence electrons are in the valence shell of a bromine atom.

44. (a) Draw the electron configuration for a calcium (Ca) atom. How many valence electrons does this atom possess? (b) Use your result from part (a) to explain why calcium forms Ca^{2+} ions.

45. Write the chemical symbol and charge for the ion that will be formed from each of the following atoms. Identify any atoms that do not form an ion.
   (a) Potassium (K)
   (b) Chlorine (Cl)
   (c) Sulfur (S)
   (d) Argon (Ar)

46. Determine the number of protons and electrons in each of the following ions:
   (a) Al^{3+}
   (b) F^-
   (c) O^{2-}
   (d) Na^+

47. Phosphorus atoms combine with hydrogen atoms to form a toxic gas called phosphine with the chemical formula PH_3. Arsenic also combines with hydrogen atoms to form a flammable and highly toxic gas called arsine. Predict the chemical formula of arsine, and explain the reasons for your choice.

48. Strontium-90 is an isotope that is produced during nuclear reactions. This isotope emits high-energy radiation that can damage human cells. When the Chernobyl nuclear reactor in the then Soviet Union exploded in 1986, large amounts of strontium-90 were released into the environment. Scientists later discovered that strontium-90 had been absorbed into the bones of exposed individuals, especially young children. (a) Find strontium (Sr) in the periodic table, and write the atomic symbol for the strontium-90 isotope. (b) How many protons and neutrons does this isotope contain? (c) Explain why strontium-90 tends to accumulate in bones.

Exploring Concepts

49. Individuals who want to follow a low-sodium diet will sometimes use sea salt instead of table salt. According to the website of the American Heart Association

(www.heart.org), are there any health advantages of consuming sea salt versus table salt? Write a two- to three-sentence explanation.

50. The inventor of the periodic table, Dimitri Mendeleev, predicted several elements from his table in addition to germanium (which was mentioned in the chapter). Use the Internet to learn the name of one of these elements. List the predictions he made about its properties, and compare them to the properties of the element that were actually found.

51. How do we know that the atom has a dense nucleus that contains nearly all of its mass? This property of the atom was discovered through a famous experiment performed in the early 20th century by Ernest Rutherford, Hans Geiger, and Ernest Marsden. Use the Internet to investigate this experiment. Write a brief description of the experiment, and explain why it demonstrates the existence of an atomic nucleus.

52. In 2012, *Consumer Reports* published articles on arsenic exposure that you can access online. According to the articles, what are the primary sources of arsenic in the foods and beverages we consume? The articles concluded that the FDA should establish standards for safe levels of arsenic in foods and beverages. Do you agree or disagree? Explain your position.

53. In this chapter, we learned that arsenic is toxic because its chemistry is similar to phosphorus, a biologically important element. Lead (Pb) is another toxic element, and exposure to this metal can occur from lead pipes and lead-based paint. Use the Internet to discover why lead is poisonous in terms of its relationship to other elements. Write a short paragraph to summarize your findings.

54. The chemistry of isotopes sometimes arises during discussions of international diplomacy. In particular, nations sometimes disagree over their ability to separate two isotopes of uranium: ^{235}U and ^{238}U. Use the Internet to investigate what is meant by "enriched uranium." Why is enriched uranium a matter of global security?

Chemical Bonding

Humans need air for survival. Depriving the body of air for only 3 to 5 minutes can be fatal. By contrast, we can survive without water for several days and without food for even longer. Why is air so essential for human life? What part of the air does our body use?

3.1 Why Do We Need Air to Survive?

Learning Objective:

Explain why breathing air is necessary for human survival.

THE KEY IDEA: The human body uses oxygen gas from the air for energy-producing reactions.

The air we breathe is a mixture of gases. Figure 3.1 shows the composition of "dry air," measured as percent by volume. We measure dry air because the amount of water in the atmosphere (humidity) varies significantly. The most abundant component of dry air is nitrogen gas, which makes up 78% of the total volume. Oxygen gas contributes 21%. Taken together, these two gases constitute almost all (99%) of the air's volume. Most of the remaining 1% is composed of argon gas. There are also tiny fractions of many other gases, including carbon dioxide.

The composition of air reveals interesting chemical principles. Argon is a noble gas and exists as isolated atoms. By contrast, oxygen, nitrogen, and carbon dioxide are all molecules. *A molecule is a combination of atoms joined by chemical bonds.* We represent molecules using a chemical formula that indicates the type and number of atoms they contain—for example, oxygen (O_2), nitrogen (N_2), and carbon dioxide (CO_2).

We can now return to our original question: Why do we need air to survive? After we inhale, the air travels through a thick tube in our throat called the *trachea*, or "windpipe," which delivers the air to our lungs (Figure 3.2). When our lungs fill with air, the gases pass through their thin lining and enter the bloodstream.

We learned in Chapter 2 that red blood cells are filled with hemoglobin, an iron-containing protein that transports oxygen throughout the bloodstream. When hemoglobin reaches the body's tissues, it discharges the oxygen molecules. The molecules enter nearby cells, where they participate in chemical reactions by combining with other molecules derived from our food (e.g., sugars). These reactions generate the molecule adenosine triphosphate (ATP), which provides chemical energy for a variety of essential cellular functions. We will examine these life-sustaining reactions in Chapter 6.

Breathing is so vital because our cells need an ongoing supply of oxygen to create ATP as an energy source. Cells cannot store ATP, so it must be continuously generated. Without this energy supply, cells quickly die. Brain cells consume large amounts of ATP, so brain damage is often the first consequence of oxygen deprivation.

What happens to the N_2 molecules that we inhale with each breath? Nothing! Nitrogen molecules do not participate in any chemical reactions within our body. Why do O_2 and N_2 molecules behave so differently?

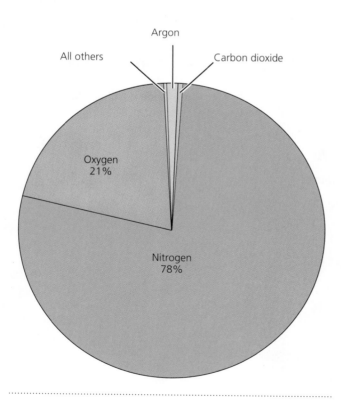

FIGURE 3.1 The composition of air. A pie chart of the gases in dry air as percent by volume.

We answer this question later in this chapter, when we investigate the types of chemical bonds that join the atoms in each molecule.

Not all chemical elements form molecules. Some elements instead form solid compounds called salts. The most familiar example is table salt—sodium chloride (NaCl). A salt contains a different type of chemical bond, which involves the transfer of an electron from one atom to another. We will examine these bonds in the final section of the chapter.

Core Concepts

- Air contains a mixture of gases. It is mostly composed of oxygen (21%) and nitrogen (78%).
- A molecule is a combination of atoms joined by chemical bonds. The composition of a molecule is represented by its chemical formula (e.g., O_2, N_2, CO_2).
- Humans use the oxygen they breathe as a component in chemical reactions that generate chemical energy (stored as ATP) within the body's cells. A constant supply of ATP is essential to sustain life.

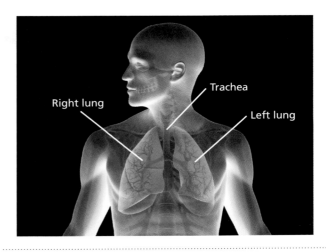

FIGURE 3.2 Breathing air. Inhaled air is transported through the trachea into the lungs. The gases then enter the bloodstream and are transported around the body.

3.2 Covalent Bonding

This section examines the type of chemical bonding that occurs when atoms share electrons. We begin by comparing two gases—hydrogen and helium.

Forming a Covalent Bond by Sharing Electrons

Imagine two balloons, one filled with helium gas and the other with hydrogen gas. Both balloons float upward because each gas is less dense than air. However, if you apply a flame to the balloons, then you immediately notice a difference. The helium balloon simply pops, allowing the helium to escape. In contrast, the flame ignites the hydrogen to produce a powerful explosion, as illustrated in Figure 3.3(a). The flammability of hydrogen led to the tragic explosion of the *Hindenburg* airship in 1937 (Figure 3.3(b)).

Why do helium and hydrogen gases behave so differently? Helium (He) is a noble gas—it is composed of individual helium atoms, which do not react chemically with other elements. Applying a flame does not to induce these atoms to give up their isolated existence. In contrast, hydrogen gas is composed of *hydrogen molecules* (H_2). Each H_2 molecule contains two H atoms linked by a chemical bond. When we ignite hydrogen gas, we provide enough energy to initiate a chemical reaction between the H_2 molecules and O_2 molecules in the air. The

Learning Objective:
Explain the formation of a covalent chemical bond.

THE KEY IDEA: A covalent bond arises from the sharing of an electron pair between two atoms.

(a)   (b)

FIGURE 3.3 Hydrogen is a flammable gas. (a) Igniting a balloon filled with hydrogen gas produces a violent explosion. (b) The flammability of hydrogen was responsible for the tragic catastrophe of the *Hindenburg* airship, which burst into flames over New Jersey in 1937. (Source: cos.fit.edu/chemistry)

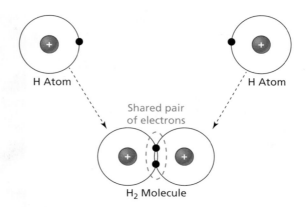

FIGURE 3.4 **Covalent bonding between two hydrogen atoms.** Two hydrogen atoms can achieve a more stable electron arrangement by sharing their electrons to form a covalent bond. The result is an H_2 molecule containing two H atoms.

reaction of hydrogen gas with oxygen releases energy as heat, which results in the explosion seen in Figure 3.3.

The distinct chemical behaviors of hydrogen and helium arise from the different electron configurations of the two elements. As we learned in Chapter 2, a helium atom is stable because its two electrons fill the first electron shell. By contrast, a hydrogen atom has an unfilled electron shell containing only one electron, so it is not stable in isolation. However, two hydrogen atoms can become more stable by sharing their electrons, as shown in Figure 3.4. Each hydrogen atom now has two electrons, which resembles the stable arrangement in helium.

The shared electron pair creates a mutual attraction between the two atoms that joins them together—this attraction is called a **covalent bond**. In general, *a covalent bond arises from the sharing of an electron pair between two atoms.* Chemists use this term because "co-" means together and "-valent" refers to the valence electrons being shared. In the case of hydrogen atoms, the single electron is the valence electron. The bonding attraction is strongest when the shared electron pair is located between the two atomic nuclei. We can think of a covalent bond as "electron glue" that holds two atoms together.

When two hydrogen atoms are joined by a covalent bond, they create a hydrogen molecule with the chemical formula H_2. We represent the covalent bond between the atoms by a pair of dots (with each dot indicating an electron) or by a straight line:

$$\mathrm{H:H \ or \ H-H}$$

Two hydrogen atoms form a covalent bond because the bonded atoms have a lower energy compared to when they exist in isolation. This general principle can be applied to all covalent bonds between atoms.

covalent bond A bond that is formed when two atoms share an electron pair.

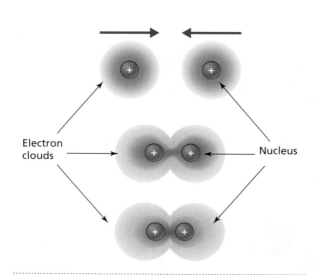

FIGURE 3.5 **Covalent bonding described in terms of electron density.** As two H atoms approach each other, their mutual interaction changes the distribution of electron density (shown as blue shading). A covalent bond arises from the increased electron density between the two positively charged nuclei.

Core Concepts

- Helium atoms are stable because they have a filled electron shell. By contrast, hydrogen atoms do not have a filled shell and are not stable in isolation.
- When two H atoms share their electrons, the electron pair creates a mutual attraction between the atoms. The sharing of an electron pair between two atoms produces a **covalent bond**.

■ A hydrogen molecule (H_2) consists of two H atoms joined by a covalent bond. When bonded together, the two H atoms have a lower energy compared to when they exist in isolation.

Quantum Mechanical Description of Covalent Bonding

Another view of covalent bonding is provided by the quantum mechanical description of atoms. As we learned in Chapter 2, quantum mechanics treats electrons as waves, and it describes the *probability* of finding an electron in a particular location. A higher probability corresponds to a greater electron density, which is represented visually by an *electron cloud*.

Consider the two hydrogen atoms at the top of Figure 3.5. As isolated atoms, they both have electron densities that are distributed symmetrically around the positively charged nucleus. As the two atoms approach each other, they begin to exert an influence on each other that distorts the electron density. This interaction generates a region of increased electron density between the two hydrogen nuclei, which bonds the atoms together. In the quantum mechanical model, *a covalent bond corresponds to increased electron density between two atomic nuclei*. We can see parallels between this description of bonding and the shared electron pair discussed above.

In most cases we will consider, shared electron pairs are sufficient to describe covalent bonding. For some situations, however, only the quantum mechanical model can provide an accurate explanation. We will encounter some examples in the next chapter.

> **THE KEY IDEA:** In the quantum mechanical description, a covalent bond corresponds to increased electron density between two atomic nuclei.

Core Concepts

■ The quantum mechanical model describes a covalent bond in terms of increased electron density in the region between two atomic nuclei.

3.3　Making Molecules

Section 3.2 examined the formation of a covalent bond between two H atoms to create a simple H_2 molecule. In general, *the principles of chemical bonding and molecular structure apply to both small and large molecules.* Figure 3.6 compares the molecular structures of H_2O and a short region of DNA. Despite the differences in their size and complexity, both molecules can be described by the same chemical principles. This chapter introduces chemical bonding and structure, with a focus on small molecules. In later chapters, we will apply these principles to larger biological molecules.

> *Learning Objective:*
> Apply the principles of chemical bonding and molecular structure.

Electrons and Bonding

The chemical bonding properties of an atom arise from its electron configuration. In our examination of atomic structure in Chapter 2, we saw that electrons in atoms are arranged within discrete shells. To recap, consider the example of carbon (Z = 6), which has six electrons. Two of these electrons fit into the first shell, but that shell then becomes filled. The other four electrons must go in the second shell, which is located farther from the nucleus (Figure 3.7). Because the first electron shell is filled and unreactive, *the chemical properties of carbon are determined by the four electrons in the second electron shell.* These are the valence electrons, which reside in the valence shell. Consequently, we can simplify our representation of a carbon atom by including only its chemical symbol and four valence electrons (Figure 3.7).

As a general rule, we can explain the bonding properties of atoms by their number of valence electrons. Figure 3.8 presents an abbreviated version of the periodic table that shows the valence electrons for the atoms of selected elements.

> **THE KEY IDEA:** The chemical bonding properties of an atom are determined by its valence electrons.

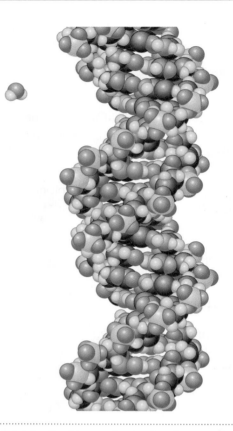

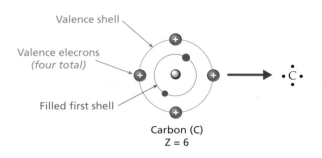

FIGURE 3.7 Arrangement of electrons in a carbon atom. A carbon atom has six electrons, two in the first shell (which is filled) and four valence electrons in the valence shell. The chemical properties of a carbon atom can be represented by depicting only its valence electrons.

FIGURE 3.6 Small and large molecules obey the same rule of chemistry. The structures of two molecules of very different size—H_2O and DNA—shown at the same size scale. A color code is used for different types of atoms: hydrogen = white, carbon = black, nitrogen = blue, oxygen = red, and phosphorus = orange. (Source for DNA structure: P. B. Kelter, J. D. Carr, and A. Scott, *Chemistry, A World of Choices*, 2nd Edition. McGraw Hill, 2003, Figure 2.27).

The older designation for the groups (e.g., 1A, 2A, and so on) provides the number of valence electrons for the atoms in each group. Because the elements within a particular group have the *same number of valence electrons*, they have similar chemical properties.

The relationship between valence electrons and chemical bonding was developed in the early 20th century by an American chemist named Gilbert Lewis (1875–1946; Figure 3.9). Lewis proposed a convenient shorthand to show how the valence electrons from different atoms pair up to form covalent bonds. He represented the electrons as dots, so this type of drawing is now called a **Lewis dot structure**. Though a simplification, these structures help us deduce how atoms combine to form stable molecules. In this section, we will use Lewis dot structures to examine how carbon, nitrogen, and oxygen atoms form covalent bonds with hydrogen atoms.

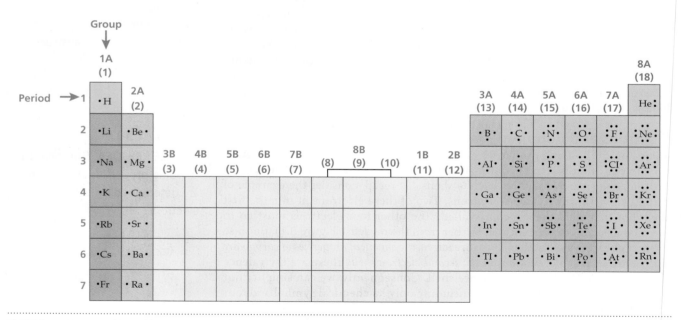

FIGURE 3.8 Representation of valence electrons for selected atoms in the periodic table. The valence electrons for each atom determine the chemical bonding properties of the element.

Lewis dot structure A representation of an atom or molecule that depicts valence electrons as dots.

Carbon

We will examine the combination of C and H atoms using a step-by-step procedure.

STEP 1: *Write the Lewis dot structures for each of the atoms in the molecule.* This step enables us to quickly identify the valence electrons. Carbon has four valence electrons, and hydrogen has one.

Carbon   Hydrogen

STEP 2: *Identify which atom should be placed at the center of the molecule.* The *central atom* is the one with the capacity to form the largest number of covalent bonds. In our example, is the central atom carbon or hydrogen? An H atom forms only one covalent bond because it has one valence electron and requires only one more to make a stable electron configuration. By contrast, carbon can form more than one covalent bond because it can fit more electrons within its second electron shell. We therefore put carbon as the central atom in the molecule.

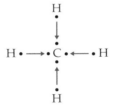

C is the central atom

STEP 3: *Determine how many additional electrons are required for carbon to achieve a stable electron configuration.* The valence electrons in carbon occupy the second electron shell, which can hold up to *eight electrons*. Because carbon has four valence electrons, it needs *four more electrons* to achieve a filled electron shell. This can be accomplished by adding four hydrogen atoms, each of which has a single valence electron.

**Add four H atoms to one C atom
to achieve a stable electron arrangement**

STEP 4: *Draw the complete Lewis dot structure for the molecule, which includes stable valence electron configurations for all the atoms.* Carbon is now surrounded by eight valence electrons—called an *octet*—which resembles the stable electron configuration of neon. The octet is composed of *four shared pairs of electrons*; for each pair, one electron is contributed by carbon and the other by hydrogen. As we learned in the previous section, a pair of electrons shared between two atoms is a covalent bond. Each hydrogen atom also has a stable bonding arrangement consisting of two electrons, which resembles the noble gas helium.

FIGURE 3.9 Gilbert Lewis. Gilbert Lewis working in his laboratory at the University of California in Berkeley. (Source: www.pbs.org)

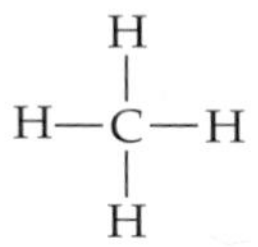

H
|
H—C—H
|
H

FIGURE 3.10 The two-dimensional structure of methane, CH₄. In this structure, each covalent chemical bond is represented as a straight line. This drawing does not convey the molecule's three-dimensional structure.

octet rule The tendency of atoms to acquire eight electrons—an octet—in their valence shell.

two-dimensional structure A flat drawing of a molecule that uses straight lines to represent covalent bonds; it is also called a Lewis structure

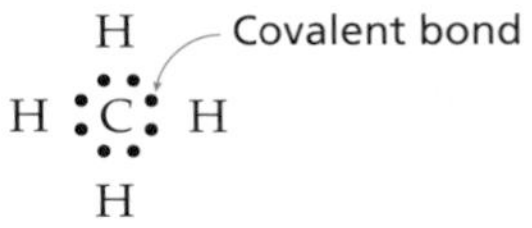

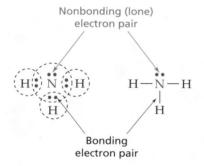

Nonbonding (lone) electron pair

Bonding electron pair

FIGURE 3.11 The Lewis dot structure and two-dimensional structure of ammonia, NH₃. Electrons from the N atom appear in blue, and electrons from the H atom appear in pink. Each shared electron pair (covalent bond) contains one electron from each atom. The N atom has a non-bonding (lone) electron pair that is not shared.

nonbonding electron pair An unshared pair of valence electrons from a single atom.

lone pair *See nonbonding electron pair.*

H
H ⫶C⫶ H   ⌐ Covalent bond
H
Lewis dot structure of CH₄

By drawing a Lewis dot structure, we have deduced that a carbon atom bonds with four hydrogen atoms to form a stable molecule with the chemical formula CH_4. This molecule is called *methane*, which is the major component of natural gas used for cooking and heating.

The structure of CH_4 illustrates a principle that Lewis called the **octet rule**. This rule states that *many atoms tend to acquire eight electrons in their valence shell*, which then resembles the electron configuration of a noble gas. This rule applies primarily to the *main-block elements* of the periodic table (i.e., elements in Groups 1A and 2A on the left, plus those in Groups 3A to 8A on the right). There are exceptions to the octet rule—for example, an He atom and an H_2 molecule are stable with only two electrons in the first electron shell. We will examine other exceptions in Section 3.5.

We can also draw a simplified version of the CH_4 molecule in which each shared pair of electrons is replaced by a straight line to indicate a covalent bond (Figure 3.10). We call this drawing a **two-dimensional structure** because all of the bonds are drawn in the flat plane of the page. This type of molecular drawing is also called a *Lewis structure*. The two-dimensional structure of CH_4 highlights the molecule's four covalent bonds. However, a two-dimensional structure does not attempt to accurately portray the molecule in three dimensions: Figure 3.10 suggests that CH_4 is a square molecule, which is incorrect. In a later section, we will learn to predict the three-dimensional structure of a molecule.

The ability of carbon atoms to form *four covalent bonds*—the maximum possible number for a second-row element—helps to explain carbon's amazing chemical diversity. Carbon is so common in biological molecules that we sometimes classify all organisms on Earth as "carbon-based" life forms. We will examine the chemical versatility of carbon in Chapter 4.

Nitrogen

We can apply the same principles to predict the stable molecule that will be formed by combining a nitrogen atom with hydrogen atoms. The nitrogen atom has *five valence electrons*, so it needs *three more electrons* for a stable octet. The nitrogen atoms can gain these extra electrons by a sharing arrangement with three hydrogen atoms. Therefore, we predict that the formula for the molecule made by nitrogen and hydrogen is NH_3. This gas is called *ammonia*, which can be dissolved in water to make cleaning solutions.

The Lewis dot structure and two-dimensional structure of NH_3 are shown side by side in Figure 3.11. Note that not all of nitrogen's valence electrons are being shared with hydrogen. Because nitrogen requires only three additional electrons for an octet, it needs to form only three covalent bonds. The two remaining valence electrons are called a **nonbonding electron pair** or a **lone pair**. The lone pair is associated *only* with the nitrogen atom, and it has no role in covalent bonding. The two-dimensional structure shows the covalent bonds in the molecule and also includes the lone pair.

PRACTICE EXERCISE 3.1

How does an oxygen atom form chemical bonds with hydrogen atoms? Draw a Lewis dot structure and a two-dimensional structure of the molecule that is formed. What is the chemical formula of the molecule, and what is its name? Does it have any nonbonding electron pairs?

In summary, the number of valence electrons in an atom determines the number of covalent chemical bonds it will form. Figure 3.12 summarizes the bonding properties of hydrogen, carbon, nitrogen, and oxygen atoms. These are the four most common atoms in biological molecules and pharmaceuticals, so learning these rules will help you analyze many different molecules.

FIGURE 3.12 Covalent bonding rules for hydrogen, carbon, nitrogen, and oxygen atoms. Hydrogen needs to share two electrons for covalent bonding. The other atoms require a stable octet of electrons.

Core Concepts

- An atom forms covalent bonds based on its number of valence electrons.
- A **Lewis dot structure** shows the valence electrons for each atom and how they pair to form stable molecule structures.
- The **octet rule** states that many atoms tend to acquire eight electrons in their valence shell, which resembles the electron configuration of a noble gas.
- A flat **two-dimensional structure** represents each covalent bond as a straight line and includes the **nonbonding electron pairs**. It does not convey the accurate three-dimensional structure of the molecule.

WORKED EXAMPLE 3.1

Question: Beginning in the 1920s, molecules known as *chlorofluorocarbons* (CFCs) were used in refrigerators and air conditioners. In the 1970s, scientists discovered that CFCs were responsible for severely depleting in the ozone layer of the atmosphere. These chemicals have since been banned under an international treaty. One popular CFC was called Freon-12, with the chemical formula CF_2Cl_2. Draw (a) the Lewis dot structure and (b) the two-dimensional structure for Freon-12.

Answer: (a) We begin by drawing the Lewis dot structure.

STEP 1: Write the Lewis dot structures for each of the atoms in the molecule. We obtain this information from the periodic table.

STEP 2: Identify which atom should be placed at the center of the molecule
The fluorine and chlorine atoms each need only one more electron to form a stable octet, so they will both form only one covalent bond. Carbon has four valence electrons, so it has the capacity to form four covalent bonds. We therefore place **C** at the center of the molecule.

STEP 3: Determine how many additional electrons are required for carbon to achieve a stable electron configuration.
Carbon can achieve a stable octet by sharing electrons with four atoms. The chemical formula of Freon-12 is CF_2Cl_2, so one **C** atom will combine with two **F** atoms and two **Cl** atoms.

STEP 4: Draw the complete Lewis dot structure for the molecule, which includes stable valence electron configurations for all the atoms. The Lewis dot structure for CF_2Cl_2 is as follows.

A Lewis dot structure shows only the configuration of valence electrons, so it does not matter where you place the **F** and **Cl** atoms. Here are two equally acceptable alternative structures:

(b) To draw the two-dimensional structure, we first identify which electron pairs form *covalent bonds* (a shared pair with one electron from each atom), and which are *lone pairs* (both electrons from the same atom). The difference can be ascertained using the color-coding in the Lewis dot structure. The two-dimensional structure of CF_2Cl_2 is shown here, with a straight line representing each covalent bond and the lone pair electrons shown as dots. The location of the **F** and **Cl** atoms relative to the **C** atom does not matter.

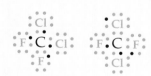

TRY IT YOURSELF 3.1

Question: Nitrogen triiodide, NI_3, is a very unstable molecule that explodes with even light contact. Draw the Lewis dot structure of NI_3.

Molecular Structure in Three Dimensions

Up to this point, we have represented molecular structures as drawings on a flat page. In reality, however, molecules exist in a three-dimensional world, just as we do. In most cases, the three-dimensional arrangement of atoms in a molecule profoundly influences its chemical or biological properties.

The three-dimensional structure of a molecule is determined by two factors: the types of atoms in the molecule and the covalent bonds that hold these atoms together. We can predict the shape of a molecule using a principle called **valence shell electron pair repulsion** (**VSEPR**), pronounced *vesper*. We need to consider two types of electron pairs within the valence shell of each atom—bonding pairs and non-bonding (lone) pairs. *VSEPR states that all valence shell electron pairs repel each other to become as far apart as possible.* Maximizing the distance between the electron pairs minimizes their repulsive interaction. The electron pair repulsion positions the atoms of the molecule in a particular spatial arrangement called the **three-dimensional structure** (or the *three-dimensional geometry*). Note that the two electrons *within* the same pair (either a bonding pair or a lone pair) *do not* repel each other. The application of VSEPR is most straightforward when the molecule has a single central atom, which is true for CH_4, NH_3, and H_2O.

How can we represent the three-dimensional structure of a molecule using a drawing on a two-dimensional surface? Chemists use **wedge-and-dash drawings** to depict three-dimensional perspective. To picture this method, imagine a person who is divided in half by a plane (Figure 3.13). The plane is a two-dimensional surface, similar to a flat sheet of paper. If the person sticks out both of his arms at an angle, one arm will extend in front of the plane and the other behind the plane. The arm that points outward from the plane, towards the viewer, is depicted by a *solid wedge*. This wedge represents three-dimensional perspective, with the thicker end being farther away from the plane. The other arm, which points behind the plane and away from the viewer, is depicted as a *dashed wedge*. As before, the thicker end of the dashed wedge is the farthest distance from the plane.

Let's apply VSEPR to predict the three-dimensional structure of methane, CH_4. This molecule has four C—H bonds organized around a central carbon atom, and each covalent bond consists of a shared electron pair. Because of mutual electron pair repulsion, the C—H bonds arrange themselves in three-dimensional space to maximize the distance between them. They do so by forming a shape called a *tetrahedron* (Figure 3.14(a)). Alternatively, we say that CH_4 has a **tetrahedral geometry**. The symmetric geometry of a

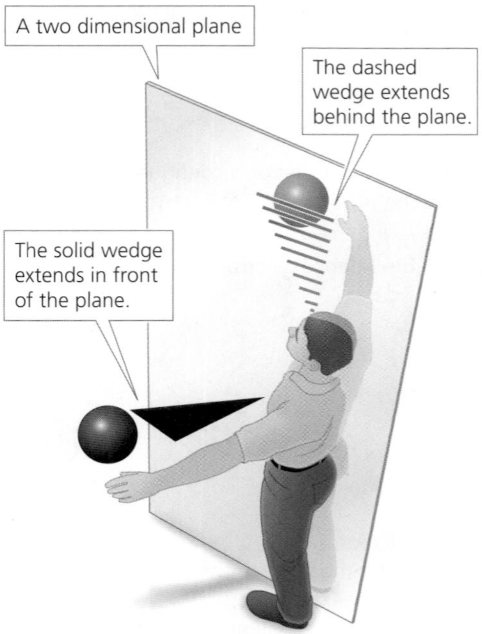

THE KEY IDEA: We can predict the three-dimensional structure of a molecule using valence shell electron pair repulsion.

▪ **valence shell electron pair repulsion (VSEPR)** A principle to predict the three-dimensional geometry of molecules; it states that all valence electron pairs in a molecule repel one another so as to maximize their separation in space.

▪ **three-dimensional structure** The spatial arrangement of atoms in a molecule.

▪ **wedge-and-dash drawing** A drawing that depicts a three-dimensional perspective for molecular structures.

▪ **tetrahedral geometry** The shape of a tetrahedron, in which all bond angles are 109.5°.

A two dimensional plane

The dashed wedge extends behind the plane.

The solid wedge extends in front of the plane.

FIGURE 3.13 A conceptual representation of a wedge-and-dash drawing for a three-dimensional structure. A person is divided in half by a two-dimensional plane. The arm that extends in front of the plane is depicted as a solid wedge, whereas the arm that extends behind the plane is depicted as a dashed wedge. The thickest edge of each type of wedge indicates the farthest distance from the plane.

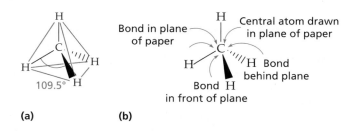

(a)   (b)

FIGURE 3.14 The three-dimensional structure of methane, CH_4. (a) Electron pair repulsion causes the four C—H bonds in CH_4 to point toward the four corners of a tetrahedron (shown as blue lines). The tetrahedral structure of CH_4 produces an H—C—H bond angle of 109.5°. (b) Use of the wedge-and-dash notation to depict the three-dimensional structure of CH_4.

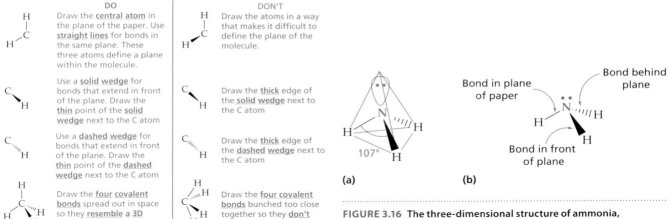

Dos and Don'ts for Drawing Wedge-and-Dash Structures

DO	DON'T
Draw the **central atom** in the plane of the paper. Use **straight lines** for bonds in the same plane. These three atoms define a plane within the molecule.	Draw the atoms in a way that makes it difficult to define the plane of the molecule.
Use a **solid wedge** for bonds that extend in front of the plane. Draw the **thin** point of the **solid wedge** next to the C atom	Draw the **thick** edge of the **solid wedge** next to the C atom
Use a **dashed wedge** for bonds that extend in front of the plane. Draw the **thin** point of the **dashed wedge** next to the C atom	Draw the **thick** edge of the **dashed wedge** next to the C atom
Draw the **four covalent bonds** spread out in space so they **resemble a 3D tetrahedron**	Draw the **four covalent bonds** bunched too close together so they **don't** resemble a 3D tetrahedron

FIGURE 3.15 Guidelines for how to use the wedge-and-dash notation to draw three-dimensional structures of molecules such as CH$_4$.

(a)   **(b)**

FIGURE 3.16 The three-dimensional structure of ammonia, NH$_3$. (a) The three-dimensional structure of NH$_3$ is trigonal pyramidal, with an H—N—H bond angle of 107°. The lone pair, shown in red, contributes to the electron pair repulsion. (b) Use of the wedge-and-dash notation to depict the three-dimensional structure of NH$_3$.

tetrahedron means that each H—C—H bond angle is exactly 109.5°. This angle is very different from the apparent 90° bond angle suggested by the two-dimensional structure in Figure 3.10.

Figure 3.14(b) illustrates how the tetrahedral geometry of CH$_4$ is represented using a wedge-and-dash drawing. The central C atom is drawn in the two-dimensional plane of the paper, and straight lines are used to represent C—H bonds that also lie in the same plane. The solid wedge indicates a C—H bond that extends in front of the plane, and the dashed wedge indicates a C—H bond that extends behind the plane.

Visualizing and drawing three-dimensional molecular structures are essential skills that require practice. Figure 3.15 summarizes the "do's and don'ts" for drawing three-dimensional molecular structures. When you view a wedge-and-dash drawing, look carefully at how each bond is represented, and use this information to imagine the molecule as a three-dimensional structure. The *Chemistry in Your Life* feature provides an analogy between perspective drawings in art and chemistry.

What is the three-dimensional structure of ammonia, NH$_3$? There are four electron pairs in the valence shell of the nitrogen atom—three bonding pairs and one lone pair. Because four electron pairs are located around the central atom, they arrange themselves in three-dimensional space in a roughly tetrahedral geometry, as demonstrated in Figure 3.16(a). However, the repulsion between the lone pair and the three bonding pairs is *slightly stronger* than the repulsion of the bonding pairs with one another. This difference arises because a lone pair occupies a slightly larger space than a bonding pair, owing to the fact that the bonding electrons are *shared* between two atomic nuclei. The strength of repulsion can be ranked from strongest to weakest:

nonbonding – nonbonding > nonbonding – bonding > bonding – bonding
 strongest   *weakest*

This extra repulsion forces the angle between the N—H bonds to close slightly from the perfect tetrahedral angle of 109.5°; in fact, the angle between each N—H bond in ammonia is 107°. The NH$_3$ molecule has a **trigonal pyramidal geometry**, in which the three N—H bonds form the shape of a three-sided pyramid. Figure 3.16(b) illustrates how to draw the three-dimensional structure of NH$_3$ using wedge-and-dash notation.

trigonal pyramidal geometry The shape of a three-sided pyramid.

CHEMISTRY IN YOUR LIFE

Three-Dimensional Perspective in Chemistry and Art

The challenge of representing a three-dimensional world on a two-dimensional surface is not unique to the study of molecules. One of the most important transformations in the history of art was the development of perspective. Medieval art depicted people, buildings, and landscapes as if they were flattened into a two-dimensional world, which distorts their spatial relationships. One example of medieval imagery is shown in Figure 1(a); the walled garden on the right side looks oddly shaped because it is missing depth perspective. We can compare two-dimensional medieval art to the two-dimensional drawing of a CH_4 molecule, also shown in Figure 1(a). The two-dimensional structure provides some useful information, such as the connectivity of the atoms. However, it erroneously implies that the molecule has the shape of a square, with H—C—H bond angles of 90°. This is a consequence of confining CH_4 to a two-dimensional world.

As an artistic contrast, consider the painting reproduced in Figure 1(b). This famous fresco is called *The School of Athens*. It

was created by Raphael from 1509 to 1510 and is considered one of the masterpieces of Renaissance art. The converging series of straight lines have been added to highlight Raphael's use of three-dimensional perspective. We can compare this fresco to the way that chemists use wedge-and-dash drawings to indicate perspective for the molecular structure of CH_4. The solid wedge represents a C—H bond extending toward you from the central carbon atom. You can see that similar wedges are formed by the perspective lines in Rafael's fresco (i.e., the lines spread farther apart as the perceived distance to the viewer become closer). The dashed wedge represents a bond extending away from you, into the plane of the paper. By this analogy, you may expect the furthest edge of the bond (connected to the H atom) to be the narrowest. However, the chemical convention is to write the dashed wedge as shown, with the thin point closest to the carbon atom.

(a)

(b)

FIGURE 1 **Representing perspective in art and chemistry.** (a) In medieval art, figures and landscapes were flattened onto a two-dimensional surface, which distorts their spatial relationship. Similarly, the two-dimensional drawing of CH_4 implies that all of the atoms exist in a flat plane, which is incorrect. (b) Rafael's Renaissance fresco—*The School of Athens*—includes a representation of depth perspective, which is highlighted by the converging geometrical lines. Chemists also include depth perspective when representing three-dimensional molecular structures using wedge-and-dash notation.

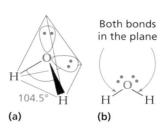

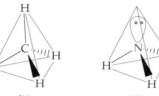

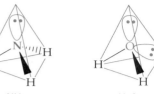

CH₄	NH₃	H₂O
Tetrahedral	Trigonal pyramidal	Bent
bond angle = 109.5°	bond angle = 107°	bond angle = 104.5°

FIGURE 3.17 **The three-dimensional structure of water, H₂O.** (a) The molecular structure of H₂O is bent, with an H—O—H bond angle of 104.5°. The two lone pairs, shown in red, contribute to the electron pair repulsion. (b) The three-dimensional structure of H₂O can be drawn with all three atoms in a two-dimensional plane.

FIGURE 3.18 **VSEPR predicts the molecular geometry of CH₄, NH₃, and H₂O.** Each molecule has four electron pairs that repel one another, producing an approximately tetrahedral arrangement. In NH₃ and H₂O, the extra repulsion from lone pairs (shown in red) slightly reduces the angles between the covalent bonds.

Finally, we turn to the H₂O molecule, which has two bonding electron pairs and two nonbonding pairs. Because this molecule contains four electron pairs, the tetrahedron again becomes a useful starting point for considering the molecular geometry. Each O—H bond and each lone pair point toward one of the four vertices of the tetrahedral structure, as shown in Figure 3.17(a). The three atoms in H₂O form an upside-down "V" shape, so the H₂O molecule is said to have a **bent geometry**. In this case, the stronger electron repulsion between the two lone pairs closes the H—O—H bond angle to 104.5°. Figure 3.17(b) illustrates the guidelines for drawing the three-dimensional structure of H₂O. The three atoms in H₂O define a plane, so they can all be drawn with straight lines within the plane of the paper.

To recap the important principles of VSEPR, Figure 3.18 compares the three-dimensional structures of CH₄, NH₃, and H₂O. In later chapters, we will see how the three-dimensional structures of biological molecules such as DNA are directly related to their biological function.

bent geometry The shape of an upside-down "V."

Core Concepts

■ The **three-dimensional structure** of simple molecules can be predicted by **VSEPR (valence shell electron pair repulsion).** According to VSEPR, all the electron pairs in the valence shell repel each other to maximize their separation in three-dimensional space.

■ Chemists use the **wedge-and-dash notation** to represent the three-dimensional structure of molecules.

■ CH₄ has a **tetrahedral geometry** with a bond angle of 109.5°. NH₃ has a **trigonal pyramidal geometry** with a bond angle of 107°. H₂O has a **bent geometry** with a bond angle of 104.5°. The bond angles in NH₃ and H₂O are smaller than the tetrahedral angle because of the extra repulsive effect of the lone pair(s).

Molecular Models

Building molecular models helps us visualize the arrangement of atoms within the three-dimensional molecular structure. In a **ball-and-stick model**, atoms are represented by "balls" and the chemical bonds between them by "sticks." Molecular models use a color code for various atoms to make them easy to identify within a molecular structure, although the atoms themselves are not really colored like this.

hydrogen = white   carbon = black   oxygen = red   nitrogen = blue

THE KEY IDEA: Molecules can be visualized using molecular models.

ball-and-stick model A molecular model that represents the atoms as balls and the bonds between them as sticks.

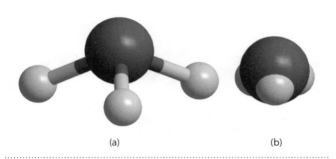

FIGURE 3.19 Two molecular models of ammonia, NH₃.
(a) Ball-and-stick model; (b) space-filling model.

■ **space-filling model** A molecular model that depicts the volume of each atom.

Figure 3.19(a) displays a ball-and-stick model for ammonia, which captures its pyramidal geometry. Note that the lone pairs are not shown explicitly in most molecular models, but the placement of the "sticks" reflects the geometrical effect of their electron pair repulsion.

Ball-and-stick modes do not accurately represent a molecule's true volume. For this purpose we use a **space-filling model**, shown in Figure 3.19(b), which presents atoms in proportion to the volume of space that they occupy.

If you examine the space-filling model of ammonia, you will notice that the nitrogen atom (blue) is shown as a larger sphere than the hydrogen atom (white). We say that nitrogen has a larger *atomic radius* than hydrogen. Why? The volume occupied by an atom depends on its total number of electrons and where they are located within the electron shells. Hydrogen has a single electron in the first shell, whereas nitrogen has a filled first electron shell plus five electrons in the second shell. Because the second electron shell is farther from the nucleus, a nitrogen atom occupies a greater volume. The space-filling volume of a molecule arises from the sum of the atomic radii of all its atoms, with modifications for the chemical bonding between them. This molecular volume is particularly important for understanding how drug molecules interact with their biological targets within a cell.

Core Concepts

■ Molecules can be represented using molecular models. A **ball-and-stick model** shows the atoms as balls and the covalent bonds as sticks. A **space-filling model** depicts the volume occupied by each atom within the molecule.

Representing Molecules

Thus far, we have learned how to represent molecules in a variety of ways. Figure 3.20 compares these representations for CH_4, NH_3, and H_2O. None of these representations is the "real" molecule; instead, each of these depictions

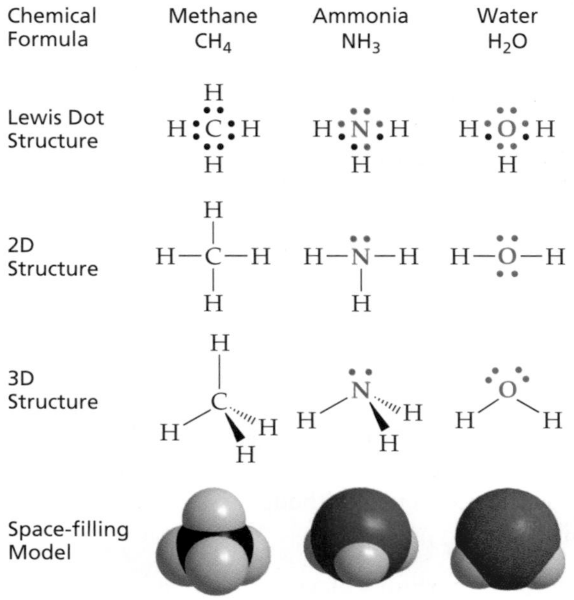

🔑 THE KEY IDEA: Different representations of molecules communicate particular features of molecular structure.

FIGURE 3.20 Representing molecules. A comparison of five different molecular representations for methane, ammonia, and water: chemical formula, Lewis dot structure, two-dimensional structure, three-dimensional structure, and space-filling model.

	Methane CH_4	Ammonia NH_3	Water H_2O
Chemical Formula	Methane CH_4	Ammonia NH_3	Water H_2O
Lewis Dot Structure	H:C:H with H above and below	H:N:H with H below	H:O:H
2D Structure	H—C—H with H above and below	H—N—H with H below	H—O—H
3D Structure	C with four H	N with three H	O with two H
Space-filling Model			

WORKED EXAMPLE 3.2

Question: Phosphine (PH_3) is a colorless, toxic, and flammable gas. Draw the following molecular structures for phosphine:

(a) Lewis dot structure

(b) Two-dimensional structure

(c) Three-dimensional structure

(d) Space-filling model

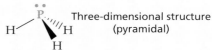

Phosphorus   Hydrogen

Answer: (a) We first draw the Lewis dot structures of the individual atoms, P and H. Using the periodic table in Figure 3.8, we see that phosphorus is in group 5A and has five valence electrons. We use color coding to distinguish between the valence electrons in phosphorus (orange) and hydrogen (pink).

It is possible to make a stable molecule that satisfies the bonding rules by combining one P atom with three H atoms. The Lewis dot structure, shown below, contains three covalent bonds and one lone pair. Note that the lone pair contains only electrons from the P atom.

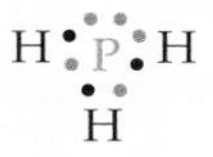

H : P : H   Lewis dot structure

H

(b) To draw the two-dimensional structure of PH_3, we replace each pair of electrons with a straight line to indicate a covalent bond. We also include the lone pair on the P atom.

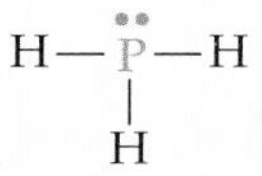

H—P̈—H

H   Two-dimensional structure

(c) We use the principle of VSEPR to predict the three-dimensional structure of PH_3. There will be mutual repulsion between the three bonding electron pairs (the covalent bonds) and the lone pair on the P atom. This repulsion will produce a trigonal pyramidal shape, similar to that of NH_3. We use the wedge-and-dash notation to represent the three-dimensional structure, and we also include the lone pair.

Three-dimensional structure (pyramidal)

(d) To draw the space-filling model of PH_3, we must represent the relative sizes of the atomic radii for the H and P atoms. An H atom has only one electron in the first electron shell, so its atomic radius is small. Phosphorus occurs in the third period of the periodic table and has electrons in the first, second, and third electron shell. Consequently, its atomic radius is much larger. The space-filling model of PH_3 is shown below; note the size difference between the P and H atoms.

Space-filling model

TRY IT YOURSELF 3.2

Question: Hydrogen sulfide (H_2S) is a colorless gas with the pungent smell of rotting eggs. Draw the following molecular structures for hydrogen sulfide:

(a) Lewis dot structure

(b) Two-dimensional structure

(c) Three-dimensional structure

(d) Space-filling model

focuses on certain aspects of the molecule. For example, the Lewis dot structure explicitly highlights the valence electrons for each atom, whereas the space-filling model indicates the volume occupied by the molecule. You should become comfortable with *all* of these representations and use the most appropriate one for the scientific question under consideration.

Core Concepts

■ Molecules can be represented in a variety of ways, each of which focuses on a different feature. All of these representations are valuable for understanding the composition and structure of molecules.

3.4 Molecules with Double and Triple Bonds

The covalent bonds that we have examined thus far are based on sharing a single pair of electrons. This type of bond is called a *single* covalent bond. Atoms may also share more than one electron pair. This section examines these bonding arrangements by focusing on the molecular gases that constitute most of our atmosphere.

Molecular Structures of Oxygen and Nitrogen Gases

At the beginning of the chapter, we examined the chemical composition of air and its two most abundant components—oxygen gas (O_2) and nitrogen gas (N_2). What are the molecular structures and bonding characteristics of these gases?

THE KEY IDEA: The O_2 molecule contains a double covalent bond, and the N_2 molecule contains a triple covalent bond.

Let us first consider the combination of two oxygen atoms, each of which has six valence electrons. If we arrange the two oxygen atoms to share one pair of electrons, we find that each oxygen atom is now surrounded by seven valence electrons—one short of a stable octet. Each oxygen atom can achieve an octet of electrons by sharing *two electron pairs* (Figure 3.21). Because each shared electron pair is a covalent chemical bond, we say that the two atoms in the O_2 molecule are joined by a **double covalent bond**; we write this bond as O=O.

We now turn to the bonding between two nitrogen atoms. Each nitrogen atom has five valence electrons and requires three more to form a stable octet. This electron configuration can be achieved if the nitrogen atoms share *three electron pairs*, producing a **triple covalent bond** that is written as N≡N. The electron-sharing arrangement in N_2 is illustrated in Figure 3.22.

double covalent bond A covalent bond formed when atoms share two pairs of electrons.

triple covalent bond A covalent bond formed when atoms share three pairs of electrons.

bond energy The energy input required to break a chemical bond.

The number of shared electron pairs between two atoms affects both the length and the strength of a covalent bond. The *bond length* is the distance between the nuclei of the two atoms joined by the bond. The strength of a bond is measured by the **bond energy,** which is the energy input required to break the bond and separate the atoms. Covalent bonds with higher bond energies are stronger than those with lower bond energies. In general, *double covalent bonds are stronger and shorter than single covalent bonds, and triple covalent bonds are stronger and shorter than double covalent bonds.* This trend occurs because of the number of electron pairs that are shared within each type of covalent bond.

This rule suggests that the bond energy for N_2 is greater than the bond energy for O_2, which is indeed the case. Nitrogen gas is mostly unreactive because a large amount of energy is required to break the triple covalent bond that joins the two N atoms. When we inhale N_2 from the atmosphere, we exhale exactly the same amount because the gas is completely inert within our body. Breaking the strong bond in N_2 usually requires high temperature and pressure within a specialized industrial facility.

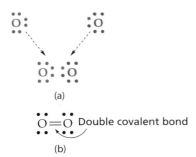

(a)

(b)

FIGURE 3.21 Covalent bonding in oxygen gas, O_2. (a) Lewis dot structure of two oxygen atoms, which combine to form an O_2 molecule. The atoms and electrons are color-coded to highlight that each oxygen atom has two shared and two unshared electron pairs. (b) The O_2 molecule has a double covalent bond, composed of two electrons from each oxygen atom.

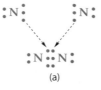

(a)

(b)

FIGURE 3.22 Covalent bonding in nitrogen gas, N_2 (a) Two nitrogen atoms can achieve a stable octet by sharing three electron pairs. (b) The two atoms in an N_2 molecule are joined by a triple covalent bond.

CONCEPT QUESTION 3.1

Usually, we pay no attention to the nitrogen gas that passes through our body because it is chemically unreactive. However, there is one situation in which nitrogen gas can affect us. When deepwater divers ascend to the water surface, they can suffer from a condition known as "the bends." Investigate this topic using the Internet. What causes the bends? How is nitrogen gas involved? What are the symptoms? What is the treatment?

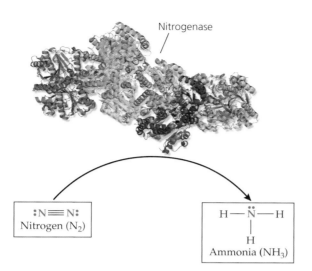

Nitrogenase

:N≡N:
Nitrogen (N₂)

H—N̈—H
|
H
Ammonia (NH₃)

FIGURE 3.23 Nitrogen fixation by bacteria. Some bacteria contain an enzyme called nitrogenase that accelerates the chemical conversion of nitrogen (N_2) to ammonia (NH_3). This reaction requires breaking the strong triple covalent bond in N_2. Nitrogenase is represented as a "ribbon diagram" that indicates the overall structure of the protein.

We now have to answer a puzzling question. Many biological molecules in our body contain nitrogen atoms. If we don't use the N_2 that we inhale from the air, where do these nitrogen atoms come from? We acquire them through our diet. The roots of plants contain bacteria that can convert nitrogen gas from the atmosphere into nitrogen-containing compounds such as ammonia (NH_3). These compounds are absorbed by the plants and then pass into our body when we eat these plants. To covert N_2 into NH_3, the bacteria must accomplish the difficult chemical task of breaking a triple covalent bond. This breakdown is achieved through the action of a protein called *nitrogenase* (Figure 3.23). This protein functions as an enzyme, a biological catalyst that makes reactions happen faster. Through the action of the nitrogenase enzyme, the bacteria are able to break the strong N_2 bond without requiring high temperature and pressure.

Core Concepts

- Molecules can achieve a stable electron arrangement by forming **double** or **triple covalent bonds**.
- The double bond in O_2 involves *two shared electron pairs*, whereas the triple bond in N_2 contains *three shared electron pairs*. Double bonds are shorter and stronger than single bonds; triple bonds are shorter and stronger than double bonds.
- The strong triple bond in N_2 is difficult to break, which makes the molecule unreactive. This is why our bodies do not use the nitrogen gas we breathe.
- The N atoms contained in the biological molecules within the human body all derive from our diet. Some bacteria produce an enzyme called *nitrogenase* that enables the conversion of N_2 gas into ammonia, which can then be utilized by plants.

Molecular Structure of Carbon Dioxide

Carbon dioxide is the fourth most abundant gas in dry air, present at a level of approximately 0.04%. Despite its low abundance, CO_2 plays an important role in life on Earth. Carbon dioxide is exhaled by humans as a product of respiration. Plants and bacteria use it during photosynthesis, a chemical reaction that utilizes light energy to synthesize sugar molecules from CO_2 and H_2O. In addition, carbon dioxide in the atmosphere has an important role in regulating the Earth's climate. The *Science in Action* feature examines the measurement of atmospheric CO_2 and what it means for the future of our planet.

The worked example presented next shows how to draw the Lewis dot structure and the two- and three-dimensional structures of CO_2. After you have reviewed this example, you can try a similar exercise with another molecule, hydrogen cyanide (HCN).

THE KEY IDEA: Carbon dioxide is a product of human respiration and a starting point for photosynthesis.

WORKED EXAMPLE 3.3

Question: Draw the following structures for carbon dioxide (CO_2).

(a) Lewis dot structure

(b) Two-dimensional structure

(c) Three-dimensional structure

Answer: (a) Draw the Lewis dot structure for the CO_2 molecule.

We begin by placing carbon in the center of the molecule, with one oxygen atom on each side. All of the atoms can achieve a stable octet of electrons if the carbon atom shares two electron pairs with each oxygen atom. In the structure provided here, the oxygen atoms and their electrons are highlighted in red.

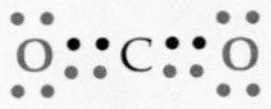

$$\ddot{O}::C::\ddot{O}$$ The C and O atoms can achieve stable electron octets by sharing two pairs of electrons

(b) Draw the two-dimensional structure, indicating whether CO_2 contains single, double, or triple covalent bonds. Include any lone pairs in the structure.

Because each shared pair of electrons constitutes a covalent bond, a *double bond* forms between the central carbon and each oxygen atom. In the two-dimensional structure, this double bond is written as two parallel lines. Each oxygen atom has two lone pairs that are not involved in chemical bonding.

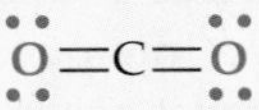

$$\ddot{O}=C=\ddot{O}$$ CO_2 has two **double covalent bonds**

(c) Use VSEPR to predict the three-dimensional geometry of CO_2.

In general, VSEPR predicts that a molecule will adopt a three-dimensional geometry that minimizes repulsion between all the electron pairs. All of carbon's electrons are involved in covalent bonding, so the carbon atom has no nonbonding electron pairs that will repel the bonding pairs The molecule will therefore adopt a *linear geometry* (i.e., all of the atoms lie in a straight line). As shown below, the linear structure minimizes the repulsion between the bonding electron pairs in the two C=O covalent bonds. The lone pairs on the O atoms extend outward and do not affect the molecule's linear geometry.

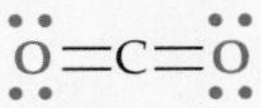

$$\ddot{O}=C=\ddot{O}$$ CO_2 is a **linear** molecule

Postscript: Why Is CO_2 not Bent like H_2O?

A common error is to draw the three-dimensional structure of CO_2 as *bent* like the structure of H_2O. You may be tempted to think that the shapes of these two molecules are the same because they both have three atoms. However, we need to remember *why* H_2O is bent—namely, the repulsive effect of the two lone pairs on the central O atom. By contrast, *the central C atom in CO_2 does not have any lone pairs, so there is no repulsive effect that bends the molecule.*

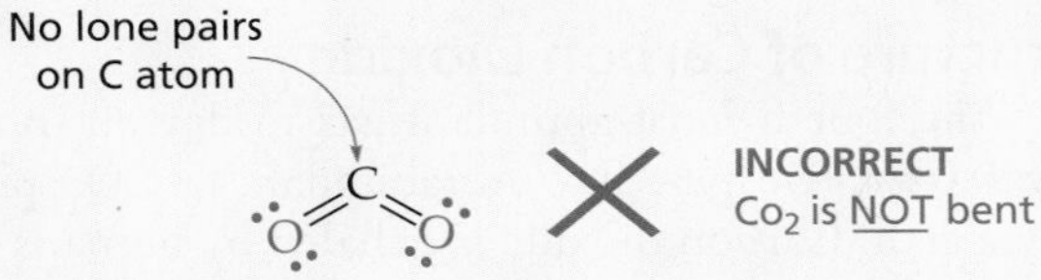

No lone pairs on C atom

$$\ddot{O}{:}{=}C{=}{:}\ddot{O}$$ ✗   **INCORRECT** Co_2 is **NOT** bent

TRY IT YOURSELF 3.3

Question: Hydrogen cyanide, HCN, is a poisonous gas. Draw the following structures for this molecule:

(a) Lewis dot structure

(b) Two-dimensional structure

(c) Three-dimensional structure

SCIENCE IN ACTION

Measuring Carbon Dioxide in the Atmosphere

Global climate change is a defining issue of the 21st century. Data records for the last century reveal an average rise in global temperature of approximately 0.75°C (3.3°F). The rate of temperature increase is getting faster. At the time of writing, the 10 warmest years on record for the planet have all happened since 1998.

What has caused this temperature increase? The Earth's temperature is controlled by its atmosphere via the greenhouse effect. Some of the heat energy emitted by the Earth is absorbed by molecules in the atmosphere, and then emitted back toward Earth. This process causes an overall warming of the planet. Without the greenhouse effect, the Earth's average temperature would be −18°C (0°F). Because this temperature is below the freezing point of water, life as we know it could not exist.

The gases that contribute to the greenhouse effect are called *greenhouse gases*. With the exception of water vapor, which is variable, the most abundant greenhouse gas is carbon dioxide (CO_2). Is there a connection between rising global temperatures and the amount of CO_2 in the atmosphere? How could we find out?

Beginning in 1958, a scientist named Charles David Keeling (1928–2005; Figure 1) began taking regular CO_2 measurements at the Mauna Loa Observatory in Hawaii. Figure 2 summarizes 50 years of atmospheric CO_2 measurements from the observatory. The concentration of CO_2 in a gas sample is quantified in units of *parts per million (ppm)*. A level of one ppm means that one-millionth of the volume of an air sample is occupied by CO_2. Before reading further, take a close look at the graph in Figure 2. What do you observe?

The data in the graph reveal two trends. First, there is an annual trend in which the CO_2 concentration peaks in May and then drops until October. These months correspond to spring and fall in the Northern Hemisphere. The variation in annual CO_2 concentration arises from *photosynthesis*, the process whereby plants utilize CO_2, H_2O, and energy from sunlight to produce sugar molecules. In the spring, seasonal trees and plants acquire more green leaves, which absorb CO_2 from the atmosphere. In the fall, this foliage begins to die, which reduces the global level of photosynthesis. The annual variation is dominated by the Northern Hemisphere because it contains a greater land area—and much more vegetation—than the Southern Hemisphere.

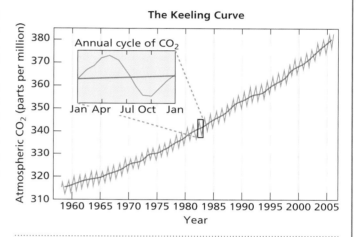

FIGURE 2 **Measurements of atmospheric carbon dioxide taken at the Mauna Loa Observatory.** The oscillating green line and the upper left inset show the annual cycle of varying CO_2 concentration. The red line shows the annual average for each year between 1958 and 2008. This graph is called the Keeling Curve in honor of Charles David Keeling.

(Source: http://en.wikipedia.org/wiki/Charles_David_Keeling)

FIGURE 1 **Charles David Keeling (1928–2005).** Keeling devoted his career to measuring levels of atmospheric CO_2. In 2002, he received the National Medal of Science, the nation's highest scientific honor. Keeling personally collected CO_2 data at the Mauna Loa Observatory for almost 50 years. Following his death, his son, Ralph Keeling, has continued this scientific project.

As a second trend, examine the red line that reports the CO_2 concentration as an *annual average*. When Keeling began his measurements, this annual average was 315 ppmv; 50 years later, it had risen to 385 ppmv, an increase of more than 20%. At this time of writing, the global CO_2 concentration has exceeded 400 ppmv.

What has caused this increase in global CO_2? All of the scientific evidence points toward CO_2 emissions from burning fossil fuels, especially coal, petroleum, and natural gas. Modern industrial society depends on a vast supply of energy, most of which is generated by fossil fuels. Because CO_2 lingers in the atmosphere for many decades, our actions have already made it inevitable that the 21st century will be warmer than at any other point in human history. How much warmer remains an uncertain question whose answer depends on the collective global response to limiting future CO_2 emissions.

3.5   Beyond the Octet Rule

Throughout this chapter, we have used the octet rule to deduce the structure of molecules. When applied to covalent bonds, this rule states that many atoms (especially C, N, and O) share electrons to achieve an octet of electrons in their valence shell, which resembles the stable electron configuration of a noble gas. However, the octet rule does not always apply, and some molecular structures violate the rule. Some of these exceptions are important for understanding biological molecules, so we explore a few of those examples here.

Radicals

As we have learned, oxygen is an essential component of human life. However, oxygen's involvement in chemical reactions within the cell can lead to the formation of dangerously reactive compounds. One of these compounds is called the *hydroxyl radical*, which has the chemical formula OH. We can understand why this compound is dangerous by drawing the Lewis dot structure (Figure 3.24). In stable molecules like H_2O, the oxygen atom is surrounded by eight electrons. However, the Lewis dot structure of the *hydroxyl radical* violates this rule because the oxygen is surrounded by only seven valence electrons. As a result of this electron deficiency, the oxygen atom has an *unpaired electron*. An unpaired electron is an inevitable consequence of having an odd number of total electrons in the molecule.

In general, *a chemical species with an unpaired electron is called a* **radical** (or a *free radical*). The unpaired electron makes a radical especially reactive. Electrons are more stable in pairs, so a radical will attack a nearby atom or molecule to obtain another electron. For example, a reactive hydroxyl radical can damage the DNA in a cell. Radicals are often written in a way that explicitly shows the unpaired electron; for example, the hydroxyl radical is written as •OH.

THE KEY IDEA: A radical contains an unpaired electron and is very reactive.

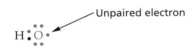
Unpaired electron

FIGURE 3.24   The Lewis dot structure of a hydroxyl radical. The O atom is surrounded by seven valence electrons, not eight, which leaves an unpaired electron.

radical   A reactive molecule that contains an unpaired electron.

Core Concepts

- A **radical**, such as a hydroxyl radical (•OH), violates the octet rule by having an electron deficiency in the valence shell, which produces an unpaired electron.
- The unpaired electron makes the radical highly reactive toward other substances.

THE KEY IDEA: Elements in the third row of the periodic table (e.g., P, S) can accommodate more than eight electrons in their valence shell.

Expanded Valence

Many biological molecules contain phosphorus and sulfur atoms. We can gain some insight into the chemical properties of these atoms by examining their position within the periodic table. Phosphorus (P) is located in Group 5B, so we expect its chemistry to resemble that of the element above it—nitrogen. Sulfur (S) occurs in Group 6B, so we expect it will behave similarly to oxygen. In some cases these predictions are borne out, but in other cases they are not.

Let's examine phosphorus in more detail, especially the types of molecules it can form with chlorine. Because phosphorus is in Group 5B, we can write a Lewis dot structure in which the P atom shares electron pairs with *three* Cl atoms to create a molecule called *phosphorus trichloride* (PCl_3). As indicated in Figure 3.25, each atom within the molecule achieves a stable octet of electrons. Using VSEPR, we predict that PCl_3 has a trigonal pyramidal geometry. So far, so good—phosphorus is behaving exactly as we would expect.

But the situation becomes more complicated because phosphorus can also form a *different* molecule called *phosphorus pentachloride*, PCl_5 (Figure 3.26). Phosphorus now shares electron pairs with *five* chlorine atoms instead of three. If each Cl atom has a stable octet, this electron-sharing arrangement can exist only if there are *10 electrons* in the valence shell of the phosphorus atom. This electron configuration is a clear violation of the octet rule.

The PCl₅ molecule is an example of **expanded valence**. We apply this term to a molecule containing a central atom (such as P) that has *more than eight electrons in its valence shell*. As a result, this atom can form more covalent bonds than we predict based on the octet rule. Applying VSEPR to PCl₅ predicts a three-dimensional geometry called *trigonal bipyramidal*. In this structure, the P atoms and the Cl atoms create a flat triangle that forms a pyramidal base. The other two Cl atoms, above and below the P atoms, are located at the apex of two pyramids (hence the term *bi*pyramidal).

expanded valence A molecule containing a central atom that has more than eight electrons in its valence shell.

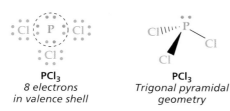

PCl₃
8 electrons in valence shell

PCl₃
Trigonal pyramidal geometry

FIGURE 3.25 The Lewis dot structure and three-dimensional structure of PCl₃. The phosphorus atom in this molecule obeys the octet rule because it has eight electrons in its valence shell. For clarity, the three-dimensional structure omits the lone pairs on the Cl atoms.

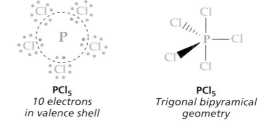

PCl₅
10 electrons in valence shell

PCl₅
Trigonal bipyramical geometry

FIGURE 3.26 The Lewis dot structure and three-dimensional structure of PCl₅. The phosphorus atom in this molecule violates the octet rule because it has 10 electrons in its valence shell. For clarity, the three-dimensional structure omits the lone pairs on the Cl atoms.

CHEMISTRY AND YOUR HEALTH

Nitric Oxide Is a Chemical Messenger

For over 100 years, physicians have been aware that the explosive nitroglycerin has a strong beneficial effect on heart patients. Alfred Nobel, who invented dynamite by blending nitroglycerin with clay, found it ironic that he needed treatment with nitroglycerin for angina, a painful condition in which the arteries of the heart are blocked. Why does nitroglycerin have potent effects on heart disease?

The answer was finally traced to a gas molecule, nitric oxide (NO). The molecule is very reactive and combines readily with many compounds in the body. This property makes it ideal for sending chemical messages because it can stimulate a rapid response but lasts only a short time after production. NO has a strong action on smooth muscle, which occurs in the walls of blood vessels, the intestines, and reproductive organs.

Why is NO so chemically reactive? We can provide the answer by drawing the Lewis dot structure for NO. When we try to construct this molecule using the octet rule, we encounter a problem. An N atom has 5 valence electrons and an O atom has 6, for a total of 11. Because this is an odd number of valence electrons, it is impossible to produce a stable octet around each atom. The best we can do is to construct the Lewis dot structure shown in Figure 1(a). This electron arrangement leaves the nitrogen atom with an unpaired electron, which is highlighted in Figure 1(b). Consequently, nitric oxide is classified as a radical and written as NO.

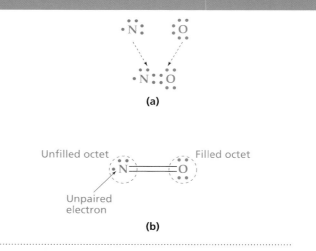

FIGURE 1 The NO molecule is a radical. (a) Lewis dot structure for nitric oxide (NO). (b) The odd number of electrons in NO leaves the N atom with an

NO molecules play an essential role in stimulating blood vessels to open wider, thereby allowing more blood to flow to various tissues. Nitric oxide is the principal regulator of human blood pressure, and it also triggers the blood flow that causes penile erections. Understanding the NO signaling pathway was a key factor in developing erectile dysfunction drugs such as Viagra, Cialis, and Levitra.

How can this violation of the octet rule occur? In Chapter 2, we learned that the third electron shell has a maximum capacity of eight electrons. However, we also explained that this number is an approximation. In addition to accommodating eight electrons, the third electron shell contains a higher-energy "subshell" that is capable of holding an additional 10 electrons. This extra electron capacity is utilized in cases of expanded valence such as PCl_5. The second electron shell does not have this additional subshell, so C, N, and O atoms cannot accommodate extra electrons. This explains why nitrogen combines with chlorine to form NCl_3 but never NCl_5.

Sulfur is another third-row element that can produce molecules with expanded valence. Because sulfur is located below oxygen in Group 6B of the periodic table, we expect it to share some of oxygen's chemical properties. In some cases, it does—for example, sulfur combines with hydrogen to form hydrogen sulfide, H_2S, a gas that is responsible for the unpleasant smell of rotting eggs. The chemical formula for hydrogen sulfide is similar to H_2O. However, sulfur can also react with fluorine to form sulfur hexafluoride, SF_6, which violates the octet rule.

PRACTICE EXERCISE 3.2

(a) Draw the Lewis dot structure for SF_6. How many electrons exist in the valence shell of the sulfur atom? (b) Apply VSEPR to predict the three-dimensional geometry of SF_6 and sketch the structure of the molecule.

Core Concepts

- **Expanded valence** occurs when the central atom in a molecule contains more than eight electrons in its valence shell, resulting in a larger number of covalent bonds. Examples of expanded valence include PCl_5 and SF_6.
- Atoms in the third row of the periodic table, such as phosphorus and sulfur, can accommodate additional electrons in a higher-energy subshell.

3.6 | Ionic Bonding

Learning Objective:

Explain the formation of ionic compounds.

🔑 THE KEY IDEA: Table salt—sodium chloride—is composed of charged sodium and chloride ions held together by ionic bonds.

■ **ionic compound** A compound composed of oppositely-charged ions ; for example, NaCl (table salt).

As we learned in Chapter 2, atoms can gain or lose electrons to form charged ions. These ions can form compounds that are held together by attractions between their electrical charges. We now examine the chemistry of these compounds, beginning with a familiar example—table salt.

The Ionic Composition of Table Salt

Recall from Chapter 2 that the chemical name for table salt is sodium chloride and its chemical formula is NaCl. If you pour a little salt into your hand and look closely, you can see that it has a solid crystalline structure. In general, a *crystal* is an arrangement of atoms, ions, or molecules that repeats in a regular pattern. When we examine salt crystals under the microscope, we can see they have the shape of a cube (Figure 3.27).

The sodium chloride crystal is an **ionic compound** composed of two types of charged ions—a sodium ion (Na^+) and a chloride ion (Cl^-). As shown in Figure 3.28, these ions are formed when a sodium atom transfers its single valence electron to a chlorine atom. The resulting Na^+ ion has a net positive charge because it has lost one electron, whereas the Cl^- ion has a net negative charge because it has gained one electron. By convention, when an element forms a positively charged ion, the name of the ion does not change (e.g., a sodium atom forms a sodium ion). However, when an element forms a negatively charged ion, the name of the ion includes *-ide* at the end (e.g., a chlorine atom forms a *chloride* ion).

Why does sodium give up its electron to chlorine? We can answer this question by considering the valence electrons for each atom. Sodium is a member of Group 1A in the periodic table, so it has a single valence electron. Chlorine is a member of Group 7A and therefore has seven valence electrons. Both atoms are very close to having the stable electron arrangement of a noble gas. When the Na atom gives up the only electron in its outer orbit, the resulting Na^+ ion has a filled electron shell like the noble gas neon (the element before it in the periodic table). Similarly, when the Cl atom adds an electron to its valence shell, the resulting Cl^- ion achieves the same stable outer octet as argon (the element that follows it in the periodic table).

After the Na^+ and Cl^- ions have been formed, they attract each other because of their opposite electrical charges. *The electrical attraction between two oppositely charged ions is called an* **ionic bond**. An ionic bond is different from a covalent bond, in which atoms *share* an electron pair.

Figure 3.29 shows the spatial arrangement of Na^+ and Cl^- ions in a crystal of NaCl. The figure shows that the NaCl crystal is composed of many small cubes. The Na^+ and Cl^- ions are located at the corners of each cube and occupy alternating positions. In other words, each Na^+ ion is closest to a neighboring Cl^- ion (and vice versa). This placement of ions serves to maximize the attraction between *opposite* charges (Na^+ and Cl^-) and minimize the repulsion between *similar* charges (two Na^+ ions or two Cl^- ions). Also shown in Figure 3.29 is a space-filling model of the NaCl crystal, which represents the different sizes of the Na^+ and Cl^- ions. The size of each ion depends on the volume occupied by its electrons.

FIGURE 3.27 Sodium chloride crystals. The structure of sodium chloride crystals as viewed through a microscope.

ionic bond The electrical attraction between two oppositely charged ions.

Na$\bullet$ $\overset{\cdot\cdot}{\underset{\cdot\cdot}{Cl}}$ $\longrightarrow$ Na$^+$ $\overset{\cdot\cdot}{\underset{\cdot\cdot}{:Cl:}}$

FIGURE 3.28 Ionization of Na and Cl atoms. The transfer of an electron (shown in red) from a sodium atom to a chlorine atom produces two ions, Na^+ and Cl^-. Each ion has a stable electron arrangement that resembles a noble gas.

CONCEPT QUESTION 3.2

The Na⁺ ion is smaller than the Na atom from which it was formed. Explain this size difference.

Core Concepts

- Sodium chloride is an **ionic compound** composed of Na^+ and Cl^- ions. These ions are formed when an Na atom transfers its valence electron to a Cl atom.
- The electrical attraction between oppositely charged ions is called an **ionic bond**. It is different from a covalent bond, in which atoms share electrons.
- A sodium chloride crystal is composed of many small cubes, with Na^+ and Cl^- ions occupying alternating positions at the corners of each cube.

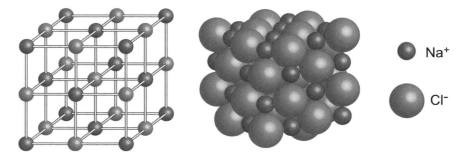

Na$^+$

Cl$^-$

FIGURE 3.29 Sodium chloride forms an ionic crystal. A crystal of sodium chloride contains many small cubes, with Na^+ and Cl^- ions positioned at the corners of the cubes. The lines in the cube indicate the geometry of the crystal and do not represent covalent bonds.

TABLE 3.1 Common Polyatomic Ions	
POLYATOMIC ION	**CHEMICAL FORMULA**
Bicarbonate	HCO_3^-
Carbonate	CO_3^{2-}
Hydroxide	OH^-
Nitrate	NO_3^-
Phosphate	PO_4^{3-}
Sulfate	SO_4^{2-}
Ammonium	NH_4^+

🔑 THE KEY IDEA: Some ions arise from a single atom (e.g., Na⁺), whereas others are polyatomic (e.g., CO_3^{2-}).

simple ion An ion that is formed when a single atom gains or loses electrons.

cation A positively charged ion; for example, Na⁺.

anion A negatively charged ion; for example, Cl⁻.

polyatomic ion An ion composed of two or more atoms that are held together by covalent bonds.

A Variety of Ions

Both Na⁺ and Cl⁻ are examples of a **simple ion**, which is formed when a *single atom* gains or loses electrons. We can predict the charge on a simple ion by examining the number of valence electrons in its parent atom. As we discussed in Chapter 2, *atoms gain or lose valence electrons to form a charged ion with the electron configuration of a noble gas.* This principle is another application of the octet rule, discussed in Section 3.3. Figure 3.30 illustrates a variety of simple ions whose charges can be predicted in this manner.

*Positively charged ions (e.g., Na⁺, Ca²⁺) are called **cations**, and negatively charged ions (e.g., Cl⁻, O²⁻) are called **anions**.* These names arise from the behavior of the ions when they are dissolved in water in the presence of metal electrodes; we will examine this topic more fully in Chapter 9. A positively charged ion in water moves toward the negative electrode, which is called the *cathode* (hence the name *cation*). By contrast, a negatively charged ion in water moves toward the positive electrode, which is the *anode* (hence the name *anion*). If you like cats, you can remember the connection by the phrase "cats are positive." Alternatively, you can remember that an anion is "a negative ion" (*a̶n̶e̶g̶a̶t̶i̶v̶e̶ion = anion*).

In contrast to simple ions, a **polyatomic ion** is a charged compound containing two or more atoms that are held together by covalent bonds. The collection of atoms in these ions must gain or lose electrons to maintain a stable bonding arrangement. The structures for four polyatomic ions are shown in Figure 3.31:

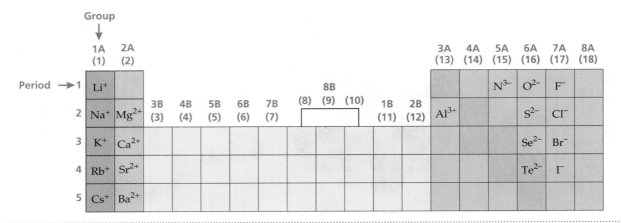

FIGURE 3.30 Simple ions formed from elements in the periodic table. Each ion has the stable electron configuration of a noble gas.

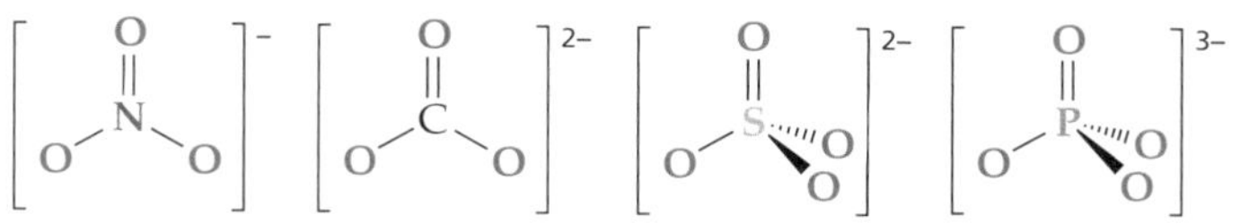

Nitrate ion NO_3^- | Carbonate ion CO_3^{2-} | Sulfate ion SO_4^{2-} | Phosphate ion PO_4^{3-}

FIGURE 3.31 The structures of four polyatomic ions. The ions shown are nitrate (NO_3^-), carbonate (CO_{32}^-), sulfate (SO_4^{2-}), and phosphate (PO_4^{3-}).

nitrate (NO_3^-), *carbonate* (CO_3^{2-}), *sulfate* (SO_4^{2-}), and *phosphate* (PO_4^{3-}). In each case, a central atom (N, C, S, or P) is covalently bonded to three or four oxygen atoms. By convention, the structure of the polyatomic ion is enclosed in square brackets, and the total ionic charge is written at the upper right. In essence, a polyatomic ion is a "charged molecule." Table 3.1 lists some polyatomic ions commonly encountered in chemistry and biology.

PRACTICE EXERCISE 3.3

Which of the polyatomic ions in Figure 3.29 exhibits expanded valence (see Section 3.5)? Specify the ion(s), identify the atom that violates the octet rule, and specify how many valence electrons surround that atom.

Core Concepts

- A **simple ion** derives from a single atom that gains or loses electrons to achieve a stable electron arrangement that resembles a noble gas. The charge on a simple ion can be predicted using the periodic table.
- A cation has a position charge (e.g., Na^+), whereas an anion has a negative charge (e.g., Cl^-).
- A **polyatomic ion** is a charged compound containing two or more atoms that are held together by covalent bonds.

Ionic Compounds

As we saw in the case of sodium chloride, ions can combine to make an ionic compound that is held together by ionic bonds. We commonly refer to sodium chloride as "salt." However, chemists describe *any* ionic compound as a **salt**. In all salts, the positive and negative electrical charges must be equally balanced so that the compound has no net charge. In NaCl, each positive charge on Na^+ is counterbalanced by a negative charge on Cl^-. We can predict the chemical formula of a salt using the principle of *charge neutralization*: *The ions in a salt combine in a ratio that neutralizes their positive and negative charges.*

THE KEY IDEA: The composition of ionic compounds is based on the principle of charge neutralization.

salt An ionic compound.

For example, calcium forms a compound with chloride ions that is held together by ionic bonds. What is the chemical formula for the calcium chloride salt? A calcium ion (Ca^{2+}) has a double positive charge, so each one needs to be balanced by two chloride ions (Cl^-). Consequently, the chemical formula for calcium chloride is $CaCl_2$. Polyatomic ions also form ionic compounds, and their chemical formulas can be deduced using the same principle. For example, when a sodium ion (Na^+) forms an ionic compound with a carbonate ion (CO_3^{2-}), its chemical formula is Na_2CO_3.

WORKED EXAMPLE 3.4

Question: What is the chemical formula of the salt made from the ions of aluminum and hydroxide?

Answer: Aluminum forms Al^{3+} ions, and the hydroxide ion is OH^-. To form a neutral salt, we need three OH^- ions to balance the charge on one Al^{3+} ion. Therefore, the chemical formula of the salt is $Al(OH)_3$. Note the position of the subscript outside the parentheses.

TRY IT YOURSELF 3.4

Question: Write the chemical formulas for the following ionic compounds:

(a) Magnesium sulfate

(b) Sodium bicarbonate (baking soda)

(c) Potassium phosphate

(d) Calcium hydroxide

(e) Ammonium nitrate

Here are some rules to follow when writing the chemical formulas of ionic compounds:

RULE 1: Every ionic compound contains a cation (positively charged ion) and an anion (negatively charged ion). *For example:* Sodium chloride contains Na^+ and Cl^- ions.

RULE 2: Chemical formulas for ionic compounds do not show the charges on the ions. *For example:* The chemical formula for sodium nitrate is $NaNO_3$, *not* $Na^+NO_3^-$

RULE 3: The chemical formula is written with the cation first and the anion second. *For example:* Write NaCl, *not* ClNa.

RULE 4: Ionic compounds are neutral. The total positive charge on the cations must balance the total negative charge on the anions. *For example:* The ions of potassium (K^+) and carbonate (CO_3^{2-}) combine to form K_2CO_3, *not* KCO_3.

RULE 5: Use parentheses ( ) in the chemical formula when more than one polyatomic ion is present. The subscripts after the parentheses refer to *all* of the atoms contained in the brackets. *For example:* The chemical formula of calcium hydroxide is $Ca(OH)_2$, *not* $CaOH_2$.

Core Concepts

- An ionic compound is composed of ions held together by ionic bonds.
- The chemical term **salt** refers to any ionic compound, not just sodium chloride.
- This principle of *charge neutralization* can be used to predict the chemical formula for an ionic compound. The ions in a salt combine in a ratio that *neutralizes their positive and negative charges.*

CHAPTER 3
VISUAL SUMMARY

3.1   Why Do We Need Air to Survive?

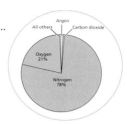

Learning Objective:

Explain why breathing air is necessary for human survival.

⊙ Air contains a mixture of gases. It is mostly composed of oxygen (21%) and nitrogen (78%).

⊙ A molecule is a combination of atoms joined by chemical bonds. The composition of a molecule is represented by its chemical formula (e.g., O_2, N_2, CO_2).

⊙ Humans use the oxygen they breathe as a component in chemical reactions that generate chemical energy (stored as ATP) within the body's cells. A constant supply of ATP is essential to sustain life.

3.2   Covalent Bonding

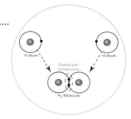

Learning Objective:

Explain the formation of a covalent chemical bond.

⊙ When two H atoms share their electrons, the electron pair creates a mutual attraction between the atoms. The sharing of an electron pair between two atoms produces a **covalent bond**.

⊙ A hydrogen molecule (H_2) consists of two H atoms joined by a covalent bond. When bonded together, the two H atoms have a lower energy compared to when they exist in isolation.

⊙ The quantum mechanical model describes a covalent bond in terms of increased electron density in the region between two atomic nuclei.

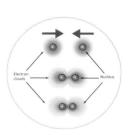

3.3   Making Molecules

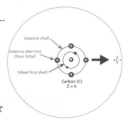

Learning Objective:

Apply the principles of chemical bonding and molecular structure.

- ⊙ An atom forms covalent bonds based on its number of valence electrons—that is, the electrons in the valence shell of the atom.

- ⊙ A **Lewis dot structure** shows the valence electrons for each atom and how they pair to form stable molecule structures. Some molecules include a **nonbonding (lone) electron pair** that is not shared and is associated with a single atom.

- ⊙ The **octet rule** states that many atoms tend to acquire eight electrons in their valence shell, which resembles the electron configuration of a noble gas.

- ⊙ A flat **two-dimensional structure** (also called a Lewis structure) represents each covalent bond as a straight line and includes the nonbonding electron pairs. It does not convey the accurate three-dimensional structure of the molecule.

- ⊙ The **three-dimensional structure** of simple molecules can be predicted by the principle of **valence shell electron pair repulsion (VSEPR)**. All electron pairs in the valence shell repel one another to maximize their separation in three-dimensional space.

- ⊙ In CH_4, which has four bonding electron pairs, this repulsion produces a **tetrahedral geometry** for the molecule with a bond angle of 109.5°. The three-dimensional structures of molecules such as CH_4 are represented on paper by **wedge-and-dash drawings**.

- ⊙ In NH_3 and H_2O, the nonbonding electron pairs exert a stronger repulsion than the bonding pairs do. The repulsive influence of the lone pair reduces the bond angle in comparison to CH_4. NH_3 has a trigonal **pyramidal geometry** with a bond angle of 107°. H_2O has a **bent geometry** with a bond angle of 104.5°.

3.4   Molecules with Double and Triple Bonds

Learning Objective:

Describe double and triple covalent bonds in molecules.

- Molecules can achieve a stable electron arrangement by forming **double** or **triple covalent bonds**.

- The double bond in O_2 involves *two shared electron pairs*, whereas the triple bond in N_2 contains *three shared electron pairs*. Double bonds are shorter and stronger than single bonds; triple bonds are shorter and stronger than double bonds.

- The strong triple bond in N_2 is difficult to break, which makes the molecule unreactive. This is why our bodies do not use the nitrogen gas we breathe.

3.5   Beyond the Octet Rule

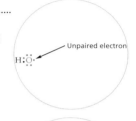

Learning Objective:

Illustrate violations of the octet rule.

- A **radical** (e.g., •OH) violates the octet rule by having an unpaired electron. Radicals are especially reactive and can damage other molecules.

- **Expanded valence** occurs when the central atom in a molecule contains more than eight electrons in its valence shell, resulting in a larger number of covalent bonds.

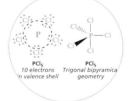

- Elements in the second row of the periodic table (e.g., P and S) can accommodate extra electrons in a subshell of the third electron shell.

3.6   Ionic Bonding

Learning Objective:

Explain the formation of ionic compounds.

- Sodium chloride (NaCl) is an **ionic compound** composed of Na^+ and Cl^- ions. These ions are formed when an Na atom transfers its valence electron to a Cl atom. The electrical attraction between oppositely charged ions is called an **ionic bond**. In a sodium chloride crystal, the Na^+ and Cl^- ions are located at the corners of a cube.

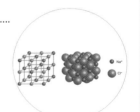

- A **simple ion** derives from a single atom that gains or loses electrons to achieve a stable electron arrangement that resembles a noble gas. The charge on a simple ion can be predicted using the periodic table.

- A **polyatomic ion** is a charged compound containing two or more atoms that are held together by covalent bonds.

- In an ionic compound (a **salt**), the ions combine in a ratio that neutralizes their positive and negative charges.

LEARNING RESOURCES

Reviewing Knowledge

3.1: Why Do We Need Air to Survive?

1. What are the two most abundant gases in the air we breathe?
2. Why do we need to breathe air to survive?
3. Describe the pathway by which inhaled oxygen gas reaches the bloodstream.

3.2: Covalent Bonding

4. Why are helium atoms stable and unreactive?
5. Describe how two H atoms interact to form an H_2 molecule.
6. What is a covalent bond?
7. How does the quantum mechanical model describe covalent bonding in an H_2 molecule?

3.3: Making Molecules

8. Why does the chemical reactivity of an atom depend only on its valence electrons and not the other electrons it contains?
9. What is the octet rule, and how does it help us to understand covalent bonding?
10. Describe the difference between a bonding electron pair and a nonbonding (lone) electron pair in terms of the atoms that contribute the electrons.
11. What general principle allows us to predict the three-dimensional geometry of molecules? Apply this principle to predict the molecular geometry of CH_4, NH_3, and H_2O.
12. Why are the bond angles slightly different in CH_4, NH_3, and H_2O? Which molecule has the largest bond angle? The smallest bond angle?
13. In a space-filling model of NH_3, why does the N atom have a larger atomic radius than the H atom?

3.4: Molecules with Double and Triple Bonds

14. What type of covalent bond is used to join (a) oxygen atoms in O_2 and (b) nitrogen atoms in N_2?
15. Which molecule has the larger bond energy –O_2 or N_2? Explain your answer.

3.5: Beyond the Octet Rule

16. What is the definition of a *radical*? Why do atoms in radicals violate the octet rule?
17. Why are radicals especially reactive toward other chemical substances?
18. Explain why PCl_5 provides an example of expanded valence. Why does the P atom in PCl_5 violate the octet rule?

3.6: Ionic Bonding

19. What is an ionic bond?
20. What is a salt?

21. What is the difference between the type of chemical bonding in a salt and a molecule?

Developing Skills

22. Chlorine (Cl_2) is a poisonous gas used as a chemical weapon during World War I. Draw the Lewis dot structure for Cl_2. How many covalent bonds does this molecule contain?
23. Hydrogen peroxide (H_2O_2) is a common disinfectant. Draw the Lewis dot structure and two-dimensional structure of this molecule.
24. Carbon tetrachloride (CCl_4) is used in dry cleaning. Draw the following structures for this molecule:
    (a) Lewis dot structure
    (b) Two-dimensional structure
    (c) Three-dimensional structure
25. A nitrogen (N) atom can form covalent bonds with one or more fluorine (F) atoms to make a stable molecule that is used in the electronics industry.
    (a) Draw the Lewis dot structure of this molecule (include all none pairs).
    (b) Use the Lewis dot structure to deduce the chemical formula of the molecule.
    (c) Draw the two-dimensional structure (include all lone pairs).
    (d) Draw the three-dimensional structure (lone pairs can be omitted for clarity).
26. Hydroxylamine, which has the chemical formula H_2N—OH, is a compound that reacts with DNA to cause mutations. Draw the Lewis dot structure and two-dimensional structure of hydroxylamine.
27. Phosphoryl chloride ($POCl_3$) is a reactive compound used in synthesizing many organic molecules. Draw the following structures for this molecule:
    (a) Lewis dot structure
    (b) Two-dimensional structure (show all lone pairs)
    (c) Three-dimensional structure (lone pairs can be omitted for clarity)
28. Formaldehyde is a molecule with formula H_2CO. Draw the Lewis structure and two-dimensional structure of formaldehyde. Use VSEPR to predict its molecular geometry.
29. Acetylene is a molecule with chemical formula C_2H_4. Draw the Lewis dot structure for this molecule. What type of covalent bond is used to join the carbon atoms?
30. For each of the chemical species below, draw the Lewis dot structure and identify whether or not it is a radical.
    (a) Hydrogen peroxide, H_2O_2
    (b) Superoxide ion, O_2^-
    (c) Hydroxide ion, OH^-
31. Boron (B) reacts with hydrogen to form a compound called borane (BH_3). (a) Draw the Lewis dot structure for BH_3 (b) Explain why BH_3 violates the octet rule. (c) Apply VSEPR to predict the three-dimensional geometry of BH_3 and draw its molecular structure.

32. Predict the chemical formulas of the salts formed by the following pairs of elements:
    (a) Na and I
    (b) Mg and O
    (c) Ca and F
    (d) Ra and O
    (e) Ar and C

33. Write the chemical formulas of the following ionic compounds that contain polyatomic ions:
    (a) Sodium carbonate
    (b) Magnesium hydroxide
    (c) Aluminum nitrate
    (d) Ammonium chloride

34. Write the chemical formulas of the following phosphate compounds:
    (a) Sodium phosphate
    (b) Magnesium phosphate
    (c) Aluminum phosphate

Exploring Concepts

35. In the late 18th century, a chemist named Joseph Priestley (1733–1804) performed some experiments that have now become famous.
    (a) In the first experiment, Priestley put a burning candle into an airtight container and observed that the flame went out. He then put a mouse into the same container and found that the mouse quickly died. Provide an explanation for these observations.
    (b) In a follow-up experiment, Priestley put both a candle and a plant into the airtight container. This time, the candle's flame continued to burn for several days. He added a mouse to the container, and it survived. Provide an explanation for these observations.

36. One difference between molecules and salts is in their melting point (m.p.), the temperature at which the solid material liquefies. Some m.p. values are shown here. Propose an explanation for these differences.

    H_2 m.p. = $-259°C$ NaI   m.p. = $662°C$
    O_2 m.p. = $-218°C$ NaCl   m.p. = $800°C$
    Cl_2 m.p. = $-101°C$ MgO   m.p. = $2850°C$

37. The noble gas xenon, Xe, has a complete octet, by definition. In a famous experiment, however, Xe was shown to form a compound with fluorine, XeF_4.
    (a) Draw a Lewis dot structure for XeF_4. How many electrons are in the valence shell of xenon?
    (b) Explain your answer to (a) using the principles in this chapter.
    (c) Use the Internet to discover the shape of the molecule and draw its structure.

38. This chapter has explained two types of chemical bonds: covalent bonds and ionic bonds. However, there is a third type called a *metallic bond*. This type of bonding is important because the majority of elements in the periodic table exist as metals. Use the Internet to investigate metallic bonding, and write a one-paragraph description that includes an explanation of why metals conduct electricity.

4

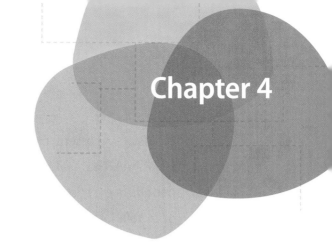

Carbon: The Element of Life

CONCEPTS AND LEARNING OBJECTIVES

FEATURES

Fats are an essential component of our diet, and those we consume are stored in our bodies as fuel. These fat molecules provide us with a source of slow-release energy for extended periods of exercise, such as long-distance running or cycling. However, excess consumption of certain fats can also cause a variety of health problems. In this chapter, we examine various types of fats and their biological effects.

4.1   What Is the Difference Between Saturated and Unsaturated Fats?

Learning Objective:

Distinguish between saturated and unsaturated fats.

THE KEY IDEA: Saturated and unsaturated fats differ because of covalent bonding between their carbon atoms.

- **fat** A class of molecules that the body uses for energy storage.

- **saturated fat** A fat that contains only single covalent bonds between its carbon atoms.

- **unsaturated fat** A fat that contains at least one double covalent bond between its carbon atoms.

The word "**fat**" describes a large class of molecules with similar chemical properties. You may be familiar with the terms **saturated fat** and **unsaturated fat** in connection with food and diet. What you may not realize is that "saturated" and "unsaturated" are terms from chemistry. The U.S. Food and Drug Administration (FDA) provides a recommended dietary allowance (RDA) for different types of nutrients, including fats, based on a 2000 calorie diet. The RDA values are 65 grams (g) of total fat, divided between 20 g of saturated fat and 45 g of unsaturated fat. By law, food labels must list the % daily value (%DV) corresponding to one "serving size" of the food. An example of a food label for a candy bar is shown in Figure 4.1.

To understand these different types of fats, we must study the chemistry of carbon. All fat molecules contain chains of carbon atoms that vary in length. The terms *saturated* and *unsaturated* refer to the types of covalent bonds that link the carbon atoms in these chains. A *saturated fat contains only single covalent bonds*

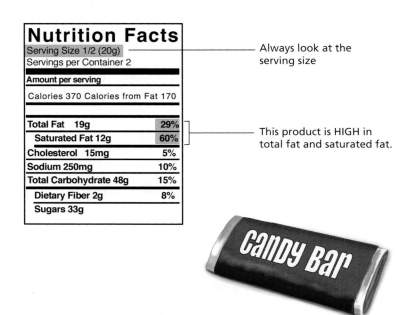

FIGURE 4.1 A food label for a candy bar. The candy contains a high amount of total fat and saturated fat—in fact, just half of this bar provides 60% of the recommended dietary allowance for saturated fat (12 g/20 g = 60 %)

Chemistry of saturated fats

$$-\overset{\displaystyle H}{\underset{\displaystyle H}{C}}-\overset{\displaystyle H}{\underset{\displaystyle H}{C}}-$$

Single covalent bond

Chemistry of unsaturated fats

$$-\overset{\displaystyle H}{C}=\overset{}{\underset{\displaystyle H}{C}}-$$

Double covalent bond

(a)

(b)

FIGURE 4.2 The chemistry of saturated and unsaturated fats. (a) Molecules of saturated fats contain only single covalent bonds between carbon atoms, whereas unsaturated fats contain at least one double covalent bond. Recall from Chapter 3 that a straight line between two atoms represents a covalent bond (a shared pair of electrons), and two lines represent a double covalent bond. The atoms at the left and right are not shown. (b) Foods that have a high content of saturated fats (left) or unsaturated fats (right).

between its carbon atoms, whereas an *unsaturated fat contains at least one double bond joining two carbon atoms.* Figure 4.2 illustrates the difference between these two bonds, along with examples of foods that are high in saturated fats or unsaturated fats. The chemical bonding within the carbon chain affects the properties of the fats. For molecules of the same size, a saturated fat more likely to be solid at room temperature than an unsaturated fat. How do the bonds between carbon atoms affect the properties of fat molecules?

We will answer this question by studying the *chemistry of carbon.* Fats are only one of the multitude of different carbon-containing molecules in living cells. In this chapter, we will examine the remarkably diverse chemistry of carbon, which provides an essential foundation for understanding the structures of most biological molecules and pharmaceuticals. As part of our investigation, we will show how carbon chemistry relates to the fats in our diet and their effect on our body's health.

Core Concepts

- A **saturated fat** contains only single covalent bonds between its carbon atoms. An **unsaturated fat** contains at least one double bond joining two carbon atoms.

4.2   Why Is Life Based on Carbon?

You may have heard the phrase "carbon-based life." In fact, the prevalence of carbon in living organisms gave rise to the term **organic chemistry** to describe the chemistry of carbon. It is often not immediately obvious that many molecules contain carbon atoms. However, you have probably seen burnt toast, as shown in Figure 4.3. Bread contains carbohydrate (sugar) molecules, which include carbon atoms. Excessive heat from the toaster decomposes the carbohydrates, which produces different carbon-containing molecules that are responsible for the black color on the surface of the burnt toast. Other carbon-containing substances are also black, including coal and the graphite used in pencils.

Why is life based on carbon rather than any other element? *Carbon has the most versatile chemistry of any element.* Carbon's capacity for covalent bonding produces more diverse molecular structures than those formed by any other element. Let's explore carbon's special properties.

Learning Objective:

Describe the unique chemical properties of carbon.

🔑 THE KEY IDEA: Life is based on carbon because it has the most versatile chemistry of any element.

organic chemistry The chemistry of molecules containing carbon.

FIGURE 4.3 Biological molecules contain carbon atoms. The black surface of burnt toast arises from carbon-containing molecules that are formed when heat decomposes the carbohydrate (sugar) molecules in the bread.

- **Carbon has four valence electrons and forms four covalent bonds.**

  A carbon atom has *four valence electrons*, and it can achieve a stable octet of electrons by forming *four covalent bonds* with other atoms. In the accompanying figure, the second atom for each covalent bond is not shown because it could be any one of many different types. Carbon's capacity to form four bonds is greater than that of any other element in the second row of the periodic table, such as nitrogen (three bonds) and oxygen (two bonds). You can imagine a carbon atom as acting like the central piece of a jigsaw puzzle, with the capability of linking to four surrounding pieces.

  4 valence electrons   4 covalent bonds

- **Carbon bonds with itself to form single, double, or triple covalent bonds.**

  Carbon atoms form carbon—carbon bonds containing one, two, or three shared pairs of electrons. This capacity to form single, double, or triple covalent bonds increases the number of bonding options available to carbon atoms. For each of the following bonding arrangements, note that all carbon atoms form four covalent bonds. Some atoms have been omitted to focus our attention on the various types of carbon—carbon bonds.

  Single bond Doule bond Triple bond

- **Carbon atoms can create numerous molecular structures.**

  Linking carbon atoms by covalent bonds can produce a variety of molecular structures in three-dimensional space. The carbon atoms can form *linear chains* (with all the atoms in a line), *branched chains* (in which one or more atoms branch off the main chain), or *rings* of various sizes. These carbon arrangements provide a structural foundation for many biological molecules, including fats, cholesterol, and steroid hormones, sugars.

  Linear chain   Branched chain   Ring

- **Carbon can form bonds with many other elements.**

  In addition to bonding with itself, *carbon can form covalent bonds with many other chemical elements*. In biological molecules and pharmaceuticals, carbon is frequently bonded to hydrogen, oxygen, nitrogen, phosphorus, and sulfur. Specific combinations of carbon with these atoms are called *functional groups*, which we will examine in the next chapter.

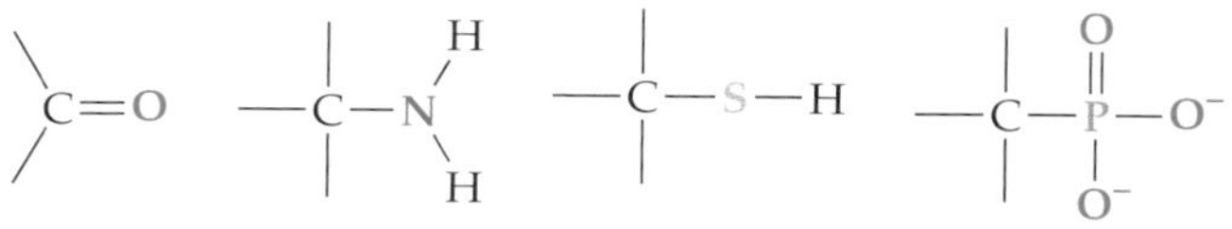

The structures shown:

$$\diagdown C=O \qquad -\overset{|}{\underset{|}{C}}-N\diagup^{H}_{\diagdown H} \qquad -\overset{|}{\underset{|}{C}}-S-H \qquad -\overset{|}{\underset{|}{C}}-\overset{O}{\underset{O^-}{\overset{\|}{P}}}-O^-$$

- **Carbon forms strong covalent bonds that are difficult to break.**

  Carbon atoms make *strong covalent bonds* that are difficult to break. The strength of the covalent bonds in a molecule determines its stability. It is vitally important that biological molecules, such as DNA, retain their structural integrity within the cell. Life could not exist if biological molecules easily fell apart.

Core Concepts

- ■ Life is based on carbon because it has the most versatile chemistry of any element.

4.3 Alkanes: Hydrocarbons with Single Bonds

In general, many biological molecules and pharmaceuticals consist of two components:

- A *structural framework* that provides a stable foundation for the molecule.
- A set of *variable molecular components* that attach to the framework and bestow specific chemical properties.

This chapter focuses on the framework, and the following chapter will examine the diversity of other molecular components.

What Are Hydrocarbons?

The molecular framework of biological and pharmaceutical molecules is often provided by a **hydrocarbon**. In general, *a hydrocarbon is a molecule that contains only hydrogen and carbon atoms.* Hydrocarbons are classified according to the types of carbon–carbon bonds they contain (Figure 4.4). The type of bonding in the molecule affects both its structure and chemical reactivity. Molecules with only single C—C bonds are called **alkanes**. Molecules with at least one C=C double bond are called **alkenes**. Molecules with one or more C≡C triple bonds are called *alkynes*, but they are less common in organic molecules. In this chapter we focus on molecules that are pure alkanes and alkenes.

Core Concepts

- ■ **Hydrocarbons** are molecules containing only hydrogen and carbon atoms.
- ■ **Akane** hydrocarbons contain only single covalent bonds.
- ■ **Alkene** hydrocarbons contain at least one double covalent bond.
- ■ Hydrocarbons frequently provide the structural framework for biological molecules and pharmaceuticals.

Methane

In Chapter 3 we encountered the simplest alkane hydrocarbon—, methane (CH_4). Although it has a single carbon atom and no carbon–carbon bonds, it is still considered an alkane because it contains only single covalent bonds. Figure 4.5 recaps various ways of representing the methane molecule. As we learned in Chapter 3, the tetrahedral structure of CH_4 is explained by valence shell electron pair repulsion.

PRACTICE EXERCISE 4.1

As a review, briefly explain what information is provided by each of the molecular representations of methane in Figure 4.5.

Learning Objective:
Characterize the bonding and structure of alkane hydrocarbons.

⊙ THE KEY IDEA: Hydrocarbon molecules contain only hydrogen and carbon atoms.

hydrocarbons Molecules that contain only hydrogen and carbon atoms.

alkanes Hydrocarbons that contain only single bonds.

alkenes Hydrocarbons that contain at least one C=C double bond.

⊙ THE KEY IDEA: Methane (CH_4) is the simplest hydrocarbon containing single covalent bonds.

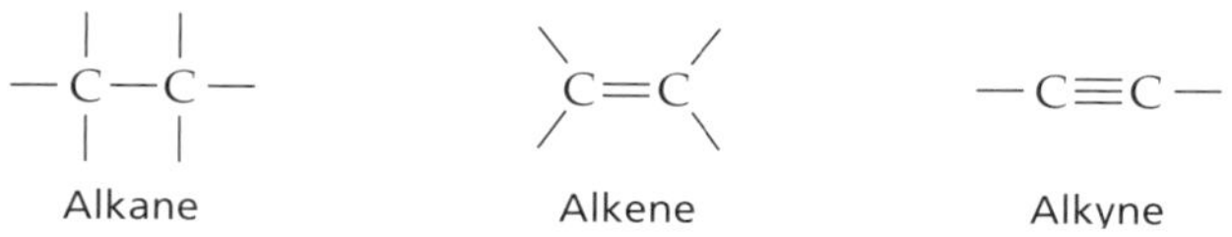

$$-\overset{|}{\underset{|}{C}}-\overset{|}{\underset{|}{C}}- \qquad \diagup^{\diagdown}C=C\diagup_{\diagdown} \qquad -C\equiv C-$$

Alkane   Alkene   Alkyne

FIGURE 4.4 Types of carbon–carbon bonds. Carbon atoms can be joined by three different types of carbon–carbon bonds, which characterize the molecule as an alkane, alkene, or alkyne.

FIGURE 4.5 Molecular representations of methane. Each representation conveys different information about the chemical bonding of molecular structure.

$$CH_4$$

Formula

$$H \overset{\overset{\displaystyle ..}{}}{:} C \overset{}{:} H$$
$$H$$

Lewis structure

2D structure

3D structure

Ball and stick model

Space-filling model

Core Concepts

■ Methane (CH_4) is the simplest alkane molecule.

Ethane

The next alkane hydrocarbon we will examine is ethane, C_2H_6 (Figure 4.6). This is the smallest molecule with a C—C single bond, which is characteristic of an alkane hydrocarbon.

How should we draw the three-dimensional structure of ethane? To begin, imagine that you take a methane molecule and remove one of its hydrogen atoms. This arrangement of atoms is called a **methyl group** and is written as —CH_3 (Figure 4.7). The line represents a bond that attaches the carbon to another atom. Note how the bonds around the carbon atom in the methyl group retain their tetrahedral geometry due to valence shell electron pair repulsion.

Imagine that we join two methyl groups together using a shared C—C bond. Each —CH_3 retains its tetrahedral structure, but the two methyl groups can be oriented in two possible ways with respect to each other. Figure 4.8 illustrates these two possibilities, which arise from how we twist the molecule around the C—C bond. Different molecular structures that arise from rotating a single bond are called *conformers*.

One option is for the two methyl groups at either end of the molecule to have the same spatial orientation and line up with each other. This structure is called the **eclipsed conformation** since each hydrogen atom from one methyl group is directly in front of a hydrogen atom from the other methyl group (think of a solar eclipse when the moon passes directly in front of the sun). The other option is for the two methyl groups to be twisted relative to each other; this arrangement is called the **staggered conformation**.

The difference between the two conformations can be observed by looking at the molecule along the direction of the C—C bond (called the "end view" in Figure 4.8). In the eclipsed conformation, the two sets of H atoms on each

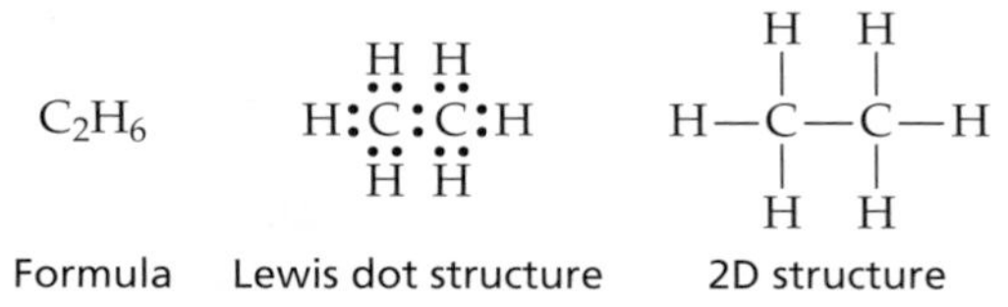

■ **methyl group** An group of atoms with the formula —CH_3.

■ **eclipsed conformation** The conformation in which two methyl groups line up with each other.

■ **staggered conformation** The conformation in which two methyl groups are twisted relative to each other.

$$C_2H_6$$

Formula

Lewis dot structure

2D structure

FIGURE 4.6 Three molecular representations of ethane.

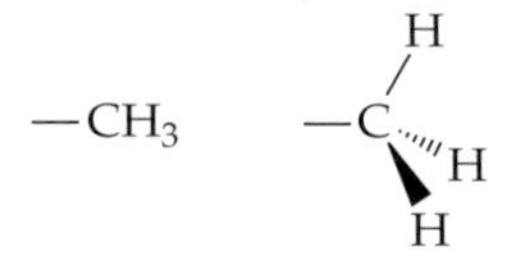

FIGURE 4.7 Chemical formula and molecular geometry of a methyl group.

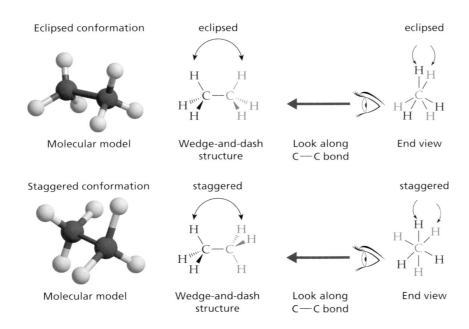

Eclipsed conformation

Molecular model

Wedge-and-dash structure

Look along C—C bond

End view

Staggered conformation

Molecular model

Wedge-and-dash structure

Look along C—C bond

End view

FIGURE 4.8 **The eclipsed (upper) and staggered (lower) conformations of ethane.** The structures on the right show the appearance of the molecule if we look at it along the C—C bond. For the eclipsed conformation, the two sets of H atoms (black and red) overlap each other, but they are shown slightly offset here so the rear H atoms are visible.

methyl group overlap each other, so the structure has three points for the positions of all the H atoms. In the staggered conformation, the H atoms from each methyl group are offset by 60°, which produces a six-pointed shape.

Which conformation—eclipsed or staggered—best describes the structure of ethane? The principle of valence shell electron pair repulsion (VSEPR; see Chapter 3) can provide a clue. In the eclipsed conformation, the C—H bonding pairs are closer to each other compared to the staggered conformation. Consequently, VSEPR predicts that *ethane adopts the staggered conformation because it minimizes the electron pair repulsion within the molecule.* This prediction is correct, although other factors are involved. The energy difference between the two conformations is relatively small, and only a small input of energy is required to rotate the methyl groups around the C—C bond. As we shall see later, the ease of rotation around a single bond is markedly different from the behavior of a C=C double bond.

Core Concepts

- Ethane can exist in an **eclipsed** or **staggered conformation,** depending on the relative orientation of its two methyl groups.
- VSEPR predicts that ethane adopts the staggered conformation because it minimizes the electron pair repulsion. However, the energy difference between the two conformations is relatively small, and rotation around the C—C bond can occur easily.

Naming Hydrocarbons

Methane and ethane are alkane hydrocarbons with one and two carbon atoms, respectively. But there exist many other alkane molecules with more carbon atoms. Chemists have devised a consistent set of rules to name these hydrocarbon molecules. The names of all alkane molecules end in the suffix *–ane*. The first part of the name, such as *meth-* or *eth-* refers to the number of carbon atoms in the molecule. Table 4.1 gives the names of the alkane hydrocarbons up to octane, which contains eight carbons (think of an "octopus").

We can use a similar naming system to describe collections of carbon and hydrogen atoms that can be attached to other molecules. For example, Figure 4.7

THE KEY IDEA: Hydrocarbons are named according to the number of carbon atoms they contain.

TABLE 4.1   Naming Rules for Alkane Hydrocarbons

NUMBER OF C ATOMS	PREFIX	ALKANE HYDROCARBON	CHEMICAL FORMULA
1	meth-	methane	CH_4
2	eth-	ethane	C_2H_6
3	prop-	propane	C_3H_8
4	but-	butane	C_4H_{10}
5	pent-	pentane	C_5H_{12}
6	hex-	hexane	C_6H_{14}
7	hept-	heptane	C_7H_{16}
8	oct-	octane	C_8H_{18}

shows a *methyl* group (—CH_3) that contains one carbon atom and an unattached covalent bond. Similarly, an *ethyl* group (—C_2H_5) contains two carbons and an unattached bond. In general, the prefix (e.g., *meth-, eth-*) describes the number of carbon atoms in the group, while the suffix *–yl* indicates that one H atom has been removed from an alkane to create an unattached bond.

Core Concepts

- Alkanes have systematic names that are based on the number of carbon atoms in the molecule.

THE KEY IDEA: Propane can be used to illustrate the principles of drawing three-dimensional structures.

Propane

The next alkane hydrocarbon we will investigate is propane, with molecular formula C_3H_8. Propane is commonly used as a fuel for home heating or camping. Its various molecular representations are shown in Figure 4.9. The three-dimensional structure of propane shows a tetrahedral geometry for the four single bonds around each carbon atom, which results in a V-shaped arrangement for the three carbon atoms. Also note the *staggered* arrangement of bonds for each adjacent carbon atom, which is caused by electron pair repulsion (recall the example of ethane). The C—H bonds of the methyl groups on each end of the propane

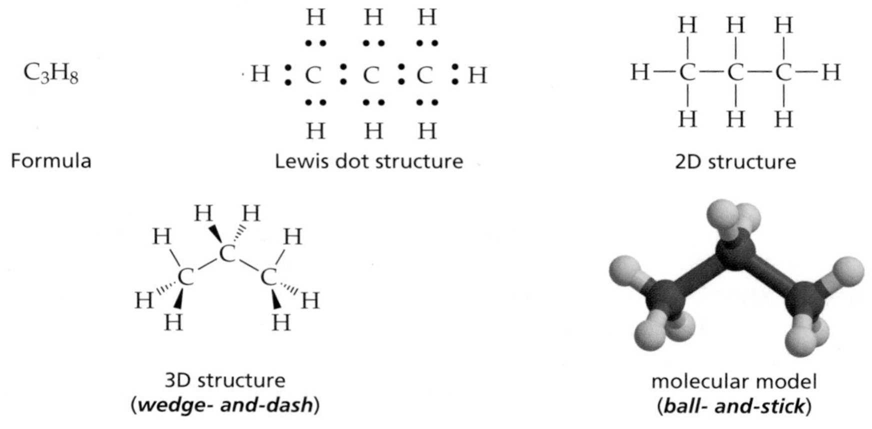

FIGURE 4.9 Molecular representations of propane.

DO
Draw the carbon atoms to show a tetrahedral geometry for the C—C bonds. In propane, this produces a V-shaped structure for the C atoms.

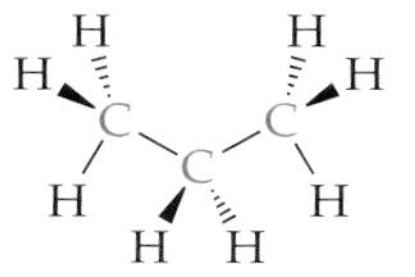

DON'T
Draw the C atoms in a straight line. This structure doesn't accurately show the tetrahedral geometry of the four bonds made by the central C atom.

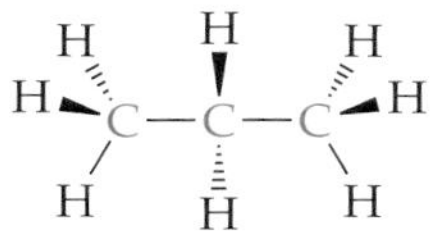

DO
Draw the C—H bonds on adjacent carbons with a staggered geometry due to electron pair repulsion.

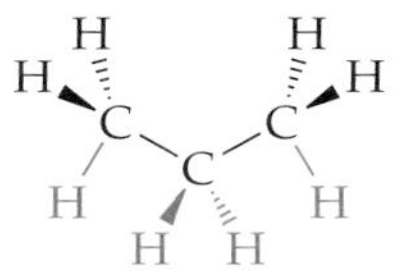

DON'T
Draw the C—H bonds on adjacent carbons with an eclipsed geometry.

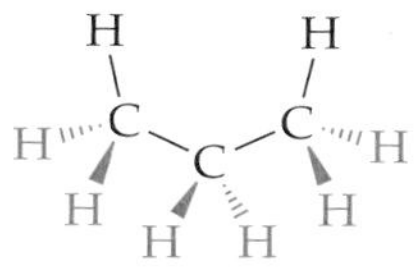

FIGURE 4.10 **Dos and don'ts for drawing the three-dimensional structure of alkanes.** These rules are applied to drawing the three-dimensional structure of propane.

molecule are eclipsed relative to each other because that is the only possible arrangement if the C—H bonds on adjacent carbons are staggered. However, the C—H bonds at each end of the molecule are farther apart (i.e., spaced by two C atoms), so there is less electron pair repulsion compared with C—H bonds on adjacent carbon atoms.

When drawing three-dimensional structures, visualize the molecule as being composed of *particular regions*, each of which obeys the chemical rules of bonding and structure. For example, each of the three carbon atoms in propane has four bonds arranged in a tetrahedral geometry, which you can see by blocking out the other regions of the molecule with your finger or a piece of paper. You can then link these regions together to generate the structure of the entire molecule. Figure 4.10 provides a summary of "dos and don'ts" for drawing three-dimensional molecular structures of alkane hydrocarbons such as propane.

Core Concepts

- Propane illustrates the three-dimensional structure of alkanes.

Butane and Structural Isomers

Butane has four carbon atoms and the chemical formula C_4H_{10}. As the size of hydrocarbon molecules increases, it becomes possible to connect the carbon and hydrogen atoms in *more than one structural arrangement*. We do not change any of the atoms in this process; rather, we rearrange the way in which the atoms are bonded together. We use the term **structural isomers** to describe various molecules that have the same chemical formula but different molecular structures. Butane is the smallest hydrocarbon for which structural isomers exist. Chemists also use *constitutional isomer* as another name for structural isomer (the words are synonyms).

THE KEY IDEA: Structural isomers are molecules with the same atomic composition but different molecular structures.

structural isomers Various molecules that have the same chemical formula but different molecular structures; they are also called constitutional isomers.

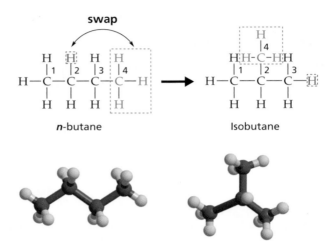

FIGURE 4.11 **Two structural isomers of butane.** These two isomers have the same chemical formula (C_4H_{10}) but different molecular structures. The isomers can be formed by rearranging the position of the methyl group (shown in red) and a hydrogen atom (shown in blue). The isomer on the left is called *n*-butane, and the isomer on the right is called isobutane. The molecular models of these isomers illustrate their different structures.

Figure 4.11 shows the two-dimensional structure of butane in which all the carbon atoms are written in a line. This form of the molecule is called *n*-butane (for "normal"). The carbon atoms have been numbered so that we can identify them. Imagine that we take a methyl group from the end of *n*-butane and swap it with the H atom on carbon #2. We have now created a new structural isomer of butane called *isobutane*. The chemical formula of isobutane is still C_4H_{10} because we have not added or removed any atoms. What differs is how the C and H atoms are connected, which is illustrated by the molecular models in Figure 4.11. We say that *n*-butane is a *linear alkane* because all the C atoms form a "straight-line" chain. By contrast, isobutane is a *branched alkane* because one of the C atoms "branches off" from the main chain of three C atoms.

Note that we use the words "linear" and "branched" to refer to how carbons are connected. Although the carbon atoms are drawn as if they make a straight line in a two-dimensional structure, the molecular model in Figure 4.11 shows that "linear" *n*-butane actually has a zig-zag molecular structure due to the tetrahedral bond geometry around each carbon atom.

Different structural isomers can be represented by *condensed chemical formulas*. For *n*-butane, we write the carbon atoms in a straight line to reflect the linear structure of the molecule. For isobutane, we can use a line to show that one methyl group branches from the main carbon chain. Alternatively, we can use parentheses to represent the branching, in which the "3" subscript outside the parentheses indicates that three methyl groups are attached to a single carbon atom.

$$CH_3CH_2CH_2CH_3 \quad CH_3CHCH_3 \quad \text{or} \quad (CH_3)_3CH$$
$$\overset{|}{CH_3}$$

n-butane   Isobutane

Structural isomers have different physical and chemical properties that arise from their different molecular structures. Although *n*-butane and isobutane are both gases at room temperature, *n*-butane boils at –1.0°C, and isobutane boils at –11.7°C. The boiling temperature of isobutane is lower because the branched structure prevents the molecules from packing together as closely, which makes it easier to convert them from a liquid into a gas.

Only two structural isomers are possible for butane, so using the iso- terminology is unambiguous. When there are more isomers, chemists employ a naming convention based on the *longest continuous carbon chain*. Using this scheme, we can still call our original molecule *n*-butane, with *n*- indicating a linear chain of four carbon atoms. But isobutane has only three carbons in its longest continuous chain, so it is designated as a type of propane. We specify exactly where the methyl group is attached by using the numbering of the carbon atoms in the longest chain (by convention, the carbon atom in the methyl group is not numbered). Since the isomer has a methyl group added to carbon #2, the complete name for the molecule is 2-methylpropane.

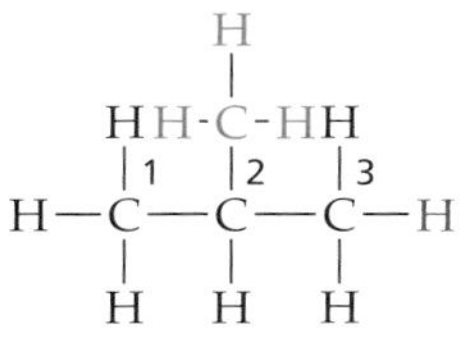

2-methylpropane

When we try to identify structural isomers, it is possible to be tricked by how we draw their molecular structures in two dimensions. For example, let's begin with *n*-butane and swap an H atom attached to carbon #1 with the —CH$_3$ group attached to carbon #3 (Figure 4.12). We may think that we have made a new structural isomer called 1-methylpropane, but we haven't. By repositioning the atoms in the drawing—but keeping all the bonds the same—we end up with *n*-butane again. *For two structural isomers to be unique, their two molecular structures cannot be transformed into each other by rotating them in space.*

PRACTICE EXERCISE 4.2

Using the isomers of butane as an example, write the condensed chemical formulas for the three structural isomers of pentane provided in Worked Example 4.1. Write these formulas on a single line using parentheses.

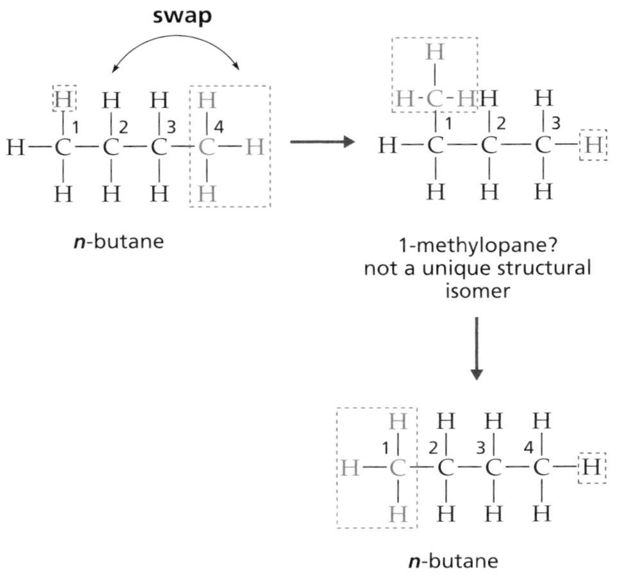

FIGURE 4.12 **Two molecules that are not structural isomers.** Modifying the structure of *n*-butane by swapping an H atom on carbon #1 with a –CH$_3$ group on carbon #3 does not produce a unique structural isomer. The rearrangement of atoms simply re-creates the original structure of *n*-butane.

WORKED EXAMPLE 4.1

Question: Pentane is a hydrocarbon with the chemical formula C_5H_{12}. Identify and name all of the unique structural isomers for this molecule.

Answer: We begin with the simplest isomer, *n*-pentane, which has a linear chain of five carbon atoms with no branching. We number the carbon atoms for reference.

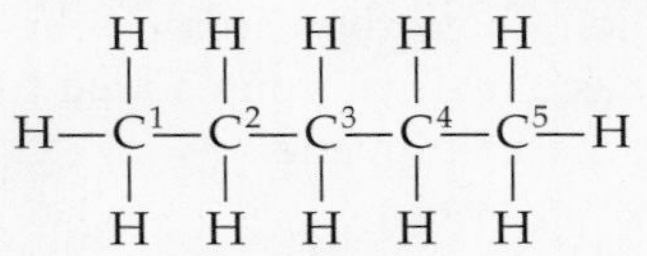

$$H-\overset{\displaystyle H}{\underset{\displaystyle H}{\overset{|}{\underset{|}{C^1}}}}-\overset{\displaystyle H}{\underset{\displaystyle H}{\overset{|}{\underset{|}{C^2}}}}-\overset{\displaystyle H}{\underset{\displaystyle H}{\overset{|}{\underset{|}{C^3}}}}-\overset{\displaystyle H}{\underset{\displaystyle H}{\overset{|}{\underset{|}{C^4}}}}-\overset{\displaystyle H}{\underset{\displaystyle H}{\overset{|}{\underset{|}{C^5}}}}-H$$

_n_-pentane (isomer #1)

As with butane, we can begin forming isomers by taking a methyl group from the end of the molecule (i.e., from carbon 4) and switching it with a hydrogen atom from another carbon atom; let's say carbon 2. This isomer has four carbon atoms in its longest continuous chain and is classified as a butane molecule. Note that we number only the carbons in the longest chain (from 1 to 4) and not the branching carbons. Since the molecule has a —CH_3 branch on carbon 2, the complete name is 2-methylbutane.

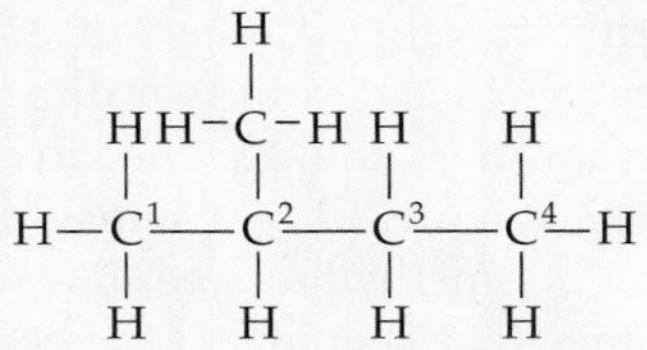

$$H-\overset{\displaystyle H}{\underset{\displaystyle H}{\overset{|}{\underset{|}{C^1}}}}-\overset{\displaystyle H-C-H}{\underset{\displaystyle H}{\overset{|}{\underset{|}{C^2}}}}-\overset{\displaystyle H}{\underset{\displaystyle H}{\overset{|}{\underset{|}{C^3}}}}-\overset{\displaystyle H}{\underset{\displaystyle H}{\overset{|}{\underset{|}{C^4}}}}-H$$

2-methylbutane (isomer #2)

Alternatively, we add the methyl group to carbon 3 instead of carbon 2, which generates the molecule shown below. Is this a unique structural isomer? If we imagine rotating the molecule by 180° (switching the left and right sides), we can see that this structure is exactly the same as 2-methylbutane; it has a chain of four carbon atoms with a methyl branch on the second carbon from the end. Consequently, this molecular arrangement is *not a unique structural isomer.* By convention, we label the carbon atoms by starting at the end that assigns the lowest possible number to the substituent carbon. That is why the correct isomer name is 2-methylbutane and not 3-methyl butane. By similar reasoning, adding a methyl group to carbon 1 does not produce an isomer since that molecular configuration is equivalent to *n*-pentane.

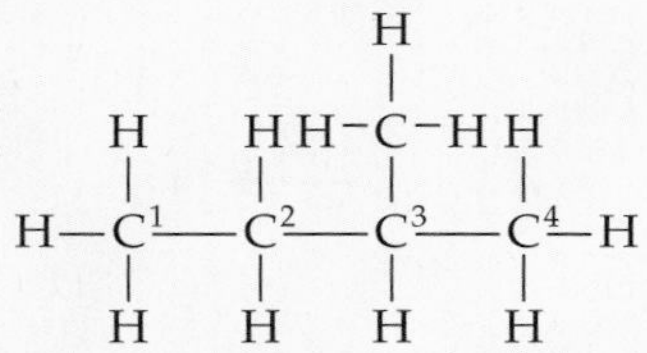

$$H-\overset{\displaystyle H}{\underset{\displaystyle H}{\overset{|}{\underset{|}{C^1}}}}-\overset{\displaystyle H}{\underset{\displaystyle H}{\overset{|}{\underset{|}{C^2}}}}-\overset{\displaystyle H-C-H}{\underset{\displaystyle H}{\overset{|}{\underset{|}{C^3}}}}-\overset{\displaystyle H}{\underset{\displaystyle H}{\overset{|}{\underset{|}{C^4}}}}-H$$

not a unique structural isomer

We can create another isomer by moving *two* methyl groups. If we take the methyl groups from each end and add them both to the central carbon atom, we obtain the following molecular structure:

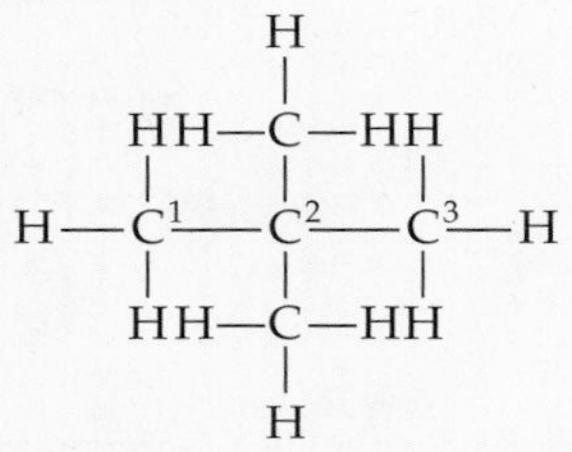

$$H-\overset{\displaystyle H-C-H}{\underset{\displaystyle H-C-H}{\overset{|}{\underset{|}{C^1}}}}... $$

2,2-dimethylpropane (isomer #3)

The longest continuous chain now has only three carbon atoms, numbered 1 through 3, making this a type of propane. Since there are two methyl groups on carbon 2, the conventional name is 2,2-dimethylpropane.

In conclusion, there are *three structural isomers* for the molecular formula C_5H_{12}: *n*-pentane, 2-methylbutane, and 2,2-dimethylpropane. The distinctive structure of each isomer is illustrated by the molecular models shown below.

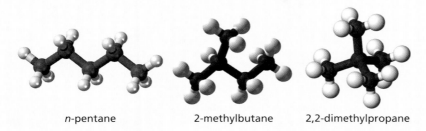

| *n*-pentane | 2-methylbutane | 2,2-dimethylpropane |

TRY IT YOURSELF 4.1

Question: Hexane is a hydrocarbon with the chemical formula C_6H_{14}. Draw and name all five structural isomers for this molecule.

Core Concepts

- **Structural isomers** are variants of a molecule that have the same number and type of atoms but different bonding connections among the atoms.
- Structural isomers exhibit different physical and chemical properties due to their different molecular structures.
- The naming convention for isomers uses the longest continuous carbon chain as a reference point and specifies the carbon atom to which the branching group is attached.

Drawing Hydrocarbons

As hydrocarbon molecules become larger, drawing their molecular structures becomes tedious. Including all of the hydrogen atoms may also obscure key structural features of how the carbon atoms are bonded together. To alleviate these problems, chemists use various shorthand ways to represent complex molecules.

Structures that we have seen up to now are called **full-atom drawings** because they include every atom in the molecule. Another representation is a **carbon-only drawing,** which shows only the carbon atoms and omits the hydrogen atoms. As examples, the carbon-only drawings for the three structural isomers of pentane are shown in Figure 4.13. This simplification clearly conveys the different structural arrangement of carbon atoms in these isomers.

As an alternative, chemists use **line-angle drawings** that include only the *bonds between the carbon atoms.* Each carbon–carbon bond is represented by a straight line. As much as possible, the line-angle drawing attempts to preserve the geometrical arrangement of the bonds.

⊙ THE KEY IDEA: Chemists use various shorthand drawings to represent hydrocarbon molecules.

full-atom drawing A representation of molecules that includes every atom in the molecule.

carbon-only drawing A simplified representation of molecules that includes only the carbon atoms and their bonds.

line-angle drawing A simplified representation of molecules that uses lines to indicate bonds between carbon atoms and the angles of those bonds.

C—C—C—C—C

```
        C
        |
  C — C — C — C
```

```
        C
        |
  C — C — C
        |
        C
```

n-pentane   1-methylbutane   2,2-dimethylpropane

FIGURE 4.13 Carbon-only drawings of the three structural isomers of pentane.

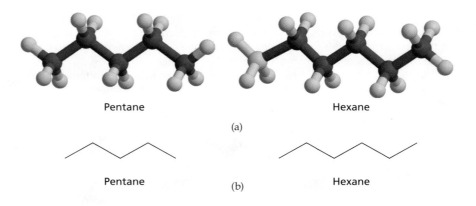

FIGURE 4.14 Line-angle drawings of hydrocarbons (a) Ball-and-stick models of *n*-pentane and *n*-hexane. (b) Line-angle drawings of the same molecules, which depict the zig-zag arrangement of the C—C bonds.

Using *n*-pentane and *n*-hexane as examples, Figure 4.14 compares ball-and-stick models with the line-angle drawings. If you take a pen and trace only the carbon–carbon bonds in the models, you will see how your pen moves up and down in a zig-zag motion that reflects the geometry of the bonds. This structural arrangement of bonds is represented in the line-angle drawing.

WORKED EXAMPLE 4.2

Question: The full-atom structure of an alkane hydrocarbon is shown here. Provide the line-angle drawing of this molecule.

Answer: The line-angle drawing of a molecule drawing omits the carbon and hydrogen atoms and preserves only the carbon–carbon bonds. There are seven C—C bonds in an eight-carbon chain, which we represent by a series of zig-zag lines. Each branching methyl group is drawn as a single line from the appropriate carbon atom. The final line-angle drawing is shown below:

TRY IT YOURSELF 4.2

Question: The full-atom structure of an alkane hydrocarbon is shown as follows. Sketch the line-angle drawing of this molecule.

WORKED EXAMPLE 4.3

Question: Octane (C_8H_{18}) has 18 structural isomers. Three of these isomers (labeled A, B, and C) appear below as line-angle drawings.

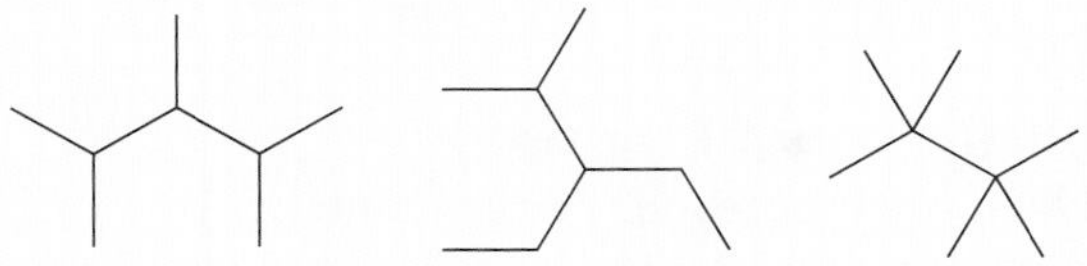

(a)   (b)   (c)

For each isomer:

1. Draw the carbon-only structure.
2. Draw the full-atom structure and confirm the chemical formula.

Answer: (a) Here are the carbon-only structures .

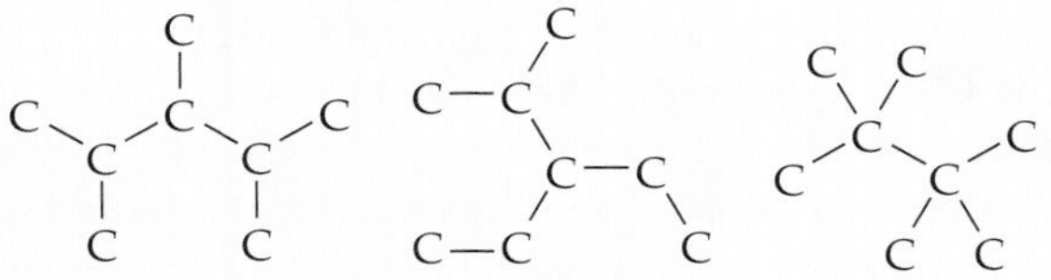

(a)   (b)   (c)

(b) To draw the full-atom structure, add the appropriate number of hydrogen atoms to each carbon atom. For an alkane hydrocarbon, each carbon atom will form *four single bonds* with other atoms. A count of the atoms confirms that each isomer has the chemical formula for octane: C_8H_{18}.

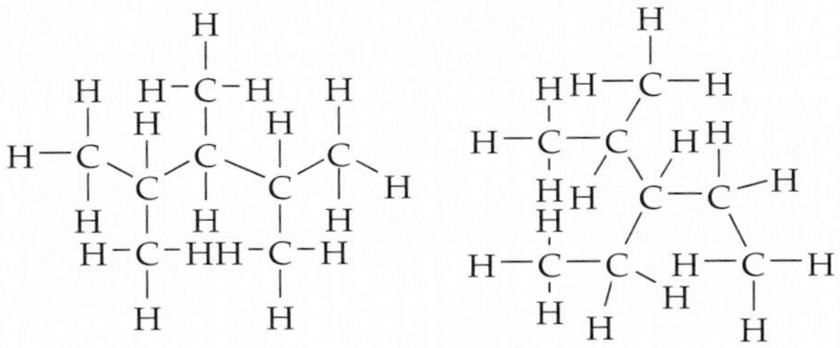

(a)   (b)   (c)
C_8H_{18} C_8H_{18} C_8H_{18}

TRY IT YOURSELF 4.3

Question: A full-atom structure of another octane isomer is provided in the following. Provide the line-angle drawing for this isomer.

PRACTICE EXERCISE 4.3

The line-angle drawing of a hydrocarbon is shown below. Draw the full-atom structure of this molecule, and write its chemical formula in the form C_xH_y.

Core Concepts

- Hydrocarbon molecules can be drawn using several types of representations.
- **Full-atom drawings** include all the atoms in the molecule, but they can become cumbersome for larger molecules.
- **Carbon-only drawings** show only the carbon atoms and do not show any hydrogen atoms.
- **Line-angle drawings** simplify the molecule even further by using lines to represent the covalent bonds that join the carbon atoms (the carbon atoms themselves are not drawn).

4.4 Alkenes: Hydrocarbons with Double Bonds

Alkenes are hydrocarbons with at least one double bond between two carbon atoms. The naming convention for alkenes resembles the one for alkanes. The prefix of the name indicates the number of carbon atoms, and the suffix *-ene* indicates an alkene

The Planar Structure of Ethene

THE KEY IDEA: Ethene is a planar molecule due to the electron density distribution within its double bond.

The simplest alkene hydrocarbon, *ethene*, contains two carbon atoms linked by a double bond. As we learned in Chapter 3, a double covalent bond consists of two shared pairs of electrons. Each carbon atom in ethene contributes two valence electrons to the C=C bond, leaving two valence electrons available to bond with hydrogen atoms. Consequently, the chemical formula for ethene is C_2H_4. Figure 4.15 compares the chemical formula, Lewis dot structure, and two-dimensional structure of ethene.

What is the three-dimensional geometry of ethene? The carbon atoms in ethene are bonded to only three other atoms. VSEPR predicts that the two electron pairs in the C=C bond repel the single electron pairs in each C—H bond. This repulsion increases the angle between the C=C and C—H bonds to approximately 120°. We may also expect the molecule to twist around the C=C bond due to electron repulsion between the adjacent C—H bonds. Surprisingly, this does not happen. It requires a large input of energy to twist a double bond because of how the electron density is distributed within the bond. As a result, *all of the atoms in ethene lie in a two-dimensional* plane (Figure 4.16). We say that ethane is a **planar** molecule.

planar A molecular geometry in which all of the atoms lie in a two-dimensional plane.

Besides their strong resistance to twisting, double covalent bonds differ from single bonds in two other ways. First, *C=C bonds are stronger than C—C bonds.* In other words, the energy required to break a C=C bond is greater compared to a C—C bond. This is because the total amount of bonding electron density joining the two carbon atoms is greater for a double bond compared to a single bond. Second, *C=C bonds are slightly shorter than C—C bonds.* In this comparison, the bond distance is measured between the nuclei of the two carbon atoms. The shorter bond length in C=C bonds maximizes the bonding electron density between the two atomic nuclei.

THE KEY IDEA: The absence of free rotation about a C=C bond produces cis- and trans- isomers.

Core Concepts

- Ethene is a **planar** molecule because of the properties of the C=C double bond.
- Rotation around a C=C bond requires a large amount of energy so molecular regions with double bonds tend to be planar and rigid.
- A C=C double bond is shorter and stronger than a C—C single bond.

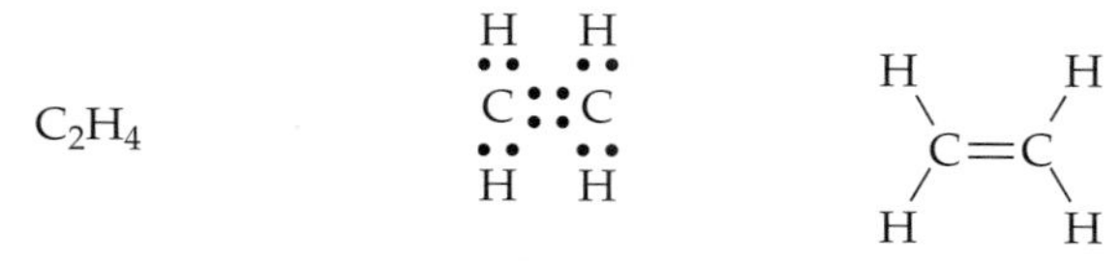

C_2H_4

Chemical formula

H H
•• ••
C :: C
•• ••
H H

Lewis dot structure

H   H
\   /
 C = C
/   \
H   H

2D structure

FIGURE 4.15 Chemical formula, Lewis structure, and two-dimensional structure for ethene.

Isomerization in Alkenes

When we studied alkanes, we saw that some molecules can exist in different shapes called structural isomers. In alkenes, a new type of structural isomerism can exist because it is very difficult to rotate the C=C bond.

For example, replacing two of the H atoms in ethene with methyl groups creates 2-butene. (the double bond begins at the second carbon). As shown in Figure 4.17, there are two possible ways to orient the methyl groups in 2-butene. Both methyl groups can be on the *same side* of the molecule, which is called the **cis isomer** (*cis* is Latin for "on this side"). Another possibility has the two methyl groups on *opposite sides* of the C=C bond. This molecule is called the **trans isomer**, from the Latin

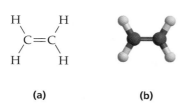

H   H
\   /
 C = C
/   \
H   H

(a)   (b)

FIGURE 4.16 Ethene is a planar molecule. (a) All of the atoms in ethene (C_2H_4) lie in a two-dimensional plane. (b) A molecular model of ethane, which illustrates its planar structure.

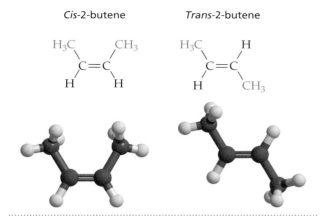

Cis-2-butene   *Trans*-2-butene

FIGURE 4.17 Molecular structures of *cis*-2-butene and *trans*-2-butene. The cis and trans isomers both have the same chemical formula, C_4H_8. In the *cis* isomer, the two methyl groups are on the same side of the C=C bond. In the *trans* isomer, the two methyl groups are on opposite sides of the C=C bond.

Cis-2-butene   *Trans*-2-butene

FIGURE 4.18 Line-angle drawings of *cis*-2-butene and *trans*-2-butene.

word for "on the other side" or "across" (think of a transatlantic flight). Butene has two isomers, *cis*-2-butene and *trans*-2-butene (by convention, the prefixes *cis*- and *trans*- are italicized). Since it is very difficult to rotate around a C=C bond, these two isomers remain locked in their respective structures. By contrast, it is not possible to have *cis*- and *trans*- isomers of butane because rotation around the C—C single bond occurs easily.

We can represent alkene molecules in shorthand form as carbon-only structures or line-angle drawings. Figure 4.18 shows the line-angle drawings for the two isomers of butene.

cis **isomer** A structural isomer in which two chemical groups are on the same side of a C=C bond.

trans **isomer** A structural isomer in which two chemical groups are on opposite sides of a C=C bond.

WORKED EXAMPLE 4.4

Question: A line-angle drawing of an alkene hydrocarbon is as follows. Draw the full-atom structure of this molecule showing all the C and H atoms. Write the chemical formula in the form of C_xH_y.

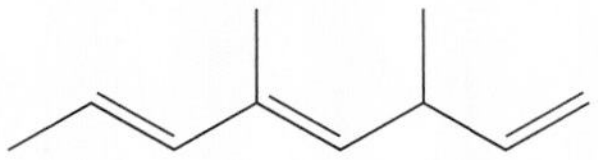

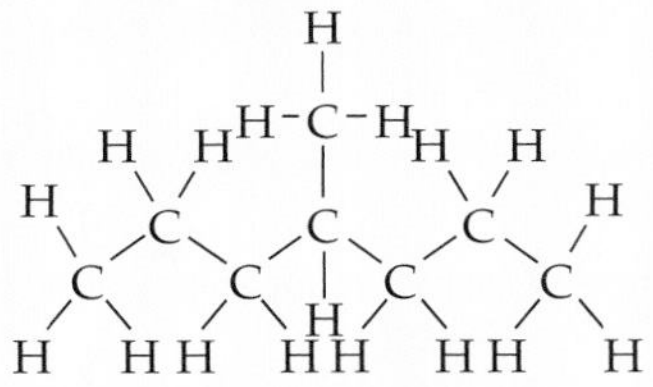

Answer: We draw the full-atom structure in two stages. First we add the carbon atoms:

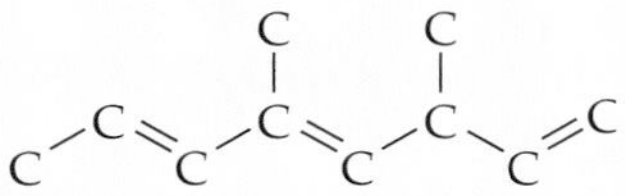

There are 10 carbon atoms in the molecular structure. We can now add the required number of hydrogen atoms to satisfy the four-bond rule for each carbon atom:

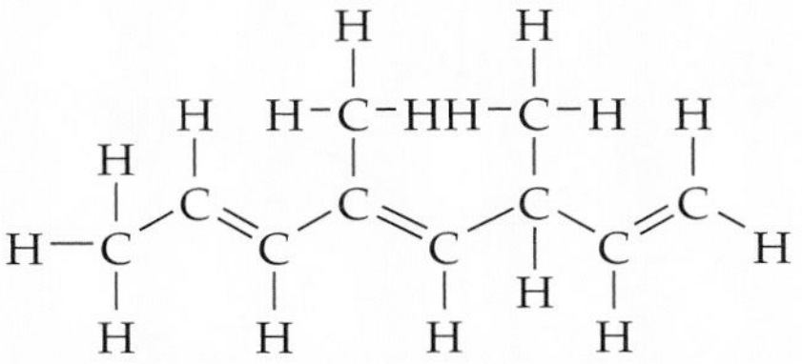

Since the molecule requires 16 hydrogen atoms to satisfy the bonding rules, the chemical formula is $C_{10}H_{16}$.

TRY IT YOURSELF 4.4

Question: The molecular starting material for the biological synthesis of cholesterol is an alkene hydrocarbon called *isoprene*. A carbon skeleton structure of this molecule is shown below. Draw the full-atom structure of isoprene showing all the C and H atoms in the molecule. Write the chemical formula in the form C_xH_y.

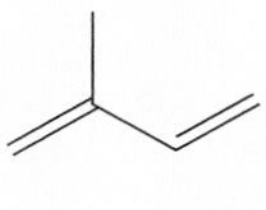

Core Concepts

- Because it is difficult to rotate a C=C bond, some alkene molecules can exist as two different isomers called *cis* and *trans*.
- In the *cis* **isomer**, the two carbon atoms are on the same side of the C=C bond. In the *trans* **isomer**, the two carbons are on opposite sides of the C=C bond.

4.5 The Chemistry of Fats

Learning Objective:

Compare and contrast saturated, unsaturated, and trans fats.

THE KEY IDEA: Triglyceride molecules—a type of dietary fat—are made from glycerol and three fatty acids.

triglyceride A dietary fat made from three fatty acids attached to a glycerol molecule.

fatty acid An acid that consists of an extended hydrocarbon chain plus a carboxylic acid group.

At the beginning of the chapter, we introduced the chemistry of carbon through the example of fats in our diet. In this section, we apply our knowledge of organic chemistry to better understand the structure, physical properties, and health effects of different types of fat.

What Are Fats?

The most important dietary fat molecule is called a **triglyceride**. These molecules are made from three **fatty acids** that are chemically attached to a glycerol molecule.

Glycerol + 3 Fatty acids ⟶ Triglyceride

Figure 4.19(a) shows the formation of a triglyceride molecule. The glycerol contains three –OH groups that are used to make chemical bonds to the fatty acids. Each fatty acid consists of an extended hydrocarbon chain containing between 10 and 20 carbon atoms (the "fatty" part), plus a chemical group called a carboxylic acid (which we discus in Chapter 5). The resulting triglyceride has a "head" group (the glycerol) plus three hydrocarbon "tails." Figure 4.19(b)

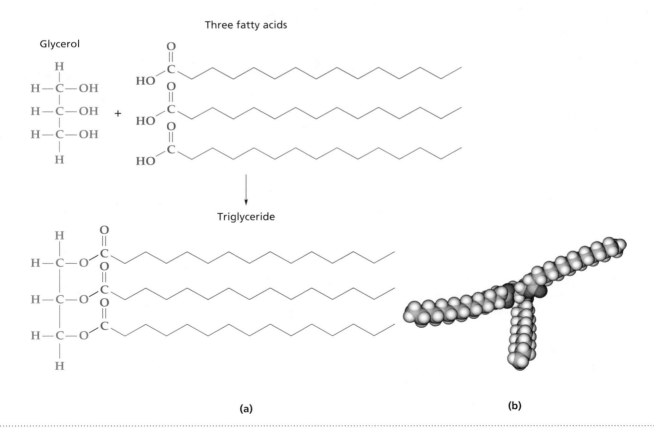

(a)   (b)

FIGURE 4.19 Molecular structure of a triglyceride. (a) A triglyceride is formed by attaching three fatty acids to a glycerol molecule. The hydrocarbon chains in the fatty acids typically contain between 10 and 20 carbon atoms. (b) A space-filling model of a triglyceride molecule, showing the "head" group and three hydrocarbon "tails." The oxygen atoms in the model are shown in red.

illustrates the shape of a triglyceride using a space-filling molecular model. All triglyceride molecules share this basic structure, so what makes a saturated fat different from an unsaturated fat?

Core Concepts

- **Triglycerides** are fat molecules that contain three **fatty acids** attached to a glycerol molecule. The fatty acids are composed of hydrocarbon chains with 10 to 20 carbon atoms.

Saturated and Unsaturated Fats

When applied to fats, the terms *saturated* and *unsaturated* refer to the chemical bonding properties of the hydrocarbon chains within the triglyceride molecule. In a saturated fat, the hydrocarbon chains contain only C—C single bonds. In an unsaturated fat, the hydrocarbon chains contain at least one C=C double bond.

Figure 4.20 compares the molecular structure and space-filling models of a saturated and unsaturated fatty acid. The saturated hydrocarbon tail forms a roughly linear zig-zag structure. The C—C bonds do permit rotation (see the discussion of ethane), but the "average" structure is the linear one illustrated in the figure. By contrast, the unsaturated fats in our bodies have C=C bonds oriented in a *cis* geometry, which places both C atoms on the same side of the double bond. This arrangement produces a permanent kink in the hydrocarbon tail, similar to bending your arm at the elbow. The unsaturated fatty acid shown in Figure 4.20 has only one C=C bond, so it is called **monounsaturated**. Other fatty acids have more than one C=C bond and are called **polyunsaturated**. *Chemistry and Your Health* describes one class of polyunsaturated fatty acid called omega-3 fatty acids.

This variation in the molecular structure of fatty acids explains the different physical properties of saturated and unsaturated fats (Figure 4.21). Animal products are richer in saturated triglycerides and tend to be solid at room temperature—examples include the fat on the edge of a steak and the fat in butter. When a triglyceride molecule contains three saturated tails, they can pack tightly together because the hydrocarbon chains all line up. By contrast, liquid oils such as olive oil, canola oil, or fish oil are richer in unsaturated fats. The kinks in the hydrocarbon tails prevent the triglycerides from packing together tightly, resulting in a fat that is less dense because there are fewer molecules in a specific volume. The lower density of fat molecules produces a liquid oil instead of a solid fat.

THE KEY IDEA: Saturated fats contain all C—C single bonds, whereas unsaturated fats contain at least one C=C double bond.

monounsaturated An unsaturated fatty acid that contains only a single C=C double bond.

polyunsaturated An unsaturated fatty acid that contains two or more C=C double bonds.

trans fat An unsaturated fatty acid containing a *trans* C=C bond; this bond causes the hydrocarbon chain to be linear.

THE KEY IDEA: Trans fats arise from hydrogenation of vegetable oils and are linked to increased risk of heart disease.

Core Concepts

- Saturated fats are triglyceride molecules that contain only C—C single bonds in their hydrocarbon chains. They are typically *solid* at room temperature because the saturated hydrocarbon tails can pack closely together to form a dense consistency.
- Unsaturated fats are triglyceride molecules that have least one C=C double bond in their hydrocarbon chains. The presence of a *cis* double bond introduces a bend in the hydrocarbon tail, which prevents the molecules from packing tightly together. For this reason, unsaturated fats are typically *liquid* oils at room temperature.

Trans Fats

Trans fats are a different type of fat molecule, which were commonly used in processed food such as pizza, popcorn, and cake (Figure 4.22(a)). In 2006, the

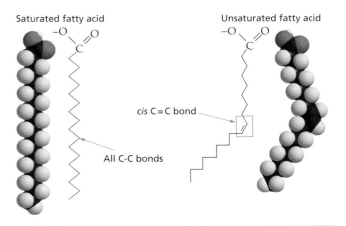

FIGURE 4.20 Chemical structures and space-filling molecular models for a saturated and unsaturated fatty acid. The hydrocarbon chain of the saturated fatty acid contains all C—C bonds. The unsaturated fatty acid contains a C=C double bond in the *cis* geometry (highlighted in pink), which creates a bend in the molecular structure.

CHEMISTRY AND YOUR HEALTH

Omega-3 Fatty Acids

In general, nutritionists recommend that you eat a higher proportion of unsaturated fats than saturated fats. However, one type of unsaturated fat has received particular acclaim. These "good" fats are made from omega-3 fatty acids and are found in fish oil and some vegetable oils like flaxseed and canola oils (Figure 1). Clinical studies show that consumption of omega-3 fatty acids lowers the amount of fat in the blood, slows the harmful hardening of arteries, and reduces the incidence of heart attacks and strokes.

FIGURE 1   Dietary sources of omega-3 fatty acids.

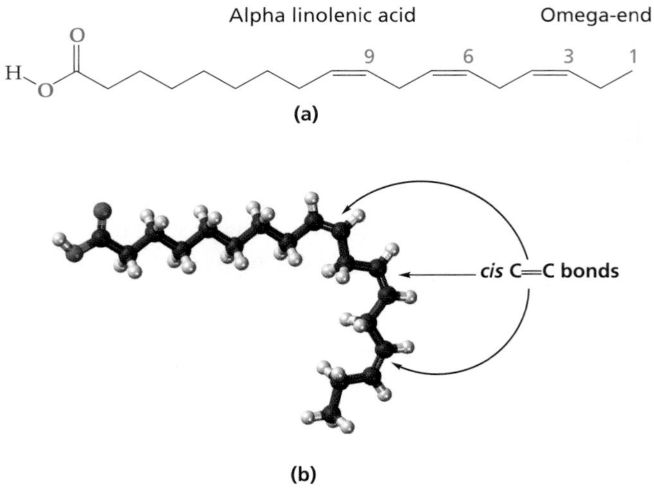

FIGURE 2   The chemical structure of alpha-linolenic acid, one type of omega-3 fatty acid. (a) The carbons are numbered beginning at the omega-end of the chain (red). (b) The fatty acid has a sharply bent structure because it contains three *cis* C=C bonds.

The term *omega-3* refers to the chemical bonding within the hydrocarbon chain of the fatty acids. Figure 2 shows the chemical structure of one example of an omega-3 fatty acid, *alpha-linolenic acid*. The presence of three double bonds means that it is a *polyunsaturated* fatty acid. A larger number of C=C bonds introduces more bends into the hydrocarbon chain, creating a hydrocarbon tail that is shaped like the letter J.

One fatty acid labeling method begins numbering carbons from the tail end of the molecule. This end is called the *omega end* since omega (ω) is the *last* letter in the Greek alphabet. Counting from the omega end, we see that the first double bond occurs between carbon atoms #3 and #4. The term *omega-3 fatty acid* means that the hydrocarbon chain has its first double bond beginning at carbon #3 as counted from the omega end.

Food and Drug Administration (FDA) ruled that nutrition labels for all packaged foods must indicate the amount of trans fats that the food contained (Figure 4.22(b)). Later the same year, New York became the first U.S. city to ban the use of trans fats by all restaurants, which was followed by similar measures in Baltimore, Boston, Cleveland, Philadelphia, and Seattle. In 2015, the FDA announced a total ban on trans fats in the food supply, to be implemented over a three-year period. This legislation reflects growing concern about the adverse effects of consuming trans fats. But what exactly are trans fats, and why are they bad for you?

Numerous medical studies have shown that unsaturated fats—especially polyunsaturated fats—are better for your health than saturated fats are. Unfortunately, the vegetable oils that provide the most convenient source of unsaturated fats have drawbacks for commercial food production. Many of these oils have a short shelf life because they react with oxygen and turn rancid. Most people in the United States prefer to spread a solid like butter or margarine on their bread, as opposed to dipping the bread in olive oil (which is the habit in Mediterranean countries). Baked goods such as pies and pastries require a fat source that is solid rather than liquid. The perfect solution would be a cooking ingredient that has fewer health risks than a saturated fat but was more convenient to use than unsaturated oils.

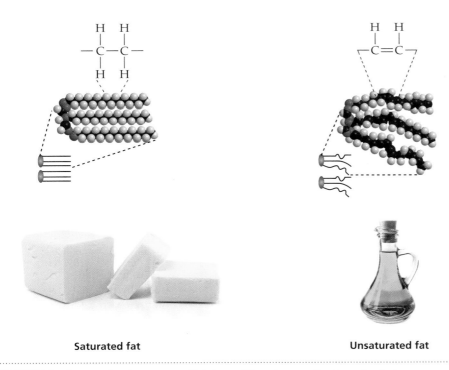

Saturated fat   **Unsaturated fat**

FIGURE 4.21 Saturated and unsaturated fats have different properties. In a saturated triglyceride, the three hydrocarbon tails are all aligned. Tight packing of saturated triglyceride molecules produces the solid consistency of saturated fats such as those found in butter. In an unsaturated triglyceride, the bent structure of hydrocarbon tails create a more disordered molecular structure. Unsaturated triglycerides cannot pack as closely together, which produces the liquid consistency of unsaturated fats (oils).

One solution is to add more hydrogen atoms to an unsaturated fat. This can be accomplished through **hydrogenation**, which involves a chemical reaction between a C=C bond and hydrogen gas. Figure 4.23 illustrates a hydrogenation reaction that converts ethene (unsaturated) into ethane (saturated). The same principle applies to the hydrogenation of longer hydrocarbon chains, which converts the C=C bonds into C—C bonds.

Food scientists used this chemical transformation to make *partially hydrogenated oils* with properties that are intermediate between those of solid saturated fats and liquid unsaturated oils. To perform the reaction,

hydrogenation A chemical reaction for adding hydrogen atoms to an unsaturated hydrocarbon.

(a)   (b)

FIGURE 4.22 Trans fats. (a) A sample of foods that commonly contained trans fats prior to the ban on their use. (b) Food and Drug Administration (FDA) guidelines state that nutrition labels on foods must specify the amount of trans fats. According to FDA regulations, an amount of trans fat that is 0.5 g or less per serving may be listed as "0 grams."

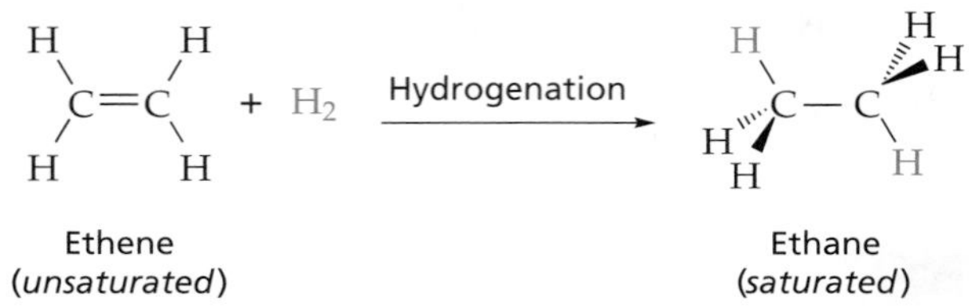

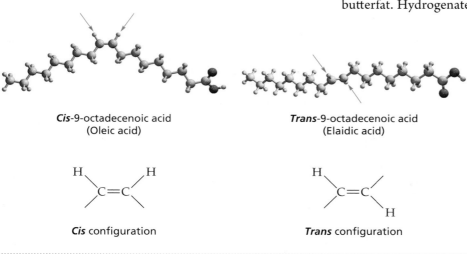

FIGURE 4.23 The hydrogenation of ethene. The addition of H_2 to ethene (an unsaturated hydrocarbon) produces ethane (a saturated hydrocarbon). During this reaction, the H atoms from the hydrogen molecule are added to the carbon atoms. The reaction is facilitated by a metal catalyst, which greatly accelerates the rate of the reaction (catalysts are discussed in Chapter 14).

FIGURE 4.24 Molecular structures of unsaturated fatty acids with *cis* and *trans* configurations of the C=C bond. The *cis* configuration of a C=C bond creates a bend in the hydrocarbon chain. By contrast, the *trans* configuration of a C=C bond produces a linear structure that resembles the shape of a saturated fatty acid.

hydrogen gas is bubbled into hot oil in the presence of a metal catalyst. Under these conditions, some of the unsaturated C=C bonds in the oil are converted into saturated C—C bonds through the addition of H atoms. The reaction is carefully controlled so that not all the C=C bonds are saturated—this is why the process is called *partial* hydrogenation. The resulting substance is a semi-solid mixture of unsaturated fats (from the original vegetable oil) and newly created saturated fats. For example, margarine is made by partially hydrogenating a selection of vegetable oils. Corn oil contains 6% saturated fats, which increases to 17% in corn oil margarine. But this amount is still much less than butter, which contains 80% saturated butterfat. Hydrogenated fats are much less prone to turning rancid than oils—they are stable, with a long shelf life.

At first glance, these partially hydrogenated vegetable oils seem to be an ideal solution. But the hydrogenation reaction also produces an unwanted by-product—*trans* unsaturated fats. Almost all of the double bonds that occur naturally in unsaturated fatty acids exist in the *cis* form. However, the high-temperature environment required for hydrogenation causes some of the fatty acids to flip the orientation of their C=C bonds and create the *trans* form. This switch from *cis* to *trans* causes the shape of the molecule to change from bent to linear, as shown in Figure 4.24. The structure of a *trans* unsaturated fatty acid now resembles that of a saturated fatty acid.

Clinical studies indicate that trans fats are even more harmful than saturated fats. Consumption of trans fats is linked to an increase in the level of blood cholesterol, which is associated with a greater risk of heart disease. We can now understand why the FDA changed the rules of nutrition labels in 2006 and why it announced a complete ban on the use of trans fats beginning in 2015.

Core Concepts

- ***Trans* fats** are typically formed during the partial **hydrogenation** of unsaturated vegetable oils. During this reaction, some of the naturally occurring *cis* double bonds are converted into the *trans* configuration.
- The conversion of the C=C bond from *cis* to *trans* orientation changes the shape of the fatty acid from bent to linear. The straightened hydrocarbon chains in *trans* fatty acids closely resemble the molecular structure of saturated fats.
- Clinical studies have linked consumption of trans fats to elevated cholesterol and an increased risk of heart disease.

4.6 Cyclic Hydrocarbons

In the previous section, we learned that consumption of trans fats is associated with an increase in the level of blood cholesterol. In fact, cholesterol is a molecule

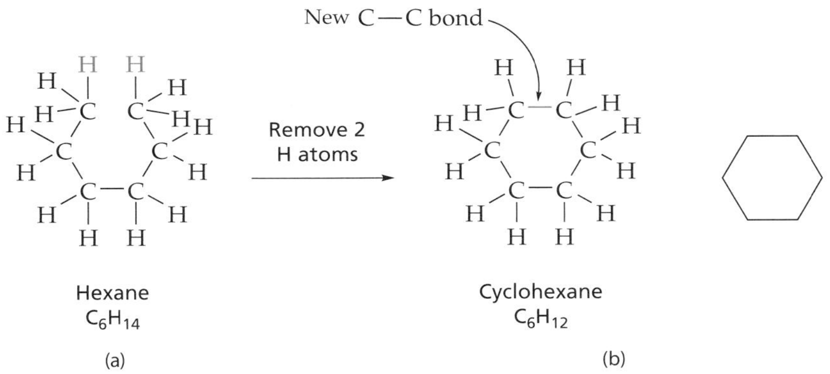

New C—C bond

Remove 2
H atoms

Hexane
C_6H_{14}

(a)

Cyclohexane
C_6H_{12}

(b)

FIGURE 4.25 The cyclic structure of cyclohexane. (a) Conversion of linear hexane (C_6H_{14}) into cyclohexane (C_6H_{12}) by removing two hydrogen atoms and forming a new C—C bond (highlighted in red). (b) Line-angle drawing of cyclohexane, showing only the bonds between the carbon atoms.

that is composed almost entirely of hydrocarbon regions. Unlike the molecules we have considered so far, which have linear carbon chains, cholesterol contains rings of carbon atoms. To understand cholesterol, we need to investigate the structure of **cyclic hydrocarbons**.

Cyclohexane

A linear alkane, such as hexane, can be converted into a cyclic alkane by removing two hydrogen atoms and forming a new carbon–carbon bond. The result is a molecule of *cyclohexane*, as shown in Figure 4.25(a). A simplified representation of cyclohexane as a line-angle drawing is provided in Figure 4.25(b).

The drawings of cyclohexane in Figure 4.25 do not capture its three-dimensional structure. Each carbon atom forms four single covalent bonds, and these bonds arrange themselves in a tetrahedral geometry due to electron pair repulsion. This effect produces a nonplanar arrangement of atoms in cyclohexane. In fact, cyclohexane can adopt two possible conformers by rotations around the C—C single bonds. The structures, illustrated in Figure 4.26, are called the *chair* and *boat.* For the chair confirmation, you can use your imagination to picture the cyclohexane molecule as a recliner chair: Four atoms provide the seat of the chair, the highest carbon (upper right) is the headrest, and the lowest carbon (bottom left) is the footrest. Cyclohexane can also form another structure called a "boat," in which the two carbon atoms at each end of the molecule both tilt *upward* to form the bow and stern of a boat. However, the boat structure is less stable than the chair structure because of electron pair repulsion between the C—H bonds for the two upward-pointing carbon atoms.

Core Concepts

- Cyclohexane (C_6H_{12}) is a **cyclic hydrocarbon** containing single carbon–carbon bonds. The tetrahedral bonding geometry around each C atom creates a nonplanar molecule that has a "chair" or a "boat" conformation.

Learning Objective:
Characterize the bonding and structure of cyclic hydrocarbons.

cyclic hydrocarbons Hydrocarbons in which the carbon atoms form a closed ring.

THE KEY IDEA: Cyclohexane is a cyclic hydrocarbon with a nonplanar structure.

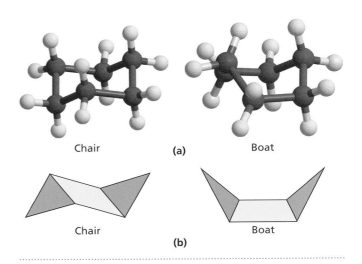

Chair   (a)   Boat

Chair   Boat

(b)

FIGURE 4.26 The chair and boat conformations of cyclohexane. (a) The two conformations are shown as ball-and-stick molecular models. (b) A simplified representation of the chair and boat structures.

○⚷ THE KEY IDEA: Chemical bonding in benzene is described by the theory of resonance.

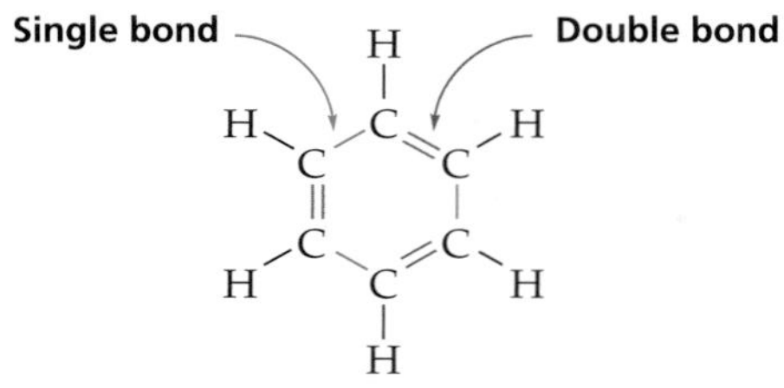

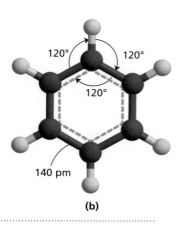

(a)

(b)

FIGURE 4.27 Deducing the molecular structure of benzene. (a) A proposed molecular structure for benzene (C_6H_6), in which a planar ring of six carbon atoms is linked by alternating single and double bonds with different lengths. Each C atom forms four covalent bonds. (b) The actual structure of benzene is a planar hexagon in which all the carbon–carbon bonds have exactly the same length. The measured bond length of 140 picometers (1 pm= 10^{-12} m) is intermediate between the typical length of a single bond (154 pm) and a double bond (133 pm).

Benzene

We now turn to the example of benzene, which has the chemical formula C_6H_6. One possible structure for benzene is a six-sided ring of carbons, with each carbon atom joined to one hydrogen atom. To satisfy the four-bond rule for carbon, we can draw a molecular structure in which the carbon atoms are joined by alternating single and double covalent bonds (Figure 4.27(a)). This looks like a reasonable molecular structure, but it has a serious flaw. Earlier in the chapter, we learned that C═C *double bonds are shorter than C—C single bonds.* If our structure is correct, the carbon–carbon bonds in the ring should alternate between longer single bonds and shorter double bonds.

However, this predicted structure does not agree with the experimental evidence. In the early 20th century, scientists developed a laboratory technique called X-ray crystallography (introduced in Chapter 1) that enabled them to determine the structures of molecules. This technique involves shining X-rays at a solid crystal of molecules and analyzing how the X-rays are deflected by the structural arrangement of atoms (see *Science in Action*). X-ray crystallography revealed that benzene's shape is planar, not buckled as in cyclohexane. In addition, it is a perfectly symmetric hexagon, with six carbon–carbon bonds of exactly the same length (Figure 4.27(b)). The length of these bonds does not fit the usual categories: They are shorter than the typical length for single bonds but longer than the typical length for double bonds. To explain these observations, we need

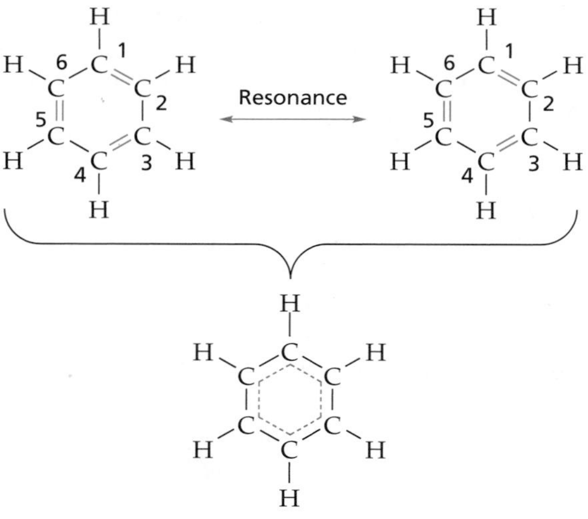

Resonance

Resonance hybrid

FIGURE 4.28 Resonance in benzene. The molecular structure of benzene can be written as two equivalent resonance structures that differ in the location of the valence electrons in the double bonds. The result is a resonance hybrid structure in which each carbon–carbon bond is intermediate between a single and double bond (indicated by the dashed lines).

a new description of the chemical bonding within benzene, which is provided by a theory called **resonance**.

As shown in Figure 4.28, there are two possibilities for placing the single and double bonds in benzene. The two bonding arrangements have the same energy because they differ only in the placement of the valence electrons in the covalent bonds. When two (or more) equivalent bonding arrangements exist in a molecule, they are called **resonance structures.** We can represent the resonance structures as being linked by a double-sided arrow, as shown in Figure 4.28. This arrow indicates the *equivalence* of the two resonance structures; it does *not* mean that the molecule is switching back and forth between them.

The theory of resonance makes a surprising prediction. *When a molecule can exist as different resonance structures, its chemical bonding is a hybrid of these structures.* For benzene, this means that each carbon–carbon bond is neither single nor double; instead, it is something in between. This prediction agrees with the experimental evidence discussed earlier, which found that the length of each carbon–carbon bond in benzene is midway between a single and double bond. The predicted structure of benzene based on resonance theory is called a **resonance hybrid**. It is represented in Figure 4.28 as a six-sided molecule with dashed lines to show that each carbon–carbon bond has properties in between those of a single and double bond.

We can explain this bonding in terms of **delocalized electrons**. Some of the bonding electrons in benzene are not associated with a specific atom or covalent bond; instead, they are delocalized around the entire six-atom ring. The benzene molecule is more stable than the alternating bond structure in Figure 4.27(a). This is attributed to the stabilizing effect of delocalized electrons rather than confining them at specific positions.

Because of its structure and stability, the planar benzene ring appears frequently in many biological molecules and pharmaceuticals. Compounds containing a benzene ring are called **aromatic** because the earliest such molecules were derived from fragrant oils and possessed a pleasant aroma. Today, this term describes any molecule with electron delocalization throughout a carbon ring structure. Because benzene rings occur so often in molecular structures, Figure 4.29 presents three common ways of drawing them.

resonance A theory that includes more than one bonding arrangement in a molecule.

resonance structures Two or more equivalent bonding arrangements for a molecule.

resonance hybrid A hybrid structure that incorporates the different resonance structures for the molecule; this structure is the most accurate representation of the chemical bonding.

delocalized electrons Electrons that are not associated with a specific atom or covalent bond.

aromatic Compounds that contain a benzene ring, or related structures.

(a)   (b)   (c)

FIGURE 4.29 Three ways to draw a benzene molecule. Drawing (a) shows alternating double and single bonds. However, we understand that the electrons in the double bonds are actually "delocalized" around the entire benzene ring. Drawings (b) and (c) denote the delocalized electrons explicitly, either as a circle inscribed within the hexagon or with the use of dotted lines.

PRACTICE EXERCISE 4.4

Naphthalene is a cyclic hydrocarbon with the following structure given as a line angle drawing:

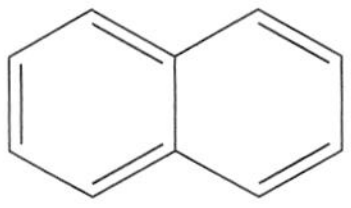

1. Deduce the chemical formula of naphthalene.

2. Draw an alternative resonance structure for naphthalene.

3. Explain why naphthalene is a planar molecule, like benzene.

SCIENCE IN ACTION

How Do We Know the Structures of Molecules?

How do we know the structures of molecules if we cannot see them? This is a central challenge in chemistry because the world of atoms and molecules is not directly observable. To solve this problem, scientists must use experimental techniques that provide information they *can* observe, which enables them to make inferences about the world that lies beyond their senses.

A widely used method for studying the structure of molecules is called *X-ray crystallography*, which is illustrated in Figure 1. The first step—and often the most difficult—is to turn a collection of molecules into a solid crystal. Within the crystal, the molecules are organized into an orderly repeating arrangement, like a group of well-trained soldiers on parade.

The next step is to shine a beam of X-rays onto the crystal. When the X-rays encounter the electrons that surround each atom, they are deflected from their original path. When these scattered X-rays reach the screen, there are certain regions where the X-ray intensity is greater, and this creates a pattern of spots on a photographic film. The collection of spots, called a diffraction pattern, contains information about the positions of the atoms in the crystal. However, extracting this information is complicated, and nowadays this task is performed by high-powered computers.

The first X-ray investigation of benzene's structure was performed in 1929 by Kathleen Lonsdale (1903–1971). To be exact, Lonsdale studied a modified form of benzene in which each of the H atoms was replaced by a methyl group (—CH₃). This molecule was used because it can be turned into a crystal at room temperature, whereas benzene is a liquid under the same conditions. Lonsdale used her X-ray diffraction data to determine that benzene has a planar and hexagonal structure, with six equal carbon–carbon bond distances. This experiment provided the first direct evidence of benzene's postulated structure.

Long before computers were available, Lonsdale pioneered the application of sophisticated mathematical methods to analyze the experimental data from X-ray crystallography experiments. Mathematical analysis of the X-ray diffraction spots generates what is called an *electron density map*, which depicts the spatial distribution of electron density within the molecule. This map reveals the location of atoms, which correspond to regions of higher electron density.

Figure 2(a) shows the electron density map obtained from Lonsdale's X-ray diffraction experiments on *anthracene*, a hydrocarbon molecule found in coal tar. This figure is a two-dimensional slice through a three-dimensional map. The regions of higher electron density are revealed by "contour lines" that are spaced closer together. This representation is similar to how the spacing of contour lines on a topographic map is used to indicate changes in elevation.

By analyzing the electron density map, we can deduce that anthracene consists of 14 carbon atoms arranged in *three fused rings* (the term *fused* refers to rings attached along their sides). The hydrogen atoms appear weakly in the electron density map because these atoms possess only one electron and thus have a very low electron density. However, we can deduce the location of the hydrogen atoms based on the position of the carbon

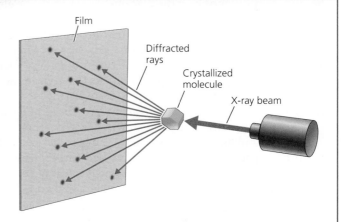

FIGURE 1 A schematic diagram of an X-ray diffraction experiment. An X-ray beam is shone onto a solid crystal of molecules. The X-rays interact with the electrons that surround the atoms to generate a diffraction pattern consisting of spots on a screen. The positions of the spots in this pattern reveal information about the location of the atoms in the crystal.

atoms and the chemical bonding rules. Combining all this information allows us to create a molecular model of anthracene, which is shown in Figure 2(b). We conclude that anthracene is a planar molecule like benzene, with chemical formula $C_{14}H_{10}$.

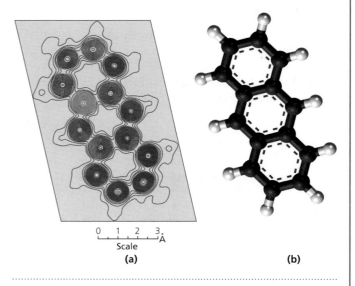

FIGURE 2 Using an electron density map to determine molecular structure. (a) An experimental electron density map for anthracene. Regions of higher electron density are indicated by contour lines that are spaced closer together. These regions correspond to the location of atoms within the molecule. The scale is in length units of Ångstroms, which are commonly used to describe the dimensions of molecules (1 Å = 10^{-10} m). (b) A molecular model of anthracene ($C_{14}H_{10}$). This model fits the electron density map in Part (a).

Core Concepts

- Chemical bonding in benzene (C_6H_6) is described by the theory of **resonance**.
- The benzene molecule can be drawn in two equivalent **resonance structures** with different placement of the electrons in the covalent bonds. The resulting **resonance hybrid** molecule contains six identical carbon–carbon bonds, with a length that is intermediate between the typical values for single and double bonds.
- The properties of benzene are explained by **delocalized electrons** in which the bonding electron density is delocalized around the molecular ring. This effect creates a planar molecule that is particularly stable.

Cholesterol

Cholesterol is an example of a biological molecule that contains cyclic hydrocarbons. Figure 4.30 shows the molecular structure of cholesterol as a line-angle drawing and a ball-and-stick model. Cholesterol contains four fused rings of carbon atoms ("fused" means that two adjacent rings share a common carbon–carbon bond). Three of these rings contain six carbon atoms, and the remaining one contains five carbons. These rings contain mostly C—C single bonds, in which each carbon atom makes four covalent bonds with a tetrahedral geometry. The nonplanar structure of cholesterol is illustrated by the molecular model. There is a C=C double bond within one ring, which causes this region of the molecule to be planar. In addition to the ring structures, cholesterol has an attached hydrocarbon chain. The only atom that is not carbon or hydrogen is a lone oxygen atom attached to one of the carbon rings.

THE KEY IDEA: Cholesterol can form waxy plaques on the walls of arteries, restricting blood flow and increasing blood pressure.

> **PRACTICE EXERCISE 4.5**
>
> Figure 4.30 shows the molecular structure of cholesterol as a line-angle drawing. Draw a full-atom structure of cholesterol, and deduce its chemical formula in the form $C_xH_yO_z$.

Because it consists almost entirely of hydrocarbon, cholesterol is not soluble in water (solubility is explained in Chapter 9). This means that cholesterol does not dissolve in blood, which is predominantly composed of water. Instead, cholesterol molecules are carried by soluble protein molecules within our bloodstream. These transport proteins play a vital role in our health. One type, called LDL (low density lipoprotein), carries cholesterol throughout the body for incorporation into cell membranes. A second type, called HDL (high density lipoprotein), transports cholesterol to the liver, where it is broken down for removal from the body.

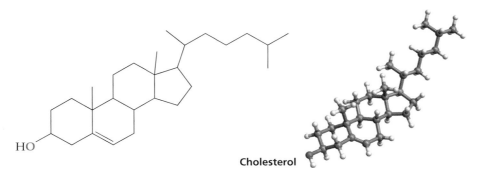

Cholesterol

FIGURE 4.30 The molecular structure of cholesterol. The molecule is depicted as a line-angle drawing and a ball-and-stick molecule model. Cholesterol is almost completely composed of hydrocarbon regions , with four fused carbon rings plus a linear carbon chain. The only exception is a single oxygen atom attached to one of the carbon rings (shown as red).

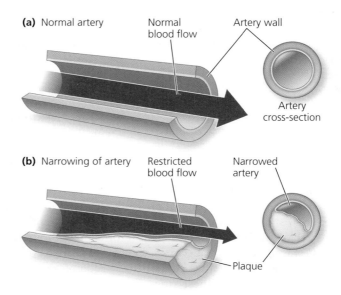

(a) Normal artery   Normal blood flow   Artery wall

Artery cross-section

(b) Narrowing of artery   Restricted blood flow   Narrowed artery

Plaque

FIGURE 4.31 **The health consequences of high cholesterol.** Excess cholesterol in the blood can contribute to the formation of a waxy deposit called plaque. The plaque accumulates on the walls of arteries, narrowing the cross-section of the artery and restricting blood flow. This medical condition, called atherosclerosis, increases blood pressure and the risk of heart disease.

A high level of LDL allows cholesterol to combine with our body's immune system cells and form waxy deposits called *plaques*. These plaques build up on the walls of arteries, which are a type of blood vessel that transports oxygen-containing blood from the heart to the rest of the body (Figure 4.31). The plaques narrow the cross-section of the artery and restrict the flow of blood, which increases blood pressure. This medical condition is called *atherosclerosis*. As an analogy, think of the pressure that builds up when you pinch a garden hose and restrict the flow of water. An increase in blood pressure raises the risk of heart disease. In addition, the plaque can rupture and allow a blood clot to form within an artery. If this happens in the heart, it can cause a heart attack; if it happens in the brain, it can produce a stroke. For these reasons, the cholesterol associated with LDL is often called "bad" cholesterol. Conversely, the cholesterol associated with HDL is called "good" cholesterol because it is removed from the bloodstream and later degraded by the liver.

Core Concepts

- Cholesterol is composed almost entirely of hydrocarbon, so it does not dissolve in the watery environment of the blood.
- Cholesterol contributes to the buildup of plaques, which are waxy deposits that form on the walls of the arteries. These plaques narrow the cross-section of the artery, restricting the blood flow and causing an increase in blood pressure. If a plaque ruptures, it can release a blood clot that causes a heart attack or stroke.

4.1   What is the Difference Between Saturated and Unsaturated Fats?

Learning Objective:

Distinguish between saturated and unsaturated fats.

⊙ A **saturated fat** contains only single covalent bonds between carbon atoms. An **unsaturated fat** contains at least one double bond joining two carbon atoms.

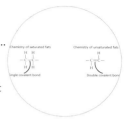

4.2   Why is Life Based on Carbon?

Learning Objective:

Describe the unique chemical properties of carbon.

⊙ Life is based on carbon because carbon has the most versatile chemistry of any element.

4.3   Alkanes: Hydrocarbons with Single Bonds

Learning Objective:

Characterize the bonding and structure of alkane hydrocarbons.

⊙ **Hydrocarbons** are molecules containing only hydrogen and carbon atoms.

⊙ Hydrocarbons often provide the structural framework for biological molecules and pharmaceuticals.

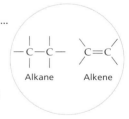

⊙ Ethane is said to exist in either of two conformations— **eclipsed** or **staggered**—depending on the relative orientation of its two **methyl groups**.

⊙ VSEPR predicts that ethane adopts the staggered conformation because it minimizes the electron pair repulsion.

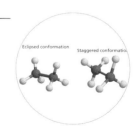

⊙ **Structural isomers** are variants of a molecule that have the same number and type of atoms but different bonding connections among the atoms.

⊙ The naming convention for isomers uses the longest continuous carbon chain as a reference point and specifies the carbon atom to which the branching group is attached.

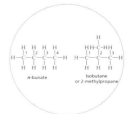

- **Full-atom drawings** include all the atoms in the molecule.

- **Carbon-only drawings** show only the carbon atoms and do not show any hydrogen atoms.

- **Line-angle drawings** use lines to represent the covalent bonds that join the carbon atoms (the carbon atoms themselves are not drawn).

4.4   Alkenes: Hydrocarbons with Double Bonds

Learning Objective:

Characterize the bonding and structure of alkene hydrocarbons.

- Ethene is a **planar** molecule because of the distribution of electron density within the $C=C$ double bond.

- Rotation around a $C=C$ bond carries a large energy penalty, so molecular regions with double bonds tend to be planar and rigid.

- Because of the rigid structure of $C=C$ bonds, some alkene molecules can exist as two different isomers called *cis* and *trans*.

- In the **cis isomer**, the two carbon atoms are on the same side of the $C=C$ bond. In the **trans isomer**, the two carbons are on opposite sides of the $C=C$ bond.

4.5   The Chemistry of Fats

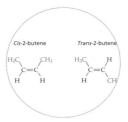

Learning Objective:

Compare and contrast saturated, unsaturated, and trans fats.

- **Triglycerides** are fat molecules that contain three **fatty acids** joined to a glycerol molecule. The fatty acids are composed of hydrocarbon chains with 10 to 20 carbon atoms.

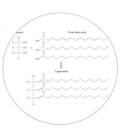

- Saturated fats are triglyceride molecules that contain only $C—C$ bonds in their hydrocarbon chains. They are typically *solid* at room temperature because the saturated hydrocarbon tails can pack closely together to form a dense consistency.

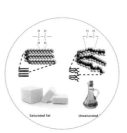

- Unsaturated fats are triglyceride molecules that have at least one $C=C$ bond in their hydrocarbon chains. The presence of a *cis* double bond introduces a bend in the hydrocarbon tail, which prevents the molecules from packing tightly together. Unsaturated fats are typically *liquid* oils at room temperature.

- **Trans fats** are typically formed during the partial **hydrogenation** of unsaturated vegetable oils. During this reaction, some of the naturally occurring *cis* double bonds are converted into the *trans* configuration.

- Clinical studies have linked consumption of trans fats to elevated cholesterol and an increased risk of heart disease.

4.6   Cyclic Hydrocarbons

Learning Objective:

Characterize the bonding and structure of cyclic hydrocarbons.

- Cyclohexane (C_6H_{12}) is a cyclic molecule containing single carbon–carbon bonds. The tetrahedral bonding geometry around each C atom creates a nonplanar molecule that forms a "chair" or "boat" structure.

- Chemical bonding in benzene (C_6H_6) is described by the theory of **resonance**.

- The **resonance hybrid** molecule contains six identical carbon–carbon bonds, with a length that is intermediate between the typical values for single and double bonds.

- The properties of benzene are explained by **delocalized electrons** within the molecular ring. This effect creates a planar molecule that is particularly stable.

- Cholesterol is composed almost entirely of hydrocarbon regions, so it does not dissolve in the watery environment of the blood.

- Cholesterol contributes to the buildup of plaques, which are waxy deposits that form on the walls of the arteries. These plaques narrow the cross-section of the artery, restricting the blood flow and causing an increase in blood pressure.

LEARNING RESOURCES

Reviewing Knowledge

4.1: What Is the Difference Between Saturated and Unsaturated Fats?

1. What type of fat has the lower RDA, saturated or unsaturated?
2. What is the chemical difference between saturated and unsaturated fats?
3. Name two foods that contain saturated fat and two foods that contain unsaturated fat.
4. Which type of fat is healthier, saturated or unsaturated?
5. Which fats are more likely to be solid at room temperature, saturated or unsaturated?

4.3: Alkanes: Hydrocarbons with Single Bonds

6. What is a hydrocarbon?
7. Why is the study of hydrocarbons important for understanding biological molecules and pharmaceuticals?
8. The two methyl groups in ethane can be staggered or eclipsed. Draw the three-dimensional structures of both molecules from the following viewpoints: (a) looking sideways at the C—C bond, (b) looking lengthwise along the C—C bond.
9. Propane is an alkane with the formula C_3H_8. Draw (a) the Lewis dot structure, (b) the two-dimensional structure, and (c) the three-dimensional structure for this molecule.
10. What are structural isomers? Use drawings of butane (C_4H_{10}) to illustrate your answer.

4.4: Alkenes: Hydrocarbons with Double Bonds

11. Explain why the presence of a C=C bond causes the geometry of ethene to be planar.
12. What are *cis* and *trans* isomers in an alkene? Draw two structures of butene to illustrate this concept.

4.5: The Chemistry of Fats

13. What is a biological role of fat in our bodies?
14. What is a fatty acid?
15. Draw the molecular structure of a triglyceride.
16. What is the difference between a saturated and an unsaturated hydrocarbon?
17. What chemical process is used to convert unsaturated hydrocarbons into saturated ones?
18. Why are saturated fats usually solid at room temperature, whereas unsaturated fats are liquid oils?
19. What are trans fats, and how are they produced? How are they similar to saturated fats? What health problems are associated with trans fats?

4.6: Cyclic Hydrocarbons

20. Draw the two-dimensional structure of cyclohexane, C_6H_{12}. What is the most stable structure for this molecule?
21. Describe how the concept of resonance explains the structure of benzene, C_6H_6. How does resonance affect the distribution of electrons and the length of each carbon–carbon bond in benzene?

Developing Skills

22. The chemical formula for a hydrocarbon is given here. Draw the full-atom structure for this molecule.

$$(CH_3)_3CCH_2CH(CH_3)C_2H_5$$

23. Candles are made from a mix of long-chain hydrocarbons. The most common component is *paraffin wax*, a linear alkane that contains 20 carbon atoms. Draw the full-atom structure for paraffin wax and use the structure to deduce its chemical formula.
24. The chemical formulas for methyl and ethyl groups are —CH$_3$ and —C$_2$H$_5$, respectively. Write the chemical formulas for (a) a propyl group and (b) a butyl group.
25. Line-angle drawings for four hydrocarbons are given below. Provide full-atom drawings for each of these structures, and give the chemical formula in the form C_xH_y.

(a)   (b)   (c)   (d)

26. Heptane is a linear hydrocarbon with the chemical formula C_7H_{16}. Use carbon-only structures to draw all nine structural isomers of heptane. [*Note: Check your structures carefully to make sure you have not written the same isomer in two different ways.*]
27. Mineral oil contains a variety of long hydrocarbon chains, some branched and some straight. Sketch the carbon-only structure of three types of hydrocarbons that each contains 18 carbon atoms: (a) a linear hydrocarbon, (b) a hydrocarbon with a single branch, and (c) a hydrocarbon with both a ring segment and a linear segment.
28. Draw a carbon-only structure for a molecule that has the chemical formula C_8H_{14}.
29. Pentene is an alkene with the chemical formula C_5H_{10}. One version of this molecule, called 2-pentene, has the C=C double bond between carbon atoms numbered 2 and 3 (counting from the left). Draw carbon-only structures for the *cis* and *trans* isomers of 2-pentene.
30. Isoprene is a molecule with the chemical formula

$$H_2C=C(CH_3)—CH=CH_2$$

that has *cis* and *trans* isomers. The two molecules have very different properties: One has a *cis* geometry at each double bond, whereas the second has a *trans* geometry. Draw the two-dimensional structures of these two isomers.

31. The reverse of hydrogenation reactions is also important in chemistry. One such reaction, shown here, converts cyclohexane into benzene. Filling in the hydrogens on the left and the right shows an imbalance in the number of H atoms. Assuming the product includes

hydrogen molecules, write a balanced chemical equation for the reaction.

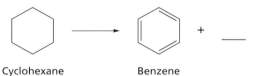

Cyclohexane   Benzene   +   ___

32. The figure that follows shows four different cyclic alkanes. Smaller angles between adjacent carbon atoms in rings are less stable (they are called "strained"). Which compound do you predict will be the least stable? The most stable?

Cydopropane

Cydobutane

Cydopentane

Cydohexane

33. Pepper contains an organic compound called phellandrene, and its molecular structure is shown here as a line-angle drawing.

(a) Draw the full-atom structure of phellandrene.
(b) Deduce the chemical formula for the molecule in the form C_xH_y.

34. Limonene is a hydrocarbon that contributes to the smell of citrus fruits, including lemons and orange. Limonene contains a cyclic hydrocarbon, and the line-angle drawing is given below.

(a) Draw the full-atom structure of limonene.
(b) Deduce the chemical formula for the molecule in the form C_xH_y.

35. Sketch the structures of linear alkane molecules with one, two, three, and four carbons. Can you write a formula that predicts the number of hydrogens for a linear alkane molecule with any number n of carbons? Write the general formula in the form C_nH?

Exploring Concepts

36. Pure carbon exists in different structural forms called *allotropes*. Two of these forms are diamond and graphite (shown below). Diamond is one of the hardest known substances, and its carbon atoms are organized into a repeating tetrahedral structure. In graphite, the carbon atoms form flat sheets with a repeating hexagonal structure.
(a) Graphite is used in pencil lead. Based on its structure, explain why graphite is able to make a mark on paper.

(b) The density of diamond is 3.53 g/cm³, and for graphite it is 2.25 g/cm³. The mass of the carbon atoms is the same in each substance. What do these density values tell you about how closely the carbon atoms are packed in diamond compared with graphite?

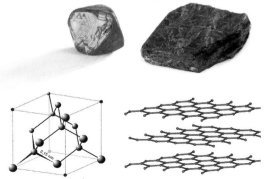

37. In 1996, three scientists won the Nobel Prize in Chemistry for the discovery of a new structure formed by carbon atoms. They called it *buckminsterfullerene*, or "buckyball" for short. Use the Internet to investigate this substance. Provide a sketch of a buckyball, and give a short description of its properties.

38. Some traditional views about fat consumption are being challenged by new medical studies. Use the Internet to investigate the latest research on the relationship between fat consumption and health. What are the health consequences of eating fats compared with eating carbohydrates? Write a short summary of this research using nontechnical language that you could share with your roommate or family.

39. Skin creams that advertise protection against wrinkles and acne often contain *retinol*. Use the Internet to investigate retinol and draw its molecular structure. Why is retinol used as an ingredient in skin creams? Using the Internet, can you find any *reputable* evidence to support the beneficial claims made by these skin creams? Write a short paragraph on your findings.

40. Androstenedione—often shortened to "andro"—has been manufactured and sold as a dietary supplement for athletes. It was commonly used by baseball players during the 1990s because it was a legal substance that was available over the counter. Androstenedione is not a steroid hormone, so why would athletes take it? Is "andro" still legal in sports? Use the Internet to research this substance. Draw its molecular structure, and write a one-paragraph report on the use of "andro" in sports.

41. The German chemist Friedrich Kekulé (1829–1896) is credited with first determining that benzene consists of a ring molecule. Later in his life, he provided an account of how he thought of this idea. Use the Internet to learn more about this story, and provide a short summary of Kekulé's reported inspiration.

42. Diesel fuel consists of a mix of hydrocarbons, but the types of hydrocarbons used are different from those found in gasoline. Use the Internet to discover what types of hydrocarbons are used in diesel fuel. How is this mix of compounds related to the operation of a diesel engine compared with a standard gasoline engine?

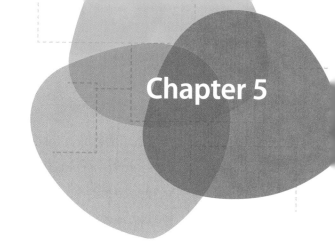

Chapter 5

Molecular Diversity

You have likely taken an over-the-counter painkiller for a headache or to relieve the symptoms of a cold. The active ingredients of these medications are small molecules that help
to reduce inflammation in the body. What do these drug molecules look like? In this chapter,
we explore and classify the various components of molecules.

5.1   What Do Drug Molecules Look Like?

Learning Objective:

Explain the role of functional
groups in molecular structure.

🔑 THE KEY IDEA: Drug mol-
ecules are often composed of a
hydrocarbon framework with attached functional groups.

The molecular structures of three common painkiller medications are shown in
Figure 5.1. Certain features of these molecules look familiar based on our study
of hydrocarbons in Chapter 4. For example, all three molecules contain a benzene
ring. Other hydrocarbon regions are also present, including the four-carbon chain
on the left side of ibuprofen. The molecules contain atoms other than carbon and
hydrogen. For example, all three drugs contain two or more oxygen atoms, and
acetaminophen contains a nitrogen atom.

These observations can be generalized to many drug molecules. Most pharmaceuticals contain hydrocarbon segments that provide a *structural framework*
for the molecule. We learned in Chapter 4 that carbon atoms are capable of forming a remarkable diversity of molecular structures. However, pure hydrocarbons
make poor drug molecules. Hydrocarbons have a limited range of chemical properties. In addition, hydrocarbons do not dissolve in water, which makes them

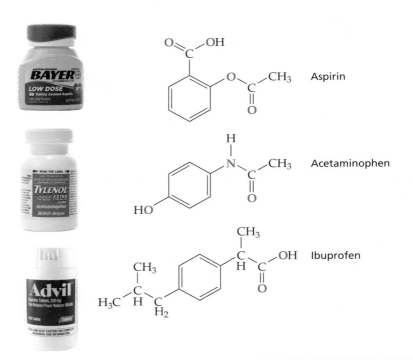

**FIGURE 5.1 Molecular structures of three common painkillers: aspirin, acetaminophen,
and ibuprofen.** Oxygen atoms are colored red, and nitrogen atoms are colored blue. The
carbon—hydrogen bonds are not shown in the ring structures.

unsuitable for the watery environments of the body's bloodstream and the interior of cells.

To function as an effective drug, hydrocarbon segments must be supplemented by other atoms—common examples include oxygen, nitrogen, sulfur, and phosphorus. These elements are called **heteroatoms** (*hetero* = different) to distinguish them from carbon and hydrogen. The addition of heteroatoms to a hydrocarbon framework dramatically alters the molecule's properties. To illustrate this point, consider the distinct properties of propane (C_3H_8) and propanol (C_3H_7OH), which differ by the addition of an oxygen atom (Figure 5.2). Propane is used as a cooking fuel; it is a gas at room temperature and does not dissolve in water. By contrast, propanol is a liquid that is very soluble in water. We see that adding a single heteroatom significantly alters the properties of the molecule. In this chapter, we examine the structure and properties of the most important heteroatoms.

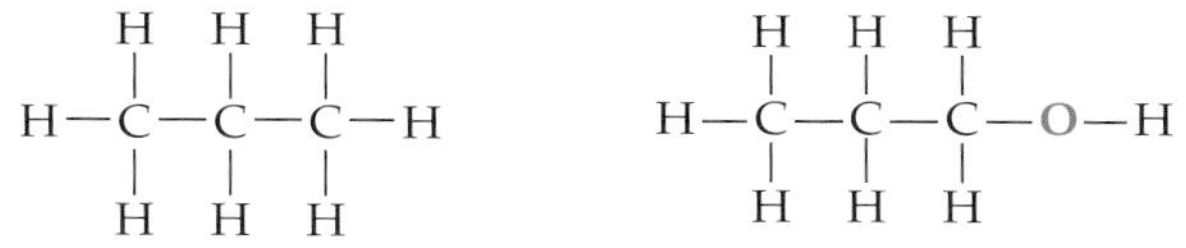

When we inspect the structures of many molecules, we observe that heteroatoms and their carbon neighbors tend to occur in specific combinations called **functional groups**. *A functional group is a collection of atoms within a molecule that has a characteristic structure and chemical behavior.* Functional groups are a useful way to categorize molecular diversity because they tend to retain their characteristic features regardless of the hydrocarbon framework to which they are attached. The presence of a functional group changes the name of molecules in a systematic way. Thus, C_3H_7OH is propanol and not propane. We will introduce the names of the different functional groups as we proceed so that we can identify the substitutions.

In general, *many drug molecules contain a hydrocarbon framework modified by functional groups.* The functional groups play a major role in the chemical properties and therapeutic effects of these molecules. In addition, most biological molecules contain functional groups that are critical for their role in the chemistry of life. For these reasons, the study of functional groups is essential for understanding the structure and function of pharmaceutical and biological molecules.

We first survey the most common functional groups, and we provide examples of molecules that contain them. In a later section, we examine ring structures that have one or more heteroatoms—these are called *heterocycles*. By the time you reach the end of this chapter, you will be able to analyze the structure of any drug molecule and identify some of its characteristic features. Understanding the richness of molecular diversity will serve as an important foundation for all of the following chapters in this book.

FIGURE 5.2 A comparison of the molecular structures of propane and propanol. The addition of a single oxygen atom in propanol dramatically changes its properties.

heteroatom An atom that is different from carbon and hydrogen; for example, O, N, P, and S.

functional group A collection of atoms within a molecule that has a characteristic structure and chemical behavior.

Core Concepts

■ A **heteroatom** is an atom that is different from carbon and hydrogen (e.g., oxygen, nitrogen, phosphorus, and sulfur). The addition of a heteroatom to a hydrocarbon molecule changes its properties (compare propane, C_3H_8, and propanol, C_3H_7OH).

■ A **functional group** is a collection of atoms within a molecule that has a characteristic structure and chemical behavior.

■ Many pharmaceuticals consist of a hydrocarbon framework modified by functional groups.

5.2 Functional Groups Containing Oxygen Atoms

The presence of oxygen heteroatoms in a molecule can generate a variety of functional groups. This section focuses on the various functional groups that are created when oxygen forms covalent bonds with carbon and hydrogen.

Learning Objective:

Illustrate functional groups containing oxygen atoms.

Alcohol

The **alcohol** functional group has the general structure shown in Figure 5.3, in which R represents any carbon-containing group of atoms. The —OH pair of atoms in an alcohol is also called a *hydroxyl* group.

Following standard practice, we do not include the nonbonding (lone) pairs of electrons in the structure of the functional group. This streamlines the molecular drawings, especially when the molecules become more complex. However, it is important to remember that *the oxygen atom has two lone pairs*, which affects the molecular geometry. Electron pair repulsion between the two lone pairs and the two bonding pairs produces the bent geometry of the R—O—H bonds (recall the bent structure of H_2O in Chapter 3).

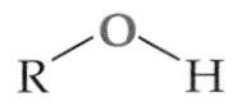

The simplest alcohol is *methanol*, CH_3OH, in which the R group is a methyl, —CH_3. This highly toxic alcohol can cause blindness and even death if consumed in excess. Figure 5.4 illustrates five molecular representations of methanol.

The three-dimensional geometry of methanol can be predicted by using the principle of electron pair repulsion (VSEPR) (see Chapter 3). When molecular structures become larger, a useful strategy is to mentally dissect the molecule into smaller regions that you can analyze more easily. To begin, the four bonds around the carbon atom (three C—H bonds and one C—O bond) have a tetrahedral geometry, which arises from the mutual repulsion of the four bonding electron pairs. We have previously seen this tetrahedral geometry in the structure of methane (CH_4). The remaining region of the molecule containing the C—O—H bonds has the bent geometry of the alcohol functional group.

■ **alcohol** A functional group with the general formula R—OH.

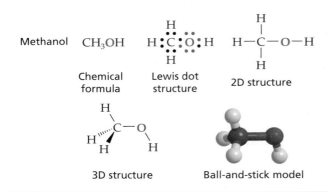

Alcohol

FIGURE 5.3 The alcohol functional group. In this general formula, R represents a carbon-containing group of atoms.

The alcohol found in alcoholic beverages is called *ethanol*, C_2H_5OH. To make these beverages, ethanol is derived from sugars by *fermentation*, a process that utilizes the biochemical properties of yeast cells. We can picture the ethanol molecule as resembling the hydrocarbon ethane, but with one hydrogen atom replaced by a hydroxyl group (—OH). Figure 5.5 presents both a three-dimensional structure and a ball-and-stick model of ethanol. As with the ethane molecule (discussed in Chapter 4), the atoms attached to each carbon in ethanol are in the *staggered* configuration to minimize bonding electron pair repulsions. The chemical formula for ethanol is sometimes written as CH_3CH_2OH to reflect the composition of the molecule (moving from left to right in Figure 5.5).

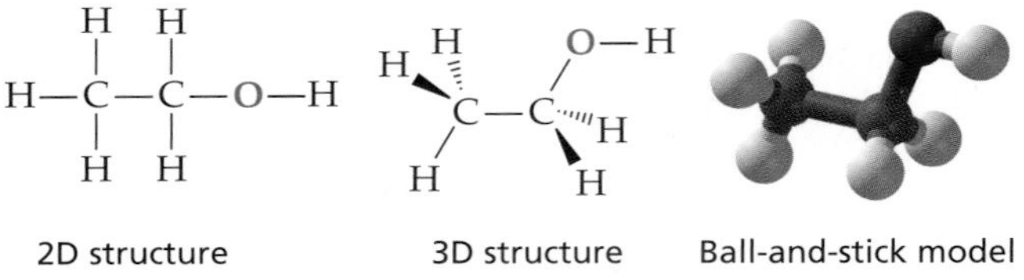

Methanol CH_3OH

Chemical formula

Lewis dot structure

2D structure

3D structure

Ball-and-stick model

FIGURE 5.4 Molecular representations of methanol, CH_3OH. The Lewis dot structure shows two lone pairs on the O atom. Lone pairs are not included in the 2D and 3D structures.

2D structure

3D structure

Ball-and-stick model

FIGURE 5.5 Molecular representations of ethanol, C_2H_5OH.

Core Concepts

■ The **alcohol** group is an —OH group attached to a carbon-containing group. One type of alcohol, ethanol, is found in alcoholic beverages.

Ether

The **ether** functional group consists of an oxygen atom that acts as a bridge between two carbon-containing groups, which can be the same or different. We use R and R′ to represent these groups when they are different. As shown in Figure 5.6, the molecular geometry of the ether group is bent because the two lone pairs on the oxygen atom contribute to electron pair repulsion.

A simple example of this functional group is *dimethyl ether*. As the name suggests, the bridging oxygen atom has a methyl group on each side. Figure 5.7 illustrates two different representations of dimethyl ether. The molecular model

■ **ether** A functional group with the general formula R—O—R′.

Ether

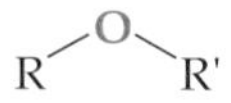

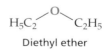

FIGURE 5.6 The ether functional group. In this general formula, R and R′ indicate two different carbon-containing groups. For a particular molecule, the two groups can be the same (R and R) or different (R and R′).

Dimethyl ether

Molecular structure Ball-and-stick model

FIGURE 5.7 Molecular structure of dimethyl ether. One of the methyl groups is drawn "backwards" (H_3C—) to indicate that the C atom (not the H atom) is attached to the central O atom.

depicts the three-dimensional structure of the molecule, with a tetrahedral geometry for each methyl group and a bent structure for the ether bridge.

When we hear the word "ether," we may think of an anesthetic. Until the mid-19th century, surgery was appallingly crude. If surgeons had to amputate an arm or a leg, the only comfort they could provide to their patient was a dose of hard liquor. Deaths during or after surgery were very common, often from the shock of the pain. This terrible situation ended with a medical breakthrough—the development of an anesthetic. On October 16, 1856, a dentist named William T. Morton (1819–1868) first administered a dose of ether to a patient in preparation for surgery. Upon awakening, the patient declared that he had not experienced any feeling during the procedure. A contemporary report of the surgery enthusiastically observed, "We have conquered pain." In 1893, this event was celebrated by a painting that is now on display at the Harvard Medical School (Figure 5.8).

The ether used by Morton and later adopted by the medical community is called *diethyl ether*, in which each R group is an ethyl group (—C_2H_5). Figure 5.9 presents the structure of this molecule. Diethyl ether is a liquid at room temperature, but it readily evaporates to form a vapor that can be inhaled. The ether molecules enter the bloodstream and cause the patient to become unconscious. However, diethyl ether does not dissolve easily in water, so it quickly leaves the bloodstream and escapes as a vapor. As a result, the unconscious state of the patient can be regulated by changing the amount of ether vapor that is inhaled. In modern-day surgery, diethyl ether has been replaced by safer anesthetics

Core Concepts

- The **ether** group contains an oxygen atom that acts as a bridge between two carbon-containing groups. An ether was used as the first anesthetic.

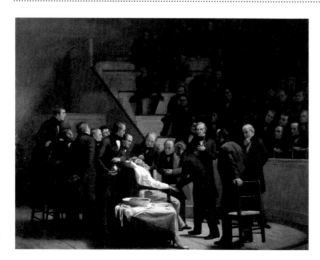

FIGURE 5.8 The use of ether as an anesthetic. This painting depicts William T. Morton's first use of ether as surgical anesthetic in 1846.

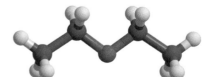

Diethyl ether

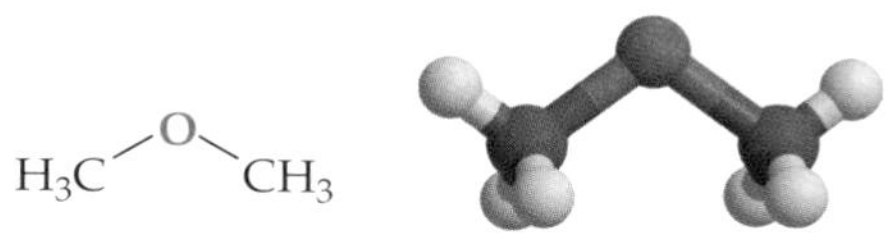

Molecular structure Ball-and-stick model

FIGURE 5.9 Molecular structure of diethyl ether. One of the ethyl groups is drawn "backwards" (H_5C_2—) to indicate that the C atom and not the H atom is bonded to the central O atom.

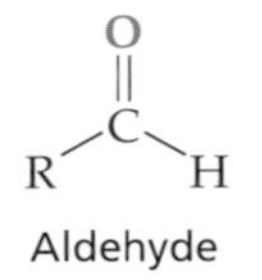

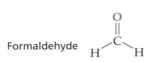

Aldehyde

FIGURE 5.10 The aldehyde functional group. In this general formula, R represents a carbon-containing group.

Formaldehyde

Molecular structure   Ball-and-stick model

FIGURE 5.11 Molecular structure of formaldehyde. This compound is shown as a molecular structure and a ball-and-stick model.

🔑 THE KEY IDEA: Aldehydes contain a C=O group attached to one carbon-containing group.

■ **aldehyde** A functional group with the general formula R—CO—H

Aldehyde

The **aldehyde** group contains a double bond between a carbon atom and an oxygen atom, plus an H atom bonded to the carbon (Figure 5.10). The bonded C=O atoms in an aldehyde are called a *carbonyl group*; this structure also appears in other functional groups.

The simplest aldehyde is called *formaldehyde,* which is illustrated in Figure 5.11. Formaldehyde is an exception to the general rule for aldehydes because it does not have a carbon-containing R group. Instead, there are two hydrogen atoms bonded to the central carbon atom. Formaldehyde is a gas that easily dissolves in water to produce a solution called *formalin,* which is used to preserve biological specimens. Any substance containing formaldehyde must be handled with extreme care because this molecule is highly toxic, causing symptoms that range from breathing problems to nausea and vomiting.

Several common food and spice flavorings are aldehydes in which the R group is a benzene ring or one of its derivatives (Figure 5.12). For example, the flavor of almonds arises from a chemical known as *benzaldehyde* (i.e., benzene + aldehyde). Another example is *cinnamaldehyde,* which produces the flavor of cinnamon.

Core Concepts

■ The **aldehyde** group contains a C=O group attached to one carbon-containing group. Some aldehydes are responsible for the flavors of foods.

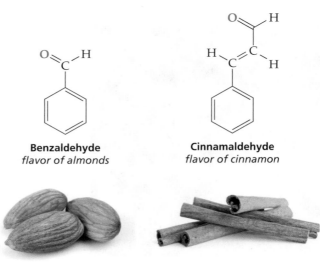

Benzaldehyde
flavor of almonds

Cinnamaldehyde
flavor of cinnamon

FIGURE 5.12 Some food and spice flavors are aldehyde molecules. Two examples are benzaldehyde (flavor of almonds) and cinnamaldehyde (flavor of cinnamon).

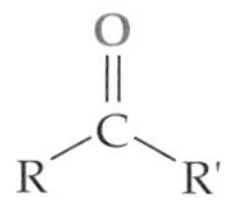

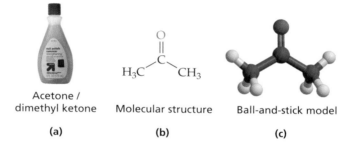

FIGURE 5.13 The ketone functional group. In this general formula, R and R′ indicate two different carbon-containing groups. For a particular molecule, the two groups can be the same (R and R) or different (R and R′).

FIGURE 5.14 Molecular structure of acetone (dimethyl ketone). (a) Acetone is a solvent commonly used in nail polish remover. The chemical name for acetone is dimethyl ketone, which is shown as (b) a molecular structure and (c) a ball-and-stick model.

Ketone

The **ketone** group is shown in Figure 5.13. It is similar to an aldehyde because it also contains a carbonyl group (C=O). However, in a ketone the carbon atom is linked to two carbon-containing groups, which can be the same or different.

The simplest example of a ketone is *acetone*, a versatile organic solvent that we frequently use to remove paint and nail polish (Figure 5.14(a)). Acetone contains two methyl groups (—CH$_3$) attached to the carbon atom, so its chemical name is *dimethyl ketone*. Figures 5.14(b) and (c) show the molecular structure and a ball-and-stick model of dimethyl ketone.

> **THE KEY IDEA:** Aldehydes contain a C=O group attached to two carbon-containing groups.

> **ketone** A functional group with the general formula R—CO—R′.

Core Concepts

■ The **ketone** group contains a C=O group attached to two carbon-containing groups. One type of ketone, acetone, is used in nail polish remover.

Carboxylic Acid

A **carboxylic acid** has the general chemical structure shown in Figure 5.15. The carbonyl group is bonded to an —OH group, which changes its chemical properties. A carboxylic acid is often written in a condensed form as R—COOH.

Why is this functional group called an "acid"? In simple terms, an *acid* is a substance that releases H$^+$ ions when it is dissolved in water. (We provide a more detailed discussion of acids in Chapter 11.) These H$^+$ ions are called *hydrogen ions* or *protons*; both terms describe what remains when a hydrogen atom loses its single electron. When a carboxylic acid is placed in water, the functional group can ionize to produce an H$^+$ ion and a negatively charged chemical group called a *carboxylate ion* (Figure 5.16).

> **THE KEY IDEA:** Carboxylic acids contain two oxygen heteroatoms and have acidic properties.

> **carboxylic acid** A functional group with the general formula R—COOH; it has the chemical properties of an acid.

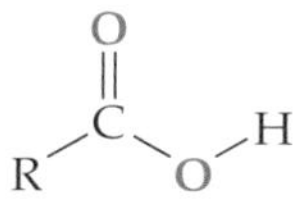

FIGURE 5.15 The carboxylic acid functional group. In this general formula, R represents a carbon-containing group.

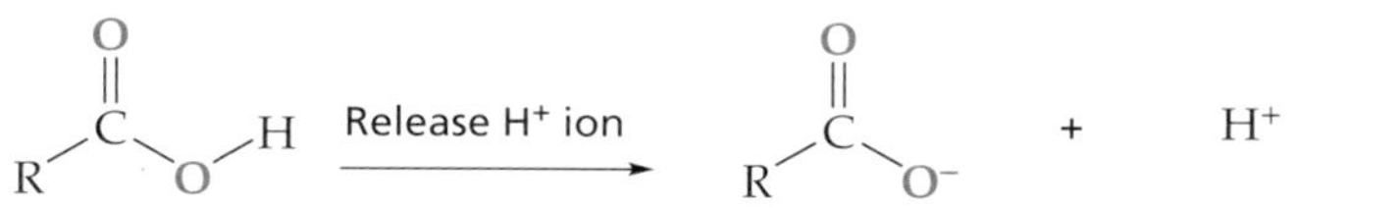

FIGURE 5.16 The acidic behavior of a carboxylic acid group. A carboxylic acid can ionize in solution to release a hydrogen ion (H$^+$) and create a negatively charged carboxylate ion. The hydrogen atom that is released in the ionization is highlighted in blue.

CHEMISTRY AND YOUR HEALTH

How Does Your Body Metabolize Alcohol?

Although humans have enjoyed alcoholic beverages for thousands of years, drinking large amounts of alcohol can be unpleasant, dangerous, and even fatal. Many of these adverse effects arise from chemical reactions that occur within living organisms—a process called *metabolism*. More specifically, alcohol metabolism refers to the chemical reactions that convert alcohol into other substances within the human body.

The type of alcohol found in beer, wine, and spirits is ethanol, C_2H_5OH. When we consume ethanol, it quickly passes from our digestive system into our bloodstream. The blood carries the alcohol to the liver, which is our primary organ for dealing with toxins that enter the body. The liver contains many types of enzymes that chemically modify these toxins. One liver enzyme, called *alcohol dehydrogenase*, facilitates the removal of two hydrogen atoms from ethanol (hence the name *de*-hydrogen-ase). This reaction, shown in Figure 1, produces a molecule called *acetaldehyde*, which contains an aldehyde functional group.

Acetaldehyde is toxic and can damage the liver. Fortunately, another liver enzyme called *aldehyde dehydrogenase* speeds the conversion of acetaldehyde into acetic acid, which is not toxic and can be fed into a cycle of energy-producing reactions within the cell. This second reaction is illustrated in Figure 2.

If a person consumes a moderate amount of alcohol over an extended period of time, then the amount of acetaldehyde in the liver does not reach damaging levels because it is converted into acetic acid. However, if one consumes a large amount of alcohol rapidly, then the excess acetaldehyde produced by alcohol dehydrogenase cannot be cleared quickly enough by aldehyde dehydrogenase. As a result, the acetaldehyde enters the bloodstream and is transported around the body and into the brain. This molecule is responsible for the unpleasant effects of drinking excess amounts of alcohol, such as nausea, impaired brain function, and loss of coordination.

The chemical process of alcohol metabolism explains the wide variation in people's responses to drinking alcohol. The instructions for making the two key enzymes are contained within *genes*, which are long stretches of DNA that encode genetic information (see Chapter 13 for a more detailed discussion). Natural variations in human DNA produce different versions of the enzymes, or they affect the amount of enzyme that is produced. For example, one version of the gene for aldehyde dehydrogenase generates very low amounts of this enzyme (less than 10% of the usual level). For individuals with this gene, drinking even modest amounts of alcohol causes a large buildup of acetaldehyde in the body, which produces the symptoms of alcohol intoxication. This genetic variant is particularly common in Asian populations and their descendants.

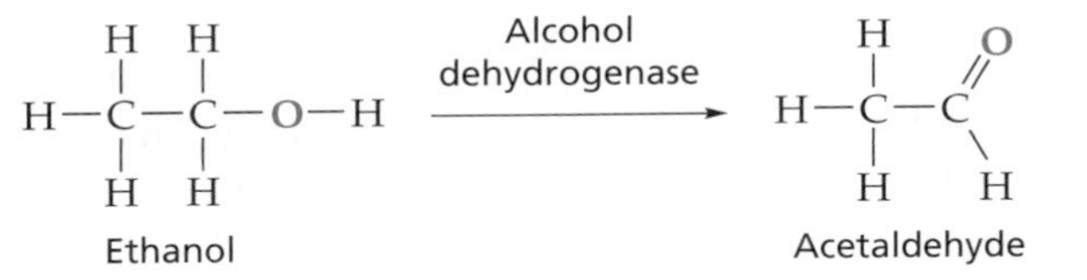

FIGURE 1 **The function of alcohol dehydrogenase.** This enzyme, which operates in the liver, facilitates the conversion of ethanol to acetaldehyde.

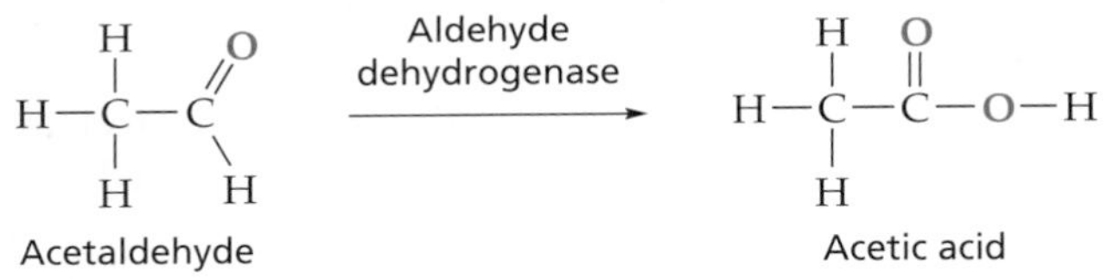

FIGURE 2 **The function of aldehyde dehydrogenase** This liver enzyme accelerates the conversion of acetaldehyde to acetic acid, which is nontoxic.

Carboxylic acids are a component of many common foods and drinks. For example, vinegar is an approximately 5% solution of *acetic acid*, CH_3COOH, dissolved in water (Figure 5.17). Another familiar example is *citric acid*, found in citrus fruits such as oranges, lemons, and limes. It is also used as a preservative in many other foods. As shown in Figure 5.18, citric acid contains three carboxylic acid groups and one alcohol group.

Functional groups are also important for understanding the chemical reactions that take place within our body. *Chemistry and Your Health* describes how the body processes alcohol by converting it into different molecular compounds.

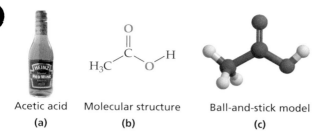

Acetic acid   Molecular structure   Ball-and-stick model
     (a)   (b)   (c)

FIGURE 5.17 Acetic acid. (a) Vinegar is an approximately 5% solution of acetic acid in water. (b) The molecular structure of acetic acid. (c) A ball-and-stick model.

Core Concepts

■ The **carboxylic acid** group contains two oxygen hereoatoms and can be written as R—COOH. It has acidic properties and is found in acetic acid and citric acid.

Ester

The terminal hydrogen in the R—COOH group can be replaced by another carbon-containing group (R or R') to form an **ester** functional group (Figure 5.19). We can write an ester group in a condensed form as R—COO—R'. Because it does not contain an H atom, an ester group is not acidic.

Esters are major contributors to the distinctive aroma of fruits such as raspberries, oranges, and pineapples, although other molecules also contribute. These ester molecules are *volatile*, which means that they easily change from a liquid into a vapor. The vapor then reaches our nostrils, and our brain processes it as a characteristic odor. Figure 5.20 shows one type of ester that contributes to the distinctive smell of pineapples.

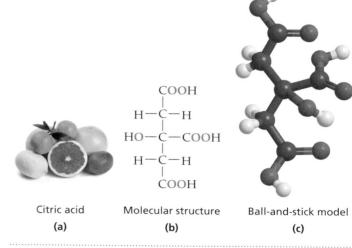

Citric acid   Molecular structure   Ball-and-stick model
    (a)   (b)   (c)

FIGURE 5.18 Citric acid. (a) Citrus fruits, such as oranges, lemons, and limes, contain citric acid. (b) The molecular structure of citric acid, which contains three carboxylic acid groups (written as —COOH) plus an alcohol group. (a) A ball-and-stick model.

🔑 **THE KEY IDEA: Esters** contains two oxygen heteroatoms and serve as a linkage between two carbon-containing groups.

ester A functional group with the general formula R—COO—R'

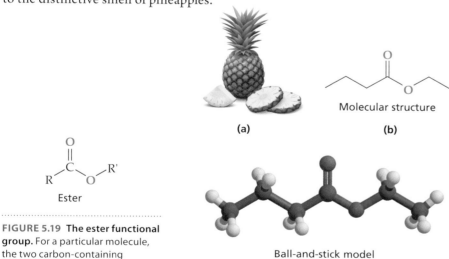

Molecular structure
      (a)   (b)

Ester

FIGURE 5.19 The ester functional group. For a particular molecule, the two carbon-containing groups can be the same (R and R) or different (R and R').

Ball-and-stick model
         (c)

FIGURE 5.20 Esters contribute to the characteristic aromas of fruits. (a) The distinctive smell of pineapples arises from volatile esters, in addition to other molecules. (b) An ester molecule that contributes to the aroma of pineapples, depicted as a line angle drawing. (c) A ball-and-stick model of the same molecule.

PRACTICE EXERCISE 5.1

The accompanying figure shows the molecular structures (line-angle drawings) of volatile esters that contribute to the characteristic aromas of different fruits.

(a) One of these esters is different in composition from the others. Identify which one, and describe the difference.

(b) Use a dotted circle to identify the ester functional group in each molecule. The circle should enclose the closest C atoms of the two carbon-containing groups (R and R′).

(c) Draw the full-atom structure of the ester that produces the aroma of bananas. Write the chemical formula for this molecule in the form $C_xH_yO_z$.

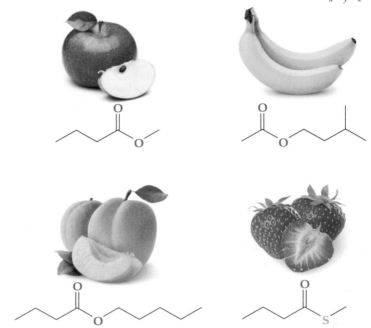

⊙ THE KEY IDEA: The presence of one or two oxygen heteroatoms can create multiple functional groups.

Functional groups containing oxygen heteroatoms

R—O—H
Alcohol

R—O—R′
Ether

$$\underset{R}{\overset{\overset{\displaystyle O}{\parallel}}{C}}\underset{H}{}$$
Aldehyde

$$\underset{R}{\overset{\overset{\displaystyle O}{\parallel}}{C}}\underset{R′}{}$$
Ketone

$$\underset{R\quad O}{\overset{\overset{\displaystyle O}{\parallel}}{C}}H$$
Carboxylic acid

$$\underset{R\quad O}{\overset{\overset{\displaystyle O}{\parallel}}{C}}R′$$
Ester

FIGURE 5.21 A summary of functional groups containing oxygen heteroatoms. An alcohol, aldehyde, and carboxylic acid each has one carbon-containing group (R). In an ether, ketone, and ester, the two carbon-containing groups can be the same (R and R) or different (R and R′).

Core Concepts

■ The **ester** group contains two oxygen heteroatoms and can be written as R—COO—R′. It serves as a linkage between two carbon-containing groups. Volatile esters contribute to the characteristic smells of fruits.

A Summary of Functional Groups Containing Oxygen Atoms

In this section, we have examined six functional groups that contain either one or two oxygen heteroatoms. Figure 5.21 shows these groups arranged in pairs to reveal their structural similarities. For example, compare the chemical structures of the alcohol (R—O—H) and ether (R—O—R′) groups. They are almost identical except that the ether has a second carbon-containing group (R′) instead of the H atom in the alcohol. The comparison is similar for both the aldehyde/ketone pair and the carboxylic acid/ester pair; that is, the H atom of the first functional group is replaced by an R′ in the partner functional group. Use these similarities to learn and identify the structures of the oxygen-containing functional groups.

Core Concepts

■ Alcohol, ether, aldehyde, ketone, carboxylic acid, and ester are functional groups that contain one or more oxygen heteroatoms.

3 Functional Groups Containing Nitrogen Atoms

Nitrogen is another common heteroatom in biological molecules and pharmaceuticals. In this section, we examine the molecular structures of functional groups that contain a nitrogen atom.

The term **amine** describes a class of functional groups that contain a nitrogen atom. Recall that nitrogen has five valence electrons, so it achieves a stable octet of electrons by forming three covalent bonds. Because there are so many amines, chemists subdivide them into three categories—**primary amine** (first order), **secondary amine** (second order), and **tertiary amine** (third order). These designations are based on whether the N atom is covalently bonded to one, two, or three carbon atoms. Figure 5.22 illustrates these three types of amines. For secondary amines and tertiary amines, the R groups can be the same or different.

Figure 5.23 provides the three-dimensional structure and a ball-and-stick model of *methylamine*, a primary amine with a single methyl as the R group. As in the case of methanol, we can understand the structure of methylamine by dissecting it into parts and applying the principle of electron pair repulsion (VSEPR). The four single bonds attached to the C atom produce a *tetrahedral structure* arising from repulsion between the four bonding electron pairs. The three single bonds attached to the central N atom produce a *trigonal pyramidal* structure because the N atom has a lone pair of electrons that repels the bonding pairs (recall the trigonal pyramidal structure of ammonia, NH_3, in Chapter 3). The two N—H bonds are *staggered* with respect to the neighboring C—H bond to minimize the electron pair repulsion.

Learning Objective:
Illustrate functional groups containing nitrogen atoms.

THE KEY IDEA: Amines and amides are functional groups that contain a nitrogen atom.

amine A functional group that contains a nitrogen atom.

primary amine An amine in which the nitrogen atom is bonded to one carbon atom.

secondary amine An amine in which the nitrogen atom is bonded to two carbon atoms.

tertiary amine An amine in which the nitrogen atom is bonded to three carbon atoms.

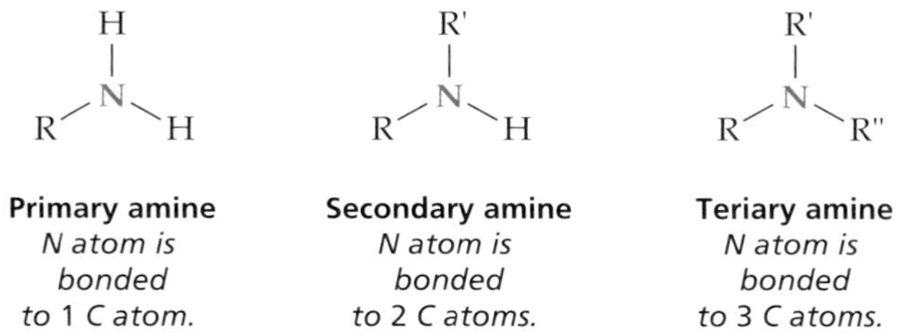

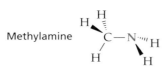

Primary amine
N atom is bonded to 1 C atom.

Secondary amine
N atom is bonded to 2 C atoms.

Teriary amine
N atom is bonded to 3 C atoms.

FIGURE 5.22 General structures for a primary, secondary, and tertiary amine. The letters R, R', and R'' represent different carbon-containing groups. In a particular molecule, these groups can be the same or different.

Methylamine

3D structure   Ball-and-stick model

FIGURE 5.23 Molecular structure of methyl amine. This compound is depicted as a 3D structure and a molecular model.

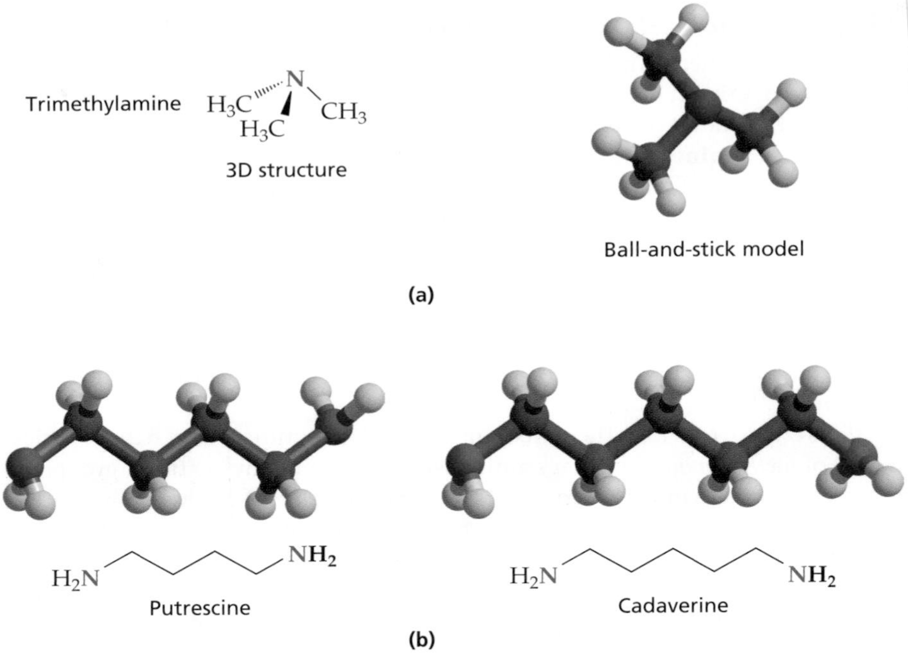

Trimethylamine

H_3C $^{\text{''''''}}N$ CH_3
H_3C

3D structure

Ball-and-stick model

(a)

H_2N NH_2

Putrescine

H_2N NH_2

Cadaverine

(b)

FIGURE 5.24 Many amines have unpleasant odors. (a) Trimethylamine is responsible for the smell of rotting fish. (b) Putrescine and cadaverine contribute to the smell of decaying flesh. In the molecular structures of these molecules, the hydrocarbon chain between the amine groups is shown as a line-angle drawing.

Some molecules containing amines have a very offensive smell. Figure 5.24(a) shows the structure of *trimethylamine*, a tertiary amine with three methyl groups attached to a central N atom. This molecule produces the odor of rotting fish, and it can contribute to bad breath. The two amines in Figure 5.24(b) have the evocative names of *putrescine* and *cadaverine* because they are partly responsible for the smell of decaying flesh.

Amines are also important for understanding the function of our brains. *Chemistry in Your Life* explains the role of dopamine molecules in transmitting chemical signals between nerve cells and how drugs can disrupt this process.

The **amide** functional group contains both nitrogen and oxygen heteroatoms. Similar to amines, which we examined earlier, amides can exist as primary, secondary, or tertiary varieties depending on whether the N atom is covalently bonded to one, two, or three carbon atoms (Figure 5.25). The N atom in an amide

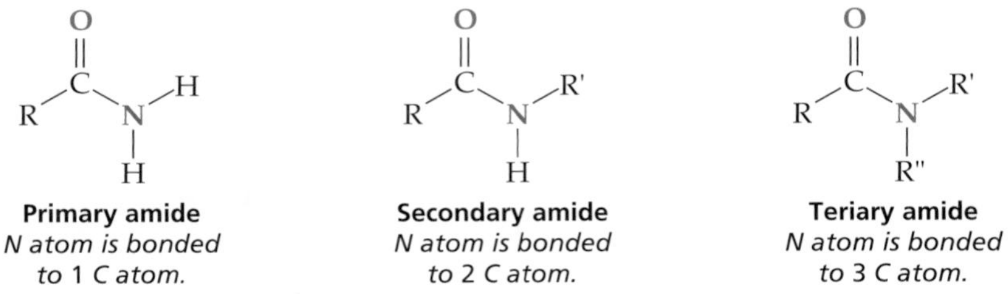

amide A functional group that contains both nitrogen and oxygen atoms.

O
‖
R C N H
|
H

Primary amide
*N atom is bonded
to 1 C atom.*

O
‖
R C N R'
|
H

Secondary amide
*N atom is bonded
to 2 C atom.*

O
‖
R C N R'
|
R''

Teriary amide
*N atom is bonded
to 3 C atom.*

FIGURE 5.25 General structures for a primary, secondary, and tertiary amide. The letters R, R', and R'' represent different carbon-containing groups. In a particular molecule, these groups can be the same or different.

CHEMISTRY IN YOUR LIFE

Chemical Signaling in the Brain

Our brain is a complex network of cells that are constantly communicating with one another. Nerve cells called *neurons* contain long extensions that come close to a neighboring neuron but do not touch. The junction between one neuron and another is called a *synapse* (Figure 1). Neurons send chemical signals across the synapse by using messenger molecules called *neurotransmitters*. The neuron that sends the neurotransmitter signal is called the *presynaptic neuron* ("pre-" means *before*), and the nerve cell that detects the signal is called the *postsynaptic neuron* ("post-" means *after*).

Neurotransmitters are stored in specialized containers in the presynaptic neuron. A pulse of electricity stimulates the containers to discharge the neurotransmitters into the synapse. The molecules then travel across the gap and attach to proteins called *receptors*, which are located on the surface of the postsynaptic neuron. When the neurotransmitter binds to the receptor, the receptor undergoes a structural change that triggers a biological response in the postsynaptic neuron. In this manner, neurotransmitters serve as the primary mode of communication between neurons.

Figure 2 shows the molecular structure of an important neurotransmitter called *dopamine*. As the name suggests, dopamine contains an amine functional group. This neurotransmitter is associated with the so-called *reward system* in the brain. When we experience pleasurable activities, certain neurons increase their release of dopamine as a neurological "reward." This chemical stimulation of the brain reinforces the behavior and prompts us to repeat it.

It is possible to overstimulate the brain's reward pathway by taking drugs. Figure 2 also shows the molecular structures of two highly addictive drugs—amphetamine and methamphetamine. The two structures are similar, but methamphetamine has an additional methyl group attached to the nitrogen, making it a secondary amine. Note that the molecular structures of both drugs are similar to that of dopamine.

Amphetamine and methamphetamine both function as chemical stimulants to the central nervous system, rapidly increasing blood pressure and pulse rate. Because they resemble dopamine, these drugs also interfere with the brain's reward system by stimulating the release of dopamine across the synapse. This flood of dopamine produces temporary feelings of energy and euphoria. These pleasurable sensations reinforce the drug-taking behavior, which makes amphetamine and methamphetamine highly addictive. Sustained use of these drugs can produce long-term brain damage.

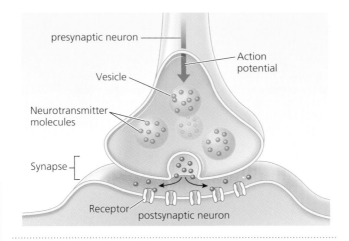

FIGURE 1 A simplified diagram of a synapse between two neurons. Communication across the synapse occurs via neurotransmitter molecules, which are released by the presynaptic neuron and bind to receptors on the surface of the postsynaptic neuron.

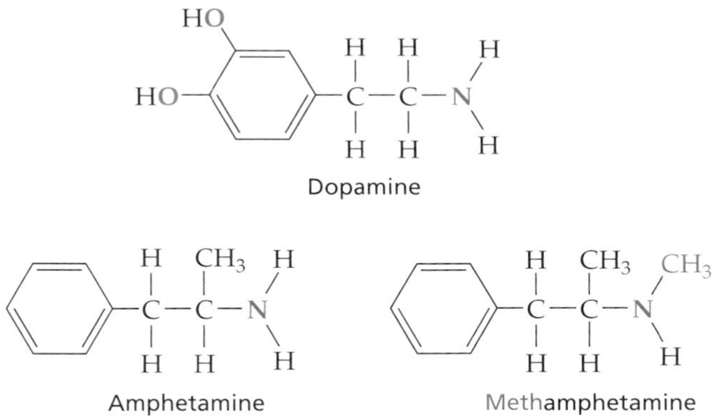

FIGURE 2 A molecular comparison of dopamine, amphetamine, and methamphetamine. Dopamine is a neurotransmitter that carries molecule signals between nerve cells. Amphetamine and methamphetamine are stimulant drugs that closely resemble the structure of dopamine. The methyl group of methamphetamine is highlighted in pink.

is always bonded to one carbon that forms part of a C=O bond. The other two bonds for the N atom make connections to H atoms or to carbon-containing R groups. For a tertiary amide, the R groups can be the same or different. The secondary amide structure has special relevance for the chemistry of biological molecules because it serves as the link between the amino acids within a protein.

Figure 5.26 illustrates two pharmaceuticals that contain an amide functional group. The first example, acetaminophen, is a popular painkiller that was introduced at the beginning of the chapter. The second example is lidocaine, which is used as a local anesthetic for minor surgery and dental procedures. Lidocaine contains a tertiary amine group, in which a nitrogen is bonded to three carbon atoms. Its molecular structure is similar to Novocain, a familiar dental anesthetic.

To recap, Figure 5.27 compares the structures of amine and amide functional groups. Amines exist in different varieties (primary, secondary, and tertiary), depending on how many carbon atoms are bonded to the N atom. Amides also exist as primary, secondary, and tertiary varieties.

Core Concepts

■ **Amines** and **amides** are functional groups that contain a nitrogen atom. Both groups exist in **primary, secondary,** and **tertiary** varieties, depending on whether the N atom is attached to one, two, or three C atoms.

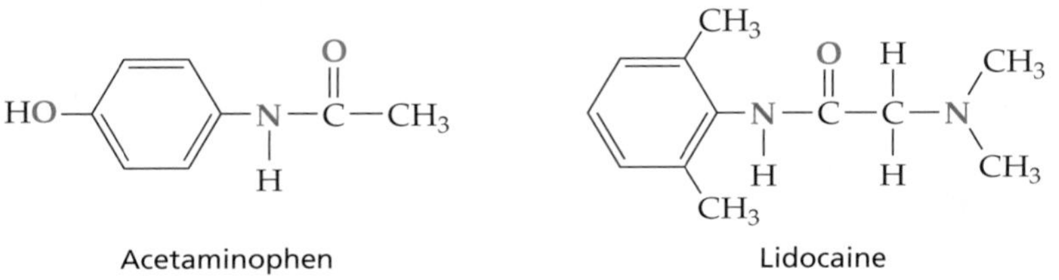

Acetaminophen   Lidocaine

FIGURE 5.26 Acetaminophen (a painkiller) and lidocaine (an anesthetic) both contain the amide functional group. The amide in these molecules is written with the N atom on the left, which is the reverse of the structure in Figure 5.25. However, the amide group is the same in both cases because it is unaffected by flipping its direction.

Functional groups containing a nitrogen heteroatom

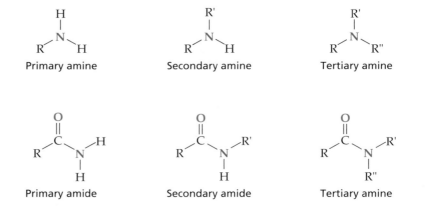

Primary amine   Secondary amine   Tertiary amine

Primary amide   Secondary amide   Tertiary amine

FIGURE 5.27 A summary of functional groups containing a nitrogen heteroatom. In these molecules, the three carbon-containing groups (R', R', and R") can be the same or different.

4 Functional Groups in Complex Molecules

Thus far, we have focused primarily on one type of functional group at a time. However, many biological molecules and pharmaceuticals have complex structures that contain multiple functional groups. The chemical and biological properties of these molecules are strongly influenced by the identity and location of these functional groups. This section gives you practice in analyzing complex molecules and identifying their functional groups.

One important example is an *amino acid*, which serves as the building block for proteins (Figure 5.28). As the name suggests, an amino acid contains both an amine and a carboxylic acid. Each functional group is bonded to a central carbon atom called the *alpha-carbon* (Cα). Also attached to this carbon atom is a variable chemical group called the *sidechain*, which is represented by the letter R. The sidechain can exist in 20 chemical varieties, which define the 20 types of amino acids that are commonly found in proteins. We examine amino acids in greater detail in Chapter 14.

Figure 5.29 provides some helpful tips for recognizing functional groups in complex molecules. Worked Example 5.1 applies these principles to identifying the functional groups in aspartame, a widely used artificial sweetener.

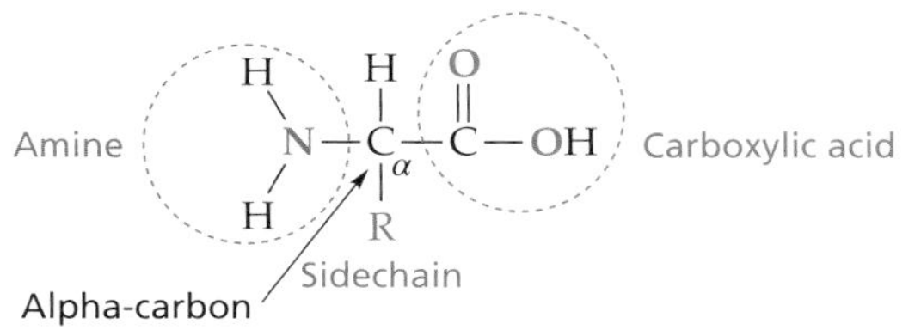

Learning Objective:

Identify a variety of functional groups in complex molecules.

THE KEY IDEA: A complex molecule can have multiple functional groups that strongly influence its chemical properties.

FIGURE 5.28 Molecular structure of an amino acid. An amino acid contains two functional groups, an amine and a carboxylic acid, that are bonded to the central alpha-carbon. The sidechain R exists in 20 different varieties that define the identity of the amino acid.

Dos and a Don't for Identifying Functional Groups in Complex Molecules

DO

When identifying the functional group, always select the largest possible grouping of heteroatoms by looking at the neighboring atoms. Simply identifying a collection of atoms like C==O is not sufficient to uniquely identify the functional group, since these atoms occur in aldehydes, ketones, esters, and amides. For example, recognize that the functional group below is an amide and NOT a ketone + an amine.

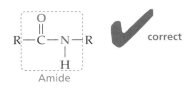

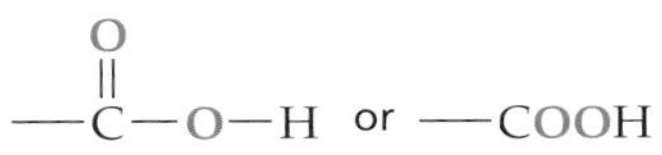

DON'T

Split apart functional groups into smaller components. For example, the molecule shown below does NOT contain a ketone and an amine.

$$R-C-N-R$$
with O double-bonded to C and H below N

Ketone amine   ✗ **incorrect**

DO

Realize that the same functional groups can be written in different ways. Sometimes all the chemical bonds are shown, but sometimes only the atoms are written and you need to infer the bonding.

e.g., Carboxylic acid

$$-\overset{\overset{\textstyle O}{\|}}{C}-O-H \quad \text{or} \quad -COOH$$

FIGURE 5.29 Do's and a don't for recognizing functional groups in complex molecules.

WORKED EXAMPLE 5.1

Question: The molecule shown here is aspartame (brand name Nutrasweet), an artificial sweetener commonly used in diet soda. Circle and identify all of the functional groups in this molecule.

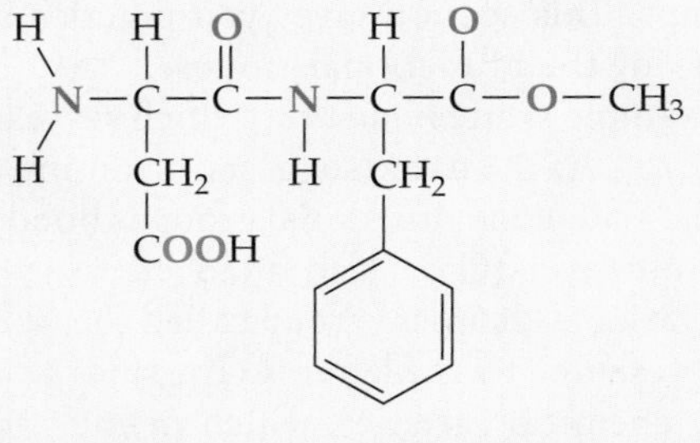

Answer: Let's begin systematically from the left end of the molecule. The first functional group that we encounter is an amine. More precisely, it is a primary amine because the nitrogen atom is attached to only one carbon atom.

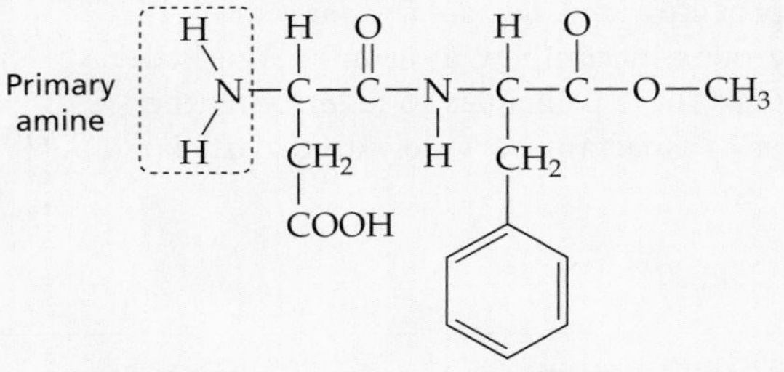

The next functional group is indicated by –COOH, which is a carboxylic acid.

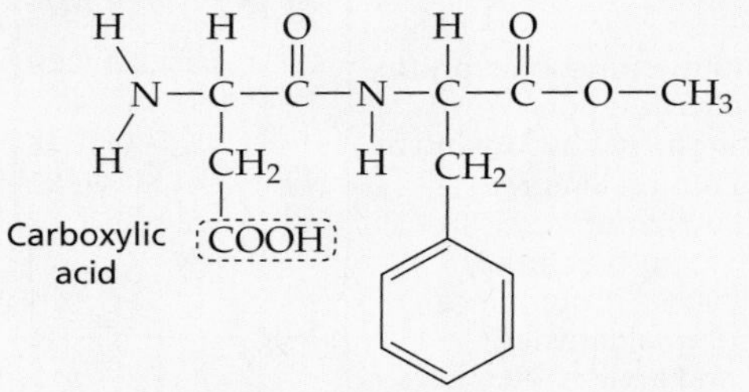

In the center of the molecule, we can identify an amide group.

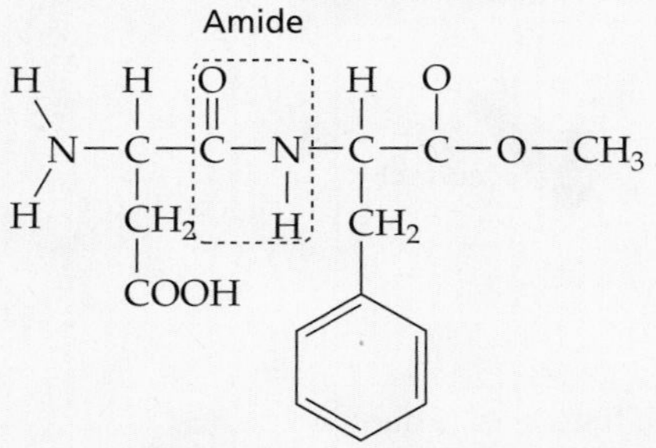

At the right end of the molecule there is an ester group.

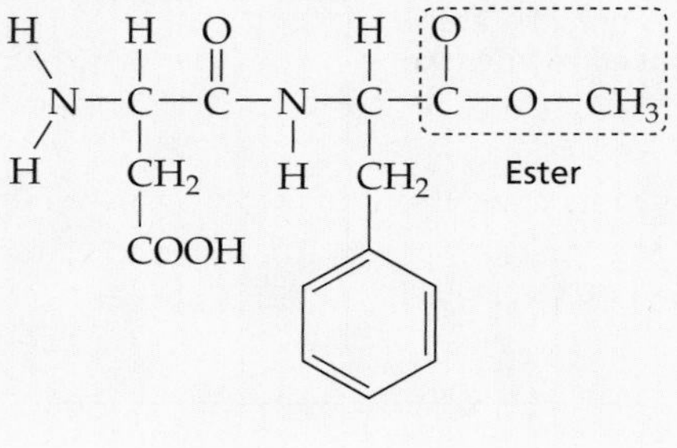

In summary, aspartame includes *four* different functional groups, as highlighted in the following figure:

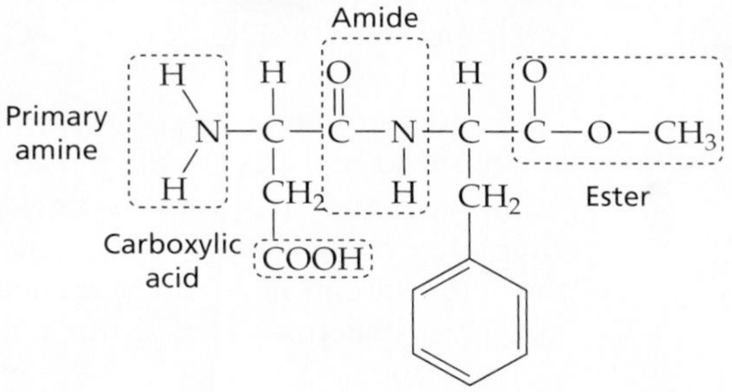

TRY IT YOURSELF 5.1

Question: Vitamin B5, also known as pantothenic acid, is found in shiitake mushrooms, cheeses, and oily fish. The structure of Vitamin B5 is shown here as a line-angle drawing. Circle and identify all of the functional groups in this molecule.

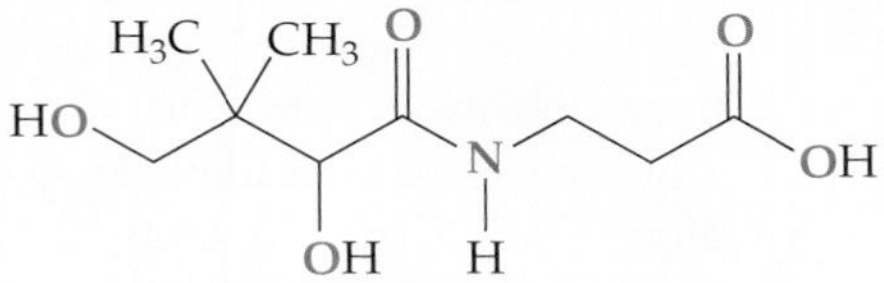

Vitamin B5

Core Concepts

- The chemical and biological properties of complex molecules are strongly influenced by their functional groups.

Learning Objective:

Illustrate functional groups containing sulfur or phosphorus atoms.

THE KEY IDEA: Sulfur and phosphorus heteroatoms are also capable of forming functional groups.

5.5 Functional Groups Containing Sulfur or Phosphorus Atoms

Although oxygen and nitrogen are the most common heteroatoms in functional groups, sulfur and phosphorus also play crucial roles in some biological molecules and pharmaceuticals. This section examines how the presence of these atoms affects a molecule's structure and properties.

Sulfur occurs in Group 6A of the periodic table, which is the same group as oxygen. Because of the similarity of these elements, the molecular structures of functional groups containing sulfur often resemble those formed by oxygen. However, the chemical properties of these sulfur-containing groups differ from those of their oxygen-containing counterparts. The presence of sulfur—especially when substituting for oxygen—is often indicated by the prefix *thio-*, which derives from the Greek word for sulfur. For example, when sulfur replaces oxygen in an alcohol group, the resulting molecule is called a **thiol**. Thiols contain a sulfur atom in anan —SH group in a similar arrangement to alcohols with their —OH group (Figure 5.30).

Perhaps the most striking feature of thiols is their terrible smell! This chemical property is used as a defensive strategy by skunks, which spray a pungent liquid at attackers. The repellent odor of skunk spray arises from two thiol compounds, each containing the –SH functional group. The more abundant of these compounds is derived from a *trans*-butene molecule in which one of the hydrogen

thiol A functional group in which sulfur replaces oxygen in an alcohol; it has the general formula R—SH.

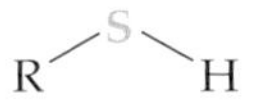

Thiol

FIGURE 5.30 The thiol functional group. This group resembles an alcohol, but with the oxygen atom replaced by a sulfur atom.

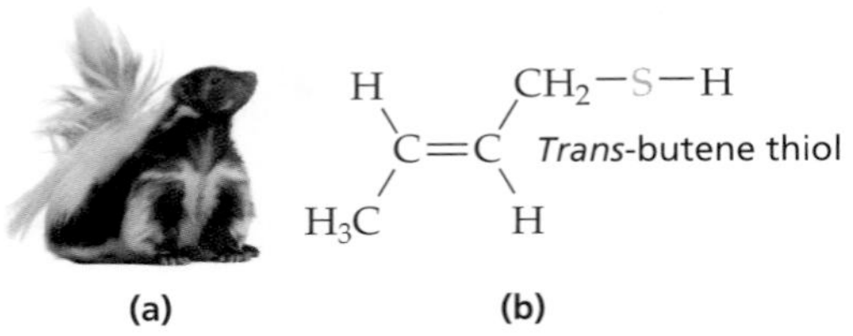

(a)   **(b)**

FIGURE 5.31 The pungent aroma of thiols. (a) Skunks defend them-selves with their spray. (b) Skunk spray contains the pungent molecule trans-butene thiol as the primary active compound.

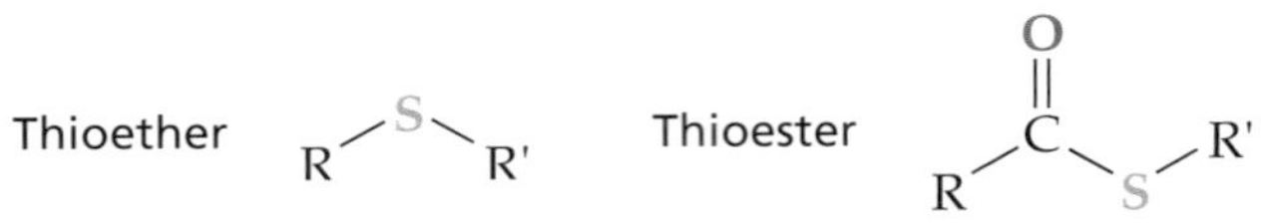

FIGURE 5.32 The thioether and thioester functional groups. In each of these groups, a sulfur atom replaces an oxygen atom. The letters R and R' represent different carbon-containing groups. For a particular molecule, these groups can be the same (R and R) or different (R and R').

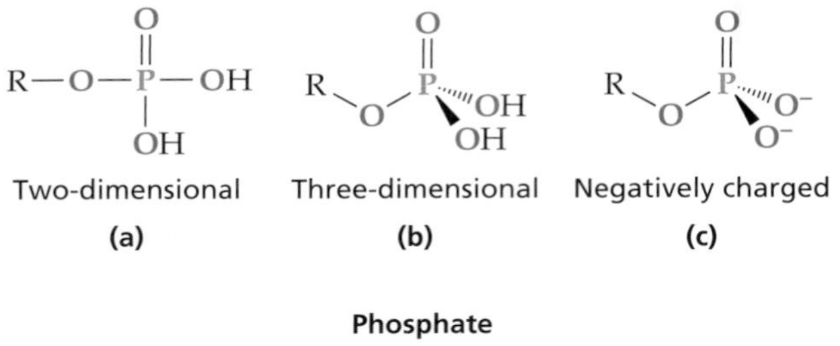

Two-dimensional   Three-dimensional   Negatively charged

(a)   **(b)**   **(c)**

Phosphate

FIGURE 5.33 The phosphate functional group. (a) A phosphate group contains a phosphorus atom bonded to four oxygen atoms. In this structure, R represents any carbon-containing group. (b) A phosphate group has an approximately tetrahedral geometry because of electron pair repulsion. (c) A negatively-charged version of the phosphate group, which arises from its acidic properties.

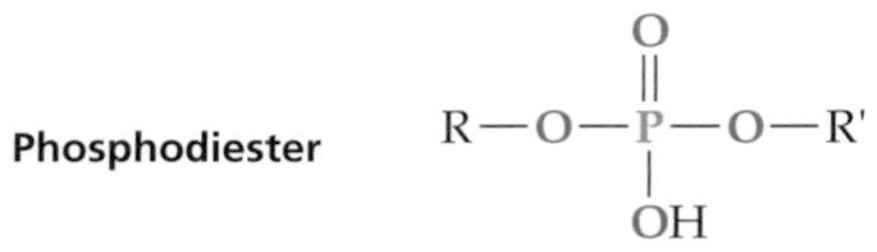

Phosphodiester

FIGURE 5.34 The phosphodiester functional group. This group resembles an ester, but with phosphorus replacing carbon plus two bridging oxygen atoms.

atoms of a methyl group has been replaced by –SH (Figure 5.31). As indicated by the *trans* designation, the two peripheral carbon atoms lie "across" from each other on opposite sides of the C=C double bond (see Chapter 4).

Sulfur can replace oxygen in other functional groups. When sulfur replaces the oxygen atom in an ether, the resulting functional group is called a **thioether** (an alternative name is a *sulfide*). Similarly, replacing one of the oxygen atoms in an ester produces a **thioester**. The molecular structures of these functional groups are pre-sented in Figure 5.32.

The most important role of phosphorus in biologi-cal molecules is forming a **phosphate group,** which is illustrated in Figure 5.33(a). In this functional group, a phosphorus atom is covalently bonded to four oxygen atoms using one double bond and three single bonds. Because of mutual repulsion between the bonding electron pairs, the phosphate group adopts an approximately tetrahedral geometry as depicted in Figure 5.33(b). However, the phosphate group is *acidic* and it can ionize to release two H^+ ions, leaving a functional group with two negative charges (recall our earlier discussion of carboxylic acid). The nega-tively charged version of the phosphate group, shown in Figure 5.33(c), closely resembles the **phosphate ion** that we introduced in Chapter 3.

Phosphate groups are a component of several im-portant biological molecules. For example, three phos-phate groups are found in adenosine triphosphate (ATP), which serves as the energy currency of the cell. Phosphates also have an important role in the molecular structure of DNA, which is the focus of Chapter 12. Within the backbone of DNA, two of the oxygen atoms in a phosphate group form covalent bonds to carbon atoms. The structure of this chemical linkage resembles an ester group, but there are two "bridging" oxygen atoms instead of one. As a result, this functional group is called a **phosphodiester** (the *di-* refers to *two* ester-type bonds). Figure 5.34 illustrates the chemical struc-ture of this phosphorus-containing group. A phospho-diester is also acidic and can release one H^+ ion from its —OH group.

Figure 5.35 summarizes the functional groups that contain sulfur or phosphorus heteroatoms.

Core Concepts

- Because sulfur and oxygen belong to the same group of the periodic table, functional groups containing sulfur often resemble those containing oxygen. Examples are **thiol**, **thioether** (sulfide), and **thioester**. However, sulfur-containing groups have different chemical properties compared to their oxygen-containing counterparts.

- The **phosphate group** contains a phosphorus atom bonded to four oxygen atoms. Phosphate groups are found in ATP and in the backbone of DNA.

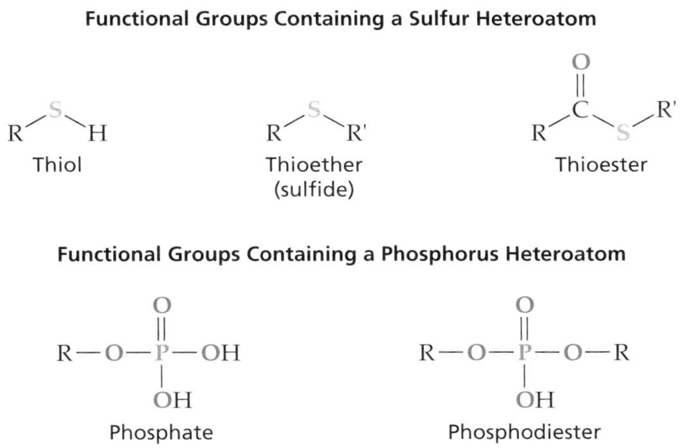

Functional Groups Containing a Sulfur Heteroatom

R—S—H
Thiol

R—S—R'
**Thioether
(sulfide)**

R—C(=O)—S—R'
Thioester

Functional Groups Containing a Phosphorus Heteroatom

R—O—P(=O)(OH)—OH
Phosphate

R—O—P(=O)(OH)—O—R
Phosphodiester

....

FIGURE 5.35 A summary of functional groups containing a sulfur or phosphorus heteroatom.

thioether A functional group in which sulfur replaces oxygen in an ether; it has general formula R—S—R'.

thioester A functional group in which a sulfur atom replaces one of the oxygens in an ester.; It has the general formula R—CO—S—R'.

Phosphate group A functional group in which a phosphorus atom is bonded to four oxygen atoms.

phosphate ion An ion with the chemical formula, PO_4^{3-}.

Phosphodiester A chemical linkage composed of a phosphorus atom and two modified ester bonds.

heterocycle A cyclic molecule in which one or more of the carbon atoms has been replaced with a heteroatom.

....

Learning Objective:

Illustrate molecules containing heterocycles.

🔑 THE KEY IDEA: A heterocycle is a cyclic molecule in which one or more of the carbon atoms has been replaced with a heteroatom.

5.6 Heterocycles

As we learned in Chapter 4, carbon atoms can arrange themselves in a cyclic (ring) structure. Heteroatoms— especially oxygen and nitrogen—can take the place of one or more carbon atoms in these cyclic molecules, and the resulting molecules are called **heterocycles**. Heterocycle structures frequently occur in biological molecules and in pharmaceuticals.

Vitamin C is probably the best-known vitamin. The human body does not make vitamin C, so we must obtain it from our diet. It occurs in abundance in citrus fruits (e.g., oranges and lemons) and in many vegetables (e.g., potatoes). The structure of vitamin C is shown in Figure 5.36 using three representations. The core of the molecule is an oxygen-containing heterocycle. In addition, the molecule contains four alcohol (—OH) groups.

Vitamin C is essential in the growth and repair of connective tissue, bones, and teeth. A lack of vitamin C can lead to scurvy, a disease that historically plagued sailors who had little access to fresh fruit on long voyages. Symptoms of this disease include bursting blood vessels, bleeding gums, and overall fatigue. There have been claims that high doses of vitamin C—up to 100 times the recommended dietary allowance—can cure the common cold or even treat cancer. However, these assertions have never been substantiated by rigorous clinical trials. Excess vitamin C, beyond what the body needs, is simply excreted in the urine.

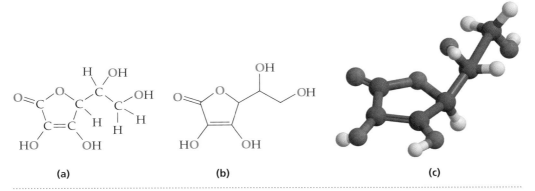

(a)   (b)   (c)

FIGURE 5.36 The molecular structure of vitamin C. (a) A molecular drawing that shows all the atoms. (b) A line-angle drawing that highlights the heteroatoms. (c) A ball-and-stick model that depicts the three-dimensional structure.

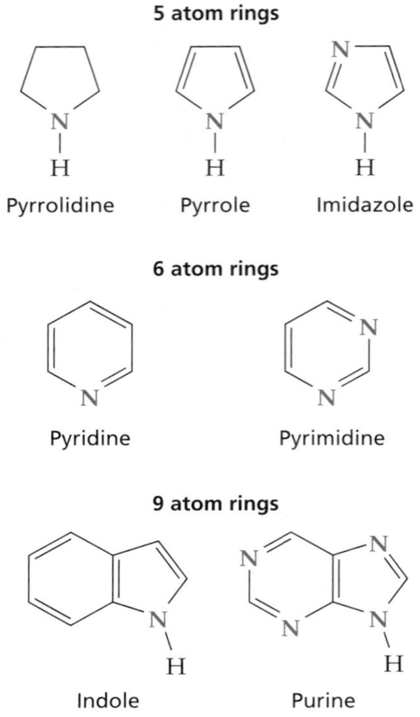

5 atom rings

Pyrrolidine   Pyrrole   Imidazole

6 atom rings

Pyridine   Pyrimidine

9 atom rings

Indole   Purine

FIGURE 5.37 Examples of heterocycles that contain nitrogen atoms. The heterocycles are organized according to the total number of atoms in the ring. Carbon atoms are not explicitly shown in order to accentuate the location of the nitrogen atoms.

Nitrogen atoms are found in a wide range of heterocycle structures. Figure 5.37 provides several examples, which are organized according to the total number of atoms in the ring. We follow the convention of not explicitly showing the carbon atoms in the ring in order to accentuate the location of the nitrogen atoms. In some of these heterocycles—such as pyridine and pyrimidine—the electrons are delocalized around the ring in a manner similar to benzene (see Chapter 4). This delocalization causes the ring to adopt a planar geometry, and it also influences the chemical properties of the nitrogen atom(s) within the ring. The purine and pyrimidine heterocycles are used as the structural foundation for making the four chemical bases in DNA; we examine this topic further in Chapter 12.

Anybody who suffers from allergies has felt the effects of *histamine*, a molecule that transmits a chemical signal between the body's cells. The molecular structure of histamine, shown in Figure 5.38, is based on an imidazole heterocycle. An allergic reaction like hay fever arises from an oversensitive immune reaction to harmless substances such as dust and pollen. As part of that response, specialized cells in the immune system secrete histamine into the bloodstream. Histamine binds to target proteins called *histamine receptors*, and this molecular interaction triggers a cascade of chemical signals that result in allergy symptoms such as sneezing and watery eyes.

Allergy medications such as Benadryl and Allegra function as *antihistamines*. They bind to the histamine receptor, which blocks histamine from attaching there. If histamine cannot access the receptor, it cannot transmit the chemical signal that stimulates an allergic response. One molecule with antihistamine properties is *diphenylhydramine* (Figure 5.39), which is the active ingredient in Benadryl and a component of over-the-counter sleep medications such as Tylenol PM. In addition to two benzene rings, diphenylhydramine contains an ether and a tertiary amine.

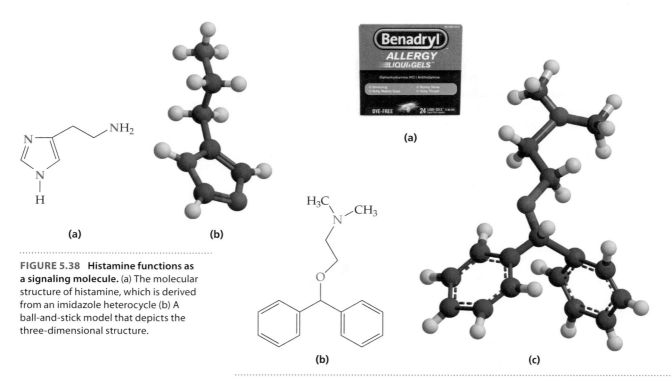

FIGURE 5.38 Histamine functions as a signaling molecule. (a) The molecular structure of histamine, which is derived from an imidazole heterocycle (b) A ball-and-stick model that depicts the three-dimensional structure.

FIGURE 5.39 Many allergy medications block the effect of histamine. (a) Benadryl is an example of an anti-histamine medication that relieves allergy symptoms. (b) The active ingredient in Benadryl is diphenylhydramine. (c) A ball-and-stick molecule that depicts the three-dimensional structure.

WORKED EXAMPLE 5.2

Question: Nicotine is an addictive compound found in tobacco. The molecular structure of nicotine, shown here, consists of two heterocycles joined by a covalent bond. Identify and name the heterocycle on the left

Nicotine

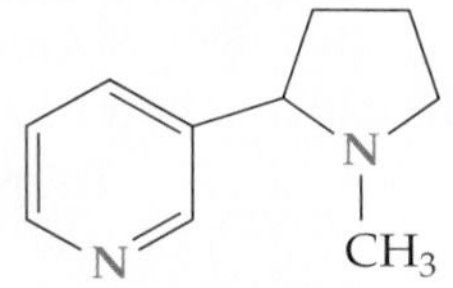

Answer: The heterocycle on the left is *pyridine*.

TRY IT YOURSELF 5.2

Question: Our body needs an adequate supply of vitamins for optimal health. One of these compounds is vitamin B3, which is also called *niacin*. The molecular structure of vitamin B3 is shown here. (a) What type of heterocycle does this vitamin contain? (b) Which functional group is attached to the heterocycle?

Vitamin B3
niacin

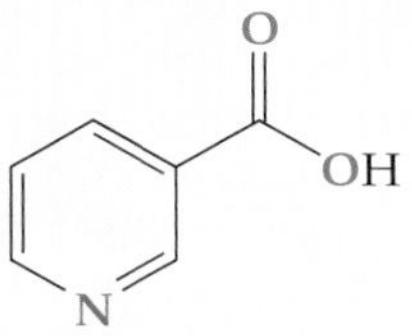

Our final example of a nitrogen-containing heterocycle is caffeine, a well-known stimulant found in coffee, tea, soda, and energy drinks. The chemical structure of caffeine, shown in Figure 5.40, is based on a purine heterocycle with the addition of extra C=O and —CH$_3$ groups. After we ingest caffeine in a drink, it enters our bloodstream and easily passes through the blood-brain barrier into our brain. Caffeine functions as a stimulant because it is chemically similar to the heterocycle in a biological molecule called *adenosine*, which is also shown in Figure 5.40. Adenosine regulates brain activity by suppressing the action of nerve cells, which produces a calming effect on the nervous system. In addition, adenosine is involved in regulating the sleep-wake cycle. Because caffeine has a similar molecular structure, it blocks the action of adenosine in the brain. This reduction of adenosine's calming effect produces the common symptoms of caffeine consumption, such as increased stimulation and wakefulness.

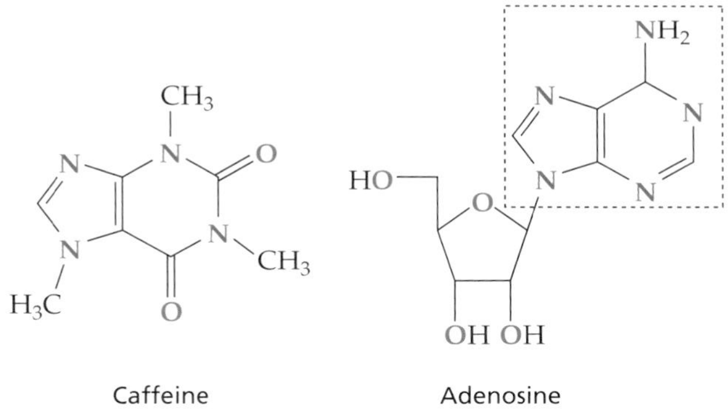

Caffeine   Adenosine

FIGURE 5.40 A comparison of caffeine and adenosine. The molecular structure of caffeine closely resembles the heterocycle in adenosine, which is highlighted by a dashed box. Caffeine achieves its stimulant effect by blocking the action of adenosine in the brain.

PRACTICE EXERCISE 5.2

Enzymes sometimes require assistance from additional molecules to perform their biological function. These accessory molecules are called *coenzymes* because they work together with the enzyme. The structure of coenzyme A is given at the right as a line-angle drawing.

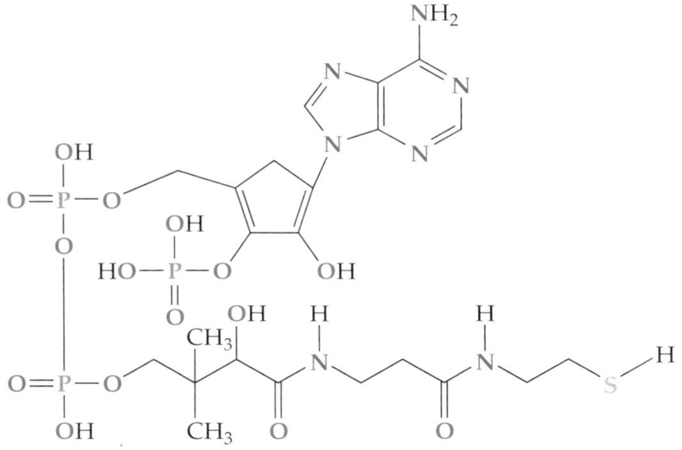

(a) The nitrogen-containing heterocycle in coenzyme A is derived from one of the heterocycles shown in Figure 5.37. Identify this heterocycle.

(b) Circle and identify three different functional groups in coenzyme A.

We conclude this chapter by connecting molecular diversity to our discussion of antibiotics and resistance in Chapter 1. *Science in Action* describes the chemical synthesis of penicillin, and it explains how this breakthrough permitted chemists to create new antibiotics such as ampicillin and methicillin.

SCIENCE IN ACTION

Chemical Synthesis of Antibiotics

In Chapter 1, we learned about the ability of penicillin to kill bacteria. This effect was first observed by Alexander Fleming in 1928. However, penicillin could not be used as a drug until it had been extracted, isolated, and purified from the *Penicillium* mold that secretes it. This difficult problem was solved by a group of scientists at Oxford University during World War II. However, their method required an enormous amount of liquid mold broth and a complicated extraction procedure to generate just a single patient dose of penicillin. During and after the war, pharmaceutical companies developed methods for producing penicillin on a large scale. However, the process still required substantial amounts of mold broth to generate sufficient amounts of the drug.

Penicillin is a *natural product*—that is, a molecule obtained from a natural source such as a *Penicillium* broth. The specific type of penicillin obtained by the Oxford scientists was given the name *benzylpenicillin* because its molecular structure contains a benzene ring. This antibiotic was later renamed *penicillin G*.

Isolating and purifying natural products can be difficult, costly, and time-intensive. Is there a better and more efficient way to make penicillin? One approach is to use *chemical synthesis*—the process of creating a complex molecule by using sequential steps to construct it from simpler components. In 1957, an American chemist named John Sheehan (1915–1992) designed a chemical synthesis that enabled him to make penicillin in the laboratory.

This laboratory synthesis of penicillin enabled chemists to introduce variations into the structure of the drug. Sheehan experimented with different chemical compounds to make new types of antibiotics that are not found in nature. One example was ampicillin, which had the same antibacterial effect as penicillin but could be taken orally rather than by injection. This development greatly enhanced the practical utility of antibiotics. In 1959, a pharmaceutical company developed another variant of penicillin called methicillin. This drug was used to treat infections by bacteria that had become resistant to penicillin.

The molecular structures of penicillin G, ampicillin, and methicillin are shown in Figure 1. The core structure of the molecule (shown in black) includes a square heterocycle that contains a nitrogen atom and a carbonyl group. The core structure also contains two functional groups, an amide and a carboxylic acid. When comparing

these three antibiotics, note that *the core structure remains the same* but the molecules differ in the molecular attachments to the left side of the molecule (which is highlighted in red). These attachments to the core structure are referred to as *substituents*.

Even though methicillin was developed to combat penicillin-resistant bacteria, many bacteria have now become resistant to methicillin as well. One example is methicillin-resistant *Staphylococcus aureus* (MRSA), which we introduced at the beginning of Chapter 1.

Today, chemists use a variety of methods to produce large-scale quantities of various pharmaceuticals. Some natural-product drugs are still isolated from their original biological source. Other drugs are produced using chemical synthesis in the laboratory. A third method, called semisynthesis, uses a natural product as a scaffold to which different chemical groups are synthetically added.

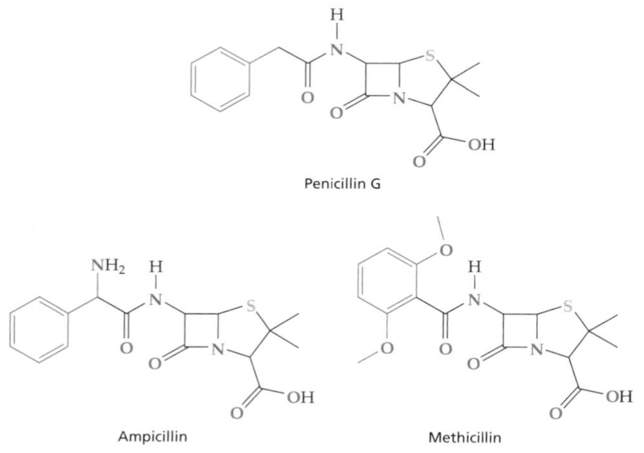

FIGURE 1 A molecular comparison of penicillin G, ampicillin, and methicillin. The core structure of all three drugs (shown in black) remains the same, but they differ in their chemical substituents (shown in pink). The hydrocarbon regions of these molecular structures are shown as line-angle drawings. Straight lines without an atom at one end terminate in a methyl group ($-CH_3$).

Core Concepts

■ A **heterocycle** is a cyclic molecule in which one or more heteroatoms (especially oxygen or nitrogen) replace the carbon atoms in the ring.

CHAPTER 5
VISUAL SUMMARY

5.1   What Do Drug Molecules Look Like?

Learning Objective:

Explain the role of functional groups in molecular structure.

⊙ A **heteroatom** is an atom that is different from carbon and hydrogen. The most common heteroatoms in molecules are O, N, S, and P.

⊙ A **functional group** is a collection of atoms within a molecule that has a characteristic structure and chemical behavior.

⊙ Many drug molecules consist of a hydrocarbon framework with attached functional groups.

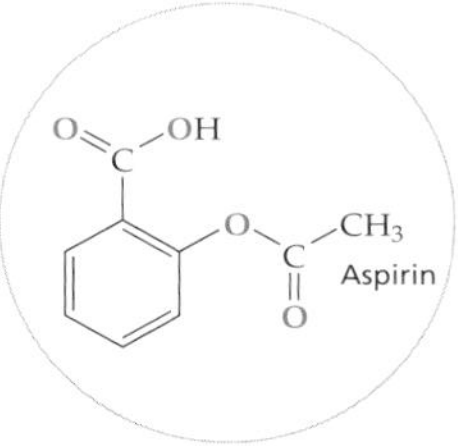

5.2   Functional Groups Containing Oxygen Atoms

Learning Objective:

Illustrate functional groups containing oxygen atoms.

⊙ The following functional groups contain one or two oxygen heteroatoms:

  ⊙ **Alcohol**

  ⊙ **Ether**

  ⊙ **Aldehyde**

  ⊙ **Ketone**

  ⊙ **Carboxylic acid**

  ⊙ **Ester**

5.3   Functional Groups Containing Nitrogen Atoms

Learning Objective:

Illustrate functional groups containing nitrogen atoms.

⊙ The following functional groups contain a nitrogen heteroatom:

  ⊙ **Amine**

  ⊙ **Amide**

⊙ Both of these functional groups exist in **primary, secondary**, and **tertiary** varieties, depending on whether the N atom is attached to one, two, or three C atoms.

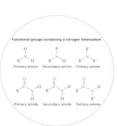

5.4   Functional Groups in Complex Molecules

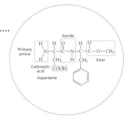

Learning Objective:

Identify a variety of functional groups in complex molecules.

⊙ Most biological molecules and pharmaceuticals contain multiple functional groups.

⊙ These functional groups strongly influence the chemical and biological properties of the molecule.

5.5   Functional Groups Containing Sulfur or Phosphorus Atoms

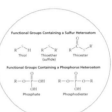

Learning Objective:

Illustrate functional groups containing sulfur or phosphorus atoms.

⊙ Because sulfur and oxygen belong to the same group of the periodic table, functional groups containing sulfur often resemble those containing oxygen. Examples include **thiol, thioether** (sulfide), and **thioester**. However, the chemical properties of sulfur-containing groups differ from those of oxygen-containing groups.

⊙ The most important biological role of phosphorus is forming the **phosphate ion** and the related **phosphate** group within molecules. Phosphate groups are found in ATP and in the structure of DNA.

5.6   Heterocycles

Learning Objective:

Illustrate molecules containing heterocycles.

⊙ A **heterocycle** is a cyclic molecule in which one or more of the carbon atoms has been replaced with a heteroatom.

LEARNING RESOURCES

Reviewing Knowledge

5.1: Introduction: What Do Drug Molecules Look Like?

1. Define the following terms: heteroatom, functional group.
2. Name four common heteroatoms.
3. Explain the importance of functional groups in pharmaceuticals and biological molecules.

5.2: Functional Groups Containing Oxygen Atoms

4. Draw the general molecular structure for the following functional groups containing one or two oxygen atoms:
     alcohol   ether   aldehyde   ketone
          carboxylic acid   ester
5. What was the first molecule to be used as a medical anesthetic?
6. Why are carboxylic acids classified as "acidic"?
7. Name one type of molecule that contributes to the fragrance of pineapples?

5.3: Functional Groups Containing Nitrogen Atoms

8. Draw the general molecular structure for primary, secondary, and tertiary amines. Explain the naming rules for these three types of amines.
9. Draw the general molecular structures of a primary, secondary, and tertiary amide.

5.4: Identifying Functional Groups in Complex Molecules

10. Summarize the dos and don'ts for identifying functional groups in complex molecules.

5.5: Functional Groups Containing Sulfur or Phosphorus Atoms

11. Draw the molecular structures for the following functional groups that contain sulfur atoms: thiol, thioester (sulfide), and thioether. Why do the structures of these functional groups resemble those containing oxygen atoms?
12. Draw the two-dimensional and three-dimensional structure of the phosphate group. Describe two important roles for phosphate groups in biological molecules.

Section 5.6: Heterocycles

13. What is a heterocycle?
14. Histamine and caffeine both contain a nitrogen-containing heterocycle. Briefly describe the biological function of each of these molecules.

Developing Skills

15. Identify each of the following functional groups, where R represents a carbon-containing collection of atoms:

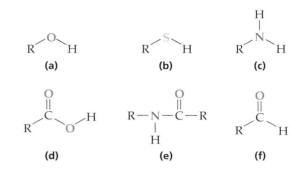

(a)   (b)   (c)

(d)   (e)   (f)

16. Aspirin is derived from salicylic acid, a natural product found in the bark of the willow tree. The two structures are compared here. In terms of their functional groups, what are the similarities and differences between salicylic acid and aspirin?

Salicyclic acid   Aspirin

17. Vanillin is a molecule in vanilla beans that produces vanilla flavor. Circle and identify all of the functional groups in this molecule.

Vanillin

18. Human urine contains urea, which is used as a waste product to remove nitrogen atoms. Urea has the chemical formula H_2NCONH_2. Draw the two-dimensional structure of this molecule.
19. Circle and identify the functional groups in these two drug molecules that are used for the treatment of pain.

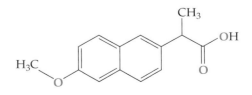

Naproxen
A painkiller used in Aleve

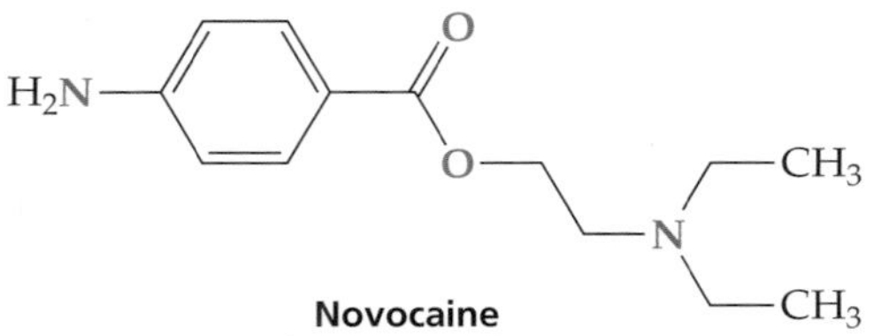

Novocaine
a dental anesthetic

20. Ginger root is sometimes used as a spice for food. The characteristic taste of ginger is produced by a molecule called *gingerol* (shown below). Circle and identify all of the functional groups in gingerol.

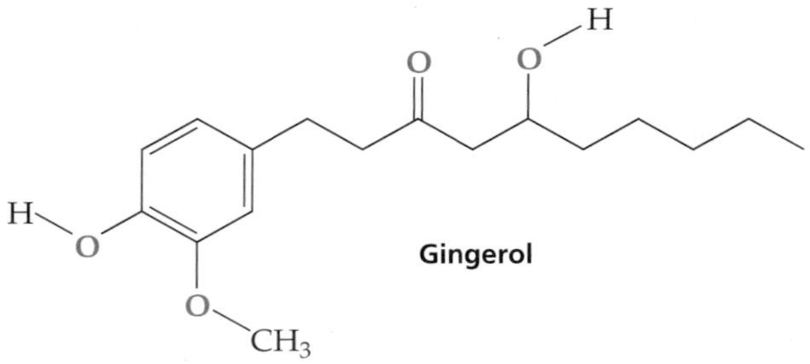

Gingerol

21. The hot taste of chili peppers is caused by a molecule called *capsaicin*. A line-angle drawing of capsaicin is provided here. Circle and identify all of the functional groups in this molecule.

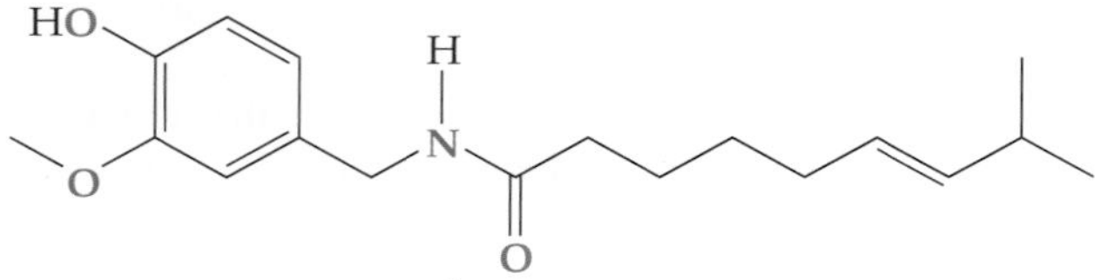

22. Ritalin is a drug used to treat attention-deficit/hyper-activity disorder (ADHD). Its molecular structure is shown here. (a) Circle and identify the functional group in Ritalin. (b) What is the name of the hetero-cycle in Ritalin?

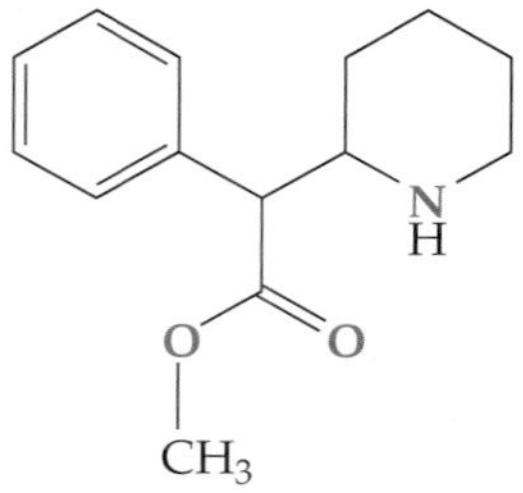

Ritalin

23. Uracil is one of the four bases in ribonucleic acid (RNA). Its molecular structure is shown here. Which heterocy-cle serves as the foundation for uracil's structure?

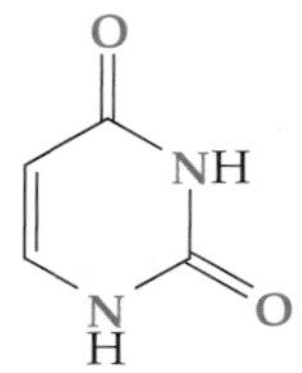

Uracil

24. Histidine and tryptophan are two of the 20 amino acids commonly found in proteins. Both contain a het-erocycle in their sidechain. Based on the following mo-lecular structures, identify the heterocycle within each sidechain.

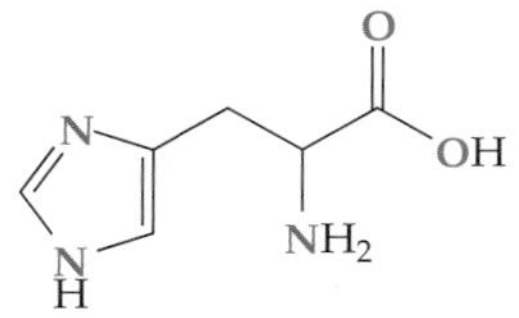

Histidine 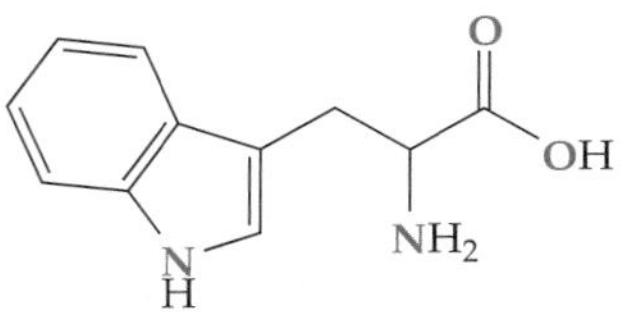   **Tryptophan**

25. The molecule shown in this exercise is called HMG coenzyme A. It serves as an accessory molecule that enables certain enzymes to perform their biological function.
    (a) The nitrogen-containing heterocycle in this mole-cule is derived from one of the heterocycles shown in Figure 5.39. Identify this heterocycle.
    (b) Circle and identify three different functional groups in HMG coenzyme A.

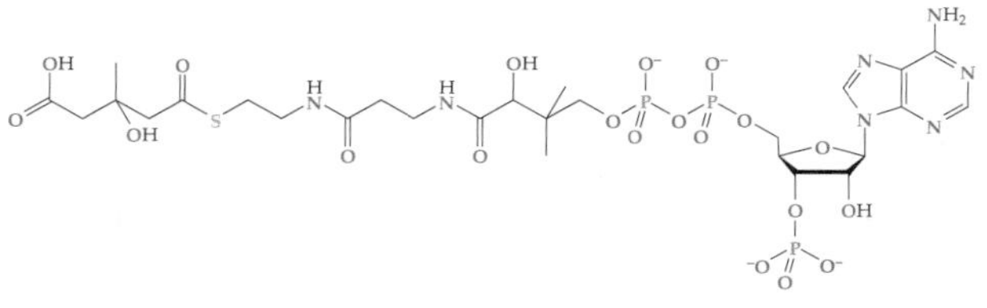

Exploring Concepts

26. Acetamide, CH_3CONH_2, has been found in space. (a) Draw the two-dimensional structure of acetamide, and circle its functional groups. (b) Use the Internet to research why the discovery of acetamide in space is important. Write a one-paragraph summary of your findings.

27. The biologically active molecule shown here is called GABA, which is an acronym for gamma-amino butyric acid. (a) Which functional groups does this molecule contain? (b) What is the biological function of GABA? Use the Internet to research this molecule, and write a paragraph about how GABA is used in the human body.

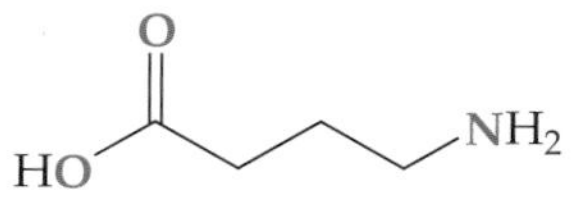

28. *Chemistry and Your Health* discussed alcohol metabolism in the human body. A molecule called disulfiram (shown below) binds to the aldehyde dehydrogenase enzyme and inhibits its function. Disulfiram (sold under the name Antabuse) has been approved by the FDA as a drug treatment for alco-hol addiction. Based on what you have learned from *Chemistry and Your Health*, explain why this treatment is used.

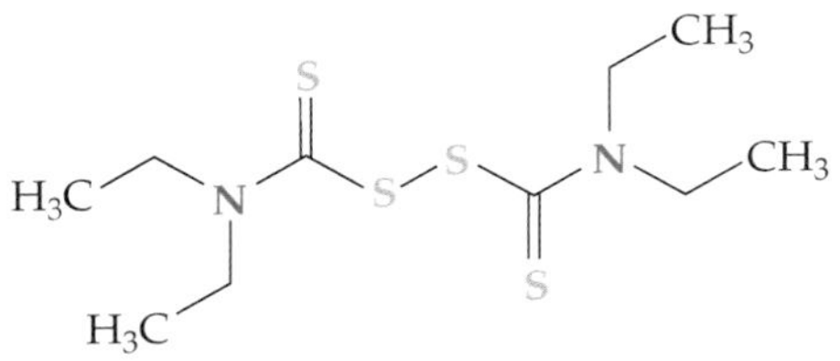

Disulfiram

29. Menthol is a molecule that contains an alcohol functional group. It is used in candy and in some types of cigarettes. (a) Use the Internet to identify the molecular structure of menthol. Draw this structure, and circle the alcohol group. (b) Why is menthol used as an additive? Use the Internet to find an explanation, and write a one- to two-sentence summary.

30. In William's Shakespeare's play *The Tempest*, one of the characters taunts another by saying that he smells like a fish:

> *What have we here? a man or a fish? dead or alive? A fish: he smells like a fish; a very ancient and fish-like smell; a kind of not of the newest Poor-John. A strange fish!*

Shakespeare may be referring to a medical condition called *fish odor syndrome*, which causes a person to smell like rotting fish. This smell is produced by one of the molecules discussed in this chapter. Use the Internet to investigate the origin of fish odor syndrome, and write a one-paragraph summary of your findings.

31. The diagram that follows compares two purine heterocycles: caffeine and theophylline. (a) Circle the region(s) of the theophylline molecule that differ(s) from caffeine. (b) Use the Internet to discover the use of theophylline, and write a two- to three-sentence summary.

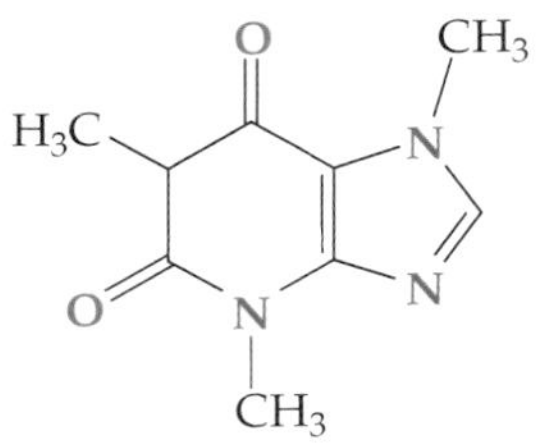

Caffeine 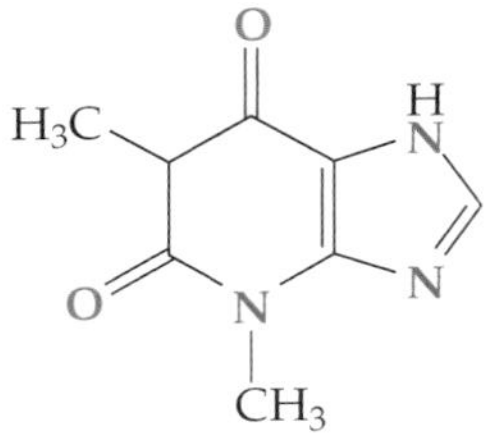   **Theophylline**

32. Eating chocolate is a pleasurable experience for humans, but it can be deadly for dogs. Use the Internet to discover which chemical component of chocolate is poisonous for dogs. Identify the name of this molecule, and draw its structure. Classify the molecule based on what you have learned in this chapter.

33. Drinking methanol (CH_3OH) is far more hazardous to your health than drinking ethanol. Review the content of *Chemistry and Your Health*, and draw the chemical structure of the molecule that is produced when alcohol dehydrogenase acts on methanol. Based on your reading of the chapter, provide the name of this molecule, and explain how its structure is related to the toxic effects of methanol consumption.

6

Chapter 6

Chemical Reactions

CONCEPTS AND LEARNING OBJECTIVES

FEATURES

When you go to the gym to exercise, you may hear people talk about "burning carbs." We use "carbs" from our diet as a fuel to generate energy. What happens inside our body when we "burn carbs"?

Learning Objective:
Relate a combustion reaction to "burning" carbs during exercise.

THE KEY IDEA: When you "burn carbs," you are using a chemical reaction to generate energy from sugar molecules.

6.1 What Happens When You "Burn Carbs" at the Gym?

The word "carb" is shorthand for **carbohydrate**. *Carbohydrates are sugar molecules that are formed from multiples of C, H, and O atoms, with the general formula* $C_x(H_2O)_y$—that is, carbon plus water ("hydrate"). However, this formula indicates the numerical ratios of the atoms in the sugar molecules, and we should not think that carbohydrates actually contain H_2O molecules.

Carbohydrates are one of the major components of our diet. When we eat carbohydrates, our digestive system breaks them down into smaller sugars. One of these sugars is **glucose** ($C_6H_{12}O_6$), which serves as a chemical fuel for living organisms. When we exercise in the gym, our cells use glucose to generate chemical energy that powers our muscles (Figure 6.1). This is what we mean by "burning carbs."

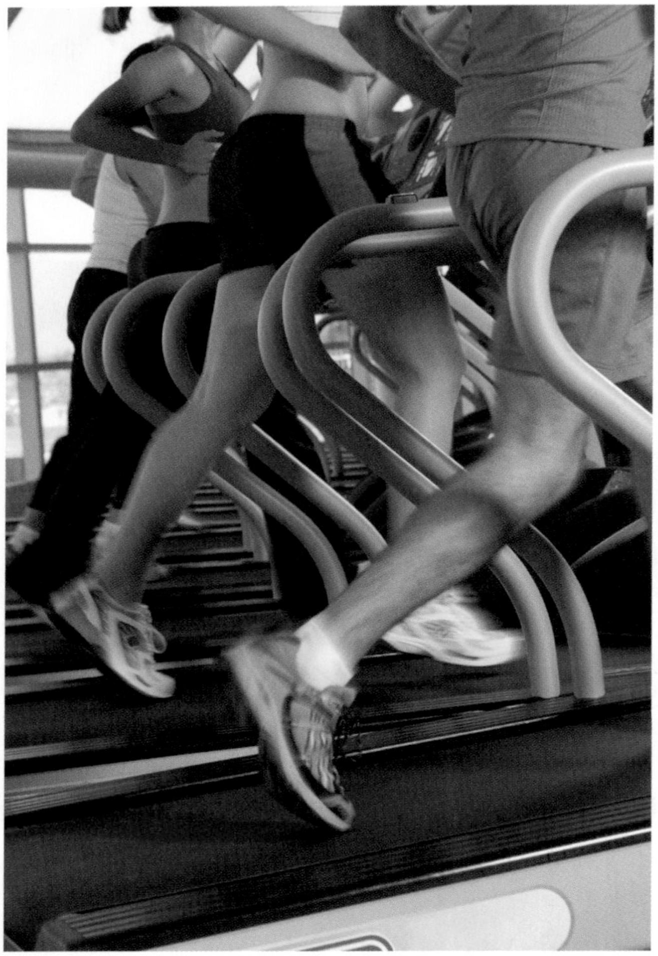

FIGURE 6.1 **"Burning carbs" at the gym.** During exercise, we use carbohydrates to generate energy.

Figure 6.2 illustrates another example of "burning carbs." Table sugar in a teaspoon is being burned in air to generate heat energy, which produces a flame. When you exercise, however, there are no tiny flames inside your body. Instead, you "burn" carbs using a different process that consumes sugars and generates energy. To understand how this happens, we need to study **chemical reactions**. In general, *a chemical reaction produces changes in matter and energy*. We begin the chapter by examining the type of reaction illustrated in Figure 6.2, which involves burning substances in air. Later in the chapter, we examine how the cells of our body extract energy from chemical reactions.

Our study of chemical reactions requires us to measure various quantities such as the mass of a chemical sample, the number of atoms or molecules that the sample contains, and the amount of energy that is released when the sample is burned in air. We will also use numerical calculations to convert between different units. Measurement and calculation are essential to understand the natural world, and no study of science would be complete without them. Please refer to Appendix B for guidelines on how to use *significant figures* in calculations.

Core Concepts:

- "Carb" is shorthand for **carbohydrate**. When you "burn carbs" at the gym, your body uses a chemical reaction to generate energy from sugar molecules.
- A **chemical reaction** produces changes in matter and energy.

2 Chemical Reactions Produce Changes in Matter

Chemical reactions are the agents of change in the physical and biological worlds. We use chemical reactions to power our vehicles and digest our food. In this section, we examine how chemical reactions generate changes at the molecular level.

Chemical Reactions Change Reactants into Products

Let's begin by examining a chemical reaction that many people use for cooking—burning natural gas (Figure 6.3). The primary component of natural gas is methane (CH_4), which reacts with oxygen gas (O_2) from the air to release heat energy and produce a flame. This type of chemical reaction is called **combustion**.

Figure 6.4 illustrates the combustion reaction between methane and oxygen using molecular models. In this reaction, one methane molecule reacts with two oxygen molecules to generate one molecule of carbon dioxide (CO_2) and two molecules of water (H_2O). The starting components of the reaction (CH_4 and O_2) are called the **reactants**; they are placed to the left of the arrow. The substances formed by the reaction (CO_2 and H_2O) are called the **products;** they are placed to the right of the arrow. In general, *a chemical reaction changes reactants into products.*

$$\text{REACTANTS} \longrightarrow \text{PRODUCTS}$$

At a more detailed level, *a combustion reaction causes the atoms in the reactant molecules to change partners and form new molecules as products.* Let's use the methane combustion reaction to illustrate this principle. In the reactants, the carbon atom is bonded to four hydrogen atoms. In the product, the same carbon atom is now bonded to two oxygen atoms. In the reactants, the oxygen atoms are joined together as oxygen molecules (O_2). In the products, the oxygen atoms have been split apart, and they are bonded to both carbon and hydrogen atoms.

You may wonder why we have included *two* molecules of O_2 (reactants) and H_2O (products) instead of just one of each. The reason is that all chemical reactions obey the law of **conservation of mass**, which states that *the total mass remains unchanged during a chemical reaction.* Because the masses of reactants and products arise from their atoms, this law also states that *atoms are neither created nor destroyed in chemical reactions.* In other words, all of the atoms that exist in the reactants must appear in the products, even though they have been

FIGURE 6.2 "Burning carbs" in a spoon. Burning table sugar generates large amounts of heat energy.

Learning Objective:
Describe chemical reactions using balanced chemical equations.

 THE KEY IDEA: Chemical reactions change matter by transforming reactants into products.

carbohydrate A sugar molecule composed of C, H, and O atoms, with the general formula $C_x(H_2O)_y$.

glucose A sugar used by living organisms to obtain energy; its chemical formula is $C_6H_{12}O_6$.

chemical reaction A process that produces changes in matter and energy.

combustion A chemical reaction that occurs when a fuel combines with oxygen, which releases heat energy.

reactants The starting components of a chemical reaction.

products The substances formed by a chemical reaction.

conservation of mass A law stating that the total mass remains unchanged during a chemical reaction.

FIGURE 6.3 Combustion of natural gas. Burning natural gas produces a flame that can be used for cooking.

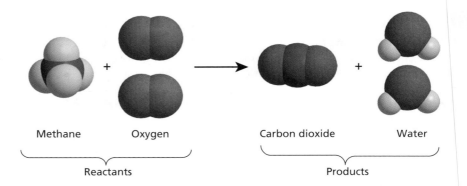

Methane   Oxygen   Carbon dioxide   Water

Reactants   Products

FIGURE 6.4 Methane combustion reaction. During a combustion reaction, one methane molecule (CH_4) reacts with two oxygen molecules (O_2) to produce one molecule of carbon dioxide (CO_2) and two molecules of water (H_2O).

rearranged into different molecules. The law of conservation of mass would be violated if there were only one O_2 molecule in the reactants because its two oxygen atoms would not be sufficient to create both of the product molecules (CO_2 and H_2O).

Core Concepts

- A chemical reaction changes **reactants** into **products**. In a **combustion** reaction, the atoms in the reactant molecules are rearranged to form the product molecules.
- The law of **conservation of mass** states that the total mass remains unchanged during a chemical reaction. This law reflects the fact that atoms are neither created nor destroyed during chemical reactions.

Chemical Equations Represent Chemical Reactions

A chemical reaction can be represented by a **chemical equation**. When writing a chemical equation, we use chemical formulas to describe the reactants and products. *All chemical equations must be balanced in order to obey the law of the conservation of mass.* The balanced equation for methane combustion is written as follows:

$$CH_4 \quad + \quad 2\,O_2 \quad \longrightarrow \quad CO_2 \quad + \quad 2\,H_2O$$

Methane   Oxygen   Carbon dioxide   Water

The numbers shown in red are called **coefficients**—they are placed in front of a chemical formula to indicate how many times that molecule must be counted. For example, the "2" in the product side of the equation tells us to count two H_2O molecules. There are three key rules for balancing chemical equations.

RULE 1: In a balanced chemical equation, *the number and types of atoms in the products must be equal to the number and types of atoms in the reactants.*

RULE 2: *Add numerical coefficients in front of the reactants and/or products to balance the number of atoms.* A coefficient placed in front of a molecule multiplies all of the atoms after the coefficient. For example, $2\,H_2O$ means $2 \times (H_2O)$, for a total of four H atoms plus two O atoms. We do not include a coefficient of "1" in front of CH_4 and CO_2. If there is no coefficient, then the molecule is counted only once.

RULE 3: *When balancing an equation, do not change the subscripts in a chemical formula.* In a chemical formula, a *subscript* placed after an atom multiplies only that atom and nothing else in the formula. To illustrate, an H_2O molecule contains two H atoms and one O atom. Changing the subscripts in a chemical formula changes the identity of the molecule. For example, H_2O is water, but H_2O_2 is hydrogen peroxide, a completely different substance.

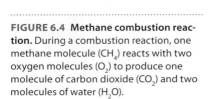 THE KEY IDEA: A chemical equation uses chemical formulas to represent a reaction.

- **chemical equation** A representation of a chemical reaction that uses chemical formulas to depict the reactants and products.

- **coefficients** Numbers placed in front of a chemical formula to indicate how many times that molecule must be counted in a balanced chemical equation.

TABLE 6.1	**An Atom Inventory for the Methane Combustion Reaction**	
	CH_4 + $2O_2$ $\longrightarrow$ CO_2 + $2H_2O$	
ATOM	**REACTANTS**	**PRODUCTS**
C	1 C from CH_4	1 C from CO_2
H	4 H from CH_4	4 H from $2H_2O$
O	4 O from $2O_2$	4 O from CO_2 + $2H_2O$

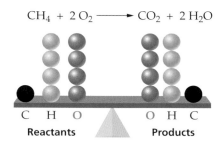

$$CH_4 + 2O_2 \longrightarrow CO_2 + 2H_2O$$

C H O   O H C

Reactants   Products

FIGURE 6.5 Balancing an equation. A balanced equation has an equal number of atoms in both reactants and products. This principle is illustrated by showing the atoms in the reactants and products on two sides of a mass balance.

The principle of a balanced equation is illustrated in Figure 6.5. We can imagine two sides of a mass balance, with one side weighing all the atoms in the reactants and the other side weighing all the atoms in the products. The two sides are balanced: Both the reactants and the products contain one C atom, four H atoms, and four O atoms. Therefore, this chemical equation obeys the principle of conservation of mass.

A helpful strategy for checking the accuracy of a chemical equation is called an *atom inventory*. This strategy involves making a table that counts the total number of each type of atom in the reactants and the products (Table 6.1). For the example of methane combustion, we have included the molecules that contribute atoms to the inventory. As you become skilled at using this tool, you will be able to omit the molecules and focus on the numbers of atoms.

Whenever you read or write a chemical equation, always verify that it is balanced. *A chemical equation that is not balanced has no scientific meaning.* Consider an analogy with a mathematical equation. The following "equation" is clearly incorrect:

3 = 6 not balanced: incorrect

However, it can be converted into a correct statement by adding a coefficient of "2":

2 × 3 = 6 balanced: correct

WORKED EXAMPLE 6.1

Question: A chemical reaction between nitrogen (N_2) and hydrogen (H_2) gases generates ammonia (NH_3). The unbalanced chemical equation for this reaction is:

$$N_2 \ + \ H_2 \ \longrightarrow \ NH_3$$

(a) Add coefficients to balance the chemical equation.
(b) Use an atom inventory to verify that the reaction is balanced.
(c) Diagram the chemical reaction using molecular models for the reactants and products.

Answer: (a) First, we add lines to the equation to indicate places for the coefficients.

$$\_\_N_2 \ + \ \_\_H_2 \ \longrightarrow \ \_\_NH_3$$

Next, we balance the reaction by focusing on one type of atom at a time. We begin by balancing the N atoms. The reactant N_2 has two N atoms, which are used to form NH_3 as a product. We can balance the N atoms by adding a coefficient of "2" in front of the NH_3, as shown below. The N atoms are highlighted in red, and the direction of the dashed arrow indicates that we have used one of the reactants to balance the product. There is no need to add a coefficient in front of N_2 in the reactants (writing "1" is not necessary).

$$\_\_N_2 \ + \ \_\_H_2 \ \longrightarrow \ \underline{2} \ NH_3$$

We now turn to the H atoms. The two NH_3 molecules in the product have a total of six H atoms, which must come from the H_2 molecules in the reactants. We can balance the H atoms by placing a coefficient of "3" in front of the H_2, as shown here. The H atoms are highlighted in red, and the direction of the dashed arrow indicates that we have used the product to balance one of the reactants.

$$\_\_N_2 \ + \ \underline{2} \ H_2 \ \longrightarrow \ \underline{2} \ NH_3$$

Both the N and H atoms are now balanced. We can therefore remove the lines and write the complete balanced equation.

$$N_2 \ + \ 3\,H_2 \ \longrightarrow \ 2\,NH_3$$

(b) We can use an atom inventory to verify that the equation is balanced:

Atom	Reactants	Products
N	2	2
H	6	6

(c) Finally, we can diagram the reaction using molecular models for the reactants and products.

TRY IT YOURSELF 6.1

Question: (a) Under conditions of high temperature, nitrogen (N_2) and oxygen (O_2) gases react to form nitric oxide (NO). Write a balanced chemical equation for this reaction. Use an atom inventory to verify that the equation is balanced. (b) Nitric oxide further reacts with oxygen gas to form nitrogen dioxide (NO_2). This gas has a brown color and is responsible for the brown haze seen in urban smog. Write a balanced chemical equation for the formation of NO_2. Use an atom inventory to verify that the equation is balanced.

PRACTICE EXERCISE 6.1

The element sulfur exists as a collection of atoms that is written as S_8. This form of sulfur reacts with oxygen gas (O_2) to produce sulfur dioxide (SO_2). Write a balanced chemical equation for this reaction.

The balanced chemical equation for a reaction can also be used to indicate the *physical state* of each reactant and product; in other words, is it a gas (*g*), a liquid (*l*), or a solid (*s*)? We add the appropriate designation (always written in *italics*)

SCIENCE IN ACTION

Antoine Lavoisier and the Origin of Modern Chemistry

What causes some substances to burn? In the 17th and 18th centuries, many chemists believed that flammable objects contained a combustible substance called *phlogiston*, a term derived from the Greek word for "burning up." According to this view, burning a piece of wood drives out the phlogiston, which can be observed as flames.

The existence of phlogiston was challenged by a French nobleman named Antoine Lavoisier (1743–1794), who is considered to be the founder of modern chemistry (Figure 1). Lavoisier lived

FIGURE 1 **Antoine Lavoisier, pioneer of modern chemistry.**
This portrait shows Antoine Lavoisier and his wife, Marie-Anne. who collaborated on his chemical investigations. Examples of Lavoisier's scientific apparatus appear on the table.

in Paris during the period before the French Revolution. He spent his wealth to purchase many types of high-quality scientific apparatus, which he used to perform accurate measurements of the substances that participate in chemical reactions.

One of Lavoisier's most important experiments involved the chemical reactions of mercury, the only metal that is liquid at room temperature. When the silvery mercury metal is heated in air, it undergoes a remarkable transformation into a red powder (Figure 2). When the English clergyman Joseph Priestley (1733–1804) visited Lavoisier in Paris, he demonstrated that heating this powder recovers the mercury metal and produces a gas that can ignite a flame. Lavoisier subsequently performed his own experiments with this reaction, and he concluded that the flammable gas was a component of normal air. He called this gas *oxygen*. Using modern notation, we can write a chemical equation for burning the red powder:

$$2\,HgO(s) \longrightarrow Hg(s) + O_2(g)$$

Lavoisier proposed that combustion reactions did not arise from the presence of phlogiston inside a flammable substance. Instead, *substances burn because they chemically react with oxygen in the air*. Using this new principle, he named the red powder *mercury oxide* (HgO).

Lavoisier also performed careful measurements of the masses of the reactants and products in this chemical reaction. These measurements enabled him to demonstrate that the combined mass of the mercury metal and the oxygen gas was the same as the mass of the mercury oxide before it was heated. These results and those from similar experiments demonstrated one of the most important principles in chemistry: *Mass is conserved during a chemical reaction.*

In 1789, Lavoisier published his experimental results and chemical theories in a groundbreaking book titled *Elementary Treatise on Chemistry*. Unfortunately for Lavoisier, this was the same year as the outbreak of the French Revolution. Because Lavoisier had served as a tax collector for the royal regime, he was guillotined in 1794. Despite his premature death, Lavoisier's legacy remains with us today in every chemistry textbook. His pioneering experiments and analysis prompted a transformation in chemists' understanding of the natural world.

FIGURE 2 **A comparison of silver mercury metal (right) and red mercury oxide (left).**

after the chemical formula for the reactant or product. For example, consider the equation for methane combustion:

$$CH_4(g) \; + \; 2\,O_2(g) \longrightarrow CO_2(g) \; + \; 2\,H_2O(g)$$

Methane   Oxygen   Carbon dioxide   Water

In this case, all of the reactants and products exist as gases. Although you may think of H_2O as a liquid, the combustion reaction produces water vapor as a product. Writing the physical states sometimes produces a cluttered chemical equation, so we will add these designations only when it is important for our discussion of the chemical changes generated by a reaction.

Core Concepts

- A **chemical equation** uses chemical formulas to describe a reaction.
- A chemical reaction must be *balanced*—the number and type of atoms must be the same in the reactants and the products. Balancing an equation involves adding numerical **coefficients** in front of the appropriate chemical formulas.
- A chemical equation can indicate the physical state of each reactant and product by using the designations for gas (*g*), liquid (*l*), and solid (*s*).

Stoichiometry: Measuring Reactants and Products

THE KEY IDEA: The stoichiometry of a reaction describes the relative amounts of reactants and products.

stoichiometry The numerical ratios of reactants and products in a chemical reaction.

After we have balanced an equation, we find that *reactants usually combine in simple ratios*. For example, the equation for methane combustion tells us that one molecule of CH_4 combines with two molecules of O_2 to produce one molecule of CO_2 and two molecules of H_2O. These ratios are called the **stoichiometry** (pronounced *stoy-key-OM-etry*), from the Greek words for "element" and "measure." *The stoichiometry of a chemical reaction describes the numerical ratios of reactants and products.*

Reaction stoichiometry may sound like an abstract chemical concept, but it can mean the difference between life and death. You may have heard news reports about individuals who have died from poisoning by *carbon monoxide* (CO). According to the Centers for Disease Control and Prevention, unintentional carbon monoxide poisoning is responsible for more than 20,000 emergency room visits each year in the United States, which lead to more than 400 fatalities annually.

Where does carbon monoxide come from? When we write the chemical equation for a combustion reaction, we usually assume that there is an abundance of oxygen gas from the atmosphere. Under these conditions, each carbon atom in the fuel can combine with two oxygen atoms to produce carbon dioxide (CO_2). However, in some situations, the supply of oxygen becomes limited. This can happen when a generator or a boiler malfunctions or when an automobile engine is not operating efficiently. If the amount of available O_2 is reduced, the combustion reaction can produce CO instead of CO_2. The scarcity of oxygen means that each carbon atom combines with only one oxygen atom instead of two.

The *unbalanced* reaction for the production of CO is written as

$$CH_4 \; + \; O_2 \longrightarrow CO \; + \; H_2O \quad \textit{unbalanced}$$

Methane   Oxygen   Carbon   Water
                                          monoxide

We now need to balance the chemical equation. First, we add lines to the unbalanced chemical equation to indicate places for the coefficients.

$$\_\_CH_4 \quad + \quad \_\_O_2 \quad \longrightarrow \quad \_\_CO \quad + \quad \_\_H_2O$$

Next, we balance the equation by focusing on one atom at a time. Which atom should we balance first? The best approach is to start with atoms in a reactant molecule that are transformed into *only one type of product molecule*. For example, the C atom in CH_4 is converted into CO and nothing else. Similarly, the H atoms in CH_4 are converted into only H_2O. By contrast, the O atoms in O_2 are converted into two different products: CO and H_2O. The O atoms are more difficult to balance, so we leave them until last.

Carbon: The reactants contain one C atom (in CH_4), and the products also contain one C atom (in CO). The C atoms are therefore balanced. We do not include a coefficient of "1" in the equation.

$$\_\_CH_4 \quad + \quad \_\_O_2 \quad \longrightarrow \quad \_\_CO \quad + \quad \_\_H_2O$$

Hydrogen: The four H atoms in CH_4 are converted into H_2O molecules. We can balance the H atoms by adding a coefficient of "2" in front of H_2O in the products. This coefficient multiplies all of the atoms in the chemical formula for H_2O.

$$\_\_CH_4 \quad + \quad \_\_O_2 \quad \longrightarrow \quad \_\_CO \quad + \quad \underline{2\,}H_2O$$

Oxygen: The final step is to balance the O atoms. The total number of O atoms in the products is three, which is an odd number. However, each O_2 molecule in the reactants contains two O atoms, and we cannot change the subscript in the chemical formula. The only way to balance the reactants and products is to use a coefficient of 1½ in front of O_2 (i.e., half of 3). In the following equation, the direction of the dashed arrow indicates that we have used the product molecules to balance one of the reactants.

$$\_\_CH_4 \quad + \quad \underline{1\frac{1}{2}\,}O_2 \quad \longrightarrow \quad \_\_CO \quad + \quad \underline{2\,}H_2O$$

What should we do about the 1½ in from of O_2? It doesn't make sense to have half of an O_2 molecule. When balancing equations, *the standard convention is to use only whole numbers and not fractions for the coefficients*. We can remove the fraction by multiplying all of the coefficients by 2, which is identical to how you would remove a ½ fractions in a mathematical equation.

The final balanced equation for methane combustion under low-oxygen conditions is as follows, with the coefficients highlighted in red and the states of matter added.

$$2\,CH_4(g) \quad + \quad 3\,O_2(g) \quad \longrightarrow \quad 2\,CO(g) \quad + \quad 4\,H_2O(g)$$

Methane	Oxygen	Carbon monoxide	Water

TABLE 6.2	An Atom Inventory for the Methane Combustion Reaction	
$2\,CH_4 + 3\,O_2 \longrightarrow 2\,CO + 4\,H_2O$		
Atom	Reactants	Products
C	2	2
H	8	8
O	6	6

Table 6.2 provides an atom inventory to verify that the equation is balanced.

You can see that the stoichiometry of the reaction has changed. When oxygen is scarce, two molecules of CH_4 react with three molecules of O_2. We say that the *combining ratio* of the reactants is 2:3. This combustion reaction produces carbon monoxide, a poisonous gas. When oxygen is plentiful and CO_2 is generated, one molecule of CH_4 reacts with two molecules of O_2 for a combining ratio of 1:2. In reality, both of these reactions can occur at the same time during methane combustion. However, the reaction producing CO will become more prevalent as the supply of O_2 decreases. *Chemistry and Your Health* describes the biological effects of carbon monoxide poisoning and the steps you can take to avoid it.

WORKED EXAMPLE 6.2

Question: Octane (C_8H_{18}) is the principal component of gasoline. The *unbalanced* equation for octane combustion is:

$$C_8H_{18} \;+\; O_2 \longrightarrow CO_2 \;+\; H_2O$$

(a) Add coefficients to balance this equation. (b) Verify that the equation is balanced by using an atom inventory.

Answer: (a) We balance the atoms one at a time. First we add lines to the equation to indicate places for the coefficients.

$$\_\_C_8H_{18} \;+\; \_\_O_2 \longrightarrow \_\_CO_2 \;+\; \_\_H_2O$$

We begin by balancing the C and H atoms because they each appear in only one product molecule.

Carbon: There are eight C atoms in octane and only one in CO_2. Therefore, there must be eight CO_2 molecules in the products, so we can add this coefficient to the equation. The C atoms are highlighted in red, and the direction of the dashed arrow indicates that we have used a reactant to balance one of the products.

$$\_\_C_8H_{18} \;+\; \_\_O_2 \longrightarrow \underline{8}\,CO_2 \;+\; \_\_H_2O$$

Hydrogen: The H atoms in C_8H_{18} are transformed into H_2O molecules. Because each H_2O molecule already includes two H atoms (indicated by the subscript), we must have a total of nine H_2O molecules. The direction of the dashed arrow indicates that we have used a reactant to balance one of the products.

$$\_\_C_8H_{18} \;+\; \_\_O_2 \longrightarrow \underline{8}\,CO_2 \;+\; \underline{9}\,H_2O$$

Oxygen: The final step is to balance the O atoms. There are a total of 25 O atoms in the products, i.e, $(8 \times 2) + (9 \times 1)$. Because the O atoms in the reactants are in the form of O_2, we need to use a coefficient of 12½ (i.e., half of 25). In this case, the dashed arrow indicates that we have used the product molecules to balance one of the reactants.

$$\_\_C_8H_{18} \; + \; \underline{12\tfrac{1}{2}\,O_2} \longrightarrow \; \underline{8}\;CO_2 \; + \; \underline{9}\;H_2O$$

We can remove the fraction by multiplying all of the numbers by a factor of 2.

$$2\,C_8H_{18} \; + \; 25\,O_2 \longrightarrow 16\,CO_2 \; + \; 18\,H_2O$$

(b) We can use an atom inventory to verify that the chemical equation is balanced.

ATOM	REACTANTS	PRODUCTS
C	16	16
H	36	36
O	50	50

TRY IT YOURSELF 6.2

Question: When the oxygen supply is limited, the combustion of octane produces carbon monoxide (CO) instead of carbon dioxide. The *unbalanced* equation for this reaction is:

$$C_8H_{18} \; + \; O_2 \longrightarrow CO \; + \; H_2O$$

(a) Add coefficients to balance this chemical equation.

(b) Use an atom inventory to verify that the equation is balanced.

(c) Compare your answer with the balanced equation for the combustion of octane when oxygen gas is plentiful (Worked Example 6.2). What differences do you notice for the coefficients used to balance the reactants and the products? Write one or two sentences to explain why the balanced equation for CO production indicates a limited supply of oxygen.

PRACTICE EXERCISE 6.2

Acetylene (C_2H_2) is a hydrocarbon with a triple covalent bond. It reacts with oxygen to generate a very hot flame, which is used to make an *oxyacetylene torch* (see the figure).

The *unbalanced* equation for acetylene combustion is:

$$C_2H_2 \; + \; O_2 \longrightarrow CO_2 \; + \; H_2O$$

(a) Add coefficients to balance this equation.

(b) Use an atom inventory to verify that the equation is balanced.

Core Concepts

- The **stoichiometry** of a reaction describes the ratios of reactants and products in a chemical reaction.
- In a combustion reaction, the availability of oxygen gas affects the stoichiometry and products of the reaction. When oxygen is plentiful, fuel combustion produces carbon dioxide (CO_2). When oxygen is scarce, the combustion reaction produces carbon monoxide (CO), a poisonous gas.

CHEMISTRY AND YOUR HEALTH

Carbon Monoxide Poisoning

Why is carbon monoxide (CO) so dangerous to humans? The answer is based on the chemistry of this molecule. The CO molecule contains a *triple covalent bond*, as shown below. The triple bond makes the molecule very stable, just as we saw with N_2 in Chapter 3. One unusual feature of the CO molecule is the presence of a lone pair of electrons attached to the C atom. This electron arrangement is uncommon because carbon does not usually have any lone pairs within stable molecular structures.

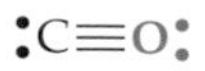

$$:C \equiv O:$$
Carbon monoxide

As discussed in the chapter, carbon monoxide is produced when fuels burn in a limited supply of oxygen. If this reaction occurs inside a closed indoor environment, then the levels of carbon monoxide begin to increase. Carbon monoxide is a particularly dangerous pollutant because it has no odor and no color, so our senses cannot detect it. Many people who are poisoned by carbon monoxide breathe the gas while they are asleep. Exposure to low levels of CO produces symptoms such as headaches, dizziness, nausea, and fatigue. High levels of CO can induce coma and death.

Carbon monoxide achieves its toxic effect by interfering with the transport of oxygen gas in our bloodstream. In Chapter 2, we learned that hemoglobin proteins in red blood cells serve as the transporters of O_2 molecules (see Figure 2.3). Hemoglobin contains four ring-shaped heme groups, each with an iron ion (Fe^{2+}) in its center. An O_2 molecule binds to each of the Fe^{2+} ions; therefore, one hemoglobin transports four O_2 molecules. After carrying O_2 molecules through the blood, hemoglobin

releases them to be used for energy-generating reactions in the body's cells.

When we inhale carbon monoxide, the gas enters our lungs and diffuses into our bloodstream. The CO molecules enter our red blood cells and attach to hemoglobin proteins. The lone pair of electrons on the C atom enables the CO molecule to bind to the Fe^{2+} ion in the heme group. In fact, a CO molecule binds to the heme iron with a strength of attraction that is approximately 250 times greater than that of oxygen (Figure 1). When a CO molecule attaches to a heme group, it prevents an O_2 molecule from binding there. Even if CO binds to only one or two of the four heme groups, it diminishes the ability of hemoglobin to release the O_2 molecules that are bound to the other heme sites.

In summary, carbon monoxide reduces the capacity of hemoglobin to transport oxygen around the bloodstream and deliver it to cells. As a result, the body's cells begin to die because they do not receive the oxygen they need to generate energy-sustaining reactions. The effect of CO poisoning is particularly pronounced in the brain, which requires a large supply of oxygen.

Because CO is undetectable by our senses, it is now standard practice to install a *carbon monoxide detector* in houses and apartments. This device alerts residents to the presence of CO molecules in the air (Figure 2). In many cases of reported fatalities from CO poisoning, a CO detector either was not present or was not functioning.

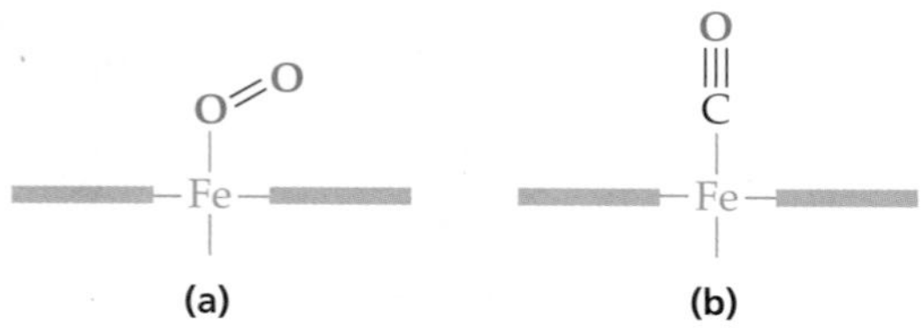

FIGURE 1 Oxygen (O_2) and carbon monoxide (CO) both bind to the iron in hemoglobin. (a) Within the hemoglobin protein, O_2 molecules bind to the iron within the heme group. The hemoglobin then transports oxygen through the bloodstream to the body's tissues. (b) CO also binds to the heme iron, with an attraction that is 250 times greater than that of oxygen. When CO is attached to the iron, it prevents the O_2 from binding.

FIGURE 2 A carbon monoxide detector. These detectors are installed indoors to warn residents of elevated levels of carbon monoxide, which is a colorless, odorless, and highly toxic gas.

3 Measuring Matter: Mass and Moles

Because molecules are so small, an observable chemical reaction involves an immensely large number of them. Consequently, we need a new scientific unit that can serve as a bridge between the invisible world of molecules and the amount of substance that we can observe and measure. This section examines chemistry's solution to this challenge.

What Is a Mole, and Why Do We Use It?

How can we measure the quantities of chemical substances? One method is to measure the mass of a sample using units of grams (g). Figure 6.6 shows how a chemical balance is used to measure 12 g of carbon powder. The significance of this mass will be explained later in the chapter.

If we burn this solid sample of carbon in air, it reacts with oxygen molecules to generate carbon dioxide gas. The reaction is shown as follows, using molecular models and a chemical equation:

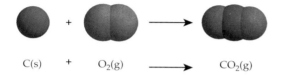

$$C(s) \quad + \quad O_2(g) \quad \longrightarrow \quad CO_2(g)$$

How many grams of carbon dioxide are produced by burning 12 g of carbon? The answer is not obvious, but the chemical equation provides an important clue. Based on the stoichiometry of the reaction, one atom of carbon reacts with one molecule of oxygen to produce one molecule of carbon dioxide. As written, the chemical reaction includes only one of each reactant and product. However, we can scale up the reaction to include more atoms and molecules. For example, one dozen (12) carbon atoms will react with 12 O_2 molecules to form 12 CO_2 molecules. What happens if we have 1 million C atoms? The answer is obvious—they will react with 1 million O_2 molecules to form 1 million CO_2 molecules. Regardless of how many atoms and molecules are involved, the ratio of $C:O_2:CO_2$ always remains 1:1:1. This principle can be summarized as:

$$C(s) \quad + \quad O_2(g) \quad \longrightarrow \quad CO_2(g)$$

1 atom	1 molecule	1 molecule
1 dozen atoms	1 dozen molecules	1 dozen molecules
1 million atoms	1 million molecules	1 million molecules

If we knew how many carbon atoms were contained in the 12-g sample, then we could deduce the number of O_2 and CO_2 molecules. We cannot count the carbon atoms directly because they are too small. Therefore, we need a strategy to relate the mass of this chemical sample, which we can measure, to the number of carbon atoms that it contains. We accomplish this task using the concept of a *mole*, a term that derives from the Latin word for "large heap." A mole tells us how many atoms or molecules are contained in a measurable "heap" of sample. As an analogy, Figure 6.7 shows a large heap of coals that is composed of many millions of smaller lumps of coal. From a distance, we can only see the large heap and not the individual coals.

Learning Objective:

Use the mole as a unit of measurement for chemical amounts.

THE KEY IDEA: A mole is used to convert between the measurable mass of a chemical sample and the number of particles in the sample.

FIGURE 6.6 Measuring the mass of a sample. A chemical balance measures the mass of a sample in grams. This sample of carbon powder has a mass of 12 g.

FIGURE 6.7 A mole analogy. A large heap of coals is composed of many smaller lumps of coal. Similarly, a measureable chemical sample is composed of a heap of atoms or molecules.

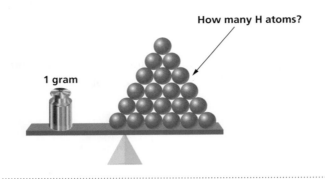

FIGURE 6.8 A thought experiment. How many hydrogen atoms are required to make a heap of atoms with a total mass of exactly 1 g?

We begin our investigation of the mole by performing a thought experiment (Figure 6.8). Imagine that we can pile hydrogen atoms on top of one another to make a heap of atoms. How many hydrogen atoms would be required to reach a mass of exactly 1 gram? We have chosen hydrogen atoms because hydrogen is the lightest element, and the gram is the standard unit of mass for measuring chemical samples.

To perform this calculation, we need to know the mass of a single hydrogen atom (m_H). This value is provided to a precision of four significant figures.

$$m_H = 0.00000000000000000000001673 \text{ g}$$

The mass of a hydrogen atom, as measured in units of grams, is incredibly small. We can use scientific notation to write this number in a more convenient form, in which the small mass of the hydrogen atom is indicated by the negative exponent.

$$m_H = 1.673 \times 10^{-24} \text{ g}$$

We can now calculate the number of hydrogen atoms in exactly 1 g (1.000 g) by using the following equation:

$$\# \text{ H atoms} = 1.00 \text{ g} \times \frac{1 \text{ H atom}}{1.673 \times 10^{-24} \text{ g}} = 5.977 \times 10^{23} \text{ H atoms}$$

Because the mass of a single H atom is so small, the number of H atoms in 1 g is enormous. For convenience, we round the number to two significant figures, which gives us 6.0×10^{23}. This number can be written in full as follows:

$$600,000,000,000,000,000,000,000$$

This number is much larger than a billion (10^9) or even a trillion (10^{12}); in fact, it is equal to six hundred thousand billion billion. That's how many H atoms are required to make a heap of atoms with a mass of only 1 g.

Although it makes conceptual sense to use the hydrogen atom as the unit of atomic mass, it is difficult to measure such a light mass. Consequently, modern chemistry bases its mass standard on the carbon-12 isotope (^{12}C), which can be measured with greater accuracy. As discussed in Chapter 2, *the atomic mass unit (1 amu) is defined as 1/12 of the mass of a ^{12}C atom.* The numerical value of the atomic mass unit is precisely known, and we provide it using six significant figures.

$$1 \text{ amu} = 1.66054 \times 10^{-24} \text{ g}$$

Note the small but important difference between the value of the amu and the mass of the hydrogen atom presented above.

We can perform a similar calculation using this new mass standard: How many atomic mass units correspond to a chemical sample of exactly 1.000 g? We divide the mass of 1 g by 1 amu to obtain an important number.

$$N_A = 6.022 \times 10^{23}$$

Avogadro's number The number of atoms in exactly 12 g of carbon-12; it is equal to 6.022 x 10²³.

This conversion factor is called **Avogadro's number,** and it is given the symbol N_A. For convenience, we provide the numerical value of N_A using four significant figures, although it has been measured to greater precision. Because ^{12}C is our reference for atomic mass, we can say that *Avogadro's number is the number of carbon atoms in exactly 12 g of* ^{12}C. As an analogy, we can think of Avogadro's number as a counting unit, like a dozen (12). We can purchase a dozen eggs or a dozen donuts; although the item changes, the number of items remains the same.

Avogadro's number provides the foundation for the **mole** as a unit of measurement. By definition, *1 mole is the amount of substance that contains Avogadro's number of atoms, molecules, ions, or other particles.* According to the international system of units, the mole is used to measure the chemical amount of a substance. Figure 6.9 illustrates 1 mole of four chemical elements—helium, sulfur, copper, and mercury. *Each chemical sample contains the same number (Avogadro's number) of atoms.* Note that each sample is an easily measurable amount.

In summary, the mole is the unit that connects the mass of a chemical sample and the number of invisible particles that the sample contains. It provides a bridge between the measurable world of chemical samples and the unseen world of atoms and molecules. In the rest of this section, we apply this principle to specific examples to illustrate its utility.

FIGURE 6.9 One mole of four elements. The elements, from left to right, are sulfur, helium (in the balloon), copper, and mercury. Each sample contains Avogadro's number of atoms.

....
CONCEPT QUESTION 6.1
....

Instead of the gram (g), suppose chemists used the kilogram (kg) as the standard unit of mass for chemical samples. What would be the new value of Avogadro's number? Explain your answer.

....

Core Concepts

- ■ **Avogadro's number** ($N_A = 6.022 \times 10^{23}$) is the number of atoms in exactly 12 g of the carbon-12 isotope.
- ■ One **mole** is the amount of substance that contains Avogadro's number of atoms, molecules, ions, or other particles. The mole measures the **chemical amount** of a substance.
- ■ The mole connects the observable mass of a chemical sample (measured in grams) and the number of invisible particles contained in the sample.

mole The chemical amount of substance that contains Avogadro's number of atoms, molecules, ions, or other particles.

chemical amount The amount of a chemical substance, measured in units of moles.

Molar Mass of Atoms

In Chapter 2, we learned that each element in the periodic table is associated with two numbers (Figure 6.10). The number above the element symbol is the *atomic number*, which is equal to the number of protons in the nucleus of the atom. It is always a *whole number* because we are counting the number of protons. The number below the element symbol is the *relative atomic mass*, written as a decimal using four significant figures. The relative atomic mass is a weighted average that reflects the natural abundance of the various isotopes of the element.

The relative atomic mass has no units because it compares the element's mass to the reference value of 1 atomic mass unit.

THE KEY IDEA: The molar mass of an atom is based on the relative atomic mass given in the periodic table.

$$\text{relative atomic mass} = \frac{\text{mass of one atom of the element (amu)}}{1\ \text{amu}}$$

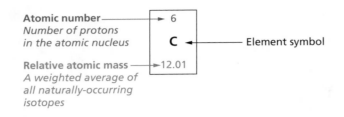

FIGURE 6.10 Atomic information. The atomic number and relative atomic mass of carbon as presented in the periodic table.

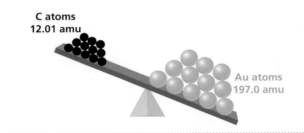

FIGURE 6.11 Comparing the mass of atoms. Because gold has a larger atomic mass than carbon, the total mass of 12 gold atoms is larger than the total mass of 12 carbon atoms. A similar principle applies to defining the molar mass of an atom, which means that the molar mass of Au atoms is larger than the molar mass of C atoms.

■ **molar mass** The mass of 1 mol (Avogadro's number) of atoms, molecules, ions, or other particles.

The mass of any atom is numerically equal to the relative atomic mass, but with the addition of amu as the unit of measurement. For example, the relative atomic mass of carbon is 12.01, so the atomic mass is 12.01 amu. In this manner, the periodic table provides the atomic masses of all the elements.

The masses of atoms vary because they contain differing numbers of protons and neutrons. For example, imagine that we could place a dozen carbon atoms on one side of a balance and the same number of gold atoms on the other side (Figure 6.11). Which collection of atoms would have the larger total mass? The answer is the dozen gold atoms because *the average mass of each gold atom (197.0 amu) is larger than the average mass of each carbon atom (12.01 amu).* The same principle holds true when we compare Avogadro's number of different atoms: The total mass of 6.022×10^{23} gold atoms is larger than the total mass of the same number of carbon atoms. This comparison enables us to introduce an important quantity called the **molar mass.** *The molar mass of an atom is defined as the mass of 1 mole (Avogadro's number) of atoms.*

Molar mass is measured in units of grams per mole (g/mol); note that the "e" in mole is generally omitted when we use it as a unit of measurement. *The molar mass of an atom is numerically equal to the relative atomic mass provided in the periodic table.* For example, the molar mass of carbon is 12.01 g/mol, and the molar mass of gold is 197.0 g/mol. To explain this numerical equivalence, recall that the mole connects the mass of atoms (measured in amu) to the mass of an observable sample (measured in grams). In other words,

$$
\underset{\substack{\text{atomic mass}\\\text{unit}}}{1.000 \text{ amu}} \times \underset{\substack{\text{Avogadro's}\\\text{number}}}{N_A} = \underset{\substack{\text{sample mass}}}{1.000 \text{ g}}
$$

We can apply this principle to the mass of a specific atom, such as carbon:

$$
\underset{\substack{\text{mass of one}\\\text{C atom}}}{12.01 \text{ amu}} \times \underset{\substack{\text{Avogadro's}\\\text{number}}}{N_A} = \underset{\substack{\text{mass of one mole}\\\text{of C atoms}}}{12.01 \text{ g}}
$$

The molar mass enables us to count the number of atoms in a sample by measuring its mass. For example, 12.01 g of carbon contains 1 mol of carbon atoms, equal to 6.022×10^{23} atoms (Avogadro's number). To obtain *twice* that number of carbon atoms, we simply double the mass of the carbon sample to 24.02 g. Similarly, we could obtain 1/10th of the number of atoms (6.022×10^{22} atoms) by measuring 1/10th of the mass of carbon (1.201 g).

If we refer to the molar mass of an atom, the meaning is unambiguous. Sometimes chemists refer to the molar mass of an element, but in this case, we must specify the chemical form of the element. For example, what is the molar mass of the element oxygen? Based on the periodic table, the molar mass of an O atom is 16.00 g/mol. However, oxygen exists naturally in the form of O_2 molecules. Therefore, the molar mass of oxygen could be 16.00 g/mol or $2 \times 16.00 = 32.00$ g/mol, depending on whether we are referring to oxygen atoms or oxygen molecules.

Use the periodic table to obtain the molar mass for the following atoms. Write your answers using the appropriate unit.

(a) Hydrogen (H)

(b) Nitrogen (N)

(c) Potassium (K)

(d) Iron (Fe)

(e) Uranium (U)

Core Concepts

■ The **molar mass** of an atom is the mass of 1 mol (Avogadro's number) of atoms. The units of molar mass are grams per mole (g/mol). The molar mass is numerically equal to the relative atomic mass provided in the periodic table.

Chemical Calculations for Atoms

Performing chemical calculation allows us to relate the measurable mass of a sample (in grams) to the number of atoms or molecules that it contains. This connection is valuable because chemical reactions typically occur via simple ratios of reactants and products. By relating the moles of reactants to moles of products, we can quantify the expected outcome of the reaction. In this section, we begin by illustrating chemical calculations for atoms.

When we perform calculations, there is a trade-off between precision and convenience. As a compromise, we will perform all of our calculations using the molar mass of atoms to one decimal place. As an example, we will use the molar mass of carbon as 12.0 g/mol and not 12.01 g/mol. By simplifying the numbers, we will be able to focus on the *reason* and the *method* for each calculation.

Our calculations will examine three important quantities: the *mass* of a sample measured in grams, its *chemical amount* measured in moles, and the *number of atoms* it contains. The relationships among these quantities are illustrated in Figure 6.12. To highlight the distinction in our calculations, we will indicate mass (g) in blue type, chemical amount (mol) in red type, and number of atoms in green type.

To convert from one quantity to another, we need to perform a *unit conversion* using a *conversion factor* (see Appendix A). To recap, a conversion factor is a ratio of numbers and associated units that is always equal to 1. For example, we can write the molar mass of carbon as

$$1 \text{ mol C} = 12.0 \text{ g C}$$

We can create two possible conversion factors from this relationship

$$\frac{1 \text{ mol C}}{12.0 \text{ g C}} = 1 \qquad \text{and} \qquad \frac{12.0 \text{ g C}}{1 \text{ mol C}} = 1$$

THE KEY IDEA: Chemical calculations allow us to relate the mass of a sample (in grams), its chemical amount (in moles), and the number of atoms it contains.

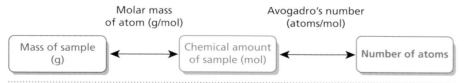

FIGURE 6.12 Chemical calculations for atoms. The relationships among three properties of a sample: mass (g), chemical amount (mol), and number of atoms. Conversion between mass and chemical amount is accomplished using the molar mass of the atom. Conversion between chemical amount and number of atoms is accomplished using Avogadro's number.

The conversion factor that we select depends on the calculation we want to perform. Suppose we are given a 15.0 g sample of carbon, and we want to calculate the chemical amount in moles. In general, we select the conversion factor that *cancels the given units (g) and introduces the new units (mol)*. In other words, the given units are on the bottom (the denominator), where they cancel, and the units for the answer are on the top (the numerator).

$$\text{chemical amount of C (mol)} = 15.0 \ \cancel{g\,C} \times \frac{1 \text{ mol C}}{12.0 \ \cancel{g\,C}} = 1.25 \text{ mol}$$

answer units (mol)

given units (g)

A similar principle works for the conversion between chemical amount and mass. Suppose we want to calculate the mass of carbon corresponding to 0.200 mol. In this case, we select the other conversion factor that cancels the given unit (mol) and introduces the new unit (g).

$$\text{mass of C} = 0.200 \ \cancel{mol\,C} \times \frac{12.0 \text{ g C}}{1 \ \cancel{mol\,C}} = 2.40 \text{ g C}$$

answer units (g)

given units (mol)

WORKED EXAMPLE 6.3

Question: In Chapter 2, we learned that arsenic (As) is a potent poison because it mimics the chemistry of phosphorus. A lethal dose of arsenic can be as small as 100 milligrams (mg). (a) For this dose, calculate the chemical amount of arsenic in units of moles. (b) How many arsenic atoms are contained in a 100 mg dose?

Answer: (a) First, we convert the mass of the sample to grams, which is the standard mass unit for chemical measurements.

$$\text{mass of As (g)} = 100 \ \cancel{mg} \times \frac{1 \text{ g}}{1000 \ \cancel{mg}} = 0.100 \text{ g As}$$

According to the periodic table, the relative atomic mass of arsenic is 74.9 (to one decimal place). Therefore, the molar mass of arsenic atoms is 74.9 g/mol.

To calculate the chemical amount of arsenic, we use a conversion factor that cancels the given units (g) in the denominator and introduces the new units (mol) in the numerator:

$$\frac{1 \text{ mol As}}{74.9 \text{ g As}} = 1$$

We now use the conversion factor to perform the calculation:

$$\text{chemical amount of As (mol)} = 0.100 \ \cancel{g\,As} \times \frac{1 \text{ mol As}}{74.9 \ \cancel{g\,As}}$$

$$= 1.34 \times 10^{-3} \text{ mol As}$$

This answer makes sense: The 100-mg dose is much smaller than the molar mass of arsenic, so the chemical amount of arsenic is much less than 1 mole.

(b) The relationship between moles and number of atoms is given by Avogadro's number. Our conversion factor must cancel the given unit (mol) and introduce the new unit (number of atoms).

$$\frac{6.022 \times 10^{23} \text{ atoms As}}{1 \text{ mol As}} = 1$$

We now use the conversion factor to perform the calculation:

$$\text{number of As atoms} = 1.34 \times 10^{-3} \text{ mol As} \times \frac{6.022 \times 10^{23} \text{ atoms As}}{1 \text{ mol As}}$$

$$= 8.07 \times 10^{20} \text{ atoms As}$$

This answer makes sense: We have less than 1 mole of arsenic, so the total number of atoms in the 100-mg dose is smaller than Avogadro's number. However, the sample still contains an enormous number of As atoms. In words, the number is more than eight hundred billion billion atoms. We can now appreciate why even a small dose of arsenic can be lethal.

TRY IT YOURSELF 6.3

Question: According to the Food and Drug Administration (FDA), the recommended dietary allowance (RDA) of iron (Fe) for adult females aged 19–50 is 18 milligrams. (a) Convert this RDA to units of moles. (b) How many iron atoms are contained in the RDA?

Core Concepts

- ◼ The mass of a sample (in grams) can be converted to chemical amount (in moles) by using the molar mass of atoms as a conversion factor.
- ◼ The chemical amount of a sample (in moles) can be converted to the number of atoms by using Avogadro's number as a conversion factor.

Molar Mass of Molecules

We can extend the concept of molar mass from atoms to molecules in the following way: *The molar mass of a molecule is equal to the sum of the molar masses of all of its constituent atoms.* The following steps illustrate how this calculation is performed for CO_2. For convenience, we will round the molar masses of the atoms to one decimal place.

🔑 THE KEY IDEA: The molar mass of a molecule is the sum of the molar masses of all of its constituent atoms.

STEP 1: *Use the periodic table to obtain the molar masses of all the atoms in the molecule.*

C 12.0 g/mol   O 16.0 g/mol

STEP 2: *Use the chemical formula to sum the molar masses of all the atoms in the molecule.*

$$\text{molar mass of } CO_4 = (1 \times \text{molar mass of } C) + (2 \times \text{molar mass of } O)$$
$$= (12.0 \text{ g/mol}) + (2 \times 16.0 \text{ g/mol})$$
$$= 44.0 \text{ g/mol}$$

Revisiting our equation for the combustion of carbon, we see that Figure 6.13 summarizes the relationships between the masses and moles of the reactants and products. The molar mass of O_2 can be calculated from the molar mass of oxygen atoms (2×16.0 g/mol $= 32$ g/mol). Note that *mass is conserved* during the reaction—the total mass of the reactants (12.0 g + 32.0 g) is equal to the mass of the products (44.0 g).

Using the stoichiometry of the chemical reaction plus the mole concept, we can predict the amount of CO_2 produced by burning any mass of carbon. For example, burning 24.0 g of carbon (2 mol) produces 2×44.0 g or 88.0 g of CO_2. Likewise, burning 3.00 g of carbon (0.25 mol) produces 0.25×44.0 or 11.0 g of CO_2. The next section explains the methods for dealing with numbers that are not as straightforward

C(s)		$O_2(g)$		$CO_2(g)$
1 mol	+	1 mol	→	1 mol
12.0 g		32.0 g		44.0 g

FIGURE 6.13 Masses and moles. This representation of a carbon combustion reaction shows the relationship between moles and masses for the reactants and products.

WORKED EXAMPLE 6.4

Question: What volume of water contains 1 mol of H_2O molecules? The density of water is 1.00 g/cm³ at 25°C.

Answer: First, use the periodic table to obtain the molar mass for each of the atoms in H_2O.

$$H \quad 1.0 \text{ g/mol} \qquad O \quad 16.0 \text{ g/mol}$$

Next, calculate the molar mass of H_2O by summing the molar masses of its constituent atoms.

$$\text{molar mass of } H_2O = (2 \times \text{molar mass of } H) + (1 \times \text{molar mass of } O)$$
$$= (2 \times 1.0 \text{ g/mol}) + (16.0 \text{ g/mol})$$
$$= 18.0 \text{ g/mol}$$

The molar mass tells us that 1 mol of H_2O has a mass of 18.0 g.

We can relate the mass of water to its volume by using the density, which is 1.00 g/cm³. Using these conversion factors, we can calculate the volume of water corresponding to 1 mol:

$$\text{volume of water} = 1 \text{ mol} \times \frac{18.0 \text{ g}}{1 \text{ mol}} \times \frac{1.00 \text{ cm}^3}{1 \text{ g}} = 18.0 \text{ cm}^3$$

The answer is given with three significant figures. As shown here, 1 mol of water occupies only 18.0 cm³, which is barely enough for a single gulp! For comparison, 1 fluid ounce (1 fl. oz.) is roughly equal to 30 cm³.

TRY IT YOURSELF 6.4

Question: Ethanol (C_2H_5OH) is the type of alcohol found in beer, wine, and spirits. What volume of ethanol contains 1 mol of molecules? The density of ethanol is 0.789 g/cm³ at 25°C.

....

Core Concepts

- The molar mass of a molecule (g/mol) is equal to the sum of the molar masses of all of its constituent atoms.
- By using the stoichiometry of a chemical reaction, it is possible to determine the mass of product generated from a given mass of reactant.

....

Chemical Calculations for Molecules

We can also perform chemical calculations using molecules instead of atoms. Many of the principles that we discussed above still apply, but we have to make an adjustment to account for the fact that a molecule contains multiple atoms. Figure 6.14 shows the relationships among three quantities and their units: the mass of the sample (g), the chemical amount of the sample (mol), and the number of molecules contained in the sample.

A particularly important type of chemical calculation aims to answer the following question: If we begin with a given mass of reactant, what mass of product is generated by the chemical reaction? For example, imagine that we have a 400.0-g sample of methane and we want to know how much carbon dioxide is produced by completely burning the gas in air. The balanced chemical equation for the reaction tells us the relationship between CH_4 and CO_2, but that relationship is provided as the *number of molecules* and not the mass of the samples. We therefore need to take a detour into the realm of chemical reactions where reactants and products exist in simple ratios. The steps of the calculation are diagrammed in Figure 6.15 and are outlined here.

THE KEY IDEA: Chemical calculations allow us to relate the mass of a sample (in grams), its chemical amount (in moles), and the number of molecules it contains.

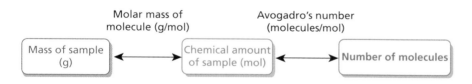

FIGURE 6.14 **Chemical calculations for molecules.** The relationships among three properties of a sample: mass (g), chemical amount (mol), and number of molecules. Conversion between mass and chemical amount is accomplished using the molar mass of the molecule. Conversion between chemical amount and number of molecules is accomplished using Avogadro's number.

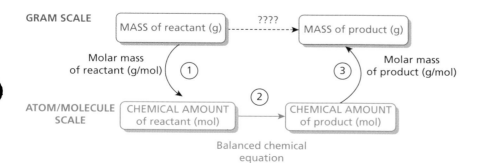

FIGURE 6.15 **The steps for calculating the mass of product obtained from a given mass of reactant.** Quantities measured in grams, which are observable amounts, are highlighted in blue. Quantities measured in moles, which refer to the number of unobservable atoms and molecules, are highlighted in red. Curved arrows indicate a transition between the two scales. The numbers in circles (such as ①) refer to the steps of the calculation.

STEP 1: *Convert the mass of the reactant (g) into the chemical amount of the reactant (mol).*

To calculate the number of moles of CH_4 in this sample, we need to determine the molar mass of CH_4.

$$\text{molar mass of } CH_4 = (1 \times \text{molar mass of C}) + (4 \times \text{molar mass of H})$$
$$= (1 \times 12.0 \text{ g/mol}) + (4 \times 1.00 \text{ g/mol})$$
$$= 16.0 \text{ g/mol}$$

The molar mass provides a conversion factor between mass and moles:

$$\frac{1 \text{ mol } CH_4}{16.0 \text{ g } CH_4} = 1$$

We can use this conversion factor to calculate the chemical amount of CH_4 in moles:

$$\text{chemical amount of } CH_4 \text{ (mol)} = 400.0 \text{ g } CH_4 \times \frac{1 \text{ mol } CH_4}{16.0 \text{ g } CH_4}$$
$$= 25.0 \text{ mol } CH_4$$

In summary, the 400-g sample of methane contains 25.0 mol of CH_4 molecules. This answer makes sense because 400 g is much larger than the mass of 1 mol of CH_4.

STEP 2: *Convert the chemical amount of the reactant (mol) into the chemical amount of the product (mol).*

The balanced chemical equation for methane combustion provides the stoichiometric relationship between the number of moles of CH_4 (reactant) and CO_2 (product):

$$CH_4 \quad + \quad 2\,O_2 \quad \longrightarrow \quad CO_2 \quad + \quad 2\,H_2O$$

| Methane | Oxygen | Carbon dioxide | Water |

In this case, the ratio is simply 1:1. However, for many other reactions, the ratio will not be as simple. To illustrate, Worked Example 6.2 showed that 1 mol of octane (C_8H_{18}) produces 8 mol of CO_2. In general, we can define a **mole ratio** that relates the moles of products to the moles of reactants, based on the stoichiometry of the balanced equation.

mole ratio The ratio of moles of product to moles of reactants in a balanced chemical equation.

$$\text{mole ratio} = \frac{\text{moles of products}}{\text{moles of reactants}}$$

In this case, the mole ratio is:

$$\text{mole ratio} = \frac{1 \text{ mol } CO_2}{1 \text{ mol } CH_4}$$

We can calculate the chemical amount of CO_2 by using this mole ratio:

$$\text{chemical amount of } CO_2 \text{ (mol)} = 25.0 \text{ mol } CH_4 \times \frac{1 \text{ mol } CO_2}{1 \text{ mol } CH_4}$$
$$= 25.0 \text{ mol } CO_2$$

STEP 3: *Convert the chemical amount of product (mol) into the mass of the product (g).*

The final step is to use the chemical amount of CO_2 to calculate the mass of CO_2 produced by the reaction. To complete this step, we need the molar mass of CO_2, which we calculated earlier as 44.0 g/mol. We now have the following conversion factor:

$$\frac{44.0\ \text{g}\ CO_2}{1\ \text{mol}\ CO_2} = 1$$

We use this conversion factor for the final step of calculating the mass of CO_2:

$$\text{mass of } CO_2\ (g) = 25.0\ \text{mol}\ CO_2 \times \frac{44.0\ \text{g}\ CO_2}{1\ \text{mol}\ CO_2} = 1100.0\ \text{g}\ CO_2$$

In summary, burning 400.0 g of CH_4 in air generates 1100.0 g of CO_2 (i.e., almost three times the starting mass of CH_4). This relationship between the observable masses of reactant and product is not obvious unless we analyze the chemical reaction in terms of the participating molecules.

All numerical conversions from mass of reactants to mass of products follow these same three steps. It is therefore possible to condense the calculation and perform all of the steps at the same time within a single equation, as shown in Figure 6.16. The units cancel to give the desired quantity—the mass of CO_2.

PRACTICE EXERCISE 6.4

Question: Propane (C_3H_8) is a hydrocarbon fuel used for gas barbecues. A typical tank, such as the one shown here, contains 20 pounds of propane. If all this propane is burned in air, what mass of CO_2 is produced? Give your answer in pounds of CO_2. Use the conversion factor: 1 kg = 2.20 pounds.

Sometimes we need to describe the amount of a compound that is not an isolated molecule. For example, sodium chloride is an ionic compound held together by ionic bonds; therefore, it would be incorrect to call it a molecule. We describe this compound in terms of its **formula unit**, which is the simplest ratio of its ionic components. The formula unit for sodium chloride is NaCl, which contains one Na^+ ion and one Cl^- ion. The same principle applies to silica, a covalent compound, which is a major component of beach sand. A single grain of silica contains many trillions of silicon and oxygen atoms, joined together by a network of covalent bonds. In every sample of silica, the Si and O atoms exist in a ratio of 1:2. Consequently, the formula unit for silica is SiO_2 (silicon dioxide).

formula unit The simplest ratio of the components in a compound.

$$\text{Mass of } CO_2\ (g) = 400.0\ \text{g}\ CH_4 \times \frac{1\ \text{mol}\ CH_4}{16.0\ \text{g}\ CH_4} \times \frac{1\ \text{mol}\ CO_2}{1\ \text{mol}\ CH_4} \times \frac{44.0\ \text{g}\ CO_2}{1\ \text{mol}\ CO_2} = 1100\ \text{g}\ CO_2$$

| ① Convert mass of CH_4 to moles of CH_4 | ② Convert moles of CH_4 to moles of CO_2 | ③ Convert moles of CO_2 to mass of CO_2. |

FIGURE 6.16 A combined calculation. A single-equation method for calculating the mass of CO_2 produced by combustion of methane.

WORKED EXAMPLE 6.5

Question: When cooking pasta, we often add a pinch of table salt for extra flavor. In culinary terms, one pinch of salt is defined as 0.36 g. How many formula units of NaCl are in a pinch of salt?

Answer: First, we need to calculate the molar mass for the formula unit NaCl. Calculating this number requires us to obtain the molar masses of Na and Cl atoms from the periodic table (to one decimal place).

$$\text{Na}\ \ 23.0\,\text{g/mol}\qquad \text{Cl}\ \ 35.5\,\text{g/mol}$$

The periodic table gives the masses of *neutral atoms*, but sodium chloride contains *charged ions* (Na^+ and Cl^-). Is this a problem? Recall from Chapter 2 that the mass of an atom arises from the masses of its protons and neutrons. The mass of an electron is so small that gaining or losing one or more electrons to form an ion has a negligible effect on the total mass. For practical purposes, then, the mass of an ion is the same as the mass of its parent atom.

$$\text{molar mass of NaCl (g/mol)} = (1 \times \text{molar mass of Na}) + (1 \times \text{molar mass of Cl})$$
$$= (23.0 + 35.5)\,\text{g/mol}$$
$$= 58.5\,\text{g/mol}$$

We now have a conversion factor between mass and moles:

$$\frac{1\,\text{mol NaCl}}{58.5\,\text{g NaCl}} = 1$$

We begin with a mass, and the final answer is the number of formula units. We therefore need to perform two conversions: mass (g) → amount (mol) and amount (mol) → formula units. We can accomplish this task by performing both conversions on a single line:

$$\text{number of formula units} = 0.36\ \text{g NaCl} \times \frac{1\ \text{mol NaCl}}{58.5\ \text{g NaCl}}$$

$$\times \frac{6.022 \times 10^{23}\ \text{NaCl}}{1\ \text{mol NaCl}} = 3.7 \times 10^{21}\ \text{NaCl}$$

We have provided the answer to two significant figures to match the number of significant figures in the mass of salt. It is remarkable that even a tiny pinch of salt contains more than three thousand billion billion NaCl formula units.

TRY IT YOURSELF 6.5

Question: A calcium dietary supplement tablet contains 1200 milligrams (mg) of calcium carbonate ($CaCO_3$), which is an ionic compound. How many formula units of $CaCO_3$ are contained in one tablet?

Core Concepts

■ Beginning with the given mass of a reactant, it is possible to calculate the mass of the product that is generated by a chemical reaction. We perform this calculation by analyzing the reaction in terms of mole quantities of reactants and products, which are related by the **mole ratio** in the balanced chemical equation.

Burning Carbs (Part 1): Changes in Matter

Now that we have studied chemical reactions, we can use this knowledge to understand what we mean by "burning carbs." When we eat carbohydrates, our digestive system breaks them down into smaller sugar units. The most important sugar for living organisms is *glucose*, which has the chemical formula $C_6H_{12}O_6$. The molecular structure of glucose is illustrated in Figure 6.17. The glucose molecule contains a ring of six atoms comprising five carbon atoms and one oxygen atom. These atoms form a nonplanar structure similar to the "chair" configuration in cyclohexane (see Chapter 4). The sixth carbon atom extends above the ring.

We will first consider burning glucose in a combustion reaction. The glucose reacts with oxygen in the air to generate carbon dioxide and water. The unbalanced chemical equation for this reaction is given here, including the physical state of the reactants and products.

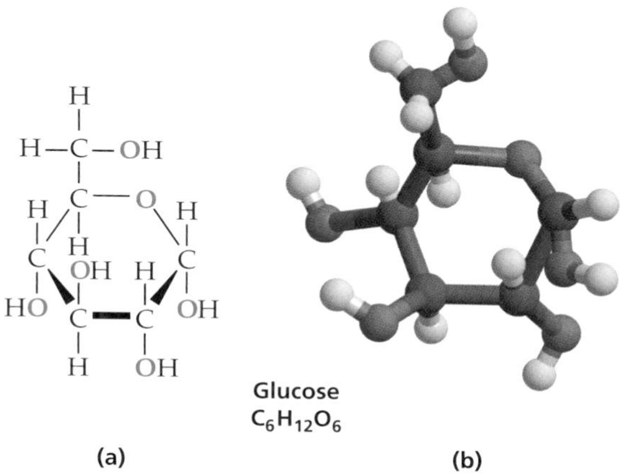

FIGURE 6.17 **The molecular structure of glucose, $C_6H_{12}O_6$.** Glucose is represented as (a) a line drawing and (b) a molecular model. In the line drawing, it is standard practice not to show the covalent bond between the O and H atoms in the –OH groups.

$$C_6H_{12}O_6(s) \ + \ O_2(g) \ \longrightarrow \ CO_2(g) \ + \ H_2O(g) \quad \textit{unbalanced}$$

Glucose   Oxygen   Carbon dioxide   Water

THE KEY IDEA: Glucose reacts with oxygen to produce carbon dioxide and water.

By applying the steps for balancing equations, we can deduce the balanced equation for burning glucose. We see that the 6 C atoms in glucose are converted into 6 CO_2 molecules, whereas the 12 H atoms in glucose generate 6 H_2O molecules.

$$C_6H_{12}O_6(s) \ + \ 6\,O_2(g) \ \longrightarrow \ 6\,CO_2(g) \ + \ 6\,H_2O(g) \ \textit{balanced}$$

How much CO_2 is produced by burning glucose? Suppose that we begin with 5.00 g of glucose, which is the amount required to fill a teaspoon. We can calculate the amount of CO_2 produced by following the steps that were diagrammed in Figure 6.15.

STEP 1: *Convert the mass of the reactant (g) into the chemical amount of the reactant (mol).*

We begin with 5.00 g of glucose, and we need to calculate the chemical amount in moles. To perform this conversion, we need the molar mass of glucose in grams per mole. We can calculate this quantity by using the molecular formula for glucose ($C_6H_{12}O_6$) and summing the molar masses of the atomic constituents.

$$\text{molar mass of glucose (g/mol)} = (6 \times \text{molar mass of C}) + (12 \times \text{molar mass of H})$$
$$+ (6 \times \text{molar mass of O})$$
$$= (6 \times 12.0\,\text{g/mol}) + (12 \times 1.0\,\text{g/mol})$$
$$+ (6 \times 16.0\,\text{g/mol})$$
$$= (72.0 + 12.0 + 96.0)\,\text{g/mol}$$
$$= 180.0\,\text{g/mol}$$

We now have a conversion factor between mass and moles of glucose:

$$\frac{1 \text{ mol } C_6H_{12}O_6}{180.0 \text{ g } C_6H_{12}O_6} = 1$$

We can use this conversion factor to calculate the chemical amount of glucose in 5.00 g:

$$\text{chemical amount of } C_6H_{12}O_6 \text{ (mol)} = 5.00 \text{ g } C_6H_{12}O_6 \times \frac{1 \text{ mol } C_6H_{12}O_6}{180.0 \text{ g } C_6H_{12}O_6}$$

$$= 0.0278 \text{ mol } C_6H_{12}O_6$$

We report the calculated number using three significant figures, following the rules in Appendix B. This answer makes sense: 5.00 g of glucose is much less than its molar mass, so it contains only a small fraction of a mole.

STEP 2: *Convert the chemical amount of reactant (mol) into the chemical amount of the product (mol).*

The balanced chemical equation for burning glucose provides the stoichiometric relationship between the number of moles of $C_6H_{12}O_6$ (reactant) and CO_2 (product):

$$C_6H_{12}O_6 \quad + \quad 6 \, O_2 \quad \longrightarrow \quad 6 \, CO_2 \quad + \quad 6 \, H_2O$$

Glucose   Oxygen   Carbon dioxide   Water

For this reaction, the mole ratio is:

$$\text{mole ratio} = \frac{\text{moles of products}}{\text{moles of reactants}} = \frac{6 \text{ mol } CO_2}{1 \text{ mol } C_6H_{12}O_6}$$

We can use this mole ratio to calculate the chemical amount of CO_2 generated by burning glucose:

$$\text{chemical amount of } CO_2 \text{ (mol)} = 0.0278 \text{ mol } C_6H_{12}O_6 \times \frac{6 \text{ mol } CO_2}{1 \text{ mol } C_6H_{12}O_6}$$

$$= 0.167 \text{ mol } CO_2$$

The stoichiometry of the chemical reaction has a multiplier effect: The number of moles of CO_2 produced as a product is equal to six times the number of moles of glucose that were present as the reactant.

STEP 3: *Convert the chemical amount of the product (mol) into the mass of the product (g).*

The final step is to use the chemical amount of CO_2 to calculate the mass of CO_2 produced by the reaction. We already know that the molar mass of CO_2 is 44.0 g/mol, so we use the conversion factor:

$$\frac{44.0 \text{ g } CO_2}{1 \text{ mol } CO_2} = 1$$

We can now calculate the mass of CO_2 produced by the reaction:

$$\text{mass of } CO_2 \text{ (g)} = 0.167 \text{ mol } CO_2 \times \frac{44.0 \text{ g } CO_2}{1 \text{ mol } CO_2} = 7.35 \text{ g } CO_2$$

$$\text{mass of CO}_2 \text{ (g)} = 5.00 \text{ g } C_6H_{12}O_6 \times \frac{1 \text{ mol } C_6H_{12}O_6}{180.0 \text{ g } C_6H_{12}O_6} \times \frac{6 \text{ mol CO}_2}{1 \text{ mol } C_6H_{12}O_6} \times \frac{44.0 \text{ g CO}_2}{1 \text{ mol CO}_2} = 7.34 \text{ CO}_2$$

① Convert mass of $C_6H_{12}O_6$ to moles of $C_6H_{12}O_6$

② Convert moles of $C_6H_{12}O_6$ to moles of CO_2.

③ Convert moles of CO_2 to mass of CO_2.

FIGURE 6.18   A combined calculation. A single-equation method for calculating the mass of CO_2 produced by combustion of glucose.

This result tells us that burning 5.00 g of glucose in air generates more than its own mass of CO_2 (almost 50% more). If we wish to perform the calculation in a single step, we can use the equation provided in Figure 6.18.

These results differ at the second decimal place because of small rounding errors in the step-by-step calculation. This discrepancy commonly occurs in calculations, and either answer is correct for our purposes.

Core Concepts

- A combustion reaction between glucose and oxygen produces carbon dioxide and water.
- We can calculate the mass of CO_2 produced by burning a given mass of glucose.

6.4   Chemical Reactions Produce Changes in Energy

Recall from the introduction that chemical reactions produce changes in matter and energy. In Chapter 2, we learned that matter is any substance that occupies space and has mass. All of the atoms and molecules that participate in chemical reactions are classified as matter. By contrast, energy is different from matter—it does not occupy space, and it does not have mass. What exactly is energy, and why does it change during a chemical reaction?

What Is Energy?

The modern science of energy emerged during the Industrial Revolution to understand the operation of steam engines. For this reason, **energy** *is defined* as the *capacity to do work*, which involves applying a force over a distance. The international unit for energy is the joule (J). In chemistry, we typically measure thousands of joules, or *kilojoules* (kJ).

Energy exists in a variety of forms, which can be interconverted. **Kinetic energy** arises from the motion of an object, and it depends on the object's mass and velocity. **Heat energy** is a type of kinetic energy because it arises from the motion of the atoms or molecules that make up matter. Heat energy can be transferred from one object to another. **Potential energy** depends on the position of an object rather than its motion. For example, water at the top of a dam has a larger potential energy than water at the dam's base. When water flows through the dam, its potential energy can be converted into kinetic energy by turning large turbines. The motion of the turbines is then used to generate electricity; this is the basis of hydroelectric power.

One type of potential energy is *bond energy*—that is, the energy stored in chemical bonds. As we saw in Chapter 3, bond energy arises from the relative positions of the electrical charges in the molecule (i.e., the electrons and atomic nuclei). Similar to the example of falling water, the bond energy is released when the bond is formed.

Learning Objective:
Relate the energy produced by a reaction to the chemical bonds in the reactants and products.

THE KEY IDEA: Energy exists in a variety of forms.

energy   The capacity to do work.

kinetic energy   A form of energy that arises from the motion of an object.

heat energy   A form of energy that can be transferred between objects.

potential energy   A form of energy that derives from an object's position.

Core Concepts

- **Energy** is the capacity to do work. The unit of energy is the joule (J). In chemistry, we typically measure thousands of joules, or *kilojoules* (kJ).
- Energy exists in a variety of forms: **kinetic energy, heat energy, potential energy,** and bond energy.

Making and Breaking Covalent Bonds

🔑 **THE KEY IDEA:** An energy input is required to break a covalent bond, and an energy output occurs when a covalent bond is formed.

In Chapter 3, we learned that two hydrogen atoms form a covalent bond. Sharing an electron pair fills the electron shells of the atoms and lowers their overall energy. Scientists often describe the bonded atoms as *more stable*, where the word "stable" means "not likely to change." For example, a pencil lying on its side is more stable than one balanced on its point. For the following discussion, "more stable" means the same thing as "having lower energy."

Figure 6.19 compares the energy of two H atoms in their separated and bonded conditions. Because the bonded H atoms are more stable, an input of energy (e.g., heat energy) is required to break the bond and detach the atoms. The amount of energy required to break a single covalent bond is very small when measured in units of joules. Therefore, it is more convenient to define this value in terms of a mole of bonds (i.e., Avogadro's number). In general, the **bond energy** *is measured as the energy required to break 1 mol of chemical bonds.* The most convenient units for bond energy are kilojoules per mole (kJ/mol). Using this unit, the bond energy for an H—H bond is 436 kJ/mol.

bond energy The energy required to break a chemical bond.

If an energy input of 436 kJ is required to break 1 mol of H—H bonds, then forming 1 mol of bonds *releases* the same amount of heat energy. Figure 6.19 shows this energy change as a *negative* number (–436 kJ/mol). By convention, we use plus and minus signs to keep track of whether energy is being absorbed (+) or released (–). This method of energy accounting is similar to keeping track of the amount of money in your bank account. When you deposit money, you *add* to your balance; when you withdraw money, you *subtract* from your balance.

Table 6.3 presents the bond energies of a selection of covalent bonds, which includes single, double, and triple bonds. In some cases, the energy of a particular covalent bond is influenced by the other atoms in the molecule. As an example, the C—H bond energy is different in methane (CH_4) and ethane (C_2H_6). For this reason, Table 6.3 reports the *average bond energy* for bonds such as C—H, O—H, and so on. These energies are obtained by averaging the energy of the same bond in a variety of molecules. In general, a larger bond energy means that it is more difficult to break the connection between the atoms; that is, the bond is more stable. We use the term *strong bond* when the bond energy is relatively large and *weak bond* when the bond energy is relatively small.

We can make some general observations based on the data in Table 6.3. First, single bonds exhibit large variations in bond energy that depend on the two atoms that are bonded together. Second, double bonds typically have a larger bond energy than single bonds, which arises from the stronger attraction of two shared electron pairs. Third, triple bonds have the largest bond energies because the atoms share three electrons pairs. For example, the large bond energy of N≡N explains why atmospheric nitrogen gas is stable and unreactive under most conditions (see Chapter 3).

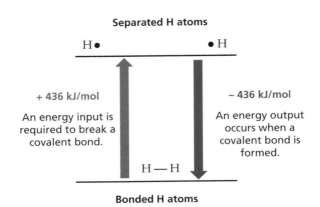

FIGURE 6.19 The bond energy of an H—H bond. An input of energy (+436 kJ/mol) is required to break 1 mole of H—H covalent bonds and separate the atoms. Conversely, energy is released (–436 kJ/mol) when all of the separated H atoms join together to form a covalent bond.

TABLE 6.3   Average Bond Energies (kJ/mol) for Covalent Bonds

SINGLE BONDS				DOUBLE BONDS		TRIPLE BONDS	
H—H	436	N—N	160	C=C	598	C≡C	813
H—F	570	N—H	391	C=O	745	C≡O	1077
		N—F	272	C=O in CO_2	803	C≡N	750
C—C	356	N—O	201	C=N	616		
C—H	416					N≡N	945
C—N	285	O—O	146	O=O	498		
C—O	336	O—H	467				
C—F	485	O—F	190	N=N	418		
				N=O	607		

Core Concepts

- The **bond energy** is measured as the amount of energy required to break 1 mol of chemical bonds. The units of bond energy are kJ/mol.
- Breaking a covalent bond requires an input of energy, and forming a covalent bond involves a release of energy (usually in the form of heat).
- The bond energy of a covalent bond depends on the atoms that are bonded together, as well as the number of electron pairs in the bond (single, double, or triple).

Energy Changes During Chemical Reactions

Earlier in this chapter, we learned that the combustion of methane produces carbon dioxide and water. This reaction also generates heat energy, as we can see from the ring of flames in Figure 6.3. What is the source of this heat energy?

We can answer this question by revisiting the chemical equation for methane combustion and explicitly showing the covalent bonds in the reactants and products (Figure 6.20). We can see that *the number and types of covalent bonds in the reactants differ from those in the products.* The chemical reaction requires an input of energy to break the covalent bonds in the reactants (four C—H bonds and two O=O bonds). This is followed by an output of energy when new covalent bonds are formed to make the products (two C=O bonds and four O—H bonds). Because each of these covalent bonds has a characteristic bond energy, it is possible to calculate the net energy change that occurs during the reaction.

The steps of this calculation are provided in Table 6.4 on page 185 and diagrammed in Figure 6.21. To begin, we set the energy of the reactants at zero to provide a convenient reference point; all energy changes are calculated relative to this baseline. This procedure is similar to using "sea level" as the reference point for "zero altitude," which allows us to measure altitudes "above sea level" and "below sea level." Next, we consider the mole quantities of each reactant and product, which enables us to use bond energy values in kJ/mol. By counting the bonds in each molecule and using the bond energies from Table 6.3, we can calculate the energy input required to break all of the bonds in the reactants (CH_4 and $2 O_2$). Similarly, we can calculate the energy output obtained when forming all of the bonds in the products (CO_2 and $2 H_2O$).

THE KEY IDEA: The energy change that occurs during a chemical reaction arises from breaking bonds in the reactants and forming bonds in the products.

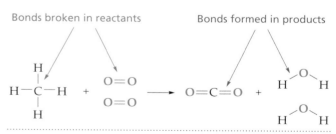

FIGURE 6.20 A chemical reaction breaks and forms covalent bonds. During methane combustion, covalent bonds in the reactants are broken (highlighted in green), and new covalent bonds are formed in the products (highlighted in blue). The arrows point to one example of each type of bond that is broken or formed.

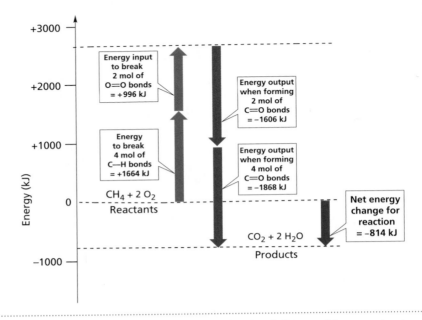

FIGURE 6.21 Energy changes during methane combustion. A diagram of the energy inputs and outputs during a methane combustion reaction.

reaction energy The net change in energy for a chemical reaction; it is the difference between the energy input to break bonds in the reactants, and the energy output from making bonds in the products.

exothermic A chemical reaction that releases energy to its surroundings.

Table 6.4 indicates that a total energy input of +2660 kJ is required to break all of the bonds in the reactants. In addition, a total energy output of –3474 kJ is obtained by forming all of the covalent bonds in the products. We calculate the overall energy change for the reaction by *adding* the energy input and output, because the energy output already contains a negative sign. This calculation yields an overall energy change of –814 kJ, which corresponds to a release of energy. This net change in energy is called the **reaction energy** of the combustion reaction. Returning to our earlier analogy of the bank account, we note that the negative sign means that the chemical reaction is giving an "energy payout."

These calculations show that methane combustion is an **exothermic** reaction. In general, *an exothermic reaction releases energy to its surroundings* (*exo* means "external"). This energy release occurs when the energy output from forming all of the bonds in the products is greater than the energy input required to break all of the bonds in the reactants. For a combustion reaction, most of the energy is released as heat. This heat output is the reason that we burn substances such as coal, methane, and octane as fuels.

We can understand why methane combustion is exothermic by comparing the numerical values of bond energies for the reactants and products. For the single bonds, the C—H bond energy in methane (416 kJ/mol) is smaller than the O—H bond energy in H_2O (467 kJ/mol). For the double bonds, the O=O bond energy in O_2 (498 kJ/mol) is significantly smaller than the C=O bond energy in CO_2 (803 kJ/mol). These differences in bond energy explain why the release of energy is greater than the input of energy. In other words, *an exothermic reaction involves the conversion of less stable reactants containing weaker bonds into more stable products containing stronger bonds.*

The heat energy released by a combustion reaction, called the *heat of combustion,* can be experimentally measured and is given as a positive number. The heat of combustion for methane is 889 kJ/mol. Our calculated energy change of 814 kJ/mol is close to the experimental value, but it is not exactly the same. There are two main sources of inaccuracy in our calculation. First, some of the bond energies used for our calculation in Table 6.4 are *average values,* so they are not an exact match for the specific molecules in the reaction. Second, we have assumed

TABLE 6.4   Calculation of Energy Changes during Methane Combustion

$$CH_4 + 2O_2 \rightarrow CO_2 + 2H_2O$$

Energy Input to Break Bonds in Reactants

Molecule	Bonds per Molecule	Moles of Reactants	Moles of Bonds	Bond Energy	Energy Input
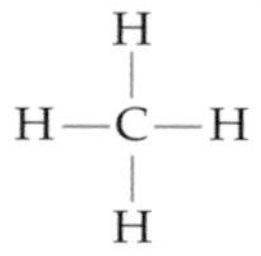 H—C—H (with H above and H below)	4	1	$4 \times 1 = 4$	416 kJ/mol	$4 \text{ mol} \times \left(\dfrac{+416 \text{ kJ}}{1 \text{ mol}} \right) = +1664 \text{ kJ}$
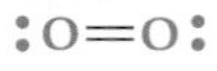 $:O=O:$	1	2	$1 \times 2 = 2$	498 kJ/mol	$2 \text{ mol} \times \left(\dfrac{+498 \text{ kJ}}{1 \text{ mol}} \right) = +996 \text{ kJ}$

Total energy input = +2660 kJ

Energy Output When Forming Bonds in Products

Molecule	Bonds per Molecule	Moles of Products	Moles of Bonds	Bond Energy	Energy Output
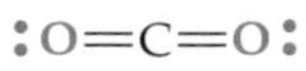 $:O=C=O:$	2	1	$2 \times 1 = 2$	803 kJ/mol	$2 \text{ mol} \times \left(\dfrac{-803 \text{ kJ}}{1 \text{ mol}} \right) = -1606 \text{ kJ}$
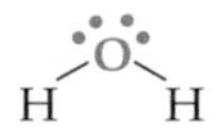 H—O—H (water)	2	2	$2 \times 2 = 4$	467 kJ/mol	$4 \text{ mol} \times \left(\dfrac{-467 \text{ kJ}}{1 \text{ mol}} \right) = -1868 \text{ kJ}$

Total energy output= −3474 kJ

Energy Change for Reaction

$$\text{energy change for reaction} = \text{energy input} + \text{energy output}$$
$$= +2660 \text{ kJ} + (-3474 \text{ kJ})$$
$$= -814 \text{ kJ}$$

that all of the bonds in the reactants are completely broken to produce isolated atoms, which then recombine to form the products. In reality, chemical reactions do not proceed in such a simple manner, as we shall see in Chapter 14. Despite these imperfections, this type of energy calculation is a valuable method for determining the approximate energy change of a chemical reaction.

We can now write a chemical equation for methane combustion that shows not just reactants and products but also the energy change that occurs during the reaction. In this equation, heat energy is one of the "products" of the exothermic reaction. We provide the experimentally determined value for the heat energy.

$$CH_4 \quad + \quad 2O_2 \quad \longrightarrow \quad CO_2 \quad + \quad 2H_2O \quad + \quad \boxed{\text{HEAT ENERGY} \atop 889 \text{ kJ/mol}}$$

Methane   Oxygen   Carbon dioxide   Water

Not all chemical reactions release heat. Some chemical reactions are **endothermic,** meaning they absorb energy from their surroundings (*endo* means "inside"). In these cases, calculating the energy difference for the reaction yields a

endothermic A chemical reaction that absorbs energy from its surroundings.

positive number—that is, the energy input (+) is greater than the energy output (–). In terms of our bank account analogy, we need to make an "energy deposit" to make the reaction occur.

PRACTICE EXERCISE 6.5

Question: The engines that power NASA's rockets obtain their energy from a chemical reaction between liquefied hydrogen (H_2) and oxygen (O_2). Calculate the amount of energy released by this reaction.

(a) Write a balanced chemical equation for the reaction between H_2 and O_2 to produce H_2O.

(b) Rewrite the balanced equation using two-dimensional structures of the reactants and products, which include all of the bonds within the molecules. Lone pairs are not required.

(c) Calculate the energy input required to break all of the bonds in the reactants.

(d) Calculate the energy output obtained when forming all of the bonds in the products.

(e) Calculate the energy change for this reaction. Explain why the result of your calculation shows that the chemical reaction is exothermic.

Core Concepts

- During a chemical reaction, covalent bonds in the reactant molecules are broken, and new bonds are formed to create the product molecules.
- An **exothermic** reaction releases energy to the external surroundings. A combustion reaction releases most of this energy as heat.
- In an exothermic reaction, the total energy released by forming all the covalent bonds in the products is greater than the total energy input required to break all the covalent bonds in the reactants.
- An **endothermic** reaction absorbs energy from its surroundings.

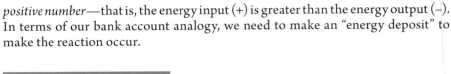 THE KEY IDEA: The combustion reaction between glucose and oxygen releases energy.

Burning Carbs (Part 2): Changes in Energy

In the previous section, we learned that the energy change for a chemical reaction arises from breaking and forming covalent bonds. Figure 6.22 shows the chemical equation for burning glucose, including the molecular structures of the reactants and products. For this reaction to occur, all the covalent bonds in the reactants must be broken. This includes the numerous bonds in glucose as well as the double bonds in the oxygen molecules. New covalent bonds must then be formed to make the reactants—namely, the C=O bonds in CO_2 and the O—H bonds in H_2O.

It is possible to calculate the energy input required to break all of the bonds in the reactants and the energy output obtained by forming all of the bonds in the products. Even without a full calculation, we can gain some insight by inspecting the types of covalent bonds in the reactants and products. For example, the energy required to break 6 mol of O=O bonds in O_2 is given by:

$$\text{energy input (kJ)} = 6 \ \cancel{\text{mol}} \times \left(\frac{+498 \text{ kJ}}{1 \ \cancel{\text{mol}}} \right) = +2988 \text{ kJ}$$

In the products, the C=O bond in CO_2 (799 kJ/mol) is stronger than the O=O bond. Each CO_2 molecule has two C=O bonds, so there are 12 moles of bonds in 6 moles of molecules. The energy output obtained by forming these bonds is:

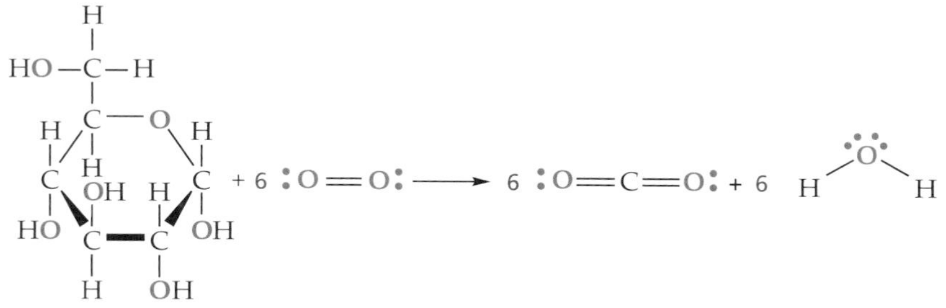

FIGURE 6.22 Combustion of glucose. The chemical equation for the combustion of glucose, showing the molecular structures of all of the reactants and products.

$$\text{energy output (kJ)} = 12 \ \cancel{\text{mol}} \times \left(\frac{-803 \ \text{kJ}}{1 \ \cancel{\text{mol}}} \right) = -9636 \ \text{kJ}$$

These calculations reveal that the amount of energy released by forming 6 moles of CO_2 is more than 3 times the amount required to break the bonds in 6 moles of O_2.

According to experimental measurement, the amount of heat energy released by burning glucose is 2805 kJ/mol. In a mole-to-mole comparison, this is more than 3 times the amount of heat released by burning methane. We can write an equation for the exothermic reaction of glucose combustion that includes the heat output:

$$\underset{\text{Glucose}}{C_6H_{12}O_6} \ + \ \underset{\text{Oxygen}}{6 \ O_2} \ \longrightarrow \ \underset{\text{Carbon dioxide}}{6 \ CO_2} \ + \ \underset{\text{Water}}{6 \ H_2O} \ + \ \boxed{\begin{array}{c} \text{HEAT ENERGY} \\ \text{2805 kJ/mol} \end{array}}$$

Core Concepts

- The combustion reaction between glucose and oxygen releases a large amount of heat energy.

6.5 Biochemical Reactions: The Basis for Life

We can now apply the principles of chemical reactions to understanding the biochemical reactions that occur in cells. We will focus on those reactions that produce the energy that cells need to function. Without these reactions, life could not exist.

Burning Carbs (Part 3): Cellular Respiration

We began this chapter by presenting the example of "burning carbs." The most common and biologically important carbohydrate is glucose. However, a combustion reaction is *not* an effective way for a biological organism to extract energy from glucose. Most cells need to function within a limited temperature range, and generating heat energy would raise the temperature of the cell to dangerous levels. In addition, the heat energy from combustion is released quickly, but our cells need to utilize energy gradually and efficiently. Finally, heat is only one form of energy, and a biological organism needs a more versatile energy supply that it can use for a variety of cellular processes.

When we "burn carbs" during exercise, the cells in our body extract energy from glucose not through combustion but through a chemical process called **cellular respiration**. Figure 6.23 compares the chemical equations for glucose combustion and cellular respiration. Note that the number and types of molecules are exactly the same for the two processes. The key difference is that *cellular*

Learning Objectives:
Explain how cells use glucose to generate energy.

THE KEY IDEA: Cellular respiration is a chemical reaction between glucose and oxygen that generates chemical energy stored in ATP molecules.

cellular respiration A chemical reaction in cells between glucose and oxygen that generates chemical energy stored in ATP molecules.

$$C_6H_{12}O_6 + 6\ O_2 \xrightarrow{\text{Combustion}} 6\ CO_2 + 6\ H_2O + \text{HEAT ENERGY}$$

$$C_6H_{12}O_6 + 6\ O_2 \xrightarrow[\text{respiration}]{\text{Cellular}} 6\ CO_2 + 6\ H_2O + \text{CHEMICAL ENERGY}$$

FIGURE 6.23 A comparison of the chemical equations for glucose combustion and cellular respiration. The combustion of glucose releases energy as heat, whereas cellular respiration generates chemical energy.

adenosine triphosphate (ATP)
A molecule containing three phosphate groups that stores chemical energy for use in cellular processes.

adenosine diphosphate (ADP) A molecule formed by the removal of one phosphate group from ATP.

respiration generates chemical energy instead of heat energy. In essence, cellular respiration extracts energy from glucose without producing a flame.

As we learned in Chapters 2 and 3, *the chemical energy used in cells is stored in the form of* **adenosine triphosphate (ATP)**. The structure of an ATP molecule, shown in Figure 6.24, consists of three regions: (1) adenine, (2) a ribose sugar, and (3) three phosphate groups. Roughly 30 ATP molecules are generated for each glucose molecule that reacts with oxygen. ATP is considered to be a "high-energy" molecule because of the three negative charges on the oxygen atoms in the phosphate groups. This is an unstable arrangement because electrical charges of the same type repel each other. Consequently, the chemical bonds in ATP that connect the phosphate groups must have a high energy to counteract the electrical repulsion.

Almost all processes within the human body require us to expend energy—moving muscles, copying DNA, and synthesizing biological molecules. ATP releases its chemical energy through a reaction that breaks off one of the phosphate groups to produce a molecule called **adenosine diphosphate (ADP)**. The ADP is a lower-energy molecule because it contains only two charged phosphate groups instead of three. Cellular respiration is used to regenerate ATP from ADP. This ATP/ADP cycle is illustrated in Figure 6.25. Our body stores very small reserves of ATP, so the energy to drive cellular processes relies on the constant regeneration of ATP from ADP.

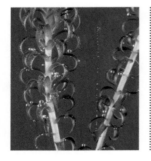

PRACTICE EXERCISE 6.6

Question: Photosynthesis is an essential biochemical reaction for life on Earth. The photograph adjacent shows bubbles of gas being released by the leaves of an underwater plant. The *unbalanced* chemical equation for photosynthesis is given below, with one product missing. (a) Balance the equation and deduce the missing product. (b) Based on your answer, identify the gas in the photograph.

$$CO_2 + H_2O \xrightarrow{\text{light energy}} C_6H_{12}O_6 + \underline{\hspace{2cm}}$$

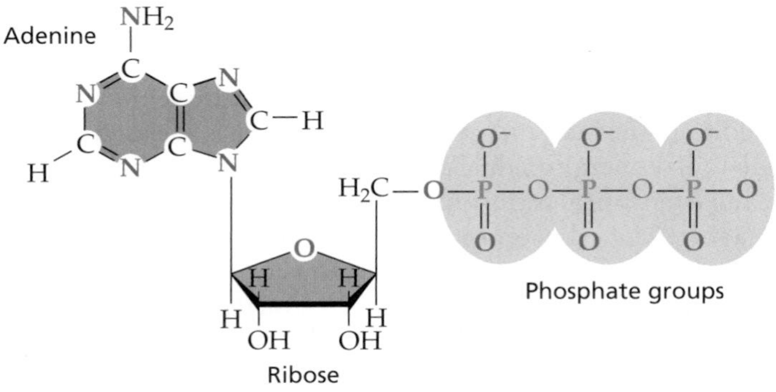

FIGURE 6.24 The structure of adenosine triphosphate (ATP). This molecule contains an adenine, a ribose sugar, and three phosphate groups. Each phosphate group contains an oxygen atom with a negative charge.

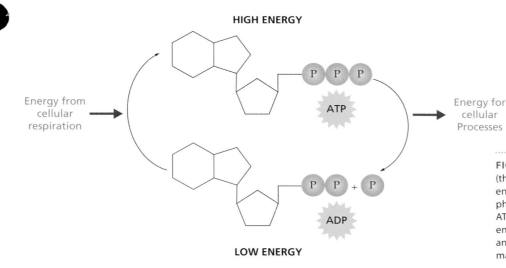

HIGH ENERGY

Energy from cellular respiration →

ATP

P P P

Energy for cellular Processes →

P P + P

ADP

LOW ENERGY

FIGURE 6.25 The ATP/ADP cycle. ATP (three phosphates) releases its chemical energy through conversion to ADP (two phosphates) plus a phosphate group. The ATP can be regenerated by an input of energy from cellular respiration. The ATP and ADP molecules are shown in schematic form without atomic details.

Core Concepts

■ "Burning carbs" in the human body involves **cellular respiration**. This reaction combines glucose and oxygen to produce carbon dioxide and water. It also generates chemical energy that is stored in the form of **adenosine triphosphate (ATP)**.

■ ATP releases its chemical energy through a reaction that breaks off one of its phosphate groups to produce a molecule of **adenosine diphosphate (ADP)**.

Exercise and Chemical Reactions

Consider the following familiar scenario. Instead of taking the elevator to your class, apartment, or dorm room, you decide to walk up the stairs. You can make it up the first one or two flights without any problem. However, the climb then starts to become more difficult. By the time you have reached the third or fourth floor, your legs feel heavy, and you start to feel a burning sensation in your muscles. Your pace begins to slow down significantly. When you eventually manage to reach the fifth or sixth floor, you are gasping for breath. You might even have to stop and rest because you have no energy left to continue.

Why do you run out of energy so quickly? We can understand this experience in terms of chemical reactions. Moving our muscles requires a supply of chemical energy derived from ATP. When you begin climbing the stairs, you rely on an available supply of ATP that is stored in your muscle cells. Unfortunately, it takes only a very short period of strenuous exercise for your muscles to use up their stored ATP. At that point your cells need a new source of this molecule.

The next available source of energy comes from the reservoir of glucose molecules that are stored in your muscle cells. A chemical reaction converts glucose to two copies of a smaller molecule called *lactic acid* (Figure 6.26); this process is called *lactic acid fermentation*. This reaction is balanced because the chemical formula for lactic acid ($C_3H_6O_3$) contains exactly half the atoms in each molecule of glucose ($C_6H_{12}O_6$).

When we compare lactic acid fermentation to cellular respiration, one reactant is conspicuously missing—oxygen. Because the production of lactic acid takes place in the absence of oxygen, we label the process as **anaerobic** ("life without air"). Lactic acid fermentation is an effective way to generate a quick burst of energy, but it is less energy efficient than cellular respiration because it produces fewer ATP molecules. In addition, the buildup of lactic acid contributes to the burning sensation you feel in your legs during strenuous exercise like climbing stairs.

THE KEY IDEA: Our bodies can generate energy from aerobic reactions (that use oxygen) and anaerobic reactions (that do not use oxygen).

anaerobic A chemical reaction that takes place in the absence of oxygen.

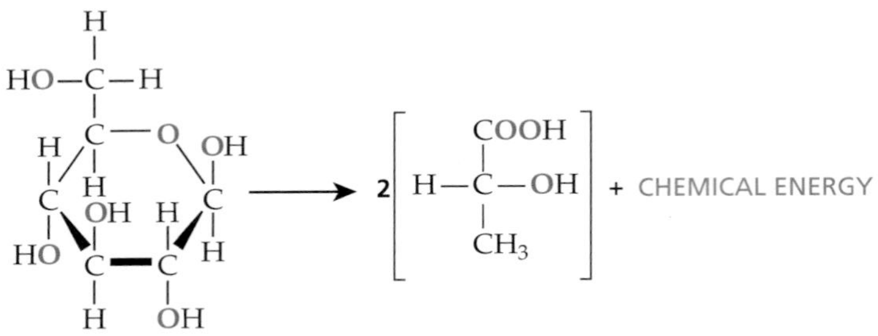

FIGURE 6.26 Lactic acid fermentation is an anaerobic reaction. During lactic acid fermentation, a glucose molecule is broken down into two lactic acid molecules. No oxygen is required for this reaction, so it is classified as anaerobic. The formation of lactic acid generates chemical energy that is stored in the form of ATP. However, the energy output from lactic acid fermentation is lower than that from cellular respiration.

Glucose
$C_6H_{12}O_6$

$2\left[\begin{array}{c} COOH \\ | \\ H-C-OH \\ | \\ CH_3 \end{array} \right]$ + CHEMICAL ENERGY

Lactic acid
$C_3H_6O_3$

aerobic A chemical reaction that uses oxygen.

The amount of ATP available from lactic acid fermentation can sustain you for roughly 20 seconds. After this time, you start breathing heavily—why? This is a reflexive response to your body's demand for more oxygen for cellular respiration, which is an **aerobic** chemical reaction. Cellular respiration is more efficient than lactic acid metabolism because it generates more ATP molecules. However, it is also much slower. As a result, your speed of climbing the stairs has to slow down after you have expended the initial surge of energy. If you are physically fit, you will be able to continue climbing the stairs based on your body's efficient delivery of oxygen from your lungs to your muscles. However, if you do not exercise regularly, then you will probably run out of energy by the fifth floor.

The need to utilize biochemical reactions efficiently explains why elite sprinters and long-distance runners have different body types. As illustrated in Figure 6.27, sprinters are muscular because they need to generate a quick burst of energy to leave the starting block, and they have to sustain their high-energy expenditure for less than 10 seconds (for a 100-meter sprint). Their large muscle mass enables them to store a large reservoir of ATP and to generate ATP anaerobically via lactic acid metabolism. By contrast, middle-distance runners use cellular respiration as their primary source of ATP. In this case, it makes sense to have a lower body mass because the body has to be moved over a longer distance. Marathon runners use yet another type of chemical reaction. After an extended period of exercise, a runner's reservoir of carbohydrates becomes depleted and additional ATP must be obtained from a chemical reaction using fats, which is called *fatty acid metabolism*.

Core Concepts

- The first burst of energy during exercise comes from stored ATP in muscle cells.
- A rapid source of ATP is obtained from lactic acid fermentation, which breaks glucose into two lactic acid molecules. This reaction is **anaerobic** because it does not use oxygen.
- Sustained exercise requires the body to generate ATP by cellular respiration, an **aerobic** reaction that uses oxygen. Cellular respiration is slower reaction that generates more ATP molecules than lactic acid fermentation.

FIGURE 6.27 Sprinters need a rapid supply of ATP. Among his many victories, Usain Bolt won the 100-meter and 200-meter sprints at the three consecutive Olympic Games in Beijing, London, and Rio de Janeiro. Sprinters have well-developed muscles that allow them to use stored ATP and anaerobic metabolism to generate a quick burst of energy for 10–20 seconds.

6.1 What Happens When You "Burn Carbs" at the Gym?

Learning Objective:

Relate a combustion reaction to "burning" carbs during exercise.

- When you "burn carbs" at the gym, your body uses a **chemical reaction** to generate energy from sugar molecules.

6.2 Chemical Reactions Produce Changes in Matter

Learning Objective:

Describe chemical reactions using balanced chemical equations.

- A chemical reaction changes **reactants** into **products**. In a combustion reaction, the atoms in the reactant molecules are rearranged to form the product molecules.

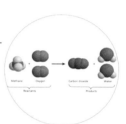

- Mass is conserved during a chemical reaction, which means that atoms are neither created nor destroyed.

- A chemical equation uses chemical formulas to describe a reaction.

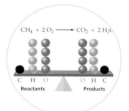

- A chemical reaction must be balanced—the number and type of atoms must be the same in the reactants and the products.

6.3   Measuring Matter: Mass and Moles

Learning Objective:

Use the mole as a unit of measurement for chemical quantities.

- ⊙ The **mol** is a unit that relates the measurable mass of a chemical sample to the number of invisible particles that it contains.

- ⊙ One mole is the **chemical amount** of substance that contains Avogadro's number of atoms, molecules, ions, or other particles.

- ⊙ **Avogadro's number** ($N_A = 6.022 \times 10^{23}$) is the number of atoms in exactly 12.00 g of the carbon-12 isotope.

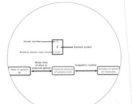

- ⊙ The **molar mass** (g/mol) of an atom is the mass of 1 mol of atoms. It is equal to the relative atomic mass provided in the periodic table.

- ⊙ The molar mass (g/mol) of a molecule is the sum of the molar masses of its constituent atoms.

- ⊙ It is possible to numerically convert between mass (g), chemical amounts (mol), and numbers of atoms or molecules.

- ⊙ The combustion reaction between glucose and oxygen produces carbon dioxide and water.

6.4   Chemical Reactions Produce Changes in Energy

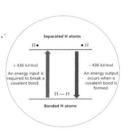

Learning Objective:

Relate the energy produced by a reaction to the chemical bonds in the reactants and products.

- ⊙ The **bond energy** is the amount of energy required to break 1 mole of chemical bonds. The units are kJ/mol.

- ⊙ Breaking a covalent bond requires an input of energy, and forming a covalent bond provides an output of energy.

- ⊙ Methane combustion is an **exothermic** reaction because it releases heat energy into the external surroundings.

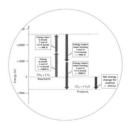

- ⊙ In an exothermic reaction, the total energy output obtained by forming the covalent bonds in the products is greater than the total energy input required to break the covalent bonds in the reactants.

⊙ An **endothermic** reaction absorbs heat energy. This type of reaction requires a greater energy input than its energy output.

⊙ The combustion reaction between glucose and oxygen releases heat energy.

6.5 Biochemical Reactions: The Basis for Life

Learning Objective:

Explain how cells use glucose to generate energy.

⊙ "Burning carbs" in the human body involves **cellular respiration**. This reaction combines glucose and oxygen to produce carbon dioxide and water. It also generates chemical energy that is stored in the form of **adenosine triphosphate (ATP)**.

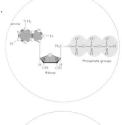

⊙ The first burst of energy during exercise comes from stored ATP in muscle cells. Further ATP production relies on lactic acid metabolism, cellular respiration, and fatty acid metabolism.

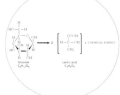

LEARNING RESOURCES

Reviewing Knowledge

6.1: What Happens When You "Burn Carbs" at the Gym?

1. What does the abbreviation "carb" mean? Why is this term used to refer to sugars?
2. Provide a definition of combustion.
3. What types of changes are produced by a chemical reaction?

6.2: Chemical Reactions Produce Changes in Matter

4. Write a balanced chemical equation for the combustion of methane. What are the reactants? What are the products?
5. What is the stoichiometry of a chemical reaction?
6. Explain the relationship between a balanced equation and the conservation of mass.
7. Write a balanced chemical equation for the combustion of butane, C_4H_{10}.

6.3: Measuring Matter: Mass and Moles

8. Explain why chemists use the mole as the unit to measure chemical quantities.
9. Define the molar mass of an atom. How is molar mass obtained from the periodic table?
10. Which method is used to calculate the molar mass of a molecule?

6.4: Chemical Reactions Produce Changes in Energy

11. Write the molecular structures of all of the reactants and products in the combustion of methane. What types of bonds exist in the products that are not present in the reactants?
12. Methane combustion is exothermic. What does this term mean?
13. Define the bond energy of a covalent bond.
14. Based on Table 6.3, which has the larger bond energy: a C—H bond or an O—H bond?

6.5: Biochemical Reactions: The Basis for Life

15. Write a balanced chemical equation for cellular respiration. In what way(s) is this reaction similar to glucose combustion? In what way(s) is it different?
16. Which molecule is used as the chemical energy source in cells? Write its abbreviation and full name. What are the chemical components of this molecule?
17. How does ATP release chemical energy? Which molecule is formed during this process? How can ATP be regenerated?
18. Write the chemical equation for lactic acid fermentation. Why is this reaction anaerobic?
19. How are the large muscles of elite sprinters related to how they generate short bursts of energy?

Developing Skills

20. When iron metal is exposed to the air for an extended period of time, the metal can form a red-orange substance called *rust*. This process occurs because iron (Fe) reacts with oxygen in the air to form an iron oxide with the chemical formula Fe_2O_3. The unbalanced equation for this reaction is given below. Add coefficients to balance the equation.

$$Fe + O_2 \rightarrow Fe_2O_3$$

21. Yeast is a single-celled organism that can utilize glucose as an energy source in the absence of oxygen. This chemical reaction is called fermentation, and it produces ethanol (C_2H_5OH) and carbon dioxide. We can write the unbalanced chemical reaction for fermentation as:

$$C_6H_{12}O_6 \rightarrow C_2H_5OH + CO_2$$

Deduce the coefficients required to balance the equation. Use an atom inventory to check that your final reaction is properly balanced.

22. Sodium azide (NaN_3) is a powder used to inflate the air bags in cars. An electrical signal triggered by a collision ignites the powder, which produces sodium metal and N_2 gas in a very rapid reaction. The release of nitrogen gas then fills the air bag. Write a balanced equation for this reaction.

23. Consult Figure 6.9, which shows 1 mol of various elements. (a) How many grams of each substance does the figure depict? (b) Comment on the relationship between the mass and the volume of these samples.

24. Which sample has the larger number of atoms, 1 mol of sodium or 1 mol of potassium?

25. Which sample has the larger number of atoms, 20 g of gold or 20 g of silver?

26. Which sample has the larger number of molecules, 50 g of water or 50 g of oxygen?

27. What is the molar mass of:
    (a) C_2H_2
    (b) O_3
    (c) $CaCO_3$?

28. How many molecules of CH_4 are in 98.0 g of CH_4?

29. If you have 1.0 g of KCl, how many formula units do you have?

30. Table sugar is composed of sucrose, which has the chemical formula $C_{12}H_{22}O_{11}$. One cup of table sugar has a mass of approximately 200 g. (a) What is the chemical amount of sucrose in one cup, expressed in units of moles? (b) How many molecules of sucrose are in the cup?

31. Mercury (Hg) is the only metal that exists as a liquid at room temperature. The density of mercury is very high, with a value of 13.5 g/cm^3. What volume of mercury contains 1 mole of Hg atoms?

32. A "standard" glass of water contains 8 fluid ounces. For reference, 1 fluid ounce = 30.0 milliliters (mL), and the density of water = 1.00 g/cm^3. (a) If you drink the entire glass, how many H_2O molecules are you consuming? (a) Express the number of molecules in words using million, billion, and/or trillion.

33. (a) Balance the following chemical equation:
    $$C_6H_{12} + O_2 \rightarrow CO_2 + H_2O$$
    (b) If you start with 25.0 g of C_6H_{12}, how many grams of CO_2 product will be produced?

34. Acetylene (C_2H_2) contains a triple covalent bond between its two carbon atoms. Acetylene reacts with oxygen to produce a very hot flame—this is the chemical basis of an oxyacetylene torch. (a) Calculate the energy change of the combustion reaction. (b) Explain how your answer to part (a) shows that the combustion reaction is exothermic.

35. Suppose that you have written a chemical reaction that is not correctly balanced. Will this affect your ability to estimate the energy change of the reaction? Explain your answer.

36. Use the balanced equation for the combustion of glucose to calculate the energy change for this reaction. You can obtain the bond energies from Table 6.3. (a) Calculate the energy input required to break all of the bonds in the reactants. (b) Calculate the energy

released when forming all of the bonds in the products. (c) Calculate the energy change for the reaction, and confirm that the reaction is exothermic (d) Compare your answer in part (c) to the experimentally measured energy change of 2805 kJ/mol. Did your calculation overestimate or underestimate the amount of energy released by the reaction?

Exploring Concepts

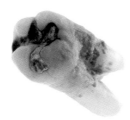

37. Tooth decay is caused by bacteria that feed on sugars in our mouth. Use the Internet to discover what types of reaction are performed by these bacteria and which chemical substances are responsible for tooth decay.

38. When oxygen is plentiful, normal cells use cellular respiration as an energy-producing reaction because it generates the largest number of ATP molecules. However, cancer cells often switch their metabolism from cellular respiration (an aerobic reaction) to lactic acid fermentation (an anaerobic reaction). This phenomenon is called the *Warburg effect,* after its discoverer, Otto Warburg (1883–1970). At first glance, this switch doesn't make sense because lactic acid fermentation produces fewer ATP molecules. Use the Internet to explore the Warburg effect, and write a one- to two-paragraph explanation of why cancer cells behave in this manner.

39. Burning any type of fossil fuel emits carbon dioxide. There is now global concern that the increase in atmospheric CO_2 is having significant effects on the Earth's climate. Most of us contribute to CO_2 emissions by driving some type of motor vehicle. Calculate the mass of CO_2 (in kilograms) that is emitted by burning 1 gallon of gasoline. We will assume that gasoline is composed of octane (C_8H_8). For reference, 1 gallon = 3.79 liters, 1 liter = 1000 cm^3, and the density of octane = 0.704 g/cm^3.

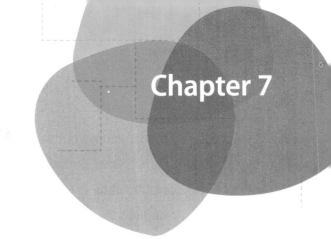

Chapter 7

Monomers and Polymers

CONCEPTS AND LEARNING OBJECTIVES

FEATURES

I n discussions about our diet, carbohydrates are classified as "simple carbs" or "complex carbs." What do these terms mean? What varieties of food contain each type of carbohydrate? Is one type of carbohydrate better for our health than the other? This chapter examines the chemistry of carbohydrates and how our bodies use them.

7.1   What Is the Difference Between Simple and Complex Carbs?

Learning Objective:

Distinguish between simple and complex carbohydrates.

THE KEY IDEA: Simple carbs are small sugars, whereas complex carbs are large sugars.

▪ **simple carbohydrate** A carbohydrate that consists of only one or two sugar molecules.

▪ **complex carbohydrate** A carbohydrate that contains hundreds or thousands of sugar units.

▪ **monomer** A single molecule that can be linked together many times to form a large polymer.

▪ **polymer** A large molecule that is constructed from many copies of a smaller monomer.

As we learned in the previous chapter, carbohydrates (often called "carbs") are sugar molecules with the general formula $C_x(H_2O)_y$. We commonly consume many types of carbohydrates in our diet. If you put table sugar into your coffee, then you are tasting *sucrose*. The sweetness of fruits is due to *fructose*. Milk from cows or goats contains *lactose*. Finally, potatoes, grains, rice, and pasta contain large amounts of *starch*.

Dietary carbohydrates are classified into two types—**simple carbohydrates** and **complex carbohydrates**. The distinction is based on the size of the sugar molecule. Glucose is a simple carbohydrate because it is a small, six-carbon sugar molecule. Sucrose is also considered a simple carbohydrate because it contains only two sugar molecules, glucose and fructose, that are joined by a chemical linkage (Figure 7.1). Lactose also contains two sugar molecules. In general, *a simple carbohydrate is small and consists of one or two sugar molecules.*

In contrast, *a complex carbohydrate is large and contains hundreds or thousands of sugar molecules.* For example, starch is composed of hundreds of glucose molecules that are linked together in long chains (Figure 7.2). Starch is an example of how a very large molecule can be constructed by repeatedly joining multiple copies of a small molecule. In chemical terms, we classify glucose as a **monomer** ("one part") and starch as a **polymer** ("many parts"). When we eat foods containing starch, our digestive system breaks down the polymer into

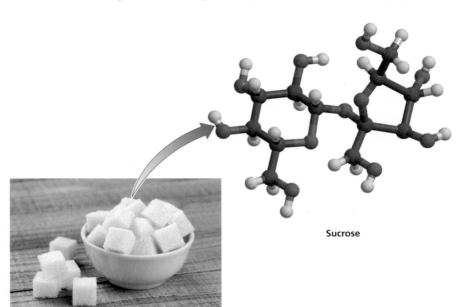

FIGURE 7.1 Table sugar is a simple carbohydrate. Sucrose (table sugar) is a small molecule made by linking two sugars, glucose and fructose. The molecule is shown as a ball-and-stick model with color-coded atoms (carbon is black, oxygen is red, and hydrogen is white).

Sucrose

glucose monomers, which are then used for cellular respiration (as discussed in Chapter 6).

In this chapter, we study the chemical principles of monomers and polymers. We begin with synthetic polymers that chemists have created in the laboratory because their simpler structures allow us to illustrate key points. We then apply these principles to the chemistry of simple and complex carbohydrates. We will learn why it is better to obtain our dietary sugar from whole grains rather than from processed foods. Although this chapter focuses on carbohydrates, there are other types of biological polymers—DNA, RNA, and proteins—that have essential roles within cells. These polymers will be examined in later chapters.

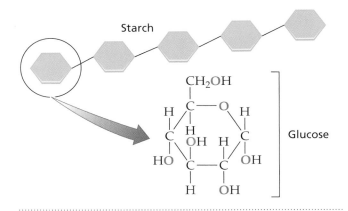

FIGURE 7.2 Starch is a complex carbohydrate. Starch is a polymer composed of many copies of a glucose monomer.

Core Concepts

- A **simple carbohydrate** is small and consists of one or two sugar molecules. A **complex carbohydrate** is large and contains hundreds or thousands of sugar molecules

- Starch is a **polymer** ("many parts") made from linking multiple copies of a glucose **monomer** ("one part").

Synthetic Polymers

Synthetic polymers are materials made by chemists and not found in nature. This section examines how these polymers are made. Our lives are touched by an enormous variety of synthetic polymers. Familiar examples include polyethylene, polystyrene, and polyester. Other polymers are sold using brand names such as Nylon, Spandex, Teflon, and Kevlar. The first synthetic polymer—Bakelite—was developed in New York in 1907 and named after its inventor, Leo Baekeland (1863–1944). Bakelite is a **plastic**, which is a term used to describe solid substances that can be molded into different shapes. Today, over a century later, the plastics industry has become the third largest manufacturing industry in the U.S. economy. In addition to their many everyday uses, plastics play an essential role in modern medicine, from surgical gloves to replacement knees and hips.

Monomers and Polymers

We can visualize a polymer as a chemical version of a metal chain—in fact, chemists often speak of *polymer chains*. As shown in Figure 7.3, a chain is constructed by joining multiple copies of a single link, which functions as the monomer. This construction makes the chain strong, yet flexible; it can be coiled into a loop or pulled taut in a straight line. A metal chain contains only one type of link. Similarly, some polymers contain only one type of monomer, and we begin with these materials.

The process of making a polymer, called **polymerization**, involves repetitively forming new covalent bonds to link the monomers. Creating these bonds modifies the chemical structure of the monomer. For example, some of the electrons that formed a covalent bond *within* the monomer must now be used to make a new bond that *joins* two monomers. We use

Learning Objective:

Illustrate examples of monomers and polymers.

plastic A solid substance that can be molded into different shapes.

polymerization The chemical process whereby multiple monomers are linked to form a polymer.

🔑 THE KEY IDEA: A large number of monomers can be chemically linked to produce a polymer.

Single link
(monomer)

Chain
(polymer)

FIGURE 7.3 A polymer chain. A chain (the polymer) is constructed by joining multiple copies of a single link (the monomer).

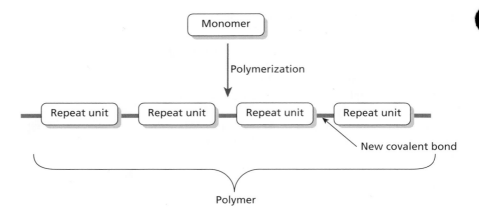

FIGURE 7.4 Monomer, polymer, and repeat unit. A polymer is formed by linking multiple copies of a monomer using new covalent bonds. This process generates a repeating arrangement of atoms called a repeat unit. A polymer can have many thousands of repeat units.

■ **repeat unit** The collection of atoms derived from the monomer that is replicated many times within the molecular structure of the polymer.

the term **repeat unit** to describe the modified collection of atoms within the polymer that is derived from the monomer. As the name suggests, the repeat unit is replicated many times to form the chemical structure of the polymer (Figure 7.4). In this manner, a chemical polymer differs from a metal chain in which the structure of the links is not altered when they are joined.

Core Concepts

- A polymer is similar to a metal chain, which is constructed by joining multiple links.
- **Polymerization** is the process by which many copies of a chemical unit (the monomer) are linked together to form a long chain-like molecule (the polymer).
- The **repeat unit** in the polymer has a different chemical formula than the monomer because new covalent bonds are formed when joining the monomers to make the polymer.

◉⚓ THE KEY IDEA: Polyethylene is formed by repeatedly linking molecules of ethylene.

■ **polyethylene** A polymer constructed from multiple copies of the monomer ethylene.

■ **addition polymerization** A chemical process in which monomers are "added together" to form a polymer with no addition or removal of atoms.

Polyethylene

As the name suggests, **polyethylene** is constructed from multiple copies of the monomer ethylene (C_2H_4); this is the older name for the molecule that chemists now call "ethene" (see Chapter 4). Ethylene contains two carbon atoms joined by a C=C double bond. Under specific reaction conditions, ethylene molecules can be joined to form a polymer (Figure 7.5(a)). This process is called **addition polymerization** because the monomers are "added together" with no addition or removal of atoms. During this addition, one pair of electrons in the double bond of ethylene is used to create a new covalent bond to a neighboring monomer. As a result, *the C=C double bond in the ethylene monomer changes into a C—C single bond in the repeat unit.* Even though the bonding has changed, the repeat unit still contains the same number and types of atoms as the monomer (two carbon atoms and four hydrogen atoms). Figure 7.5(b) illustrates a molecular model of a small region of polyethylene.

Because polymers are so large, their chemical formulas are often written in terms of the repeat unit. The subscript n indicates that the collection of atoms is repeated n times, and the size of n is possibly in the thousands.

$$\left[\begin{array}{cc} H & H \\ | & | \\ -C & -C- \\ | & | \\ H & H \end{array}\right]_n$$

Repeat unit for polyethylene

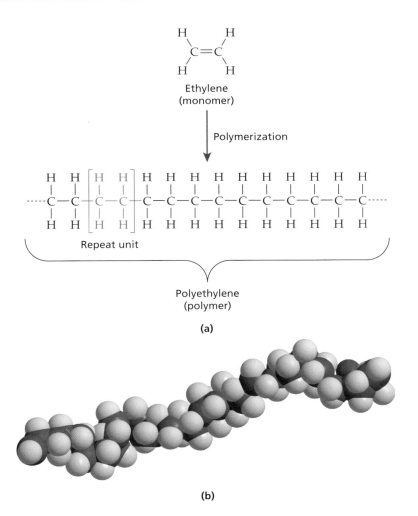

Ethylene
(monomer)

Polymerization

Repeat unit

Polyethylene
(polymer)

(a)

(b)

FIGURE 7.5 Making polyethylene.
(a) Polyethylene (a polymer) is made by joining multiple copies of ethylene (a monomer). The repeat unit in the polymer, highlighted in square brackets, contains a C—C single bond, whereas the monomer contains a C=C double bond. The dashed lines at either end indicate that the polymer extends beyond the length shown in the figure. (b) A short section of polyethylene represented as a molecular model. The carbon atoms are shown as black and the hydrogen atoms as white.

CONCEPT QUESTION 7.1

Your classmate, a business major, approaches you with a proposition for a new start-up company. The company will polymerize ethane *to create a new polymer called* polyethane. *Is this a good investment? Explain your answer using chemical principles.*

FIGURE 7.6 High-density and low-density polyethylene. Plastic containers are commonly labeled to indicate the type of polymer they contain. HDPE (high-density polyethylene) is used to make milk containers, shown on the left, whereas LDPE (low-density polyethylene) is used to make trash bags, shown on the right.

Polyethylene is just one of the many types of plastic that we use in our daily lives. To aid with recycling, plastics are commonly labeled with a code that identifies the type of polymer they contain. Figure 7.6 displays two of these labels that identify different types of polyethylene. HDPE stands for *high-density polyethylene*, a tough and rigid plastic that is used to make milk containers. LDPE stands for *low-density polyethylene*, a thinner and more flexible plastic that is used to make trash bags.

How can two types of polyethylene possess such different properties? The difference arises from the structure of the polymer chains that are used to make the plastic (Figure 7.7). High-density polyethylene is made from a **linear polymer** in which the ethylene monomers are attached end-to-end. An HDPE milk container contains many such polymer chains, which pack closely against one another along their sides. This structure creates a material that is very dense because many atoms can be crowded within a particular volume. In addition, HDPE is hard and tough because the position of its polymer chains cannot be moved except by exerting considerable force.

linear polymer A polymer in which the monomers are attached end-to-end.

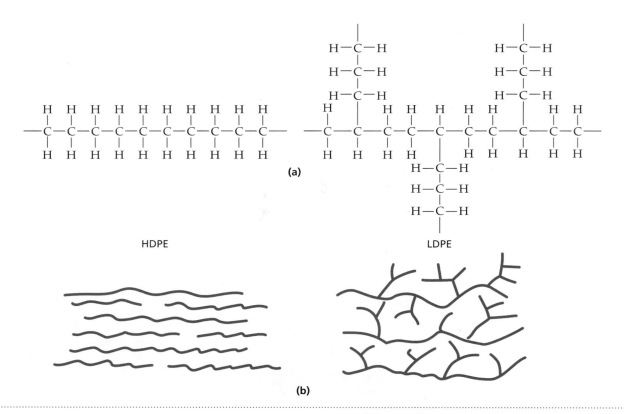

HDPE

LDPE

(a)

(b)

FIGURE 7.7 Linear and branched polyethylene. HDPE is constructed from linear polymers, which can pack closely together to form a dense material (i.e., more atoms within a particular volume). LDPE is constructed from branched polymers, which have a more irregular structure and cannot pack so closely to one another. Consequently, LDPE is less dense and more flexible than HDPE.

branched polymer A polymer in which some of the monomers branch sideways from the primary polymer chain.

It is also possible to attach ethylene molecules *sideways* onto the main polymer chain. This process is called *branching* because it is similar to how the branches of a tree extend outward from the trunk. The resulting molecule is called a **branched polymer**. The branches prevent the polymer chains from packing tightly together, which results in the formation of low-density polyethylene. The LDPE material made from these polymers is thin and flexible. Because the polymer chains are only loosely packed, little effort is required to shift their position. You can easily demonstrate this weaker association by stretching a piece of LDPE plastic in a garbage bag. A slight pull is sufficient to make the branched polymer molecules realign, and a stronger effort will cause the plastic to break. When this happens, you have *not* broken the polymer chains themselves, which are tightly linked by covalent chemical bonds. Instead, you have disrupted the interactions among the polymer molecules within the plastic.

CONCEPT QUESTION 7.2

Here is a summary of all the plastic recycling symbols:

PETE   HDPE   V   LDPE   PP   PS   Other

(a)   *Use information from this chapter and the Internet to find the full names for all of the polymers used to make these different types of plastics (e.g., what does "PP" mean?).*

(b)   *Perform a "plastic inventory" of your personal use of plastics during one week. Each time you use any type of plastic, examine the recycling label to see which type of polymer it contains. Do you use certain types of plastic more than others? Do you use all of the types of plastic listed in the summary?*

WORKED EXAMPLE 7.1

Question: Polystyrene is a polymer made from joining monomer units of styrene. It is commonly used in the form of Styrofoam to make coffee cups.

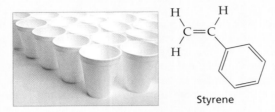

Styrene

(a) Draw a segment of polystyrene made from three styrene monomers.

(b) Use square brackets to highlight the repeat unit in polystyrene.

Answer: The chemical structure of the polymer is shown below, with the repeat unit highlighted in red. The dashed lines at either end indicate that a typical polymer has many more repeat units.

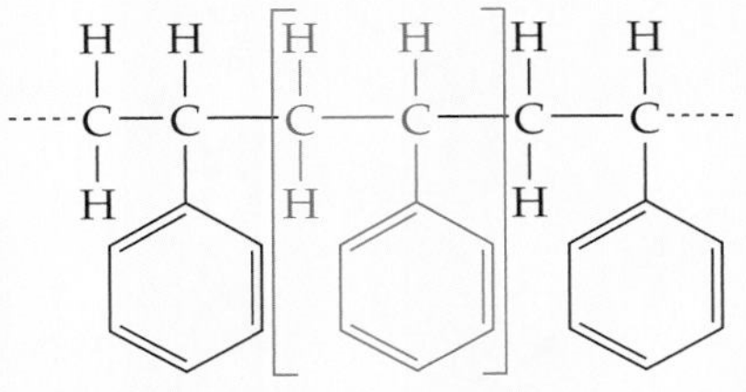

Repeat unit

Note the similarities between this example and the formation of polyethylene as shown in Figure 7.5. The monomer (styrene) contains a C=C double bond, but the repeat unit contains a C—C single bond.

TRY IT YOURSELF 7.1

Question: Polypropylene is a hard polymer that is used to make plastic water bottles. The structure of the propylene monomer is:

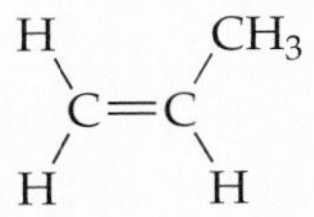

Draw a segment of polypropylene made from four monomers. Use square brackets to identify the repeat unit.

PRACTICE EXERCISE 7.1

Teflon is a heat-resistant polymer used for nonstick cookware. A region of Teflon polymer is shown below. Draw the structure of the *monomer* that is used to make Teflon.

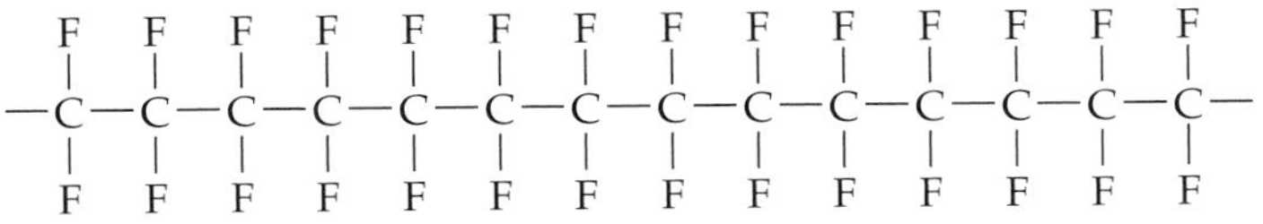

Teflon polymer (segment)

SCIENCE IN ACTION

Measuring the Sizes of Polymers

In the early 20th century, a major challenge for chemists was how to measure the size of polymers. Molecular size is a factor in a process called *sedimentation*. In general, *sedimentation is the motion of particles through a liquid because of the influence of forces that act on those particles*. We can illustrate sedimentation by considering wet sand in a bucket. This type of sample is called a *suspension* because solid grains of sand are suspended in liquid water. Over time, the sand particles will settle at the bottom of the bucket, leaving a layer of clear water at the top. This effect occurs because the *force of gravity* drags the sand particles downward. The gravitational force depends on the mass of the particle, so large particles will move downward more rapidly than small particles.

Because molecules are much smaller than sand particles, the influence of gravity on them is miniscule. Is it possible to amplify the force acting on molecules? This can be accomplished by spinning a sample in a circle, which subjects the molecules to a centrifugal force. You may have experienced this effect when riding on a roller coaster. When you accelerate around a sharp bend, your body feels the influence of the centrifugal force pushing outward. This effect is measured as the *g-force*, where 1 *g* corresponds to the force of normal gravity. During an intense roller coaster ride, you may feel a centrifugal force of 5 *g*.

A device that spins a sample is called a *centrifuge*. However, the earliest generation of centrifuges did not spin fast enough to study molecules. In the 1920s, a scientist named Theodor Svedberg (1884–1971) built a new type of machine called an *ultracentrifuge*. It was capable of spinning samples very rapidly, thereby generating a centrifugal force that was many thousand times greater than gravity. Figure 1(a) shows Svedberg in his ultracentrifuge room at the University of Uppsala in Sweden. Figure 1(b) illustrates the operation of the ultracentrifuge. The samples to be analyzed are contained in tubes, which are then placed in a rotor that is quickly spun.

Figure 2 shows what happens over time to a sample of molecules in an ultracentrifuge. Initially, molecules of different sizes are evenly mixed together. The spinning rotor subjects the molecules to a very large centrifugal force—a typical value is 10,000 *g*—that causes the molecules to move via sedimentation. The larger molecules move more quickly through the liquid and are the first to reach the bottom of the tube. The smaller molecules move more slowly and remain near the middle of the tube. Scientists can determine the size of a molecule by measuring the rate at which it moves within the tube.

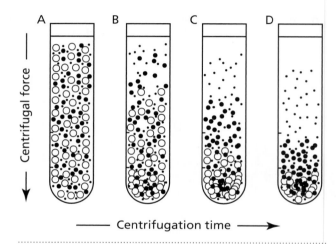

FIGURE 2 An ultracentrifuge separates molecules according to their sizes. (A) At the beginning of the experiment, molecules of different sizes are evenly mixed. As the centrifugation time increases (B and C), the larger molecules move more rapidly toward the bottom of the tube. After an extended time (D), the large molecules have all collected at the bottom of the tube, whereas the smaller molecules remain in the center.

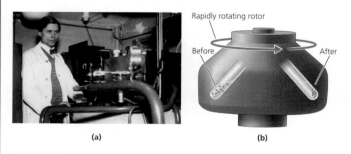

(a)   (b)

FIGURE 1 Operation of an ultracentrifuge. (a) Theodor Svedberg invented the ultracentrifuge. (b) An ultracentrifuge uses a rotor to spin samples very rapidly, subjecting them to forces that are many thousand times greater than gravity.

The ultracentrifuge has a special importance for the study of biological molecules. In the 1920s, many scientists were skeptical about the existence of large biological polymers. Instead, they believed that proteins were based on a loose association of small molecules, called a *colloid*. Svedberg used the ultracentrifuge to measure the rates of movement of molecules through a solution, which enabled him to determine the size of the molecules. The results were astonishing—proteins were much larger than any other molecule that chemists had previously measured. In 1926, Svedberg was awarded the Nobel Prize in Chemistry for his invention of the ultracentrifuge and his measurements of molecular sizes. Ultracentrifuges are still routinely used in scientific and medical laboratories.

Core Concepts

- **Polyethylene** is a polymer made by joining multiple copies of ethylene (the monomer). The reaction is called **addition polymerization** because the monomers are "added together" with no addition or removal of atoms.
- Ethylene contains a C=C bond, whereas the repeat unit in polyethylene contains a C—C bond. One electron pair from the C=C creates new covalent bonds that join the repeat units within the polymer.
- High-density polyethylene (HDPE) is made from a **linear polymer**. These polymer chains pack tightly together to form a tough, rigid plastic. Low-density polyethylene (LDPE) is made from a **branched polymer**. These polymers cannot pack closely together, so they produce a thin, flexible plastic.

Polyesters and Polyamides

Thus far, all of the polymers that we have examined are constructed from a single monomer. It is possible to make other types of polymers using two different monomers—these molecules are called *copolymers*. Using more than one monomer increases the possibilities for chemical variety in the polymer.

Polyethylene is named for its monomer unit. By contrast, **polyesters** and **polyamides** are named for the types of chemical linkages that connect the repeat units. These polymers are widely used to make synthetic fabrics. We will focus on how the chemical linkages in synthetic polymers are formed between various types of monomers. In this discussion, we refer back to some of the functional groups that you studied in Chapter 5.

Consider the two different monomer units (A and B) depicted in Figure 7.8(a). Monomer A contains two alcohol groups, and monomer B contains two carboxylic acid groups. The circle and square represent two different carbon-containing molecules that are unreactive. The two monomers can be joined by

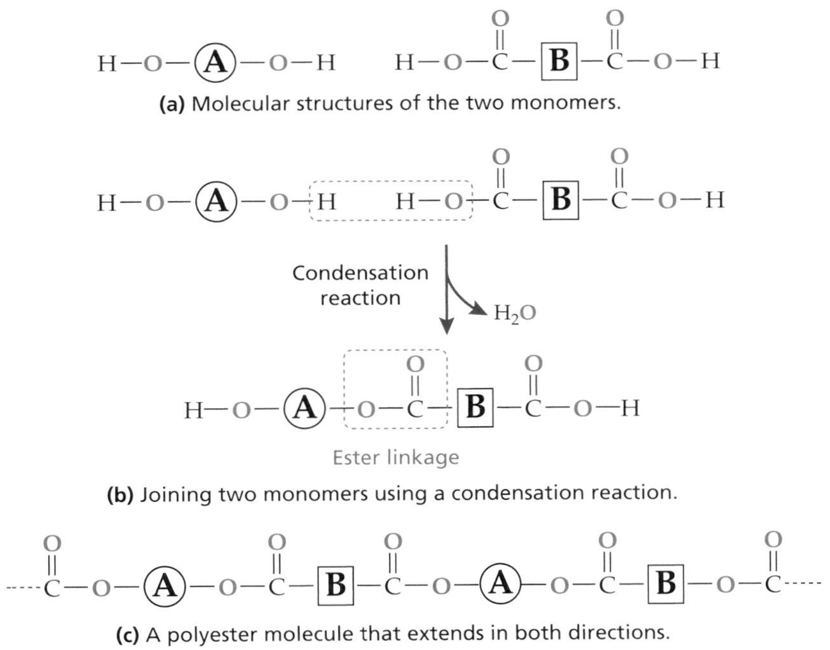

THE KEY IDEA: Polyesters and polyamides are formed by condensation reactions.

polyester A polymer in which the monomers are joined by ester linkages.

polyamide A polymer in which the monomers are joined by amide linkages.

(a) Molecular structures of the two monomers.

(b) Joining two monomers using a condensation reaction.

Ester linkage

(c) A polyester molecule that extends in both directions.

FIGURE 7.8 Making a polyester by condensation reactions. The two monomers (A and B) are joined by a condensation reaction, which removes an H_2O molecule and creates an ester linkage.

a chemical reaction between the —OH group of monomer A and the —COOH group of monomer B. As shown in Figure 7.8(b), the reaction removes three atoms to form a water molecule (H_2O). The oxygen atom in H_2O comes from the carboxylic acid, not the alcohol. This type of chemical reaction is called a **condensation reaction**. In general, *a condensation reaction joins two molecules to form a larger molecule while releasing a smaller molecule* (which is often H_2O). We indicate the release of H_2O by using a side arrow that curves away from the main reaction arrow.

condensation reaction A chemical reaction that joins two molecules to form a larger product while releasing a smaller molecule such as H_2O.

The remaining atoms join together to create a *new ester linkage* in the product molecule. Note that neither monomer A nor monomer B contains an ester group. This process is repeated many times to generate a long polyester molecule, as illustrated in Figure 7.8(c). Polyester is an example of a condensation polymer because it is formed via condensation reactions. This type of reaction is different from the addition reaction used to form polyethylene, in which no atoms are gained or lost.

We can employ a similar approach to study the formation of polyamides, as shown in Figure 7.9. Once again we use monomer B, but now we include a different monomer (C) that contains primary amine groups at either end of an unreactive core component. The two monomers can be joined by a condensation reaction involving the —COOH group of monomer B and the —NH$_2$ group of monomer C. After H_2O is removed, the remaining atoms become joined by a new amide linkage. Note again the absence of an amide in either monomer B or monomer C. Repeating this reaction generates a polyamide, which is another example of a condensation polymer.

(a) Molecular structures of the two monomers.

(b) Joining two monomers using a condensation reaction.

(c) A polyamide molecule that extends in both directions.

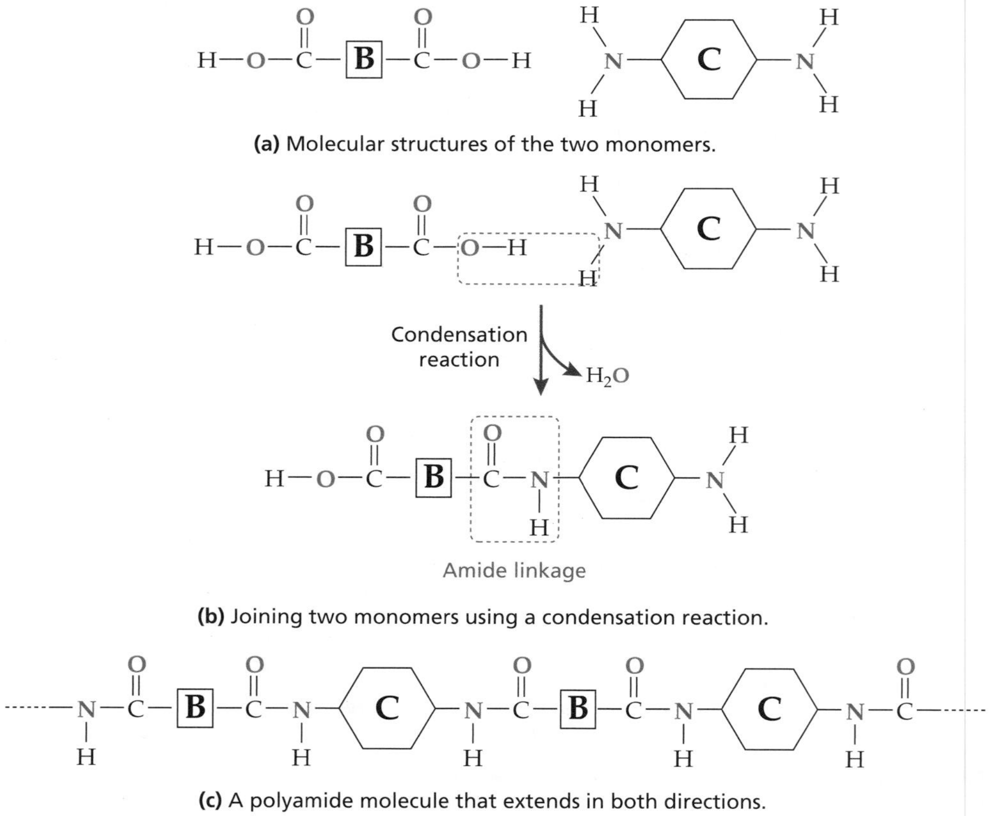

FIGURE 7.9 Making a polyamide by condensation reactions. Two monomers (B and C) are joined by a condensation reaction, which removes an H_2O molecule and creates an amide linkage.

One familiar polyamide is *nylon*, which was invented in 1935 by Wallace Carothers (1896–1937) at the DuPont Experimental Station in Delaware. The goal was to create a synthetic polymer to replace silk, a natural polymer made by silkworms that was costly and time consuming to produce. Nylon fabric was introduced to the public at the 1939 World's Fair in New York, and it immediately became a sensation. During World War II, when silk was scarce, nylon was used to make such diverse products as women's stockings and military parachutes. Today, we use nylon in many substances, including clothing, ropes, machine parts, and tennis racket strings.

The chemical reaction that produces nylon is illustrated in Figure 7.10(a). Two monomers are required, which we designate as X and Y. Monomer X has a carboxylic acid at both ends, and monomer Y has an amine at both ends. In each monomer, a short hydrocarbon chain lies between the two functional groups. The carboxylic acid in monomer X undergoes a condensation reaction with the amine group of monomer Y, releasing an H_2O molecule and creating a new amide linkage. There are still reactive functional groups at either end, so the polymerization reaction can continue repeatedly to join multiple monomers. Figure 7.10(b) shows the repeat unit of this nylon polymer, which is given the name *nylon 6,6*. This name reflects the fact that each monomer contains 6 carbon atoms, giving a total of 12 carbons in the repeat unit.

Many types of monomers can be linked via condensation reactions to create a multitude of **condensation polymer**s. A key point is that the monomer must have *a reactive functional group at each end*. Two reactive functional groups are necessary for the monomer to participate repeatedly in the condensation reactions that form the polymer. We can observe this principle in the formation of the condensation polymers illustrated in figures 7.8 through 7.10. If the monomer has only one reactive functional group, then the possible number of linking reactions is limited, and a long polymer chain cannot be created.

condensation polymer A polymer formed via condensation reactions.

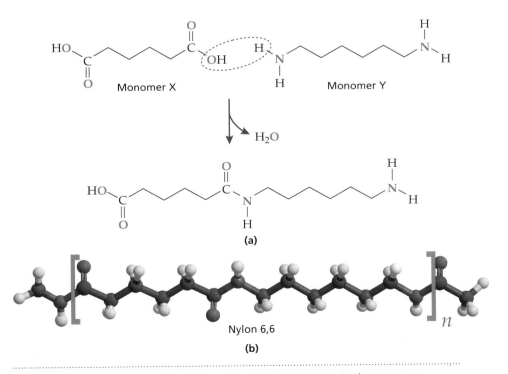

(a)

Nylon 6,6

(b)

FIGURE 7.10 Making nylon. (a) Nylon is formed by a condensation reaction between two monomers (X and Y), producing an amide linkage in the product. (b) The repeat unit of a nylon 6,6 polymer.

WORKED EXAMPLE 7.2

Question: *Kevlar* is an extremely strong polymer used for making bulletproof vests. It is constructed from two different types of monomers, which are shown below.

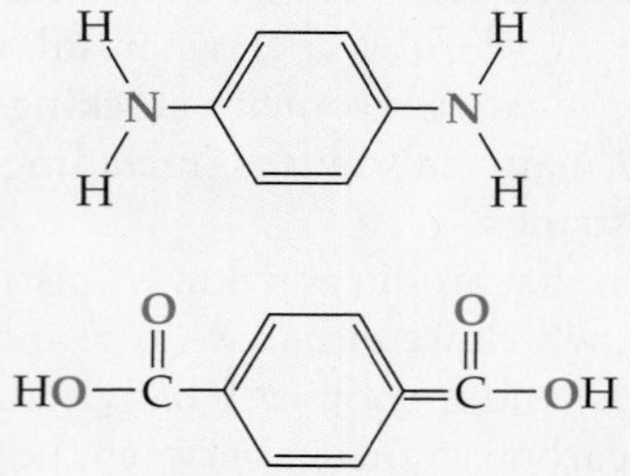

(a) Draw a diagram to show how these two monomers are joined by a condensation reaction (include the molecular structures of all the reactants and the products).

(b) What type of chemical linkage is produced? Highlight and label this linkage in your drawing.

(c) Add two more monomers to form a short region of a polymer chain.

Answer: (a) The condensation reaction between the two monomers is

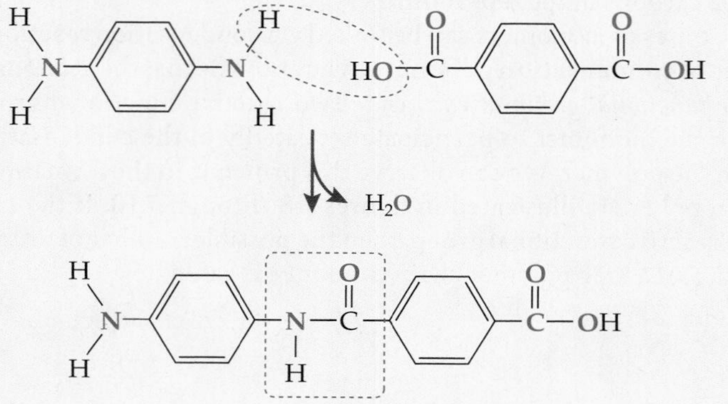

H_2O

Amide linkage

(b) The reaction creates an amide linkage in the product molecule, which is highlighted by the dashed box.

(c) Adding two additional monomers creates a short region of a polyamide. This process can be repeated multiple times to create a much longer polymer chain.

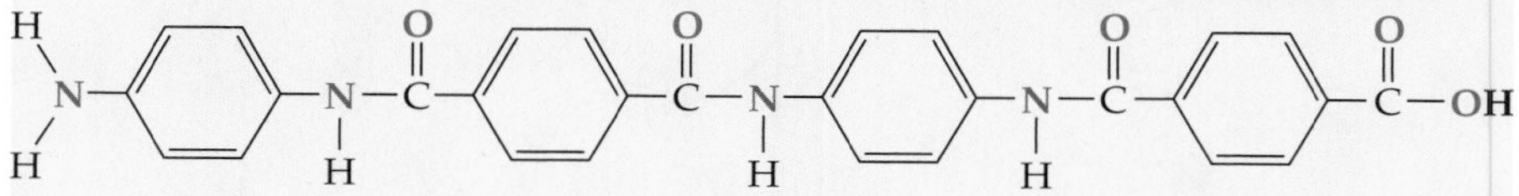

TRY IT YOURSELF 7.2

Question: PETE is the shorthand name for a polymer that is commonly used to create plastic bottles for soda beverages. Examples of these bottles and the recycling symbol for PETE appear below, along with the two monomer units that are used to make PETE.

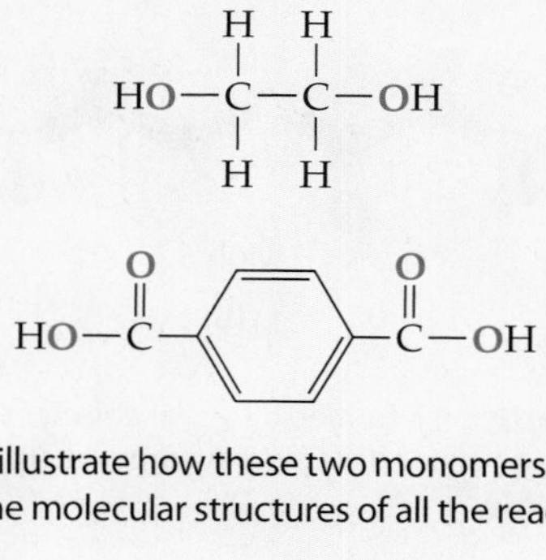

PETE

(a) Draw a diagram to illustrate how these two monomers are joined by a condensation reaction (include the molecular structures of all the reactants and products).

(b) What type of chemical linkage is produced by joining the monomers? Circle and label this linkage in your drawing.

(c) Add two more monomers to form a short region of a polymer chain.

WORKED EXAMPLE 7.3

Question: *Lactic acid* is a small biological molecule that can be used to make a condensation polymer called *polylactic acid* (PLA). Because PLA can be decomposed by composting, it is being investigated as a "green" alternative to conventional plastics. The molecular structure of lactic acid is shown here.

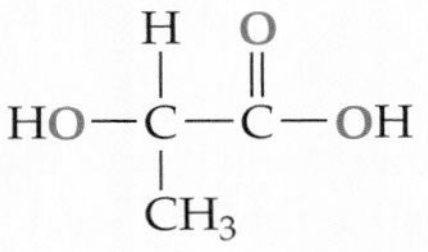

Lactic acid

(a) Which components of the lactic acid molecule enable it to form a condensation polymer? Identify and label them on the structure of PLA.

(b) Draw a diagram to show condensation reactions among *four* lactic acid monomers to create a short region of PLA.

(c) Which type of linkage is used to join the repeat units in PLA? Circle and label this linkage in your drawing.

Answer: (a) Lactic acid contains two reactive functional groups—an alcohol group and a carboxylic acid group (see the figure below). These reactive groups are located on opposite ends of the molecule. Therefore, it is possible for lactic acid to participate in repeated condensation reactions to form a condensation polymer.

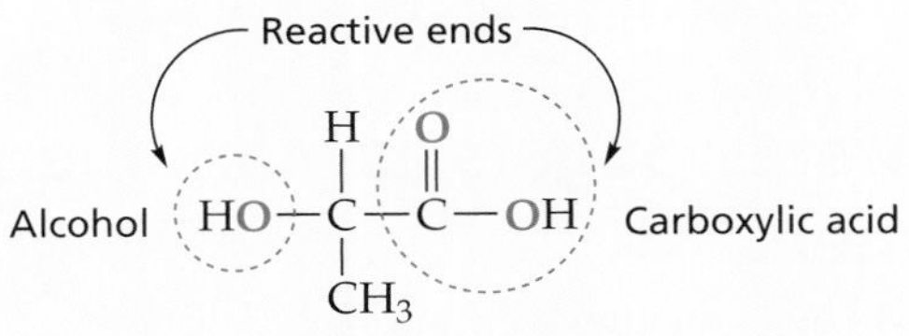

(b) The condensation reactions among four lactic acid monomers are illustrated below. Note how the alcohol group of one monomer reacts with the carboxylic acid group of an adjacent monomer.

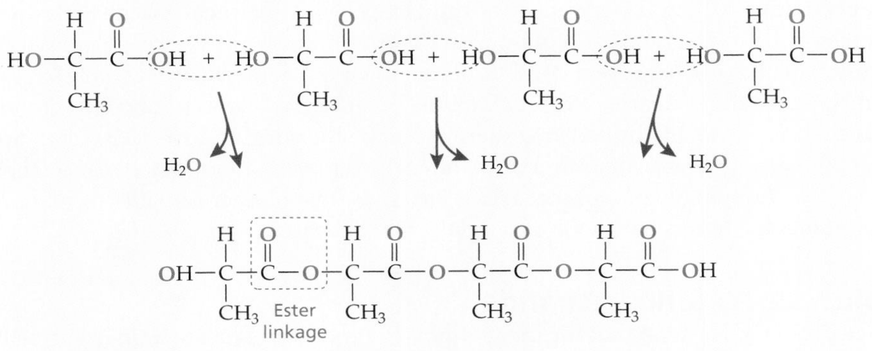

Polylactic acid

(c) In polylactic acid, the repeat units are joined by an ester linkage.

TRY IT YOURSELF 7.3

Question: The molecular structure of acetic acid is as shown here. Is it possible to use acetic acid as a monomer to make a condensation polymer called *polyacetic acid*? Explain your answer.

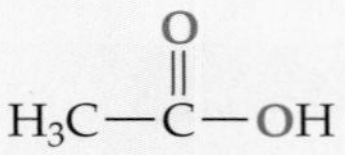

$$H_3C-\overset{\overset{\displaystyle O}{\|}}{C}-OH$$

Acetic acid

PRACTICE EXERCISE 7.2

Silicone is a water-resistant polymer made from a monomer unit (shown at right) that contains a silicon atom. The alcohol groups in the monomer can participate in a condensation reaction. Draw a diagram to show condensation reactions among four monomers to create a short region of silicone polymer.

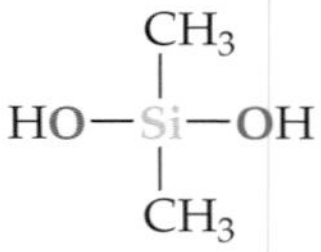

$$HO-\overset{\overset{\displaystyle CH_3}{|}}{\underset{\underset{\displaystyle CH_3}{|}}{Si}}-OH$$

Core Concepts

■ Two monomers containing reactive functional groups can be linked by a **condensation reaction**, which generates H_2O as a product.

■ A **polyester** is formed by condensation reactions between two monomers, one containing alcohol groups and the other containing carboxylic acid groups.

■ A **polyamide** is formed by condensation reactions between two monomers, one containing carboxylic acid groups and the other containing amine groups.

7.3 Carbohydrates: Sugars as Monomers and Polymers

Learning Objective:
Relate the principles of monomers and polymers to various sugars.

■ **saccharide** A chemical term for a sugar.

■ **monosaccharide** A sugar that consists of a single molecule.

■ **disaccharide** A sugar that consists of two linked monosaccharides.

■ **polysaccharide** A sugar made from many monosaccharides.

🔑 THE KEY IDEA: Glucose is the most important and abundant biological sugar.

We can now apply the principles of monomers and polymers to understanding carbohydrates. Sugar molecules can be described using various terms, which are summarized in Figure 7.11. In the chapter introduction, we saw that carbohydrates can be classified as simple or complex, depending on their size. These names are commonly used in connection to our diet. The chemical term for sugars is **saccharides**, which is derived from the Greek word for sugar. Saccharides are categorized by the number of monomers they contain. For example, glucose is a **monosaccharide** because it consists of one sugar molecule. Sucrose (table sugar) and lactose (milk sugar) are **disaccharides** because they each contain two monosaccharides that are linked together. Monosaccharides and disaccharides are simple carbohydrates. Starch and cellulose are **polysaccharides** because they are made from many monosaccharides, just like any polymer is made from many monomers. Polysaccharides are complex carbohydrates.

Glucose: A Monosaccharide

Recall from Chapter 6 that the most important and abundant sugar in living cells is glucose ($C_6H_{12}O_6$). Glucose can exist as different *isomers*—molecules with the same atoms but different structures (recall our discussion in Chapter 4). Two common isomers of glucose are *α-glucose* and *β-glucose* (α and β are the Greek

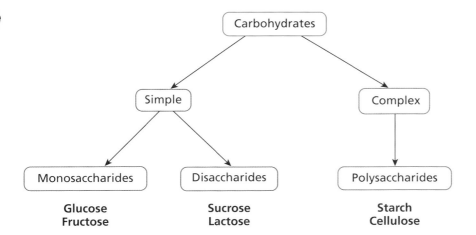

FIGURE 7.11 Describing sugars. Carbohydrates can be classified as simple or complex. Simple carbohydrates include monosaccharides (one sugar molecule) and disaccharides (made from two sugar molecules). Complex carbohydrates are polysaccharides that contain many monosaccharides.

letters *alpha* and *beta*). Figure 7.12 compares their structures using full-atom drawings and line-angle drawings. The carbons atoms are numbered 1 to 6 using the standard convention.

Both isomers of glucose contain a six-atom ring composed of five carbon atoms and one oxygen atom; the sixth carbon atom projects above the ring. Glucose also contains five —OH groups attached to specific carbon atoms. The structural difference between α-glucose and β-glucose is subtle but important. In α-glucose, the —OH group on C_1 points *below the ring* in the *opposite direction* to the sixth carbon atom (C_6). In β-glucose, the —OH group points *above the ring* in the *same direction* as C_6. The type of isomer affects how the glucose monomers are linked to form various sugar molecules.

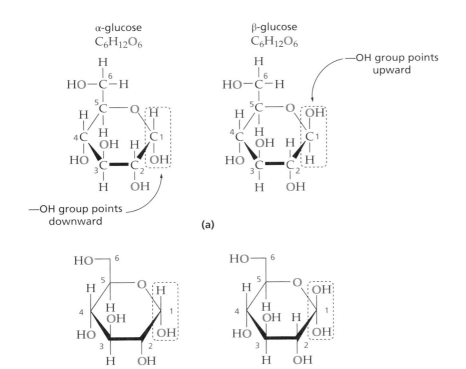

FIGURE 7.12 Two isomers of glucose. (a) The molecular structures of α-glucose and β-glucose are shown as full-atom structures. The carbon atoms are numbered 1–6 for identification. The two isomers differ in the orientation of the —OH group on C_1. (b) Simplified drawings of α-glucose and β-glucose that omit the carbon atoms. These simpler drawings enable us to distinguish the two isomers more clearly.

Core Concepts

- Glucose ($C_6H_{12}O_6$) is the most important and abundant biological sugar.
- Glucose can exist as two possible isomers: α-glucose and β-glucose. In α-glucose, the —OH group on C_1 points below the ring, whereas in β-glucose, it points above the ring,

🔑 THE KEY IDEA: Sucrose (table sugar) and lactose (found in milk) are disaccharides made by joining two monomer units.

sucrose The disaccharide commonly known as table sugar; it is made by joining α-glucose and β-fructose.

Sucrose and Lactose: Two Important Disaccharides

When we talk about "sugar" in our food, we are usually referring to *sucrose*. We obtain sucrose from plant sources, either sugarcane (which grows in warm climates) or sugar beets (which grow in cooler environments). Figure 7.13 shows a picture of sucrose alongside its molecular structure. **Sucrose** is a disaccharide because it is formed by joining two monosaccharides—*glucose* and *fructose*.

Fructose is often called "fruit sugar" because it is abundant in fruits such as grapes, pears, and apples. You may have heard of fructose in connection with *high fructose corn syrup*, which is added as a sweetener to many processed foods. Fructose has the same chemical formula as glucose ($C_6H_{12}O_6$), but it has a different molecular structure. Like glucose, fructose can exist as two isomers, *α-fructose* and *β-fructose*. The β-fructose isomer is used to make sucrose. Glucose and fructose are joined by a condensation reaction between an —OH group on α-glucose and another —OH group on β-fructose. The reaction removes a water molecule and creates a new C—O—C linkage between the sugars.

How do our bodies digest sucrose? These small sugars pass through the stomach and into the small intestine, which contains specialized enzymes that decompose sugars. As shown in Figure 7.14, sucrose is digested by an enzyme

α-glucose
$C_6H_{12}O_6$

β-fructose
$C_6H_{12}O_6$

Condensation reaction $\rightarrow$ H_2O

New linkage

Sucrose
$C_{22}H_{22}O_{11}$

FIGURE 7.13 The molecular structure of sucrose (table sugar). Sucrose is formed by a condensation reaction between α-glucose and β-fructose, which removes an H_2O molecule and creates a new chemical linkage.

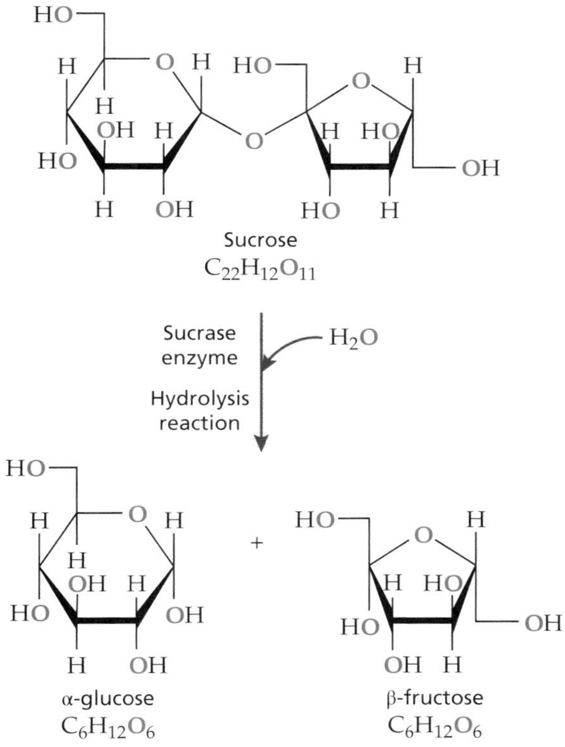

Sucrose
$C_{22}H_{12}O_{11}$

Sucrase enzyme

H_2O

Hydrolysis reaction

α-glucose
$C_6H_{12}O_6$

+

β-fructose
$C_6H_{12}O_6$

FIGURE 7.14 Digestion of sucrose. Sucrase (an enzyme) facilitates the digestion of sucrose (a disaccharide) into glucose and fructose (monosaccharides). The connecting bond in sucrose is broken by the addition of a water molecule (hydrolysis).

called *sucrase* (the suffix "–ase" refers to an enzyme). Sucrase catalyzes a reaction in which a water molecule breaks the chemical linkage in sucrose, producing glucose and fructose as monosaccharides. This reaction is called a **hydrolysis reaction** (or, simply, *hydrolysis*), where *hydro* refers to water and *lysis* means "to break." In essence, hydrolysis is the *reverse* of the condensation reaction that created the linkage. The glucose and fructose molecules pass through the walls of the small intestine into the bloodstream, which distributes them throughout the body.

A different disaccharide, **lactose**, is present in animal milk (Figure 7.15). Lactose is formed by linking β-glucose to a different sugar monomer called β-*galactose*. Almost everyone can digest sucrose, but the inability to digest lactose is very common—this condition is called *lactose intolerance*. After drinking milk, affected individuals often experience unpleasant symptoms, which include stomach cramps, bloating, gas, and diarrhea. According to the National Institutes of Health, approximately 65 percent of the world's adult population has lactose intolerance. However, other people can drink milk without suffering any of these adverse effects. What causes these different reactions to lactose? We answer this question in Chapter 14.

hydrolysis reaction A chemical reaction in which water is used to break the bond between two molecules.

lactose The disaccharide found in milk; it is made by joining *a*-glucose and β-galactose.

Core Concepts

- **Sucrose** (table sugar) is a disaccharide formed by joining two monosaccahrides—α-glucose and β-fructose.
- Sucrose is digested within the small intestine by an enzyme called sucrase. This enzyme catalyzes a **hydrolysis reaction** in which water is used to break the bond joining the two sugars, releasing glucose and fructose as products.
- **Lactose** (milk sugar) is a disaccharide formed by a condensation reaction between β-galactose and β-glucose.

FIGURE 7.15 **The molecular structure of lactose (milk sugar).** Lactose is formed by a condensation reaction between two sugars, β-glucose and β-galactose.

Starch: A Polysaccharide

THE KEY IDEA: Starch is a polysaccharide made from glucose monomers.

starch A polysaccharide composed of many *a*-glucose molecules, linked together as linear or branched chains.

Starch is an example of a sugar polymer—a polysaccharide. It is also classified as a complex carbohydrate. Starch functions as the energy reservoir of plant cells, and it serves as the primary source of nutrition for billions of people worldwide. Figure 7.16 shows some foods that contain starch, together with a photograph of pure starch.

Starch consists of two types of sugar polysaccharides—*amylose* and *amylopectin*. A typical sample of starch contains approximately 30% amylose and 70% amylopectin by mass. Both molecules are constructed from glucose monomers, but they have different structures (Figure 7.17). Amylose is composed of a linear glucose chain in which the number of monomers ranges from several hundred to a few thousand. Amylopectin is a larger molecule with a more complex branched structure. A single amylopectin molecule may have as many as 1 million glucose monomers. These linear and branched molecules in starch are comparable to the linear and branched versions of polyethylene that we examined in Section 7.2.

The assembly of both polymer types uses condensation reactions between —OH groups within the glucose molecules. In this case, we need to be more

(a)   (b)

FIGURE 7.16 **Starch is a major component of our diet.** (a) A sample of common foods that contain large amounts of starch—wheat, bread, pasta, potatoes, and rice. (b) Pure starch is a white powder.

specific about which —OH groups are involved. Consequently, we refer to the standard numbering of the carbon atoms in glucose, which was introduced in Figure 7.12.

The formation of starch polymers is based on the —OH groups within the glucose molecules. Consider two α-glucose molecules placed side by side (Figure 7.18). These molecules can be joined by a condensation reaction between an —OH on C_1 of one glucose and an —OH on C_4 of its neighbor. Removing an H_2O molecule creates a new connection between the glucose molecules. This connection is called an α*(1-4) linkage*; the name derives from the isomer of glucose and the numbering of the carbon atoms. The condensation reaction is repeated multiple times to generate an amylose polymer. Figure 7.19 displays a short region of amylose; note the presence of multiple α(1-4) linkages that join the glucose molecules. Amylose is a linear polymer, and it curls into a compact helical structure that contains six glucose molecules within each complete turn.

The helical structure of amylose is the basis for a chemical test for starch. This test uses a solution containing a mixture of iodine (I_2) and potassium iodide (KI), which has a light orange color. When we add a few drops of I_2/KI solution to a potato, which is rich

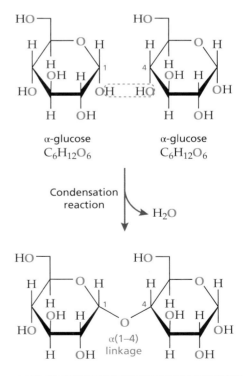

α-glucose
$C_6H_{12}O_6$

α-glucose
$C_6H_{12}O_6$

Condensation
reaction H_2O

α(1–4)
linkage

FIGURE 7.18 **Formation of an α(1-4) linkage between two glucose molecules.** A condensation reaction between —OH groups in neighboring glucose molecules creates an α(1-4) linkage. These linkages are used to make linear chains of glucose molecules in polysaccharides.

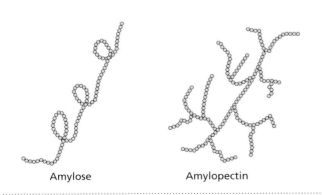

Amylose   Amylopectin

FIGURE 7.17 **Starch consists of amylose and amylopectin.** Amylose is a smaller linear polymer, whereas amylopectin is a larger polymer containing multiple branches. Each of the small circles represents a glucose monomer.

FIGURE 7.19 Molecular structure of amylose. (a) Amylose is a linear polymer of glucose monomers. (b) The amylose chain adopts a helical structure with six glucose molecules per turn.

FIGURE 7.20 A chemical test for starch. (a) A starch test using I_2/KI solution shows a blue-black color for a potato (right), which contains large amounts of starch, but no color change for an apple (left), which has little starch. (b) The blue-black color arises from the tri-iodide ion (I_3^-), which fits into the space in the middle of the amylose helix.

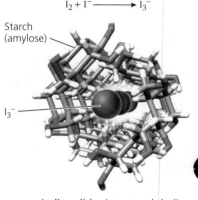

in starch, the potato turns a blue-black color (Figure 7.20(a)). However, if we add the same solution to an apple, which contains little starch, there is no color change. What is the explanation for this difference? A reaction takes place in the I_2/KI solution, between an iodine molecule (I_2) and an iodine ion (I^-) to form a tri-iodide ion (I_3^-).

$$I_2 + I^- \rightarrow I_3^-$$

The structure of the tri-iodide ion is the right size and shape to slip into the hollow center within the helical structure of amylose (Figure 7.20(b)). When this happens, the I_3^- ion generates the intense blue-black color that is characteristic of the starch test. In the absence of amylose, the I_2/KI solution does not change from its original color.

Because glucose contains five —OH groups, the monomers can be linked in other ways. Figure 7.21 illustrates a condensation reaction between an —OH on carbon-1 and an —OH on carbon-6, which creates an α(1-6) linkage. The structure of amylopectin uses both α(1-4) and α(1-6) linkages. The α(1-4) linkages are used to make linear glucose chains, like those found in amylose, while α(1-6) linkages are used to make branches that turn off from the main chain. An example of a branching structure is shown in Figure 7.22(a). Many such branches are connected to make the structure of amylopectin, which is illustrated in Figure 7.22(b).

The flexible branched structure of amylopectin has an important biological purpose—it provides quick and convenient access to stored carbohydrate energy. When our body utilizes these polysaccharides as an energy source, enzymes have to remove the glucose monomers from the *end of a chain*. The large number of branches in amylopectin generates many chain ends where the enzymes can access the stored glucose.

How does our body digest starch to release the glucose monomers as chemical fuel? The first step in this process is accomplished by an enzyme called *salivary amylase*. As the name suggests, this enzyme is found in the saliva in our mouth, and it acts on both the amylose and amylopectin components of starch. Figure 7.23 illustrates the action of salivary amylase on amylose. The polysaccharide chain is shown as curved to indicate the polymer's helical structure. Salivary amylase targets the sugar linkages in amylose, and it catalyzes a hydrolysis reaction that breaks these linkages. This reaction can be repeated again and again to cut a long amylose or amylopectin polymer into smaller sugar molecules. These fragments then travel to the small intestine, where they are broken down into glucose by other enzymes. The glucose can then be transported across the walls of the small intestine and into the bloodstream.

Core Concepts

- **Starch** is composed of two types of polysaccharides constructed from glucose monomers: amylose and amylopectin. Amylose is a smaller, linear molecule; amylopectin is a larger, branched molecule.
- In amylose, the glucose monomers are linked by α(1-4) linkages. These bonds create a linear molecule that curls into a helix with six glucose units per turn.
- In amylopectin, the glucose monomers are joined by α(1-4) linkages or α(1-6) linkages The α(1-4) linkages make linear glucose chains, whereas the α(1-6) linkages create branches.
- The branched structure of amylopectin is a rapidly accessible source of energy because it provides many chain ends where enzymes can remove glucose monomers.
- The first stage of starch digestion occurs in our mouth, where the salivary amylase enzyme cuts the polysaccharide into smaller sugar molecules.

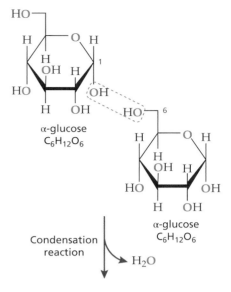

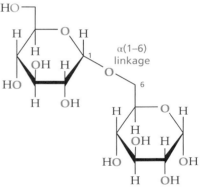

FIGURE 7.21 Formation of an α(1-6) linkage between two glucose molecules. These linkages are used to create branches in the structure of amylopectin and other polysaccharides.

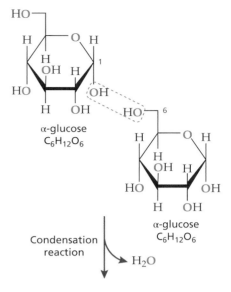

(a)

(b)

FIGURE 7.22 Molecular structure of amylopectin. (a) Amylopectin contains glucose molecules connected via α(1-4) and α(1-6) linkages. The α(1-4) linkages create linear glucose chains, whereas the α(1-6) linkages create branches from the main polymer chain. (b) The structure of the amylopectin polymer contains multiple branches. The end of each branch provides an access point for enzymes to remove glucose monomers, which can then be used for energy-generating reactions.

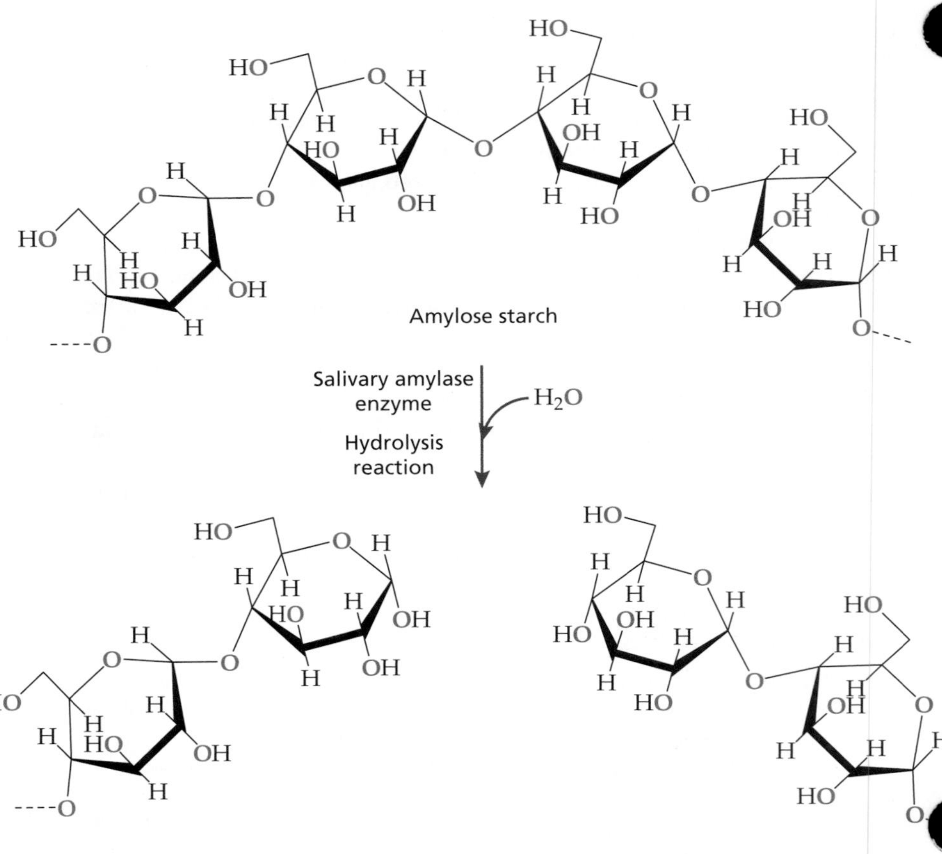

FIGURE 7.23 The first step in the digestion of starch. Salivary amylase, an enzyme, breaks down starch polymers into smaller sugars by catalyzing a hydrolysis reaction. The amylose polymer is shown as curved to indicate its helical structure.

Amylose starch

Salivary amylase enzyme

H_2O

Hydrolysis reaction

Blood Sugar and Glycemic Index

🔑 THE KEY IDEA: The glycemic index of a food indicates blood glucose levels after eating.

glycemic index   A number that indicates the effect of a food on a person's blood glucose level.

After we eat any carbohydrates, our digestive system breaks down the sugars into simple monomers. These monomers enter the bloodstream and then pass from the blood into the interior of cells. The most important monomer is glucose because it serves as the fuel for cellular respiration and the generation of chemical energy in the form of ATP.

To understand the relationship between diet and blood sugar, nutritionists have developed a measurement called the **glycemic index**. *The glycemic index of a food is a number that indicates the effect of the food on a person's blood glucose level.* As a reference, pure glucose has a glycemic index of 100, which defines the upper limit of the scale. The glycemic index can be applied to the sucrose in processed foods and also to the complex carbohydrates in our diet.

Figure 7.24 displays a graph that compares the change in blood glucose levels over time after a person eats a food with either a high or a low glycemic index. Food with a high glycemic index is digested rapidly, producing a rapid surge in the concentration of blood glucose after 30 minutes. After reaching a high peak value, the blood glucose quickly begins to fall. After only one hour, it is close to its initial baseline level. As more time passes, the cells absorb all of the available glucose, and the blood glucose level begins to dip below the baseline level. This is the cause of the "sugar crash" that we sometimes experience after eating sugary foods. The pattern is very different for a food with a low glycemic index, which is digested more slowly. As a result, the blood glucose level takes longer to

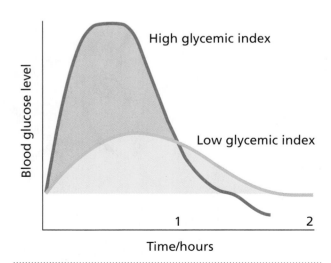

FIGURE 7.24 A graph of blood glucose levels over time for a food with a high glycemic index (red) and one with a low glycemic index (orange).

rise, the peak value is much smaller, and the blood sugar decreases gradually over time. These foods sustain our energy level for longer periods of time, and they do not lead to a sugar crash.

Figure 7.25 shows a collection of foods that have a high glycemic index. Table 7.1 provides examples of the glycemic index for a selection of foods. As a general rule, a "high glycemic index" refers to values that are 70 or above, and a "low glycemic index" refers to values equal to or lower than 55. When comparing various foods, the *total amount* of carbohydrate consumed must be the same in each case (50 g of carbohydrate is typically used).

What causes the large difference in glycemic index between bread made with flour and with whole grains? Flour is made from pulverized grains, such as wheat. This process creates tiny particles that can be easily accessed by enzymes that digest the starch within the grain. By contrast, whole grains still contain the hard outer layer—called the *bran*—that surrounds the starchy interior. This covering slows the ability of enzymes to access the starch, which results in slower digestion of the carbohydrate. The same principle explains the difference in glycemic index between brown rice (which retains the bran coating) and white rice, from which the coating is removed.

The glycemic index uses a standard amount of carbohydrate. In reality, the various foods we eat contain *different amounts* of carbohydrates. A meal with a large quantity of carbohydrates will generate more blood glucose than a low-carbohydrate meal will. It would be helpful to have a measurement of blood glucose that is more aligned with our real-world eating habits. This quantity is provided by the **glycemic load** of a food, which is calculated by the following equation:

FIGURE 7.25 Examples of foods with a high glycemic index.

glycemic load A measure of how a food affects blood glucose; it includes the glycemic index and the amount of carbohydrates consumed.

$$\text{glycemic load} = \frac{(\text{glycemic index}) \times (\text{grams of carbohydrates})}{100}$$

Let us interpret this equation. The glycemic index for a particular food remains constant; for example, white rice always has a value of 89 (see Table 7.1). However, the equation tells us that the glycemic load also depends on the *amount* that you eat, which is indicated by the grams of carbohydrates. For example, two cups of white rice have a glycemic load that is twice the amount of only one cup.

We can use white rice for an example of calculating glycemic load. One cup of cooked white rice contains 45 g of carbohydrates. Consequently,

$$\text{glycemic load (1 cup white rice)} = \frac{89 \times 45}{100} = 40$$

We can also compare the glycemic load of brown rice. One cup of cooked brown rice also contains 45 g of carbohydrates, but the glycemic index is 50. Therefore,

$$\text{glycemic load (1 cup brown rice)} = \frac{50 \times 45}{100} = 22.5$$

TABLE 7.1	The Glycemic Index of Selected Foods[1]
FOOD	**GLYCEMIC INDEX**
BREADS	
White flour bread	71
Whole wheat flour bread	71
100% whole grain bread	51
GRAINS	
White rice	89
Brown rice	50
Quinoa	53

[1]*Data provided by Harvard Medical School. The reported glycemic index is an average value calculated from several measurements of different food samples.*

To illustrate the impact of portion size on glycemic load, we can calculate the glycemic load of two cups of brown rice (a total of 90 g of carbohydrates) as:

$$\text{glycemic load (2 cups brown rice)} = \frac{50 \times 90}{100} = 45$$

Our calculation shows that eating two cups of brown rice will generate more blood glucose than eating one cup of white rice. When controlling blood sugar, it is therefore important to consider both the glycemic index and the total amount of carbohydrates. For one serving of food, a glycemic load greater than 20 is considered high, values from 11 to 19 are considered medium, and a value of 10 or less is considered low.

These dietary considerations are important because high levels of blood sugar can have detrimental effects on our health. If the amount of blood glucose exceeds the capacity of the cells to absorb it, then the glucose remains in the blood supply. High blood sugar is connected to serious health problems such as loss of vision, damage to blood vessels, and an increased risk for heart disease, stroke, and amputations. The medical term for an excess amount of sugar in the blood is *hyperglycemia* (pronounced as *hi-per-gly-SEE-me-uh*); *hyper* means "over," as in "hyperactive." *Chemistry and Your Health* explains how cells control the influx of sugar from the bloodstream into the cell's interior using a hormone called *insulin* and how this process is connected to diabetes.

PRACTICE EXERCISE 7.3

Cornflakes breakfast cereal has a glycemic index of 93. A serving size of cornflakes is 30 g, of which 26 g is carbohydrates. What is the glycemic load for a serving of cornflakes? Round your answer to the nearest whole number.

Core Concepts

- The **glycemic index** of a food measures the food's effect on a person's blood glucose level.
- Foods with a high glycemic index produce a rapid surge in blood glucose to reach a high peak level, followed by a quick drop to below the baseline. By contrast, foods with a low glycemic index cause a slower increase in blood glucose, a smaller peak value, and a gradual decline.
- High concentrations of blood sugar can lead to unhealthy consequences.
- The **glycemic load** is a measurement of blood glucose that includes the glycemic index and the amount of carbohydrates consumed.

Cellulose: A Polysaccharide We Cannot Digest

Cellulose is the most abundant natural polymer on Earth. It is used to construct the tough walls of plant cells, and it also provides a major component of tree bark. Like starch, cellulose is a polysaccharide that is constructed from glucose monomers. Unlike starch, however, our body cannot digest cellulose, and it passes through our digestive system as fiber. In general, *fiber is the indigestible part of food that comes from plants.* If starch and cellulose are both made from glucose, then why can we digest one but not the other?

The answer is based on the different chemical structures of the two polymers. Starch is made from α-glucose monomers, whereas cellulose is constructed from β-glucose monomers. The β-glucose molecules are joined by a β(1-4) linkage, as shown in Figure 7.26. The repeat unit within the polymer consists of two linked β-glucose molecules, with one flipped upside-down with respect to the other.

THE KEY IDEA: The bonding of glucose monomers in cellulose creates a tough polymer that is resistant to human digestion.

cellulose An indigestible polysaccharide composed of linear chains of β-glucose molecules; it provides fiber in our diet.

CHEMISTRY AND YOUR HEALTH

Insulin and Diabetes

When we digest carbohydrates, glucose molecules pass from the intestines into the bloodstream. As we learned in Chapter 6, glucose is the primary fuel for cellular respiration. In order to participate in this energy-generating reaction, the glucose molecules must leave the bloodstream and enter the cell.

The passage of glucose into cells is controlled by a hormone called *insulin*. When our body senses an increase in the blood glucose level after a meal, an organ called the *pancreas* releases insulin into the bloodstream. As shown in Figure 1, insulin binds to a specialized protein called the *insulin receptor* that is embedded within the cell membrane. The membrane of a cell creates a boundary between the interior of the cell and the external environment.

One region of the insulin receptor extends outside the cell and provides a site for binding the insulin hormone. To visualize the binding between insulin and its receptor, imagine a baseball

(insulin) being caught by a baseball glove (the receptor). Insulin binding activates the receptor, which transmits a chemical signal to the other regions of the receptor protein that extend through the membrane into the cell's interior. The activated receptor triggers another set of chemical signals that is passed in a relay between multiple proteins—this process is known as *signal transduction*. At the end point of this pathway, a signal activates a glucose transporter protein (known as GLUT4), which resides within the cell membrane. This glucose transporter acts like a tunnel that allows glucose molecules to pass through the cell membrane and enter the cell. This is a greatly simplified description of a complex biochemical process.

Malfunctions in this insulin signaling system can cause *diabetes*, which is characterized by an elevated level of glucose in the blood. In Type 1 diabetes, the body's immune system attacks and destroys the insulin-producing cells in the pancreas. As a result, the body cannot produce enough insulin to control the sugar supply. There are no insulin molecules to bind to the receptor and trigger the movement of sugar from the bloodstream into the cell. Type 1 diabetes is the most common type in children. Individuals with this disorder can be treated with insulin injections, which replace the missing insulin that is not produced naturally by the pancreas.

Type 2 diabetes usually affects adults in later life, and it accounts for 90% of all cases of diabetes. There are two main causes of Type 2 diabetes. First, the cells of the pancreas begin to produce lower quantities of insulin. Second, the cells in a person's body become less sensitive to the effect of insulin— this behavior is called *insulin resistance*. Refer again to Figure 1. In Type 2 diabetes, insulin still binds to its receptor, but there is some malfunction in the signal transduction pathway between the insulin receptor and the glucose transporter. As a result, the glucose transporter does not receive its instructions to allow glucose molecules to enter the cell. The causes and treatment of Type 2 diabetes are subjects of ongoing research. However, diet and exercise play important roles in treating this disorder.

FIGURE 1 Insulin is a protein that controls the passage of glucose into cells. The first step occurs when insulin binds to its receptor, which triggers a sequence of chemical signals. As a result of this signal transduction pathway, the glucose transporter protein (GLUT4) enables glucose molecules to enter the cell.

In effect, the β(1-4) linkage causes the glucose molecules to have an *alternating orientation* within the polymer. As a consequence of this molecular structure, the linear polymer chains in cellulose have a stiff structure. This property is very different from the flexible polymer chains found in starch.

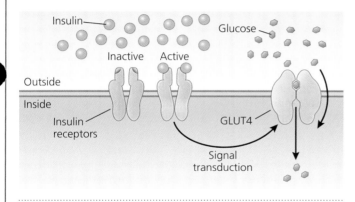

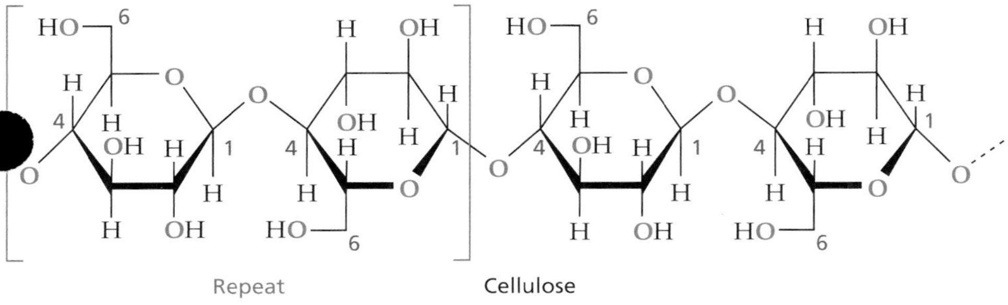

Repeat   Cellulose

FIGURE 7.26 A short region of a cellulose polymer. Cellulose is composed of β-glucose monomers that are joined by a β(1-4) linkages. The repeat unit consists of two monomers with alternating orientations; the C_6 carbon points upward in one monomer and downward in the other.

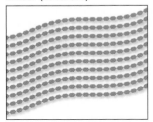

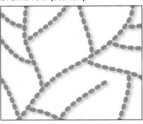

FIGURE 7.27 **Structural comparison of cellulose and starch.** Cellulose is composed of rigid linear polymers that pack together to form tough fibers. This structure makes cellulose resistant to human digestion. The amylopectin version of starch has a branched structure that is easily digested.

FIGURE 7.28 **Cellulose fibers in paper.** This photograph shows a highly magnified view of cellulose fibers at the cut end of a piece of paper. These fibers are constructed from many cellulose polymers that are tightly packed together. The fibers have been artificially colored to make them more visible.

Figure 7.27 provides a schematic comparison between cellulose and starch (amylopectin).

The linear polymers in cellulose cause them to stick together to form tough fibers, as shown in Figure 7.28. Because of this tight packing, the digestive enzymes in our body cannot cut the chemical linkages in the polymer and release the glucose monomers. As a result, the cellulose passes through our body undigested. By contrast, the branched structure of starch provides digestive enzymes with easy access to the chemical linkages between the glucose monomers. The comparison of cellulose and starch demonstrates how changes in molecular structure can affect the biological and nutritional properties of molecules.

CONCEPT QUESTION 7.3

Humans cannot digest cellulose, but other animals can. Cows can break down the cellulose in grass, and termites can digest wood. Propose a hypothesis to explain how these animals are capable of digesting cellulose.

Core Concepts

- **Cellulose** is a polymer composed of β-glucose monomers that are joined by β(1-4) linkages. The glucose monomers have an alternating orientation within the polymer, which creates a stiff, linear structure.
- The cellulose polymers pack together tightly to form tough cellulose fibers. These fibers are resistant to digestion by enzymes in our digestive system.

7.1 What Is the Difference Between Simple and Complex Carbs?

Learning Objective:

Distinguish between simple and complex carbohydrates.

⊙ A **simple carbohydrate** is small and consists of one or two sugar molecules. One example is sucrose (table sugar).

⊙ A **complex carbohydrate** is large and contains hundreds or thousands of sugar molecules. For example, starch is a **polymer** made from many copies of a glucose **monomer**.

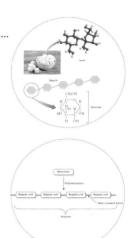

7.2 Synthetic Polymers

Learning Objective:

Illustrate examples of monomers and polymers.

⊙ Multiple copies of a monomer are joined by a **polymerization** reaction to form a polymer.

⊙ Creating new bonds in the polymer requires chemical modification of the monomers. The group of atoms that repeats within the polymer is called the **repeat unit**.

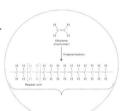

⊙ **Polyethylene** is made from ethylene monomers by **addition polymerization**. No atoms are gained or lost during polymerization.

⊙ The C=C double bond in the ethylene monomer changes into a C—C single bond in the repeat unit.

⊙ **Polyester** is a **condensation polymer**. One monomer contains —OH groups, and the other monomer contains —COOH groups. These monomers react via a **condensation reaction**, which removes an H₂O molecule. The reaction creates a new ester linkage between the monomers.

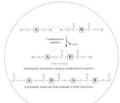

⊙ A **polyamide** is also a condensation polymer. **One monomer** contains —COOH groups, and the other monomer contains —NH₂ groups. These monomers react via a condensation reaction, which creates a new amide linkage between the monomers.

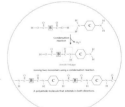

⊙ Nylon is an example of a polyamide.

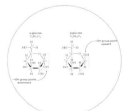

7.3   Carbohydrates: Sugars as Monomers and Polymers

Learning Objective:

Relate the principles of monomers and polymers to various sugars.

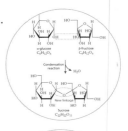

⊙ Glucose is a **monosaccharide** because it consists of one sugar molecule.

⊙ Two common isomers of glucose are α-glucose and β-glucose. These isomers differ in the orientation of the —OH group attached to the C_1 atom.

⊙ **Sucrose** (table sugar) is a **disaccharide** made from joining two monosaccharides: α-glucose and β-fructose.

⊙ **Lactose** (milk sugar) is another example of a disaccharide. It is made by joining β-galactose and β-glucose.

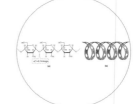

⊙ **Starch** consists of two types of **polysaccharides**—amylose and amylopectin—that are both made from glucose monomers. Amylose is a smaller polymer with a linear structure. Amylopectin is a larger polymer containing multiple branches.

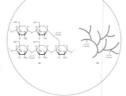

⊙ The glucose monomers in amylose are joined by α(1-4) linkages. The linear polysaccharide chain forms a helical structure. This structure is detected by a chemical test for starch using an I_2/KI solution.

⊙ Amylopectin contains α(1-4) linkages to make regions of linear polymer. In addition, it uses α(1-6) linkages to create branches that split from the main polymer chain.

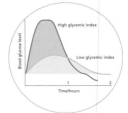

⊙ The **glycemic index** of a food indicates the effect of the food on a person's blood glucose level.

⊙ The **glycemic load** of a food includes both the glycemic index and the amount of carbohydrate.

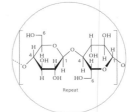

⊙ **Cellulose** is a glucose polymer that we cannot digest and acts as fiber in our diet. In cellulose, the β-glucose monomers are joined by β(1-4) linkages. This linkage produces rigid chains that attach tightly to one another, forming tough fibers.

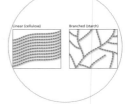

LEARNING RESOURCES

Reviewing Knowledge

7.1: What Is the Difference Between Simple and Complex Carbs?

1. What is the difference between simple carbohydrates and complex carbohydrates?
2. Define the terms *monomer* and *polymer*.
3. Describe the relationship between glucose and starch.
4. Which polymer provides the fiber in our diet?

7.2: Synthetic Polymers

5. Explain why the repeat unit within a polymer has a different chemical structure compared to the monomer that is used for the polymerization reaction.
6. Draw the chemical structures of (a) ethylene and (b) the repeat unit in polyethylene. How do these structures differ?
7. Although they are made from the same monomer, high-density polyethylene (HDPE) is hard, whereas low-density polyethylene (LDPE) is soft and flexible. Explain this difference in terms of the structures of the polymers.
8. What is a condensation reaction?
9. How do polyesters and polyamides get their names?

7.3: Carbohydrates: Sugars as Monomers and Polymers

10. What is the most important and abundant sugar in cells?
11. Draw the molecular structures of α-glucose and β-glucose. Use a box to highlight the structural difference(s) between these two isomers.
12. Which two monosaccharides are used to make sucrose (table sugar)? What type of chemical reaction is used to join these monosaccharides? What is the name of the linkage that is formed?
13. Which two monosaccharides are used to make lactose (milk sugar)? What type of chemical reaction is used to join these monosaccharides? What is the name of the bond that is formed?
14. Describe how sucrose is digested. Which enzyme is involved, and what reaction does it catalyze?
15. Draw a diagram to show how two glucose monomers are joined by an α(1-4) linkage.
16. What type of molecular structure is formed by the polysaccharide chain in amylose?
17. Explain how the structure of amylose is responsible for the blue-black color test for starch using I_2/KI solution.
18. Draw a diagram to show how two glucose monomers are joined by an α(1-6) linkage. Explain how this linkage creates a branch in the amylopectin molecule.
19. What does a food's glycemic index measure? Sketch a graph of blood glucose levels versus time to show the body's response to eating a high-glycemic-index food and a low-glycemic-index food.

20. You are preparing to eat before taking a final exam. Is it best to choose a food with a high glycemic index or one with a low glycemic index? Explain your choice.
21. Compare and contrast the use of glycemic index and glycemic load to measure the effect of a food on blood glucose levels.
22. Sketch the β(1-4) linkage between two glucose molecules in cellulose.
23. Explain why humans cannot digest cellulose.

Developing Skills

24. Polyvinyl chloride (PVC) is a polymer constructed from the vinyl chlorine monomer shown here.

Vinyl chloride

(a) Draw the structure of a PVC polymer that is constructed from four monomers added end-to-end.
(b) Unlike ethylene, the vinyl chloride molecule is not symmetric. Therefore, it is possible to add the monomers with different orientations to create polymers with various chemical structures. Use this principle to draw the structure of a polymer that is different from the one that you drew in part (a). Write a sentence to explain the difference.

25. Suppose you are measuring the mass of a polyethylene sample using an ultracentrifuge. You determine that the average mass of the polymer is 100,000 amu (amu = atomic mass unit). The mass of an ethylene monomer is 28 amu. How many monomers are contained within an average polyethylene molecule? Round your answer to the closest whole number.

26. The mass of a specific polystyrene molecule is 120,000 amu (amu = atomic mass unit). The mass of a styrene monomer is 104 amu. How many monomers are contained within the polystyrene molecule? Round your answer to the closest whole number.

27. Polyoxymethylene is a hard polymer that is used in materials for the automobile industry. The repeat unit for the polymer is given below. (a) Draw the *two-dimensional* structure of a region of polyoxymethylene that contains three monomers. (b) Draw the *three-dimensional* structure of the same region of polymer. Is the polymer planar (i.e., flat) or nonplanar?

28. This chapter describes the formation of a nylon 6,6 polymer using two different monomers. It is possible to make another type of nylon, called nylon 6, using a single monomer. The structure of this monomer is shown below.

(a) Explain why this monomer is suitable for forming a condensation polymer.
(b) Draw a region of a nylon 6 polymer made from three monomers.

29. The chemical structures of acetic acid and methyl amine are shown below.

Acetic acid   Methylamine

(a) These two molecules can join together to form an amide linkage. Draw the chemical reaction that must occur for this to happen. What is this type of reaction called?
(b) Once these two molecules are joined, it is *not* possible to repeat the reaction to form a long polyamide chain. Why not?
(c) How would you modify the structures of acetic acid and methyl amine to obtain molecules that could be used to make a polyamide chain? Draw the chemical structures of these modified molecules, and explain your choice.

30. When grain is allowed to soften in water, it produces a sugar called *maltose*. It is found in cereal, pasta, potatoes, and beer. Maltose is formed by condensation reaction between two α-glucose molecules, which produces a disaccharide with an α(1-4) bond. (a) Draw the molecular structure of maltose. (b) Circle and label the α(1-4) bond.

31. Elastomers are rubbery materials made from soft plastics. The classical rubber polymer is polybutadiene, with the repeat unit shown below. From the structure of this repeat unit, can you identify why this polymer is more flexible than polyethylene or polystyrene? (*Hint:* Consider the number and type of covalent bonds in the repeat unit. Can chains of this polymer pack as closely as those of the other polymers?)

Exploring Concepts

32. Our body uses a molecule called *glycogen* as a form of energy storage. Glycogen is a branched polysaccharide made from glucose monomers, and it is stockpiled within cells of the liver and muscles. The structure of glycogen is shown below. Explain how the structure of glycogen provides a rapidly accessible source of glucose.

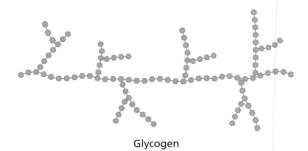

Glycogen

33. Despite the fact that we cannot digest cellulose, fiber is recommended for a healthy diet. Use the Internet to investigate the role of fiber in our diet, and write a brief report. What problems are associated with diets that are low in fiber?

34. Many processed foods contain *high fructose corn syrup* as a sweetener. (a) Use the Internet to research the source of this food additive and how it is produced. Write a short report on your findings. (b) There is disagreement about whether adding high fructose corn syrup to sweeten food is better or worse than adding sucrose. Use the Internet to investigate this issue, and write an evaluation of the pros and cons of high fructose corn syrup. Base your response on the most reliable scientific evidence you can find.

35. Patients who suffer from Type 2 diabetes often are resistant to insulin, and a drug called metformin is used to reduce their glucose levels. Use the Internet to find out how metformin accomplishes this effect and what risks the drug presents in doing so.

36. There is great interest in the possibility of using cellulose to make ethanol, which can serve as a transportation fuel. This approach is appealing because corn stalks and other sources of agricultural cellulose are currently discarded and have no practical use. Use the Internet to research this topic, and write a two- to three-paragraph summary that addresses the following questions:
(a) How can ethanol be produced from cellulose using chemical methods?
(b) How can ethanol be produced from cellulose using biological methods (i.e., enzymes)?
(c) What are the current limitations on large-scale production of ethanol from cellulose?
(d) What possible innovations could overcome these limitations?

37. Celluloid is a polymer that was commonly used for many decades in the entertainment industry. There is a close resemblance between the words "celluloid" and "cellulose." Use the Internet to investigate celluloid. How is the polymer made, and what was it used for? Write a one-paragraph report on your findings.

38. The disaccharide trehalose consists of two glucose molecules connected by an $\alpha(1\text{-}1)$ linkage. (a) Draw the structure of this molecule. (b) Use the Internet to investigate why trehalose is used as a food preservative. Write two or three sentences on your findings.

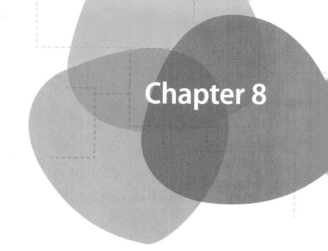

Chapter 8

The Unusual Nature of Water

E very living organism on Earth requires water. Humans can survive only three days without water, compared to more than 3 weeks without food. Many types of bacteria can survive without oxygen, but all of them need water. Why is water so important for life?

8.1 Why Is Water Essential for Life?

Learning Objective:

Characterize the ways in which water's properties sustain life on Earth.

🔑 THE KEY IDEA: Liquid water has unusual properties that are essential for sustaining life on Earth.

boiling point The temperature at which a liquid changes into a vapor.

vapor A gas that arises from a liquid or a solid.

After a workout, one of the first things you want to do is drink water (Figure 8.1). This urge to drink is your body's reminder that you need to remain sufficiently hydrated. Water is the most abundant substance in our body and accounts for approximately 60% of our total mass (which includes the mass of our bones). To appreciate this fact, think of the differences in size, shape, and consistency between a grape and a raisin (a dehydrated grape).

Water is necessary for all life on Earth. When NASA launches satellites to explore planets and moons, they usually search for liquid water as the first step in determining whether the environment could sustain life. Detection of salty water on the surface of Mars has stimulated excitement about the possibility that life once existed or may even exist today on that planet.

Why is water essential for life? The answer to this question can be found in liquid water's unusual properties. To understand this point, Figure 8.2 compares the boiling points of three molecules of similar size: methane (CH_4), ammonia (NH_3), and water (H_2O). The **boiling point** is the temperature at which a liquid changes into a **vapor**, which is a gas that arises from a liquid or solid. The boiling point of H_2O is much higher than that of either CH_4 or NH_3. As a result, H_2O exists as a liquid at room temperature (approximately 25°C), whereas CH_4 and NH_3 are both gases. In addition, H_2O remains a liquid over a large temperature range, from 0°C to 100°C.

This liquid property of water has a special significance for living organisms. The chemical reactions that sustain life require molecules to collide with one another before the reaction can occur. A gas is a poor medium for sustaining

FIGURE 8.1 Water is essential for life. We need to supply our body with water to stay alive.

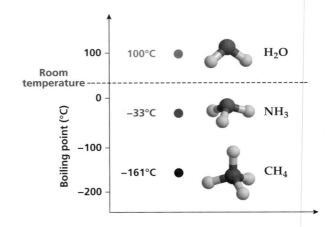

FIGURE 8.2 A comparison of the boiling points for CH_4, NH_3, and H_2O. The boiling point of each molecule is indicated by a colored dot. Of these molecules, only H_2O is a liquid at room temperature (25°C).

SCIENCE IN ACTION

Does Liquid Water Exist on Other Planets?

Because water is fundamental to life on Earth, scientists searching for life on other planets first try to locate sources of liquid water. Our close planetary neighbor, Mars, shows evidence for the presence of water in its ancient geological past because its surface is marked with deep canyons. One theory proposes that these canyons were formed by flows of liquid water during the planet's ancient past. However, Mars lost its water several billion years ago. Today, the average global temperature on Mars is –55°C, much below the freezing point of water. Images taken by the Mars Rover showed the surface of the planet as rocky, dry, and barren (Figure 1).

In 2015, NASA scientists announced surprising new measurements taken by a satellite—the Mars Reconnaissance Orbiter. The instruments on this satellite detect characteristic signatures from specific ions that bind tightly to H_2O molecules. The measurements indicated that very salty water makes a seasonal appearance on the surface of Mars. The existence of liquid water is revealed by increases in streaks on the walls on a Martian crater (Figure 2). These streaks increase in the Martian summer, when the crater is heated by the sun. Water that contains a high concentration of salt freezes at a lower temperature than pure water; this is why we put salt on roads in the winter. The streaks disappear in winter, when the temperature of the planet's surface drops to the point where even salty water freezes.

This is the strongest evidence to date that Mars has an environment that may be hospitable to life. Even though the concentration of salt is very high, there are organisms on Earth—called *extremophiles*—that can survive in very salty water. As yet, however, there is no scientific evidence that life exists on Mars today or has existed in the past.

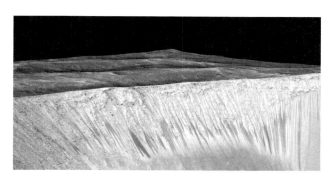

FIGURE 2 **Evidence for seasonal salty water on Mars.** NASA's Mars Reconnaissance Orbiter detected seasonal dark streaks on the walls of a crater on Mars, which are caused by water molecules bound to minerals. The streaks increase in darkness during the summer and disappear in winter.

FIGURE 1 **The dry and barren surface of Mars.** This photograph was taken by the Mars Rover.

life because the molecules are far apart, and they collide only fleetingly. A solid medium is also unsuitable, but for the opposite reason. The components of a solid are fixed in place, which prevents the molecular motion necessary for life. By contrast, *a liquid provides an environment in which different types of molecules can interact easily.* The molecules in a liquid are always moving and bumping into one another, which is the type of interaction best suited for the molecular processes of life.

H_2O is not the only molecule that exists as a liquid on Earth. Many different types of hydrocarbon molecules are liquids, such as octane and benzene. However, water has another unique property: *It dissolves a wide variety of molecules and ions.* All organisms rely on a diversity of dissolved substances within the cell to generate the chemical reactions that life requires.

Water has other unusual properties that we will explore in this chapter. Because water is so familiar to us, we might also assume it is simple to understand. In fact, explaining the properties of water remains an ongoing scientific challenge. As Felix Franks, one of the world's leading experts on water, has remarked: "Of all the known liquids, water is probably the most studied and least understood."

8.2 Chemical Bonding in H_2O

The remarkable properties of water arise from the chemical characteristics of the H_2O molecule. To understand this molecule, we need to extend our earlier description of covalent bonding by examining how the bond is affected by the particular atoms that it joins. Throughout this chapter, we will use "H_2O" to refer to an isolated molecule and "water" to indicate the familiar liquid. This distinction is important because H_2O molecules can exist in solid and gaseous states as well as in their liquid form.

H_2O: A Review of Bonding and Structure

An H_2O molecule is formed when an oxygen atom achieves a stable octet of outer electrons by forming covalent bonds with two hydrogen atoms. This arrangement results in two bonding pairs and two nonbonding pairs (lone pairs).

The repulsion between the four electron pairs (two bonding pairs and two lone pairs) generates an approximately tetrahedral orientation of the electron pairs (Figure 8.3(a)). Valence shell electron pair repulsion forces the two O—H bonds into a bent geometry with an H—O—H bond angle of 104.5°, which is slightly smaller than the ideal tetrahedral angle of 109.5° (Figure 8.3(b)). The difference occurs because repulsion between the two nonbonding electron pairs is slightly greater than repulsion between the two bonding electron pairs, which pushes the two O—H bonds closer together to create a smaller bond angle.

Polar and Nonpolar Covalent Bonds

Until this point, we have treated all covalent bonds as being alike. In reality, *the types of atoms that are joined by a covalent bond affect how the electron pair is shared.* In some covalent bonds, the electron pair is shared equally between two atoms. In other cases, one atom attracts a larger share of the electron pair. This difference has important consequences for the molecular properties of H_2O.

Consider first the example of a covalent bond between two hydrogen atoms to form an H_2 molecule. *When a covalent bond joins two identical atoms, the atoms share the electron pair equally.* However, the situation is different for the covalent bond between the H and F atoms in an HF molecule. *When a covalent bond is formed between two different atoms, the electron pair is not shared equally between the atoms.* In the HF molecule, the fluorine atom exerts a stronger attraction on the bonding electron pair than the hydrogen atom does.

This effect is best illustrated using the quantum mechanical model of covalent bonding, which describes the spatial distribution of electron density within a molecule (see Chapter 3). Figure 8.4 compares the electron density for the H_2 and HF molecules. In H_2, the electron density is distributed equally because each atom has the same attraction for the shared electron pair. By contrast, the electron density in HF is not distributed equally because the

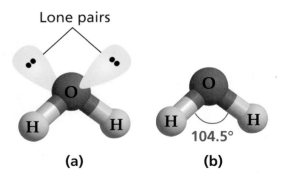

Lone pairs

(a)   (b)

104.5°

FIGURE 8.3 Structure of an H_2O molecule. (a) An H_2O molecule contains four electron pairs—two bonding pairs and two nonbonding pairs—that have an approximately tetrahedral orientation. The lone pairs are shown as yellow spheres. (b) The H_2O molecule is bent with a bond angle of 104.5°.

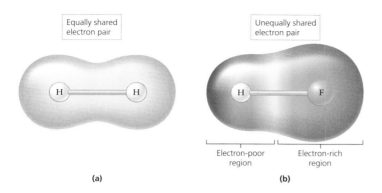

Equally shared electron pair

Unequally shared electron pair

H — H

H — F

Electron-poor region   Electron-rich region

(a)   (b)

FIGURE 8.4 Covalent bonding and electron density in H₂ and HF. (a) In H₂, the electron density is distributed equally between the two H atoms. (b) In HF, the electron density is not distributed equally because the F atom exerts a stronger attraction for the shared electron pair. Because of the unequal distribution of electron density, the F atom becomes slightly more negative (indicated by red) and the H atom becomes slightly more positive (indicated by blue).

F atom has a stronger attraction for the shared electron pair. This attraction results in a greater electron density for the F atoms as compared to the H atom.

The allocation of electron density within a molecule affects the net electrical charge on each atom. All *isolated* atoms are electrically neutral because the total negative charge of the electrons exactly balances the total positive charge of the protons in the atomic nucleus. However, this charge balance is upset in covalent bonds with an unequally shared electron pair. When the F atom in an HF molecule attracts the shared electron pair, it is "stealing" some of the negative charge that belongs to the H atom. The F atom therefore acquires a *net negative charge* because the total negative charge surrounding the F atom, which arises from the increased electron density, is *slightly larger* than the positive charge on the fluorine nucleus. On the other side of the molecule, the H atom has "lost" some of its negative charge because of the unequal sharing. Consequently, the H atom acquires a *net positive charge* because the negative charge due to the electron density is *slightly less* than the positive charge on the hydrogen nucleus.

In Figure 8.4(b), the electron density for the F atom is colored red to indicate that it has a net negative charge, whereas the electron density for the H atom is colored blue to indicate that it has a net positive charge. The small electrical charges acquired by the H and F atoms are called **partial charges**. The F atom acquires a *partial negative charge*, which is written $\delta-$ (the lowercase Greek letter delta δ means "small"). Similarly, the H atom acquires a *partial positive charge*, which is written as $\delta+$. The magnitude of the partial charge is typically less than a full electrical charge of ±1 (using the relative charge units introduced in Chapter 2).

A key point is that *the partial charges on the atoms arise because they are sharing electrons in a covalent chemical bond*. We can illustrate this concept with an analogy. Suppose you and your friend both have $1 in the form of four quarters. You decide to combine your money, so you both put your quarters on a table. At this point, your greedy friend slides one of your quarters over to his side of the table, so he now has $1.25 and you have 75 cents. The *total amount of money on the table remains the same*, but your friend is now richer and you are poorer because the sharing is now unequal. This scenario is similar to the unequal sharing of the electron pair in the H—F covalent bond. Because fluorine is "electron greedy," it attracts a larger share of the electron pair. As a result, the F atom becomes "electron rich" ($\delta-$) while the H atom becomes "electron poor" ($\delta+$). As with our money analogy, the total amount of electrical charge on the molecule remains the same, but it is now unequally distributed between the atoms.

The $\delta+$ and $\delta-$ charges on each end of an H—F covalent bond form a **dipole**. In general, *a dipole is formed when positive and negative charges are separated by a short distance*. Because of this dipole, the H—F bond is called a **polar covalent bond**. By contrast, there are no partial charges on the H atoms in an H—H bond because the electron pair is shared equally. The H—H bond does not form a dipole and is therefore called a **nonpolar covalent bond**. Figure 8.5 compares the polar vs. nonpolar properties of H—H and H—F bonds. By convention, a dipole is drawn as an arrow that points from the positive charge toward the negative charge.

partial charge A small electrical charge acquired by an atom as a result of unequal sharing of the electron density in a covalent bond.

dipole A pair of positive and negative charges separated by a short distance.

polar covalent bond A covalent bond that forms a dipole.

nonpolar covalent bond A covalent bond that does not form a dipole.

No dipole

H — H

H — H is a
nonpolar covalent bond

Dipole

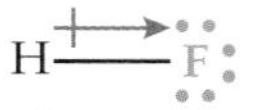

H — F
$\delta+$ $\delta-$

H — F is a
polar covalent bond

FIGURE 8.5 A comparison of a nonpolar covalent bond (H—H) and a polar covalent bond (H—F). For the H—F bond, the separation of partial charges produces a dipole.

Core Concepts

- When two identical atoms form a covalent bond, such as H—H, the electron pair is shared equally between the atoms. However, when a covalent bond is formed between two different atoms, such as H—F, the electron pair is not shared equally.
- Unequal electron sharing in a covalent bond (H—F) creates **partial charges** on the bonded atoms. The F atom acquires a small excess of negative charge ($\delta-$), while the H atom has a small deficiency of negative charge ($\delta+$).
- The separation of partial charges within the H—F bond creates a **dipole**. Consequently, the H—F bond is called a **polar covalent bond**. The H—H bond does not form a dipole, so it is called a **nonpolar covalent bond**.

Electronegativity: Attracting a Shared Electron Pair

The unequal sharing of electrons in a covalent bond can be explained by a property of atoms called *electronegativity*. **Electronegativity** *measures the ability of an atom to attract the shared pair of electrons in a chemical bond.* Atoms with a *higher* electronegativity exert a *stronger* attraction for the shared electron pair. In the previous example of HF, fluorine is more electronegative than hydrogen. The principle of electronegativity applies only when an atom is chemically bonded to another atom.

A scale for measuring electronegativity values was devised in the early 20th century by the chemist Linus Pauling (1901–1994). The *Pauling electronegativity scale* assigns every atom a numerical electronegativity value between 0 and 4. Figure 8.6 shows the electronegativity values for elements in the periodic table, which reveals two trends. First, *the electronegativity increases from left to right across a period.* Second, *the electronegativity within a group decreases from top to bottom.* According to these trends, *atoms in the top right corner of the periodic table have the highest electronegativity.* Fluorine has the highest electronegativity of any atom, with a value of 4.0 on the Pauling scale. The next highest is oxygen, with an electronegativity value of 3.5. As we shall see, this number is important for understanding the chemical properties of H_2O.

CONCEPT QUESTION 8.1

The noble gases (Group 8A) are not included in Figure 8.6. Why are electronegativity values not given for the noble gases?

We can imagine a covalent bond between two atoms as being like a tug-of-war for the shared electron pair. Each atom in the bond exerts a pull on the electron pair with a strength that is determined by its electronegativity value. The *difference* between the electronegativity values tells us if the atoms are evenly matched or if one is stronger than the other. We can quantify this difference by using an equation. Consider a covalent bond A—B between two atoms, A and B, which have electronegativity values of EN_A and EN_B. The electronegativity difference

---

THE KEY IDEA: When two atoms form a chemical bond, the difference in their electronegativity values determines whether the bond is nonpolar covalent, polar covalent, or ionic.

electronegativity The ability of an atom to attract the shared pair of electrons in a chemical bond.

FIGURE 8.6 Electronegativity values of the elements. The elements with highest electronegativity values are located in the top right corner of the periodic table.

Increasing electronegativity →

1A	2A											3A	4A	5A	6A	7A
H 2.1																
Li 1.0	Be 1.5											B 2.0	C 2.5	N 3.0	O 3.5	F 4.0
Na 0.9	Mg 1.2			Transition elements								Al 1.5	Si 1.8	P 2.1	S 2.5	Cl 3.0
K 0.8	Ca 1.0	Sc 1.4	Ti 1.5	V 1.6	Cr 1.7	Mn 1.6	Fe 1.8	Co 1.9	Ni 1.9	Cu 2.0	Zn 1.6	Ga 1.8	Ge 2.0	As 2.2	Se 2.6	Br 2.8
Rb 0.8	Sr 1.0	Y 1.2	Zr 1.3	Nb 1.6	Mo 2.2	Tc —	Ru 2.2	Rh 2.3	Pd 2.2	Ag 1.9	Cd 1.7	In 1.8	Sn 1.8	Sb 2.0	Te 2.1	I 2.5
Cs 0.79	Ba 0.9								Pt 2.3	Au 2.5	Hg 2.0	Tl 2.0	Pb 2.3	Bi 2.0	Po —	

Increasing electronegativity

within the bond is designated by ΔEN_{AB} where the uppercase Greek letter delta (Δ) denotes a difference. We calculate the magnitude of ΔEN_{AB} as the difference between the electronegativity values of each atom in the bond.

$$\Delta EN_{AB} = EN_A - EN_B$$

When performing these calculations, we choose atom A as the one with the higher electronegativity value to ensure that the difference is always positive.

The numerical value of ΔEN_{AB} allows us to classify covalent bonds according to their **bond polarity**—the degree to which they are polar or nonpolar. We can use the following rules:

> **Rule 1: If EN$_{AB}$ is less than 0.5, then the bond is a *nonpolar covalent bond*.**
>
> **Rule 2: If ΔEN_{AB} is between 0.5 and 2.0, then the bond is a *polar covalent bond*.**

> **bond polarity** The degree to which a covalent bond is polar or nonpolar; it is determined by the electronegativity difference between the two bonded atoms.

Although these are useful guidelines, the cutoffs between categories are somewhat arbitrary. In reality, electronegativity differences have a continuous range that produces variation in the bond polarity. Within the class of polar covalent bonds, a bond with a larger numerical value of ΔEN_{AB} has greater polarity than a bond with a smaller value. Bonds with greater polarity generate larger partial charges on their atoms. These bonds also have a larger dipole, because the strength of the dipole depends on the size of the partial charges.

We can apply this equation to determine the polarity for H₂ and H—F. In the hydrogen molecule, we calculate the ΔEN value as follows:

$$\Delta EN_{HH} = EN_H - EN_H = 2.1 - 2.1 = 0$$

This calculation makes sense because the two hydrogen atoms are identical, so the H—H bond is nonpolar. The ΔEN value does not need to be exactly zero for a bond to qualify as nonpolar. A small ΔEN value means that the electron pair is shared *almost* equally, so the partial charges on the atoms are very small.

Now, let's consider the ΔEN for hydrogen fluoride.

$$\Delta EN_{HF} = EN_F - EN_H = 4.0 - 2.1 = 1.9$$

The electronegativity value for the fluorine atom is almost double that of the hydrogen atom, so fluorine attracts the shared electrons much more strongly. The value of ΔEN is large, so we classify H—F as a strongly polar bond with large partial charges and a large dipole.

Now, let's consider the bond formed between sodium (Na) and fluorine (F). We can calculate the electronegativity difference based on the values in Figure 8.7.

$$\Delta EN_{NaF} = EN_F - EN_{Na} = 4.0 - 0.9 = 3.1$$

The electronegativity difference is so large that it far exceeds the threshold for a polar covalent bond ($\Delta EN_{AB} = 2.0$). Because of this large difference, the fluorine atom acquires *both* of the shared electrons. The sodium atom loses one of its electrons to the fluorine atom, creating a positively charged sodium ion (Na^+) and a negatively charged fluoride ion (F^-) (Figure 8.7). The oppositely charged ions attract each other to form an *ionic bond*, which we discussed in Chapter 3. Instead of the partial charges ($\delta+$ and $\delta-$) observed for polar covalent bonds, the atoms in an ionic bond have full charges (+ or –) because the electron has been completely transferred from one atom to another.

We can now add a third category of bonding based on the electronegativity difference of two atoms.

> **Rule 3: If ΔEN_{AB} is greater than 2.0, the bond is an *ionic bond*.**

Like many rules, this one has exceptions. There are some cases in which ΔEN_{AB} values in the range of 1.6 to 2.0 produce an ionic bond instead of

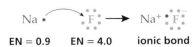

EN = 0.9   EN = 4.0   ionic bond

FIGURE 8.7 A very large difference in electronegativity produces an ionic bond. The large electronegativity difference between sodium and fluorine causes the transfer of an electron to create an Na⁺ ion and an F⁻ ion. The charged ions then form an ionic bond.

WORKED EXAMPLE 8.1

Question: Consider the following three covalent bonds that are commonly found in biological molecules and pharmaceuticals:

C—H   N—H   O—H

For each of these bonds:
   (a) Calculate the electronegativity difference (ΔEN), and classify the bond as polar or nonpolar.
   (b) For the polar bonds, assign partial charges to the atoms, and draw the direction of the dipole.
   (c) Identify which of the three bonds is the most polar.

Answer:
   (a) C—H $\Delta EN = 2.5 - 2.1 = 0.4$ nonpolar
        N—H $\Delta EN = 3.0 - 2.1 = 0.9$ polar
        O—H $\Delta EN = 3.5 - 2.1 = 1.4$ polar
   (b) $^{\delta-}N \leftarrow H^{\delta+}$ $^{\delta-}O \leftarrow H^{\delta+}$
   (c) The O—H bond is the most polar.

TRY IT YOURSELF 8.1

Question: Hydrogen atoms form molecules with the elements in Group 7A of the periodic table (called the halogens). Classify the following covalent bonds as polar or nonpolar. What trend do you notice regarding the polarity of the bond as the atomic number of the halogen increases?

   (a) H—F   (b) H—Cl   (c) H—Br

a polar covalent bond. For example, sodium bromide (NaBr) is an ionic solid. Using the electronegativity values from Figure 8.6, we calculate

$$\Delta EN_{NaBr} = EN_{Br} - EN_{Na} = 2.8 - 0.9 = 1.9$$

This is the same electronegativity difference that we calculated for the H—F molecule. To account for this "in-between" region, we use a fourth rule.

Rule 4: If $\Delta EN_{AB} = 1.6–2.0$, the bond is (a) *polar covalent* **if only nonmetals are bonded or (b)** *ionic* **if a metal is involved.**

In the case of NaBr, sodium (Na) is a metal from Group 1A of the periodic table. Consequently, we classify the bond in NaBr as an ionic bond.

In summary, electronegativity explains why specific pairs of atoms form non-polar covalent bonds, polar covalent bonds, or ionic bonds. We compare these three types of bonds in Figure 8.8, although the demarcations between the different types of bonds are only approximate.

FIGURE 8.8 Electronegativity difference affects the type of chemical bond. When atoms A and B form a bond, the difference in electronegativity (ΔEN_{AB}) affects whether the bond will be nonpolar covalent, polar covalent, or ionic. In this figure, we show only polar covalent bonds in the region between 1.6 and 2.0, although ionic bonds will form if one of the atoms is a metal.

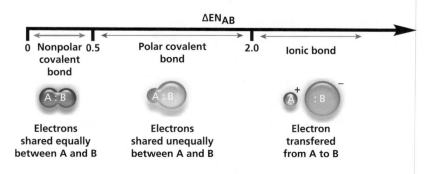

WORKED EXAMPLE 8.2

Question: Chapter 3 used table salt (sodium chloride) to introduce the topic of ionic bonding. Use the principle of electronegativity to explain why sodium chloride has ionic bonds.

Answer: We use Figure 8.6 to obtain the electronegativity values of sodium (0.9) and chlorine (3.0).

The electronegativity difference is calculated as

$$\Delta EN_{NaCl} = EN_{Cl} - EN_{Na} = 3.0 - 0.9 = 2.1$$

This electronegativity difference is greater than 2.0, which is above the threshold for an ionic bond.

TRY IT YOURSELF 8.2

Question: For the following pairs of atoms, classify the bond between them as nonpolar covalent, polar covalent, or ionic.

   (a) Hydrogen (H) and iodine (I)

   (b) Lithium (Li) and fluorine (F)

   (c) Lithium (Li) and iodine (I)

Core Concepts

- **Electronegativity** measures the ability of an atom to attract the shared pair of electrons in a chemical bond.
- The Pauling electronegativity scale provides a numerical value for the electronegativity of the elements. Within the periodic table, electronegativity increases from left to right across a period and decreases from top to bottom within a group.
- The **bond polarity** of a covalent bond between two atoms (A and B) can be calculated from the difference in their electronegativity values (ΔEN_{AB}). Covalent bonds are classified as nonpolar ($\Delta EN_{AB} < 0.50$) or polar ($\Delta EN_{AB} = 0.5$–2.0).
- An ionic bond forms when the electronegativity difference between two atoms is very high ($\Delta EN_{AB} > 2.0$). In this case, an electron is transferred from the less electronegative atom to the more electronegative atom to form two charged ions (e.g., Na^+ and F^-). When metals are involved, ionic bonds can also form when $\Delta EN_{AB} = 1.6$–2.0

H_2O Is a Strongly Polar Molecule

We can now apply our knowledge of electronegativity and polar covalent bonds to understand the properties of H_2O. As shown in Figure 8.9, the O—H bonds in H_2O are *strongly polar* because of the large electronegativity difference between O and H ($\Delta EN_{OH} = 1.4$). Because the O atom is more electronegative, it acquires a partial negative charge ($\delta-$), while the H atom gains a partial positive charge ($\delta+$).

In an H_2O molecule, each O—H bond acts as a **bond dipole**. The two O—H bond dipoles are oriented at an angle because of the bent geometry of the molecule. Because both bond dipoles point toward the O atom, they *add together* in this direction to create an even larger dipole for the H_2O molecule. The total dipole of a molecule is called the **molecular dipole**. A molecule such as H_2O with a molecular dipole is called a **polar molecule**. Figure 8.10 illustrates the distribution of electrical charges within an H_2O molecule, which has a negative charge on the O atom and a positive charge on each H atom.

In general, dipoles are sensitive to the presence of an electrical charge. As an example, rubbing the surface of a balloon generates static electrical charges. If

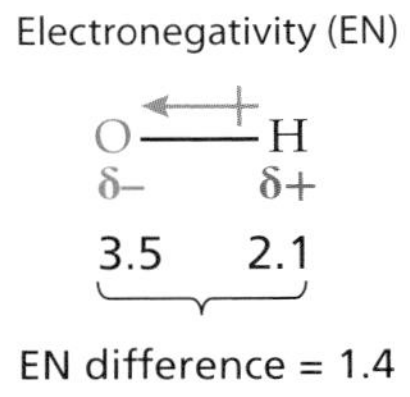

THE KEY IDEA: The large electronegativity difference between O and H creates a strongly polar O—H bond, and the two O—H bonds in H_2O generate a large molecular dipole.

Electronegativity (EN)

$$\underset{\substack{\delta- \\ 3.5}}{O} \xrightarrow{\hspace{1.5cm}} \underset{\substack{\delta+ \\ 2.1}}{H}$$

EN difference = 1.4

FIGURE 8.9 The O—H bond is strongly polar. The large electronegativity difference between the O and H atoms creates a strongly polar O—H bond. The dipole formed is shown in pink.

bond dipole The dipole of an individual covalent bond within a molecule.

molecular dipole The directional sum of all the bond dipoles in a molecule.

polar molecule A molecule that contains a molecular dipole (e.g., H_2).

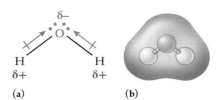

FIGURE 8.10 Distribution of electrical charges in an H_2O molecule. (a) The H_2O molecule has two strongly polar O—H bonds that create relatively large partial charges on the O atom ($\delta-$) and the H atoms ($\delta+$). (b) Distribution of electrical charge within H_2O, where red indicates negative charge and blue indicates positive charge.

FIGURE 8.11 Polar H_2O molecules can be deflected by electrical charges. The flow of the water stream from a faucet can be changed by a rubbed balloon. This effect is caused by an interaction between the static electrical charges on the balloon and the dipoles of the H_2O molecules.

the charged balloon is held next to a stream of water from a faucet, the interaction of the electrical charges with the H_2O dipoles is strong enough to deflect the direction of the water (Figure 8.11). This simple experiment, which you can try at home, is an observable consequence of the polar character of H_2O molecules.

We must be careful to distinguish between a *bond dipole* (which refers to an individual bond) and the *molecular dipole,* which arises from the net effect of adding all the bond dipoles within the molecule. When we add bond dipoles together, the *direction* of the dipoles is particularly important. Although the idea sounds counterintuitive, it is possible to have a molecule that contains individual bond dipoles but has no molecular dipole (i.e., the molecule is *nonpolar*). An example is provided in Concept Question 8.2.

CONCEPT QUESTION 8.2

As we learned in Chapter 3, carbon dioxide (CO_2) is a linear molecule with double covalent bonds between the central C atom and the two O atoms.

Carbon dioxide (CO_2)

(a) *Redraw the CO_2 molecule to show the partial charges on the atoms and the bond dipoles.*

(b) *Surprisingly, CO_2 has no molecular dipole. Propose an explanation for this observation.*

Core Concepts

- The two O—H bonds in H_2O are *strongly polar* ($\Delta EN_{OH} = 1.4$). This polarity generates significant partial charges on the atoms: $\delta-$ for the oxygen atom and $\delta+$ for each hydrogen atom. Each O—H bond forms a dipole.
- Because H_2O has a bent geometry, the two **bond dipoles** formed by each O—H bond add together to form a large **molecular dipole.** Consequently, H_2O is a strongly **polar molecule.**

From H$_2$O Molecules to Liquid Water

Previously in this chapter, we have examined the properties of a *single* H$_2$O molecule. However, a sample of water contains many billions of H$_2$O molecules. Instead of thinking about individual molecules, we now must consider a *collection* of molecules and how they interact. The unusual properties of liquid water emerge from these interactions.

Hydrogen Bonding Between H$_2$O Molecules

We have learned that H$_2$O molecules have large partial charges on the O atom ($\delta-$) and the H atoms ($\delta+$). When two H$_2$O molecules are in close proximity, the $\delta-$ charge on the O atom of one molecule attracts the $\delta+$ charge on the H atom of its neighboring molecule (and vice versa). This type of attraction is called a **hydrogen bond**. In general, *a hydrogen bond is an electrical attraction between a hydrogen atom in a polar bond and another highly electronegative atom (e.g., N, O, F)*. This bonding interaction is oriented toward a nonbonding electron pair on the electronegative atom.

A hydrogen bond is distinct from a covalent bond. As we learned in Chapter 3, a covalent bond is formed when two atoms share a pair of electrons. It is called an **intramolecular bond** because it occurs within a single molecule (*intra* means "within"). By contrast, *a hydrogen bond does not involve a shared pair of electrons*. In other words, the H atom in a hydrogen bond does not share an electron pair with the electronegative atom to which it is bonded. A hydrogen bond between two H$_2$O molecules is called an **intermolecular bond** (*inter* means "between"). Table 8.1 summarizes the terms used to describe different types of bonds.

Figure 8.12 shows a central H$_2$O molecule surrounded by four neighboring molecules. This arrangement represents the maximum possible number of hydrogen bonds, which are depicted as dotted lines. For the central H$_2$O molecule, two hydrogen bonds are formed by mutual attraction between the $\delta+$ charge on each H atom and the $\delta-$ charge on the O atoms of two neighboring molecules. The two H atoms on the central H$_2$O molecule act as **hydrogen bond donors** because each donates hydrogen bonds to another molecule. The other two hydrogen bonds arise when $\delta+$ charges on the H atoms of two neighboring molecules are mutually attracted to the $\delta-$ charge on the O atom of the central H$_2$O molecule. The O atom serves as the **hydrogen bond acceptor**, and the two hydrogen bonds are directed towards the two nonbonding electron pairs on the O atom. In summary, a single H$_2$O molecule is capable of donating two hydrogen bonds and accepting two hydrogen bonds.

Hydrogen bonds are much weaker than covalent bonds; the strength of each intermolecular hydrogen bond in water is only about 5% of the strength of an intramolecular O—H bond. In addition, the distance between two atoms that participate in a hydrogen bond is longer than the distance between the two atoms in a covalent bond.

Although Figure 8.12 illustrates the origin of hydrogen-bonding interactions, it does not capture the three-dimensional arrangement of the H$_2$O molecules. A more accurate depiction is provided in Figure 8.13. The four neighboring molecules surround the central H$_2$O with an *approximately tetrahedral geometry*. This

Learning Objective:

Explain why H$_2$O is a liquid at room temperature.

THE KEY IDEA: A hydrogen bond is an electrical attraction between a hydrogen atom of one molecule and a highly electronegative atom of a neighboring molecule.

hydrogen bond An electrical attraction between a hydrogen atom in a polar bond and another highly electronegative atom (e.g., N, O, F).

intramolecular bond A chemical bond within a single molecule.

intermolecular bond A chemical bond between two molecules.

hydrogen bond donor The H atom that forms a hydrogen bond; this H atom is attached to a highly electronegative atom.

hydrogen bond acceptor A highly electronegative atom with a pair of nonbonding electrons that attracts the hydrogen bond donor to form a hydrogen bond.

TABLE 8.1	Bonding Within and Between Molecules
BOND TYPE	**DESCRIPTION**
*Intra*molecular bond	A chemical bond connecting atoms *within* a single molecule.
*Inter*molecular bond	A chemical bond *between two* molecules.
Hydrogen bond	A type of chemical bond based on electrical attraction between polar molecules.

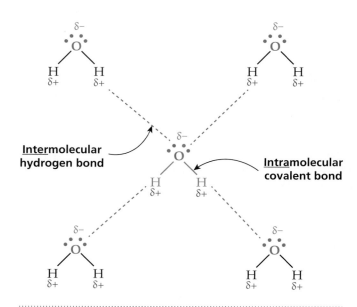

FIGURE 8.12 Hydrogen bonding among H₂O molecules. A central H₂O molecule (highlighted in red) forms hydrogen bonds with four neighboring H₂O molecules. The intermolecular hydrogen bonds arise from a mutual electrical attraction between the δ+ charge on the H atoms and the δ– charge on the O atoms. The central H₂O molecule donates two hydrogen bonds and accepts two hydrogen bonds, creating a total of four hydrogen bonding interactions.

structure arises from the tetrahedral orientation of the four electron pairs in H_2O (two bonding pairs and two lone pairs), which imposes a directionality on the hydrogen-bonding interactions with its neighbors.

We can now understand why water is a liquid at room temperature. Hydrogen bonds make the H_2O molecules more "sticky" in their interactions with one another. Rather than flying apart to form a gas, they stick together to form a liquid. Even though each hydrogen bond is weak, the combined effect of many trillions of these bonds holds the molecules together in a liquid state. To summarize, *water is a liquid at room temperature because of hydrogen-bonding interactions among its highly polar H_2O molecules.*

By contrast, both CH_4 and NH_3 are gases at room temperature. Why? The C—H bonds in methane are nonpolar ($\Delta EN_{CH} = 0.4$), so the partial charges on the atoms are only small. Attraction among the CH_4 molecules is very weak, so the molecules act independently and form a gas. The N—H bond in ammonia is moderately polar ($\Delta EN_{NH} = 0.9$), which generates small partial charges on the atoms. There are weak hydrogen bonds among NH_3 molecules, but their attraction is not strong enough to stick the molecules together as a liquid at room temperature. In addition, NH_3 has only one lone pair, as opposed to the two lone pairs for H_2O. Consequently, an NH_3 molecule has only one site to accept a hydrogen bond from a neighboring molecule. This characteristic limits its ability to form an extensive hydrogen-bonding network like that found in water.

CONCEPT QUESTION 8.3

Water is essential for life on Earth, but life might exist elsewhere in the universe under different conditions. Consider a newly discovered planet that has the same average temperature as Earth but an abundance of hydrogen sulfide, H_2S, instead of H_2O. Would you expect H_2S on this planet to exist as a liquid, just as H_2O exists as liquid water on Earth? Explain the chemical basis for your answer.

Many molecules are capable of hydrogen-bonding interactions. This is especially true for molecules containing a polar functional group that includes an H atom, such as an alcohol (—OH) or an amine (—NH₂). The capacity to form hydrogen bonds significantly affects the properties of molecules. As one example,

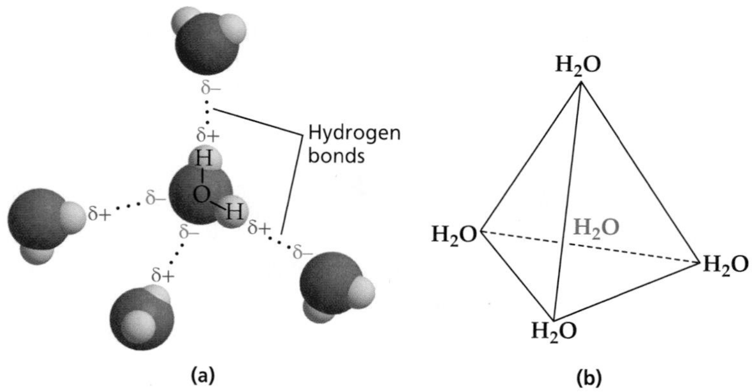

FIGURE 8.13 The spatial orientation of hydrogen bonds. (a) The four hydrogen bonds around a central H₂O molecule are oriented in a tetrahedral geometry. (b) The shape of a tetrahedron highlights the spatial arrangement of the four outer H₂O molecules with respect to the central H₂O molecule (highlighted in red).

methane (CH$_4$) is a gas at room temperature, whereas methanol (CH$_3$OH) is a liquid because of the hydrogen-bonding interactions of the —OH group. Hydrogen bonds also stabilize the three-dimensional structures of large biological molecules such as DNA and proteins. Hydrogen bonds also allow these molecules to be flexible, because these weak bonds can easily be broken when a molecule changes shape. In this manner, hydrogen bonding allows biological molecules to achieve both stability and flexibility.

Core Concepts

- A **hydrogen bond** is an electrical attraction between a hydrogen atom in a polar bond and another highly electronegative atom (e.g., N, O, F).
- A covalent bond within a single H$_2$O molecule is a strong **intramolecular bond**. By contrast, a hydrogen bond between two molecules is a weak **intermolecular bond.**
- One H$_2$O molecule is capable of forming four hydrogen bonds with neighboring H$_2$O molecules. The two H atoms in H$_2$O serve as **hydrogen bond donors**, and the O atom (with two lone pairs) serves as a **hydrogen bond acceptor** for two hydrogen bonds.
- In liquid water, the four hydrogen bonds that form among neighboring H$_2$O molecules create an approximately tetrahedral geometry.
- H$_2$O molecules form a liquid at room temperature because the combined effect of many hydrogen-bonding interactions cause the molecules to stick together.

FIGURE 8.14 An iceberg floating in liquid water with a cloud overhead. The H$_2$O molecule is unique in its ability to exist as solid, liquid, and gas at temperatures and pressures found at the Earth's surface.

Molecular Organization of Ice, Water, and Vapor

One unique property of H$_2$O is its ability to exist in all three states of matter—solid, liquid, and gas—under conditions of temperature and pressure that are commonly encountered on Earth. Consider the photograph of an iceberg in Figure 8.14. The iceberg is composed of H$_2$O arranged in a *solid* structure, the water in which it floats is *liquid* H$_2$O, and the clouds overhead contain H$_2$O as water *vapor*, which is the gaseous form of H$_2$O. Although water vapor is colorless, we can see clouds because the vapor condenses on dust particles in the air to form microscopic water droplets. The physical characteristics of ice, water, and vapor are very different. For example, water will easily flow around in your hand, but ice is remarkably hard. How can the same molecule produce substances with such distinctive properties?

The differences among ice, water, and vapor arise from changes in the hydrogen-bonding interactions among H$_2$O molecules that occur at different temperatures (Figure 8.15). We begin with ice, the solid form of H$_2$O that exists at or below 0°C, which is the *freezing point* of water (Figure 8.15(a)). At this low temperature, the hydrogen bonds lock the H$_2$O molecules into a fixed location to make a solid structure called an *ice crystal*. Why is the structure of ice so hard? Each H$_2$O

THE KEY IDEA: Ice, liquid water, and water vapor differ because of changes in the hydrogen-bonding interactions among H$_2$O molecules that occur at different temperatures.

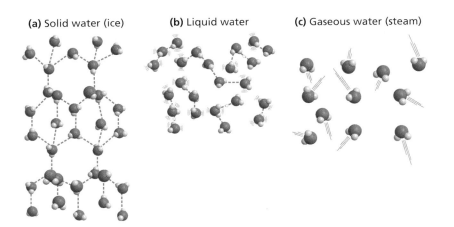

(a) Solid water (ice) **(b)** Liquid water **(c)** Gaseous water (steam)

FIGURE 8.15 A comparison of ice, water, and vapor. A molecular view of H$_2$O molecules in (a) solid ice, (b) liquid water, and (c) gaseous vapor. Oxygen atoms are displayed as red circles, and hydrogen atoms as smaller white circles. The dashed lines around the atoms in liquid water and water vapor represent the speed of motion of the molecules.

FIGURE 8.16 The geometry of a snowflake. The six-sided shape of snowflakes arises from the six-fold symmetry of hydrogen-bonded H_2O molecules in ice crystals.

molecule is hydrogen-bonded to four neighbors in a tetrahedral geometry (refer back to Figure 8.13). This configuration creates a very strong three-dimensional network of hydrogen bonds. The geometry of the hydrogen-bonding interactions in ice creates a six-fold symmetry, which you can observe by looking at the positions of the oxygen atoms in Figure 8.15(a). This symmetry is responsible for the six-sided shape of snowflakes, as illustrated in Figure 8.16.

Why does ice melt when the temperature rises above 0°C? The answer lies in the competition between two opposing effects: the energy of the hydrogen bonds, which holds the H_2O molecules together, and the energy of molecular motion, which pulls them apart. This energy of motion—called *kinetic energy*—is related to temperature. Higher temperatures correspond to faster speeds of molecular motion and greater kinetic energy.

When the temperature is low enough to form solid ice, the kinetic energy of the H_2O molecules is relatively small. The strength of the hydrogen bonds is therefore sufficient to hold the molecules in a rigid crystal. As heat energy is added, the kinetic energy of the H_2O molecules begins to increase. At first, a few H_2O molecules begin to distort the rigid hydrogen-bonding network in ice. As the temperature continues to rise, the amount of molecular motion becomes sufficient to break more of the hydrogen bonds among the H_2O molecules. The solid structure of ice begins to crumble, eventually producing the fluid form of liquid water (Figure 8.15(b)).

The H_2O molecules in liquid water are constantly moving, in contrast to the fixed molecular structure of in ice. Each H_2O molecule rapidly makes and breaks hydrogen bonds with various neighboring molecules, like a molecular dance with constantly changing partners. At the molecular level, liquid water is a random, jostling arrangement of H_2O molecules held together by a dynamic network of hydrogen bonds. Because of the molecular motion, the hydrogen-bonding interactions can deviate from the ideal tetrahedral geometry. For a fleeting moment, an H_2O molecule may form only three hydrogen bonds as it switches from one partner to another. The configuration of H_2O molecules in one region of water is different from that in other regions, reflecting the absence of any long-range structural order. If we were able to take a sequence of quick molecular snapshots, of liquid water, each picture would show the H_2O molecules in slightly different positions.

What happens during the transition from liquid water to water vapor? As we continue to heat the water, the molecules begin to move even faster. This additional kinetic energy causes the H_2O molecules to pull and distort the hydrogen bonds within liquid water. At 100°C, the *boiling point* of water, some H_2O molecules have sufficient kinetic energy to break free of the hydrogen-bonding network and form water vapor. As we continue to heat the water, even more molecules escape. After enough time, all of the liquid water will turn into a vapor. In water vapor, the H_2O molecules no longer have any hydrogen bonding with one another. Instead, each H_2O molecule behaves independently, only rarely colliding with its neighbors (Figure 8.15(c)). As a result, water vapor occupies a much larger volume than the same amount of H_2O molecules in liquid water.

PRACTICE EXERCISE 8.1

Question: Methanol (CH_3OH) is a liquid at room temperature because of intermolecular hydrogen-bonding interactions. The molecular structure of methanol is shown below.

(a) Which of the covalent bonds in methanol are polar? Add partial charges to the atoms that participate in polar bonds.

(b) The C—H bonds in methanol do *not* form hydrogen bonds. Why not?

(c) Draw a diagram that indicates the hydrogen-bonding interactions among four methanol molecules. Use dashed lines to represent the hydrogen bonds. What is the maximum number of hydrogen bonds that each methanol molecule can form?

(d) Do you predict that the boiling point of methanol will be higher than, lower than, or the same as that of liquid water? Explain your answer.

CHEMISTRY IN YOUR LIFE

Chemical and Biological Antifreeze

If you drive a car in a climate with very cold winters, then you need to use *antifreeze* in your car's radiator. The most common antifreeze is a small molecule called ethylene glycol, which is illustrated in Figure 1. The usual practice is to add a mixture of water and ethylene glycol to the radiator. The presence of ethylene glycol lowers the freezing point of the mixture below that of pure water. For example, a 50:50 mixture of ethylene glycol and water has a freezing point of –37°C. Adding an antifreeze prevents the radiator contents from turning into ice when the temperature drops below 0°C. Ethylene glycol also raises the boiling point of the mixture, so it will not turn into vapor (steam) in hot desert temperatures.

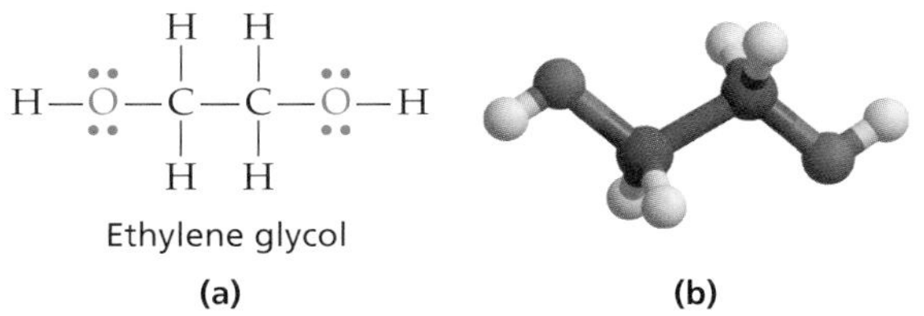

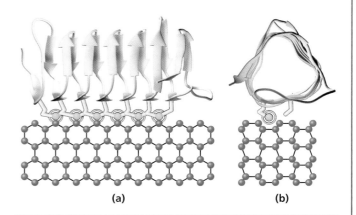

FIGURE 1 Ethylene glycol acts as an antifreeze. (a) The chemical structure and (b) a molecular model of ethylene glycol, the most commonly used antifreeze in automobiles.

How does ethylene glycol inhibit the freezing of water? In effect, ethylene glycol lowers the freezing point by preventing the formation of ice. The molecular structure of ethylene glycol contains two alcohol functional groups with polar O—H bonds. These alcohol groups enable ethylene glycol to form hydrogen bonds with H$_2$O molecules. Because the shapes of the two molecules are different, ethylene glycol interferes with the ability of H$_2$O to form the tetrahedral network of hydrogen bonds that is found in solid ice. In this manner, the presence of ethylene glycol prevents the H$_2$O molecules from freezing. However, ethylene glycol is very toxic, so it must be kept away from children and pets.

Biological organisms also confront the danger of freezing. Some insects exist in temperatures that should freeze their internal fluids. Some types of fish swim in ocean water that is below 0°C (due to the presence of salt). These organisms contain a particular type of protein called an *antifreeze protein*. Unlike ethylene glycol,

antifreeze proteins do not lower the freezing point of liquid water. Instead, they interfere with the molecular process by which ice begins to form.

The freezing transition from water into ice does not take place uniformly throughout the liquid. Instead, it begins at specific points that serve as the "nucleus" for ice formation; for this reason, the process is called *nucleation*. The first step in nucleation is the formation of a tiny ice crystal, which then grows by the successive addition of H$_2$O molecules to form bulk ice. The antifreeze proteins bind tightly to the surface of these nascent ice crystals, preventing them from growing (Figure 2). Remarkably, insects and fish can survive with these tiny ice crystals within their blood.

FIGURE 2 An antifreeze protein. (a) The structure of an insect antifreeze protein bound to a surface of ice. (b) A side view of the same protein showing how chemical groups from the protein insert into the ice layer, preventing growth of ice crystals.

Antifreeze proteins potentially have many beneficial applications. Unlike ethylene glycol, these proteins are not toxic. In addition, they can inhibit ice formation at far lower concentrations than conventional antifreeze chemicals. Scientists are currently studying antifreeze proteins as a way to prevent ice damage to frozen food and to improve the preservation of organs at low temperatures for surgical transplants.

Core Concepts

- The differences among solid ice, liquid water, and water vapor arise from changes in hydrogen-bonding interactions that occur at different temperatures.
- In *solid ice*, which forms at or below 0°C, the H_2O molecules form a rigid hydrogen-bonded structure. Each H_2O molecule participates in a tetrahedral bonding arrangement with four neighboring molecules. The H_2O molecules have low kinetic energy, so the hydrogen bonds are able to lock them into a fixed position.
- In *liquid water*, which exists at room temperature, the molecules have increased kinetic energy. The result is a dynamic hydrogen-bonding network in which the interactions among neighboring molecules are constantly changing.
- When the temperature is raised to 100°C, the H_2O molecules gain sufficient kinetic energy to break the hydrogen bonds and form *water vapor*. This vapor contains free H_2O molecules, with no intermolecular hydrogen bonding.

8.4 The Unusual Properties of Water

To conclude the chapter, we will examine four distinctive properties of water:

- Liquid water is denser than solid ice.
- Water has an unusually high boiling point.
- Water has a large capacity for absorbing heat energy.
- Boiling water requires a large input of energy.

Liquid Water Is Denser than Solid Ice

What happens if you put an ice cube in water? The ice cube becomes partially submerged, but it floats on the surface. This behavior of ice and water is quite unusual. If we make a solid by cooling another liquid such as methanol or benzene, the solid sinks when placed in the liquid.

Ice floats in water because *the density of solid ice is slightly less than the density of liquid water*. As discussed in Chapter 2, the density of a substance is defined as the mass per unit volume.

$$\text{density} \left(g/cm^3 \right) = \frac{\text{mass} \left(g \right)}{\text{volume} \left(cm^3 \right)}$$

The density of water is 1.00 g/cm³. If we put a solid cube of iron in water, the iron will sink because its density (7.60 g/cm³) is much greater than that of water. By contrast, a cork from a wine bottle will float in water because its density (0.25 g/cm³) is less than that of water. Ice has a density of 0.92 g/cm³, so 1 cm³ of ice has a lower mass of 0.92 g compared to 1.00 g for 1 cm³ of liquid water.

The different densities of ice and water arise from the arrangement of H_2O molecules in these two substances. Figure 8.17 compares computer models of ice and water that show the H_2O molecules as space-filling models. As you can see, the rigid hydrogen-bonding structure in ice leaves large empty cavities among the H_2O molecules. By contrast, H_2O molecules in water are mobile and can squeeze closer together, so there are no cavities in liquid water like those in solid ice. Because of this tight packing, 1 cm³ of water contains more H_2O molecules than the same volume of ice. Because more molecules correspond to a larger mass, and the volume remains the same, the density (g/cm³) of water is greater than that of ice.

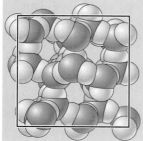
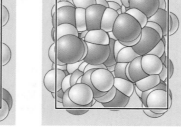

Ice   Water

FIGURE 8.17 Solid ice is less dense than liquid water. Computer models of solid ice and liquid show the H_2O molecules as space-filling models. The rigid hydrogen-bonding arrangement in ice produces empty cavities among the H_2O molecules. In water, the H_2O molecules are mobile and can squeeze close to one another. As a result, a higher number of H_2O molecules can fit within in a given volume of liquid water.

This chemical feature of water has important biological consequences for aquatic organisms such as fish. When the temperature falls below 0°C, the surface water in ponds and lakes freezes. The solid ice, which is less dense, remains on the surface, while the denser water remains liquid underneath. Consequently, fish can swim within the liquid water even though the surface is frozen solid. The sport of ice fishing depends on this unusual property of water (Figure 8.18).

Core Concepts

- The rigid, hydrogen-bonded structure of solid ice contains large, empty cavities among the H_2O molecules. By contrast, there are no cavities in liquid water because the moving H_2O molecules can get closer to one another. As a result, the density of water is greater than that of ice.

Water Has an Unusually High Boiling Point

This chapter has stressed the essential role of liquid water in sustaining life on Earth. Yet the very fact that water is a liquid under the conditions found on Earth is a chemical anomaly. We can appreciate this unusual behavior by comparing the properties of H_2O with similar molecules.

In Chapter 2, we learned that elements in the periodic table are arranged in *groups* that have similar chemical properties. Elements in Groups 4A and 6A of the periodic table form chemical compounds with hydrogen, which are known as *hydrides*. Figure 8.19 compares the boiling point temperatures for eight hydrides under atmospheric pressure. The graph also shows room temperature (25°C) for comparison.

Let's begin by examining the hydrides of the Group 4A elements, which include carbon (C), silicon (Si), germanium (Ge), and tin (Sn). You can observe an obvious trend in this graph: The boiling point for the liquid is lower for lower atomic numbers. For example, methane (CH_4) boils at –164°C, whereas stannane (SnH_4) boils at the higher temperature of –52°C. All of these compounds are gases at room temperature because the attraction among the molecules is not strong enough to stick them together and form a liquid. They turn into liquids only at low temperatures, where the kinetic energy of the molecules is greatly reduced.

Moving to Group 6A, we observe a similar trend for the boiling points of the hydrides of sulfur (S), selenium (Se), and tellurium (Te). Although these values are somewhat higher than those for Group 4A, they all fall below 0°C, and they decrease as the atomic number becomes smaller. If we extrapolate this linear trend line to H_2O, we predict a boiling point of approximately –100°C. Based on this prediction, all H_2O on Earth should exist as water vapor—an unsuitable environment for the development of life. Fortunately for us, the prediction is not correct. The actual boiling point of water is 100°C, which deviates from the predicted value by a large margin of 200°C. This unexpectedly high boiling point arises from extensive intermolecular hydrogen bonding among H_2O molecules, which is not present for the other hydrides of Group 6A.

Core Concepts

- Periodic trends for hydride compounds predict that water should boil at –100°C, but it actually boils at +100°C. This large discrepancy arises from extensive intermolecular hydrogen bonding among H_2O molecules in liquid water.

Water Has a Large Capacity for Absorbing Heat Energy

When boiling water for cooking, you may have noticed that the metal pan becomes hot to the touch much more quickly than the water inside does. This observation reveals another unusual property of water—it has a particularly

FIGURE 8.18 Ice fishing. Even though the surface of the lake is frozen, fish survive within the denser water underneath.

THE KEY IDEA: Water's unusually high boiling point arises from extensive hydrogen bonding among H_2O molecules.

THE KEY IDEA: A large amount of heat energy is required to raise the temperature of liquid water.

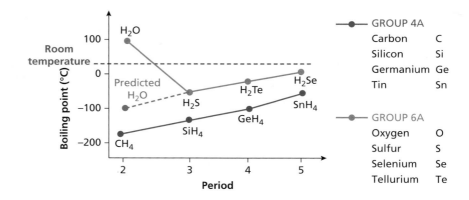

specific heat capacity The amount of heat energy required to raise the temperature of 1 gram of a substance by 1 degree Celsius.

large capacity for absorbing heat energy. The ability of a substance to absorb heat energy is measured by its *specific heat capacity*. The **specific heat capacity** of a substance is the amount of heat energy required to raise the temperature of 1 gram of the substance by 1 degree Celsius.

The relationship between the heat energy input and temperature rise of a substance can be expressed as the equation

$$Q = m \cdot C \cdot \Delta T$$

that contains the following quantities (with units):

$$Q = \text{heat energy added (joules, J)}$$
$$m = \text{mass of substance (grams, g)}$$
$$c = \text{specific heat capacity (joules per gram per degree Celsius, J/g°C)}$$
$$\Delta T = \text{temperature change (degrees Celsius, °C)}$$

Table 8.2 compares the specific heat capacities of three metals (copper, iron, and aluminum) plus two liquids (ethanol and water). Note that the specific heat capacity of water is more than 10 times larger than that of copper (which is used for many cooking pans). In fact, *liquid water has the highest specific heat capacities of any substance that we encounter in everyday life.*

We can illustrate different heat capacities by plotting a graph that compares the temperature rise for 1 g samples of aluminum and water when heat energy is added (Figure 8.20). From reading the graph, we see that it takes 90 J of energy to raise the temperature of aluminum by 100°C. For the same input of heat energy, water rises in temperature by only 21.5°C. Even with an energy input of 200 J, water's temperature increase does not reach 50°C.

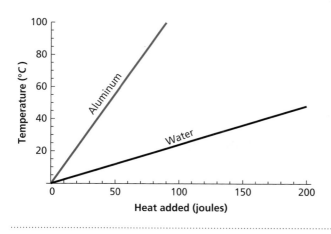

FIGURE 8.20 **A comparison of specific heat capacities.** The graph shows the temperature increase (°C) that occurs when heat energy (J) is added to a 1-g sample of aluminum and water. A comparison of the two graph lines shows that the same amount of heat input produces a smaller temperature rise for water compared to aluminum. This difference occurs because water has a greater heat-absorbing capacity compared to aluminum.

CONCEPT QUESTION 8.4

(a) On the graph in Figure 8.20, add a line that depicts the temperature rise of 1 g of copper when heat energy is added. (b) Similarly, add a second line that depicts the temperature rise of 1 g of ethanol. (c) For a specific input of energy, such as 100 J, which substance (copper or ethanol) will exhibit the larger increase in temperature? Refer to Table 8.2 for specific heat capacities.

TABLE 8.2	Examples of Specific Heat Capacities
SUBSTANCE	**SPECIFIC HEAT CAPACITY (J/g°C)**
Copper (Cu)	0.386
Iron (Fe)	0.450
Aluminum (Al)	0.900
Ethanol (C_2H_5OH)	2.460
Water (H_2O)	4.184

WORKED EXAMPLE 8.3

Question: Calculate the amount of heat energy required to raise the temperature of 50.0 g of iron by 10.0°C.

Answer: We use the equation

$$Q = m \cdot C \cdot \Delta T$$

and the following quantities for the sample of iron:

$$m = 50.0 \text{ g} \qquad c = 0.450 \text{ J/g°C (from Table 8.2)} \qquad \Delta T = 10.0°C$$

Using these quantities, we can calculate the heat input, Q.

$$Q = 50.0 \, \cancel{g} \times 0.450 \, \frac{J}{\cancel{g} \, \cancel{°C}} \times 10.0 \, \cancel{°C} = 225 \text{ J}$$

Note that the units of g and °C cancel within the equation to produce a final result in joules (J).

TRY IT YOURSELF 8.3

Question: In the introduction to this section, we stated that a metal cooking pan becomes hotter to the touch much faster than the water it contains. We can test this statement using a calculation. (a) Suppose that a cooking pan is made of pure copper and has a mass of 250 g. Calculate the increase in temperature (ΔT) that occurs when the copper pan is supplied with 10,000 J of heat energy. (b) Calculate the increase in temperature (ΔT) that occurs when the same mass of water (250 g) is supplied with the same amount of heat energy (10,000 J). (c) Write a sentence to relate these two numbers to the statement at the beginning of the question.

PRACTICE EXERCISE 8.2

Question: In an experiment on a solid material, 480 J of heat energy was required to raise the temperature of a 20.0 g sample by 40.0°C. Calculate the specific heat capacity of the material, and provide the correct units.

Why does water have such a large specific heat capacity? We can provide an answer by visualizing what happens at the molecular level as we begin to supply heat energy to the liquid. A small fraction of this energy is used to speed up the motion of the molecules, increasing their kinetic energy and producing a rise in temperature. However, most of the heat energy input is used to break the hydrogen bonds among the H_2O molecules. Breaking these hydrogen bonds separates the H_2O molecules, but it does *not* increase their kinetic energy and thus does not affect the temperature of the water As a result of these molecular changes, only a fraction of the heat input is transferred to the kinetic energy of the H_2O molecules. Consequently, a relatively large amount of heat energy is needed to raise the temperature of a gram of water by 1°C. If the hydrogen bonds were not present, then all of the heat input would be converted to kinetic energy, and the temperature would increase by a greater amount.

The large specific heat capacity of water has an important consequence for life. Most complex organisms, including human, can survive only within a narrow range of temperatures. Normal human body temperature is 37°C (98.6°F), based on an oral thermometer measurement. Even small deviations from this average can result in serious medical conditions. For example, *heat-stroke* is defined as having an elevated body temperature of 40°C (104°F), and *hypothermia* occurs when the body temperature falls to 35°C (95°F). The large amount of water in our bodies (roughly 60% by mass) regulates our body temperature, enabling us to gain or lose a large amount of heat energy without a dangerous change in body temperature.

CONCEPT QUESTION 8.5

Do you predict that the specific heat capacity of water vapor will be larger, smaller, or the same as the specific heat capacity of liquid water? Provide a chemical explanation for your prediction. Use the Internet to find the specific heat of water vapor. Was your prediction correct?

Core Concepts

- The ability of a substance to absorb heat energy is measured by its **specific heat capacity**. Water has a very high specific heat capacity, which means that it can absorb a large amount of heat with only a small rise in temperature.

⊙╌ THE KEY IDEA: Phase changes between ice, water, and vapor require an input of energy that does not change the temperature.

phase change A transition between two phases of matter; for example, solid ice melts to form liquid water.

latent heat The amount of heat energy required to produce a phase change without an increase in temperature.

Boiling Water Requires a Large Input of Energy

Suppose that you want to cool a cup of soda. Which would work best: adding an amount of cold water, or adding the same mass of ice? Based on our experience, we typically add ice. We also know that the ice melts over a period of time. Why is ice so effective at cooling our drink?

Earlier in the chapter, we examined the molecular differences between solid ice, liquid water, and water vapor. These three states of matter are called *phases*. The transition between two phases is called a **phase change**. We are already familiar with several phase changes: ice to water (melting), water to ice (freezing), water to vapor (boiling), and vapor to water (condensation).

Figure 8.21 provides a graph that illustrates these phase changes. In particular, it shows the response of ice, water, and vapor to the addition of heat energy (Q). Let's begin with ice at the lower left region of the graph (gray line). When ice below 0°C is heated, the temperature of the ice begins to rise. The amount of the temperature increase depends on the specific heat capacity of ice (as discussed in the previous section). When the ice reaches 0°C, something interesting happens: *adding more heat energy does not increase the temperature of the ice*. This effect is represented by the flat region of the graph at 0°C. What is happening?

In fact, *the heat energy causes a phase change from solid ice to liquid water (melting)*. At a molecular level, the heat energy breaks some of the hydrogen bonds between H_2O molecules in ice. This change enables the H_2O molecules to move more freely, which causes the solid structure of ice to melt and produce the fluid structure of water. We use the term **latent heat** to describe the amount of heat energy required to produce a phase change without an increase in temperature (*latent* means "hidden"). In particular, the energy required to change ice to water is called the *latest heat of fusion* (*fusion* is another word for "melting").

After all the ice has been converted to water, adding more heat energy causes the water temperature to rise (the blue line in the graph). This process continues until the water reaches 100°C. At this point, the graph exhibits another flat region with no temperature increase (shown in red). In this region, the input of heat energy produces a phase change from liquid water to water vapor. The heat energy required to cause this transition is called the *latent heat of vaporization*. In this context, vaporization is another word for what we commonly call boiling. The heat energy is used to break all the hydrogen bonds among the H_2O molecules in liquid water, which enables the molecules to escape from the liquid and form a vapor. Once no liquid water remains, adding further heat energy increases the temperature of the water vapor (the green line in the graph).

The amount of energy required to convert water into vapor is very large and much greater than the amount of energy required to turn ice into water. This observation is represented in Figure 8.21 by the greater length of the red line compared to the black line. The reason for the difference is that the ice-to-water transition requires only some of the hydrogen bonds to be broken because hydrogen bonding still exists in liquid water. By contrast, the water-to-vapor transition involves breaking *all* the hydrogen bonds among the H_2O molecules in liquid water, which requires a much greater input of heat energy.

Some important practical consequences arise from these phase changes. Consider, for example, why we add ice to cool our soda drink instead of cold water. It is not the temperature of the ice that cools the drink but rather the *melting* of the ice. The heat input to melt the ice comes from the liquid in the cup that surrounds the ice. As a result, the liquid loses some heat energy and its temperature is lowered.

Emergency caregivers know that burns caused by steam are more severe than those caused by boiling water. When water vapor touches the skin, some of it turns into liquid water. This process of condensation is the reverse of boiling, and it releases heat energy instead of absorbing energy. Because the latent heat of vaporization is so large, the condensation of steam releases enough energy to greatly increase the severity of a burn.

In addition, H_2O molecules in the Earth's atmosphere provide the energy needed to drive hurricanes. When water condenses, it releases the energy stored in the vapor. As long as hurricanes are above the ocean, the vapor above the ocean's surface powers the storm through condensation. But once the hurricane moves over land, the supply of energy is interrupted and the storm's intensity begins to dissipate.

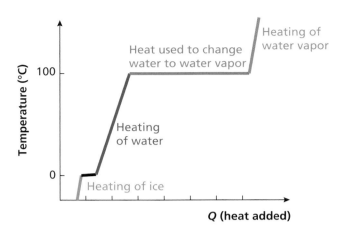

FIGURE 8.21 A graphical illustration of phase changes. In the sloped regions of the graph, adding heat energy to ice (gray line), water (blue line), or water vapor (green line) causes an increase in their temperature. The flat regions of the graph indicate phase changes from ice to water (black line) and from water to vapor (red line). During these phase changes, adding heat energy does not change the temperature.

Core Concepts

- A **phase change** is a transition between two different phases of matter. Examples include the conversion of solid ice into liquid water (melting) or the transformation of liquid water into water vapor (boiling).
- Phase changes require an input of heat energy, but there is no change in temperature while the phase change is occurring. The amount of energy required to cause a phase change is called the **latent heat**.
- The quantity of energy that is necessary to boil water (the latent heat of vaporization) is particularly large. This heat energy is required to break all the hydrogen bonds among H_2O molecules in liquid water.

CHEMISTRY AND YOUR HEALTH

Why Does Perspiring Make You Cooler?

When you exercise, your internal body temperature rises. This causes you to perspire, which releases water onto the surface of your skin (Figure 1). Why does this happen? Perspiration cools your body because of the unusual properties of water.

We know that water turns into a vapor at its boiling point of 100°C. Yet water can also turn into vapor *below the boiling point*—this process is called *evaporation*. It is the reason that a puddle of rainwater disappears during a sunny day, even though the temperature remains far below the boiling point.

Evaporation occurs at the *surface* of liquid water (Figure 2). Because of their random motion, some H_2O molecules have a higher kinetic energy than others. These high-energy molecules can escape from the surface of the water and become water vapor. Molecules with lower kinetic energy are left behind in the liquid. As a result, the loss of higher-energy molecules has a *cooling effect* on the liquid because a lower kinetic energy corresponds to lower temperature.

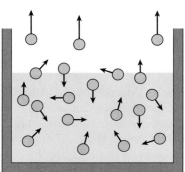

High energy:
Evaporating

Medium energy:
Pulled back into wat(

Lower energy:
Remain as liquid

FIGURE 2 Evaporation occurs at the surface of a liquid.
High-energy molecules escape from the surface and turn into a gas, leaving the less energetic molecules behind.

The evaporation of perspiration from the surface of our skin is a very effective cooling mechanism for our body. Turning liquid water into water vapor requires a large amount of energy, because the H_2O molecules need to break free of their hydrogen bonds to escape the liquid surface. As a result, the H_2O molecules that manage to escape have a significant cooling effect on the remaining liquid.

On a hot, muggy summer day, you may notice that you perspire but the water remains on your skin without evaporating. This occurs when there is high *humidity*, which corresponds to more water vapor in the air. If the air already contains a significant amount of water vapor, then it is less capable of absorbing additional water vapor from your perspiring skin. You can think of the air as being like a kitchen sponge that is already saturated with water and cannot absorb any more. Perspiring without evaporation doesn't cool our skin, which is why we feel so unpleasantly hot and damp on humid days.

FIGURE 1 Perspiration releases water onto the surface of the skin as a method for cooling the body.

8.1   Why Is Water Essential for Life?

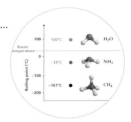

Learning Objective:

Characterize the ways in which water's properties sustain life on Earth.

⊚ Every living organism on Earth requires water. Water is the most abundant substance in the human body, accounting for 60% of its mass.

⊚ H_2O is a liquid at room temperature (approximately 25°C), whereas CH_4 and NH_3 are both gases. Water remains a liquid over a large temperature range, from 0°C to 100°C.

⊚ Liquid water provides a good environment to sustain life. It allows different types of molecules to interact and produce the chemical reactions that are necessary for life. Water also dissolves a wide variety of molecules and ions.

8.2   Chemical Bonding in H_2O

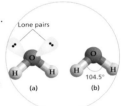

Learning Objective:

Compare and contrast polar and nonpolar covalent bonds.

⊚ An H_2O molecule has two covalent O—H bonds and two lone pairs of electrons. Electron repulsion orients the four electron pairs (two bonding pairs and two lone pairs) in an approximately tetrahedral geometry.

⊚ The structure of the H_2O molecule is bent with an H—O—H bond angle of 104.5°.

⊚ When a covalent bond forms between two identical atoms (e.g., H—H), they share the electron pair equally.

⊚ When a covalent bond forms between two different atoms (e.g., H—F), the electron pair is not shared equally. The F atom exerts a stronger attraction on the bonding electron pair than the H atom does.

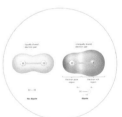

⊚ In H—F, the unequal sharing of the electron pair creates **partial charges** on the atoms. The F atom gains a net negative charge ($\delta-$), whereas the H atom is left with a net positive charge ($\delta+$).

⊚ The separation of partial charges in H—F creates a **dipole**; therefore, the bond is called a **polar covalent bond**. There are no partial charges and no dipole in H—H, so it contains a **nonpolar covalent bond**.

⊚ **Electronegativity** measures the ability of an atom to attract the shared pair of electrons in a chemical bond Atoms with a higher electronegativity exert a stronger attraction for the electron pair.

- Electronegativity is measured using the Pauling scale (0–4). The most electronegative element is fluorine (4.0), followed by oxygen (3.5).

- The type of bond formed between two atoms (A and B) depends on the difference in their electronegativity.

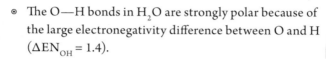

$$\Delta EN_{AB} = EN_A - EN_B$$

$\Delta EN_{AB} < 0.5$ nonpolar covalent bond

$\Delta EN_{AB} = 0.5\text{–}2.0$ polar covalent bond

$\Delta EN_{AB} > 2.0$ ionic bond

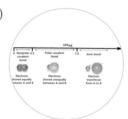

- The O—H bonds in H_2O are strongly polar because of the large electronegativity difference between O and H ($\Delta EN_{OH} = 1.4$).

- Each O—H bond forms a dipole. Because of the bent geometry of H_2O, the two **bond dipoles** add together to create a larger **molecular dipole**. As a result, H_2O is a strongly polar molecule.

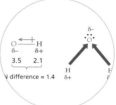

8.3   From H_2O Molecules to Liquid Water

Learning Objective:

Explain why H_2O is a liquid at room temperature.

- H_2O molecules form a liquid at room temperature because of intermolecular hydrogen bonds.

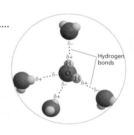

- A **hydrogen bond** is an electrical attraction between a hydrogen atom in a polar bond and another highly electronegative atom (e.g., N, O, F). The bond is oriented toward a nonbonding electron pair on the electronegative atom.

- One H_2O molecule is capable of forming four hydrogen bonds with neighboring H_2O molecules. The central H_2O molecule serves as a **hydrogen bond donor** for two bonds, and a **hydrogen bond acceptor** for another two bonds.

- Hydrogen bonding among neighboring H_2O molecules occurs with an approximately tetrahedral geometry.

- Ice (solid), water (liquid), and water vapor (gas) are all composed of H_2O molecules.

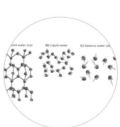

- In ice, the kinetic energy of the H_2O molecules is low, so the hydrogen bonds can lock the molecules into a solid crystal.

- In water, increased kinetic energy produces dynamic hydrogen-bonding interactions in which moving H_2O molecules are constantly changing partners.

- In water vapor, the kinetic energy is sufficiently large to break all the hydrogen bonds among H_2O molecules.

8.4   The Unsual Properties of Water

Learning Objective:

Illustrate how water's unusual properties are related to hydrogen bonding.

⊚ Solid ice is less dense than liquid water, but most substances are more dense as solids than they are as liquids.

⊚ In ice, the hydrogen-bonded structure of H_2O molecules contains large empty cavities. In water, the motion of H_2O molecules fills these cavities, so more H_2O molecules are packed within the same volume.

⊚ Based on trends from the periodic table, we predict that water has a boiling point of $-100°C$. In fact, the boiling point of water is $+100°C$.

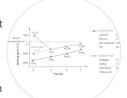

⊚ The exceptionally high boiling point of water arises from extensive hydrogen bonding among H_2O molecules.

⊚ Water has an unusually high **specific heat capacity**, the ability to absorb heat energy. It has the highest value for any substance that we encounter in everyday life.

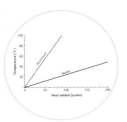

⊚ When water absorbs heat, most of the energy is used to break hydrogen bonds among the H_2O molecules. Only a small fraction is converted into kinetic energy of the molecules, which produces a rise in temperature.

⊚ **Phase changes** are transitions between different phases of matter (e.g., ice and water). Phase changes require an input of heat energy, but there is no change in temperature. The amount of energy required to cause a phase change is called the **latent heat**.

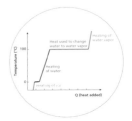

⊚ The quantity of energy necessary to boil water (the latent heat of vaporization) is particularly large. This heat energy is used to break all the hydrogen bonds among H_2O molecules in liquid water, which releases the H_2O molecules as a vapor.

LEARNING RESOURCES

Reviewing Knowledge

8.1: Why Is Water Essential for Life?

1. What properties of liquid water are important for sustaining life?
2. Why is H_2O unique among all molecules?

8.2: Chemical Bonding in H_2O

3. What causes the bent geometry of the H_2O molecule?
4. Why does the bond angle in H_2O ($104.5°$) differ slightly from the ideal tetrahedral angle of $109.5°$?

5. Describe how electron sharing in an H—H covalent bond differs from that in an H—F bond.
6. Sketch the electron density for the covalent bonds in H—H and H—F. Explain why the electron density distribution is different for these two molecules.
7. What is electronegativity? What scale is used to assign electronegativity values to the elements?
8. Explain how the unequal sharing of electrons in an H—F molecule generates partial charges on the H and F. Which atom becomes more negatively charged ($\delta-$), and which becomes more positively charged ($\delta+$)?
9. What is a dipole? Why does H—F have a dipole, whereas H—H does not?

10. Define the following terms:
    polar covalent bond   nonpolar covalent bond
    polar molecule   nonpolar molecule
11. What two trends are observed for the electronegativity values of elements in the periodic table?
12. Describe how the electronegativity difference (ΔEN) is used to classify covalent bonds as nonpolar, polar, and ionic.
13. Sketch the structure of an H_2O molecule, and assign partial charges ($\delta-$ or $\delta+$) to all the atoms.

8.3: From H_2O Molecules to Liquid Water

14. What is the difference between an intramolecular bond and an intermolecular bond?
15. Describe a hydrogen bond, and explain how it differs from a covalent bond.
16. Sketch the hydrogen-bonding interactions that occur among a central H_2O molecule and four neighboring H_2O molecules.
17. Methane (CH_4) is a gas at room temperature, whereas methanol (CH_3OH) is a liquid. Why?
18. Describe how the H_2O molecules are arranged in solid ice. How does this arrangement explain why ice is so hard?
19. What changes occur at the molecular level when ice melts to become liquid water? What type of interaction exists among the H_2O molecules in water?
20. What changes occur at the molecular level when water boils to become a vapor? What is the role of kinetic energy in this process?
21. Explain why water is a liquid at room temperature, whereas other molecules with similar sizes (CH_4, NH_3) are gases.

8.4: The Unusual Properties of Water

22. What is unusual about the density of solid ice in comparison with the density of liquid water? How can this density difference be explained in terms of the organization of the H_2O molecules in these two phases?
23. Why is the boiling point of H_2O considered unusual in comparison with the hydrides of oxygen's chemical neighbors in Group 6 of the periodic table? What property of water accounts for this anomaly?
24. Explain why water has an unusually large capacity to absorb heat energy.
25. When the temperature of water reaches 100°C, adding extra heat energy does not increase the temperature further. Explain this observation.

Developing Skills

26. Which of the following elements does not have an electronegativity value? Explain your answer.
    sulfur (S)   carbon (C)   neon (Ne)   chlorine (Cl)
27. For each covalent bond below, (a) assign the appropriate partial charge ($\delta-$ or $\delta+$) for each atom, and (b) draw an arrow to indicate the direction of the dipole for the bond.
    O—N   N—H   Cl—O   S—N

28. For each covalent bond below, (a) assign the appropriate partial charge ($\delta-$ or $\delta+$) for each atom, and (b) draw an arrow to indicate the direction of the dipole for the bond
    N—P   S—O   F—Cl
29. Consider the chemical bonds between the following pairs of elements. Predict whether the bond is nonpolar covalent, polar covalent, or ionic.
    (a) Na and O
    (b) N and Cl
    (c) Ca and F
    (d) S and O
30. Silane (SiH_4) is a molecule that is analogous to methane (CH_4). Compare the properties of Si—H and C—H bonds by (a) calculating the electronegativity difference (ΔEN) and (b) assigning partial charges to each atom. What do you notice when comparing the partial charge on the H atoms in each bond?
31. Ammonia has the chemical formula NH_3. (a) Is the N—H bond classified as polar or nonpolar? (b) Draw the three-dimensional structure of ammonia using the principle of VSEPR. (c) Does NH_3 have a molecular dipole? Explain your answer.
32. Ammonia (NH_3) and methane (CH_4) both exist as gases at room temperature, but we can turn them into liquids by cooling them. If we lower the temperature for samples of ammonia and methane gases, predict which one will turn into a liquid first. Provide a molecular explanation for your prediction.
33. Why do homeowners in cold climates worry about their water pipes bursting in winter?
34. Suppose that you put three ice cubes in a glass and then fill it with water to the very top of the rim. Will the water overflow the glass when the ice cubes melt? Explain your answer.
35. On a hot summer day, which would warm up more quickly—a concrete sidewalk or water in a swimming pool? Explain the chemical reason for your answer.
36. The specific heat capacity of ice at –10 °C is 2.108 J/g°C. The value is approximately half of the specific heat capacity of water (4.184 J/g°C). Propose an explanation for why the heat capacity of ice is less than that of water.
37. The graph below shows the boiling points for the hydrides of elements in Groups 6A and 7A of the periodic table. Use chemical principles to explain why HF has a higher boiling point than HCl by nearly 100°C.

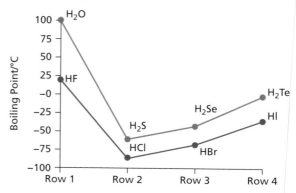

38. The following table gives the boiling points for the hydrides of elements in Group 5A of the periodic table. (a) Using Figure 8.19 as a guide, plot a graph of these boiling points. Use a data point for each hydride, and join the data points by straight lines. (b) What do you notice about the trend of the boiling points for PH_3, AsH_3, and SbH_3? (c) Does the boiling point for NH_3 fit this trend? Explain why or why not. (d) Provide a chemical explanation for the boiling point of NH_3.

PERIOD	ELEMENT	HYDRIDE	BOILING POINT (°C)
2	Nitrogen (N)	NH_3	−33.3
3	Phosphorus (P)	PH_3	−87.7
4	Arsenic (As)	AsH_3	−62.5
5	Antimony (Sb)	SbH_3	−17.0

39. (a) How much heat energy (in joules) is required to heat 10.0 g of water from room temperature (25.0°C) to 65.0°C? (b) How much heat energy is required to heat the same mass of ethanol by the same amount? (c) Which amount of heat is larger? Use the data in Table 8.2.

40. A standard gold bar held by the U.S. Treasury has a mass of 12.4 kg. In an experiment, 50 kJ of heat energy was used to heat a gold bar, resulting in a temperature change of 31°C. (a) Calculate the specific heat capacity of gold (assuming that the bar is 100% gold). (b) Compare your calculated results to the data in Table 8.2. Write a one- to two sentence description of whether the specific heat capacity of gold is higher, lower, or approximately the same as that of the substances in the table.

Exploring Concepts

41. A human who weighs 70 kg contains about 42 L of water $(1 L = 1000 cm^3)$. Use the known density of water $(1 g/cm^3)$ to convert this volume into mass. Based on these numbers, what percentage of our total body weight is water? Does your calculation agree (approximately) with the percentage of water in the human body, as discussed in Section 8.1?

42. Consider an experiment using two cups of water at the same temperature. You place a cube of ice in one cup. In the second cup, you place a cube of iron with the same volume and the same temperature. Predict which liquid will show the largest temperature decrease, and explain your answer.

43. In Section 8.2, we explained that the electronegativity of elements within a group of the periodic table tends to decrease as we move from the top to the bottom within the group. This trend can be explained by an effect called *shielding*. Use the Internet to investigate shielding. Write a short paragraph about this effect and how it explains the electronegativity trend.

44. Figure 8.6 provides the electronegativity values of elements in the periodic table, but it does not include the noble gases (Group 8A). Helium and neon do not bond with any other elements. However, xenon and krypton do form chemical compounds under some conditions. Use the Internet to find the electronegativity values of xenon and krypton, and investigate what types of compounds these elements form. Write a paragraph about your discoveries.

45. Two H—F molecules can form a strong intermolecular hydrogen bond because of the large partial charges on the H and F atoms. Why is hydrogen fluoride (H—F) a *gas* at room temperature, not a liquid like H_2O?

46. Use the Internet to discover the molecular component in "dry ice". This substance freezes at around −70°C. Why is the freezing point of dry ice much lower than the freezing point of "water ice"?

47. Although ethylene glycol is an effective antifreeze, it is also poisonous to animals and humans. A nontoxic antifreeze molecule is used in some frozen foods such as ice cream, where it reduces the formation of ice crystals. Use the Internet to discover the name of this molecule, and draw its molecular structure.

48. We sometimes say that a cup of coffee is "steaming hot." What we are seeing are droplets of water that form from the steam (water vapor). After the coffee has cooled, we no longer see the "steam." Why not?

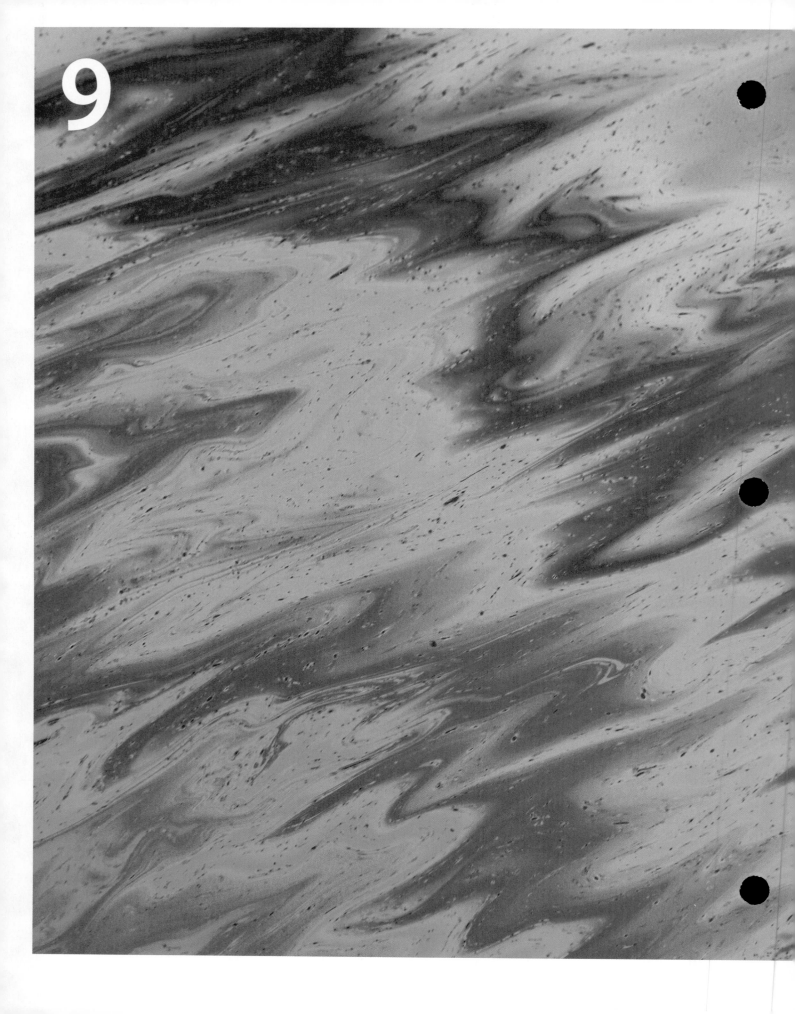

Chapter 9

Molecules and Ions in Solution

Y ou may be one of the many people who take daily vitamin supplements. You probably know the names of several vitamins, such as vitamin C or vitamin D. What are vitamins, and how do they contribute to our health?

9.1 How Do Water-Soluble and Fat-Soluble Vitamins Differ?

Learning Objective:

Distinguish between water-soluble and fat-soluble vitamins.

THE KEY IDEA: Vitamins are classified according to their solubility in water or fat.

vitamin An organic molecule that plays an essential role in many biological processes.

water-soluble vitamin A vitamin that dissolves in any watery environment within the body.

fat-soluble vitamin A vitamin that dissolves in regions of the body that store fat molecules.

Vitamins are organic molecules that play an essential role in many biological functions within an organism. The name is based on "vita" (meaning "life") and "amine" (the functional group). We know now that not all vitamins contain an amine, but the name is still used. Many vitamins are "helper molecules" that assist an enzyme in performing its role as a biological catalyst. Other vitamins enable cells to synthesize important molecules that are necessary for cells to function.

With few exceptions, our body cannot make vitamins, so we must obtain them from our diet. An inadequate supply of a vitamin can sometimes lead to disease. For example, a deficiency of vitamin C causes a disease called *scurvy*, which was common among sailors in the 18th and 19th centuries. This problem was remedied by providing citrus fruits to be eaten during long sea voyages. Today, many of us take vitamin supplements to ensure an adequate supply of vitamins (Figure 9.1). According to *The Wall Street Journal*, sales of vitamins and other dietary supplements in the United States totaled nearly $23 billion in 2013.

Vitamins are grouped into two categories: **water-soluble vitamins** and **fat-soluble vitamins.** Water-soluble vitamins dissolve in any watery environment within the body, including the blood and the interior of cells. Fat-soluble vitamins dissolve in regions of the body that store fat molecules such as triglycerides (see Chapter 4). Table 9.1 provides examples of vitamins and their classification, whereas Figure 9.2 illustrates the molecular structures of vitamins A and C. Vitamin C is a water-soluble vitamin present in citrus fruits, and vitamin A is a fat-soluble vitamin that is important for vision. Why are some vitamins soluble in water, whereas others are soluble in fat? We will answer this question later in the chapter. As a prelude, Concept Question 9.1 asks you to compare the structures of vitamins C and A.

FIGURE 9.1 A selection of vitamin supplements. Vitamin supplements are popular additions to our daily diet.

TABLE 9.1 Classification of Selected Vitamins according to Solubility	
WATER-SOLUBLE VITAMINS	**FAT-SOLUBLE VITAMINS**
Vitamin C	Vitamin A
Vitamin B6 (riboflavin)	Vitamin D
Niacin	Vitamin E
Folic acid	Vitamin K

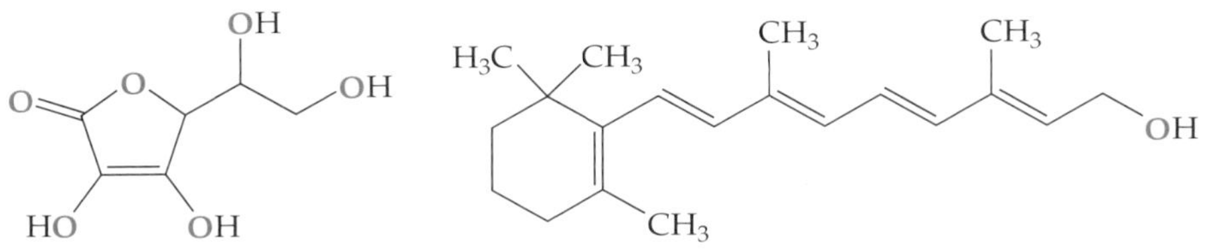

Vitamin C
water-soluble

Vitamin A
fat-soluble

FIGURE 9.2 Molecular structures of two vitamins. Vitamin C is a water-soluble vitamin, whereas vitamin A is a fat-soluble vitamin.

CONCEPT QUESTION 9.1

Compare the molecular structures of vitamin C and vitamin A as shown in Figure 9.2. What similarities and differences do you observe?

The relationship between molecular structure and solubility is important for all biological molecules. For example, the structure of cell membranes is based on the solubility properties of phospholipid molecules. Solubility is also important for ions (e.g., Na^+), which play a role in many biological processes. In this chapter, we will examine the chemistry of solutions to understand why molecules and ions tend to dissolve in some environments but not in others.

Core Concepts

- **Vitamins** are organic molecules that play an essential role in various biological functions. Our body cannot make most vitamins, so we must obtain them from our diet.
- Vitamins are grouped into two categories: **water-soluble vitamins** and **fat-soluble vitamins**.

9.2 Molecules in Solution

Learning Objective:
Analyze polar and nonpolar molecules to predict their solubilities in different solvents.

Everything we drink is a solution, with the primary component being liquid water. For example, coffee contains a range of molecules—including caffeine—that are dissolved in water. Sodas contain dissolved sugar or a sugar substitute to provide a sweet taste, along with dissolved carbon dioxide to provide the "fizz." Even bottled water is not "pure" H_2O because it contains dissolved ions such as sodium, magnesium, and calcium. In this section, we examine how solutions are formed and why substances dissolve (or don't) in different chemical environments.

THE KEY IDEA: A solution is a homogeneous mixture of a solute dissolved in a solvent.

What Is a Solution?

*A **solution** consists of a **solute** that is dissolved in a **solvent**.* The solute can be a molecule or an ion. The solvent is the chemical substance that dissolves the solute; it is present in a larger amount than the solute. We focus on liquid solvents because they are most common in both chemistry and biology. Solutions that use liquid water as the solvent are called **aqueous solutions**. Water is a very effective solvent that can dissolve a wide variety of substances.

A solution is a *homogeneous mixture* in which the solute is uniformly distributed throughout the solution (*homo* = same). A sample taken from one location within the solution is identical to a sample taken from any other location. Figure 9.3(a) shows a solution of copper sulfate in water. Note how the blue color, caused by the dissolved copper sulfate is spread evenly throughout the solution. By contrast, Figure 9.3(b) shows a *heterogeneous mixture* (*hetero* = different) of oil, water, and sand. This mixture is *not* a solution because the components are distinct and are not mixed together.

solution A homogeneous mixture that consists of a solute dissolved in a solvent.

solute The component of a solution that is dissolved in the solvent.

solvent The component of a solution that dissolves the solute.

aqueous solution A solution in which water is the solvent.

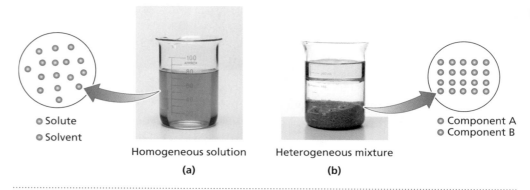

FIGURE 9.3 What is a solution? (a) A solution is a homogeneous mixture in which the solute particles (represented by blue dots) are distributed uniformly throughout the solvent. (b) In a heterogeneous mixture, the different components are distinct from each other. Note that the particle size is not to scale in these diagrams.

■ **solubility** The ability of a solute to dissolve in a solvent.

■ **soluble** The relative ease of dissolving a solute in a solvent.

■ **insoluble** The relative difficulty of dissolving a solute in a solvent.

Although liquid water is the most common solvent in biological systems, it is possible to make solutions using other solvents such as liquid hexane. Substances that dissolve easily in water do not dissolve in liquid hexane, and vice versa. We use the term **solubility** to describe *the ability of a solute to dissolve in a solvent*. In general, the solubility depends on the chemical properties of both the solute and the solvent. We will use the terms **soluble** and **insoluble** to indicate the relative ease or difficulty of dissolving a solute in a solvent. However, these terms are not meant to indicate an "all or none" distinction. Even an "insoluble" substance can dissolve to a very limited degree. It is more accurate to view solubility as varying between very high and very low.

Core Concepts

■ A **solution** is a homogeneous mixture that consists of a solute dissolved in a solvent.
■ A solution in which water is the solvent is called an **aqueous solution**.
■ **Solubility** describes the ability of a solute to dissolve in a solvent. The solubility depends on the properties of both the solute and the solvent. **Soluble** and **insoluble** are terms that describe the relative ease or difficulty of dissolving a solute in a solvent.

Nonpolar Molecules in Water

THE KEY IDEA: A nonpolar methane (CH$_4$) molecule in water cannot form hydrogen bonds with polar H$_2$O molecules.

You have probably heard the saying that "oil and water don't mix." This principle is illustrated in Figure 9.3. What is the chemical basis for this lack of mixing? "Oil" is a generic term that refers to a liquid composed of hydrocarbon molecules (see Chapter 4). Before examining the behavior of oils, we will begin with the simplest hydrocarbon—methane (CH$_4$). This molecule illustrates the general principle that hydrocarbon molecules do not mix with water.

Methane is a gas at room temperature, but we can use pressure to inject it into liquid water. As we learned in Chapter 8, liquid water is composed of polar H$_2$O molecules that are held together via intermolecular hydrogen bonds. However, the *C—H bonds in CH$_4$ are nonpolar and cannot participate in hydrogen bonding*. Recall that the polarity of a chemical bond depends on the difference in electronegativity (EN) values between the two bonded atoms. This difference indicates the degree to which the electron pair is shared equally or unequally. The electronegativity value of carbon is 2.5 and that of hydrogen is 2.1, so the difference (ΔEN) is only 0.4. Because the C—H bonds are nonpolar, the C and H atoms lack the significant partial charges that are required for hydrogen bonding.

As a consequence of their polarity difference, the nonpolar CH$_4$ molecules and the polar H$_2$O molecules want nothing to do with each other. When CH$_4$ is forced into liquid water, the H$_2$O molecules form a cage-like structure called a **clathrate** that surrounds and isolates the nonpolar solute (Figure 9.4). The clathrate structure enables the water molecules to accommodate the unwelcome

■ **clathrate** A rigid, cage-like structure of H$_2$O molecules that completely surrounds a nonpolar solute molecule.

solute while maintaining the maximum possible number of intermolecular hydrogen bonds. This behavior is sometimes called an "iceberg effect" because the arrangement of H_2O molecules in the clathrate bears some resemblance to the rigid structure of ice.

We must expend energy to force methane into water. Why? The explanation is based on how the CH_4 molecule disturbs the interaction among H_2O molecules in the solvent. The CH_4 molecule has little effect on the hydrogen bond energies because both liquid water and the clathrate cage contain intermolecular hydrogen bonds. Instead, we must consider a property called **entropy**. In general, *entropy measures the degree of disorder or randomness within a system, such as a collection of atoms or molecules.* One fundamental law of nature is that *any system tends to progress toward greater entropy (or disorder).* Any attempt to act against this trend and "organize" a system requires an input of energy.

We can illustrate the principle of entropy with an analogy, which is depicted in Figure 9.5. When you open a new deck of playing cards, the cards are divided into suits (hearts, spades, and so on) and arranged sequentially within each suit (Figure 9.5(a)). This is an organized pattern of playing cards, which corresponds to a state of low entropy. Imagine that you hold the deck of cards and then let them fall (Figure 9.5(b)). The cards that are now scattered on the floor are in a random, disorganized arrangement that corresponds to a state of high entropy. The increase in randomness (entropy) is spontaneous, and no input of energy is required. If you want to reorganize the deck of cards, you have to invest effort (energy) to transform a disordered arrangement of cards into their ordered suits. Similarly, an input of energy is required to change a system from a state of high entropy (more disorder) to one of low entropy (less disorder).

The principle of entropy can explain why methane does not dissolve easily in water. The H_2O molecules in liquid water are randomly arranged and constantly moving, which is a state of high entropy. When methane is added to water, the nonpolar CH_4 molecules force the nearby H_2O molecules to form an ice-like clathrate structure. Within the clathrate, the H_2O molecules are forced into fixed positions within a well-organized structure, which represents a state of low entropy. Consequently, the CH_4 molecules cause the neighboring H_2O molecules to change from a state of high entropy to one of low entropy. This *decrease in entropy* runs counter to the general tendency for entropy to increase. As a result, there is an *entropy penalty* to pay for dissolving methane in water, which must be overcome by an input of energy. This is why pressure must be used to force CH_4 into water.

A clathrate structure can form around a CH_4 molecule because the molecule is small. When the hydrocarbon solute becomes larger, such as hexane (C_6H_{14}), the H_2O molecules cannot form the same type of well-organized structure. The nonpolar hexane molecules still cause the nearby H_2O water molecules to become more structurally organized, but they do not achieve the ordered ice-like geometry of a clathrate. Nevertheless, the entropy penalty

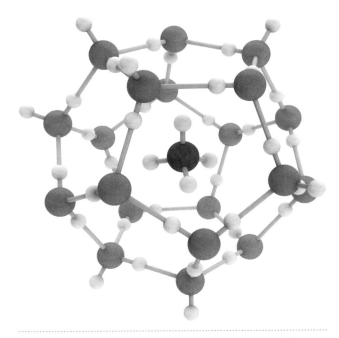

FIGURE 9.4 **A water clathrate surrounds a CH$_4$ molecule.** Water molecules cannot form hydrogen bonds with a nonpolar solute like methane, so they create a cage-like structure called a clathrate that surrounds the CH_4 molecule. The CH_4 molecule is located in the center, and hydrogen bonds between H_2O molecules are shown as gray lines.

entropy A measure of the degree of disorder or randomness within a system.

(a)   (b)

FIGURE 9.5 **An analogy for entropy.** (a) In a new pack of playing cards, the cards are organized sequentially within four suits. This organized arrangement corresponds to a state of low entropy. (b) After the pack of cards is dropped to the floor, the cards become random and disorganized. This arrangement corresponds to a state of high entropy.

still applies because the dissolved hexane molecule is causing the entropy of the neighboring H_2O molecules to decrease. In fact, hexane molecules tend to avoid water by clustering together, as described in the following section.

Core Concepts

- Forcing nonpolar methane (CH_4) molecules into water causes the polar H_2O molecules to form a cage-like structure called a **clathrate**.
- The presence of CH_4 molecules in water changes the **entropy**. In general, entropy is a measure of randomness or disorder in a system, such as a collection of molecules. Entropy naturally tends to increase.
- When H_2O molecules form a rigid clathrate structure, the entropy of the molecules decreases in comparison to their random arrangement in liquid water. As a result, forcing methane into water requires an expenditure of energy to overcome the decrease in entropy.

The Hydrophobic Effect

The example of methane in water illustrates how polar and nonpolar molecules avoid interacting. Molecules that do not mix with water are called **hydrophobic**, which means "water fearing." In general, hydrophobic molecules are nonpolar and cannot form hydrogen bonds with H_2O molecules in water. For example, liquid oils contain a variety of hydrocarbon molecules, many of which exist as long chains. These hydrocarbon chains contain a combination of C—C, C=C, and C—H bonds, all of which are nonpolar bonds. Figure 9.6 shows a close-up view of olive oil in water. The oil forms large droplets that separate the oil molecules from the surrounding water. Why does this happen?

When two or more nonpolar molecules are dissolved in water, *they aggregate to minimize their interaction with the solvent.* This principle is called the **hydrophobic effect.** In the first panel of Figure 9.7, two nonpolar molecules are each surrounded by water. The exposed surfaces of these molecules induce the nearby H_2O molecules to form ordered structures, a process that causes an unfavorable decrease in entropy. However, if the two hydrophobic molecules huddle closely together, as illustrated in the second panel of Figure 9.7, then some of their nonpolar surfaces become shielded from the surrounding water. This *hydrophobic interaction* reduces the exposed surface area, which decreases the entropy penalty incurred by forcing the solvent water into a more ordered state. When the two hydrophobic molecules are in close proximity, they are further stabilized by attractive forces that we describe later in the chapter. The aggregation of nonpolar molecules can lead to the formation of two distinct layers, which occurs when oil rests on the surface of water.

> **⊙→ THE KEY IDEA:** Nonpolar molecules aggregate in water to reduce the exposure of their hydrophobic surfaces.

> **hydrophobic** The inability to mix with water.

> **hydrophobic effect** A principle that describes the association of nonpolar solutes in water to minimize their interaction with the solvent.

> **⊙→ THE KEY IDEA:** Methanol (CH_3OH) is soluble in water because its polar –OH group forms hydrogen bonds with H_2O.

Core Concepts

- Molecules that do not mix with water are called **hydrophobic** ("water fearing"). Hydrophobic molecules are nonpolar and cannot form hydrogen bonds with H_2O molecules in water.
- When placed in water, nonpolar molecules aggregate to minimize their interaction with the solvent. This behavior is called the **hydrophobic effect**.

Polar Molecules in Water

Why is methanol (CH_3OH) soluble in water, whereas methane (CH_4) is not? The answer is related to the reason why methanol is a liquid at room temperature, whereas methane is a gas. The molecular structures of H_2O, CH_4, and CH_3OH are compared in Figure 9.8. Methanol has a nonpolar —CH_3 group in common with methane, but it also has a polar —OH group in common with H_2O.

FIGURE 9.6 Oil in water. Olive oil in water forms large oil droplets in which the oil molecules cluster together.

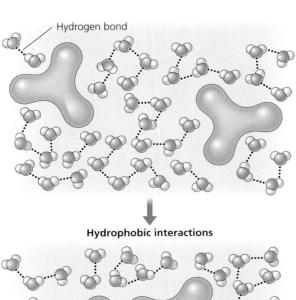

Hydrogen bond

Hydrophobic interactions

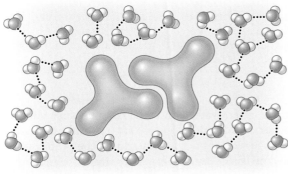

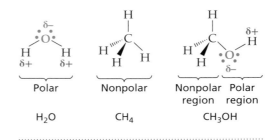

Polar
H_2O

Nonpolar
CH_4

Nonpolar   Polar
region   region
CH_3OH

FIGURE 9.8 Structural comparison of H_2O, CH_4 (methane) and CH_3OH (methanol). The polar and nonpolar regions are highlighted for each molecule.

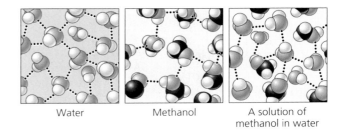

Water   Methanol   A solution of
                                    methanol in water

FIGURE 9.7 The hydrophobic effect. The top panel shows two nonpolar hydrocarbon molecules that are each surrounded by liquid water molecules. The hydrophobic surface of each molecule induces the ordering of nearby H_2O molecules, which carries an entropy penalty. The bottom panel shows a hydrophobic interaction between the two molecules, which reduces the total exposed surface and lessens the entropy penalty.

FIGURE 9.9 A molecular view of water, methanol, and a solution of methanol in water. The polar—OH group in methanol is readily incorporated into water's hydrogen-bonding network, in contrast to the nonpolar—CH_3 group.

The—OH groups in methanol can form intermolecular hydrogen bonds with one another, similar to the way that H_2O molecules interact in liquid water. These hydrogen-bonding interactions "stick" the methanol molecules together as a liquid at room temperature. By contrast, methane does not form hydrogen bonds, so it exists as a gas at room temperature (see Chapter 8).

Methanol dissolves easily in water because its polar—OH group can form hydrogen bonds with neighboring H_2O molecules For this reason, methanol is classified as **hydrophilic** ("water loving"). Figure 9.9 illustrates molecular views of liquid water, liquid methanol, and a solution of methanol in water. Note that the—CH_3 group in methanol does not participate in hydrogen bonding because its C—H bonds are nonpolar.

hydrophilic The ability to mix with water.

PRACTICE EXERCISE 9.1

For each molecule given below, draw the two-dimensional structure (including any lone pairs), and predict whether or not it is soluble in water. Explain the reason for your prediction.

(a) Ethanol (C_2H_5OH)

(b) Propane (C_3H_8)

(c) Methylamine (CH_3NH_2)

CONCEPT QUESTION 9.2

The boiling point is the temperature at which a liquid turns into a vapor. Explain why the boiling point of methanol (65°C) is substantially lower than that of water (100°C).

CONCEPT QUESTION 9.3

The photograph to the right shows a sugar cube dissolving in water. The fluid flowing beneath the sugar cube contains dissolved sugar (sucrose) molecules. After a short time, the entire sugar cube disappears. How does the entropy of the sugar molecules in the sugar cube compare to the entropy of the same molecules when they are dissolved in water?

Core Concepts

- Methanol (CH_3OH) is a **hydrophilic** molecule because it dissolves easily in water.
- Methanol dissolves in water because its polar —OH group forms hydrogen bonds with H_2O molecules in the solvent. However, the –CH_3 group in methanol does not form hydrogen bonds because its C—H bonds are nonpolar.

🔑 THE KEY IDEA: Nonpolar molecules such as hexane form liquids due to intermolecular dispersion forces.

Nonpolar Solvents

As we explained earlier in the chapter, water is not the only possible solvent. Liquid hydrocarbons such as hexane can also dissolve solutes and make solutions. These **nonpolar solvents** are relevant for understanding the properties of fat-soluble vitamins. Liquid hexane has chemical properties similar to regions of cells that are composed primarily of hydrocarbons, such as cellular stores of fat molecules and the hydrophobic interior of cell membranes (see Section 9.3).

We first must confront an apparent paradox: Why can *any* hydrocarbon exist as a liquid at room temperature? Recall that methane (CH_4) is a gas at room temperature because it cannot form hydrogen bonds. A larger hydrocarbon such as hexane (C_6H_{14}) also cannot form hydrogen bonds; nevertheless, it exists as a liquid. Why?

The liquid state of certain hydrocarbons is *not* due to hydrogen bonding. Rather, it arises from another type of intermolecular attraction called **dispersion forces**, which are illustrated in Figure 9.10. They are also called *Van der Waals forces*, in recognition of the Dutch chemist who first investigated them. As we saw in Chapters 2 and 3, quantum mechanics tells us that electrons do not have a precisely defined location and instead must be described as "clouds" of electron density. In nonpolar molecules, like hydrocarbons, the distribution of electron density produces a neutral molecule that has no regions of positive or negative charge (Figure 9.10(a)). However, the electron density can fluctuate momentarily to produce a slightly uneven distribution of electrical charge. This flickering change makes one region of the molecule slightly more negative (increased electron density) and another region slightly more positive (reduced electron density). For a fleeting moment, there is now an *instantaneous dipole* within the molecule. Recall from Chapter 8 that a dipole is a separation of positive and negative charges.

If two molecules are close together, then a momentary change in the electron distribution of one molecule can induce a response in its neighbor. For example, if a region of one molecule becomes more negative, then it can influence a region of a nearby molecule to become more positive. The outcome of this effect is called an *induced dipole* (Figure 9.10(b)). As a result, the neighboring hydrocarbon molecules

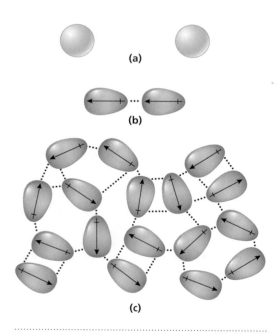

FIGURE 9.10 Dispersion forces between nonpolar molecules. (a) Two nonpolar molecules that are far apart do not interact. (b) As the molecules approach each other, a momentary change in the electron density of one molecule (an instantaneous dipole) can generate a complementary change in a neighboring molecule (an induced dipole). This interaction creates an electrical attraction between the two nonpolar molecules. (c) A sufficient number of these dispersion force attractions produces a liquid state.

have acquired some "polar" character due to mutual interactions between their electron densities. A collection of these "polar" molecules can become electrically attracted to one another through the interaction of their induced dipoles (Figure 9.10(c)). Although each intermolecular attraction is very weak, the aggregate effect of multiple attractions among many hydrocarbon molecules is sufficient to make them a liquid at room temperature.

The dispersion forces increase in strength as the molecule becomes larger because bigger molecules have a greater number of electrons that can fluctuate in their positions. This is why larger hydrocarbons such as hexane and octane are liquids, whereas smaller ones such as methane, ethane, propane, and butane are gases. Molecules with delocalized electrons—such as benzene (C_6H_6)—also tend to form liquids via dispersion forces. This occurs because the delocalized electron density is more easily perturbed by a nearby instantaneous dipole.

nonpolar solvent A solvent in which the molecules are nonpolar, such as liquid hexane.

dispersion forces Forces that produce mutual attractions among nonpolar molecules.

CONCEPT QUESTION 9.4

Hydrocarbons with short carbon chains (e.g., propane, C_3H_8) are gases, whereas those with medium-length chains (e.g., hexane C_6H_{14}) are liquids. When the carbon chain reaches 20 atoms and beyond, the hydrocarbons exist as solid waxes—for example, paraffin wax, which is used to make candles. Use chemical principles to explain why long-chain hydrocarbons form waxes. Why do you think these waxes are so easy to mark or cut compared to other solids?

Core Concepts

- Some hydrocarbon molecules such as hexane are liquid at room temperature and function as nonpolar solvents.
- The attraction among hexane molecules comes from **dispersion forces**. A momentary fluctuation in the electron density of a nonpolar molecule produces an instantaneous dipole. This distribution of charge creates an induced dipole in a nearby molecule. The dispersion forces arise from mutual attraction among the induced dipoles in nonpolar molecules.

Comparing Solubilities in Polar and Nonpolar Solvents

Let us consider the solubility of molecules in polar and nonpolar solvents. Figure 9.11 compares two alcohol molecules—methanol and hexanol. Both molecules contain a hydrocarbon region terminated with an alcohol group. Which of these molecules do you predict to be more soluble in water? Which would be more soluble in hexane, a nonpolar solvent?

To answer this question, we compare the *relative sizes* of the polar and nonpolar regions in the two molecules. Methanol has a small nonpolar region (one $-CH_3$ group) attached to its polar—OH group. It is relatively easy to incorporate methanol molecules into the hydrogen-bonding network of liquid water, so methanol is highly soluble in water. By contrast, the polar—OH group interferes with the dispersion force attractions among hexane molecules. Consequently, methanol dissolves only slightly in hexane.

By contrast, hexanol has a much longer hydrocarbon chain consisting of six carbon atoms. Although the —OH group is still present, liquid water has difficulty accommodating the large nonpolar region within its hydrogen-bonded network. Interaction between polar H_2O molecules and hydrocarbons is unfavorable because stabilizing hydrogen bonds cannot be formed. As a consequence, hexanol is very insoluble in water despite having an—OH group. By contrast, hexanol is highly soluble in liquid hexane because of

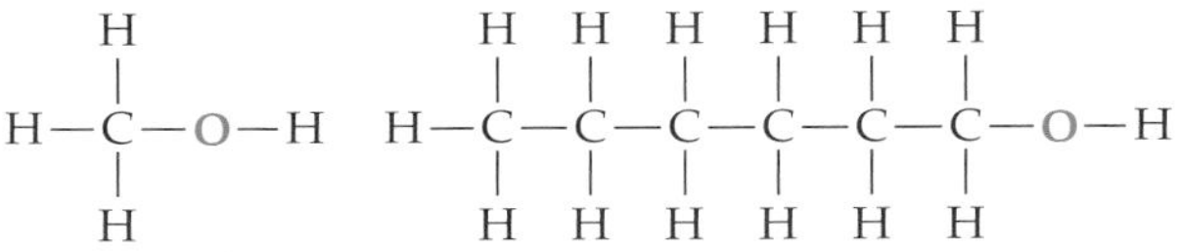

🔑 THE KEY IDEA: Polar molecules are soluble in polar solvents, and nonpolar molecules are soluble in nonpolar solvents.

FIGURE 9.11 Molecular structures of methanol and hexanol. Both molecules contain an alcohol group (—OH), but hexanol contains more carbon atoms.

favorable dispersion forces that exist between the nonpolar regions of the alcohol and the hydrocarbon solvent. Table 9.2 summarizes these solubility behaviors.

The solubility of molecules in polar and nonpolar solvents can be expressed by a simple rule: *Like dissolves like.* Polar solvents such as water easily dissolve other polar molecules such as methanol. Similarly, nonpolar solvents such as hexane dissolve other less polar molecules such as hexanol.

TABLE 9.2	Comparison of Solubilities in Water and Hexane	
	SOLUBILITY IN WATER	**SOLUBILITY IN HEXANE**
Methanol	Very high	Very low
Hexanol	Very low	Very high

PRACTICE EXERCISE 9.2

The structures of four molecules are given as follows:

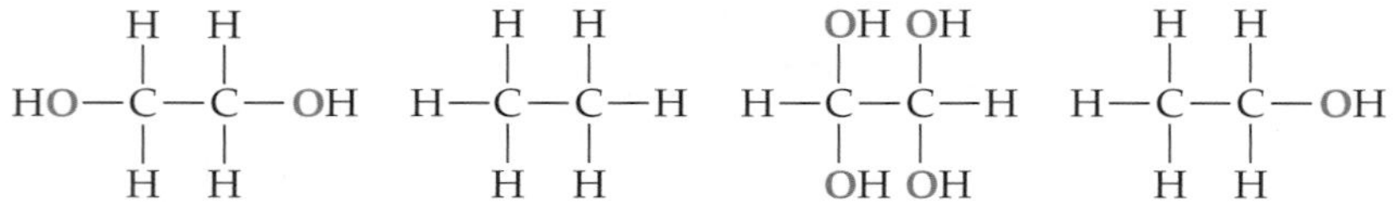

(a)   (b)   (c)   (d)

(a) Order these molecules from most to least soluble in water.

(b) Order these molecules from most to least soluble in hexane.

Core Concepts

■ Polar molecules dissolve readily in polar solvents such as water, whereas nonpolar molecules dissolve easily in nonpolar solvents such as hexane. These results are summarized as "like dissolves like."

THE KEY IDEA: Water-soluble vitamins have multiple polar functional groups, whereas fat-soluble vitamins are primarily composed of hydrocarbons.

Water-Soluble and Fat-Soluble Vitamins

We can now return to the question that began this chapter: What is the difference between water-soluble and fat-soluble vitamins? The term "water-soluble" is obvious, but what does "fat-soluble" mean?

Chapter 4 introduced the molecular structure of fat molecules. Most fats in our body are triglycerides, composed of a glycerol "head" group attached to three fatty acids (Figure 9.12). Each fatty acid has a "tail" made from a long hydrocarbon chain. When many triglycerides aggregate, their hydrocarbon regions provide a nonpolar environment for dissolving other molecules. The term *fat soluble* is equivalent to being soluble in a nonpolar solvent such as hexane.

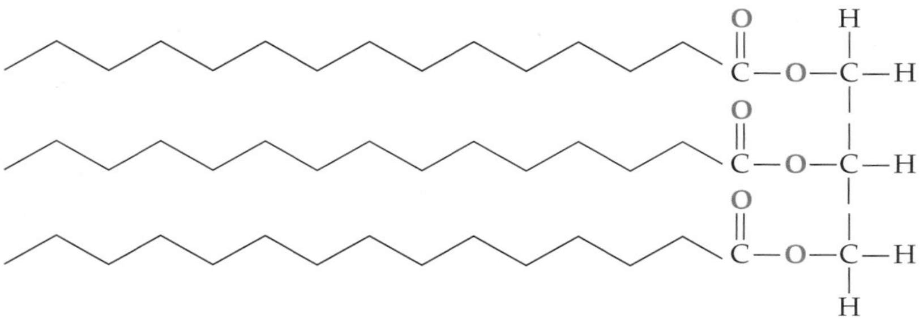

FIGURE 9.12 Molecular structure of a triglyceride. A triglyceride molecule is made from a glycerol molecule and three fatty acids.

3 Fatty acids   +   Glycerol

Triglycerides are stored as fat droplets within special-ized cells. A collection of these cells forms *adipose tissue*, which typically accounts for 20% of body weight in adult males and 25% in adult females. Figure 9.13 displays a thin slice of adipose tissue that has been magnified using a microscope. The tissue looks mostly "white," which cor-responds to the regions where the fat molecules are stored. The thin purple regions are watery interiors of cells that have been stained by a dye. The triglyceride molecules in adipose tissue serve as a reserve of stored energy. When this energy is needed, an enzyme liberates the fatty acids by cutting the bonds that connect them to the glycerol molecule. The fatty acids are then transported to the mi-tochondria—the cell's "energy generators"—where they are converted to carbon dioxide and water using a series of chemical reaction that generates energy stored in the form of ATP.

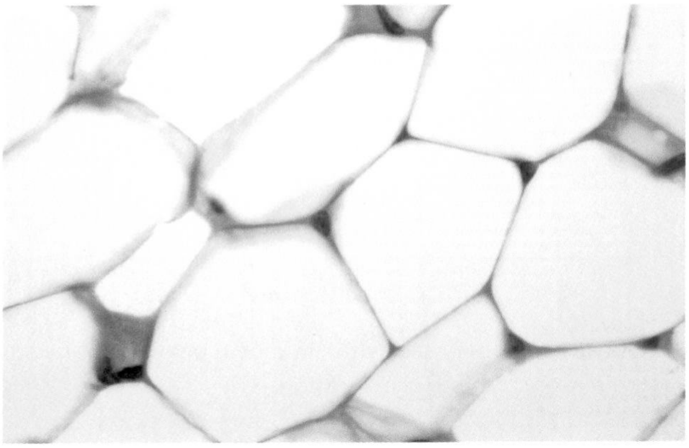

FIGURE 9.13 A magnified microscope image of adipose tissue. The "white" regions are where triglycerides are stored as fat droplets. The watery regions of the cell, which have been stained purple with a dye, constitute only a small fraction of the tissue volume.

We can now classify vitamins using the solubility rule—like dissolves like. Because water is a polar sol-vent, we predict that water-soluble vitamins are *hydrophilic polar molecules*. Conversely, we predict that fat-soluble vitamins are *hydrophobic nonpolar mol-ecules* because they dissolve in nonpolar environments. Figure 9.14 illustrates the structures of two water-soluble vitamins (C and B6) and two fat-soluble vi-tamins (A and E). Both water-soluble vitamins are constructed from a hetero-cycle plus several polar —OH groups (refer to Chapter 5). These polar groups enable the water-soluble vitamins to form hydrogen bonds with solvent H_2O molecules. By contrast, the fat-soluble vitamins are almost completely hydro-carbon. The lone —OH group in each of these two vitamins cannot offset their strongly hydrophobic properties.

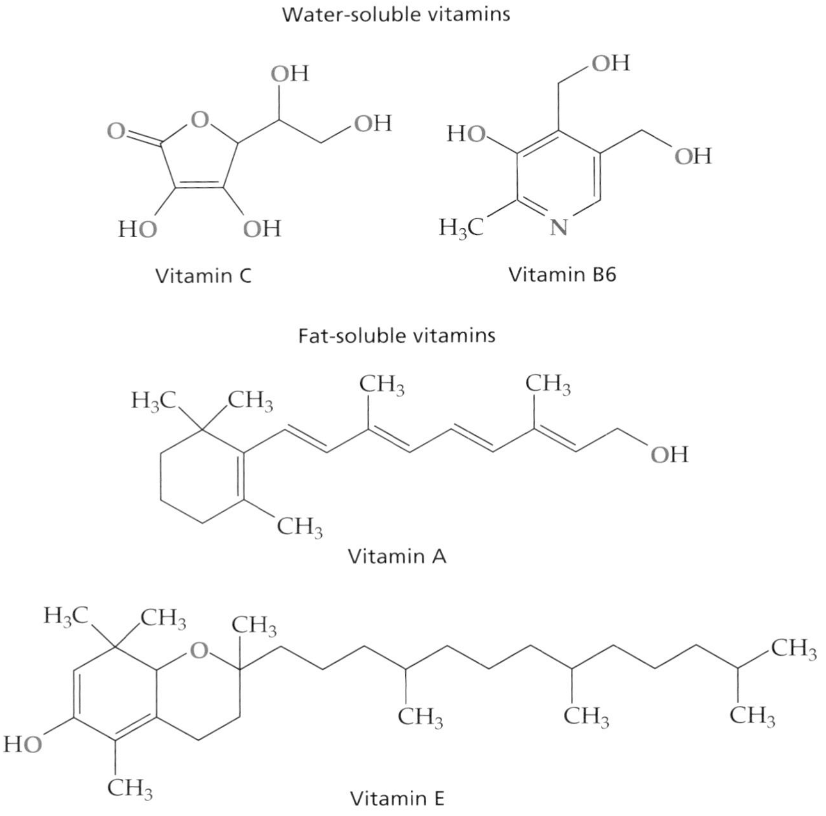

FIGURE 9.14 Molecular structures of water-soluble and fat-soluble vitamins. The water-soluble vitamins contain several polar functional groups, whereas the fat-soluble vitamins primarily consist of nonpolar hydrocarbons.

Although we are using the designations "water soluble" and "fat soluble," there is a *range* of solubility that depends on the vitamin's molecular structure. For example, vitamin C is approximately twice as soluble in water as vitamin B6. In addition, vitamin E is more soluble in nonpolar solvents (fat soluble) than is vitamin A.

WORKED EXAMPLE 9.1

Question: Classify each vitamin shown below as water soluble or fat soluble. Explain the reasons for your classification.

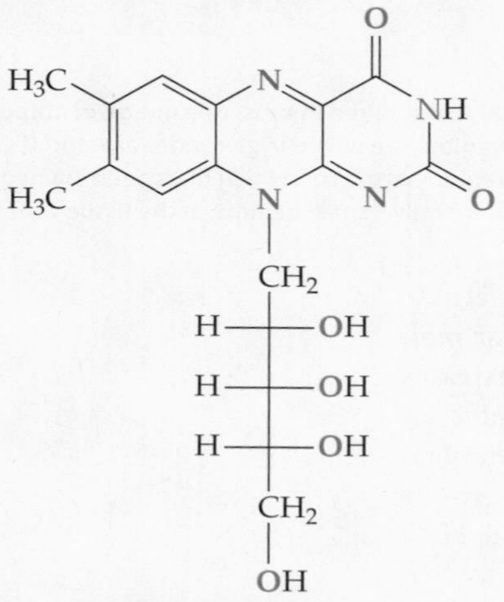

Vitamin B2 (riboflavin)

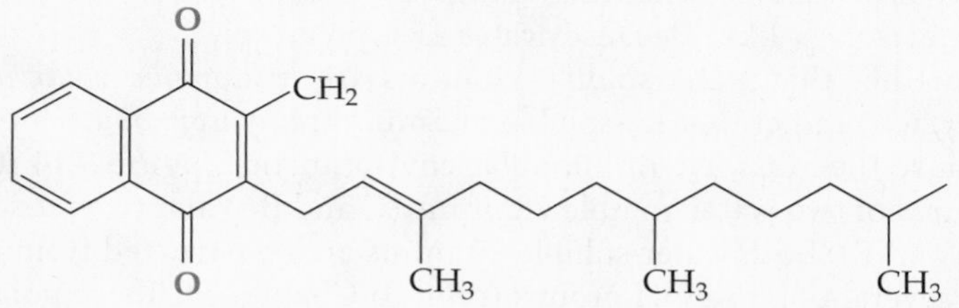

Vitamin B1

Answer: We can classify each vitamin using the solubility rule that "like dissolves like."

Vitamin B2 (riboflavin) contains four polar —OH groups. In addition, the three-ring heterocycle has an N—H group plus two C═O groups, which are also polar. Vitamin B2 will dissolve in water, a polar solvent, so we classify it as *water soluble*.

Vitamin K1 is composed almost completely of hydrocarbon regions. It contains a two-ring cyclic hydrocarbon, plus a long hydrocarbon chain. There are two polar C═O groups attached to the ring, but this is not sufficient to offset the dominant contribution from the hydrocarbon regions. Vitamin K1 will dissolve in nonpolar solvents, so we classify it as *fat soluble*.

TRY IT YOURSELF 9.1

Question: Classify each vitamin shown here as water soluble or fat soluble. Explain the reasons for your classification.

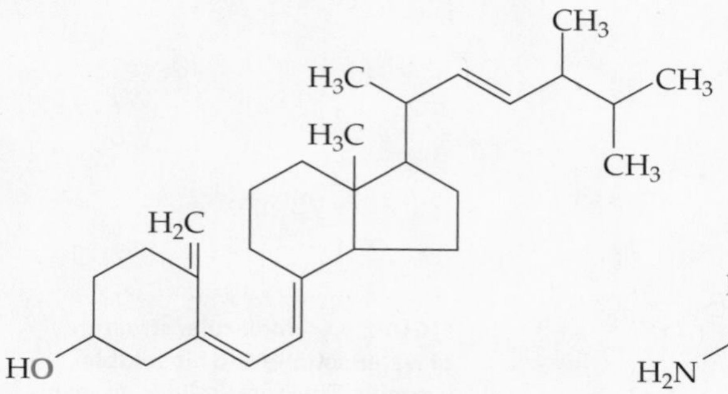

Vitamin D3

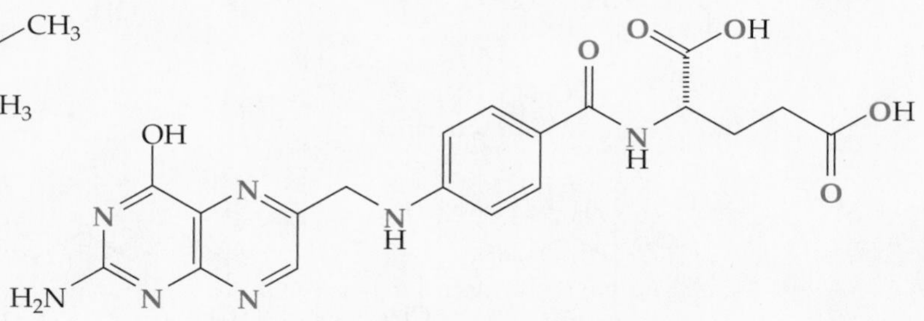

Vitamin B9 folic acid

Propose an explanation for the greater water solubility of vitamin C compared to vitamin B6. Similarly, propose an explanation for the greater fat solubility of vitamin E compared to vitamin A.

Whether a vitamin is water soluble or fat soluble affects the daily dose that we need. The U.S. Institute of Medicine publishes a Recommended Dietary Allowance (RDA) for each of the major vitamins. The RDA for water-soluble vitamins is typically much higher than the RDA for fat-soluble vitamins. For example, the RDA for vitamin C (water soluble) is given in units of *milligrams* per day (1 mg = 10^{-3} g), whereas the RDA for vitamin A (fat soluble) is given in units of *micrograms* per day (1 μg = 10^{-6} g). Why is there a thousand-fold difference in the recommended doses?

The solubility of a vitamin affects how it is synthesized and transported in the body. All water-soluble vitamins can travel freely within the bloodstream, and excess amounts of these vitamins are easily removed by excretion in the urine. By contrast, fat-soluble vitamins avoid these watery regions, and their circulation around the body often requires assistance by transport molecules. These vitamins settle in adipose tissue and the liver, where they can reside for a long time and are released only when needed. Unlike water-soluble vitamins, high levels of fat-soluble vitamins can become toxic because they accumulate in the body and cannot be removed in the urine.

Core Concepts

- The term "fat soluble" refers to the ability of a molecule to dissolve within a nonpolar environment.
- Triglycerides (glycerol plus three fatty acids) are the most common type of fat molecule in the human body. These molecules are stored within adipose tissue, and they serve as an energy storage reserve that can generate ATP.
- The solubility rule of "like dissolves like" can be applied to water-soluble and fat-soluble vitamins. Water-soluble vitamins (e.g., vitamins C and B6) have polar groups that interact with solvent H_2O molecules. Fat-soluble vitamins (e.g., vitamins A and E) are composed primarily of nonpolar hydrocarbon regions that dissolve in nonpolar environments.

9.3 Molecular Self-Assembly

Some collections of molecules spontaneously arrange themselves into an organized structure. This process, called **molecular self-assembly,** is driven by the solubility properties of the molecules. Self-assembly occurs when molecules have a "split personality"—part of the molecule is polar (hydrophilic), and another part is nonpolar (hydrophobic). This section presents two examples of self-assembly.

Self-Assembly of Detergents

Each time you wash your dirty dishes in the sink, you are utilizing the chemical principles of solubility. Many types of foods leave a residue of grease on plates and cooking pans (the term *grease* refers to any type of fatty or oily substance). Trying to clean the dishes with water alone will not work because grease is not soluble in water. For this reason, we use dishwashing liquid that *solubilizes* the grease (i.e., makes it soluble). How does this process work?

Dishwashing liquid contains a type of molecule called a **detergent** (Figure 9.15). Detergent molecules contain a hydrophilic head group plus a long hydrocarbon tail, which is hydrophobic. Detergents are similar in structure to the fatty acids that we examined in Chapter 4. However, detergents differ from fatty acids in the type of head group they contain. For example, several types of

Learning Objective:
Apply solubility principles to illustrate examples of molecular self-assembly.

THE KEY IDEA: Detergent molecules self-assemble in water because they contain both a hydrophilic region and a hydrophobic region.

molecular self-assembly The spontaneous assembly of a collection of molecules to form an organized structure.

detergent A molecule used for cleaning; it contains a hydrophilic head group plus be long hydrocarbon tail.

CHEMISTRY IN YOUR LIFE

Why Do Chili Peppers Taste Hot?

If you enjoy spicy food, then you have probably eaten a chili pepper. If so, you will recall a burning sensation in your mouth. Perhaps the hot taste was too much, and you tried to quench the effect by drinking a glass of water. Unfortunately, that didn't help. Why do chili peppers burn our mouth even though they don't have a hot temperature? Why does water not alleviate the burning sensation?

The molecule responsible for the hot taste of chili is called *capsaicin*, which is depicted in Figure 1 as a line-angle drawing and a space-filling model. Capsaicin is composed primarily of hydrocarbon regions plus a few functional groups. As a result, it is a hydrophobic molecule that is very soluble in oily environments but much less in water. This lack of solubility explains why drinking a glass of water does not wash away the capsaicin.

Capsaicin tastes hot because it triggers the same biochemical pathway that our cells use to detect high temperatures. A cell membrane contains *ion channel proteins* that control the flow of ions into and out of the cell. One of these proteins is sensitive to temperature: If the environment becomes too hot, it allows positively charged ions (Na^+ and Ca^{2+}) to flow into the cell (Figure 2). This increased ion flow triggers a series of cellular responses that our brain interprets as a burning sensation. Capsaicin binds to this temperature-sensitive protein and triggers the same flow of ions. When this happens inside our mouth, our brain tells us that our mouth is burning even though the temperature is normal.

What is the remedy for the burning sensation produced by capsaicin? Water doesn't work, but milk products will reduce the effect—this is why spicy dishes are often served with yogurt. Milk contains large amounts of a protein called *casein*, which has an unusual structure. Most proteins fold into a complex three-dimensional structure in which the hydrophobic chemical groups are buried within the interior, shielding them from exposure to water. By contrast, casein has a floppy, unfolded structure that exposes its hydrophobic regions. The unfolded protein provides an ideal nonpolar environment for "dissolving" the nonpolar capsaicin molecule. Once capsaicin is attached to a casein protein, it is no longer available to trigger the activity of the heat-sensitive ion channel protein.

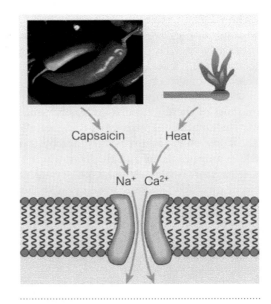

FIGURE 2 Heat and capsaicin both trigger the same response from a temperature-sensitive protein that is embedded within the cell membrane. The increased flow of Na^+ and Ca^{2+} ions into the cell triggers a response that is interpreted by our brain as a burning sensation.

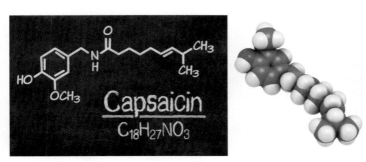

FIGURE 1 Capsaicin gives chili peppers their hot taste.

detergents contain a *sulfate* ion (SO_4^{2-}) that is attached to the hydrocarbon chain. The sulfate ion is highly soluble in water.

Figure 9.16 illustrates what happens when a collection of detergent molecules is placed in water. Initially, the molecules are randomly spread out. However, this is not a stable situation because the hydrophobic tails are exposed to the aqueous solvent. Consequently, a collection of molecules spontaneously self-assembles into a roughly spherical structure called a **micelle**. In the figure,

■ **micelle** A structure formed by the self-assembly of molecules with both hydrophilic and hydrophobic regions.

we show a two-dimensional cross-section through the micelle to illustrate its molecular organization. The hydrocarbon tails pack closely together within the center of the micelle, providing a localized nonpolar environment that shields them from water. This behavior is an example of the hydrophobic effect, which was discussed earlier. In contrast, the hydrophilic head groups face outward because they are soluble in the aqueous solvent. The micelle structure enables both regions of the molecule— hydrophilic and hydrophobic—to be in its preferred environment.

The self-assembly of a micelle may appear to violate the principle of entropy increase, which favors systems that are less organized and more random. It is true that the organization of detergent molecules into a structured micelle produces a decrease in entropy. However, a counteracting positive entropy change arises from shielding the hydrophobic tails from water, which enables nearby H_2O molecules to become more disordered. Overall, the entropy gain from increasing randomness in the solvent is greater than the entropy loss from organizing the micelle structure. In addition, packing the hydrocarbon chains together generates favorable dispersion forces that contributes to the stability of the micelle.

The solubility behavior of detergent molecules explains how they can clean the grease from your dishes. Figure 9.17 shows the stages in the cleaning process, proceeding from left to right. The hydrophobic grease that coats the surface of the pan is not soluble in water. However, the hydrophobic tails of the detergent molecules can dissolve the grease by providing a nonpolar environment. The detergent molecules form a micelle that surrounds the grease. The nonpolar grease is sequestered in the interior of the micelle, while the hydrophilic head groups create an exterior surface that is soluble in water. The grease is now enclosed within the water-soluble micelles and can be discarded in the dirty water.

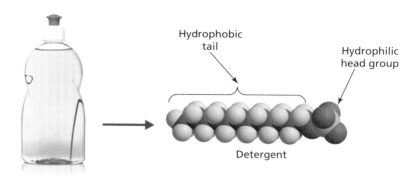

FIGURE 9.15 The structure of a detergent molecule. Dishwashing liquid contains detergent molecules, which contain a hydrophilic head group attached to a long hydrophobic tail. In this detergent, the head group is sulfate ion (SO_4^{2-}).

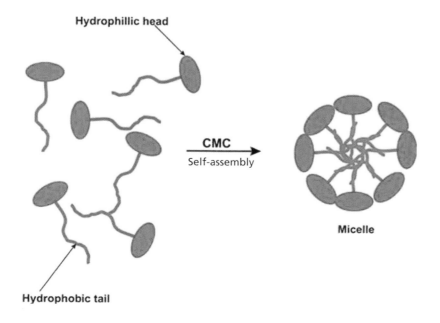

FIGURE 9.16 Self-assembly of detergent molecules to form a micelle. When dissolved in water, detergent molecules self-assemble to form a spherical micelle (shown here in cross-section). The detergent is represented schematically as a hydrophilic head (red) and a hydrophobic tail (green).

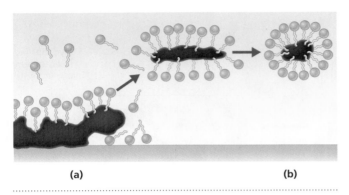

(a)   **(b)**

FIGURE 9.17 How detergent works to clean grease. (a) The hydrocarbon tails of the detergent provide a nonpolar environment for dissolving the hydrophobic grease (shown in black). (b) The detergent molecules form a water-soluble micelle, with the grease sequestered within its hydrophobic interior.

Soap contains molecules of stearic acid (shown below). Use solubility principles to explain why this molecule works as an effective cleaning agent.

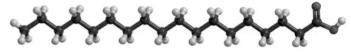

Stearic acid

CONCEPT QUESTION 9.5

Consider a collection of detergent molecules that is dissolved in liquid hexane. How will these molecules organize via self-assembly? Draw a sketch of the predicted structure, and explain your answer.

Core Concepts

- A detergent molecule acid contains a hydrophilic "head" group and a hydrophobic "tail" composed of a hydrocarbon chain.
- When a collection of detergent molecules is placed in water, they exhibit **molecular self-assembly** by spontaneously organizing to form a roughly spherical **micelle**. The hydrocarbon tails are buried within the interior of the micelle, and the polar head groups are exposed to the solvent.
- Detergent molecules clean dishes by sequestering grease within the nonpolar interior of micelles.

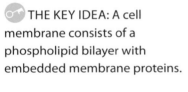 THE KEY IDEA: A cell membrane consists of a phospholipid bilayer with embedded membrane proteins.

- **phospholipid** A molecule composed of two fatty acids, a glycerol unit, a phosphate group, and a head group with polar components.

Self-Assembly of Cell Membranes

The principles of solubility and self-assembly also apply to the molecular organization of cell membranes. These membranes are constructed from molecules called **phospholipids** (Figure 9.18). A phospholipid bears some resemblance to a triglyceride molecule because it contains two fatty acids that are chemically linked to a glycerol. However, in phospholipids the glycerol's third –OH group links to a phosphate group (hence the use of "phospho" in the lipid's name). The

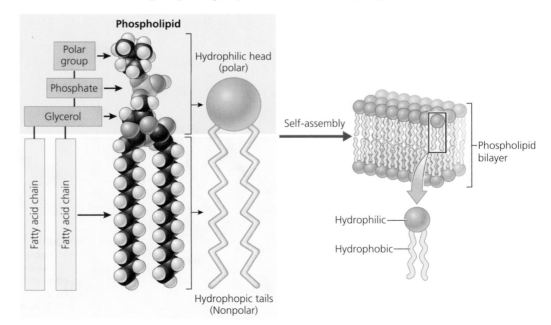

FIGURE 9.18 Phospholipids assemble into a bilayer structure. A phospholipid molecule contains a polar hydrophilic head group and two nonpolar hydrophobic tails. A collection of phospholipids self-assembles to form a phospholipid bilayer.

FIGURE 9.19 Schematic structure of a cell membrane. A cell membrane is composed of a phospholipid bilayer with embedded membrane proteins. The proteins are shown in purple, and some of them have attached carbohydrate chains.

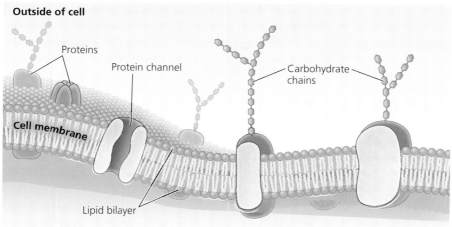

final component is the head group, which typically has polar or charged components. Chemical variations in the head group produce different types of phospholipids, and biological membranes include a mixture of various kinds.

Phospholipids are significantly larger than detergent molecules, but they still possess hydrophilic and hydrophobic regions. What happens to these molecules in water? Like detergents, they self-assemble into a structure that shields the hydrophobic regions from water. Instead of creating a micelle, however, the phospholipids form a **phospholipid bilayer** (often shortened to "lipid bilayer"). This difference arises because the phospholipid molecules have two long hydrocarbon tails, which are too bulky to fit within the interior of a micelle. Figure 9.18 shows the structure of a phospholipid and the organization of a bilayer. The bilayer resembles a sandwich in which the hydrophilic head groups form two thin layers that are exposed to water, while the hydrophobic tails create a much thicker nonpolar region within.

phospholipid bilayer A double layer of phospholipids that forms the structural foundation of the cell membrane.

The phospholipid bilayer forms the structural foundation of the cell membrane, but it is not the only component. As Figure 9.19 illustrates, various *membrane proteins* are embedded within the lipid bilayer. Some of these proteins have attached carbohydrate chains composed of linked sugar molecules.

Membrane proteins perform many essential biological tasks. For example, consider the challenge of getting H_2O molecules into cells. The presence of water inside cells is essential for their survival and function. However, H_2O is a very polar molecule, and it is not soluble within the hydrophobic interior of the lipid bilayer. How do H_2O molecules pass through the cell membrane?

PRACTICE EXERCISE 9.5

Cholesterol is a component of our cell membranes. The structure of cholesterol is provided below as a line-angle drawing and molecular model. Explain why cholesterol molecules are found within the hydrophobic interior of the lipid bilayer. What region of the cholesterol molecule would you predict will interact with the polar head groups of the phospholipids?

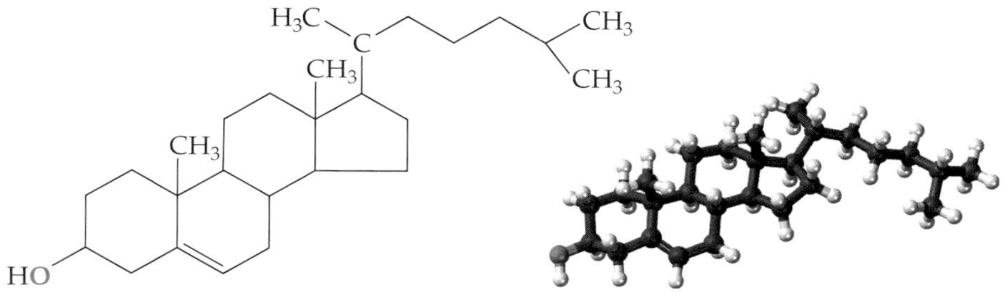

Cholesterol

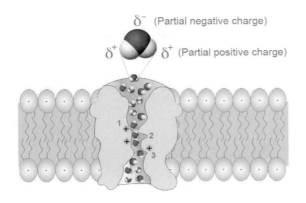

δ^- (Partial negative charge)

δ^+ δ^+ (Partial positive charge)

FIGURE 9.20 Aquaporin transports H$_2$O molecules across the cell membrane. Aquaporin is a membrane protein located within the phospholipid bilayer. It contains a channel that enables H$_2$O molecules to cross the membrane without encountering the nonpolar interior. The narrow region in the middle of the channel enables H$_2$O molecules to travel in single file, while preventing larger molecules from passing through.

The solution is provided by a transport protein called *aquaporin* ("water pore") that spans the cell membrane (Figure 9.20). Aquaporin contains a water channel that enables H$_2$O molecules to travel through the lipid bilayer without encountering the nonpolar tails of the phospholipids. The interior of the channel is coated with hydrophilic chemical groups that provide a hospitable environment for the H$_2$O molecules. A constriction in the middle of the channel creates a pathway so narrow that H$_2$O molecules have to pass through in single file. The positive charges located at this region prevent the H$_2$O molecules from forming hydrogen bonds, which would slow their progress through the channel. The narrowness of the channel allows small H$_2$O molecules to flow through but prevents the passage of larger molecules. In general, membrane transport proteins allow only a specific type of molecule or ion to cross the cell membrane.

SCIENCE IN ACTION

Using Solubility Principles for Drug Discovery

To be effective, a drug must be able to reach its biological target in the body. The relative ease by which an orally administered drug is absorbed into the bloodstream is called its *oral bioavailability*. To dissolve in the stomach and travel within the bloodstream, the molecule must be soluble in an aqueous environment. However, to pass through cell membranes—which have a hydrophobic interior—the drug must also be soluble in a nonpolar environment.

Is it possible to examine the structure of a potential drug molecule and predict its oral bioavailability? Christopher Lipinski, a senior research scientist at Pfizer Pharmaceuticals, proposed guidelines that are now called Lipinski's Rule of Five because they all contain a multiple of 5. The rule states that absorption of a drug into the bloodstream is *likely* when the following criteria are met:

(1) The molecular mass is *less than 500 amu.*

(2) The number of hydrogen bond donors is *no more than 5.*

(3) The number of hydrogen bond acceptors is *no more than 10.*

(4) The partition coefficient between nonpolar/polar solvents is *less than 10^5.*

Let's apply Lipinski's Rule of Five to acetaminophen, the active ingredient in Tylenol. Acetaminophen has a molecular mass of 151 amu, which satisfies point (1). As a reminder, amu stands for *atomic mass unit*, and 1 amu is approximately the mass of one hydrogen atom.

Points (2) and (3) relate to hydrogen bonds. In general, a molecule that is capable of forming hydrogen bonds is soluble in water. However, too many hydrogen bonds will make the molecule insoluble in nonpolar environments, which means that it cannot pass easily through cell membranes.

A *hydrogen bond donor* is typically a polar O—H or N—H bond. These bonds can "donate" their terminal hydrogen atom by forming a hydrogen bond with an H$_2$O molecule in the solvent. Conversely, a *hydrogen bond acceptor* is an atom that can accept a hydrogen bond from an H$_2$O molecule. A typical hydrogen bond acceptor is an atom with a lone pair (like O or N) that also has a partial charge as a result of participating in a polar covalent bond. As shown in Figure 1, acetaminophen has *two* hydrogen bond donors and *three* hydrogen bond acceptors. Points (2) and (3) are therefore satisfied.

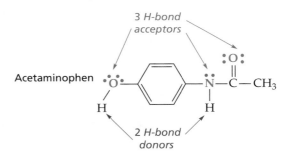

Acetaminophen

3 *H-bond acceptors*

2 *H-bond donors*

FIGURE 1 The molecular structure of acetaminophen contains two hydrogen bond donors and three hydrogen bond acceptors.

Point (4) refers to a *partition coefficient*, which indicates the *relative solubility* of a molecule in a nonpolar solvent compared to a polar solvent. We use octanol as the nonpolar solvent because it mimics the hydrophobic interior of cell membranes. Water is used as the polar solvent. These two solvents do not mix, so they can be layered on top of each other, as shown in Figure 2. To measure the partition coefficient, we add a sample of the drug and then shake the container to disperse the molecules. Depending on their solubility, the drug molecules will partition themselves between the two solvents (recall that "partition" means "divide into parts").

We then measure the relative concentration of solute in both the nonpolar and polar layers. The partition coefficient (P) of the solute molecule is defined as the ratio of nonpolar/polar concentrations. For example, if the concentration of solute is 100 times greater in n-octanol than in water, then the partition coefficient is equal to 100 (10^2). A higher number for P indicates that the solute has a stronger preference for nonpolar solvents, and a lower number signifies an affinity for polar solvents. As examples, the partition coefficient is 0.17 for methanol (more soluble in water) and 10,000 (10^5) for hexane (more soluble in octanol). The partition coefficient for acetaminophen is roughly equal to 3, which satisfies Point (4). This low P value indicates that acetaminophen has a very mild preference for octanol over water.

Because the values of P vary over a wide range, the partition coefficient is often reported as a logarithm in the form of $\log P$. Using the principle of logarithms, $\log P$ reflects the power of 10 in the numerical value of P (see Appendix C). For hexane, $P=10^5$ so $\log P = 5$. A larger, positive value of $\log P$ signifies that the solute prefers nonpolar solvents. If $\log P$ is a smaller or a negative number, then the solute has greater affinity for polar solvents. For methanol, which is highly water soluble, $P=0.17$ and $\log P = -0.77$.

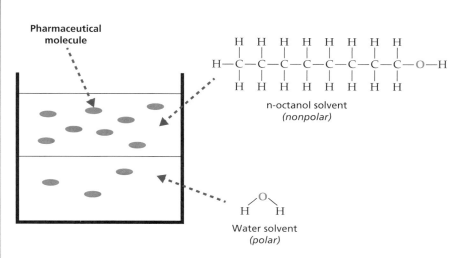

Pharmaceutical molecule

n-octanol solvent
(nonpolar)

Water solvent
(polar)

FIGURE 2 **The partition coefficient measures the relative solubility of a solute molecule in a nonpolar solvent (n-octanol) compared to a polar solvent (water).** This diagram illustrates a drug molecule (shown in red) that is more soluble in octanol than in water, resulting in a higher concentration of molecules in the nonpolar solvent.

Core Concepts

- Cell membranes are constructed from **phospholipid** molecules, which self-assemble into a **phospholipid bilayer**. This sandwich structure has two external hydrophilic surfaces and a thick hydrophobic interior.
- Cell membranes in living organisms contain membrane proteins that perform various tasks. For example, aquaporin is a transport protein that enables H_2O molecules to pass through the cell membrane.

9.4 Ions in Solution

The chemical principles of solubility apply not only to molecules but also to ions. Because our cells and blood contain multiple types of ions, their solubility is important for the chemistry of life. We examined ions and ionic compounds in earlier chapters. As a reminder, a simple ion (such as K^+ or F^-) is formed when an atom gains or loses one or more electrons. A polyatomic ion (such as NH_4^+ or SO_4^{2-}) is one in which a bonded collection of atoms has gained or lost one or more electrons. All ions possess a net electrical charge, either positive or negative.

Learning Objective:

Characterize the solubility of ions and ionic compounds.

Ions in Aqueous Solution

We begin our examination of ions in solution by performing a simple experiment to measure the flow of electrical current in a liquid. Figure 9.21 presents the experimental apparatus, which consists of two electrodes that are immersed

FIGURE 9.21 Measuring the flow of electrical current in a liquid. (a) When the beaker contains distilled water, the bulb does not light. (b) Adding sodium chloride to the water causes the bulb to light, which indicates a flow of electrical current.

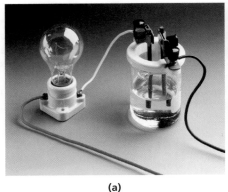 

(a)   (b)

FIGURE 9.21 Measuring the flow of electrical current in a liquid. (a) When the beaker contains distilled water, the bulb does not light. (b) Adding sodium chloride to the water causes the bulb to light, which indicates a flow of electrical current.

🔑 THE KEY IDEA: Solid sodium chloride dissolves in water to produce aqueous Na+ and Cl- ions.

in a beaker of liquid, plus a light bulb to detect the flow of current. The device is powered by a battery, which is not shown. An electric current is produced by the movement of electrons in a circuit, but current will flow *only* if the circuit is complete (i.e., a closed loop).

The beaker in Figure 9.21(a) contains distilled water, which is (almost) pure H_2O. Because the bulb is not lit, we conclude that distilled water does not conduct an electrical current. We now add a spoonful of table salt (NaCl) to the distilled water and let it dissolve. The result is shown in Figure 9.21(b)—the bulb is now lit, which indicates a flow of electrical current in the circuit. What has happened in the water to produce this change?

Recall that NaCl is an ionic compound that consists of two oppositely charged ions. The Na atom has lost an electron to form Na^+, whereas the Cl atom has gained an electron to form Cl^-. Solid NaCl is held together by ionic bonds that arise from the electrical attraction between oppositely charged ions (See Chapter 3). In its solid form, NaCl does not conduct electricity. However, when NaCl is added to water, the salt crystals dissolve to produce free-floating Na^+ and Cl^- ions. Because the ions are dissolved in water, they are called **aqueous ions**. The process whereby an ionic compound falls apart is called *dissociation*, which is the opposite of association (coming together). The dissociation of solid NaCl in solution can be written as a chemical equation in which (*s*) means *solid* and (*aq*) means *aqueous*.

■ **aqueous ion** An ion dissolved in water.

$$NaCl(s) \xrightarrow{H_2O} Na^+(aq) + Cl^-(aq)$$

The aqueous ions are mobile within the solvent—in fact, the name *ion* means "wanderer." The flow of $Na^+(aq)$ and $Cl^-(aq)$ ions conducts an electrical current through the aqueous solution. The $Na^+(aq)$ ions are attracted to the negative electrode (the cathode), so they are named **cations**. Conversely, the $Cl^-(aq)$ ions flow toward the positive electrode (the anode) and are named **anions**. Any dissolved solute that conducts electricity in this manner is called an **electrolyte**. You may have heard of *electrolyte imbalance*, which can be caused by profuse sweating during exercise.

■ **cation** A positively charged ion.

■ **anion** A negatively charged ion; for example, Cl⁻.

■ **electrolyte** A dissolved solute that conducts electricity.

Why do NaCl crystals dissolve in water? When the crystals enter the water, the partial charges on the H_2O molecules are attracted to the charged ions on the surface of the crystals. The ions now have a choice between two different environments—the solid or the solvent. Because the H_2O molecules in the solvent are so abundant, the attraction between the ions and H_2O is sufficient to overcome the ionic attractions within the NaCl crystal. As a result, the Na^+ and Cl^- ions break away from the crystal lattice and become aqueous ions in solution. Finally, all of the ions have entered the solvent and the salt crystal completely dissolves.

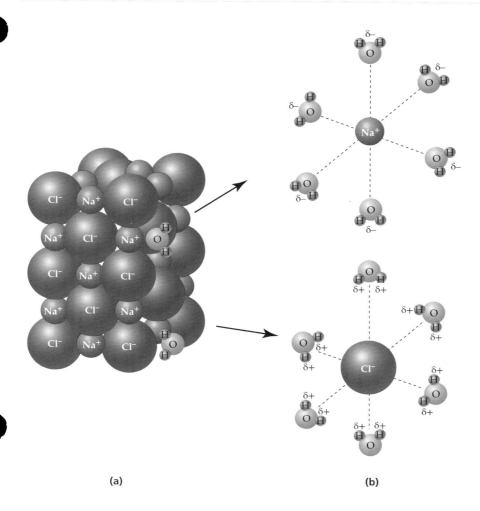

FIGURE 9.22 NaCl dissociates in water. (a) A crystal of NaCl just after being placed in water. In an aqueous environment, the H_2O molecules begin to dissolve the crystal lattice structure. (b) The arrangement of H_2O molecules around sodium and chloride ions in an aqueous solution. The $O^{\delta-}$ atoms are oriented toward the Na^+ ions, and the $H^{\delta+}$ atoms are oriented toward the Cl^- ion.

(a)   (b)

The $Na^+(aq)$ and $Cl^-(aq)$ ions are now surrounded by H_2O molecules, as shown in Figure 9.22. Note that the molecular orientation of H_2O molecules is different for the two ions. For the Na^+ ion, the H_2O molecules orient themselves so that the partial negative charge on the oxygen (δ^-) points toward the ion's positive charge. By contrast, the negatively charged Cl^- ion is surrounded by the partial positive charges (δ^+) on the hydrogen atoms in H_2O.

CONCEPT QUESTION 9.6

Imagine that we repeat the experiment shown in Figure 9.21, but we use liquid hexane as the solvent instead of water. We add solid NaCl to the solvent and observe the light bulb. Will the bulb light or not? Explain your prediction using chemical principles.

Although electrolytes dissolve in water, not all chemical substances that dissolve in water are electrolytes. For example, sucrose is a type of sugar molecule that is highly soluble in water (recall the structure of sucrose from Chapter 7). However, if we add sugar to distilled water in our conductivity apparatus, the bulb will not light. Sucrose is soluble in water because it has many—OH groups that form intermolecular bonds with H_2O molecules in the solvent. However, the atoms in the sucrose molecule are joined by strong covalent bonds that do not break in solution. As a result, sucrose does not generate any ions, so there is no conduction of electricity. Consequently, *sucrose is not an electrolyte.*

CHEMISTRY AND YOUR HEALTH

Electrolyte Imbalance

As we exercise, the movement of our muscles generates heat that is transferred to the blood and the body's core. As a result, our internal body temperature increases, and we begin to perspire. Perspiration is a natural mechanism designed to cool our body through the evaporation of water from the surface of the skin (see *Chemistry and Your Health* in Chapter 8). The sweat released by our skin is composed of water that contains dissolved ions such as sodium and chloride. If you lick the sweat on your skin, it will taste salty. In some cases of strenuous and prolonged exercise, excessive loss of ions through the skin can lead to *electrolyte imbalance*. Symptoms of this condition include muscle cramps, nausea, and fluid retention.

A simple solution for mild electrolyte imbalance due to exercise is to consume an electrolyte sports drink. One example is Gatorade, which was developed in 1965 by researchers in the University of Florida to help the members of the school football team—the Gators (hence the name of the drink). Nowadays, many types of sports drinks are available. To determine the quantity of electrolytes in the drink, consult the Nutrition Facts label on the bottle. The amount of electrolytes that you need to replace depends on how much sweat you produce while exercising.

All sports drinks contain dissolved ions to replace the electrolytes that are lost through sweating. Many of these drinks also contain sugars, both as an energy source and as a way to facilitate our body's uptake of the ions. When we consume a sports drink, the sodium must move from the intestine into the cellular environment. A transporter protein pumps both glucose and sodium across the membranes of the intestinal cells in a concerted action. The presence of sugars in sports drinks serves to stimulate the uptake of sodium ions into our cells.

Nutrition Facts

Serving Size 8 fl 0z (240 mL)
Servings Per Container 4

Amount Per Serving

Calories 50

	% Daily Value**
Total Fat 0g	0%
Sodium 110mg	5%
Potassium 30mg	1%
Total Carbohydrate 14g	5%
Sugars 14g	
Protein 0g	

Not a significant source of Calories from Fat, Saturated Fat, Cholesterol, Dietary Fiber, Vitamin A, Vitamin C, Calcium, Iron.

* Percent Daily Values are based on a 2,000 calorie diet.

← **Electrolytes**

FIGURE 1 **Sports drinks contain dissolved ions (sodium and potassium) to replace the electrolytes that are lost by sweating during exercise.** The amounts of electrolytes are listed on the nutrition label for the drink.

Core Concepts

- Solid crystals of NaCl dissolve in water to produce aqueous ions: Na$^+$(aq) and Cl$^-$(aq). These ions are mobile in water, and they conduct an electrical current.
- The Na$^+$(aq) ions are called **cations** because they are attracted to the cathode (the negative electrode). The Cl$^-$(aq) ions are called **anions** because they are attracted to the anode (the positive electrode).
- A dissolved solute that conducts electricity is called an electrolyte. Na$^+$(aq) and Cl$^-$(aq) ions are electrolytes.
- NaCl crystals dissolve because of the attraction between the charged ions and the partial charges on H$_2$O molecules in the solvent. Na$^+$(aq) and Cl$^-$(aq) ions in solution are surrounded by H$_2$O molecules; the orientation depends on the ion's electrical charge.
- Sucrose dissolves easily in water, but it does not dissociate to produce ions. Therefore, sucrose is not an electrolyte.

TABLE 9.3	Solubility Rules for Ionic Compounds in Water at 25°C[1]	
IONS THAT FORM SOLUBLE COMPOUNDS	**INSOLUBLE EXCEPTIONS**	
Group 1A (alkali metals): lithium (Li$^+$), sodium (Na$^+$), potassium (K$^+$)	None	
Ammonium (NH$_4^+$)	None	
Nitrate (NO$_3^-$)	None	
Bicarbonate (HCO$_3^-$)	None	
Chlorate (ClO$_3^-$)	None	
Group 7A (halides): chloride (Cl$^-$), bromide (Br$^-$), iodide (I$^-$).	Compounds formed with silver (Ag$^+$), mercury (Hg^{2+}), lead (Pb^{2+})	
Sulfate (SO$_4^{2-}$)	Compounds formed with silver (Ag$^+$), calcium (Ca^{2+}), strontium (Sr^{2+}), barium (Ba^{2+}), mercury (Hg^{2+}), lead (Pb^{2+})	
IONS THAT FORM INSOLUBLE COMPOUNDS	**SOLUBLE EXCEPTIONS**	
Carbonate (CO$_3^{2-}$), phosphate (PO$_4^{3-}$), sulfide (S^{2-})	Compounds formed with sodium (Na$^+$), potassium (K$^+$), ammonium (NH$_4^+$)	
Hydroxide (OH$^-$), oxide (O^{2-})	Compounds formed with sodium (Na$^+$), potassium (K$^+$), barium (Ba^{2+})	

[1] Soluble compounds are defined as those that dissolve 1 g or more in 100 g of water. Insoluble compounds are defined as those that dissolve less than 0.01 g per 100 g of water.

precipitate An insoluble ionic solid formed by the combination of an aqueous cation and anion.

precipitation reaction A chemical reaction that forms an insoluble precipitate.

Solubility of Ionic Compounds

Although some ionic compounds dissolve easily in water, others are insoluble or dissolve only to a tiny extent. Based on experiments, chemists have developed solubility rules that specify which ionic compounds are soluble in water. Table 9.3 provides a selection of solubility rules, along with some exceptions. Refer to Chapters 2 and 3 to refresh your memory on the varieties of ions and ionic compounds.

THE KEY IDEA: Solubility rules indicate which ionic compounds are soluble in water and which are not.

The solubility rules in Table 9.3 enable us to understand a type of chemical reaction in which aqueous cations and anions combine to form an insoluble ionic solid called a **precipitate**. This type of reaction is called a **precipitation reaction**. Figure 9.23 illustrates one example of a precipitation reaction. In Figure 9.23(a), a colorless solution of potassium iodide (KI) is about to be poured into a beaker containing a colorless solution of lead nitrate (PbNO$_3$). According to the solubility rules, both of these ionic compounds are soluble in water. Figure 9.23(b) shows the result of adding the two solutions—the formation of a bright yellow precipitate of lead iodide (PbI$_2$). The solubility rules tell us that most compounds containing iodide (I$^-$) ions are soluble, but there is an insoluble exception when I$^-$ combines with a lead ion (Pb^{2+}).

We can represent the precipitation reaction by a chemical equation. The ionic compounds are color-coded so that we can track their behavior in the reaction. The designation (*aq*) means that the ionic compound is soluble in water and has dissociated into aqueous ions. The formation of a solid precipitate is indicated by (*s*).

(a)   (b)

FIGURE 9.23 A precipitation reaction. (a) Preparing to add a colorless solution of potassium iodide (KI) to a colorless solution of lead nitrate (PbNO$_3$). (b) Mixing these solutions produces an insoluble yellow precipitate of lead iodide (PbI$_2$).

$$2\,KI(aq) \;+\; Pb(NO_3)_2(aq) \;\longrightarrow\; PbI_2(s) \;+\; 2\,KNO_3(aq)$$

Potassium Lead nitrate   Lead Potassium
iodide   iodide nitrate

Precipitate

We can write a *complete ionic equation* that represents all of the aqueous ions in the reactants and products. The solid precipitate of $PbI_2(s)$ does not contribute to the aqueous ions.

$$\underbrace{2\,K^+(aq) + 2\,I^-(aq)}_{\text{Potassium iodide}} \;+\; \underbrace{Pb^{2+}(aq) + 2\,NO_3^-(aq)}_{\text{Lead nitrate}} \;\longrightarrow\; PbI_2(s)$$

Lead iodide

Precipitate

$$+\; \underbrace{2\,K^+(aq) + 2\,NO_3^-(aq)}_{\text{Potassium nitrate}}$$

spectator ion An ion that appears on both sides of a chemical equation and does not participate in the chemical reaction.

The $K^+(aq)$ and $NO_3^-(aq)$ ions appear on both sides of this equation, which means that they have not participated in the chemical reaction. We call them **spectator ions** because they only "watch" the reaction. In general, *a spectator ion is not changed as a result of the chemical reaction, so it appears on both the reactant and product sides of the chemical equation.* We can eliminate the spectator ions and write an equation that contains only those ions that participate in the reaction. This type of equation is called a *net ionic equation,* and it is written as follows:

$$2\,I^-(aq) \;+\; Pb^{2+}(aq) \;\longrightarrow\; PbI_2(s)$$

Lead iodide

Precipitate

WORKED EXAMPLE 9.2

Question: Barium nitrate, $Ba(NO_3)_2$, and potassium sulfate, K_2SO_4, are ionic solids that dissolve in water. Mixing these two solutions produces a white precipitate as shown in the photograph.

(a) Write chemical equations to show which aqueous ions are present in the solutions of $Ba(NO_3)_2$ and K_2SO_4.

(b) Identify the name and chemical formula of the precipitate.

(c) Write a complete ionic equation for the precipitation reaction, and identify the spectator ions.

(d) Write the net ionic equation for the reaction.

Answer: (a) The equations below describe how each ionic compound dissociates in water to form a solution of aqueous ions:

$$Ba(NO_3)_2(s) \;\xrightarrow{\;H_2O\;}\; Ba^{2+}(aq) + 2\,NO_3^-(aq)$$

$$K_2SO_4(s) \;\xrightarrow{\;H_2O\;}\; 2\,K^+(aq) + SO_4^{2-}(aq)$$

(b) When the two solutions are mixed together, we have a combination of all four ions shown on the product side of the equations. The next step is to examine which pair of cation and anion could combine to form a solid precipitate. In essence, we are investigating what happens when the ions "swap places" compared to the original clear solutions. Using the solubility rules, we see that $Ba^{2+}(aq)$ and $SO_4^{2-}(aq)$ can combine to form a solid precipitate of *barium sulfate*, $BaSO_4$. This prediction uses the solubility rule that compounds containing sulfate (SO_4^{2-}) ions are usually soluble, *except* when paired with barium ions (Ba^{2+}).

(c) We can write the complete ionic equation for the precipitation reaction to show the aqueous ions.

$$Ba^{2+}(aq) + 2\,NO_3^-(aq) + 2\,K^+(aq) + SO_4^{2-}(aq)$$

$$\longrightarrow BaSO_4(s) + 2\,NO_3^-(aq) + 2\,K^+(aq)$$

Precipitate

This form of the equation enables us to identify the spectator ions as $Na^+(aq)$ and $NO_3^-(aq)$. These ions appear on both the reactant and product sides of the chemical equation.

(d) The net ionic equation eliminates the spectator ions and includes only the ions that participate in the chemical reaction.

$$Ba^{2+}(aq) + SO_4^{2-}(aq) \longrightarrow BaSO_4(s)$$

Precipitate

TRY IT YOURSELF 9.2

Question: Potassium chloride, KCl, and silver nitrate, $AgNO_3$, are ionic solids that dissolve in water. When these solutions are mixed together, will a precipitate be formed? If so, identify the name and chemical formula of the precipitate.

PRACTICE EXERCISE 9.6

You have samples of two solutions: barium chloride, $BaCl_2$, and ammonium sulfate, $(NH_4)_2SO_4$. If you mix the solutions, will a precipitate form? Answer this question by writing a chemical equation and providing an explanation.

The solubility rules have practical consequences for living organisms. Within the aqueous environment of living cells, it would be detrimental for two ions to combine and form a precipitate. In other contexts, however, an insoluble ionic compound can be useful. As one example, calcium (Ca^+) ions combine with carbonate ions to form an insoluble solid known as *calcium carbonate*. You know this substance as blackboard chalk (Figure 9.24(a)). Calcium carbonate is also used by marine organisms to build hard shells, which we sometimes find on the beach (Figure 9.24(b)). Birds also use calcium carbonate to make the outer shells of their eggs (Figure 9.24(c)).

Another insoluble salt, calcium phosphate, is the major form of calcium found in cow's milk. The calcium ion is Ca^{2+} and the phosphate ion is PO_4^{3-}, so the chemical formula for calcium phosphate is $Ca_3(PO_4)_2$. The calcium phosphate attaches to a milk protein called casein, which prevents the ionic compound from forming a precipitate. Calcium phosphate is absorbed into our bones and teeth, where it is

(a)

(b)

(c)

FIGURE 9.24 Three examples of calcium carbonate, an insoluble ionic compound. (a) Blackboard chalk. (b) A shell from a marine organism. (c) The outer shell of a bird's egg.

used to make a more complex ionic solid called *hydroxylapatite*, $Ca_5(PO_4)_3(OH)$. Hydroxylapatite makes up 70% of our bone mass and almost 90% of our tooth enamel. This is why drinking milk or obtaining calcium from another source contributes to healthy bones and teeth.

Core Concepts

- Solubility rules indicate which ionic compounds are soluble in water and which are not.
- A **precipitation reaction** occurs when dissolved cations and anions in aqueous solution combine to form an insoluble ionic solid called a **precipitate**.
- A **spectator ion** is not changed as a result of a chemical reaction. Therefore, it appears in both the reactant and product side of the chemical equation.
- Soluble ionic compounds are used in the aqueous interiors of cells. Insoluble calcium carbonate is used to build the shells of sea organisms and the outer shells of bird eggs.

Ion Transport Across Cell Membranes

THE KEY IDEA: Charged ions require the assistance of ion channel proteins to cross the nonpolar interior of the cell membrane.

Ions serve many important biological functions, such as transmitting chemical signals, stimulating muscle contraction, and conducting electrical signals along nerve cells. In many cases, these functions involve the flow of ions into or out of cells. As we learned in Section 9.3, cells are enclosed by a cell membrane that is composed of a phospholipid bilayer with embedded membrane proteins.

Aqueous ions are highly soluble in water, where they interact favorably with the polar H_2O molecules. However, the electrical charges on ions make them *very insoluble in nonpolar environments*. Consequently, the hydrocarbon interior of a phospholipid bilayer forms an impenetrable barrier to ions. To cross the cell membrane, ions receive assistance from membrane proteins called *ion channels*. These channels operate in a similar way to aquaporin, discussed earlier in the chapter. The interior of the ion channel provides a tunnel containing polar and charged chemical groups, which offer a more favorable environment for the transport of ions.

Some ion channels can be selectively opened or closed based on electrical or chemical signals. Figure 9.25 illustrates an ion channel in these two states. When the channel is closed, ions cannot cross the membrane. Conversely, when it is open, ions can flow into or out of the cell. Ion channels often discriminate among different types of ions. For example, one type of channel admits potassium ions (K^+) but not sodium (Na^+) ions, even though they possess the same electrical charge.

A disruption of ion transport is the primary cause of *cystic fibrosis*, the most common genetic disorder in the United States. Normal lung cells contain a protein channel that enables the transport of chloride (Cl^-) ions across the cell membrane. The instructions for making this ion channel are encoded within a gene, which is a specific region of DNA. If a child inherits two mutated versions of the gene—one from each parent—then a malfunctioning protein is generated. The reduction in functioning chloride channels disrupts the normal flow of chloride ions across the cell membrane. This disruption generates the characteristic symptoms of the disease—very salty sweat and, more seriously, a buildup of thick mucus in the lungs that interferes with breathing and is prone to bacterial infection.

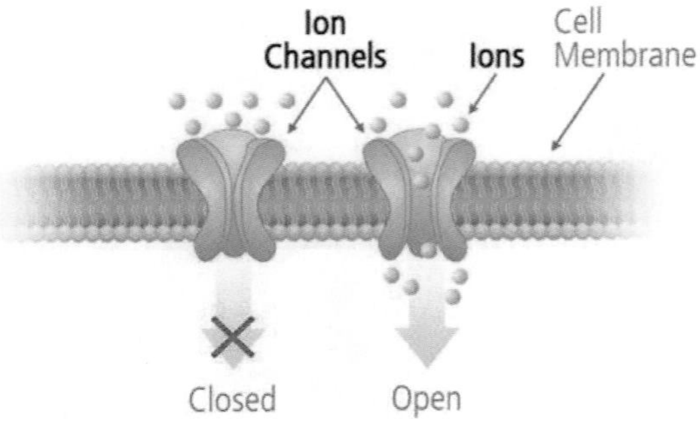

FIGURE 9.25 Ion channels are membrane proteins that permit the selective transport of charged ions across the nonpolar interior of the cell membrane. When the channel is closed, no ions can pass through the membrane. The protein can be triggered by a chemical or an electrical stimulus, which opens the channel and allows ions to flow into the cell.

Core Concepts

- The phospholipid bilayer forms an impenetrable barrier to ions because they are not soluble within the nonpolar interior of the bilayer.
- Ion channel proteins provide a pathway for ions to cross the cell membrane.

VISUAL SUMMARY

9.1   How Do Water-Soluble and Fat-Soluble Vitamins Differ?

Learning Objective:

Distinguish between water-soluble and fat-soluble vitamins.

⊙ **Vitamins** are organic molecules that play an essential role in various biological functions. Our body cannot make most vitamins, so we must obtain them from our diet.

⊙ Vitamins are grouped into two categories: **water-soluble vitamins** and **fat-soluble vitamins.**

9.2   Molecules in Solution

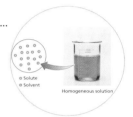

Learning Objective:

Analyze polar and nonpolar molecules to predict their solubilities in different solvents.

⊙ A **solution** is a homogeneous mixture that consists of a **solute** dissolved in a **solvent**.

⊙ A solution that uses liquid water as the solvent is called an **aqueous solution**.

⊙ When a nonpolar methane (CH_4) molecule is forced into liquid water, the H_2O molecules form a rigid, cage-like structure called a **clathrate** that surrounds the nonpolar solute.

⊙ Forming a clathrate requires the H_2O molecules to become more structurally organized compared to liquid water. This process results in a decrease in **entropy**, which must be overcome by an input of energy.

⊙ Molecules (like oils) that do not mix with water are called **hydrophobic** ("water fearing").

⊙ When oil is poured into water, it forms large oil droplets that segregate the oil molecules from the surrounding water.

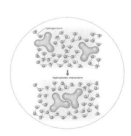

⊙ This behavior can be explained by the **hydrophobic effect**: If two or more nonpolar molecules are dissolved in water, they aggregate to minimize their interaction with the solvent.

⊙ Molecules that mix easily with water are called **hydrophilic** ("water loving").

⊙ Methanol (CH_3OH) contains a polar—OH group. It is soluble in water because the—OH group in methanol forms hydrogen bonds with H_2O molecules.

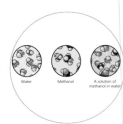

⊙ The hydrocarbon hexane (C_6H_{12}) is a liquid at room temperature because of molecular attractions called **dispersion forces**.

⊙ Dispersion forces are produced when fluctuations in the electron density of one hexane molecule (an instantaneous dipole) cause a change in the distribution of electron density in a neighboring molecule (an induced dipole).

⊙ Methanol is more soluble in water than in liquid hexane. Conversely, hexanol (with a larger hydrocarbon chain) is more soluble in hexane than in water.

⊙ In general, polar molecules are soluble in polar solvents, and nonpolar molecules are soluble in nonpolar solvents ("like dissolves like").

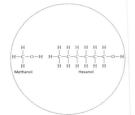

⊙ Most of the stored fat in our body is composed of triglycerides, which contain three long hydrocarbon chains. "Fat soluble" is equivalent to being soluble in a nonpolar solvent such as hexane.

⊙ Water-soluble vitamins have several polar—OH groups that can form hydrogen bonds to solvent H_2O molecules.

⊙ Fat-soluble vitamins are composed primarily of hydrocarbon regions, which dissolve in the nonpolar environment of triglyceride molecules.

9.3 Molecular Self-Assembly

Learning Objective:

Apply solubility principles to illustrate examples of molecular self-assembly.

⊙ **Molecular self-assembly** is the process by which molecules spontaneously arrange themselves into an organized structure.

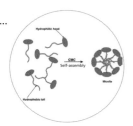

⊙ Self-assembly occurs when molecules have a polar (hydrophilic) region and a nonpolar (hydrophobic) region.

⊙ A collection of fatty acids in water self-assembles to form a spherical **micelle**. The polar carboxylic acid groups are exposed to the solvent, and the nonpolar hydrocarbon tails are buried within the interior.

- Cell membranes are constructed from **phospholipids**. These molecules contain a polar head group (hydrophilic) and two hydrocarbon tails (hydrophobic).

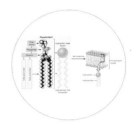

- A collection of phospholipids self-assembles to form a **phospholipid bilayer**. The hydrophilic head groups create two thin layers that are exposed to water, while the hydrophobic tails create a much thicker nonpolar interior.

- A cell membrane contains membrane proteins that are embedded within the phospholipid bilayer.

9.4   Ions in Solution

Learning Objective:

Characterize the solubility of ions and ionic compounds.

- Solid NaCl dissolves in water to produce an aqueous solution of Na$^+$(aq) ions and Cl$^-$(aq) ions.

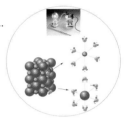

$$NaCl(s) \xrightarrow{\text{H}_2\text{O}} Na^+(aq) + Cl^-(aq)$$

- These aqueous ions flow through the solution and conduct electricity—they are **electrolytes**. Positively charged ions are called **cations,** and negatively charged ions are called **anions**.

- Solid NaCl dissolves in water because the polar H$_2$O molecules surround the aqueous ions. The partial charges on the O and H atoms in H$_2$O form an electrical attraction with the charges on the Na$^+$ and Cl$^-$ ions.

- Solubility rules indicate which ionic compounds are soluble in water and which are not.

- These rules can be used to explain a **precipitation reaction**, which occurs when dissolved cations and anions combine to form an insoluble ionic solid called a **precipitate**.

- Aqueous ions that do not participate in the precipitation reaction are called **spectator ions**.

- Because ions are charged, they are not soluble in the hydrophobic interior of the cell membrane. Transport of ions across the membrane requires ion channel proteins. Ion channels can be selectively opened or closed, based on electrical and chemical signals.

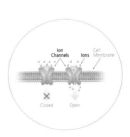

LEARNING RESOURCES

Reviewing Knowledge

9.1: How Do Water-Soluble and Fat-Soluble Vitamins Differ?

1. What are the biological functions of vitamins?

9.2: Molecules in Solution

2. Define the following terms: *solute, solvent, solution*.
3. Explain the difference between a solution and a heterogeneous mixture.
4. What is an aqueous solution?
5. Describe what happens when methane (CH_4) is forced into liquid water.
6. Describe the concept of entropy, and explain why it is relevant to the solubility of methane in water.
7. Explain why hydrophobic molecules tend to aggregate in water.
8. Methanol (CH_3OH) is a liquid at room temperature, whereas methane (CH_4) is a gas. Why?
9. Explain how dispersion forces cause hexane molecules to form a liquid at room temperature.
10. Explain the chemical principles underlying the solubility rule: "like dissolves like."
11. Which molecular characteristics can be used to distinguish water-soluble from fat-soluble vitamins?

9.3: Molecular Self-Assembly

12. What is molecular self-assembly? What types of molecules exhibit self-assembly?
13. What are the two regions of a detergent molecule, and how do their solubility properties differ?
14. Explain why fatty acids in water self-assemble to form a micelle.
15. What are the molecular components of a phospholipid?
16. What type of biological structure is formed by the self-assembly of phospholipids?
17. Describe the components of a cell membrane.
18. Why is it not possible for H_2O molecules to cross the cell membrane without assistance? How do cells facilitate the transport of H_2O molecule across the membrane?

9.4: Ions in Solution

19. Explain why a solution of table salt in water conducts electricity but purified water does not.
20. What is an electrolyte?
21. Write a chemical equation to describe how solid sodium chloride dissociates in water.
22. How are H_2O molecules oriented when they surround a sodium (Na^+) ion? Why does the orientation change when H_2O molecules surround a chloride (Cl^-) ion?
23. Although sucrose (a sugar) is very soluble in water, it is not an electrolyte. Why not?

24. What happens during a precipitation reaction?
25. Why do cells need ion transport proteins?

Developing Skills

26. Entropy applies to all situations, not only to solutions of molecules. Consider the figures presented here, which shows a volume of gas in a container. In Figure A, the gas particles (shown in red) are held in one half of the container by a piston, and there is a vacuum in the other half. In Figure B, the piston has been quickly withdrawn, and the gas has expanded to fill the entire container.

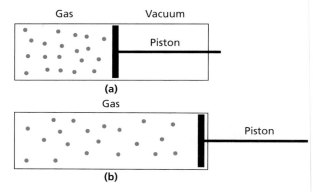

(a) How has the entropy of the gas particles changed in Figure B compared to Figure A?
(b) What is the likelihood that the gas particles in Figure B will spontaneously return to occupying only the left half of the container?
(c) What could you do to reverse the expansion of the gas and return the gas particles to occupying only the left half of the container?

27. It is possible to dissolve a very small amount of hexane in pure water. Describe how the presence of a nonpolar hexane molecule affects the surrounding H_2O molecules.
28. Methylamine (CH_3NH_2) is soluble in water. Sketch a diagram showing the hydrogen-bonding interactions that occur between the methylamine molecule and the solvent H_2O molecules.
29. Predict whether helium gas is more soluble in water or hexane. Explain your prediction.
30. The molecular structure of glycerol is shown below. Explain why glycerol is soluble in water.

31. Classify the following vitamins as either water soluble or fat soluble. Explain your choices.

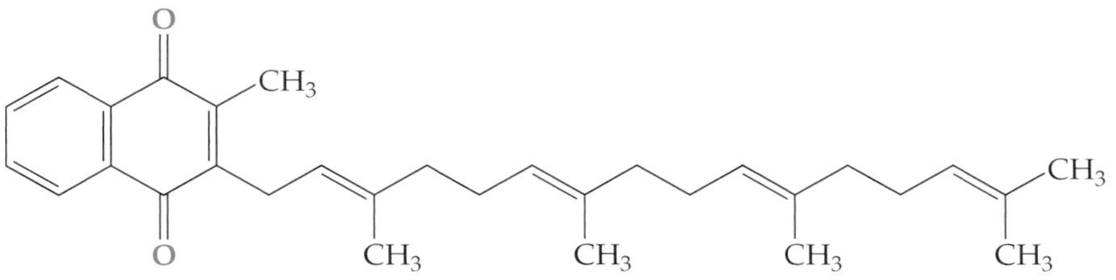

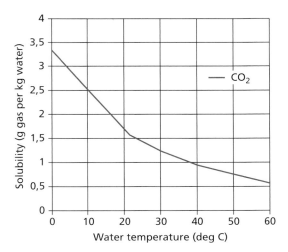

Vitamin K2 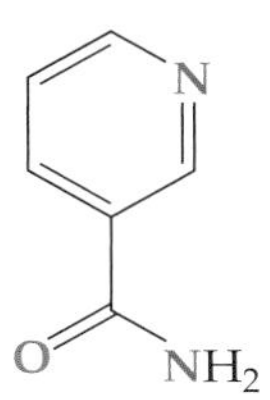   *Vitamin B3*

32. Fatty acids contain a long hydrocarbon chain linked to a carboxylic acid. Common fatty acids contain 16, 18, and 20 carbon atoms in their "tails." Will their solubility in water increase or decrease with the length of the tails? Explain your answer.

33. Table salt (sodium chloride) is more soluble in liquid water than in liquid methanol (CH_3OH). Explain this observation using chemical principles.

34. Which of the molecules below (A, B, and C) will self-assemble into a micelle, and which will not? Explain your prediction for each molecule.

(a)

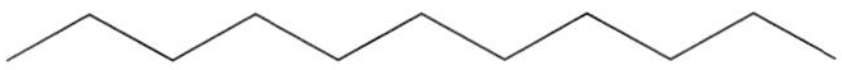

(b)

(c)

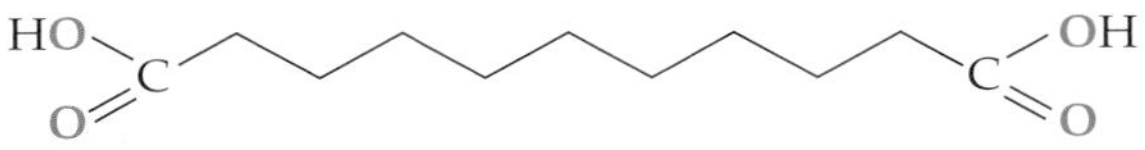

35. The solubility of a substance measures the maximum amount of solute that can dissolve in a particular volume of solvent. The graph shows how the solubilities of salt (NaCl) and table sugar (sucrose) change with temperature. The units of the vertical axis are grams of solute that dissolve in 100 milliliters (mL) of water (1 mL = 1/1000th of a liter).

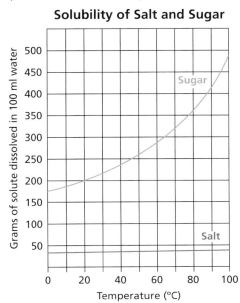

Solubility of Salt and Sugar

Compare how the solubilities of NaCl and sucrose change with increasing temperature.

Using the graph, explain why it is easier to dissolve more sugar in hot coffee than in cold coffee.

Calculate the percent increase in the solubility of sucrose when the temperature rises from 20°C to 100°C.

36. The graph shows the solubility of carbon dioxide (CO_2) in water. The units of the vertical axis indicate how many grams of gas dissolve in 1 kilogram (kg) of water.
    (a) Using the graph, describe how the solubility of CO_2 changes with increasing temperature.
    (b) Calculate the percentage change in solubility from 10°C to 40°C.
    (c) Imagine that you take a bottle of carbonated soda from the refrigerator and place it on a counter at room temperature. After 30 minutes, you remove the bottle cap, and the soda spills over the counter. Use the graph to explain why this happens.

37. You have been given a solution of an unlabeled salt sample to analyze. You perform the following tests:
    Test 1. Add $CaCl_2$ and observe a precipitate.
    Test 2. Add K_2CO_3 and observe no precipitate.
    Is the sample more likely to be Na_2SO_4 or $Hg(NO_3)_2$? Explain your reasoning.

Exploring Concepts

38. Titan, a moon of Jupiter, has an average surface temperature of –178°C. Scientists speculate that Titan may possess oceans of hydrocarbons. No evidence for life has been observed on Titan, but we can imagine what types of constraints the planetary environment would impose. Suggest what type of membrane structure could exist for a cell that must survive in a liquid hydrocarbon ocean on Titan.

39. The formation of clathrates is favored by conditions of low temperature and/or high pressure. Under such conditions, methane gas can get trapped within clathrates to produce an ice-like structure that burns with a violet flame (see the figure). Use the Internet to research methane clathrates. Identify several places in the world where they are found, and explain how they are connected to concerns about global warming. Write a one- to two- paragraph report on your findings.

40. Elements in Groups VIIA of the periodic table (the halogens) form molecules that contain two atoms (i.e., they are *diatomic*). Fluorine (F_2) and chlorine (Cl_2) are gases at room temperature, whereas bromine (Br_2) is a liquid, and iodine (I_2) is a solid. The differences among these states can be explained by dispersion forces. Explain why the dispersion forces among molecules are strongest in I_2, somewhat weaker in Br_2, and weakest in F_2 and Cl_2.

41. The U.S. Department of Agriculture (USDA) has compiled information on the recommended dietary allowance (RDA) for various vitamins. Select *three vitamins*, and use the USDA website (https://fnic.nal.usda.gov/food-composition/vitamins-and-minerals) to obtain the following information for each:

    (a) What is the recommended dietary allowance? Which units are used for this value?

    (b) What function(s) does this vitamin have in our bodies?

    (c) What are the health consequences of a deficiency in this vitamin?

42. You may have heard that high doses of vitamin C—much larger than the recommended dietary allowance—can prevent colds. Use the Internet to investigate whether vitamin C can prevent or treat colds. Write a two-paragraph summary that either supports or refutes this claim based on the evidence that you have discovered.

43. Cholera is a disease caused by an infectious bacterium that is particularly prevalent in Africa and South Asia. The characteristic symptom of cholera is watery diarrhea, which often leads to severe dehydration and electrolyte imbalance. One strategy for combating these effects is called *oral rehydration therapy*. Use the Internet to research oral rehydration therapy. Write a one- to two-paragraph report that explains this therapy in terms of the chemical concepts discussed in this chapter.

44. Cholesterol is a biological hydrocarbon that is very insoluble in water. Consequently, the transport of cholesterol in the blood requires a class of proteins called *lipoproteins*. Two variants of these proteins are *high density lipoprotein* (HDL) and *low density lipoprotein* (LDL). Use the Internet to investigate the biological roles of HDL and LDL. How are measurements of HDL and LDL used in medical cholesterol tests? Write a two-paragraph summary of your findings.

45. For many years the pesticide DDT (shown) was sprayed to control mosquitos and other insects. However, the use of DDT was restricted after scientists discovered that it was poisoning fish and birds. DDT is very insoluble in water. Use the Internet to investigate DDT. Based on your findings, explain how DDT accumulates in wildlife.

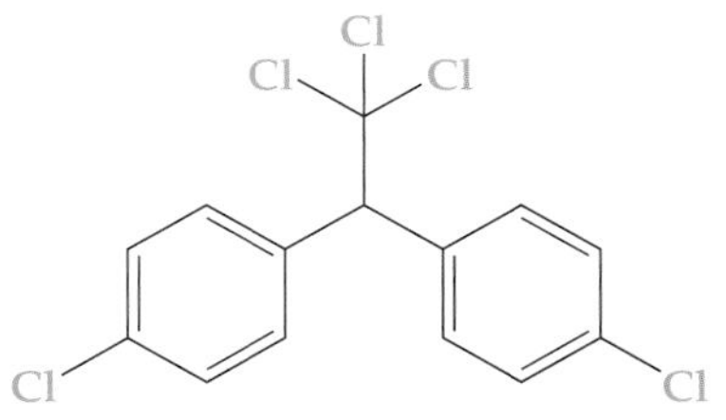

46. Polychlorinated biphenyls (PCBs) and dioxins are two classes of chemicals that have been linked to harmful environmental effects. One example molecule from each class appears below.
    (a) The harmful effects of these molecules are related to their solubility properties. Use the Internet to investigate the following questions, and write a short report on your findings.
    (b) What are the industrial uses of these molecules?

(c) How do they escape into the environment?
(d) What are their environmental impacts?

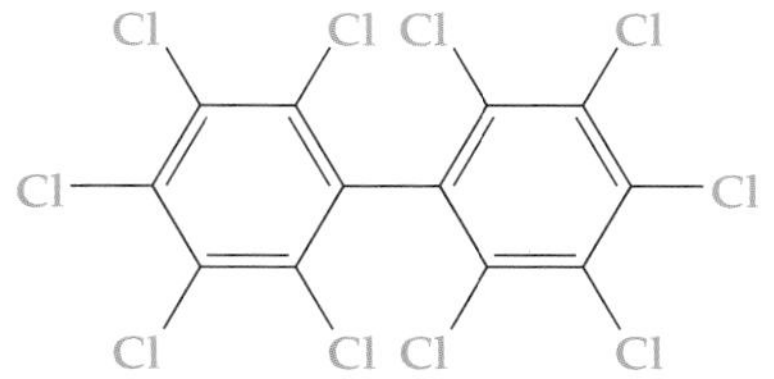

PCB

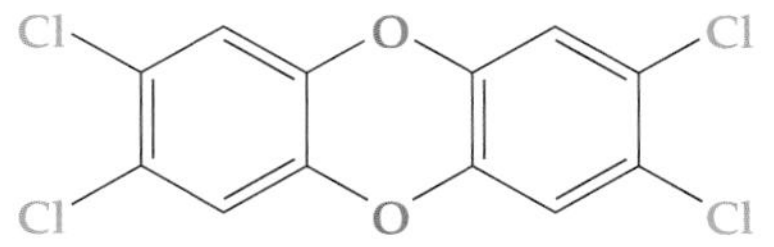

Dioxin

10

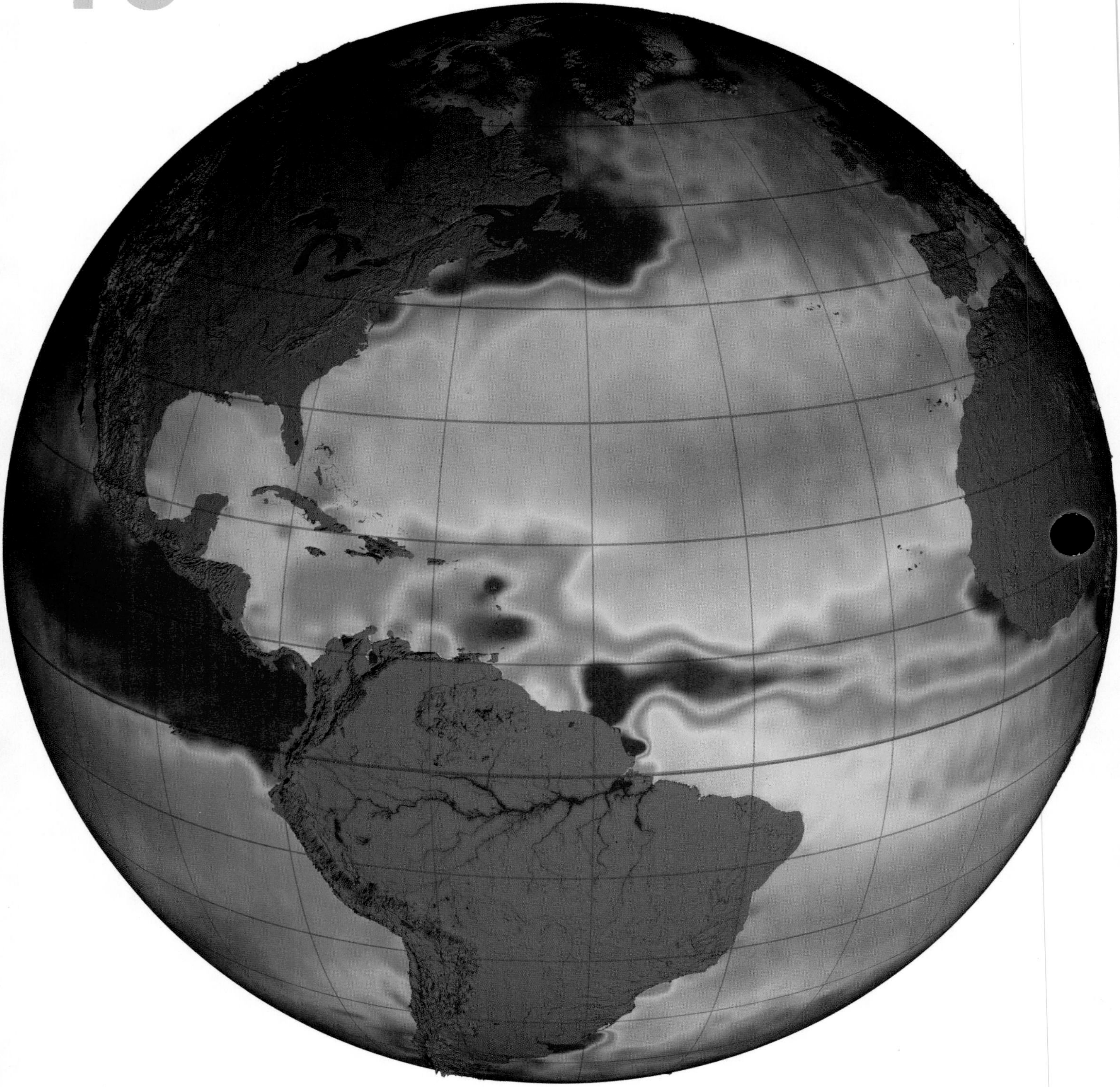

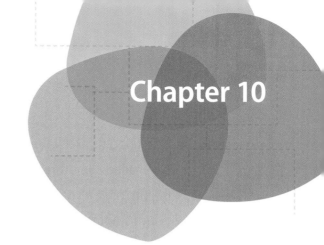

Measuring Concentration

CONCEPTS AND LEARNING OBJECTIVES

FEATURES

D uring the 2002 Boston Marathon, a young woman reached a region of the course called "Heartbreak Hill." She had been running for five hours and had consumed large quantities of a sports drink. After struggling to the top of the hill, she drank more liquid. Within minutes she collapsed and was rushed to the hospital, where she later died. What caused the runner's death?

10.1 Why Can Drinking Too Much Water be Harmful?

Learning Objective:

Explain the importance of maintaining an appropriate concentration of sodium ions in the blood.

🔑 THE KEY IDEA: Drinking an excessive amount of water can lead to a condition called hyponatremia.

hyponatremia A health-threatening condition caused by a lowered amount of sodium ions in the blood.

concentration A measure of the amount of substance dissolved in a specific volume of solution.

When we exercise, we usually drink water or a sports drink to avoid becoming dehydrated (Figure 10.1). In an endurance event like a marathon, we might be tempted to drink a lot of fluid. But is it possible to drink *too much*? If a person drinks a large amount of water or a sports drink (which is mostly water), the extra fluid lowers the level of sodium ions in the blood. If the amount of sodium becomes too low, the condition is called **hyponatremia**. This name is derived from "hypo" (meaning "below") and "natrium" (the historical name for sodium, Na).

The marathon runner described earlier drank so much fluid during the race that she developed severe hyponatremia, which caused the cells in her brain to swell. She was not the only runner affected. A follow-up report in the *New England Journal of Medicine* provided further details about 488 participants in the same marathon race. Based on blood sample measurements, 13% of the runners were diagnosed with mild hyponatremia, and 0.6% had sodium levels so low that they posed an immediate health risk.

The example of hyponatremia illustrates the importance of **concentration**, which is a quantitative property of a solution. In Chapter 9, we learned that a

FIGURE 10.1 Too much water? Drinking a moderate amount of water during exercise will prevent dehydration. However, drinking a very large amount of water during a marathon race can lead to a dangerously low salt concentration in the blood

solution is composed of a solute dissolved in a solvent. In general, *concentration measures the amount of solute dissolved in a specific volume of solution*. We can gain an intuitive understanding of solution concentration by considering the task of making a pitcher of orange juice from a can of frozen concentrate (Figure 10.2). The can contains concentrated orange extract that is convenient to store because the volume is small. To make juice that is suitable for drinking, we pour the contents of the can into a pitcher, add water, and stir to mix everything. After these steps, the total amount of orange extract in the pitcher remains the same, but it is now *less concentrated* because the total volume is larger.

The clinical diagnosis of hyponatremia is based on a reduced concentration of sodium ions in the *blood serum* (the fluid component of blood). Concentration is also an important factor in the therapeutic effect of pharmaceuticals. Most drugs that we take orally pass through our digestive system and enter the bloodstream, where they become a solute in a solution. If the concentration of the drug is too small, then it will not provide any health benefit. Conversely, if the concentration is too high, then the drug could become toxic. For this reason, it is important to know *how much* of the drug is present in a solution. In this chapter, we examine how concentration is defined, calculated, and measured.

FIGURE 10.2 Concentration of orange juice. In a can of orange "concentrate," the orange extract is confined within a small volume. In the pitcher of juice, the total amount of orange extract remains the same, but it is now less concentrated because the addition of water has produced a larger volume.

Core Concepts

- Drinking too much fluid can lower the level of sodium ions in the blood. This potentially dangerous condition is called **hyponatremia**.
- **Concentration** measures the amount of solute dissolved in a particular volume of solution.

10.2 Measuring Concentration

Concentration can be measured using a variety of units. Some of these measurements may be familiar, like the concentration of alcohol in beer, wine, and spirits. Chemists use a measurement of concentration based on moles because it provides information about the number of solute particles in the solution. This section introduces various measurements and units of concentration.

What is Concentration?

The concentration of a solution can be defined using an equation:

$$\text{concentration} = \frac{\text{amount of solute}}{\text{volume of solution}}$$

Note that the volume measurement used in this definition is the volume of the *solution* and not the volume of the solvent used to make the solution. This distinction is important because the amount of solute also contributes to the total volume of the solution. Using the terminology from Chapter 2, we note that concentration is an *intensive property*—it remains the same regardless of how much solution we have. Using the earlier example of orange juice, we find that the concentration of orange extract remains the same even after we have consumed half of the juice in the pitcher.

Learning Objective:
Use various measurements and units of concentration.

THE KEY IDEA: A quantitative definition of concentration is based on the amount of solute and the volume of the solution.

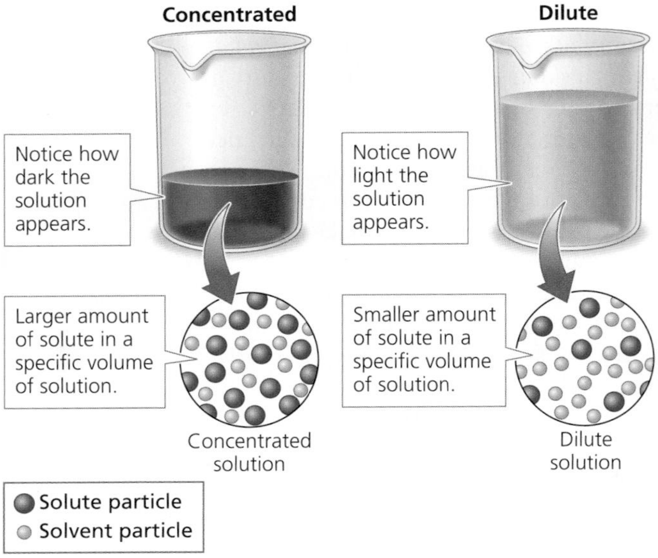

Concentrated

Notice how dark the solution appears.

Larger amount of solute in a specific volume of solution.

Concentrated solution

Dilute

Notice how light the solution appears.

Smaller amount of solute in a specific volume of solution.

Dilute solution

● Solute particle
● Solvent particle

FIGURE 10.3 Concentrated and dilute solutions. The solution on the left is more concentrated because it has a large number of solute molecules (blue dye) in a small volume of solution. Adding water to the beaker creates a more dilute solution because the amount of solute is the same, but the volume of the solution has increased. The dilute solution has fewer solute particles in a specified volume compared to concentrated solution.

dilution The addition of solvent to a solution to lower the concentration of the solution.

liter The standard unit for measuring the volume of a solution; $1 L = 10^3 cm^3$.

Figure 10.3 illustrates concentration using a blue dye in water. The solution on the left is *more concentrated* because the blue dye molecules are contained in a smaller volume of solution. The higher concentration of dye molecules causes the dark blue color of the solution. We can add more water to increase the volume of the solution, as shown on the right side of the figure. Significantly, the *number of dye molecules remains the same*. However, they are now dispersed throughout a larger volume of solution. As a result, *the concentration of the solute is reduced*. Referring to the equation for concentration, we see that keeping the amount of solute the same (in the numerator) and increasing the volume of the solution (in the denominator) will *lower* the concentration. The reduced concentration of dye molecules accounts for the lighter blue color of the solution. The addition of solvent to lower the concentration is called **dilution**; we say that the resulting solution is more *dilute* when compared to the original.

The circles in Figure 10.3 provide a molecular-level view of each solution. In the concentrated solution on the left, there are a greater number of solute particles within a particular volume. In the dilute solution on the right, there are fewer solute particles within the same volume. Referring again to the equation, we see that a larger amount of solute in a fixed volume corresponds to a higher concentration.

To measure the concentration of a solution, we need to know two quantities—the amount of the solute and the volume of the solution. The amount of solute can be measured using volume, mass, or moles. We will see examples of all these measurements within the chapter. The standard unit for measuring the volume of solution is the **liter**. *One liter is a volume that corresponds to a cube measuring 10 cm along each side* (Figure 10.4). For reference, the small bottles of water that you buy from the store are usually 0.5 L. We can write a liter in terms of cubic centimeters:

$$1 L = 10 cm \times 10 cm \times 10 cm = 1000 cm^3 = 10^3 cm^3$$

Samples of solution are often smaller than 1 L, especially in biological or medical contexts. In these cases, the more convenient unit is a cubic centimeter (1 cm^3). Because this volume corresponds to one-thousandth of a liter, it is called a *milliliter* (mL). For even smaller samples, such as a forensic DNA sample in a

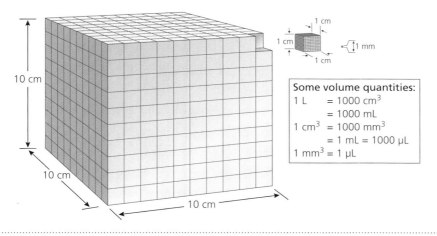

FIGURE 10. 4 Volume definitions. A volume of 1 liter (1 L) corresponds to a cube measuring 10 cm along each side (i.e., 1000 cm³). The smaller cube at the bottom right has a volume of 1 cm³ or 1 milliliter (1 mL). The smallest cube shown has a volume of 1 mm³, which is equal to 1 microliter (1 μL).

crime lab, scientists work with volumes of 1 *microliter* (μL). This volume corresponds to a volume of 1 cubic millimeter (1 mm³), which is one-millionth of a liter.

> **PRACTICE EXERCISE 10.1**
>
> Use powers of 10 to prove that 1 μL is equal to one-millionth of a liter.

Core Concepts

- The quantitative definition of concentration is based on the amount of solute and the volume of the solution.
- The amount of solute can be measured using *volume, mass,* or *moles.*
- The volume of solution can be measured in **liters** (L), *milliliters* (mL), or *microliters* (μL).

Measuring Concentration—Percent by Volume or Mass

If you read the label on a bottle of beer, you will see a number such as "5% ABV." This number tells you that the beer contains 5% "alcohol by volume." Wine contains around 13% ABV, and a distilled spirit like vodka might be 40% ABV "Alcohol by volume" is a measurement of concentration. In this case, the solute is ethanol (C_2H_5OH), the type of alcohol found in alcoholic drinks. Because pure ethanol is a liquid, it makes sense to specify the volume of ethanol as a percentage of the total volume of the beverage (the solution). The concentration measurement is a ratio called the **volume/volume percent**; it is written as % (v/v), where *v* refers to volume.

Under current regulations, manufacturers have wide latitude in how they display the alcohol content of a beverage. The U.S. Treasury Department has proposed the use of a standard label, like the one shown in Figure 10.5, which lists the alcohol content by volume, in addition to the amount of alcohol per serving and the serving size. At the current time, these descriptive labels are not mandatory.

⊙ **THE KEY IDEA:** Concentration can be measured as a percentage ratio of solute to solution.

Alcohol Facts	
 Contains **5** Servings	**Calories per Serving:** **98** Alcohol by Volume: 13% Alcohol per serving: 0.5 oz
Servings Size: 5 fl oz	

U.S. Dietary Guidelines advice on moderate drinking: no more than two drinks per day for men, one drink per day for women.

Ingredients: Grapes, yeast, sulfiting agents, and sorbates.

FIGURE 10.5 Labeling the alcohol content of drinks. The U.S. Treasury Department has proposed a standard label for alcoholic beverages. This label is for a hypothetical bottle of wine.

volume/volume percent A measure of concentration that uses a percentage ratio of the volume of solute compared to the volume of solution.

We can write an equation for % (v/v), beginning with our definition of concentration and quantifying the amount of solute in terms of its volume in milliliters (mL):

$$\text{concentration} = \frac{\text{volume of solute (mL)}}{\text{volume of solution (mL)}}$$

This equation can be converted to a percentage by multiplying the ratio by 100:

$$\% \ (v/v) = \frac{\text{volume of solute (mL)}}{\text{volume of solution (mL)}} \times 100\%$$

Using the definition of percentage, we can deduce that a 1% (v/v) solution corresponds to 1 mL of solute dissolved in 100 mL of solution. Therefore, a beer with 5% ABV contains 5 mL of ethanol in 100 mL of total volume. In the following equation, the volume units of mL cancel in the numerator and denominator of the ratio.

$$\% \ (v/v) = \frac{5 \ \cancel{mL} \ \text{ethanol}}{100 \ \cancel{mL} \ \text{solution}} \times 100\% = 5\%$$

In contrast to beer, the rubbing alcohol you buy in the drugstore as a disinfectant has a much higher alcohol concentration. It typically contains 70% (v/v) isopropanol, a molecule that contains three carbon atoms and an alcohol functional group.

WORKED EXAMPLE 10.1

Question: A standard bottle of wine has a volume of 750 mL. What volume of ethanol is contained in a bottle of wine with an alcohol content of 13% (v/v)?

Answer: We apply the formula for measuring concentration in % (v/v). The alcohol content of the wine, 13% (v/v), can be defined in terms of the volume of ethanol and the total volume of the solution (i.e., the 750 mL bottle of wine).

$$13\% = \frac{\text{volume of ethanol (mL)}}{750 \ \text{mL}} \times 100\%$$

$$\text{volume of ethanol (mL)} = \frac{13 \ \cancel{\%} \times 750 \ \text{mL}}{100 \ \cancel{\%}} = 98 \ \text{mL}$$

TRY IT YOURSELF 10.1

Question: Your local drugstore sells aqueous solutions of hydrogen peroxide (H_2O_2) as a disinfectant. Hydrogen peroxide is a liquid, and a typical concentration is 3.0% (v/v). What volume of hydrogen peroxide is contained in a bottle with a volume of 946 mL (8 fluid ounces)?

FIGURE 10.6 A bag of physiological saline. The intravenous fluid used in hospitals is an aqueous solution of sodium chloride with a concentration of 0.9% (m/v). This concentration is chosen to match the amount of dissolved salts in human blood serum.

As another example of concentration measurement, consider the use of an intravenous (IV) drip in a hospital to rehydrate a patient or deliver a medicine. Figure 10.6 illustrates a bag of IV solution with the label "0.9% sodium chloride." This concentration is chosen to match the concentration of salts that are dissolved in human blood serum. The medical term for this solution is *physiological saline*, where *saline* refers to a salt solution.

Sodium chloride is a solid, not a liquid, so it is more appropriate to measure the amount of solute as a mass rather than a volume. The concentration measurement used for IV fluid is called the **mass/volume percent;** it is written as % *(m/v).* Sometimes the older term *weight/volume percent* is used. We can derive an equation for this measurement by starting with our general definition of concentration and adding appropriate units:

$$\text{concentration} = \frac{\text{mass of solute (g)}}{\text{volume of solution (mL)}}$$

This equation can be converted to a percentage by multiplying the ratio by 100:

$$\% \text{ (m/v)} = \frac{\text{mass of solute (g)}}{\text{volume of solution (mL)}} \times 100\%$$

A 1% (m/v) solution corresponds to 1 g of solute dissolved in 100 mL of solution. We can make a 0.9% solution of physiological saline by dissolving 0.9 g of NaCl in 100 mL of solution.

$$\% \text{ (m/v)} = \frac{0.9 \text{ g NaCl}}{100 \text{ mL}} \times 100\% = 0.9\%$$

As a technical note, the measurement of % (m/v) has a mathematical inconsistency. Strictly speaking, a percentage is a ratio of two quantities that are *measured in the same units*. In the equation for % (m/v), mass (g) and volume (mL) have different units, so it is not possible to cancel the units in the ratio. However, we can still use this equation as a convenient way to express the concentration of a solid solute dissolved in a solution.

> **mass/volume percent** A measure of concentration that uses a percentage ratio of the mass of solute compared to the volume of solution.

WORKED EXAMPLE 10.2

Question: A sugar solution contains 750 mg of sucrose in a volume of 1.5 L. What is the concentration of sucrose in % (m/v)?

Answer: The equation for % (m/v) specifies mass in grams (mg) and volume in milliliters (mL). Therefore, we need to convert these given quantities to the appropriate units:

$$\text{mass of sucrose (g)} = 750 \text{ mg} \times \frac{1 \text{ g}}{1000 \text{ mg}} = 0.750 \text{ g}$$

$$\text{volume of solution (mL)} = 1.50 \text{ L} \times \frac{1000 \text{ mL}}{1 \text{ L}} = 1500 \text{ mL}$$

We can now apply the equation for % (m/v):

$$\% \text{ (m/v)} = \frac{0.750 \text{ g}}{1500 \text{ mL}} \times 100\% = 0.050\%$$

TRY IT YOURSELF 10.2

Question: What mass of glucose is required to make 0.500 L of a 5.0% (m/v) solution?

Core Concepts

■ Concentration can be measured as a percentage ratio of solute and solution. We use **volume/volume percent,** % (v/v), if the solute is a liquid, such as ethanol. We use **mass/volume percent,** % (m/v), if the solute is a solid, such as sodium chloride.

Measuring Concentration—Parts per Million or Billion

The previous section explains how to measure concentrations in percentages. This calculation works for concentrated solutions, but it is not a convenient measurement for very dilute solutions. When a solute is present in very small amounts, we use concentration units called **parts per million** (ppm) and **parts per billion** (ppb). For example, the U.S. Environmental Protection Agency (EPA) uses ppm to regulate the concentration of potentially harmful contaminants in drinking water.

To understand these units, let's begin with something familiar—percentage. The term *percent* means "per hundred," so 1% is the same as "one part per hundred." We can express this as a ratio with 1 in the numerator and 100 in the denominator:

$$1 \text{ part per hundred } (\%) = \frac{1}{100} = \frac{1}{10^2}$$

Similarly, we can define 1 ppm as a ratio with one million in the denominator

$$1 \text{ ppm} = \frac{1}{1,000,000} = \frac{1}{10^6}$$

Finally, 1 ppb is defined as a ratio with one billion in the denominator:

$$1 \text{ ppb} = \frac{1}{1,000,000,000} = \frac{1}{10^9}$$

We can apply these definitions to solutions, which typically have water as a solvent. We begin by defining 1 ppm as a ratio of masses measured in grams:

$$1 \text{ ppm} = \frac{1 \text{ g solute}}{10^6 \text{ g solution}}$$

For dilute solutions, a typical mass of solute is measured in milligrams. We can substitute $1 \text{ g} = 10^3 \text{ mg}$ in the equation, which then allows us to cancel a factor of 10^3 in the numerator and denominator:

$$1 \text{ ppm} = \frac{10^3 \text{ mg solute}}{10^3 \times 10^3 \text{ g solution}} = \frac{1 \text{ mg solute}}{10^3 \text{ g solution}}$$

We typically measure the volume of a solution, not its mass. It is possible to convert from mass to volume using *density* (see Chapter 2). For concentrations measured in ppm, we assume that the solution is very dilute. As a result, the density of the solution will be almost exactly the same as the density of pure water (1 g/cm^3 at 25°C). Therefore,

$$10^3 \text{ g solution} \times \frac{1 \text{ cm}^3}{1 \text{ g}} = 10^3 \text{ cm}^3 \text{ solution}$$

By definition, $1 \text{ L} = 10^3 \text{ cm}^3$, so we can replace the denominator by 1 L. We now have a practical equation for measuring solution concentration: *1 ppm is equal to 1 mg of solute in 1 L of solution.*

$$1 \text{ ppm} = \frac{1 \text{ mg solute}}{1 \text{ L solution}}$$

As an example of this unit, the U.S. Department of Health and Human Services (HHS) recommends that municipal water supplies should add 0.7 mg/L (0.7 ppm) of fluoride ions (F^-) to reduce tooth decay.

Similarly, we can define ppb using the mass unit of micrograms ($1\ g = 10^6\ \mu g$).
1 ppb is equal to 1 μg of solute in 1 L of solution.

$$1\ \text{ppb} = \frac{1\mu g\ \text{solute}}{1\ \text{L solution}}$$

PRACTICE EXERCISE 10.2

Following the method used to define ppm, derive the equation that defines ppb.

Table 10.1 provides a selection of water contaminants that are regulated by the EPA. The agency uses two types of concentration standard for each contaminant. The first is the maximum contaminant level goal (MCLG), which is defined as *the maximum concentration of contaminant that produces no known adverse health effects.* For example, the MCLG for arsenic is zero because any exposure to this element is harmful. The MCLG standards are public health goals, but they are not legally enforced. The EPA has a second standard called the maximum contaminant level (MCL), which is the *legal limit* for a contaminant in municipal water supplies. The EPA sets the enforceable MCL as close as possible to the ideal MCLG. However, the agency must also consider practical challenges, such as the availability and cost of water purification treatments to remove the contaminant. For this reason, the MCL for arsenic is 0.01 ppm and not zero.

TABLE 10.1   EPA Limits for Drinking Water Contaminants[1]

CONTAMINANT	MCLG (PPM)	MCL (PPM)	SOURCES OF CONTAMINANT	ADVERSE HEALTH EFFECTS
Arsenic (As)	0	0.01	Erosion of natural deposits; waste discharge from glass and electronics manufacturing	Skin damage, problems with blood circulation
Cadmium (Cd^{2+})	0.005	0.005	Corrosion of pipes; erosion of natural deposits; waste discharge from battery manufacturing	Kidney damage
Lead (Pb^{2+})	0	0.015	Corrosion of household plumbing; old types of paint; erosion of natural deposits	Children: Delays in physical and cognitive development. Adults: Kidney problems, high blood pressure.
Mercury (Hg)	0.002	0.002	Erosion of natural deposits; discharge from factories and refineries; runoff from landfills and croplands	Kidney damage

[1]*Data obtained from the U.S. Environmental Protection Agency at http://water.epa.gov/drink/contaminants/*

WORKED EXAMPLE 10.3

Question: When the city of Flint, Michigan, switched its water source in 2014, residents began complaining about discolored, odorous, and foul-tasting water. When scientists from Virginia Tech tested the tap water in various homes, they frequently found highly elevated levels of lead that were much greater than the EPA's MCL of 15 ppb.

Suppose that a resident of Flint drinks a standard glass of tap water with a volume of 8.0 fluid ounces (fl. oz.). How much lead is contained in this glass of water if the lead concentration is 2000 ppb? As an approximate conversion, 1 fl. oz. = 30 mL.

Answer: A concentration of 1 ppb is defined as 1 μg of solute in 1 L of solution. Therefore, a lead concentration of 2000 ppb can be written as:

$$2000 \text{ ppb} = \frac{2000 \ \mu\text{g lead}}{1 \text{ L solution}}$$

Next, we use a conversion factor to convert the volume of the tap water from fluid ounces to milliliters.

$$\text{volume of tap water (mL)} = 8.0 \ \cancel{\text{fl. oz.}} \times \frac{30 \text{ mL}}{1 \ \cancel{\text{fl. oz.}}} = 240 \text{ mL}$$

Since the definition of ppb uses a volume of 1 L, we convert the volume of tap water to liters:

$$\text{volume of tap water (L)} = 240 \ \cancel{\text{mL}} \times \frac{1 \text{ L}}{1000 \ \cancel{\text{mL}}} = 0.240 \text{ L}$$

In this case, the tap water is the solution that contains the lead. To find the total mass of lead, we need to multiply the lead concentration (ppb) by the volume of the drinking water (L):

$$\text{mass of lead } (\mu\text{g}) = \frac{2000 \ \mu\text{g lead}}{1 \ \cancel{\text{L solution}}} \times 0.240 \ \cancel{\text{L solution}} = 480 \ \mu\text{g lead}$$

To conclude, we calculate that a single glass of contaminated tap water contains 480 μg of lead.

TRY IT YOURSELF 10.3

Question: In Chapter 2, we learned that arsenic is a poisonous element. In some regions of Bangladesh, the arsenic concentration in groundwater measures as high as 100 ppb. (a) Convert this concentration to units of parts per million (ppm). (b) How many times higher is this concentration than the MCL for arsenic listed in Table 10.1?

PRACTICE EXERCISE 10.3

The salinity ("saltiness") of water is expressed in *parts per thousand* (ppt), which measures the mass of salt (in grams) found in 1000 g of solution. A typical value for salinity is 35 ppt. Chapter 2 described how salts can be extracted from sea water by evaporation. Suppose that we put one metric ton (1,000 kg) of sea water in a shallow pan and allowed the water to evaporate. What mass of salt will remain as a solid after evaporation?

Core Concepts

- A concentration of 1 **part per million** (ppm) is equal to 1 mg of solute in 1 L of solution.
- A concentration of 1 **part per billion** (ppb) is equal to 1 μg of solute in 1 L of solution.

THE KEY IDEA: Molarity measures the number of moles of solute dissolved in 1 liter of solution.

Measuring Concentration–Molarity

Up to this point, we have measured solute in terms of volume and mass. However, these measurements do not tell us how many particles are present in the solution. For example, a 1 ppb solution of arsenic and a 1 ppb solution of mercury do not contain the same number of arsenic or mercury atoms. Why not? The reason is that the mass of a mercury atom (200 amu) is much larger than the mass of an arsenic atom (75 amu). As a result, 1 mg of mercury contains fewer atoms than 1 mg of arsenic. This inconsistency causes a problem when we want to understand the poisonous effects of these elements. The toxicity of a dose of mercury or of arsenic depends on the *number of atoms* and not on the mass of the atoms. The number of atoms is critical because these elements achieve their toxic effects by participating in chemical reactions within the human body. As we saw in Chapter 6, chemical reactions are best described in terms of the relative numbers of reactants and products.

For this reason, chemists prefer a measurement of concentration that is based on the mole. Chapter 6 introduced the *mole* as a unit to measure the *chemical amount* of a substance. *One mole of a substance contains Avogadro's number (6.022 × 10²³) of atoms, molecules, ions, or other entities that make up the substance.* Using the mole allows us to relate the mass of the solute to the number of atoms, molecules, or ions that are present in the solution.

We use the term **molarity** to define the *molar concentration* of a solution. The *molarity (M) of a solution is equal to the chemical amount of solute (measured in moles) that is present in 1 L of solution.* This definition can be written as an equation:

molarity The molar concentration of a solution; it is equal to the chemical amount of solute, measured in moles, that is present in 1 liter of solution.

$$\text{molarity (M)} = \frac{\text{chemical amount of solute (mol)}}{\text{volume of solution (L)}}$$

The units of molarity are *moles per liter* (mol/L). A solution that contains 1 mole of solute in 1 liter of solution is said to have a concentration of "one molar" (1 M). The advantage of using molarity is that solutions with the same molarity contain the same number of dissolved molecules (or ions) per unit volume. For example, if we make 1 L of a 1 M solution of glucose and a 1 M solution of ethanol, then both solutions will contain the *same number of solute molecules* (either glucose or ethanol). This is true even though 1 mole of glucose has a different mass than 1 mole of ethanol.

Earlier in the chapter, we introduced physiological saline as a 0.9% (m/v) solution of NaCl. However, this concentration measurement does not tell us *how many* Na^+ and Cl^- ions are present in the saline solution. We can gain this information by calculating the molar concentration of the solution. To be more precise, we assume that physiological saline contains 0.90 g of NaCl dissolved in 100 mL of solution. As with Chapter 6, we use color coding to indicate different types of chemical quantity. We will indicate mass (g) in blue type, chemical amount (mol) in red type, and number of atoms in green type. We now add the volume of the solution (L) in purple type. We will use the molar mass of atoms to one decimal place.

> **STEP 1:** First, we must convert the mass of the solute (in grams) into the chemical amount of solute (in moles). Because sodium chloride is an ionic compound and not a molecule, we deal with *formula units* of NaCl (see Chapter 6). The relative atomic masses listed in the periodic table are for neutral atoms, not for ions such as Na^+ and Cl^-. However, the mass difference between an atom and its ion is negligible because the mass of an electron is so small compared to the masses of the protons and neutrons.
>
> Using the periodic table, we find that the molar masses of sodium and chlorine are
>
> Na 23.0 g/mol   Cl 35.5 g/mol
>
> Therefore, we can calculate the molar mass of NaCl as:

$$\text{molar mass of NaCl (g/mol)} = (1 \times \text{molar mass of Na}) + (1 \times \text{molar mass of Cl})$$
$$= (23.0 \text{ g/mol}) + (35.5 \text{ g/mol})$$
$$= 58.5 \text{ g/mol}$$

The molar mass provides a conversion factor between mass and moles:

$$\frac{1 \text{ mol NaCl}}{58.5 \text{ g NaCl}} = 1$$

The physiological saline solution contains 0.90 g NaCl. This quantity can be converted to the chemical amount as follows:

$$\text{chemical amount of NaCl (mol)} = 0.90 \ \cancel{\text{g NaCl}} \times \frac{1 \text{ mol NaCl}}{58.5 \ \cancel{\text{g NaCl}}} = 0.015 \text{ mol NaCl}$$

STEP 2: Next, we can introduce the volume of the solution. We must convert 100 mL into units of liters:

$$100 \ \cancel{\text{mL}} \times \frac{1 \text{ L}}{10^3 \ \cancel{\text{mL}}} = 0.100 \text{ L}$$

STEP 3: We now have all the necessary quantities to calculate the molarity of the solution:

$$\text{molarity of NaCl (M)} = \frac{\text{chemical amount of NaCl (mol)}}{\text{volume of solution (L)}} = \frac{0.015 \text{ mol}}{0.100 \text{ L}}$$

$$= 0.15 \text{ M NaCl}$$

This result tells us that physiological saline contains 0.15 M NaCl; that is, 0.15 moles per liter. Alternatively, we can write the molar concentration as 150 millimolar (150 mM).

Solid sodium chloride dissociates in water according to the following equation:

$$\text{NaCl}(s) \xrightarrow{\text{H}_2\text{O}} \text{Na}^+(aq) + \text{Cl}^-(aq)$$

Each formula unit of NaCl dissociates into one $\text{Na}^+(aq)$ ion and one $\text{Cl}^-(aq)$ ion. Similarly, 0.15 mol NaCl in solution produces 0.15 mol $\text{Na}^+(aq)$ and 0.15 mol $\text{Cl}^-(aq)$. Because the initial concentration of NaCl is 0.15 M and the volume remains constant, the molar concentrations of $\text{Na}^+(aq)$ and $\text{Cl}^-(aq)$ are also 0.15 M (150 mM). This result will be important when we return to our discussion of hyponatremia.

Figure 10.7 shows the relationships among the quantities that are used to characterize a solution. The solute can be specified in terms of mass (g), chemical amount (mol), or number of molecules or formula units. To calculate the molarity, we also include the volume of the solution.

Because molarity is such a common measure of concentration, chemists employ a shorthand convention using square brackets. *[X] is a shorthand for the molarity of X expressed as moles per liter.* For example, the molarity of NaCl in physiological saline is [NaCl] = 0.15 M.

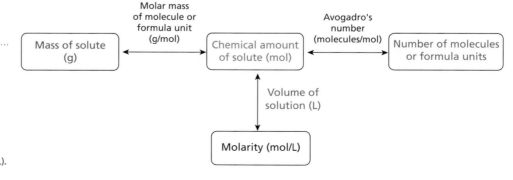

FIGURE 10.7 The relationships among quantities used to characterize a solution. The amount of solute in a solution can be described by the mass (g), the chemical amount (mol), or the number of molecules or formula units. The concentration can be characterized by the molarity (mol/L).

In summary, we can use the following equations for calculations involving mass, moles, and molarity.

To calculate the molarity of a solution:

$$\text{molarity (M)} = \frac{\text{chemical amount of solute (mol)}}{\text{volume of solution (L)}}$$

To calculate the chemical amount of solute in a given solution:

$$\text{chemical amount of solute (mol)} = \text{molarity (mol/L)} \times \text{volume (L)}$$

To calculate the mass of solute in a given solution:

$$\text{mass of solute (g)} = \text{chemical amount of solute (mol)}$$
$$\times \text{molar mass of solute (g/mol)}$$

WORKED EXAMPLE 10.4

Question: What mass of methanol (CH_3OH) is required to make 250 mL of a 0.10 M aqueous solution?

Answer: First, we calculate the chemical amount of methanol (in moles) required to make this solution:

$$\text{chemical amount of } CH_3OH \text{ (mol)} = 0.250 \; \cancel{L} \times \frac{0.10 \text{ mol}}{1.00 \; \cancel{L}} = 0.025 \text{ mol } CH_3OH$$

To convert from moles to mass, we use the molar mass of CH_3OH.

$$\text{molar mass of } CH_3OH \text{ (g/mol)} = (1 \times \text{molar mass of C})$$
$$+ (4 \times \text{molar mass of H})$$
$$+ (1 \times \text{molar mass of O})$$
$$= (1 \times 12.0 \text{ g/mol}) + (4 \times 1.0 \text{ g/mol})$$
$$+ (1 \times 16.0 \text{ g/mol})$$
$$= 32.0 \text{ g/mol}$$

The molar mass provides a conversion factor between moles and mass:

$$\frac{32.0 \text{ g } CH_3OH}{1 \text{ mol } CH_3OH} = 1$$

We can now convert between moles and mass of the solute:

$$\text{mass of } CH_3OH \text{ (g)} = 0.025 \; \cancel{\text{mol } CH_3OH} \times \frac{32.0 \text{ g } CH_3OH}{1 \; \cancel{\text{mol } CH_3OH}} = 0.80 \text{ g } CH_3OH$$

The calculation tells us that 0.80 g of CH_3OH in 250 mL of solution will produce a solution concentration of 0.10 M.

TRY IT YOURSELF 10.4

Question: You dissolve 9.0g of glucose ($C_6H_{12}O_6$) in water to make 200 mL of aqueous solution. What is the molar concentration of glucose?

Solution concentrations of 1 M are common in the laboratory. However, the concentrations of some substances are much lower, such as biological molecules in cells or contaminants in a water supply. In these cases, we express the concentration using units of *millimolar (mM), micromolar (μM),* or even *nanomolar (nM).*

The following equations define these different units using chemical amounts expressed in millimoles ($1 \text{ mmol} = 10^{-3} \text{ mol}$), micromole ($1 \text{ μmol} = 10^{-6} \text{ mol}$), and nanomoles ($1 \text{ nm} = 10^{-9} \text{ mol}$).

$$\text{millimolar} \qquad 1 \text{ mM} = 10^{-3} \text{ M} = \frac{1 \text{ mmol of solute}}{1 \text{ L of solution}}$$

$$\text{micromolar} \qquad 1 \text{ μM} = 10^{-6} \text{ M} = \frac{1 \text{ μmol of solute}}{1 \text{ L of solution}}$$

$$\text{nanomolar} \qquad 1 \text{ nM} = 10^{-9} \text{ M} = \frac{1 \text{ nmol of solute}}{1 \text{ L of solution}}$$

WORKED EXAMPLE 10.5

Question: (a) Calculate the number of mercury (Hg) atoms in an aqueous solution with a volume of 1.00 L and a concentration 1.00 ppb. Similarly, calculate the number of arsenic atoms in an aqueous solution with the same volume and concentration. Are these two numbers the same? (b) Calculate the number of Hg atoms in an aqueous solution with a volume of 1.00 L and a concentration of 1.00 μM. Similarly, calculate the number of As atoms in an aqueous solution with the same volume and concentration. Are these two numbers the same?

Use the following quantities and approximations for this calculation:
- The relative atomic mass of Hg and As can be obtained from the periodic table. Use the values to one decimal place.
- For the atomic mass unit, use $1 \text{ amu} = 1.66 \times 10^{-24} \text{ g}$ (to three significant figures).
- For Avogadro's number, use $N_A = 6.02 \times 10^{23}$ (to three significant figures)

Answer: (a) We will begin by calculating the total mass of Hg atoms in the aqueous solution. By definition, 1 ppb = 1 mg/L, and we have 1.00 L of solution. Because the definition of amu is given in units of grams, we need to convert from μg to g.

$$\text{mass of Hg (g)} = 1.00 \; \cancel{L} \times \frac{1 \; \cancel{\text{μg}} \; \text{Hg}}{1 \; \cancel{L}} \times \frac{10^{-6} \; \text{g}}{1 \; \cancel{\text{μg}}} = 10^{-6} \; \text{g Hg}$$

According to the periodic table, an Hg atom has a relative atomic mass of 200.6 (to one decimal place). As we learned in Chapter 2, this is numerically equal to the mass of Hg atoms in amu (atomic mass units). We now need a conversion factor to convert the mass from amu to grams. By definition,

$$1 \text{ amu} = 1.66 \times 10^{-24} \text{ g} \qquad \text{or} \qquad \frac{1 \text{ amu}}{1.66 \times 10^{-24} \text{ g}} = 1$$

We can now calculate the number of Hg atoms contained in 10^{-6} g.

$$\text{number of Hg atoms} = 10^{-6} \; \cancel{g} \times \frac{1 \; \cancel{\text{amu}}}{1.66 \times 10^{-24} \; \cancel{g}} \times \frac{1 \text{ Hg atom}}{200.6 \; \cancel{\text{amu}}}$$

$$= 3.00 \times 10^{15} \text{ Hg atoms}$$

(b) We can use the same method to calculate the number of As atoms. The mass, volume, and definition of amu remain the same. The only difference is the relative atomic mass of As, which is given in the periodic table as 74.9 (to three significant figures). To calculate the number of As atoms, we need only to replace the final term in the previous equation.

$$\text{number of As atoms} = 10^{-6} \ \cancel{g} \ \times \ \frac{1 \ \text{amu}}{1.66 \times 10^{-24} \ \cancel{g}} \times \frac{1 \ \text{As atom}}{74.9 \ \text{amu}}$$

$$= 8.04 \times 10^{15} \ \text{As atoms}$$

These calculations show that two aqueous samples of Hg and As with the same concentration in ppb *do not contain the same number of atoms*. In fact, there are over 2½ times more As atoms compared to the number of Hg atoms. This difference arises from the differing atomic masses of the two elements. The smaller atomic mass of As means that there are more atoms in a 1 ppb sample.

(b) We will begin with the Hg atoms. By definition, 1 μM = 1 μmol / 1 L = 10^{-6} mol / 1 L. The definition of a mole and Avogadro's number allows us to proceed directly to calculating the number of Hg atoms.

$$\text{number of Hg atoms} = 1.00 \ \cancel{L} \ \times \ \frac{10^{-6} \ \cancel{\text{mol Hg}}}{1 \ \cancel{L}} \times \frac{6.02 \times 10^{23} \ \text{Hg atoms}}{1 \ \cancel{\text{mol Hg}}}$$

$$= 6.02 \times 10^{17} \ \text{Hg atoms}$$

The definition of the mole means that we can perform exactly the same calculation for the As atoms.

$$\text{number of As atoms} = 1.00 \ \cancel{L} \ \times \ \frac{10^{-6} \ \cancel{\text{mol As}}}{1 \ \cancel{L}} \times \frac{6.02 \times 10^{23} \ \text{Ag atoms}}{1 \ \cancel{\text{mol As}}}$$

$$= 6.02 \times 10^{17} \ \text{As atoms}$$

These calculations show that the two aqueous samples of Hs and As with the same concentration in μM *do contain the same number of atoms*. The equality arises from the definition of a mole in terms of Avogadro's number, which counts atoms rather than measuring their mass.

TRY IT YOURSELF 10.5

Question: (a) Calculate the number of fluoride ions (F^-) in an aqueous solution with a volume of 50.0 mL and a concentration of 1.00 ppm. Similarly, calculate the number of lead ions (Pb^{2+}) in an aqueous solution with the same volume and concentration. Are these two numbers the same? (b) Calculate the number of F^- ions in an aqueous solution with a volume of 50.0 mL and a concentration of 1.00 mM. Similarly, calculate the number of Pb^{2+} ions in an aqueous solution with the same volume and concentration. Are these two numbers the same?

Use the following quantities and approximations for this calculation:
- The relative atomic masses of F and Pb can be obtained from the period table. Use the values to one decimal place. Gaining or losing electrons to form an ion has a negligible effect on the mass.
- For the atomic mass unit, use 1 amu = 1.66×10^{-24} g (to three significant figures).
- For Avogadro's number, use $N^A = 6.02 \times 10^{23}$ (to three significant figures)

Core Concepts

- The **molarity** (M) is the *molar concentration* of a solution. The molarity is the chemical amount of solute (measured in moles) that is present in 1 L of solution. The units are moles per liter (mol/L).
- A 1 molar (1 M) solution of any substance contains the same number (Avogadro's number, N_A) of atoms, molecules, ions, or other entities.
- For any solution, it is possible to calculate the relationships among mass, moles, molarity, and number of entities (atoms, molecules, ions, etc.).

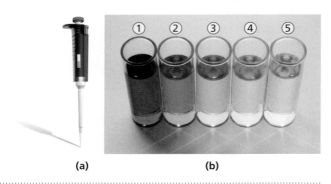

THE KEY IDEA: A solution of required concentration and volume can be made by diluting a stock solution.

Diluting a Concentrated Solution

When making solutions in the laboratory, chemists usually begin by preparing a solution that is more concentrated than they need. This concentrated solution is called a *stock solution*, where the word "stock" refers to something that can be stored for repeated use (think of the phrase "in stock"). Next, the chemist dilutes the stock solution to the desired volume and concentration. This dilution method enables the chemist to create multiple solutions of different volumes and concentrations from a single stock solution. For example, a solution of physiological saline can be made by creating a concentrated stock solution of NaCl and then diluting the stock to achieve the desired concentration of 0.15 M.

Figure 10.8 illustrates the process of preparing different solutions by diluting a stock solution. In this case, the stock solution is made with a red dye so we can observe the effects of the dilution. The dilution requires the use of a pipette, shown in Figure 10.8(a), which is precisely calibrated in units of volume and allows us to withdraw a known volume from the stock solution. Water is then added to the sample to reach the desired concentration. Beginning with a 1.00 M solution in Figure 10.8(b), each subsequent dilution reduces the concentration by a factor of 10.

Let's use the dilution method to make physiological saline from a 1.00 M stock solution of NaCl. The first step is to make the 1.00 M stock solution, which contains 1 mole of NaCl in 1 L of solution (Figure 10.9).

FIGURE 10.8 Preparing solutions by diluting a concentrated stock solution. (a) Calibrated pipettes allow the delivery of known volumes from a stock solution. (b) The stock solution ① of 1.00 M dye is shown on the left with the darkest red color. Solution ② is a 1:10 dilution of the stock solution to produce a solution concentration of 0.10 M. Solutions ③, ④, and ⑤ are each 1:10 dilutions of the previous solution. The final solution has a concentration of only 0.0001 M and is almost colorless.

STEP 1: Based on our earlier calculation, we know that the molar mass of NaCl is 58.5 g/mol. We therefore measure a mass of 58.5 g on a chemical balance, which corresponds to 1 mol NaCl.

STEP 2: We add the solid NaCl to a *volumetric flask*. These flasks are accurately calibrated to a volume of exactly 1.00 L, indicated by a mark on the neck.

STEP 3: We dissolve the NaCl in approximately half the volume of the volumetric flask. We do not add 1.00 L of water because the dissolved NaCl will affect the final volume. Recall that molarity is measured using 1.00 L of *solution* and not 1.00 L of *solvent*.

STEP 4: We carefully add more water until the volume of the solution exactly reaches the 1.00 L mark.

FIGURE 10.9 Preparing a 1M solution of NaCl. We use 58.5 g of NaCl, which corresponds to 1 mol of solute. The NaCl is dissolved in water within a calibrated volumetric flask, which has a volume of exactly 1.00 L.

We now need to determine what volume of the 1.00 M stock solution is required to provide the appropriate number of moles of solute in the solution of physiological saline that we want to prepare. In this case, suppose that we want to prepare 0.100 L of 0.150 M NaCl solution. We accomplish this task by using a *dilution equation* that compares two solutions, the stock solution (solution 1) and the solution to prepare (solution 2) (Figure 10.10).

Both sides of the equation correspond to the same chemical amount of solute expressed in moles. We can verify this statement by multiplying the concentration of a solution by its volume, which produces the moles of solute.

$$\underbrace{\text{Concentration }(C_1) \times \text{Volume }(V_1)}_{\textbf{Stock solution}} = \underbrace{\text{Concentration }(C_2) \times \text{Volume }(V_2)}_{\textbf{Solution to prepare}}$$

FIGURE 10.10 A dilution equation. The dilution equation relates the concentration and volume of the stock solution (C_1 and V_1) to the concentration and volume of the solution that we wish to prepare (C_2 and V_2). The chemical amount of solute, measured in moles, is the same for both sides of the equation.

$$\text{concentration}\left(\frac{\text{mol}}{\cancel{L}}\right) \times \text{volume }(\cancel{L}) = \text{moles of solute (mol)}$$

For convenience, we can write the dilution equation in an abbreviated form as:

$$\mathbf{C_1 V_1 = C_2 V_2}$$

We can use this equation with the following information:

Stock solution: $C_1 = 1.00\text{ M}$ V_1 is the unknown volume we need to use

Solution to prepare: $C_2 = 0.150\text{ M}$ $V_2 = 0.100\text{ L}$

Our equation now reads:

$$1.00\text{ M} \times V_1(\text{L}) = 0.150\text{ M} \times 0.100\text{ L}$$

Solving the equation for V_1 gives:

$$V_1\ (\text{L}) = \frac{0.150\ \cancel{M} \times 0.100\text{ L}}{1.00\ \cancel{M}} = 0.0150\text{ L}$$

For convenience we convert this calculated volume to milliliters (mL):

$$0.0150\ \cancel{L} \times \frac{1000\text{ mL}}{1\ \cancel{L}} = 15.0\text{ mL}$$

To make our desired solution, we take 15.0 mL of the stock solution and add water to obtain a total volume of 0.100 L. This type of dilution is an everyday procedure in laboratories and clinics.

For these dilution calculations, it is also possible to use volume units of milliliters (mL). The dilution equation remains valid because the conversion factor from liters to milliliters (1 L = 1000 mL) will cancel on both sides of the equation. We can revisit the calculation above using the following quantities:

Stock solution: $C_1 = 1.00\text{ M}$ V_1 is the unknown volume we need to use

Solution to prepare: $C_2 = 0.150\text{ M}$ $V_2 = 100\text{ mL}$

Our equation now reads:

$$1.00\text{ M} \times V_1\ (\text{mL}) = 0.150\text{ M} \times 100\text{ mL}$$

Solving the equation for V_1 gives:

$$V_1\ (\text{mL}) = \frac{0.150\ \cancel{M} \times 100\text{ mL}}{1.00\ \cancel{M}} = 15.0\text{ mL}$$

This answer is the same as we obtained using volume units of liters. In summary, it is possible to apply the dilution equation using volume units of liters or milliliters, but *we must always use the same volume units on both sides of the equation.*

Core Concepts

- It is often convenient to use a stock solution to prepare a variety of solutions with specific concentrations and volumes.
- The required dilution of the stock solution can be calculated using a dilution equation, $C_1V_1 = C_2V_2$.
- The volumes in the dilution equation can be measured in liters (L) or milliliters (mL), provided that the same volume units are used on both sides of the equation.

WORKED EXAMPLE 10.6

Question: Aqueous sodium hydroxide (NaOH) can be purchased at a standard concentration of 1.00 M. How much of this solution is required to make 250 mL of 0.200 M NaOH?

Answer: We use the dilution equation, $C_1V_1 = C_2V_2$, in which solution 1 is the stock solution and solution 2 is the one we wish to prepare. Wall volumes are specified in units of liters.

Stock solution: $C_1 = 1.00\,M$ V_1 is the unknown volume we need to use

Solution to prepare: $C_2 = 0.200\,M$ $V_2 = 0.250\,L$

Applying the equation gives:

$$1.00\,M \times V_1\,(L) = 0.200\,M \times 0.250\,L$$

Solving the equation for V_1 gives:

$$V_1\,(L) = \frac{0.200\,\cancel{M} \times 0.250\,L}{1.00\,\cancel{M}} = 0.0500\,L$$

For convenience, we convert this calculated volume to milliliters (mL):

$$0.0500\,\cancel{L} \times \frac{1000\,mL}{1\,\cancel{L}} = 50.0\,mL$$

To make the desired solution, we take 50.0 mL of the stock solution and add water to make a total volume of 250 mL.

As a check, we can also perform this calculation using volumes in milliliters.

Stock solution: $C_1 = 1.00\,M$ V_1 is the unknown volume we need to use

Solution to prepare: $C_2 = 0.200\,M$ $V_2 = 250\,mL$

Applying the equation gives:

$$1.00\,M \times V_1\,(mL) = 0.200\,M \times 250\,mL$$

Solving the equation for V_1 gives:

$$V_1 = \frac{0.200\,\cancel{M} \times 250\,mL}{1.00\,\cancel{M}} = 50.0\,mL$$

This is the same volume as we obtained using liters as the volume unit in the dilution equation.

TRY IT YOURSELF 10.6

Question: You are given 100 mL of a 3.00 M glucose solution. You now dilute the solution by adding water to increase the solution volume to 500 mL. What is the new concentration of glucose?

SCIENCE IN ACTION

Measuring Unknown Concentrations Using Light

In this chapter, we have explained how to make a solution of known concentration by specifying the quantity of the solute and the volume of the solution. However, scientists frequently need to measure the unknown concentration of a novel solution. How is this done?

A common technique to measure concentrations uses the properties of light. Light is a wave that is composed of oscillating electric and magnetic fields—for this reason, light is called an *electromagnetic wave*. When white light is passed through a glass prism, the prism disperses the light into a range of colors called the *visible spectrum* (Figure 1(a)). The various colors in the visible spectrum correspond to different *wavelengths* of light. The wavelength is the distance between successive peaks of the wave (Figure 1(b)). For the visible spectrum, the wavelength is measured in units of nanometers (1 nm = 10^{-9} m or one-billionth of a meter). Figure 1(c) illustrates the wavelengths that correspond to different colors in the spectrum. Light at the violet edge of the spectrum has a wavelength of 400 nm. As we scan through other colors in the spectrum—such as blue, green, and yellow—the wavelength increases. Red light has the longest wavelength, with a value of 700 nm at the outer edge of the visible spectrum.

We can use light to measure concentration because *the amount of light absorbed by a solution— called the absorbance—is directly proportional to the molar concentration of solute molecules*. This relationship exists because a solution with a higher molarity of solute contains a greater number of light-absorbing molecules or ions. As an example, Figure 2(a) shows five solutions of copper nitrate in water. At low concentration (0.1 M), the solution has no discernible color because it absorbs very little light. As the solute concentration increases, the copper nitrate solute appears blue because more light is being absorbed. At the highest concentration (0.5 M), the solution is dark blue because the absorbance is high.

Figure 2(b) plots the concentration (horizontal axis) and absorbance (vertical axis) for each of the copper nitrate samples. The data points fall on a straight line, such that higher concentrations of solution have a higher absorbance. In general, the absorbance of a solution is related to its concentration by an equation called *Beer's law*.

$$A = \varepsilon\, l\, c$$

- *A* is the absorbance of light at a particular wavelength (a number with no units).

- ε is the molar absorption coefficient of the molecule at the same wavelength (measured in units of cm^{-1}M^{-1}).

- *l* is the path length through which the light travels (measured in centimeters, cm).

- *c* is the concentration of the absorbing substance (measured in moles per liter, M).

The graph in Figure 2(b) is called a Beer's Law plot. We can use this graph to determine the concentration of an unknown solution. Suppose that we are given a new solution of copper nitrate, but we do not know its concentration. We measure the absorbance and find that *A* = 0.5. According to Figure 2(b), this absorbance corresponds to *c* = 0.25 M.

In addition to its application in chemistry, Beer's Law is widely used for measuring unknown concentrations in medicine (e.g., the concentration of hemoglobin in a blood sample) and environmental science (e.g., the concentration of a water pollutant).

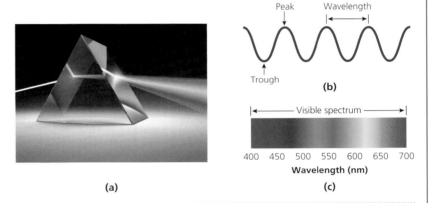

(a)

(b)

(c)

FIGURE 1 The wave properties of light. (a) A glass prism disperses white light into a visible spectrum of colors that correspond to different wavelengths. (b) The wavelength of light is measured as the distance between successive peaks of the wave. (c) The visible spectrum consists of colors with wavelengths that range from 400 nm (violet) to 700 nm (red). Ultraviolet and infrared light have wavelengths that fall outside the range of the visible spectrum.

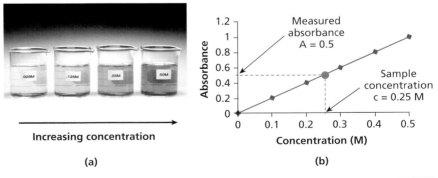

(a)

(b)

FIGURE 2 Measuring concentration by absorbance of light. (a) As the solution concentration of copper nitrate increases, the samples appear darker because they absorb more light. (b) A graph of absorbance versus concentration (molarity, M) produces data points that fall on a straight line. This relationship allows us to determine the unknown concentration of a solution by measuring its absorbance. In this example, an absorbance of 0.5 corresponds to a concentration of 0.25 M.

10.3 # Biological Applications of Concentration

In this section, we examine the importance of concentration for living organisms. In particular, we explain why a low concentration of sodium in the blood plasma can be harmful by causing the cells of the body to swell.

Osmosis: The Flow of Water Across a Membrane

Osmosis occurs when H_2O molecules flow across a membrane that is *selectively permeable*. This type of membrane allows H_2O molecules to cross, but it blocks the passage of solutes such as sugars and ions. Figure 10.11 shows two different sugar solutions in either side of a U-shaped tube, separated by a selectively permeable membrane. The sugar solution on the left side of the tube has a lower concentration of sugar with a smaller number of solute molecules. By comparison, the sugar solution on the right side has a higher concentration of solute molecules. If the volumes of these two solutions are initially the same, then there must be fewer water molecules in the concentrated solution because it contains a higher number of solute molecules. Therefore, *the concentration of H_2O molecules in the left side of the tube is greater than in the right side.* We are using the term *concentration* loosely here because it typically refers to the solute and not the solvent.

In general, molecules spontaneously flow from a region of higher concentration to a region of lower concentration. This process is called *diffusion*. The difference in H_2O concentrations causes the H_2O molecules to flow from the left side (where the H_2O concentration is high) to the right side (where the H_2O concentration is low). Because the membrane is selectively permeable, it allows the H_2O molecules to flow across, but it prevents the passage of glucose molecules. The net flow of water molecules from left to right decreases the volume in the left-hand compartment and increases the volume in the right-hand compartment. After some time, these changes in volume cause the concentration of solute molecules and H_2O molecules to become the same on both sides of the membrane. At this point, the flow of water stops because the H_2O concentration is balanced. In

Learning Objective:

Relate concentration to osmosis and hyponatremia.

⚷ THE KEY IDEA: Osmosis is the flow of water from high to low concentration across a selectively permeable membrane.

■ **osmosis** The net flow of water through a selectively permeable membrane, from a region of higher solvent concentration to one of lower solvent concentration.

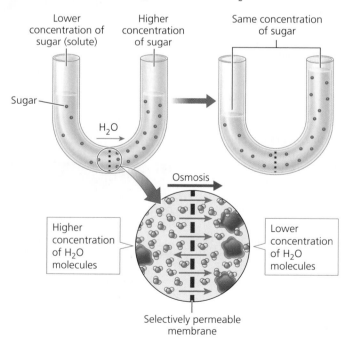

FIGURE 10.11 Osmosis is the net flow of H_2O molecules across a selectively permeable membrane. Two halves of a U-tube are separated by a selectively permeable membrane, which permits H_2O molecules to cross but blocks the passage of glucose solute molecules. Osmosis is the net flow of H_2O molecules across the membrane, from a region in which their concentration is higher to a region of lower concentration.

summary, *osmosis is the net flow of water through a selectively permeable membrane, from a region of higher solvent concentration to a region of lower solvent concentration.*

Core Concepts

- **Osmosis** is the net flow of H_2O molecules from a region of higher solvent concentration to a region of lower solvent concentration. This spontaneous flow occurs through a selectively permeable membrane, which allows H_2O molecules to cross but blocks the passage of solutes.

Why Is Hyponatremia Harmful?

In the chapter introduction, we discussed how blood sodium concentration can be reduced by drinking an excessive amount of water. Exercise can also stimulate the release of *antidiuretic hormone*, which signals our body to stop urinating. This hormone leads to water retention, which can also dilute the sodium concentration. In this section, we explain how the harmful effects of hyponatremia arise from osmosis.

Osmosis is relevant to cells because the cell membrane is selectively permeable, permitting H_2O molecules to cross easily while impeding the passage of solute molecules and ions. Transport of H_2O molecules across a cell membrane is facilitated by *aquaporin*, a membrane protein that provides a "water channel" (see Chapter 9). Other membrane proteins permit the passage of sugar, ions, and other solutes across the membrane. However, opening the channels in these proteins often requires activation by a biological trigger such as a chemical signal or an electrical voltage. As a result, the passage of solute molecules and ions across the cell membrane is much less than the flow of H_2O molecules.

In general, osmosis depends on the *tonicity* ("strength") of a solution, which could be the interior of a cell or the outside environment. Tonicity arises from solutes that are *not* able to cross the selectively permeable membrane. The tonicity of a solution depends on the total molar concentration of *all* such solutes, regardless of their identity. By contrast, molecules that can cross the selectively permeable membrane, such as water, do not contribute to the tonicity. In the example shown in Figure 10.11, the tonicity of the solutions arises from the dissolved sugar molecules because they cannot cross the membrane.

Figure 10.12 shows photographs and sketches of red blood cells that have been placed in three solutions made with different molar concentrations of sodium chloride. The first solution is called *isotonic* because the concentration of ions in the solution is the same as the concentration of solutes within the cell (*iso* means "same" and *tonic* means "strength"). There is no difference in the concentration of H_2O molecules in these two regions, so the rate at which H_2O molecules flow into the cell is exactly balanced by the rate at which they flow out of the cell.

The second solution is called *hypertonic* (*hyper* means "above") because the concentration of ions in the solution is *greater* than the ion concentration in the cell. In a hypertonic solution, the concentration of H_2O molecules *outside* the cell is less than the concentration *inside* the cell. Consequently, H_2O molecules flow from the inside (higher H_2O concentration) to the outside (lower H_2O concentration). This loss of water causes the cell to shrivel and shrink. The third solution is called *hypotonic* (*hypo* means "below") because the concentration of ions in the solution is *less* than the solute concentration in the cell. The H_2O concentration is higher *outside* the cell, so water flows *into* the cell. In this case, osmosis causes the cell to swell and possibly burst.

The terminology used to describe these different solution conditions can be confusing. Although the terms *isotonic, hypertonic,* and *hypotonic* refer to the concentration of *solute molecules,* it is the relative concentration of H_2O molecules that determines the direction of water flow during osmosis.

THE KEY IDEA: A low blood concentration of sodium ions causes water to flow into cells and produce swelling.

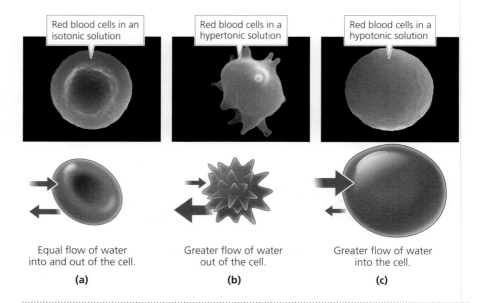

FIGURE 10.12 The effects of placing red blood cells into solutions that are isotonic, hypertonic, and hypotonic. (a) The ion concentration of an isotonic solution is the same as the solute concentration within the cell. Therefore, the net flow of water into and out of the cell are the same. (b) In a hypertonic solution, the ion concentration in the solution is greater than the solute concentration inside the cell. As a result, water flows out of the cell and it shrivels. (c) In a hypotonic solution, the solute concentration in the solution is less than the solute concentration inside the cell. Consequently, water flows into the cell, causing it to swell.

Avoiding the problems of osmosis is the reason that bags of IV fluid, like the one shown in Figure 10.6, need to contain a specific concentration of salt that is approximately *isotonic* to the concentration of ions in the patient's blood serum. If the ion concentration delivered from the IV bag is too high or too low, then the patient's red blood cells will shrivel or swell. In extreme cases, the cells could be destroyed.

The normal concentration of sodium ions in human blood serum ranges between 135 and 145 mM (millimolar). The medical diagnosis of hyponatremia is defined as having a blood sodium concentration of *135 mM or less*. This reduced concentration can induce nausea and grogginess, but the condition is not life threatening. However, if the blood sodium concentration drops to *120 mM or less*, then the person is diagnosed with *critical hyponatremia*, which can cause coma or death.

A decrease in concentration of sodium ions in the blood serum means that the blood is now *hypotonic* when compared to the solute concentration within cells. If the blood is hypotonic, there is a *net flow of water from the blood into the body's cells*, which causes the cells to swell. Many cells have room to expand because they are embedded in flexible tissue such as fat and muscle. However, the cells in our brain are packed tightly together, and the total volume is constrained by a rigid bony skull. Critical hyponatremia can be fatal because it results in swelling of the brain cells. The swelling creates additional pressure within the brain, which can prevent the flow of blood to the brain and deprive its cells of life-sustaining oxygen. Brain swelling induced by hyponatremia caused the death of the marathon runner discussed in the introduction.

Hyponatremia is an extreme and uncommon condition that typically occurs for long-distance running events such as a marathon. When you exercise at the gym or play sports, it is still important to drink a sufficient amount of water to avoid dehydration. How much water should you drink? Your body is very sensitive to what it needs, so experts provide a simple guideline: Drink when you feel thirsty.

CONCEPT QUESTION 10.1

Imagine you are stranded in a lifeboat after a shipwreck. You are surrounded by ocean with no land in sight. Because you have no supply of fresh water, you consider drinking sea water. On average, the concentration of dissolved ions in the ocean is approximately 600 mM. Explain what would happen to your red blood cells if you drank a large quantity of sea water.

WORKED EXAMPLE 10.7

Question: The nutrition label for a sports drink is given below. Does the concentration of sodium in this drink match the normal concentration of sodium (140 mM) in a person's blood plasma?

Answer: We first need to calculate the concentration of sodium in the sports drink and then compare it with the normal concentration of sodium in a person's blood plasma.

The label tells us that the drink contains 110 mg of sodium in a volume of 240 mL. We first convert the mass of sodium to the number of moles by using the molar mass. The mass of the sodium atom (Na) from the periodic table is the same as the mass of the sodium ion (Na^+) because the mass of the lost electron is negligible.

$$\text{chemical amount of Na}^+ \text{ (mol)} = 0.110 \text{ g Na}^+ \times \frac{1 \text{ mol Na}^+}{23.0 \text{ g Na}^+} = 4.78 \times 10^{-3} \text{ mol Na}^+$$

We can now calculate the molarity of Na^+ ions in the sports drink:

$$\text{concentration of Na}^+ (M) = \frac{4.78 \times 10^{-3} \text{ mol Na}^+}{0.240 \text{ L}} = 0.0199 \text{ M Na}^+$$

The concentration of sodium ions in a sports drink is 19.9 mM, which is much lower than the typical concentration in blood serum (135–145 mM). Consuming a sports drink does replace some electrolytes that we lose via perspiration but not at the same concentration.

TRY IT YOURSELF 10.7

Question: What mass of sodium would be required in the sports drink to produce a concentration of Na^+ ions that is isotonic to our blood serum? Why do you think that companies do not make sports drinks with this concentration of sodium ions?

Nutrition Facts

Serving Size 8 fl oz (240 mL)
Servings Per Container 4

Amount Per Serving

Calories 50

	% Daily Value*
Total Fat 0g	0%
Sodium 110mg	5%
Potassium 30mg	1%
Total Carbohydrate 14g	5%
Sugars 14g	
Protein 0g	

Not a significant source of Calories from Fat, Saturated Fat, Cholesterol, Dietary Fiber, Vitamin A, Vitamin C, Calcium, Iron.

* Percent Daily Values are based on a 2,000 calorie diet.

Core Concepts

- The tonicity of a solution arises from solutes that cannot cross the selectively permeable membrane. Tonicity depends on the total molar concentration of all such solutes, regardless of their identity.
- When cells are in an isotonic solution, there is an equal flow of water into and out of the cell. When cells are in a hypertonic solution, there is a net flow of water out of the cell. When cells are in a hypotonic solution, there is a net flow of water into the cell.
- The normal concentration of sodium ions in blood serum ranges from 135–145 mM. Hyponatremia is defined as a sodium ion concentration of 135 mM or less. Critical hyponatremia occurs when the sodium ion concentrations falls to 120 mM or less.
- Hyponatremia causes the blood to become hypotonic. As a result of osmosis, water flows from the blood into the body's cells. This increased water content causes the cells to swell, which is particularly dangerous in the brain.

CHEMISTRY AND YOUR HEALTH

How Much Salt Should We Eat?

All of us need salt in our daily diet. Table salt (NaCl) is essential for vital processes such as nerve signaling, fluid balance, and the proper functioning of organs such as the kidneys. How much sodium do we need in our diet? U.S. dietary guidelines, provided by the FDA, recommend that we consume no more than 2300 milligrams of salt per day, which is approximately the amount in a single teaspoon. How closely do Americans follow these guidelines? A 2010 report by the Institute of Medicine revealed that Americans consume an average of more than 3400 milligrams of salt per day, which is close to 50% higher than the recommended amount. Excessive salt consumption elevates blood pressure, which in turn increases the risk for heart attack and stroke—the leading causes of death in the United States. Figure 1 identifies the sources of the salt that Americans consume.

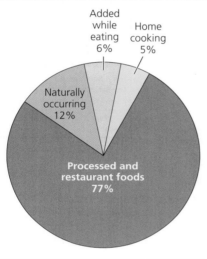

FIGURE 1 Relative contributions of salt in the American diet.

Read some product labels in your local market to see how much salt is added to processed foods. Figure 2 shows a chart of salt content in different varieties of food, which was compiled by the Centers for Disease Control and Prevention (CDC). Note that each food type has a wide range of salt content, so we advise you to investigate the specific foods that you eat.

If you eat many of your meals in your school cafeteria or in fast-food restaurants, then controlling your sodium intake presents a major challenge. You are living largely on prepared or processed foods. Fortunately, many university cafeterias have responded to the excess sodium problem by offering low-salt alternatives to the standard, highly salted foods.

Food	Sodium Range (in miligrams)
1 slice white bread	80-230
3 oz turkey breast, deli or pre-packaged luncheon meat	450-1,050
4 oz slice frozen pizza, plain cheese, regular crust	370-730
4 oz slice restaurant pizza, plain cheese, regular crust	510-760
4 oz boneless, skinless chicken breast, fresh	40-330
3 oz chicken strips, restaurant, breaded	430-900
3 oz chicken nuggets, frozen, breaded	200-570
1 cup chicken noodle soup, canned prepared	100-940
1 corn dog, regular	350-620
1 cheeseburger, fast food restaurant	710-1,690
1 oz slice American cheese, processed (packaged or deli)	330-460
1 cup canned pasta with meat sauce	530-980
5 oz pork with barbecue sauce (packaged)	600-1,120
1 oz potato chips, plain	50-200

FIGURE 2 Salt content in a variety of foods. The amount of salt is measured in milligrams.

THE KEY IDEA: The concentration of molecules in the body changes over time due to biological processes.

Changing Concentrations In the Human Body

When we make and store a solution in the laboratory, its concentration remains constant over time. By contrast, the concentration of substances in the human body typically changes as time passes. When you drink coffee, you don't notice the effect of the coffee immediately. It takes around 15 minutes for the caffeine to reach your brain and produce its stimulating effect. The blood concentration of caffeine peaks around 30 to 45 minutes after your first drink. The amount of caffeine then begins to decrease because the caffeine molecules are converted into other substances. In addition, some of the caffeine can be removed through urination. The decrease in blood caffeine concentration happens slowly; after eight hours, you still have significant levels of caffeine in your body.

The rise and fall of concentration over time is particularly important for understanding the effectiveness of pharmaceuticals. The study of how drugs move through the body is called **pharmacokinetics**, where the word *kinetic* refers to motion. A typical pharmacokinetic profile for a drug is shown as a graph in Figure 10.13. In this graph, the horizontal axis measures time and the vertical axis measures the blood (plasma) concentration of the drug. Time zero refers to the point when the drug is taken orally as a tablet or liquid.

pharmacokinetics The analysis of how drugs move through the body.

The factors that affect the concentration form the acronym ADME, which stands for the following biological processes:

A   Absorption

D   Distribution

M   Metabolism

E   Excretion

The initial rise of drug concentration in the blood plasma is called the *absorption phase*. Absorption measures the degree to which a drug reaches the bloodstream. For example, any tablet taken orally has to pass through the stomach and intestines before it enters the bloodstream. The ability of the drug to make this journey depends on its solubility in two different environments: the aqueous environment of the digestive system and the hydrophobic interior of cell membranes (we discussed this process in Chapter 9 in the feature *Science in Action*). For the drug represented in Figure 10.13, the maximum blood concentration (C_{max}) is reached one hour after the patient ingests the initial dose. The second phase involves the *distribution* of the drug throughout the body. To be effective, the drug must reach its biological target, which is usually an enzyme or a receptor protein. Distribution causes the plasma concentration to decrease because the drug molecules are moving from the bloodstream into the body's organs, tissues, and cells.

Another pathway for decreasing drug concentration is *metabolism*, which breaks down drug molecules into other substances. Drug metabolism often takes place through the action of enzymes in the liver, which is the organ that clears

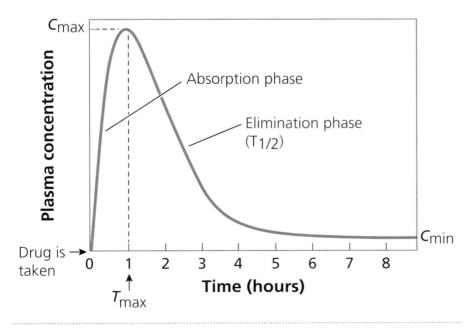

FIGURE 10.13 The pharmacokinetic profile of a drug. The pharmacokinetic profile shows how the plasma concentration of a drug changes over time. The drug concentration rises during the absorption phase, reaches a maximum concentration (C_{max}), and decreases during the elimination phase. The half-life ($T_{1/2}$) of the elimination phase is the time required for half of the drug to be eliminated from the body.

toxins from the body. The other mechanism for reducing drug concentration is *excretion* of the drug in the urine. Taken together, metabolism and excretion constitute the *elimination phase*, which slowly removes the drug from the body. The elimination phase can be characterized by its *half-life* $(T_{1/2})$, which is the time required to reduce the concentration of the drug to half of its original amount. Drugs with a short half-life are eliminated quickly, whereas those with a long half-life linger in the body for a long time. The elimination phase and the half-life are important factors to consider when recommending the time duration between doses of the drug.

PRACTICE EXERCISE 10.4

Estimate the half-life of the drug depicted in Figure 10.13.

CONCEPT QUESTION 10.2

Some drugs are taken orally, whereas other drugs are administered intravenously. How does the mode of delivery affect the pharmacokinetic profile of the drug? Using Figure 10.13 as a guide, sketch the pharmacokinetic profile for an intravenously administered drug. For convenience, you can assume that the C_{max} and the elimination half-life remain the same as in Figure 10.13.

As another example, Figure 10.14 shows the experimentally measured pharmacokinetic profile of alcohol in the human body. In this study, subjects were given one type of alcoholic drink—spirits, wine, or beer—which they consumed after fasting overnight. Blood samples were taken from each participant at regular intervals and were then analyzed to determine the *blood alcohol concentration* (BAC). The concentration of ethanol in the graph is presented in units of milligrams per deciliter (mg/dL); one deciliter (dL) is equal to 100 mL (1/10th of a liter).

Figure 10.14 shows that spirits are absorbed into the bloodstream more quickly than wine or beer. The ethanol from spirits reached its peak concentration after 30 minutes, compared with 45 minutes for wine and beer. The subsequent decrease of BAC is caused by the metabolism of ethanol in the liver. As

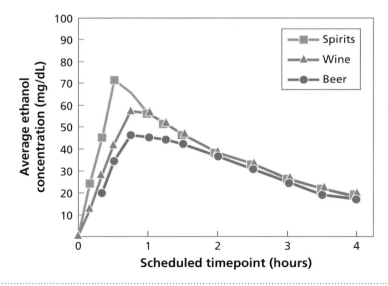

FIGURE 10.14 The pharmacokinetic profile of alcohol in the human body. Blood samples were collected at various timepoints, and blood alcohol concentration was measured in mg/dL. The pharmacokinetic profile shows a rapid rise in blood alcohol level within the first 30 to 45 minutes, flowed by a slow decline that extends over several hours.

we discussed in Chapter 5, a liver enzyme called *alcohol dehydrogenase* catalyzes the conversion of ethanol into acetaldehyde. A second liver enzyme, *acetaldehyde dehydrogenase*, subsequently converts acetaldehyde into acetic acid, which is a commonly used molecule in biochemical reactions. Note that the blood alcohol content remains at a significant level even four hours after a single drink. For beer, the four-hour blood alcohol level is slightly less than half of the peak concentration. In this study, there was no excretion of alcohol by urination.

We can relate this study to the legal limits for BAC while driving. BAC is measured as a concentration in units of mass per volume, % (m/v), which we discussed in Section 10.2. A BAC of 1% is defined as a concentration of 1 g of ethanol in 100 mL (1 dL) of blood. We can express this measurement as a ratio:

$$\%BAC = \frac{\text{mass of ethanol (g)}}{100 \text{ mL of blood}} \times 100$$

A BAC of more than 0.5% causes serious and potentially fatal alcohol poisoning, so typical BAC values are expressed as a fraction of a percent.

What are the BAC values shown in Figure 10.14? The mass units for the alcohol measurement are in milligrams (mg), and we can convert this mass to grams using a conversion factor:

$$\frac{1 \text{ g}}{1000 \text{ mg}} = 1$$

For the sample of beer, the peak alcohol concentration is 45 mg/dL. Using our conversion factor:

$$\frac{45 \text{ mg ethanol}}{100 \text{ mL}} \times \frac{1 \text{ g}}{1000 \text{ mg}} = \frac{0.045 \text{ g ethanol}}{100 \text{ mL}}$$

We can now insert this quantity into the equation for %BAC.

$$\%BAC = \frac{0.045 \text{ g ethanol}}{100 \text{ mL of blood}} \times 100 = 0.045\%$$

This amount of blood alcohol does not qualify a person as being legally impaired. In the United States, a national standard states that it is illegal to drive with a BAC of 0.08% or higher. However, the National Transportation Safety Board has proposed that the limit be lowered to 0.05%. This proposal is based on research showing that even drivers with a BAC below the legal limit still have slower reaction times and diminished decision-making capacity. In practice, a person's blood alcohol content depends on several factors in addition to the amount of alcohol consumed. These factors include body weight, gender, stomach contents, and human variability in the amount of enzymes available to metabolize alcohol in the liver.

Core Concepts

- **Pharmacokinetics** is the analysis of how drugs move through the body.
- The blood concentration of a drug is influenced by four factors: absorption, distribution, metabolism, and excretion.
- Alcohol has a pharmacokinetic profile in the human body. Blood alcohol content (%BAC) is measured in units of g/dL. Based on a national standard, it is illegal to drive with %BAC of 0.08% or higher.

CHAPTER 10
VISUAL SUMMARY

10.1 Why Can Drinking Too Much Water Be Harmful?

Learning Objective:

Explain the importance of maintaining an appropriate concentration of sodium ions in the blood.

⊙ **Concentration** measures the amount of solute in a particular volume of solution.

⊙ Drinking too much fluid can cause **hyponatremia**, which is a reduction of sodium concentration in the blood serum.

10.2 Measuring Concentration

Learning Objective:

Use various measurements and units of concentration.

⊙ Concentration can be defined using an equation:

$$\text{concentration} = \frac{\text{amount of solute}}{\text{volume of solution}}$$

⊙ The amount of solute can be measured using *volume, mass,* or *moles.*

⊙ The standard unit of volume for a solution is the **liter**. Additional volume units are the milliliter $(1 \text{ mL} = 10^{-3} \text{ L})$ and the microliter $(1 \text{ μl} = 10^{-6} \text{ L})$.

⊙ Various concentration units are expressed as a ratio of solute to solution.

$$\% \text{ (v/v)} = \frac{\text{volume of solute (mL)}}{\text{volume of solution (mL)}} \times 100$$

$$\% \text{ (m/v)} = \frac{\text{mass of solute (g)}}{\text{volume of solution (mL)}} \times 100$$

$$1 \text{ ppm} = \frac{1 \text{ mg solute}}{1 \text{ L solution}}$$

$$1 \text{ ppb} = \frac{1 \text{ μg solute}}{1 \text{ L solution}}$$

⊙ The **molarity** (M) of a solution is equal to the chemical amount of solute (measured in moles) that is present in 1 L of solution.

⊙ The mass of solute (g) is converted to the chemical amount (mol) by using the molar mass (g/mol).

- Solutions are commonly made by diluting a stock solution to obtain the desired concentration and volume.

- The **dilution** amount is calculated using the equation

$$C_1V_1 = C_2V_2$$

- where C is concentration, V is volume, solution$_1$ refers to the stock solution, and solution$_2$ is the solution being made.

10.3   Biological Applications of Concentration

Learning Objective:

Relate concentration to osmosis and hyponatremia.

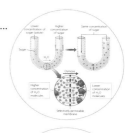

- **Osmosis** is the net flow of water from a region of higher solvent concentration to a region of lower solvent concentration. This spontaneous flow occurs across a selectively permeable membrane, which allows H_2O molecules to cross but blocks the passage of solutes

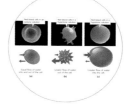

- The cell membrane is selectively permeable to the flow of water. The flow of water into or out of a cell depends on whether a cell is placed in an isotonic, hypertonic, or hypotonic solution.

- Hyponatremia occurs when the concentration of sodium ions in blood plasma falls below the normal range (135–145 mM). Osmosis produces a net flow of water from the blood into the body's cells, which causes the cells to swell. Swelling of cells in the brain can be fatal.

- **Pharmacokinetics** is the analysis of how drugs move through the body.

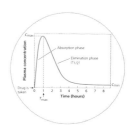

- The blood concentration of a drug is influenced by four factors: absorption, distribution, metabolism, and excretion.

LEARNING RESOURCES

Reviewing Knowledge

10.1: Why Can Drinking Too Much Water Be Harmful?

1. What was the cause of a runner's death in the 2002 Boston Marathon?

10.2: Measuring Concentration

2. What is the definition of concentration?
3. What is the standard unit for measuring the volume of concentrations? How is this unit defined?
4. Which equations define concentration measurements in (a) % (v/v) and (b) % (m/v)?
5. Which equations are used to define measurements of solution concentration in parts per million (ppm) and parts per billion (ppb)?
6. The EPA classifies drinking water pollutants using two standards: the maximum contaminant level goal (MCLG) and the maximum contaminant level (MCL). (a) Explain the difference between these two standards. (b) Why are the two standards not the same for some pollutants?
7. Define the molarity of a solution, and write an equation for it.
8. Which equation is used to calculate the dilution of a stock solution to make a solution of desired volume and concentration?

10.3: Biological Applications of Concentration

9. Explain the process of osmosis.
10. Explain what happens to the flow of water when a cell is placed in a solution that is (a) isotonic, (b) hypertonic, and (c) hypotonic.
11. What are the definitions of (a) hyponatremia and (b) critical hyponatremia?
12. Sketch a diagram to show the pharmacokinetics of a drug in the human body. Describe how the plasma concentration of the drug is affected by absorption, distribution, metabolism, and excretion.

Developing Skills

13. You are collecting water samples as part of a water quality survey. For each of the water samples given below, (a) calculate the contaminant concentration in ppm, and (b) determine whether the contaminant level exceeds the MCL. Refer to Table 10.1 for MCL values.
    Sample A: 0.12 mg of cadmium in 15 L of sample
    Sample B: 1.5 mg of lead in 5.0 L of sample
14. What is the concentration, in millimolar (mM), of a solution made from 1 mg of solid glucose dissolved in 50 mL of water?
15. The molar mass of NaCl is 58.5 g. What are the concentrations of the following solutions expressed in molarity?
    (a) 5.85 g NaCl dissolved in a total volume of 1.00 L
    (b) 5.85 g NaCl dissolved in a total volume of 1.00 L

(c) 5.85 g NaCl dissolved in a total volume of 0.100 L
(d) 58.5 g NaCl dissolved in a total volume of 10.00 L

16. Calculate the mass of solute (in grams) that would be required to make up the following solutions.
    (a) 1 L of a 0.5 M solution of sodium chloride (NaCl)
    (b) 5 L of 0.001 M acetic acid (CH_3COOH)
    (c) 250 mL of 0.02 M methanol (CH_3OH)

17. This question involves estimating the concentration of drug molecules in the body. Suppose that you swallow two aspirin tablets, each with a dose of 325 mg. When you take aspirin orally, the proportion that reaches your bloodstream is 68%. Assume the aspirin is distributed uniformly within your blood, which has a total volume of 5.5 L. Calculate the concentration of aspirin in your blood in units of (a) % (m/v), (b) ppm, and (c) molarity. (*Note: Your answer will be an overestimate because we have not accounted for any factors that remove aspirin from the bloodstream.*)

18. Consider a U-shaped tube containing two solutions that are separated by a selectively permeable membrane (see Figure 10.11). The left side of the U-tube contains physiological saline, and the right side contains one of the solutions given below. For each solution, predict whether there would be a flow of water, and, if so, the direction of the flow (left to right or right to left).
    (a) 0.20 M NaCl
    (b) 0.15 M KCl
    (c) 0.30 M glucose

19. NaCl is soluble in water, but adding more and more salt finally reaches a limit. A solution that cannot dissolve any additional solute is called a saturated solution. The amount of salt in the saturated solution defines the solubility of the salt. The solubility of NaCl is measured as 359 g/L at room temperature. What is the molarity of the saturated solution?

20. The solubility of plastics in water is very low, but it is not zero. This is significant because millions of tons of plastics have accumulated in the Pacific Ocean, raising concerns about the safety of marine life. Suppose the solubility of polyethylene is 20 ppm. For each metric ton of plastic, how much polyethylene would be dissolved in the area? For a million metric tons? (A metric ton is 1000 kg.)

21. A bottle of a certain liquor is labeled "80 proof." The proof unit is defined as twice the alcohol by volume. (a) If the volume of the bottle is 750 mL, then what volume of alcohol is contained in the bottle? (b) One standard "shot" of alcohol is 1.5 fluid ounces. What volume of alcohol is contained in one "shot" of this liquor? One fluid ounce is approximately equal to 30 mL.

22. Ethylene glycol, shown here, is very soluble in water. It is added to automobile radiators to prevent freezing.

$$H-\overset{..}{\underset{..}{O}}-\overset{\overset{H}{|}}{\underset{\underset{H}{|}}{C}}-\overset{\overset{H}{|}}{\underset{\underset{H}{|}}{C}}-\overset{..}{\underset{..}{O}}-H$$

As shown in the graph below, the freezing point of ethylene glycol–water solutions drops as a function of concentration. Use this graph to determine how much ethylene glycol is needed to prevent freezing down to $-10°C$ in winter.

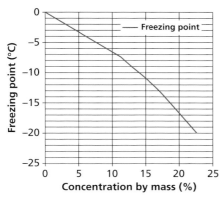

23. A bottle of concentrated HCl has a concentration of 12.0 M. How would you prepare 100 mL of a 0.300 M solution of HCl?

24. The Dead Sea is among the saltiest bodies of water on earth. Its water contains 31.5 g of salt in 100 g of solution, by volume. Its density is 1.24 g/cm^3. What is the molarity of this solution? If the density of an average human body is 1.01g/cm^3, can you sink in the Dead Sea?

25. Arsenic contaminates groundwater in many locations within the United States. For example, consider a sample of water from a well in Nevada, which contains 50 mg of arsenic in 1 L of water. Does this exceed the MCL for arsenic?

26. An important principle regulates the solubility of all salt. Let's take KCl as an example. If you add a different salt that contains either K^+ ions or Cl^- ions to a solution of KCl, then its solubility decreases. So, for example, if you add KNO_3 or NaCl to the original solution, then KCl will become less soluble than before. However, if you add a salt such as $LiNO_3$ to the KCl solution, then there is no effect. This phenomenon is called the *common ion effect*, and it is a general chemistry principle. Predict which of the following solutions will show the common ion effect:
    (a) $LiCl + KNO_3$
    (b) $NaCl + Na_2SO_4$
    (c) $CaCl_2 + HCl$

27. According to the nutrition label, a can of soup contains 880 mg of "sodium" (which refers to the amount of Na^+ ions). What mass of NaCl was added to the can to produce this quantity of sodium? Recall that the dissociation of 1 mole of NaCl produces 1 mole of Na^+ and 1 mole of Cl^-.

28. The average composition of cow's milk in the United States is 87.7% water and 4.9% lactose. The remainder is composed of protein, fat, and ions. How much lactose is contained in a pint (473 mL) of milk, if the density of milk is 1.033 g/mL?

29. Polymers that are very large in size are conveniently measured on a scale of *micromolar*, abbreviated as mM. A concentration of 1 mM corresponds to one micromole of solute in 1 L of solution. For example, a certain virus contains a DNA molecule with a molar mass of 2.5×10^7 g/mol.

In the laboratory, you are asked to make up a stock solution of this very large polymer. What mass of DNA do you need to make up 10 mL of a 1.0 mM solution?

30. The *Science in Action* feature described how to measure unknown concentrations using a Beer's Law plot. The figure below shows a Beer's Law plot of a solute molecule. You are given a solution with an unknown concentration, and you measure an absorbance of $A = 0.9$. What is the molar concentration of the solute?

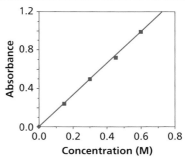

Exploring Concepts

31. Suppose you need to take aspirin tablets over three days to relieve the pain from a muscle injury. Identify and describe the processes that affect the concentration of aspirin molecules in your body over this time.

32. Our kidneys are important organs that filter our blood and separate waste products, which are removed from the body in urine. If a person experiences kidney failure, it is common medical practice to provide *kidney dialysis* using a machine. Kidney dialysis uses the principle of osmosis. Use the Internet to research kidney dialysis and explain how it works using chemical principles. Write a one- to two-paragraph report on your discoveries, and illustrate your report with a sketch.

33. Alcohol content can be measured using a *breathalyzer*. This device detects the alcohol in your breath and converts it into a measurement of blood alcohol content (BAC). Use the Internet to investigate the chemistry of how a breathalyzer works. Write a one-paragraph report on your findings, and include some relevant chemical equations.

34. A biochemist named Peter Agre shared the 2003 Nobel Prize in Chemistry for his discovery and characterization of the aquaporin water channel. One of his key experiments was to compare the osmotic response of cells containing aquaporin to cells that lack this protein. These figures show the condition of cells at various times after being placed in a medium of pure water. What difference do you observe between the behaviors of the two types of cells? Explain this observation.

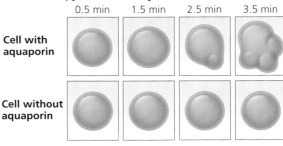

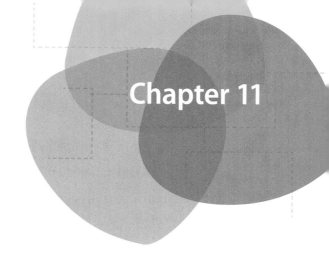

Acids and Bases

CONCEPTS AND LEARNING OBJECTIVES

FEATURES

You have probably experienced the unpleasant symptoms of "heartburn," which is also called acid indigestion. You may have taken an antacid to alleviate the symptoms. If your heartburn occurs frequently, you may have used an over-the-counter medication. What is an acid? What causes acid indigestion? How do antacids and heartburn medications provide relief?

11.1 What Causes Acid Reflux Disease?

Learning Objective:

Explain the origin of acid reflux disease.

🔑 THE KEY IDEA: Acid reflux is caused by the backflow of stomach acid into the esophagus.

■ **acid reflux** The backflow of stomach acid into the esophagus.

Acid indigestion occurs when the contents of the stomach flow into the *esophagus*, the muscular tube that connects the mouth to the stomach (Figure 11.1). The stomach contains acid that helps us digest food. The interior wall of the stomach is coated by a layer of mucus that protects it from damage by the acid. However, the tissues lining the esophagus lack this protection, so the acid creates a burning sensation in the chest and throat. Because the esophagus passes near the heart, this symptom was given the misleading name of "heartburn."

In medical terminology, this condition is called **acid reflux**. *Flux* means "flow" and *reflux* means "backflow", so *acid reflux* describes the backflow of acid from the stomach into the esophagus. For some individuals, acid reflux occurs frequently, perhaps several times every week. This condition is termed *gastroesophageal reflux disease* (GERD) but is commonly known as *acid reflux disease*. Over time, repeated acid reflux can damage the esophagus lining, which makes swallowing painful and can increase the risk of developing cancer in the esophagus.

For decades, the only remedy for heartburn or acid reflux disease was to take an antacid such as TUMS, Rolaids, or Alka-Seltzer. Antacids contain a chemical compound that acts as a *base*, which neutralizes some of the stomach acid. However, antacids provide only temporary relief. Newer classes of pharmaceuticals (e.g., Pepcid AC, Prilosec) treat GERD by reducing the production of stomach acid. Figure 11.2 illustrates three treatments for acid reflux disease.

This chapter describes the chemistry of acids and bases. How are acids and bases defined, and what are their properties? Why are some acids stronger than others? How do we measure the degree of acidity of a solution? We also examine why acids and bases are fundamental in biological processes, including the origin and treatment of acid reflux.

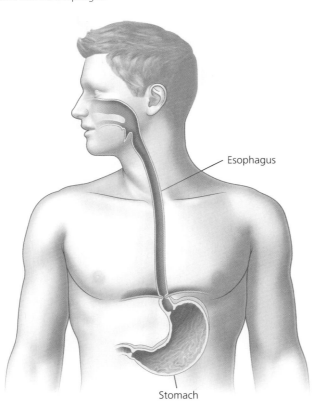

Esophagus

Stomach

FIGURE 11.1 The cause of acid indigestion. The esophagus is a muscular tube that connects the mouth to the stomach. The flow of acid from the stomach into the esophagus is responsible for acid indigestion, or "heartburn."

Core Concepts

■ **Acid reflux** (also called acid indigestions or heartburn) is the backflow of stomach acid into the esophagus, where it can cause a painful burning sensation.

■ Individuals who experience frequent acid reflux have a condition called gastroesophageal reflux disease (GERD).

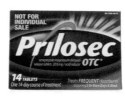

FIGURE 11.2 **Three types of treatment for acid reflux disease.** Antacids like TUMS neutralize stomach acid, whereas drugs like Pepcid AC and Prilosec reduce the production of stomach acid.

11.2 What Are Acids and Bases?

Acids and bases are part of many chemical and biological processes. This section explains the classification of acids and bases according to their chemical properties. Later in the chapter, we will apply these concepts in a biological context.

Examples of Acids and Bases

The terms **acid** and **base** refer to the chemical properties of substances. Concept Question 11.1 asks you to think about acids and bases that you already know.

CONCEPT QUESTION 11.1

Name all the chemical substances you know that are acids. Name all the chemical substances you know that are bases. Can you think of any characteristics of acids and/or of bases?

You likely have named several acids. Certain acids are commonly used in the chemical laboratory (e.g., *hydrochloric acid*, *nitric acid*). Others are part of everyday life—for example, a car battery contains *sulfuric acid*. The term *acid* derives from the Latin word for "sour," and acids are responsible for the sour taste of certain foods and beverages. The sour taste of yogurt comes from *lactic acid*, and the tartness of vinegar comes from *acetic acid*. Citrus fruits such as oranges, lemons, and limes contain two types of acid: *citric acid* and *ascorbic acid* (better known as vitamin C).

Figure 11.3 illustrates a selection of these acids. Sulfuric acid (H_2SO_4) is an *inorganic acid* because it does not contain any carbon atoms. In contrast, acetic acid, lactic acid, and citric acid are *organic acids* that are found in biological cells. These molecules derive their acidic properties from the *carboxylic acid* functional group (see Chapter 5).

Common examples of bases have less obvious names. In the chemical laboratory, the most commonly used base is *sodium hydroxide*. This substance is also found in household cleaning products such as Lysol, Drano, and Clorox. Another common base, *ammonia*, is found in glass cleaners like Windex. In the kitchen, one familiar base is baking soda (*sodium bicarbonate*), which causes batter to "rise." In contrast to acids, bases are not frequently found in food because they taste bitter. All of the antacids used for heartburn are bases; for example, TUMS and Maalox both contain *calcium carbonate* as their active ingredient. Figure 11.4 illustrates some of these bases.

Core Concepts

- Examples of **acids** you encounter in everyday life are *sulfuric acid* (car battery), *acetic acid* (vinegar), *lactic acid* (yogurt), and *citric acid* (citrus fruits). Organic acids derive their acidic properties from the *carboxylic acid* functional group.
- Examples of **bases** you encounter in everyday life are *sodium hydroxide* (drain cleaner), *ammonia* (window cleaner), *sodium bicarbonate* (baking soda), and *calcium carbonate* (antacid tablets).

Learning Objective:
Apply the Brønsted-Lowry theory of acids and bases.

THE KEY IDEA: Acids and bases are found in many common substances.

acid A substance that can donate a hydrogen ion (H⁺) to another substance.

base A substance that can accept a hydrogen ion (H⁺) from another substance.

FIGURE 11.3 A selection of common acids. Acetic acid, lactic acid, and citric acids are organic acids, and their molecular structures contain the carboxylic acid functional group.

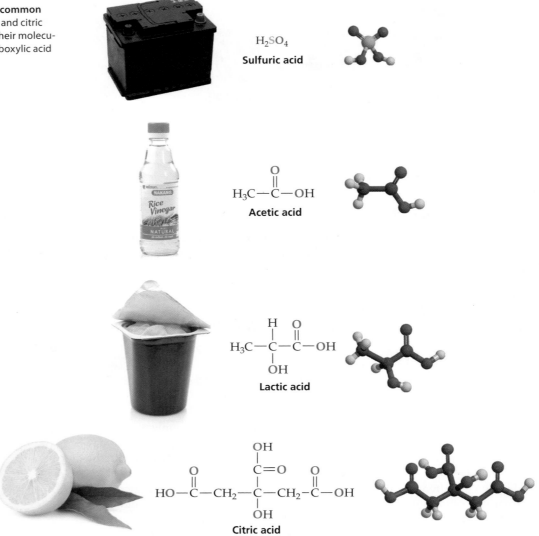

H$_2$SO$_4$
Sulfuric acid

H$_3$C—C(=O)—OH
Acetic acid

H$_3$C—C(H)(OH)—C(=O)—OH
Lactic acid

HO—C(=O)—CH$_2$—C(OH)(C=O)—CH$_2$—C(=O)—OH
Citric acid

FIGURE 11.4 A selection of common bases. The uses of these bases vary from cleaning solutions to antacids and baking products.

NaOH
Sodium hydroxide

NaHCO$_3$
Sodium bicarbonate

CaCO$_3$
Calcium carbonate

NH$_3$
Ammonia

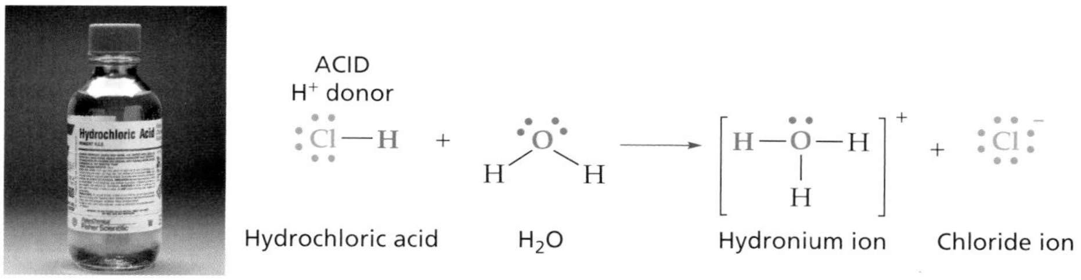

FIGURE 11.5 **Hydrochloric acid (HCl) is a proton donor.** When HCl is dissolved in water, it donates an H^+ ion to a nearby H_2O molecule. The acidic hydrogen atom in HCl is highlighted in pink. In this reaction, the H_2O molecule acts as a base. The proton transfer produces a hydronium ion (H_3O^+) and a chloride ion. By convention, a polyatomic ion like H_3O^+ is enclosed on square brackets. In this equation, acids are shown in red and bases in blue.

Brønsted-Lowry Theory of Acids and Bases

All acids share a distinctive set of chemical properties, and all bases share a different set. We need a *theory* of acids and bases that explains these general properties. Recall from Chapter 1 that a theory provides a general explanation of an observable phenomenon in the natural world.

In 1923, two chemists, J. N. Brønsted (1879–1947) and T. M. Lowry (1874–1936), independently proposed a general theory of acids and bases. In the Brønsted-Lowry theory, *an acid is any substance that can donate a hydrogen ion (H⁺) to another substance.* Conversely, *a base is any substance that can accept an H⁺ ion from another substance.* In essence, acid-base chemistry is the transfer of H^+ ions from one chemical substance to another. An H^+ ion is a proton—that is, a hydrogen atom that has lost its electron. Therefore, we use *hydrogen ion* and *proton* as equivalent terms in the chemistry of acids and bases. Throughout this chapter, we will mostly consider acid–base reactions in water because it is the most common solvent. However, the Brønsted-Lowry theory is not limited to reactions in water.

To illustrate the Brønsted-Lowry theory, consider the chemical properties of hydrochloric acid (HCl). This acid is used as stomach acid in our body. As shown in Figure 11.5, an HCl molecule in water behaves as an acid by donating an H^+ ion to a nearby H_2O molecule. The H atom in HCl that produces the H^+ ion is called an *acidic hydrogen atom*. The H^+ ion is picked up by one of the lone pairs of electrons on the H_2O molecule, which forms a new O—H covalent bond. In this reaction, the H_2O molecule is acting as a base by accepting a proton. The result of the proton transfer is the formation of a **hydronium ion** (H_3O^+). The other product of the reaction is a chloride ion (Cl^-); it has a negative charge because the electron that belonged to the H atom in HCl is left behind when the proton is transferred. As we shall see later, the products of this reaction can also behave as an acid or a base.

The acid–base reaction in Figure 11.5 can be written in a simplified form as shown below. We use *(aq)* to designate an aqueous solution and *(l)* to indicate a liquid. As discussed in Chapter 8, we use "water" to refer to the liquid and "H_2O" to refer to the individual molecules that make up the liquid.

$$HCl(aq) + H_2O(l) \longrightarrow H_3O^+(aq) + Cl^-(aq)$$

| Hydrochloric acid | Water | Hydronium ion | Chloride ion |

We can predict the structure of the H_3O^+ ion using the principle of valence shell electron pair repulsion (VSEPR) (see Chapter 3). The valence shell of H_3O^+ contains three bonding pairs and one lone pair, so electron repulsion will produce a *trigonal pyramidal* shape similar to the structure of ammonia (NH_3). The three-dimensional structure of H_3O^+ is shown in Figure 11.6. In reality, the hydronium ion

⚙ THE KEY IDEA: The Brønsted-Lowry theory states that acids donate H^+ ions and that bases accept H^+ ions.

hydronium ion An ion that is formed when a water molecule accepts a proton; its chemical formula is H_3O^+.

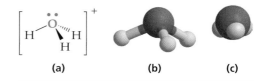

(a)   (b)   (c)

FIGURE 11.6 **The H_3O^+ ion has a trigonal pyramidal structure.** (a) Three-dimensional structure of the H_3O^+ ion using the wedge-and-dash notation. (b) Ball-and-stick and space-filling molecular models of the H_3O^+ ion.

is a simplification of how protons behave in water. Experiments and computer simulations have revealed that a single proton is surrounded by more than one H_2O molecule.

WORKED EXAMPLE 11.1

Question: Apply the Brønsted-Lowry theory to explain why nitric acid (HNO_3) behaves like an acid when it is dissolved in water. Use a chemical equation in your answer.

Answer: When HNO_3 is dissolved in water, it donates an H^+ ion to a nearby H_2O molecule. This chemical behavior matches the definition of an acid according to the Brønsted-Lowry theory. The result of the proton transfer is the formation of a hydronium ion (H_3O^+) and a nitrate ion (NO_3^-). The chemical equation for the reaction is

$$\boxed{\begin{array}{c} \text{ACID} \\ H^+ \text{ donor} \end{array}}$$

$$HNO_3(aq) \ + \ H_2O(I) \ \longrightarrow \ H_3O^+(aq) \ + \ NO_3^-(aq)$$

Nitric acid   Water   Hydronium   Nitrate
                                        ion   ion

TRY IT YOURSELF 11.1

Question: Consider a compound with the chemical formula $HClO_4$. According to the Brønsted-Lowry theory, do you predict that the compound is an acid or a base? Use a chemical equation in your answer.

PRACTICE EXERCISE 11.1

Write a chemical equation to show how hydrobromic acid (HBr) transfers a proton to H_2O in aqueous solution.

To examine the chemical properties of bases, we use sodium hydroxide (NaOH) as an example. When this solid compound is dissolved in water, it ionizes to produce an aqueous solution of $Na^+(aq)$ ions and $OH^-(aq)$ ions (Figure 11.7).

Although we conventionally write the hydroxide ion as OH^-, the negative charge resides on the electronegative oxygen atom and *not* on the hydrogen atom. We can clarify this point by drawing the Lewis dot structure of the OH^- ion, as shown in Figure 11.8. The O atom acquires an extra electron to achieve a stable octet. In the case of NaOH, this extra electron comes from ionization of an Na atom to produce an Na^+ ion.

We commonly think of sodium hydroxide as being the base. However, *in the Brønsted-Lowry theory, it is the hydroxide ion, OH^-, that acts as the base.* In aqueous solution, the OH^- ion accepts a proton from a nearby H_2O molecule. By donating this H^+ ion, the H_2O molecule functions as an *acid.* In fact, *H_2O can behave as both an acid and a base,* depending on the circumstances. Figure 11.9 provides a chemical equation for the proton

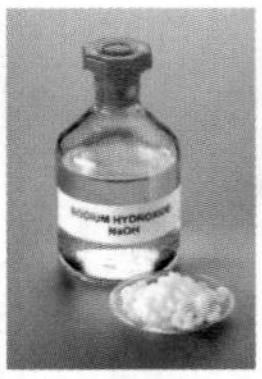

$$NaOH(s) \xrightarrow{\ H_2O\ } Na^+(aq) \ + \ OH^-(aq)$$

FIGURE 11.7 Ionization of sodium hydroxide. Solid sodium hydroxide ionizes in water to produce an aqueous solution of $Na^+(aq)$ ions and $OH^-(aq)$ ions.

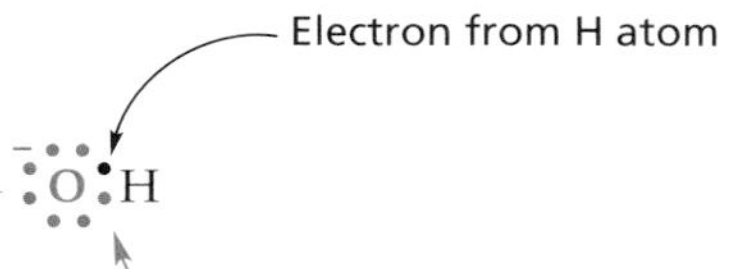

Electron from H atom

Extra electron to
achieve a stable octet

Electron from O atom

FIGURE 11.8 The Lewis dot structure of the hydroxide ion (OH⁻). The oxygen atom acquires an extra electron to achieve a stable octet, which produces the negative charge on the ion.

BASE
H⁺ acceptor

H_2O Hydroxide ion   Hydroxide ion H_2O

FIGURE 11.9 The hydroxide ion (OH⁻) acts as a Brønsted-Lowry base. In aqueous solution, the hydroxide ion acts as a base by accepting a proton from a nearby H_2O molecule, which serves as the acid. The acidic hydrogen is highlighted in pink. This equation represents the multiple proton transfer reactions that occur within aqueous solution

transfer reaction, in which an H_2O molecule donates an H^+ ion to an OH^- ion. This equation looks unusual because the products are the same as the reactants. However, the proton transfer reaction still indicates a chemical change because a specific H_2O molecule is being converted into an OH^- ion (and vice versa). In aqueous solution, these proton transfer reactions are constantly occurring, and they involve multiple H_2O molecules and OH^- ions.

The Brønsted-Lowry theory can also explain why ammonia acts as a base, even though it does not dissociate to produce an OH^- ion. As we discussed earlier, aqueous solutions of ammonia are commonly used as household cleaners. When NH_3 dissolves in water, the lone pair on the N atom accepts an H^+ ion from a nearby H_2O molecule to create an ammonium ion (NH_4^+) and a hydroxide ion (OH^-). The proton transfer reaction is illustrated in Figure 11.10. This reaction can also be written in a simpler form.

$$H_2O(l) + NH_3(aq) \longrightarrow OH^-(aq) + NH_4^+(aq)$$

Water   Ammonia   Hydroxide   Ammonium
ion   ion

$$H_2O(l) + NH_3(aq) \longrightarrow OH^-(aq) + NH_4^+(aq)$$

Water   Ammonia   Hydroxide ion   Ammonium ion

BASE
H⁺ acceptor

H_2O Ammonia   Hydroxide ion   Ammonium ion

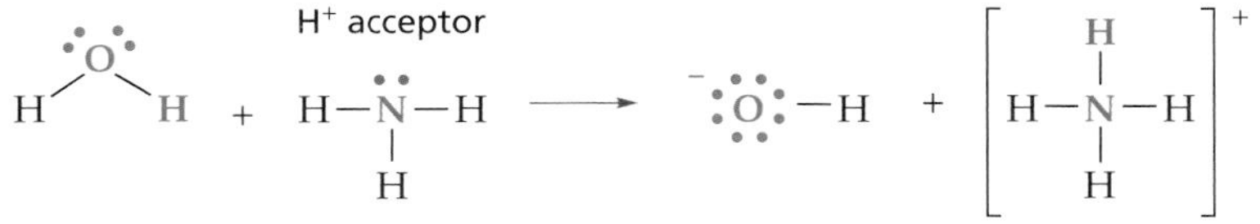

FIGURE 11.10 Ammonia (NH₃) is a proton acceptor. When NH_3 is dissolved in water, it acts as a base by accepting an H^+ ion from a nearby H_2O molecule. The H_2O molecule acts as an acid by donating an H^+ ion (the acidic hydrogen atom is highlighted in pink). The proton transfer produces a hydroxide ion (OH^-) and an ammonium ion (NH_4^+). By convention, the polyatomic NH_4^+ ion is enclosed in square brackets. In this equation, acids are shown in red and bases in blue.

PRACTICE EXERCISE 11.2

Use VSEPR to predict the three-dimensional structure of the ammonium ion (NH_4^+). Draw the structure using wedge-and-dash notation.

CONCEPT QUESTION 11.2

Ammonia (NH_3) and water (H_2O) both act as Brønsted-Lowry bases. Do you predict that methane (CH_4) can also act as a base? Explain your reasoning using chemical principles.

Just as acids have an acidic hydrogen atom, many bases have a basic nitrogen atom. These nitrogen atoms have a lone pair of electrons that can accept an H^+ ion, forming a new N—H bond and acquiring a net positive charge. For example, the amine functional group acts as a base because of its nitrogen atom (see Chapter 5). One interesting class of biological molecules is called *alkaloids*. This group includes morphine, caffeine, and nicotine. The name derives from "alkaline," which is an older term used to describe the chemical property of bases. Alkaloids behave as bases because they contain one or more basic nitrogen atoms.

WORKED EXAMPLE 11.2

Question: Methylamine (CH_3NH_2) acts as a Brønsted-Lowry base. (a) Draw the two-dimensional structure of methylamine (including lone pars), and use an arrow to label the basic nitrogen atom. (b) Using two-dimensional molecular structures (including lone pairs), write a chemical equation to show the proton transfer reaction between an H_2O molecule and methylamine. Write one to two sentences to describe this reaction.

Answer: (a) The two-dimensional structure of methylamine is shown below, with a label for the basic N atom.

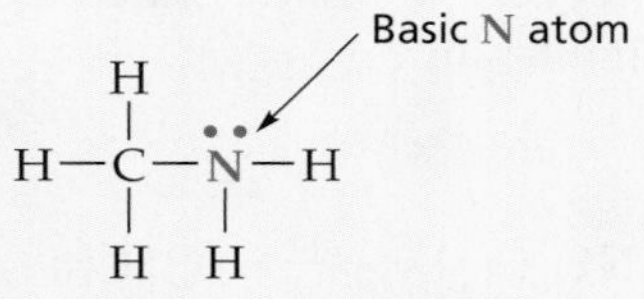

(b) The proton transfer reaction between an H_2O molecule and methylamine is shown below. The H_2O molecule functions as an acid by donating an H^+ ion to the basic N atom in methylamine. The product molecule contains a positively charged group of atoms called an *ammonium* group ($-NH_3^+$).

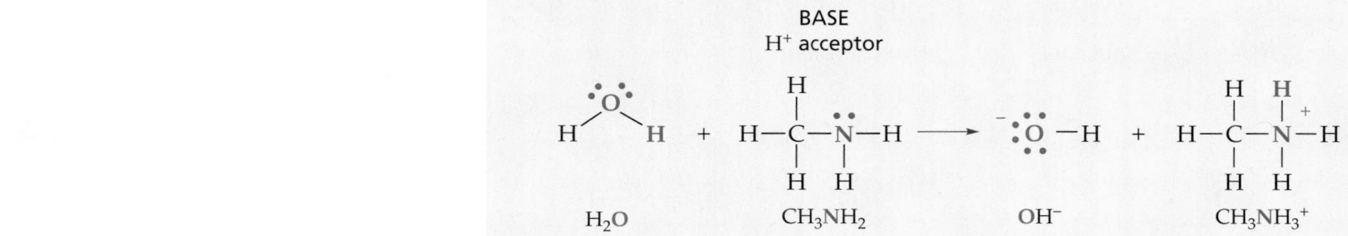

TRY IT YOURSELF 11.2

Question: (a) Using the Brønsted-Lowry theory, predict whether an ammonium ion (NH_4^+) in water acts as an acid or a base. (b) Using two-dimensional structures (including lone pairs), write a chemical equation to show the proton transfer reaction between an NH_4^+ ion and an H_2O molecule. (c) In this reaction, does the H_2O molecule act as an acid or a base?

CHEMISTRY AND YOUR HEALTH

What Is "Freebase" Cocaine?

Cocaine is a powerful and addictive stimulant drug. The 2014 National Survey on Drug Use and Health estimated that 1.5 million individuals in the United States were active users of cocaine (i.e., they had used the drug within the past month). Among young adults aged 18 to 25, an estimated 1.5% are cocaine users.

Figure 1 illustrates the molecular structure of cocaine. It is a bitter-tasting alkaloid, with a single basic nitrogen atom. This form of the drug, called "freebase" cocaine, is an oily liquid that is not very soluble in water. As a result, it is difficult to extract and purify. However, the freebase can be neutralized by reacting it with hydrochloric acid to form an ionic salt called *cocaine hydrochloride*. This version of cocaine is soluble in water and is easier to extract. Cocaine hydrochloride is the primary ingredient of powdered cocaine, which is injected or snorted. It is possible to convert the salt back to freebase cocaine by "cooking" it with

sodium bicarbonate ($NaHCO_3$) and water. This process is an acid-base reaction.

Cocaine achieves its effect by artificially elevating the level of dopamine in the brain. Dopamine is a type of neurotransmitter, a substance that transfers chemical signals between nerve cells (see Chapter 5). Dopamine is particularly active in the regions of the brain that produce pleasurable sensations. Injecting or smoking cocaine provides an initial euphoria because the drug temporarily boosts the levels of dopamine. However, this effect quickly wears off, leaving the user feeling depressed. Restoring the euphoria requires more cocaine. After repeated exposure to the drug, the brain becomes less sensitive to dopamine. This change is associated with addiction because the user must take more of the drug to achieve the same effect. The medical consequences of long-term cocaine use can be severe and include seizures, strokes, and heart attacks.

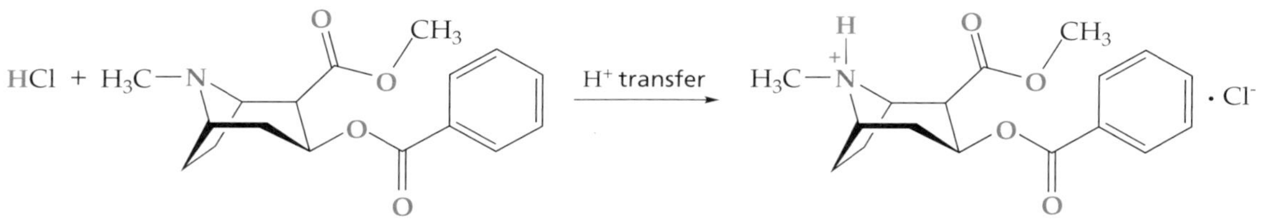

FIGURE 1 Pure cocaine contains a nitrogen atom that acts as a base—hence the name "freebase cocaine." The lone pair on the nitrogen atom is not shown, which is common practice for large molecules. The freebase form of the drug can react with hydrochloric acid to generate cocaine hydrochloride, an ionic salt found in powdered cocaine. The • symbol before the Cl^- indicates a connection between the positively and negatively charged ions.

Core Concepts

- According to the Brønsted-Lowry theory of acids and bases:
  - An **acid** *donates* a hydrogen ion (H^+) to another substance.
  - A base *accepts* a hydrogen ion (H^+) from another substance.
- When hydrochloric acid (HCl) is dissolved in water, it acts as an acid by donating an H^+ ion to a water molecule to generate a **hydronium ion** (H_3O^+).
- When sodium hydroxide (NaOH) is dissolved in water, it ionizes to produce $Na^+(aq)$ and $OH^-(aq)$ ions. The aqueous hydroxide ion acts as a base by accepting an H^+ ion from an H_2O molecule.
- When ammonia (NH_3) is dissolved in water, it acts as a base by accepting an H^+ ion from a water molecule to generate an ammonium ion (NH_4^+).

neutralization A chemical reaction
between an acid and a base.

Acid-Base Neutralization

Recall from the beginning of the chapter that antacids provided the first type of
treatment for acid indigestion. All antacids are bases ("anti-acids"). When an acid
and a base are combined, they undergo a chemical reaction called **neutralization**.
After we eat a meal, there is a large quantity of acid in our stomach. As we learned
in Section 11.1, some of this stomach acid can backflow into the esophagus.
Taking an antacid can neutralize *some* of the stomach acid and alleviate the unpleasant symptoms of acid reflux. In this section, we examine the chemistry of
acid-base neutralization.

Consider the neutralization reaction between aqueous solutions of hydrochloric acid and sodium hydroxide.

$$HCl(aq) \ + \ NaOH(aq) \ \xrightarrow{\text{neutralization}} \ NaCl(aq) \ + \ H_2O(l)$$

The products of the reaction are sodium chloride (table salt) and liquid water,
which is written as $H_2O(l)$. If equal amounts of HCl and NaOH are mixed, the
acid and base are completely consumed by the neutralization reaction, so neither
remains in the products.

In general, any acid-base neutralization reaction in aqueous solution can be
described by the following equation:

$$acid \ + \ base \ \xrightarrow{\text{neutralization}} \ salt \ + \ water$$

salt An ionic compoundw.

In Chapter 3, we mentioned that "salt" can refer to any ionic compound, not only
sodium chloride (table salt). In fact, the chemical definition of a salt comes from
acid-base chemistry. *A **salt** is an ionic compound that forms as the result of a neutralization reaction between an acid and a base.*

We can explain what happens during a neutralization reaction by examining
the dissolved ions. When HCl is dissolved in water, it produces $H^+(aq)$ and Cl^-
(aq) ions. This process is called **ionization** (an alternative term is *dissociation*).

ionization The decomposition of a
neutral compound into charged ions; it is
sometimes called *dissociation*.

$$HCl(aq) \ \xrightarrow{\text{ionization}} \ H^+(aq) \ + \ Cl^-(aq)$$

Similarly, NaOH ionizes in water to produce $Na^+(aq)$ and $OH^-(aq)$ ions.

$$NaOH(aq) \ \xrightarrow{\text{ionization}} \ Na^+(aq) \ + \ OH^-(aq)$$

In aqueous solution, *neutralization is a chemical reaction between $H^+(aq)$ and
$OH^-(aq)$ ions to form $H_2O(l)$.* In effect, this reaction "neutralizes" the chemical
properties of the acid and the base.

$$H^+(aq) \ + \ OH^-(aq) \ \xrightarrow{\text{neutralization}} \ H_2O(l)$$

We can rewrite the original neutralization reaction between HCl(aq) and
NaOH(aq) to show all of the dissolved ions.

$$\underbrace{H^+(aq) + Cl^-(aq)}_{HCl(aq)} \ + \ \underbrace{Na^+(aq) + OH^-(aq)}_{NaOH(aq)} \ \xrightarrow{\text{neutralization}} \ \underbrace{Na^+(aq) + Cl^-(aq)}_{NaCl(aq)} \ + \ H_2O(l)$$

(a)   (b)   (c)

FIGURE 11.11 Milk of Magnesia neutralizes acid. (a) Milk of Magnesia is a white suspension of $Mg(OH)_2$ particles that resembles milk. (b) When hydrochloric acid (HCl) is added, it reacts with $Mg(OH)_2$ and dissolves the particles. (c) A completed neutralization reaction between HCl and $Mg(OH)_2$ produces a clear solution.

Note that the $Na^+(aq)$ and $Cl^-(aq)$ ions, appear on both sides of the equation, and they play no role in the neutralization reaction. They are *spectator ions* because they only "watch" the chemical reaction. (Recall the discussion of precipitation reactions in Chapter 9.) The spectator ions are the components of the salt that is formed by the neutralization reaction.

We can apply our understanding of neutralization reactions to explain how antacids work. One antacid is Milk of Magnesia, which contains magnesium hydroxide, $Mg(OH)_2$; this formula can be deduced using the rules for ionic compounds introduced in Chapter 3. Magnesium hydroxide is not very soluble in water, so it does not form a proper solution. Instead, small white particles of $Mg(OH)_2$ are suspended in water to produce a thick liquid that resembles milk (Figure 11.11(a)).

When you drink Milk of Magnesia, it reacts with hydrochloric acid in your stomach. The result is a neutralization reaction that is illustrated in Figures 11.11(b) and 11.11(c). This neutralization reaction reduces the amount of stomach acid. However, the dose of an antacid is not sufficient to neutralize *all* the acid in the stomach. In fact, that would not be a desirable outcome because we need stomach acid for digestion.

The chemical equation for the neutralization reaction is written below.

$$2\,HCl(aq) \;+\; Mg(OH)_2(aq) \;\longrightarrow\; MgCl_2(aq) \;+\; 2\,H_2O(l)$$

      Acid   Base   Salt   Water

In this example, the chemical formula $Mg(OH)^2$ tells us that magnesium hydroxide can produce two $OH^-(aq)$ ions. In a neutralization reaction, each of these $OH^-(aq)$ ions reacts with an $H^+(aq)$ ion. This is why we need a coefficient of "2" in front of the $HCl(aq)$ in order to obtain a balanced equation. The neutralization reaction produces magnesium chloride, $MgCl_2$, as the salt.

WORKED EXAMPLE 11.3

Question: Sulfuric acid, H_2SO_4, can donate two H^+ ions. Write the balanced chemical equation for the neutralization reaction between sulfuric acid and magnesium hydroxide, $Mg(OH)_2$.

Answer: The balanced chemical equation is written as.

$$H_2SO_4(aq) \;+\; Mg(OH)_2(aq) \;\longrightarrow\; MgSO_4(aq) \;+\; 2\,H_2O(l)$$

      Acid   Base   Salt   Water

Each $H^+(aq)$ from H_2SO_4 neutralizes one $OH^-(aq)$ from $Mg(OH)_2$. Consequently, we do not need a coefficient of "2" in front of the H_2SO_4. However, we do need a coefficient of "2" in front of the H_2O.

TRY IT YOURSELF 11.3

Question: Phosphoric acid, H_3PO_4, can donate three H^+ ions. Write the balanced chemical equation for the neutralization reaction between phosphoric acid and magnesium hydroxide, $Mg(OH)_2$.

WORKED EXAMPLE 11.4

Question: (a) Write the chemical equation for the neutralization reaction between aqueous solutions of nitric acid (HNO_3) and potassium hydroxide (KOH). (b) Identify the spectator ions.

Answer: (a) The chemical equation presented here follows the general rule for a neutralization reaction in aqueous solution.

$$HNO_3(aq) \ + \ KOH(aq) \ \xrightarrow{\text{neutralization}} \ KNO_3(aq) \ + \ H_2O(l)$$

Acid   Base   Salt   Water

(b) To identify the spectator ions, we rewrite the equation showing all the dissolved ions.

$$\underbrace{H^+(aq) + NO_3^-(aq)}_{HNO_3(aq)} \ + \ \underbrace{K^+(aq) + OH^-(aq)}_{KOH(aq)} \ \xrightarrow{\text{neutralization}}$$

$$\underbrace{K^+(aq) + NO_3^-(aq)}_{KNO_3(aq)} \ + \ H_2O(l)$$

The spectator ions are potassium (K^+) and nitrate (NO_3^-). Alternatively, we can deduce the spectator ions from the ionic composition of the salt in part (a).

TRY IT YOURSELF 11.4

Question: (a) Write the chemical equation for the neutralization reaction between aqueous solutions of hydrobromic acid (HBr) and lithium hydroxide (LiOH). (b) Identify the spectator ions.

PRACTICE EXERCISE 11.3

Baking soda contains sodium bicarbonate, ($NaHCO_3$). When baking soda reacts with an acid, it generates bubbles of gas that cause batter to rise during cooking. (a) Write a chemical equation for the neutralization reaction between sodium bicarbonate and hydrochloric acid. (b) Name the gas produced by the reaction.

PRACTICE EXERCISE 11.4

Acid-base neutralization is not limited to aqueous solutions. The photograph shows a neutralization reaction between two gases: hydrochloric acid (HCl) and ammonia (NH_3). Mixing these gases produces a dense white cloud of ammonium chloride (NH_4Cl). (a) Write a chemical reaction for this neutralization reaction. (b) Briefly describe the proton transfer reaction that occurs during this neutralization.

Core Concepts

- A chemical reaction between an acid and a base is called a **neutralization reaction**. The products of the reaction are a **salt** and water.
- Acid-base neutralization in aqueous solution involves a reaction between H⁺(aq) and OH⁻(aq) to produce liquid water.

$$H_3O^+(aq) + OH^-(aq) \rightarrow H_2O(l)$$

- The other aqueous ions in the acid-base reaction are called spectator ions because they do not participate in the neutralization reaction.

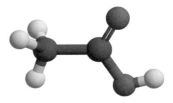

Hydrochloric acid (HCl)
strong acid

Acetic acid (CH₃COOH)
weak acid

FIGURE 11.12 A strong and weak acid. Hydrochloric acid is a strong acid because it ionizes completely in an aqueous solution. Acetic acid is a weak acid because it ionizes only partially in an aqueous solution.

11.3 Relative Strengths of Acids and Bases

Many people enjoy the tart taste of acids in vinegar, yogurt, and orange juice. However, we would be ill-advised to taste hydrochloric acid at the same concentration. The reason is that acids differ in their relative strength—hydrochloric acid is a **strong acid,** whereas acetic acid is a **weak acid** (Figure 11.12). *The terms "strong acid" and "weak acid" describe the extent to which an acid ionizes in solution.* As a very good approximation, *a strong acid ionizes completely (100%) in aqueous solution.* By contrast, *a weak acid ionizes to a small extent in aqueous solution.* For example, only 1% of the acetic acid in vinegar is ionized.

The difference between these two acids is illustrated by the electrical conductivity apparatus displayed in Figure 11.13. We introduced this apparatus in Chapter 9, when we discussed ions in aqueous solution. The bulb will light only if there are enough ions in the liquid to complete the electrical circuit. If we use distilled water as the liquid, then the bulb does not light because pure water contains very few ions. If we use a solution of 0.10 M hydrochloric acid (HCl), then the bulb shines brightly. However, if we use a solution of 0.10 M acetic acid (CH₃COOH), the bulb is lit only dimly. What is the explanation for this observed difference?

Let's first consider the 0.10 M solution of HCl (Figure 11.14). Because HCl is a strong acid, each molecule ionizes to produce one H₃O⁺ ion and one Cl⁻ ion. By the same argument, 0.10 mole of HCl produces 0.10 mole of H₃O⁺ and 0.10 mole of Cl⁻. As a result of the ionization, the solution contains concentrations of 0.10 M H₃O⁺ and 0.10 M Cl⁻. Using our approximation of 100% ionization, we can say that *no intact HCl molecules remain in solution because all of them have ionized.* A larger number of ions in solution increases the conduction of electricity, which results in the brightly lit bulb in Figure 11.13.

We now consider the 0.10 M solution of acetic acid. When dissolved in water, the CH₃COOH molecule ionizes to produce a hydronium ion and an acetate ion (CH₃COO⁻). The acetate ion differs from acetic acid by only an H⁺ ion.

Learning Objective:

Compare and contrast strong and weak acids and bases.

THE KEY IDEA: Strong acids and bases ionize completely in aqueous solutions, whereas weak acids and bases ionize only partially.

strong acid An acid that ionizes almost completely in aqueous solution.

weak acid An acid that ionizes to a limited extent in aqueous solution.

Pure water **Hydrochloric acid** **Acetic acid**

FIGURE 11.13 Demonstrating the strength of an acid by electrical conductivity. An electrical conductivity apparatus contains a battery, a light bulb, and two electrodes immersed in a liquid. This device can be used to compare the electrical conductivity of distilled water, hydrochloric acid, and acetic acid. The two solutions of acids were made with identical concentrations (0.10 M).

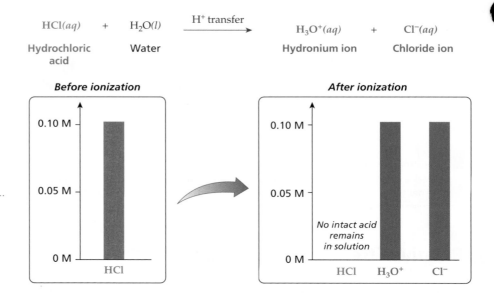

FIGURE 11.14 A strong acid ionizes almost completely in aqueous solution. HCl is a strong acid that ionizes almost completely in aqueous solution to produce $H_3O^+(aq)$ and $Cl^-(aq)$. After ionization, virtually no intact HCl molecules remain in solution.

$$CH_3COOH(aq) \; + \; H_2O(l) \xrightarrow{\text{H}^+ \text{ transfer}} H_3O^+(aq) \; + \; CH_3COO^-(aq)$$

Acetic acid   Water   Hydronium   Acetate ion
                                               ion

In solution, *the acetate ion can act as a base* and accept an H^+ ion from a nearby hydronium ion. This reaction produces acetic acid and water.

$$H_3O^+(aq) \; + \; CH_3COO^-(aq) \xrightarrow{\text{H}^+ \text{ transfer}} CH_3COOH(aq) \; + \; H_2O(l)$$

Hydronium   Acetate ion   Acetic acid   Water
ion

Note that the second reaction is the reverse of the first reaction. We can therefore combine the two reactions using double arrows ⇌, where the top arrow describes the forward (left-to-right) reaction and the bottom arrow describes the reverse (right-to-left) reaction. This two-way type of chemical reaction is called a **reversible reaction** (see Figure 11.15). Because the reaction is reversible, *all three chemical species (acetic acid, acetate ion, and hydrogen ion) are present in the solution.*

This behavior of acids in solution is an example of **chemical equilibrium**. Any reversible reaction has a particular equilibrium. For some reactions, the forward reaction proceeds easily and the reverse reaction is very difficult. In these cases, the equilibrium favors the products. This situation occurs for strong acids like HCl. Although the ionization of HCl is reversible to a very small degree, the equilibrium point strongly favors the ionized products (H_3O^+ and Cl^-) and not the intact acid (HCl). For this reason, we say that HCl ionizes almost completely in aqueous solution, such that almost no HCl molecules remain.

In other cases, the reverse reaction occurs more readily than the forward reaction, and the equilibrium favors the reactants. For a solution of acetic acid in water, the equilibrium favors the reactant (the intact CH_3COOH) rather than the products. As a result, only 1% of CH_3COOH molecules are ionized to make $CH_3COO^-(aq)$ and $H_3O^+(aq)$. The remaining 99% of the $CH_3COOH(aq)$ molecules do not ionize. Figure 11.16 illustrates the relative concentrations of CH_3COOH, H_3O^+, and CH_3COO^- in aqueous solution of acetic acid. The most abundant chemical species is the intact acetic acid molecule. The concentration

reversible reaction A two-way chemical reaction in which some of the products are transformed back into the reactants.

chemical equilibrium The balance point of a reversible reaction that describes the relative amounts of reactants and products.

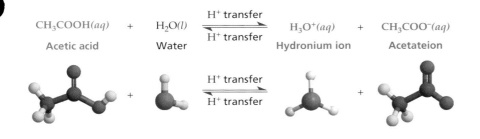

$$CH_3COOH(aq) \quad + \quad H_2O(l) \xrightleftharpoons[\text{H}^+\text{ transfer}]{\text{H}^+\text{ transfer}} H_3O^+(aq) \quad + \quad CH_3COO^-(aq)$$

Acetic acid   Water   Hydronium ion   Acetate ion

FIGURE 11.15 A reversible reaction. The ionization of acetic acid (CH_3COOH) in aqueous solution is a reversible reaction. The forward and reverse reactions each involve the transfer of an H⁺ ion.

$$CH_3COOH(aq) \quad + \quad H_2O(l) \xrightleftharpoons[\text{H}^+\text{ transfer}]{\text{H}^+\text{ transfer}} H_3O^+(aq) \quad + \quad CH_3COO^-(aq)$$

Acetic acid   Water   Hydronium ion   Acetate ion

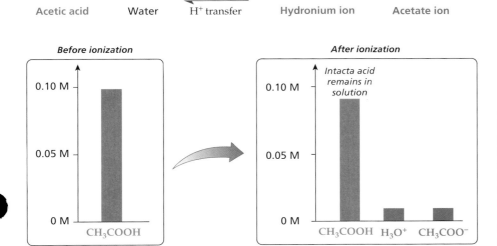

Before ionization   **After ionization**

FIGURE 11.16 A weak acid ionizes partially in aqueous solution. CH_3COOH is a weak acid that ionizes to a small extent in aqueous solution to produce $H_3O^+(aq)$ and $CH_3COO^-(aq)$. This reaction is reversible, so it can proceed in the opposite direction to regenerate CH_3COOH molecules. At equilibrium, all three chemical species are present in solution.

of $H_3O^+(aq)$ and $Cl^-(aq)$ ions is very small, which explains why the bulb of the conductivity apparatus is lit only dimly by a solution of acetic acid.

Because acetic acid and acetate are closely connected, they are called a **conjugate acid-base pair** (*conjugate* means "together"). In general, *the two chemical substances in a conjugate acid-base pair differ by only an H⁺ ion.* Acetic acid behaves as an acid by donating an H⁺ ion, and acetate behaves as the *conjugate base* by accepting an H⁺ ion.

It is a common misconception that CH_3COOH molecules contribute to the acidity of the solution. *In fact, the acidic properties of a solution arise from the concentration of $H_3O^+(aq)$ ions, not from the concentration of intact acid molecules.* Because weak acids dissociate only partially, they generate a relatively small concentration of H_3O^+ ions in solution.

We can summarize this comparison of strong and weak acids as follows:

strong acid → more dissociation → higher concentration of H_3O^+
weak acid → less dissociation → lower concentration of H_3O^+

Figure 11.17 provides a molecular-scale comparison of aqueous solutions of hydrochloric acid and acetic acid. For HCl, a strong acid, there are almost no intact HCl molecules, and the solution contains only $H_3O^+(aq)$ and $Cl^-(aq)$ (to a good approximation). By contrast, a solution of acetic acid, a weak acid, contains many intact CH_3COOH molecules and only a small number of $H_3O^+(aq)$ and $CH_3COO-(aq)$ ions.

It is no accident that the organic acids found in cells—such as acetic acid, lactic acid, and citric acid—are all weak acids. These substances can be present in typical cellular concentrations ranging from 10 to 100 mM without making

conjugate acid-base pair An acid-base pair in which the two substances differ only by an H⁺ ion.

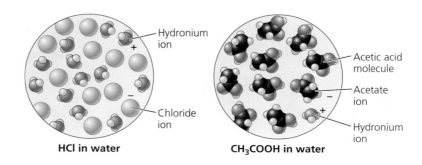

FIGURE 11.17 A molecular-level comparison of a strong and a weak acid. In the aqueous solution of HCl, essentially all of the acid molecules have ionized to produce $H_3O^+(aq)$ and $Cl^-(aq)$ ions. In the aqueous solution of CH_3COOH, most of the acid molecules remain intact, and the concentrations of $H_3O^+(aq)$ and $CH_3COO^-(aq)$ ions are very small. This reaction is reversible, so it can proceed in the opposite direction.

the cell too acidic. Most biological molecules are very sensitive to the acidity of their environment. For example, the three-dimensional structure of proteins can unfold in strongly acidic solutions. In fact, the human body contains only one example of a strong acid, which is the solution of hydrochloric acid found in the stomach. As we learned at the beginning of the chapter, the cells in the stomach lining are protected from this acid by a thick layer of mucus.

The behavior of strong and weak bases parallels our description of strong and weak acids. *A strong base ionizes almost completely in aqueous solution, whereas a weak base ionizes to a small extent.* For example, sodium hydroxide (NaOH) is a strong base that ionizes almost completely into $Na^+(aq)$ and $OH^-(aq)$ ions when it is dissolved in water. By contrast, ammonia (NH_3) is a weak base that ionizes very little in aqueous solution. When NH_3 ionizes, it accepts an H^+ ion to form NH_4^+. We can write the ionization of a weak base as a reversible chemical reaction.

$$NH_3(aq) \ + \ H_2O(l) \ \rightleftharpoons \ NH_4^+(aq) \ + \ OH^-(aq)$$

Similar to weak acids, the equilibrium for this reaction favors the reactants and not the products. As a result, the concentration of OH^- ions in the solution is very small compared to the concentration of NH_3 molecules. In summary,

strong base → more dissociation → higher concentration of OH^-
weak base   → less dissociation   → lower concentration of OH^-

We conclude by connecting several topics that we have examined in this section. Figure 11.18 presents a selection of acids and their conjugate bases. As we learned earlier, the two substances in a conjugate acid-base pair differ by an H^+ ion. For this reason, it is common to designate a generic acid as HA and its conjugate base as A^-. As the table in the figure shows, *there is an inverse relationship between the strength of an acid and the strength of its conjugate base.* Strong acids are shown at the top of the table. HCl is a strong acid, so its conjugate base, the chloride ion (Cl^-), is a weak base. $HCl(aq)$ ionizes almost completely in aqueous solution, but $Cl^-(aq)$ shows little tendency to accept an H^+ ion and re-create HCl. The difficulty of this reverse reaction explains why HCl is a strong acid and remains ionized.

In the middle of the table are weak acids that have corresponding weak bases. For example, acetic acid (CH_3COOH) is a weak acid with a weak conjugate base (CH_3COO^-). A similar relationship exists between hydrofluoric acid (HF) and its conjugate base, the fluoride ion (F^-). For these conjugate acid-base pairs, the loss or gain of a proton is reversible; consequently, an aqueous solution contains a mixture of the acid (HA), the base (A^-), and hydronium ions (H_3O^+). Finally, the bottom of the table lists very weak acids that have correspondingly strong bases. For example, the strong base $NH_2^-(aq)$ has a very pronounced tendency to accept a proton and form $NH_3(aq)$. However, $NH_3(aq)$ is a very weak acid that shows little inclination to lose a proton and form $NH_2^-(aq)$. This means that the ionization strong bases is almost 100%, similar to the behavior of strong acids.

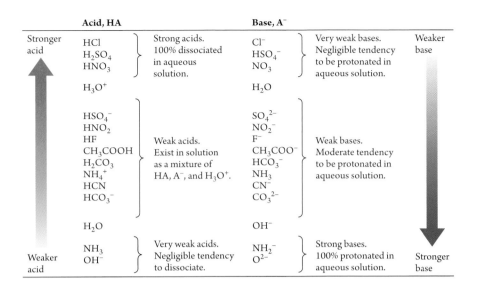

Acid, HA		Base, A⁻		
Stronger acid	HCl H_2SO_4 HNO_3	Strong acids. 100% dissociated in aqueous solution.	Cl^- HSO_4^- NO_3^-	Very weak bases. Negligible tendency to be protonated in aqueous solution.	Weaker base
	H_3O^+		H_2O		
	HSO_4^- HNO_2 HF CH_3COOH H_2CO_3 NH_4^+ HCN HCO_3^-	Weak acids. Exist in solution as a mixture of HA, A⁻, and H_3O^+.	SO_4^{2-} NO_2^- F^- CH_3COO^- HCO_3^- NH_3 CN^- CO_3^{2-}	Weak bases. Moderate tendency to be protonated in aqueous solution.	
	H_2O		OH^-		
Weaker acid	NH_3 OH^-	Very weak acids. Negligible tendency to dissociate.	NH_2^- O^{2-}	Strong bases. 100% protonated in aqueous solution.	Stronger base

FIGURE 11.18 Relative strengths of conjugate acid-base pairs.

WORKED EXAMPLE 11.5

Question: Lactic acid is a weak acid found in yogurt and sourdough bread. The reversible chemical equation for the ionization of lactic acid in aqueous solution is:

$$\underset{\text{Lactic acid}}{H_3C\!-\!\overset{\overset{\textstyle OH}{|}}{\underset{\underset{\textstyle H}{|}}{C}}\!-\!COOH} + \underset{\text{Water}}{H_2O(l)} \rightleftharpoons \underset{\text{Hydronium ion}}{H_3O^+(aq)} + \underset{\text{Lactate ion}}{H_3C\!-\!\overset{\overset{\textstyle OH}{|}}{\underset{\underset{\textstyle H}{|}}{C}}\!-\!COO^-}$$

(a) Use an arrow to label the acidic hydrogen atom in lactic acid.

(b) What is the conjugate base in this reaction?

Answer: (a) The acidic hydrogen atom belongs to the carboxylic acid functional group.

$$H_3C\!-\!\overset{\overset{\textstyle OH}{|}}{\underset{\underset{\textstyle H}{|}}{C}}\!-\!COOH \quad \overset{\text{Acidic}}{\text{H atom}}$$

(b) The conjugate base in this reaction is the lactate ion. It differs from lactic acid by a single H^+ ion.

TRY IT YOURSELF 11.5

Question: The bicarbonate ion (HCO_3^-) is a weak acid that is found in many natural water supplies. The following equation describes the reversible ionization of bicarbonate ions in aqueous solution. Add the missing conjugate base to this equation, and provide its name.

$$HCO_3^- + H_2O(l) \rightleftharpoons H_3O^+(aq) + \underline{\hspace{2cm}}$$

PRACTICE EXERCISE 11.5

Phosphoric acid (H_3PO_4) is a weak acid that is used to give a tart taste to sodas. (a) Write the reversible chemical equation for the reaction of phosphoric acid with water, which results in the transfer of 1 proton. (b) Circle the conjugate base in the equation.

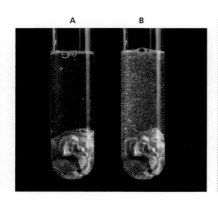

A　　B

PRACTICE EXERCISE 11.6

When a strip of zinc metal is placed in an acidic solution, a chemical reaction occurs that releases bubbles of hydrogen gas (H_2). The equation for this reaction is.

$$Zn(s) + 2\,H_3O^+(aq) \rightarrow Zn^{2+}(aq) + 2\,H_2O(l) + H_2(g)$$

The photographs show the reaction of zinc strips with a 1.0 M solution of two different acids: acetic acid and hydrochloric acid. Which photograph (A or B) corresponds to which acid? Provide the chemical reason for your answer.

Core Concepts

- A **strong acid** ionizes almost completely (100%) in aqueous solution, whereas a **weak acid** ionizes to a small extent.
- The ionization of a weak acid in water is a **reversible reaction** that can proceed in both directions. In a reversible reaction, all of the reactants and products are present in solution.
- The **chemical equilibrium** for a reversible reaction is a balance point that describes the relative amounts of reactants and products. For acetic acid in water, the equilibrium condition is 99% intact acid and only 1% ionization.
- The acidic properties of a solution arise from the concentration of H_3O^+ ions, not the concentration of intact acid molecules.
- Acetic acid (CH_3COOH) and the acetate ion (CH_3COO^-) form a **conjugate acid-base pair**. The two chemical substances in a conjugate acid-base pair differ by only an H^+ ion.
- A strong base ionizes almost completely (100%) in aqueous solution, whereas a weak base ionizes to a very limited extent
- There is an inverse relationship between the relative strength of an acid and its conjugate base. A strong acid has a very weak conjugate base, and vice versa.

11.4 Measuring Acidity: The pH Scale

Learning Objective:
Use pH as a measurement of acidity.

Because acids produce H_3O^+ ions in solution, a logical measure of acidity is the concentration of $H_3O^+(aq)$ ions. As we saw in Chapter 10, square brackets [ ] act as shorthand to denote concentration in moles per liter (molarity). The acidity of a solution is therefore described by $[H_3O^+]$, with the (aq) omitted. While we can report the hydronium ion concentration directly, the numbers are very small and inconvenient to write. As an alternative, we can measure acidity using the pH scale. This section explains the pH scale and how it is used to measure the acidity of a solution.

Ionization of Pure Water

THE KEY IDEA: Water ionizes to a very small extent to produce aqueous H_3O^+ ions and OH^- ions.

To create a scale of acidity, it is logical to begin with pure water as a reference point. Liquid water is composed primarily of intact H_2O molecules. However, a very small fraction of these molecules ionize to form aqueous hydronium ions $H_3O^+(aq)$ and hydroxide ions $OH^-(aq)$. The ionization reaction occurs via the transfer of a proton from one H_2O molecule to another. This reaction can also occur in reverse: An $H_3O^+(aq)$ ion can transfer a proton to a nearby $OH^-(aq)$ ion to regenerate two H_2O molecules. Because the reaction can proceed in both directions, *the ionization of pure water is a reversible reaction* that can be written as follows:

$$H_2O(l) \;+\; H_2O(l) \;\rightleftharpoons\; H_3O^+(aq) \;+\; OH^-(aq)$$

Because it is a reversible reaction, the ionization of pure water is another example of chemical equilibrium. In this case, the equilibrium point strongly favors the intact H_2O molecules on the left side of the equation. As a result, the molar concentrations of $H_3O^+(aq)$ and $OH^-(aq)$ in pure water are very small. On average, only 1 in every 550 million molecules dissociates into a hydronium and hydroxide ion; all of the other molecules remain as intact H_2O.

Experimental measurements of pure water reveal that *the product of the molar concentrations of $H_3O^+(aq)$ ions and $OH^-(aq)$ ions is constant* (at a fixed temperature). This constant is so important in acid-base chemistry that it is given a special name and symbol: the *ion product constant of water* (K_w). At room temperature (25°C), K_w is given by the following equation:

$$K_w = [H_3O^+][OH^-] = 10^{-14}$$

The numerical value of K_w indicates that the amount of water ionization is very small. Each ionization reaction in water produces one $H_3O^+(aq)$ ion and one $OH^-(aq)$ ion. As a result, the molar concentration (molarity) of the ions must be equal.

$$[H_3O^+] = [OH^-]$$

Using the equation for K_w, we deduce that

$$[H_3O^+] = 10^{-7} \text{ M and } [OH^-] = 10^{-7} \text{ M} \quad (\text{because } 10^{-7} \times 10^{-7} = 10^{-14})$$

This is a very low concentration, which can be written as 0.0000001 M. These small concentrations of $H_3O^+(aq)$ and $OH^-(aq)$ ions explain why pure water does not conduct electricity, as illustrated in Figure 11.13.

Core Concepts

- The ionization of pure water is a reversible reaction that generates very small amounts of aqueous hydronium ions $H_3O^+(aq)$ and hydroxide ions $OH^-(aq)$.
- The product of the concentrations of $H_3O^+(aq)$ and $OH^-(aq)$ ions is a constant called the *ion product constant of water*. At room temperature (25°C), $K_w = [H_3O^+][OH^-] = 10^{-14}$.
- In pure water, the concentrations of $H_3O^+(aq)$ and $OH^-(aq)$ ions are equal: $[H_3O^+] = [OH^-] = 10^{-7}$ M.

Neutral, Acidic, and Basic Solutions

Solutions can be classified as **neutral**, **acidic**, or **basic** based on their relative concentration of $H_3O^+(aq)$ and $OH^-(aq)$ ions. For dilute concentrations of acids and bases in water, the value of K_w remains constant. If we add an acid to pure water, $[H_3O^+]$ *increases* because the acid generates $H_3O^+(aq)$ ions. To counterbalance this effect, $[OH^-]$ must *decrease* to maintain the constant value of K_w. Conversely, adding a base to pure water increases $[OH^-]$ and decreases $[H_3O^+]$.

Figure 11.19 provides a seesaw analogy for neutral, acidic, and basic solutions. In a neutral solution,

$$[H_3O^+] = [OH^-] \quad \text{which means that} \quad [H_3O^+] = 10^{-7} \text{ M and } [OH^-] = 10^{-7} \text{ M}$$

In the seesaw analogy, the concentrations of $H_3O^+(aq)$ and $OH^-(aq)$ are balanced. As we saw earlier, this relationship also applies to pure water.

In an acidic solution,

$$[H_3O^+] > [OH^-] \quad \text{which means that} \quad [H_3O^+] > 10^{-7} \text{ M and } [OH^-] < 10^{-7} \text{ M}$$

In this definition we use > to mean "greater than" and < to mean "less than."

In a basic solution,

$$[H_3O^+] < [OH^-] \quad \text{which means that} \quad [H_3O^+] < 10^{-7} \text{ M and } [OH^-] > 10^{-7} \text{ M}$$

THE KEY IDEA: Neutral, acidic, and basic solutions are defined by the relative concentrations of H_3O^+ ions and OH^- ions.

neutral solution A solution in which the concentrations of H_3O^+ and OH^- ions are equal.

acidic solution A solution in which the concentration of H_3O^+ ions is greater than that of OH^- ions.

basic solution A solution in which the concentration of OH^- ions is greater than that of H_3O^+ ions.

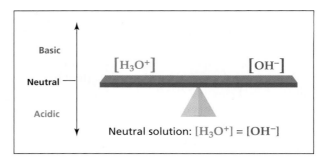

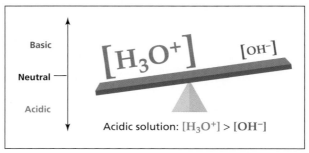

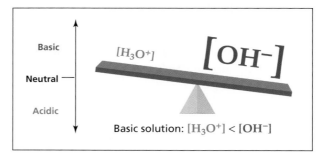

FIGURE 11.19 Comparing neutral, acidic, and basic solutions. A seesaw analogy for the relative concentrations of $H_3O^+(aq)$ and $OH^-(aq)$ concentrations in neutral, acidic, and basic solutions. The size of the square brackets [ ] corresponds to the magnitude of the molar concentration.

WORKED EXAMPLE 11.6

Question: In a dilute solution at 25°C, the concentration of $H_3O^+(aq)$ ions is 10^{-4} M. Calculate the concentration of $OH^-(aq)$ ions in the solution.

Answer: We can use the definition of K_w, which connects $[H_3O^+]$ and $[OH^-]$.

$$K_w = [H_3O^+][OH^-] = 10^{-14}$$

If $[H_3O^+] = 10^{-4}$ M, then

$$10^{-4} \times [OH^-] = 10^{-14}$$

We can rearrange this equation to calculate $[OH^-]$.

$$[OH^-] = \frac{10^{-14}}{10^{-4}} = 10^{-10} \text{ M}$$

TRY IT YOURSELF 11.6

Question: In a dilute solution at 25°C, the concentration of $OH^-(aq)$ ions is 10^{-8} M. Calculate the concentration of $H_3O^+(aq)$ ions in the solution.

WORKED EXAMPLE 11.7

Question: Classify each of the following solutions as acidic, basic, or neutral. Explain your choice.

(a) $[H_3O^+] = 10^{-3}$ M   (b) $[H_3O^+] = 10^{-9}$ M   (c) $[OH^-] = 10^{-3}$ M

Answer:

(a) $[H_3O^+] = 10^{-3}$ M is an *acidic solution* because $[H_3O^+] > 10^{-7}$ M.

(b) $[H_3O^+] = 10^{-9}$ M is a *basic solution* because $[H_3O^+] < 10^{-7}$ M.

(c) $[OH^-] = 10^{-3}$ M is a *basic solution* because $[OH^-] > 10^{-7}$ M.

For this last example, we can check our answer by calculating $[H_3O^+]$.

$$[H_3O^+] = \frac{10^{-14}}{[OH^-]} = \frac{10^{-14}}{10^{-3}} = 10^{-11} \text{ M}$$

We confirm that the solution is basic because $[H_3O^+] < 10^{-7}$ M.

TRY IT YOURSELF 11.7

Question: Classify each of the following solutions as acidic, basic, or neutral. Explain your choice.

(a) $[H_3O^+] = 10^{-12}$ M   (b) $[H_3O^+] = 10^{-7}$ M   (c) $[OH^-] = 10^{-10}$ M

Core Concepts

- In dilute solutions of acids and bases, the concentrations of $H_3O^+(aq)$ and $OH^-(aq)$ ions are connected by the ion product constant of water (K_w).
- The definitions of **neutral**, **acidic**, and **basic** solutions are based on the relative concentrations of $H_3O^+(aq)$ and $OH^-(aq)$:
  - neutral solution: $[H_3O^+] = [OH^-]$
  - acidic solution: $[H_3O^+] > [OH^-]$
  - basic solution: $[H_3O^+] < [OH^-]$

Defining the pH Scale

In pure water, the numerical value of $[H_3O^+]$ and $[OH^-]$ is 10^{-7} M. In a dilute solution of acid, a typical hydronium ion concentration might be $[H_3O^+] = 10^{-4}$ M. Writing negative powers of 10 becomes awkward, and it would be more convenient to measure the acidity of a solution using a simple number.

The **pH scale** was devised precisely for this purpose. As shown in Equation 1, the pH value replaces the negative power of 10 in the measurement of $[H_3O^+]$.

$$[H_3O^+] = 10^{-pH} \qquad\qquad \text{Equation 1}$$

To write pH as a regular number, we use the property of logarithms (see Appendix C). The relationship between pH and $[H_3O^+]$ is given by Equation 2.

$$pH = -\log_{10}[H_3O^+] \qquad\qquad \text{Equation 2}$$

Equations 1 and 2 are two equivalent ways of showing the numerical relationship between $[H_3O^+]$ and pH. The minus sign in Equation 2 means that *an increase in* $[H_3O^+]$ *will cause a decrease in pH, and vice versa*.

We can now use the pH scale to classify neutral, acidic, and basic solutions.

THE KEY IDEA: The pH scale is a logarithmic measure of acidity based on the concentration of H_3O^+ ions.

pH scale A logarithmic scale for measuring the acidity of a solution that is based on the concentration of $H_3O^+(aq)$ ions.

- **pH = 7 for pure water and neutral solutions**

  For pure water, $[H_3O^+] = 10^{-7}$ M and pH = 7. The same relationship holds true for neutral solutions.

- **pH < 7 for acidic solutions**

  By definition, *an acidic solution has a pH less than 7.* For an acidic solution in which $[H_3O^+] = 10^{-4}$ M, then pH = 4.

- **pH > 7 for basic solutions**

  By definition, *a basic solution has a pH greater than 7.* For a basic solution in which $[H_3O^+] = 10^{-10}$ M, then pH = 10.

Remember that the pH scale is logarithmic and is based on powers of 10. Consider a solution in which $[H_3O^+] = 10^{-5}$ M and pH = 5. Suppose that we add more acid to make the hydronium ion concentration 10 times greater. Now, $[H_3O^+] = 10^{-4}$ M and pH = 4. We see that *increasing $[H_3O^+]$ by a factor of 10 reduces the solution pH by one unit (more acidic). Conversely, decreasing $[H_3O^+]$ by a factor of 10 increases the solution pH by one unit (less acidic).* One consequence of the logarithmic pH scale is that substantial changes in hydronium ion concentration produce relatively small changes in pH.

CONCEPT QUESTION 11.3

The pH scale is based on the ionization of pure water, in which $K_w = [H_3O^+][OH^-] = 10^{-14}$. Suppose that the K_w value for water on a different planet is measured as $K_w = 10^{-10}$. What would be the pH of pure water on the planet? Explain your reasoning.

Figure 11.20 illustrates the relationship between pH and $[H_3O^+]$ for common solutions. Note that more acidic solutions (e.g., stomach acid) have *higher* concentrations of H_3O^+ ions and *lower* pH values.

WORKED EXAMPLE 11.8

Question: (a) Nitric acid (HNO_3) is a strong acid. What is the pH of a 0.00010 M aqueous solution of HNO_3? (b) Extra HNO_3 is added to the solution to double the hydronium ion concentration. What is the new pH of the solution?

Answer: (a) Because HNO_3 is a strong acid, it ionizes completely in aqueous solution.

$$HNO_3(aq) + H_2O(l) \longrightarrow H_3O^+(aq) + NO_3^-(l)$$

Consequently, an aqueous solution of HNO_3 generates an *equal molar concentration* of $H_3O^+(aq)$ ions. We can now write the hydronium ion concentration in scientific notation:

$$[H_3O^+] = 0.00010 \text{ M} = 1.0 \times 10^{-4} \text{ M}$$

We can use Equation 1 to deduce the pH of the solution.

$$[H_3O^+] = 10^{-pH} \quad \text{therefore} \quad pH = 4.00$$

Alternatively, we can use Equation 2 to calculate the pH.

$$pH = -\log_{10}[H_3O^+] = -\log_{10}(1.0 \times 10^{-4}) = -(-4.00) = 4.00$$

(b) If the hydronium concentration is doubled, the new concentration is

$$[H_3O^+] = 2 \times (1.0 \times 10^{-4} \text{ M}) = 2.0 \times 10^{-4} \text{ M}$$

In this case, it is not possible to deduce the pH directly using Equation 1, so we need to use Equation 2:

$$pH = -\log_{10}[H_3O^+] = -\log_{10}(2.0 \times 10^{-4}) = -(-3.70) = 3.70$$

We can see that doubling the hydronium ion concentration produces only a small decrease in pH of 0.3 unit.

TRY IT YOURSELF 11.8

Question: (a) Hydrochloric acid (HCl) is a strong acid. What is the pH of a 0.010 M solution of HCl? (b) Extra HCl is added to the solution to increase the hydronium ion concentration to 5 times its original value. What is the new pH of the solution? Has the pH increased or decreased?

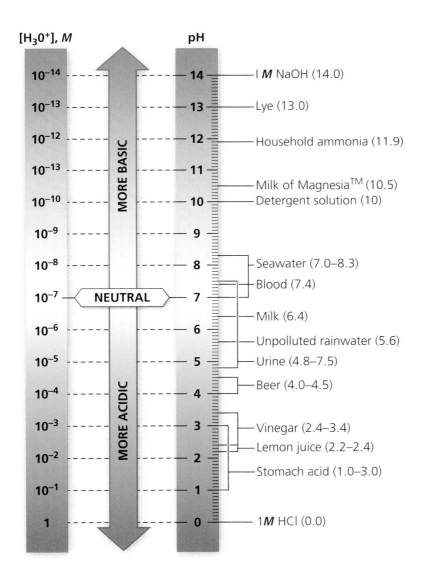

FIGURE 11.20 The relationship between [H_3O^+] and pH for a variety of aqueous solutions.

WORKED EXAMPLE 11.9

Question: Vinegar is made by dissolving acetic acid in water. If the pH of white vinegar is 2.50, then what is the concentration of hydronium ions in the solution?

Answer: To answer this question, we use Equation 1, which tells us how to calculate $[H_3O^+]$ for a known pH value.

$$[H_3O^+] = 10^{-pH}$$

We now substitute the pH of the white vinegar solution. This calculation can be performed by using the 10^x button on your calculator.

$$[H_3O^+] = 10^{-2.50} = 3.2 \times 10^{-3} \, M \quad or \quad 3.2 \, mM$$

TRY IT YOURSELF 11.9

Question: The typical pH of human blood is 7.40.

(a) Based on its pH, is blood neutral, acidic, or basic?

(b) *Without using a calculator*, predict the approximate concentration of H_3O^+ ions in blood. Explain the chemical basis for your prediction. (*Hint*: Refer to the earlier discussion of K_w.)

(c) Calculate the concentration of H_3O^+ ions in blood.

(d) Was your predicted concentration close to the calculated value? Write one sentence to explain the comparison.

PRACTICE EXERCISE 11.7

What is the pH of a very concentrated, 10.0 M solution of hydrochloric acid? What is unusual about the pH you have calculated?

Core Concepts

- The **pH scale** is a logarithmic measurement of acidity based on the molar concentration of hydronium ions in solution. The pH value is related to $[H_3O^+]$ by two equivalent equations:
$$[H_3O^+] = 10^{-pH} \quad or \quad pH = -\log_{10}[H_3O^+]$$
- Because of the negative sign in the pH equation, an increase in $[H_3O^+]$ causes a decrease in pH. Conversely, a decrease in $[H_3O^+]$ causes an increase in pH
- The pH scale can be applied to classify neutral, acidic, and basic solutions:
  - neutral solution: pH = 7.0
  - acidic solution: pH < 7.0
  - basic solution: pH > 7.0
- Because the pH scale is logarithmic, changing the pH by one unit corresponds to a 10-fold change in $[H_3O^+]$.

Learning Objective:

Illustrate two biological applications of acid-base chemistry.

11.5 Biological Applications of Acid-Base Chemistry

The chemistry of acids and bases has widespread importance in biological organisms. In this section, we describe the physiology of stomach acid and the development of pharmaceutical treatments for acid reflux. We also discuss how the components of blood stabilize its pH within a narrow range.

Origin and Treatment of Acid Reflux Disease

The inside of our stomach is the most acidic environment in our body. Lining the interior of the stomach wall are glands that secrete a mixture of hydrochloric acid, enzymes, and mucus. The hydrochloric acid is responsible for the acidity of the stomach. The pH of our stomach contents varies, depending on the time of day and when we eat. Our stomach contents have a resting pH of 4 to 5. After a meal, more hydrochloric acid is released into the stomach to aid with digestion, which lowers the pH to between 1 and 2. As we learned in the previous section, a change in pH by one unit corresponds to a 10-fold change in the concentrations of $H_3O^+(aq)$ ions. Consequently, a post-meal decrease in stomach pH from 4 to 2 corresponds to an $H_3O^+(aq)$ concentration *that is 100 times greater.*

> ⊙ THE KEY IDEA: Two classes of acid reflux drugs are histamine blockers and proton pump inhibitors.

PRACTICE EXERCISE 11.8

If the contents of the stomach have a pH of 2.0, then what is the concentration of $H_3O^+(aq)$ ions?

The connection between the stomach and the esophagus is controlled by a circular muscle called a *sphincter* (Figure 11.21). When we are not eating or drinking, the sphincter remains closed to keep the acidic contents of the stomach inside. However, the sphincter must open to allow food to pass from the esophagus into the stomach. When this happens, some of the stomach acid can escape and flow back ("reflux") into the esophagus. Because the inside of the esophagus lacks a protective lining of mucus, the acid creates a burning sensation. For some individuals, the sphincter does not close all the way, or it opens too often. These individuals are prone to frequent acid indigestion, which is classified as gastroesophageal reflux disease, or GERD.

We are all familiar with the experience that acid indigestion is more pronounced during or shortly after a meal. One reason is that the sphincter must open and close to allow food to pass into the stomach, which can allow the stomach acid to escape. The second reason, mentioned above, is that the contents of the stomach are particularly acidic at mealtime, which increases the corrosive effect of the stomach acid within the esophagus.

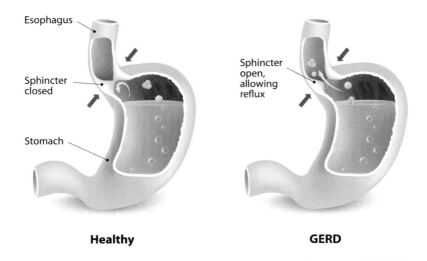

Healthy   **GERD**

FIGURE 11.21 The origin of acid reflux disease. The connection between the esophagus and the stomach is controlled by a sphincter muscle. A closed sphincter confines the acidic contents of the stomach. When the sphincter is open, stomach acid can flow into the esophagus and create a burning sensation.

Although an antacid can provide temporary relief, newer medications for acid reflux *reduce the production of stomach acid at its source.* To understand how these new drugs work, we need to examine how stomach acid is produced. An important molecule in this process is *histamine,* which contains an amine functional group. As we discussed in Chapter 5, histamine is used to transit chemical signals between cells. You probably associate histamine with allergies because it is one of the molecules responsible for triggering an allergic response (e.g., sneezing).

Histamine stimulates the production of stomach acid by binding to a receptor protein located within the outer membrane of cells that line the interior of the stomach. This protein is called the H2 receptor to distinguish it from a similar histamine receptor (H1) that is involved in the allergic response. When histamine binds to its receptor, the interaction triggers a sequence of chemical signals that activates another type of membrane protein called a *proton pump.* As the name suggests, this protein pumps H⁺ ions across the cell membrane, which generates acidity within the stomach. The role of the histamine receptor and proton pump is illustrated in Figure 11.22(a).

One class of drugs reduces the production of stomach acid by interfering with histamine signaling. The drug attaches to the H2 receptor and prevents histamine from binding (Figure 11.22(b)). Because these drugs utilize this mechanism, they are called *H2 receptor blockers.* Figure 11.23 compares the molecular structure of histamine with the structures of two H2 receptor blockers: cimetidine (Tagamet HB) and famotidine (Pepcid AC). Each drug molecule contains a ring of five atoms (a heterocycle), which closely resembles the nitrogen-containing ring in histamine (see Chapter 5). *These drugs mimic the structure of a biological molecule, which is a common strategy in drug design.* Because of their close molecular similarity, these drugs bind to the same H2 receptor that normally binds histamine.

PRACTICE EXERCISE 11.9

Based on other molecules we have examined, predict whether histamine acts as an acid or a base. In addition, predict whether it is strong or weak. Explain your predictions.

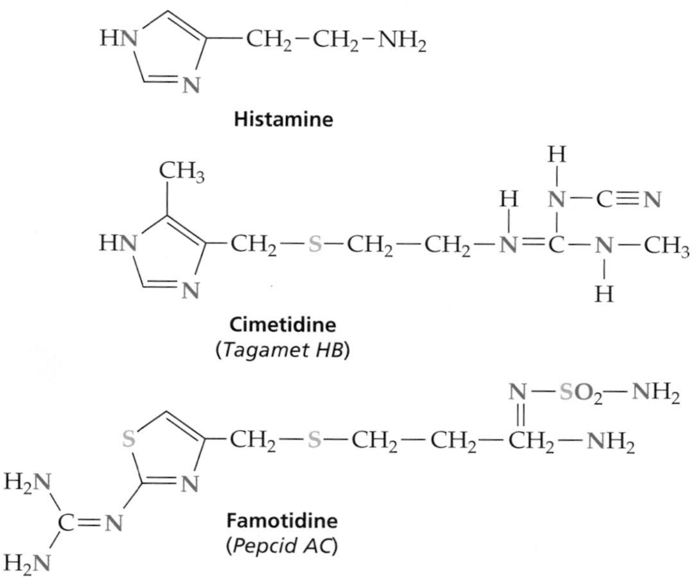

FIGURE 11.23 Drugs that function as histamine receptor blockers. Comparison of histamine's structure with two pharmaceutical analogs that function as H2 receptor blockers.

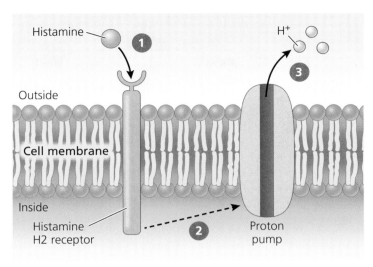

(a) Histamine signaling generates H⁺

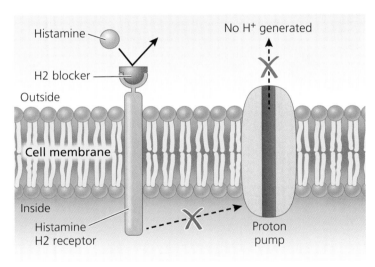

(b) Histamine receptor blocker

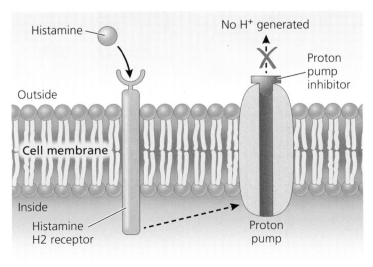

(c) Proton pump inhibitor

FIGURE 11.22 Reducing the production of stomach acid. (a) The first step in the production of stomach acid occurs when histamine binds to its receptor. This binding triggers a sequence of chemical signals to activate a proton pump ②, which then produces H⁺ ions ③. (b) A histamine receptor blocker binds to the H2 histamine receptor and prevents histamine from reaching the receptor. These drugs prevent the transmission of chemical signals to the proton pump. (c) A proton pump inhibitor binds directly to the membrane protein that produces the H⁺ ions, which prevents it from releasing acid into the stomach.

SCIENCE IN ACTION

Don't Try This at Home: Discovering the Cause of Stomach Ulcers

How far would you go to test a scientific hypothesis? In the early 1980s, the Australian physician Barry Marshall (b. 1951; Figure 1) was investigating the cause of stomach ulcers, which are painful sores in the stomach lining. The established medical theory was that ulcers were caused by stress and/or diet, which increased stomach acidity. However, Marshall and his mentor, Robin Warren (b. 1937), discovered that most patients with stomach ulcers were infected with a previously unknown type of bacterium. They cultivated this slow-growing bacterium in a Petri dish, and they named it *Helicobacter pylori* (usually shortened to *H. pylori*). Based on these studies, Marshall and Warren proposed a radically different hypothesis: The cause of stomach ulcers is *H. pylori* infection, not stress or diet.

FIGURE 1 Barry Marshall challenged medical convention. Marshall proposed that infection by *H. pylori* bacteria, not stress and diet, is the cause of stomach inflammation and ulcers. He tested his hypothesis on himself by drinking a culture of the bacteria.

Their idea was almost universally rejected by the medical community. Frustrated by this response, Marshall took a radical step—he drank a culture of *H. pylori* bacteria. Within a few days he became nauseous and started vomiting. After a week, his stomach lining was severely inflamed, and it was possible to extract colonies of *H. pylori* bacteria. After two weeks, Marshall began taking antibiotics to kill the bacteria, and his symptoms rapidly disappeared. Although Marshall did not develop a stomach ulcer (which takes longer), his dramatic demonstration of *H. pylori*'s effects began to turn the tide of medical opinion. Medical science now accepts Marshall's theory that *H. pylori* infection can cause many stomach problems, including inflammation, ulcers, and cancer. In 2005, Marshall and Warren received the Nobel Prize in Physiology or Medicine.

H. pylori is a spiral-shaped bacterium that swims using thin flagella (Figure 2(a)). Most bacteria cannot survive in the highly acidic environment of the stomach, but *H. pylori* has adapted to these conditions. The bacteria use their flagella to burrow through the protective layer of mucus that lines the inside of the stomach, enabling them to reach the less acidic tissue that lies underneath. In addition, *H. pylori* produces large amounts of an enzyme called *urease*, which breaks down urea (present in stomach contents) into carbon dioxide and ammonia. The ammonia acts as a base and neutralizes some of the stomach acid in the bacteria's local environment.

$$\underset{\textbf{Urea}}{H_2N-\overset{\overset{\displaystyle O}{\|}}{C}-NN_2} \; + \; H_2O \; \xrightarrow{\text{Urease enzyme}} \; 2\,NH_3 \; + \; CO_2$$

After *H. pylori* has colonized a region of the stomach lining, the infection triggers an immune response that causes tissue inflammation. This process disrupts the mucus layer, allowing stomach acid to leak through to the sensitive tissue underneath. Over time, the acid erodes the stomach lining and produces a crater-shaped ulcer, as shown in Figure 2(b). If the erosion is deep enough, it can rupture blood vessels and create a bleeding ulcer. Fortunately, new antibiotic therapies are effective at treating *H. pylori* infections.

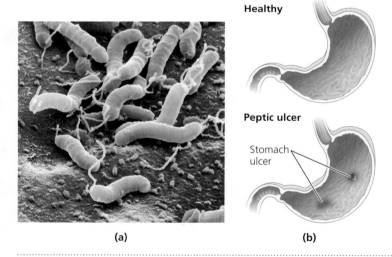

(a)   (b)

FIGURE 2 *H. pylori* bacteria cause stomach ulcers. (a) A magnified microscope image of *H. pylori* bacteria. These bacteria use their thin flagella for swimming and burrowing through the stomach mucus. (b) Infection of the stomach lining by *H. pylori* can lead to the formation of a crater-like ulcer.

The next generation of drugs, called *proton pump inhibitors*, is even more effective at reducing stomach acid. *A proton pump inhibitor binds to the proton pump protein and prevents it from releasing acid into the stomach* (Figure 11.22(c)). Examples of these drugs are Prilosec, Nexium, and Prevacid. Because they target a different biological molecule, these drugs have different types of molecular structures than H2 receptor blockers.

Core Concepts

- Acid reflux disease (GERD) is caused by the flow of stomach acid into the esophagus. This flow occurs when the sphincter muscle that normally constrains the acid within the stomach does not close properly or opens too frequently.
- The production of stomach acid is initiated by the binding of histamine to its H2 receptor protein. This binding triggers a sequence of chemical signals that activates a proton pump to produce H^+ ions.
- Acid reflux drugs employ two possible mechanisms to block the production of stomach acid. Histamine receptor blockers attach to the H2 receptor and prevent histamine binding. Proton pump inhibitors bind directly to the proton pump and prevent it from producing H^+ ions.

FIGURE 11.24 Molecular structure of carbonic acid.

Blood as a Buffer System

If a person is having a panic attack, the recommended procedure is for that person to breathe into a paper bag. Why would this odd suggestion help? To answer this question, we apply our knowledge of acid-base chemistry to the properties of blood.

Most biological molecules are very sensitive to the pH of their surroundings and function best within a narrow pH range. For example, the hemoglobin protein in red blood cells, which transports oxygen from the lungs to the other organs, is very sensitive to changes in acidity. The pH of human blood is 7.4, slightly above neutral. Even deviations as tiny as a few tenths of a pH unit can be dangerous or even fatal. If the blood pH falls below 7.35 (i.e., becomes more acidic), the clinical condition is called *acidosis*. If the pH rises above 7.45, the condition is called *alkalosis* (derived from the term *alkaline*).

To prevent these harmful conditions, the pH of our blood is stabilized by a **buffer**. In general, *a buffer is a mixture of chemical substances in solution that resists changes in pH*. In addition to their important role in cells, buffer mixtures are commonly used in the chemical laboratory to stabilize the pH of a solution.

Human blood is a complex mixture of cells, large molecules, and small molecules. One important contributor to blood chemistry is the gas carbon dioxide (CO_2), which is generated as a product of respiration. Carbon dioxide is soluble in water, as you know from drinking carbonated beverages. Dissolved CO_2 reacts with water to form carbonic acid (H_2CO_3), shown in Figure 11.24.

Dissolving CO_2 in water is a reversible process that can be written as a chemical equation,

$$CO_2(g) \;+\; H_2O(l) \; \rightleftharpoons \; H_2CO_3(aq)$$

Carbon dioxide   Carbonic acid

Carbonic acid partially ionizes in water to produce aqueous hydronium and bicarbonate ions. Using terminology from earlier in the chapter, we note that $H_2CO_3(aq)$ is a *weak acid* and that the bicarbonate ion, $HCO_3^-(aq)$, is the *conjugate base*. The weak acid and its conjugate base differ by only an H^+ ion.

$$H_2CO_3(aq) \;+\; H_2O(l) \; \rightleftharpoons \; H_3O^+(aq) \;+\; HCO_3^-(aq)$$

Carbonic acid   Bicarbonate ion

| Weak acid | | Conjugate base |

One type of buffer solution contains a mixture of a weak acid and its conjugate base. In blood, the mixture of $H_2CO_3(aq)$ and $HCO_3^-(aq)$ functions as a buffer. Figure 11.25 illustrates what happens when a strong acid or base is added to a buffer. For simplicity, we will consider an example when the concentrations of weak acid and conjugate base are equal. However, the concentrations are not equal in human blood.

THE KEY IDEA: A buffer is a mixture of chemical substances in solution that resists changes in pH.

buffer A mixture of chemical substances in solution that resists changes in pH; in most cases, it consists of a weak acid and its conjugate base.

When a strong acid is added to the buffer, the acid generates additional $H_3O^+(aq)$ ions. Many of these $H_3O^+(aq)$ ions will react with the conjugate base, $HCO_3^-(aq)$, which accepts an H^+ ion to generate $H_2CO_3(aq)$. However, carbonic acid is a weak acid that ionizes only partially in solution. Remember that the acidity of a solution arises from H_3O^+ ions and not from the weak acid itself. In effect, the conjugate base acts like an "acid sponge," soaking up the acidic protons from the $H_3O^+(aq)$ ions. A large fraction of the acidic protons then become "locked up" within the molecular structure of H_2CO_3, which reduces the number of free $H_3O^+(aq)$ ions in solution. As a result, the pH of the buffer solution changes only slightly.

Suppose that we add a strong base to the original buffer, which increases the number of $OH^-(aq)$ ions. The extra $OH^-(aq)$ ions react with $H_2CO_3(aq)$ to generate more of the conjugate base, HCO_3^-. In this case, the carbonic acid acts like a "base sponge" to soak up the extra $OH^-(aq)$ ions. As a result, the pH of the buffer solution undergoes only a slight change.

The situation in blood is more complicated, but the same buffer principle applies. In effect, the chemical partnership between $H_2CO_3(aq)$ and $HCO_3^-(aq)$ creates a solution that resists changes in pH. This pH stability is essential for the biological molecules that reside within our blood to function effectively.

PRACTICE EXERCISE 11.10

Solutions A and B each contain a mixture of two substances dissolved in water.

> Solution A: Hydrofluoric acid (HF) and sodium fluoride (NaF).
> Solution B: Nitrous acid (HNO_2) and sodium chloride (NaCl)

Can either or both of these solutions form a buffer? Explain your answers. You can refer to Figure 11.8 for relevant information.

CONCEPT QUESTION 11.4

Panic attacks often cause a person to hyperventilate-that is, exhale excess amounts of CO_2. One common remedy is to have the person inhale and exhale slowly and deeply into a paper bag. What is the purpose of this treatment, and what effect will it have on the person's blood pH value? Explain your answers in terms of the blood's acid–base chemistry.

Core Concepts

- Human blood has a pH of 7.4, and even small changes can result in serious illness or death. As a result, blood pH is stabilized by a buffer.
- A **buffer** is a mixture of chemical substances in solution that resists changes in pH.
- One type of buffer solution contains a mixture of a weak acid and its conjugate base. For example, a mixture of carbonic acid, H_2CO_3, and the bicarbonate ion, HCO_3^-, contributes to the buffering capacity of blood.

FIGURE 11.25 Response of a buffer to addition of a strong acid or base. The initial buffer solution contains equal amounts of $H_2CO_3(aq)$, the weak acid, and $HCO_3^-(aq)$, the conjugate base. Adding a strong acid to the buffer generates extra $H_3O^+(aq)$ ions, which react with HCO_3^- (aq) to create $H_2CO_3(aq)$. The increase in $H_2CO_3(aq)$ is indicated by the larger height of the bar. Conversely, adding a strong base generates extra $OH^-(aq)$ ions, which react with $H_2CO_3(aq)$ to create an increased amount of $HCO_3^-(aq)$. In each case, addition of the strong acid or base changes the pH only slightly.

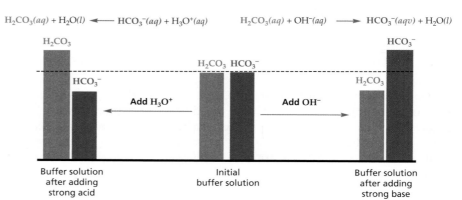

CHAPTER 11
VISUAL SUMMARY

11.1 What Causes Acid Reflux Disease?

Learning Objective:

Explain the origin of acid reflux disease.

⊙ **Acid reflux** is the backflow of stomach acid into the esophagus, where it can cause a painful burning sensation.

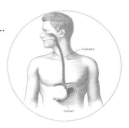

11.2 What Are Acids and Bases?

Learning Objective:

Apply the Brønsted-Lowry theory of acids and bases.

⊙ According to the Brønsted-Lowry theory:

⊙ An **acid** is a substance that can donate a hydrogen ion (H^+) to another substance (e.g., hydrochloric acid).

⊙ A **base** is a substance that can accept an H^+ ion from another substance (e.g., ammonia).

⊙ In a **neutralization** reaction, an acid reacts with a base to produce a **salt** plus water.

⊙ In aqueous solution, neutralization is a chemical reaction between $H^+(aq)$ and $OH^-(aq)$ to form $H_2O(l)$.

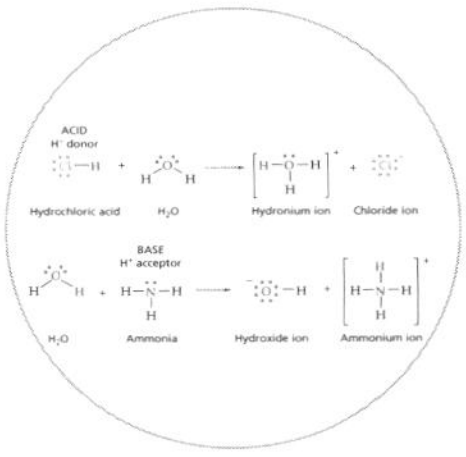

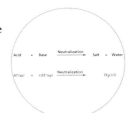

11.3 Relative Strengths of Acids and Bases

Learning Objective:

Compare and contrast strong and weak acids and bases.

⊙ As a very good approximation, a **strong acid** ionizes almost completely (100%) in aqueous solution. By contrast, a **weak acid** ionizes to a small extent.

⊙ HCl is a strong acid. The ionization of HCl in water produces hydronium ions (H_3O^+) and chloride ions (Cl^-). Almost no intact HCl molecules remain.

⊙ Acetic acid, CH_3COOH, is a weak acid. The ionization of CH_3COOH in water is a **reversible reaction** that results in all reactants and products being present in solution This reaction generates a low concentration of $H_3O^+(aq)$ ions.

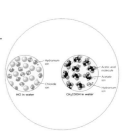

11.4   Measuring Acidity: The pH Scale

Learning Objective:

Use pH as a measurement of acidity.

⊚ The H_2O molecules in liquid water ionize to a very small extent, producing $H_3O^+(aq)$ and $OH^-(aq)$ ions. This is a reversible reaction.

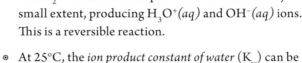

⊚ At 25°C, the *ion product constant of water* (K_w) can be written as

$$K_w = [H_3O^+][OH^-] = 10^{-14}$$

⊚ From this result, we derive that $[H_3O^+] = 10^{-7}$ M and $[OH^-] = 10^{-7}$ M in pure water.

⊚ A **neutral solution** is one in which:

$$[H_3O^+] = [OH^-]$$

⊚ An **acidic solution** is one in which:

$$[H_3O^+] > [OH^-]$$

⊚ A **basic solution** is one in which:

$$[H_3O^+] < [OH^-]$$

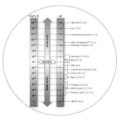

⊚ The **pH scale** is a logarithmic scale used to measure acidity. pH is related to $[H_3O^+]$ by two equations:

$$[H_3O^+] = 10^{-pH} \text{ and } pH = -\log_{10}[H_3O^+]$$

⊚ The pH can be used to classify a solution:

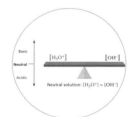

  ⊚ pH = 7 for pure water and neutral solutions.
  ⊚ pH < 7 for acidic solutions.
  ⊚ pH > 7 for basic solutions.

⊚ Changing $[H_3O^+]$ by a factor of 10 changes the pH by one unit.

11.5   Biological Applications of Acid-Base Chemistry

Learning Objective:

Illustrate two biological applications of acid-base chemistry.

⊚ Acid reflux disease occurs when the sphincter muscle at the top of the stomach does not close properly, which allows stomach acid to flow back into the esophagus.

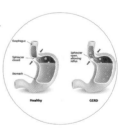

⊚ Acid reflux drugs reduce the production of stomach acid. One class of drugs blocks the histamine receptor, and another class inhibits the proton pump protein.

⊚ The pH of blood is stabilized to avoid harmful consequences.

⊚ A **buffer** is a mixture of chemical substances in solution that resists changes in pH. One type of buffer solution contains a mixture of a weak acid and its conjugate base.

⊚ One contribution to blood's buffering capacity comes from dissolved CO_2 gas, which produces carbonic acid (H_2CO_3), a weak acid. The acid ionizes to generate a bicarbonate ion (HCO_3^-), which is its conjugate base.

LEARNING RESOURCES

Reviewing Knowledge

11.1: What Causes Acid Reflux Disease?

1. Describe the cause and symptoms of acid reflux disease.

11.2: What Are Acids and Bases?

2. How are acids and bases defined according to the Brønsted-Lowry theory?
3. Write a chemical equation to show how hydrochloric acid (HCl) behaves as an acid when it is dissolved in water.
4. Write the chemical formula of the hydronium ion, and draw its three-dimensional structure.
5. Describe what happens when sodium hydroxide (NaOH) dissolves in water. What chemical substance acts as a base?
6. Use the Brønsted-Lowry theory to explain why aqueous ammonia acts as a base.
7. Write the chemical equation for the neutralization reaction between HCl(aq) and NaOH(aq). What are the spectator ions in this reaction, and why is this term used to describe them?

11.3: Relative Strengths of Acids and Bases

8. What is the difference between a strong and a weak acid?
9. Write a chemical equation for the ionization of acetic acid (CH_3COOH) in water. Why is it called a reversible reaction?
10. What is the relationship between the two components of a conjugate acid-base pair? Which chemical substances make up the conjugate acid-base pair for the ionization of acetic acid?
11. Why does a 0.1 M solution of HCl exhibit stronger acidic properties than a 0.1 M solution of CH_3COOH?

11.4: Measuring Acidity: The pH Scale

12. Write a chemical equation to show how H_2O molecules ionize in liquid water.
13. Use the ion product constant of water to derive the molar concentrations of H_3O^+ and OH^- in pure water.
14. Define a neutral solution, an acidic solution, and a basic solution in terms of $[H_3O^+]$ and $[OH^-]$.
15. Write two equations to show the relationship between pH and $[H_3O^+]$.
16. Use pH to define a neutral solution, an acidic solution, and a basic solution.

11.5: Biological Applications of Acid-Base Chemistry

17. What is the biological role of stomach acid?
18. Describe the role of histamine in the production of stomach acid.
19. Modern pharmaceutical treatments for acid reflux disease include histamine receptor blockers and proton pump inhibitors. Describe how each type of drug works.

20. What is the pH of human blood? What are acidosis and alkalosis?
21. Write chemical equations for (a) the formation of carbonic acid in blood and (b) the ionization of the carbonic acid. (c) How do these equations explain why blood is a buffer?

Developing Skills

22. The chemical reaction between hydrobromic acid (HBr) and H_2O can be written as.

$$HBr(aq) + H_2O(l) \rightarrow H_3O^+(aq) + Br^-(aq)$$

Use the Brønsted-Lowry theory to identify the acid and base within the reactants.

23. Strong acids neutralize strong bases to produce a salt plus water. Write the neutralization reaction for nitric acid, HNO3(aq), and sodium hydroxide, NaOH(aq).
24. Aluminum hydroxide is only slightly soluble in water, but the fraction that dissolves acts as a base.
    (a) Write the chemical formula for aluminum hydroxide. Refer to Chapter 3 for the rules for writing the formulas of ionic compounds.
    (b) Write a balanced chemical equation for the neutralization reaction between dissolved aluminum hydroxide and hydrochloric acid (HCl).
25. (a) Write a chemical equation for the neutralization reaction between carbonic acid, $H_2CO_3(aq)$, and sodium hydroxide, NaOH(aq). (b) Identify the spectator ions in this reaction.
26. Classify the following solutions as acidic, basic, or neutral:
    (a) $[H_3O^+] = 10^{-6}$ M
    (b) $[H_3O^+] = 10^{-12}$ M
    (c) $[OH^-] = 10^{-4}$ M
    (d) $[H_3O^+] = 1.0$ M
    (e) $[OH^-] = 0.1$ M
27. For solutions with the following pH values, calculate the molar concentration of $H_3O^+(aq)$ ions and $OH^-(aq)$ ions:
    (a) pH = 10
    (b) pH = 8
    (c) pH = 6
    (d) pH = 3
    (e) pH = 0
28. You are given a solution of acid with a pH value of 4. If you dilute the acid with water by a factor of 100, what will be the new pH value?
29. Calculate the pH of a solution of 10^{-5} M NaOH.
30. Water ionizes to produce both hydronium and hydroxide ions. Is water a weak acid? Is it a weak base? Or is it both? Explain your answer.
31. Consider a buffer solution that contains acetic acid (CH_3COOH), H_3O^+ ions, and acetate ions (CH_3COO^-). If you add a small amount of a strong base such as NaOH to this solution, what changes will occur in each of the three chemical substances?

32. The antacid sold as TUMS contains calcium carbonate, $CaCO_3$. When hydrochloric acid (HCl) is added to $CaCO_3$, the chemical reaction releases a gas. (a) Write a chemical equation for this reaction, and identify the gas that is released. (b) What type of chemical reaction is this?

33. Compare two solution mixtures, one containing 1 mM each of weak acid and conjugate base and the other 1 M each of the two components. Which solution do you expect to have the greater capacity to buffer pH? Why?

34. The pH scale measures acidity using a logarithmic scale. The accompanying table gives the first two data points of the relationship between $[H_3O^+]$ and pH.
    (a) Complete the table for the remaining values of $[H_3O^+]$. Enter the pH value using three significant figures.
    (b) Use all the data points to plot a graph showing how pH varies with $[H_3O^+]$. Use $[H_3O^+]$ as the horizontal axis and pH as the vertical axis.
    (c) Explain how the graph you have plotted differs from a linear graph (i.e., a straight line).

$[H_3O^+]$	pH
1×10^{-4} M	4.00
1×10^{-4} M	3.70
3×10^{-4} M	
4×10^{-4} M	
5×10^{-4} M	
6×10^{-4} M	
7×10^{-4} M	
8×10^{-4} M	
9×10^{-4} M	
1×10^{-3} M	

35. Consider four solutions: (a) 1.0 M NH_3, (b) 1.0 M NaOH, (c) 1.0 M HCl, and (d) 1.0 M CH_3COOH. Which solution has the highest pH? Which has the lowest pH?

36. You are given two mixtures with the same concentrations of an acid and a base:
    $$HCl(aq) + NaOH(aq)$$
    $$H_2CO_3(aq) + NaHCO_3(aq)$$
    Will either mixture act as a buffer? Explain why or why not.

37. Coal naturally contains a small amount of sulfur. Burning coal is one cause of acid rain. Write balanced chemical equations for the following reactions:
    (a) Sulfur atoms react with oxygen gas to produce sulfur dioxide (SO_2).
    (b) Sulfur dioxide reacts with oxygen gas to produce sulfur trioxide (SO_3).
    (c) Sulfur trioxide reacts with water to form an acid.
    (d) What is the name of the acid that causes the acid rain?

38. Sourdough bread is traditionally made in San Francisco. The sour taste of this bread comes from the presence of bacteria in the mix that is used to prepare the dough. These bacteria produce acetic acid and lactic acid. The pH of a 0.1 M solution of acetic acid is close to 3, and the pH of the same concentration of lactic acid is around 2.4. Which of the two acids is stronger? Explain your answer.

Exploring Concepts

39. The ion-product constant K_w measures the equilibrium ionization of water. However, the equilibrium point for this ionization varies with temperature. We learned that $K_w = 10^{-14}$ at room temperature (25°C). However, K_w becomes lower as the temperature increases. How will the pH of a neutral solution change as the temperature increases?

40. Some salts change the solution pH when they dissolve in water. For example, a certain solution of sodium acetate has pH = 9. What is the $[H_3O^+]$ for this solution? Why do you think the sodium acetate changes the pH from that of pure water? Use the Internet to investigate "hydrolysis," and use this concept as the basis for your explanation.

41. If a person hyperventilates, he or she could develop respiratory alkalosis. Under what conditions could a person develop acidosis?

42. Section 11.5 describes two classes of acid reflux drugs: histamine receptor blockers and proton pump inhibitors. Select one of these drugs, and use the Internet to investigate its effectiveness and side effects.

43. As part of a science fair project, a student compared the ability of two tablets to neutralize a sample of dilute HCl that was similar in concentration to that in stomach acid. She found that TUMS (calcium carbonate) was more potent than Tagamet (cimetidine). Is she justified in her conclusion that TUMS is a more effective antacid than Tagamet? Explain why or why not.

44. Oven cleaners contain high concentrations of sodium hydroxide (NaOH). These cleaners work by a process called *saponification*. Use the Internet to investigate this term, and write a one-paragraph chemical description for why NaOH is an effective oven cleaner.

45. Use the Internet to investigate two proton pump inhibitor drugs: omeprazole (Prilosec) and lansoprazole (Prevacid). Draw the molecular structure of these drugs, and write a one-paragraph description of how they block the action of the proton pump.

46. Gardeners know that the color of hydrangea plants depends on the pH of the soil in which they grow. The relationship between pH and plant color is shown here. Use the Internet to discover which molecules in hydrangea plants are responsible for this sensitivity to pH. Draw the molecular structure of one of these molecules, and provide its name.

DEEP BLUE			PURPLE -PINK			DEEP PINK
4.5	5	5.5	6	6.5	6.8	7

47. Many bacteria cannot grow in acidic solutions. A product called "sour salt" is used to preserve canned tomatoes, for example. Sour salt is not a salt at all. Rather, it consists of crystals of citric acid, an ingredient of citrus fruits. The proteins that speed up biological reactions are sensitive to pH, and they often can be inhibited by low pH. Citric acid prevents the browning of apples and potatoes, for example. Use the Internet to identify the enzyme involved in the browning reaction. How is its activity blocked by low pH?

12

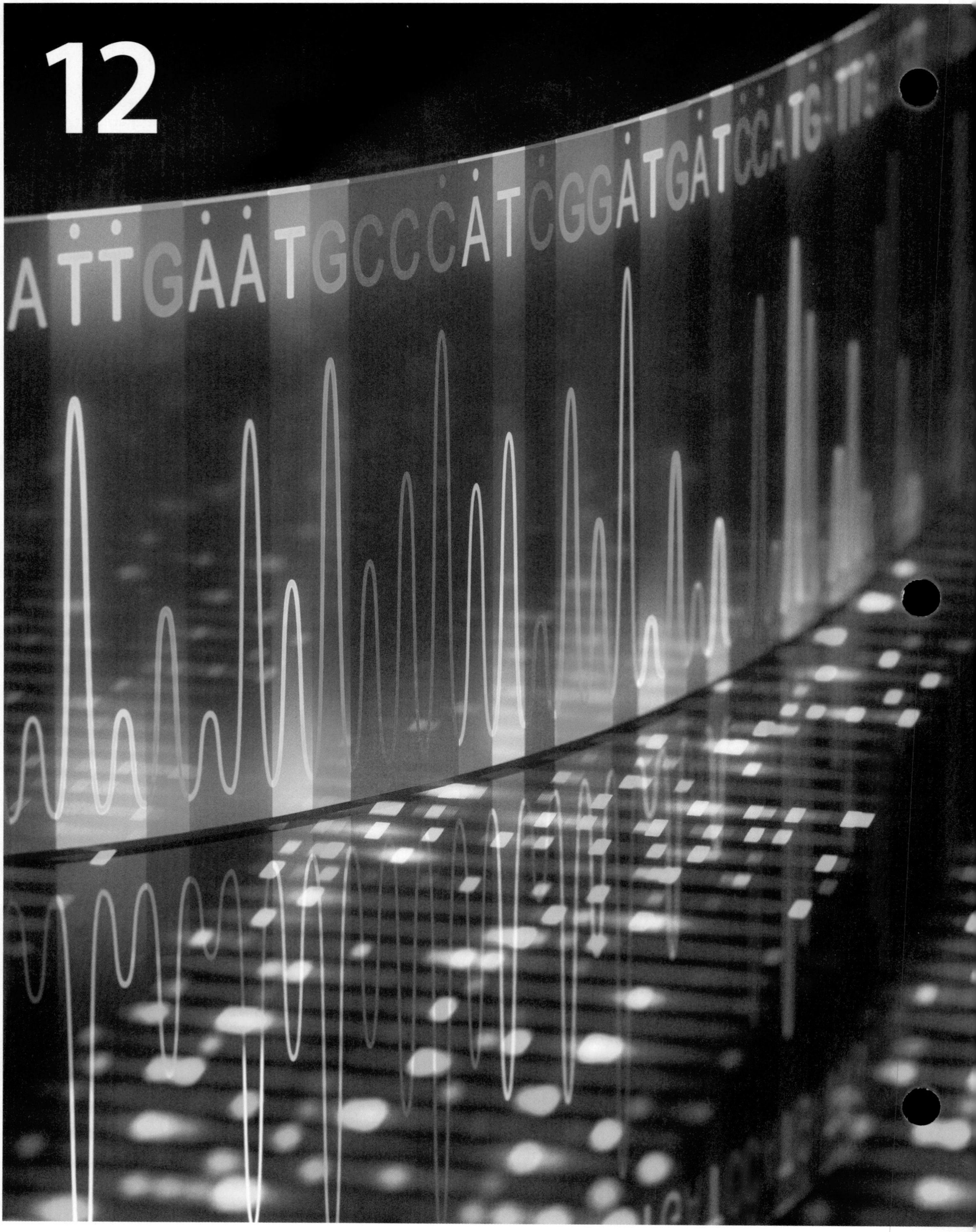

DNA: The Molecule of Heredity

D NA is the most famous biological molecule. It contains the genetic information that we inherit from our parents, and it also contains directions to make proteins that perform essential biological tasks within our body. The shape of DNA is a double helix, consisting of two spirals that twist around each other. How was the double helix discovered, and how is DNA's structure related to its biological function?

12.1 How Was the DNA Double Helix Discovered?

Learning Objective:

Explain why the DNA double helix was an important scientific discovery.

THE KEY IDEA: All cell-based organisms use DNA to store genetic information.

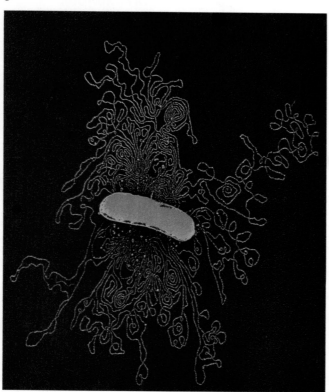

FIGURE 12.1 DNA is the molecule of heredity for all cell-based organisms. In this photograph, DNA bursts from a treated *E. coli* bacterial cell (center).

Every cell-based organism on Earth contains **DNA**, which is an abbreviation for <u>d</u>eoxyribo<u>n</u>ucleic <u>a</u>cid. Figure 12.1 shows DNA spilling out of an *E. coli* bacterial cell. What distinguishes bacteria from humans is the different genetic information stored within the millions or billions of chemical building blocks contained within DNA. Scientists have determined the entire sequence of these building blocks in human DNA, which is called the *human genome.* They have also identified many genetic diseases—including various types of cancer—that arise from changes, or *mutations,* in DNA. Medical researchers are now investigating how small variations in our DNA can influence our susceptibility to diabetes and heart disease, as well as our varying responses to pharmaceuticals. We are entering a new era of medicine that will be based on connections between our DNA and our health.

None of these modern advances in science would be possible without knowing the molecular structure of DNA. The famous DNA double helix was discovered more than 60 years ago by James Watson (b. 1928) and Francis Crick (1916–2004), who are pictured together in Figure 12.2. When Watson and Crick first met in the Cavendish Laboratory of Cambridge University, they could hardly have been less alike. Watson was a child prodigy: He entered the University of Chicago at the age of 15, and he earned his Ph.D. in zoology at age 22. When Watson arrived in Cambridge in 1951, following his postdoctoral studies, he was less than 25 years old. In contrast, Francis Crick was 12 years older, and he was still pursuing his Ph.D. research after working for the British Navy during World War II. Crick was very bright, but he was also brash and unfocused.

Despite their different backgrounds, Watson and Crick instantly clicked. Their scientific skills were also complementary. Watson, the biologist, was obsessed with discovering the structure of DNA, which he believed to be the "secret of life." Crick, who was trained in physics, had technical and mathematical skills that Watson lacked. Their work together became one of the most famous collaborations in science. In 1953, Watson and Crick integrated several strands of scientific evidence to deduce the double-helix structure of DNA. In a short publication the same year, they described the shape of DNA and hinted at a mechanism by which DNA can copy itself. This paper sparked a revolution in science, enabling scientists to explain biological processes in terms of the structure and function of molecules. The legacy of Watson and Crick's discovery remains with us today in the form of modern genetic science and medicine.

The structure of DNA provides a valuable case study of scientific discovery. This chapter examines the scientific techniques, evidence, and insights that this breakthrough required. In the next chapter, we examine how the genetic information stored in DNA provides instructions for making proteins.

FIGURE 12.2 A famous partnership. Francis Crick (left) and James Watson at Cambridge University.

Core Concepts

■ All biological organisms use **DNA** (deoxyribonucleic acid) to store genetic information.

■ James Watson and Francis Crick deduced that the structure of DNA is a double helix. This breakthrough provides a case study of scientific discovery.

12.2 Nucleotides: The Building Blocks of DNA

The introduction to this chapter stressed the biological importance of DNA. However, we should remember that DNA is a molecule that can be examined using the principles of chemistry. It is possible to extract DNA from cells using simple chemical procedures, which produces a white precipitate of string-like molecules (Figure 12.3).

DNA is a very long molecule that is an example of a *biological polymer*. In Chapter 7, we explained how a polymer chain can be constructed by the repeated addition of small monomer units. For DNA, the monomer is called a *nucleotide*, and the polymer is a *polynucleotide*. A DNA double helix consists of two polynucleotide chains that are twisted around each other. This section provides an overview of their molecular structures.

Molecular Structure of a Nucleotide

Nucleotides serve as the molecular building blocks of DNA. A nucleotide has three components: a *sugar*, a *phosphate group*, and a *base*. The sugar is called **deoxyribose**; the prefix *deoxy-*indicates that the molecule has one fewer oxygen atom than a related sugar called ribose. We introduced the structure of the phosphate group in Chapter 5; it is composed of a phosphorus atom bonded to four oxygen atoms. The **nucleotide base** exists in four varieties, which we examine in the next section.

Figure 12.4 shows the molecular structure of deoxyribose. The molecule contains a five-membered ring with four carbon atoms and one oxygen atom. These five atoms lie in a plane that is perpendicular to the paper. A fifth carbon is not in the ring; instead, it extends above its plane. The carbon atoms in deoxyribose are labeled 1′, 2′, 3′, 4′, and 5′, which are pronounced "one prime, "two prime," and so on. We use primes (′) because the numbers 1, 2, and so on are reserved for labeling the carbon atoms in the DNA bases. These labels make it easier to refer to specific carbon atoms in the molecule. For example, we say that the 5′-carbon extends above the ring and that the 2′-carbon is where deoxyribose is missing an oxygen atom compared to ribose.

Learning Objective:

Analyze the structures of nucleotides and polynucleotides.

DNA Deoxyribonucleic acid.

THE KEY IDEA: Nucleotides are composed of three components: a deoxyribose sugar, a phosphate group, and a base.

nucleotide The molecular building block of DNA; it consists of a deoxyribose sugar, a phosphate group, and a nitrogen-containing base.

deoxyribose The sugar contained in a DNA nucleotide.

nucleotide base The nitrogen-containing base in in a nucleotide; it exists in four varieties: adenine, thymine, guanine, and cytosine.

FIGURE 12.3 A DNA precipitate. DNA can be extracted from cells, producing a white precipitate that can be pulled out of the solution by a metal wire.

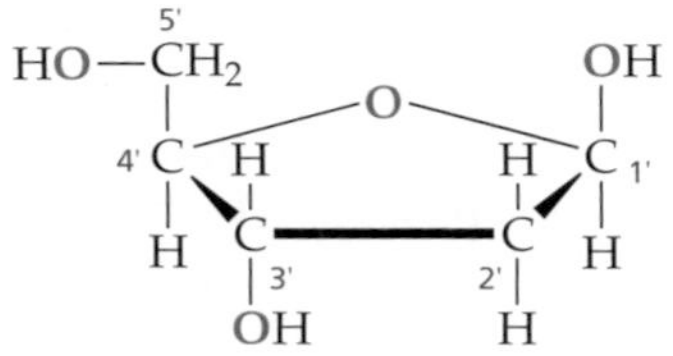

HO—CH₂ ... OH (deoxyribose structure)

FIGURE 12.4 Molecular structure of deoxyribose. The carbon atoms in this sugar molecule are labeled from 1′ to 5′. The solid wedges and thick line indicate that the 2′ and 3′ carbon atoms are closer to the viewer than the other atoms in the ring are. The –OH group on the 5′-carbon is written "backwards" (HO–) to indicate a carbon–oxygen bond.

Figure 12.5 illustrates the molecular structure of a complete nucleotide. The base is attached to the 1′-carbon of deoxyribose, and the phosphate is attached to the 5′-carbon. All nucleotides contain the same deoxyribose sugar and phosphate components. Variation in the base produces four possible nucleotides, and this small set of molecules is all that is required to construct a DNA double helix.

Core Concepts

- **Nucleotides** are the molecular building blocks of DNA.
- A nucleotide contains three components: a **deoxyribose** sugar, a phosphate, and a **nucleotide base**. The phosphate is attached to the 5′-carbon of deoxyribose, and the base is attached to the 1′-carbon.

🔑 **THE KEY IDEA:** DNA nucleotides contain four bases: adenine, thymine, guanine, and cytosine.

- **adenine, thymine, guanine, and cytosine** The four nucleotide bases in DNA.

- **pyrimidine** A nitrogen-containing heterocycle comprised of a single ring; it serves as the foundation for cytosine and thymine.

- **purine** A nitrogen-containing heterocycle comprised of two fused rings; it serve as the foundation for adenine and guanine.

The Four Varieties of DNA Bases

The four varieties of DNA base are called **adenine, thymine, guanine,** and **cytosine** (often abbreviated as A, T, G, and C, respectively). Despite the vast diversity of life on Earth, *all biological organisms use the same four DNA bases.* We use the term *base* because these molecules have the chemical properties of bases in acid–base reactions. In terms of the molecular structure, the bases are classified as *heterocycles.* As we discussed in Chapter 5, a heterocycle is a cyclic molecular structure in which at least one atom is a *heteroatom* that differs from carbon and hydrogen. All of the nucleotide bases contain multiple nitrogen atoms, so they are also called *nitrogenous bases.*

As illustrated in Figure 12.6, two of the bases—cytosine and thymine—are derived from a type of heterocycle called a **pyrimidine**, which contains one ring of atoms. In contrast, guanine and adenine are derived from a different heterocycle called a **purine**. A purine is a larger molecule composed of two rings of atoms that are connected along one carbon—carbon bond; these are called *fused rings.* Remember that the longer name ("pyrimidine") is associated with the smaller base, and vice versa.

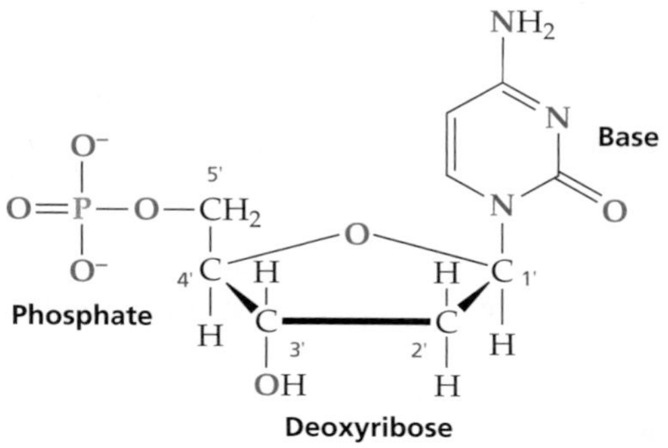

FIGURE 12.5 Molecular structure of a nucleotide. A DNA nucleotide contains a deoxyribose sugar, a phosphate, and a nucleotide base. The base is attached to the 1′-carbon of deoxyribose, and the phosphate is attached to the 5′-carbon.

Pyrimidine DNA bases

Pyrimidine

NH₂ → Cytosine

O → Thymine

Purine DNA bases

Purine

O → Guanine

NH₂ → Adenine

FIGURE 12.6 The four types of DNA base. These bases are derived from nitrogen-containing heterocycles. Cytosine and thymine are pyrimidine bases, whereas guanine and adenine are purine bases.

WORKED EXAMPLE 12.1

Question: For the two pyrimidine bases, cytosine and thymine, circle the regions within each molecule that differ from the original pyrimidine heterocycle.

Answer: The regions are circled in the structures shown below.

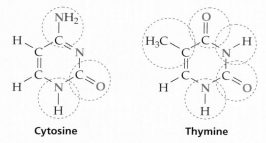

Cytosine   Thymine

TRY IT YOURSELF 12.1

Question: For the two purine bases, guanine and adenine, circle the regions within each molecule that differ from the original purine heterocycle.

PRACTICE EXERCISE 12.1

Gout is a medical condition caused by excess *uric acid*. A high level of uric acid in the blood leads to the formation of solid crystals within the joints of the feet, which causes redness, swelling, and severe pain. Based on the structure of uric acid shown here, what type of base is used as the source for making this molecule—purine or pyrimidine? Explain your answer.

Uric acid

Core Concepts

- The four DNA bases are called **adenine** (A), **thymine** (T), **guanine** (G), and **cytosine** (C).
- All of the DNA bases are derived from nitrogen-containing heterocycles. Thymine and cytosine are **pyrimidine** bases, whereas guanine and adenine are **purine** bases.

Making a Polynucleotide Chain

DNA consists of two **polynucleotides**, which are long chains constructed from the addition of thousands or millions of nucleotides. We also use the term *DNA strand* to refer to a long sequence of attached nucleotides. To begin, we examine how two nucleotides are joined together (Figure 12.7). The phosphate group on the lower nucleotide attaches to the deoxyribose sugar on the upper nucleotide. To be specific, a new connection forms between the phosphate and the O atom that is connected to the 3′-carbon of deoxyribose. This new chemical

 THE KEY IDEA: Nucleotides can be joined together to make long polynucleotide chains.

polynucleotide A polymer containing many linked nucleotides.

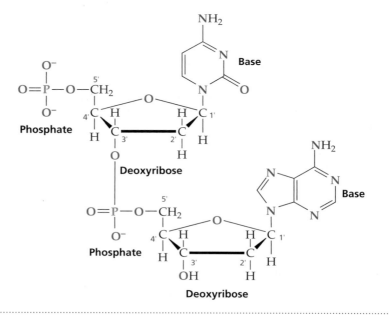

FIGURE 12.7 **Linking two nucleotides.** The phosphate group of the lower nucleotide attaches to the deoxyribose sugar of the upper nucleotide. This connection occurs through the O atom that is attached to the 3′-carbon of deoxyribose. The complete chemical linkage is called a phosphodiester bond.

linkage between the two deoxyribose molecules is called a *phosphodiester bond* (see Chapter 5). We use this term because the chemical linkage resembles an ester group but with two modifications. First, the carbon atom in the ester is replaced by a phosphorus atom (*phospho-*). Second, there are two bridging oxygen atoms instead of one (*-diester*).

This process of joining nucleotides can be repeated to create a polynucleotide chain. Figure 12.8 shows a short polynucleotide containing four nucleotides,

Phosphodiester bond

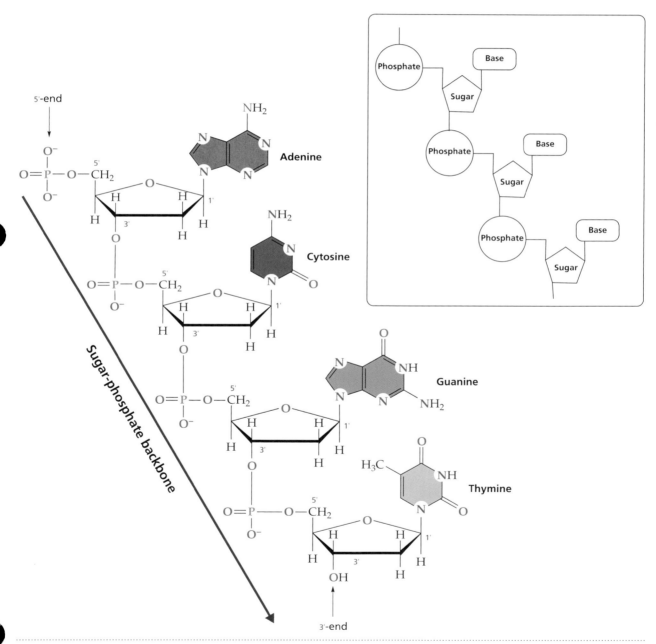

FIGURE 12.8 The molecular structure of a polynucleotide chain. The diagram shows a short polypeptide chain made from joining four different nucleotides. The two ends of the chain are labeled as the 5′-end and the 3′-end. The sugar-phosphate backbone is an alternating arrangement of sugar and phosphate groups that connects the nucleotides and supports the bases. The sequence of bases (in the 5′-to-3′ direction) encodes genetic information within DNA. The box shows a simplified representation of the components of a polynucleotide chain.

5′-end The end of the polynucleotide at which a phosphate group is attached to the 5′-carbon of deoxyribose.

3′-end The end of the polynucleotide at which an OH group is attached to the 3′-carbon of deoxyribose.

directionality The orientation of a polynucleotide strand; it is defined in the 5′-to-3′ direction.

sugar-phosphate backbone An alternating sequence of sugars and phosphates; it links the nucleotides in a polynucleotide chain and supports the DNA bases.

each of which has a different DNA base. The two ends of the polynucleotide chain are defined in terms of which component of the nucleotide is left *unattached*. Examine the top of the chain in Figure 12.8, and you will see that no other chemical groups are attached to the phosphate that is connected to the 5′-carbon. For this reason, this end of the polynucleotide is called the **5′-end**. At the other end, the —OH group on the 3′-carbon is not attached to anything else; therefore, it is called the **3′-end**. This is analogous to the two distinct ends of a pencil, with a writing point at one end and an eraser at the other end.

The distinct structures of the 3′-end and 5′-end enable us to define the **directionality** of a polynucleotide strand. This is similar to defining the direction in which a pencil is pointing. *The directionality of a polynucleotide strand is defined in the 5′-to-3′ direction;* it is often written with an arrow as 5′ → 3′. This orientation is chosen because the synthesis of a polynucleotide chain involves the addition of new nucleotides to the 3′-end, so the chain always lengthens in the 5′ → 3′ direction. DNA base sequences are written in the 5′ → 3′ direction; for example, the DNA sequence in Figure 12.8 is written as 5′- A C G T -3′. The sequence of bases in a polynucleotide strand encodes the genetic information that is stored within DNA.

Use your finger to trace a continuous line from the 5′-end to the 3′-end. Your finger will pass over an alternating sequence of sugar—phosphate— sugar—phosphate— and so on. This component of the polynucleotide is called the **sugar-phosphate backbone.** It links the nucleotides, and it provides the structural framework to support the DNA bases. The sugar-phosphate backbone remains the same throughout the entire polynucleotide chain, regardless of the specific bases attached to the deoxyribose sugars.

The chemistry of the phosphate groups makes DNA an acid (recall the meaning of "A" in DNA). Within a cell, the phosphate group donates an H^+ ion and thereby becomes ionized with a net negative charge. This chemistry has an important influence on the structure of DNA. The double helix is stabilized by positive ions (cations) such as sodium (Na^+) and magnesium (Mg^{2+}), which neutralize the negative charges on the phosphate groups. Without the presence of these cations, DNA would be unable to maintain its structure due to the repulsive effect of the phosphate's negative charges.

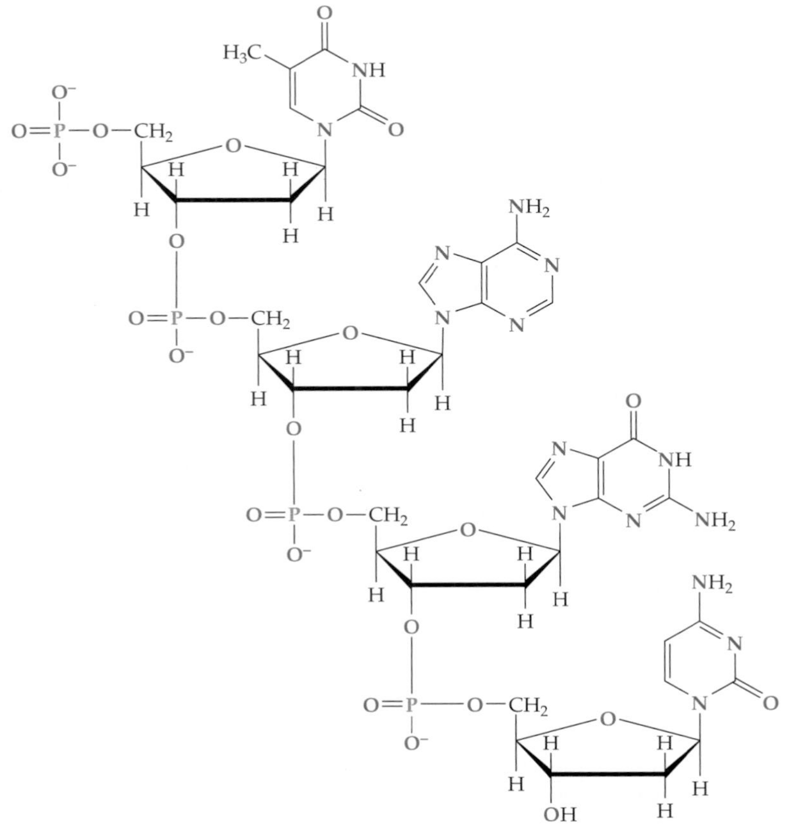

PRACTICE EXERCISE 12.2

The adjacent diagram shows a short region of a polynucleotide chain.

(a) For one deoxyribose sugar, label the carbon atoms from 1′ to 5′.

(b) Identify and label the 5′-end and the 3′-end of the polynucleotide chain.

(c) Identify the name of each base, and classify the base as a purine or a pyrimidine.

(d) Write the sequence of DNA bases for this region of the polynucleotide chain.

Core Concepts

- Two nucleotides are joined together by a chemical reaction between the phosphate group of one nucleotide and the —OH group on the 3'-carbon of a neighboring nucleotide.
- A **polynucleotide** chain is formed by joining multiple nucleotides. The chain has two distinct ends, the **5'-end** and the **3'-end**.
- The **directionality** of a polynucleotide is defined in the 5' → 3' direction. Genetic information is encoded as a sequence of DNA bases that is read in the 5' → 3' direction
- The **sugar-phosphate backbone** links the nucleotides and provides a structural support for the DNA bases, which are attached to the deoxyribose sugars.

12.3 Discovering the Double Helix

This section provides a brief history of the discovery of the DNA double helix. This account illustrates how scientific discovery works in practice, which is often more convoluted than the simplified steps presented in the scientific method (see Chapter 1). We will focus on the scientific techniques and evidence that were necessary to deduce the double-helix structure of DNA. We will also introduce the cast of characters who contributed to this groundbreaking discovery.

Deducing the structure of DNA provides an example of how scientists approach the study of biological molecules. This discovery emerged from a convergence of scientific perspectives derived from three disciplines—biology, chemistry, and physics—all of which were necessary to solve the puzzle (Figure 12.9). This multidisciplinary perspective is widely utilized in modern biomolecular science.

Biology: DNA Is the Molecule of Heredity

During the first half of the 20th century, many scientists were convinced that DNA was not capable of storing genetic information. One reason for this belief was that isolating DNA from cells often caused the molecule to break down into small fragments. In addition, scientists thought that the chemical structure of DNA, a polynucleotide with only four different bases, was too simple to encode the complex genetic instructions for a living organism. Instead, many scientists believed that protein molecules—which are more diverse than DNA in form and function—were the molecules of heredity.

Learning Objective:

Characterize the scientific methods and evidence that led to the discovery of the DNA double helix.

THE KEY IDEA: Experiments on two strains of bacteria demonstrated that DNA, not protein, is the molecule of heredity.

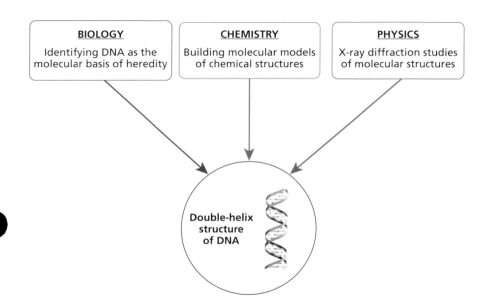

FIGURE 12.9 **Pathways of scientific discovery.** Three scientific disciplines (biology, chemistry, and physics) were all crucial to deducing the double-helix structure of DNA.

FIGURE 12.10   A microscope image of two strains of *Pneumococcus* bacteria. The rough (R) strain on the left is harmless, whereas the smooth (S) strain causes pneumonia. The R strain can be transformed into the smooth strain.

An important clue about DNA's biological function came from three scientists, led by Oswald Avery (1877–1955), who worked at the Rockefeller Institute for Medical Research in New York City. In the 1940s, these scientists were investigating two strains of the *Pneumococcus* bacterium, the organism that causes pneumonia infections. These strains are called rough (R) and smooth (S), based on how they appear under the microscope (Figure 12.10). The R strain is harmless when injected into test animals, but the S strain causes pneumonia. Interestingly, it is possible to transform the harmless R strain into the infectious S strain. How does this happen?

Scientists hypothesized that the change was caused by some substance—called the "transforming principle"—that was transmitted from the S strain into the R strain. Avery and his colleagues wanted to discover the identity of the transforming principle. They used different enzymes to break down the structures of different types of biological molecules, such as proteins, fats, and DNA. Destroying the proteins and fats did not prevent the transformation. However, destroying the DNA stopped the transformation from occurring. To confirm these results, the team purified the intact DNA from the S strain and discovered that it was capable of changing the R strain into the S strain. Based on these results, they proposed that DNA must be the "transforming principle."

These results convinced the young James Watson that DNA held the "secret of life." However, not everyone was persuaded by the experiments. In fact, some skeptics believed that the observations were caused by contamination of the experiment. At the beginning of the 1950s, a significant number of scientists continued to believe that proteins, and not DNA, held the key to understanding genetic information.

Core Concepts

- *Pneumococcus* bacteria can exist as two strains: The rough (R) strain is harmless, whereas the smooth (S) strain causes pneumonia.
- Oswald Avery and his colleagues identified DNA as the "transforming principle" that changed the R strain into the S strain.

THE KEY IDEA: Molecular models enable scientists to visualize the structure of biological molecules.

Chemistry: Building Molecular Models

Chemistry played an essential role in the discovery of DNA structure. In the 1950s, the dominant figure in chemistry was Linus Pauling (1901–2004), a professor at the California Institute of Technology (Caltech). Early in his career, Pauling determined the molecular structures of amino acids and small fragments of proteins. As a result of this work, he realized that scientists could obtain important chemical insights from building *molecular models* that accurately represent the precise structural arrangement of atoms. As a groundbreaking example of this approach, Pauling constructed a model of one type of structure found in fiber-like proteins such as keratin, which is an important structural component of skin, hair, and nails. He designed the structure as a twisted arrangement of atoms that he called the *alpha helix*, or *a-helix*. This helical structure later proved to be a fundamental component of many proteins. Figure 12.11 shows Pauling posing next to a molecular model of the α-helix.

Watson and Crick were greatly inspired by the success of Pauling's model-building method. They asked a machinist to construct a set of metal models that resembled the chemical components of DNA. Following in Pauling's footsteps, they attempted to piece together the models in a three-dimensional molecular jigsaw. Unfortunately, Watson and Crick's first model of DNA was an embarrassing failure. They proposed that DNA was a *triple helix* consisting of three polynucleotide chains. Because of their limited knowledge of chemistry, Watson and Crick placed the phosphates on the inside of the helix and the bases on the outside. They mistakenly thought that (Mg^{2+}) ions would neutralize the negative charges on the phosphate groups. In addition, they failed to account for the fact that DNA must be surrounded by a large amount of water because it is located within the aqueous environment of the cell. Their superior at Cavendish Laboratory, Sir Lawrence Bragg (1890–1971), was so humiliated by this scientific fiasco that he banned Watson and Crick from conducting any further work on DNA. The failure of this first DNA model illustrates how scientific research often involves missteps and mistakes.

Core Concepts

- Linus Pauling pioneered the use of molecular models to represent molecular structures. His success with applying these models to proteins inspired Watson and Crick to utilize this method to deduce the structure of DNA.

FIGURE 12.11 Linus Pauling pioneered the use of molecular models. Linus Pauling, the 20th century's most influential chemist, points to his molecular model of the α-helix, which is an important structural element in proteins.

Physics: X-ray Diffraction Studies of Molecular Structures

How can scientists determine the structure of molecules that are too small to be observed directly, even with the most powerful microscope? In the early years of the 20th century, physicists discovered that they could use a beam of X-rays to probe the atomic structure of crystals, in which individual atoms are fixed in a regular and repeating arrangement. The electrons surrounding the atoms scatter the incoming X-rays, which are then deflected onto a photographic plate. This process is called *X-ray diffraction* because it is similar to the diffraction of visible light. In 1915, William Bragg (1862–1942) and his son Lawrence used this method to deduce the cubic arrangement of sodium and chloride ions in crystals of table salt. When the two Braggs received the Nobel Prize in Physics for this research, Lawrence was only 25 years old. After a distinguished scientific career and a knighthood, Sir Lawrence Bragg became the head of the Cavendish Laboratory.

Scientists quickly realized that X-ray diffraction was a powerful technique to study the structural arrangement of atoms in molecules. In Chapter 4, the *Science in Action* feature describes how this technique was used to prove the planar structure of benzene. By the middle of the 20th century, scientists had begun applying X-ray diffraction to examine the structures of large biological molecules such as proteins. In some cases, protein molecules could be induced to form crystals, which is the ideal sample for analysis. Other proteins, however, had elongated structures that were more difficult to crystallize. In these cases, scientists stretched the proteins into long fibers that also diffracted X-rays.

Scientists also began applying X-ray diffraction to the study of DNA. In the early 1950s, this research was centered at King's College in the University of London. X-ray studies of DNA had been pioneered by a physicist, Maurice Wilkins (1916–2004; Figure 12.12), who obtained the first X-ray photographs. Because DNA is a long, thin molecule, it was impossible to obtain crystals for analysis. Instead, Wilkins used fibers of DNA for his X-ray diffraction experiments, employing methods similar to those used in the study of protein fibers.

Figure 12.13(a) illustrates a simplified X-ray diffraction experiment, A DNA fiber is placed in the path of the X-ray beam, and the X-rays are scattered by the

THE KEY IDEA: X-ray diffraction provides information about the structure of biological molecules.

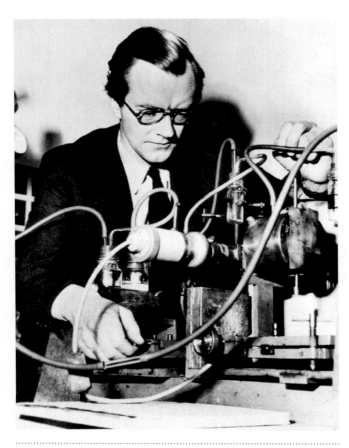

FIGURE 12.12 **Maurice Wilkins.** Wilkins, who worked at King's College in London, obtained the first X-ray diffraction photographs of DNA. This photograph shows Wilkins using his X-ray diffraction equipment.

electrons surrounding each atom within the DNA molecule. Because the X-rays are waves, they *interfere* with one another. Two waves can add together, a process called *constructive interference*. This type of interference produces a high X-ray intensity. By contrast, two waves can also cancel each other, a process that is called *destructive interference*. In this case, the X-ray intensity vanishes. The regions of constructive and destructive interference can be detected by a photographic plate. The result is a *diffraction pattern* of dark regions caused by strong X-ray intensity (constructive interference). Figure 12.13(b) shows a diffraction pattern for one of Wilkins's early X-ray studies of DNA.

The next step is to use the X-ray diffraction pattern to infer the three-dimensional structure of the molecule. This is a very difficult task, requiring the application of sophisticated mathematical techniques. The analysis is especially challenging for a large molecule such as DNA, which contains many atoms. Before the widespread availability of computers, a detailed interpretation of a single DNA diffraction pattern could take months to complete.

The X-ray diffraction technique illustrates a general principle in scientific research. It is rare that scientists can directly observe the entities or processes they are investigating. Instead, science typically involves using an experimental technique to collect *indirect evidence*, such as the spots in a diffraction pattern, which then has to be interpreted to produce scientific insights, such as the structure of a molecule. Making the intellectual leap from evidence to interpretation is an important and creative component of scientific investigation.

In 1951, the head of the King's College laboratory hired a young scientist named Rosalind Franklin (1920–1958) to extend the X-ray studies of DNA (Figure 12.14). Franklin had studied chemistry in college and then obtained her Ph.D. from Cambridge University, where she became an expert in X-ray crystallography. After working in Paris for several years, she was eager to tackle a new scientific project at King's College. Unfortunately, a misunderstanding of responsibilities, as well as their very different personalities, prevented Wilkins and Franklin from working together. This lack of partnership was a distinct contrast to the collaborative research style of Watson and Crick.

FIGURE 12.13 **An X-ray diffraction experiment on DNA.** (a) X-rays from a source are directed toward a DNA fiber. The X-rays are scattered by the electron density surrounding the atoms within the molecule. The scattered X-rays produce dark regions on a photographic plate, corresponding to regions of high X-ray intensity. (b) An X-ray diffraction pattern from a DNA fiber, obtained by Maurice Wilkins.

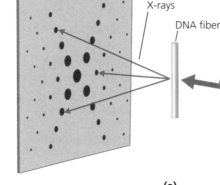

Film

Diffracted X-rays

DNA fiber

X-ray beam

(a)

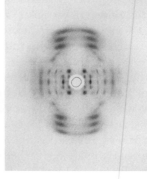

(b)

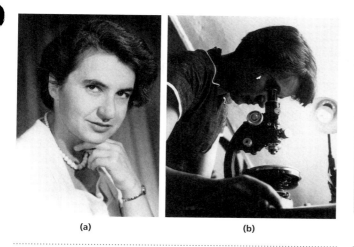

(a)   (b)

FIGURE 12.14 **Rosalind Franklin.** (a) Franklin was recruited to King's College to study the structure of DNA. (b) Franklin prepares an X-ray diffraction sample using (a) microscope.

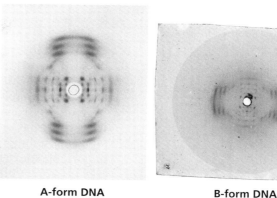

A-form DNA   **B-form DNA**

FIGURE 12.15 **X-ray diffraction photographs of A-form and B-form DNA.** The A-form occurs when DNA is dehydrated, whereas the B-form is DNA's structure when it is surrounded by water.

Within a short time, Franklin made major advances in the structural study of DNA. Using X-ray diffraction, she discovered that DNA can exist in two different structures, which she called the *A-form* and the *B-form*. Figure 12.15 compares the distinct X-ray diffraction photographs of these two structures. The A-form is produced when DNA is dehydrated to remove much of the surrounding water. Under these conditions, the fibers pack together more tightly, so they behave more like a solid crystal. This regular arrangement of atoms produces a more defined set of diffraction spots. The B-form of DNA occurs when it is surrounded by water. Under these conditions, the orientation of the DNA fibers is less organized, and the diffraction pattern is more blurred.

Which type of DNA is more conducive to studying in detail? If you are a biologist, you may think that B-form DNA is more interesting because it is the type that occurs in the aqueous environment of cells. Franklin, however, was trained in the technique of X-ray diffraction. Therefore, she decided to focus on A-form DNA because the X-ray photographs provided a greater wealth of diffraction data.

Before Franklin concentrated on the A-form, she took a final set of X-ray photographs of B-form DNA. One of these photographs was the high-quality image shown in Figure 12.16. Franklin labeled the picture "Photo 51" and put it in her research files. Without Franklin's knowledge or consent, Wilkins showed this photograph to Watson and Crick. The valuable information contained in Photo 51 was immediately obvious to Crick, who was an expert in X-ray diffraction. In fact, Crick had made important contributions to the development of sophisticated mathematical theories to interpret the X-ray diffraction patterns generated by helical protein fibers.

The X-pattern of spots in Photo 51 is the characteristic signature of a helix. Watson and Crick already suspected that long threads of DNA formed some kind of helix because helices (plural of *helix*) had already been discovered within the structure of fiber-like proteins. Photo 51 provided definitive evidence that these hunches were correct. Other features of Photo 51

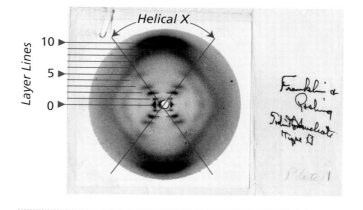

FIGURE 12.16 **Photo 51 provided information about the helical structure of DNA.** This high-quality X-ray photograph of B-form DNA was taken by Rosalind Franklin and Raymond Gosling, who was Franklin's graduate student. Their handwritten names appear to the right of the photograph. The X-shaped diffraction pattern is characteristic of a helix. The angle between the two arms of the "X" and the spacing between the spots (layer lines) provide information about the spatial dimensions of the helix.

provided important clues about the specific structure of the DNA helix. The angle between the two lines of the X reveals the *pitch* of the helix, which describes whether the helix is tightly or loosely coiled. The arrangements of spots in each arm of the X are called *layer lines*. The presence of 10 layer lines between the middle and the edge of the photograph indicates that DNA contains 10 nucleotides within each complete turn of the helix.

Equipped with this new information, Watson and Crick became convinced they could now solve the structure of DNA. They would use Franklin's X-ray data to guide their construction of a new molecular model. At this time, Watson and Crick learned that Pauling had also proposed a structure for DNA. Fortunately for them, Pauling's model was a triple helix that contained the same errors as their original blunder. However, it would not take long for Pauling to realize his mistake. Faced with the prospect of being beaten by Pauling, Sir Lawrence Bragg relented and allowed Watson and Crick to resume their work on DNA.

Core Concepts

- X-ray diffraction is an experimental technique for studying the structures of molecules, which cannot be observed directly. X-rays are directed at a sample (a crystal or a fiber) and are then scattered by the electrons surrounding the atoms. The scattered X-rays interfere with one another and create a pattern of dark spots on a photographic plate.
- Maurice Wilkins took the first X-ray diffraction photographs of DNA fibers. Rosalind Franklin used the same technique to discover that DNA exists in two structural varieties: the A-form and B-form. She also obtained the best-quality X-ray photograph of B-form DNA, known as Photo 51. Watson and Crick used information from this Photo 51 to construct a new molecular model of DNA.

Constructing the DNA Double Helix

Watson and Crick returned to building their DNA model, but this time they had the structural data from Photo 51 to guide them. The photograph revealed that DNA has *two* polynucleotide chains—a double helix—with the phosphate groups on the outside and the bases on the inside. However, one piece of the puzzle was missing: How did the bases fit within the structural framework of the polynucleotide chains?

Watson prepared replicas of the four DNA bases and tried to fit them together as pairs. However, each pair he tried didn't fit within the framework of their model. At this point, Watson had a fortuitous conversation with a visiting research scientist, who said that the chemical structures of the bases printed in textbooks were incorrect. After Watson learned the correct structures, he made a major scientific breakthrough.

Watson deduced that the bases in the DNA can fit within the dimensions of the double helix if they are paired in a specific way. These **base pairs** are:

Adenine—Thymine

Guanine—Cytosine

Purine—Pyrimidine

Figure 12.17 illustrates the molecular structures of these base pairs. The two bases in a pair are linked by *hydrogen bonds:* The A—T pair has two hydrogen bonds, and the G—C pair has three hydrogen bonds. Bases that form pairs are said to be *complementary* to each other.

Each complementary pair includes *one purine base* (adenine or guanine) and *one pyrimidine base* (thymine or cytosine). This purine–pyrimidine pairing of a large and a small base means that *the total width of the base pair remains constant*, regardless of the particular bases that are present or their relative order (i.e., purine–pyrimidine vs. pyrimidine–purine). As a result, the DNA double

THE KEY IDEA: James Watson and Francis Crick combined various types of scientific evidence to deduce the structure of DNA.

base pairs Specific pairings of the DNA bases within the double helix; adenine pairs with thymine, and guanine pairs with cytosine.

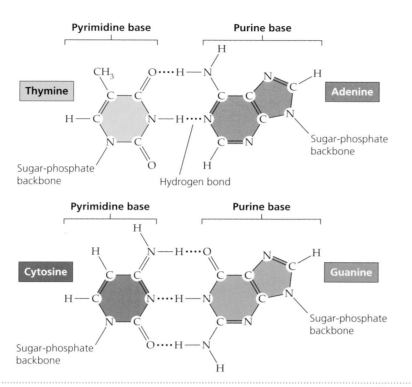

FIGURE 12.17 Base pairing in DNA. In the structure of DNA, the four bases form characteristic pairs. Each pair contains one purine base and one pyrimidine base: Adenine pairs with thymine, and guanine pairs with cytosine. The two bases within a pair are held together via hydrogen bonds.

antiparallel The spatial arrangement of two polynucleotide chains that are oriented in opposite directions.

helix can encode genetic information within a changing base-pair sequence, while still preserving the structural regularity of the sugar-phosphate backbone.

The symmetry of Photo 51 revealed that the two polynucleotide chains in a DNA helix are **antiparallel**; that is, *they are oriented in opposite directions*. As an analogy, Figure 12.18(a) illustrates an antiparallel orientation of two colored pencils. For DNA, recall that the directionality of a polynucleotide is always defined in the 5′ → 3′ direction. The antiparallel orientation of two polynucleotide chains is illustrated schematically in Figure 12.18(b). Note how the arrows representing the 5′ → 3′ orientation of each chain are pointing in opposite directions.

Based on this information, Watson and Crick constructed a new molecular model of DNA. In this model, the two nucleotide strands form a double helix with an antiparallel orientation. The helical twist was determined by the layer lines in Photo 51. The phosphate groups are located on the outside of the double helix, where they are soluble in water and provide DNA with the chemical properties of an acid. This location of the phosphates also explains how their negative charges can be neutralized by aqueous cations in solution (like Na⁺). The DNA bases are not very soluble in water, so they are

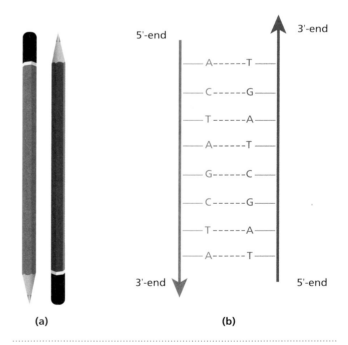

FIGURE 12.18 The two DNA strands are antiparallel. (a) Antiparallel orientation of two colored pencils, where the "direction" of the pencil is defined by the position of the writing point and the eraser. (b) Antiparallel orientation of two polynucleotide strands in a DNA molecule. The sugar-phosphate backbones are represented by thick lines, and the arrow represents the orientation of each strand in the 5′ → 3′ direction. The DNA bases are represented by single letters, and hydrogen bonding between complementary bases is indicated by dotted lines.

FIGURE 12.19 A model of the DNA double helix. James Watson (left) and Francis Crick pose by their model of the DNA double helix.

buried within the interior of double helix, where they interact according to the base-pairing rules (A with T, and G with C). When Watson and Crick had completed their work, they proudly posed for a photograph with their model of the DNA double helix (Figure 12.19).

Watson and Crick announced their discovery in a short paper that was published on April 6, 1953, in the pages of *Nature*, a prestigious scientific journal. Following this paper, in the same issue, was an experimental X-ray report by Wilkins. In third place was another X-ray paper by Franklin. Watson and Crick made no mention of Franklin's experimental data in their paper, and they did not acknowledge its valuable role in their deduction of the double-helix structure of DNA.

Since 1953, decades of research on DNA have corroborated and refined the double-helix structure originally proposed by Watson and Crick. Figure 12.20(a) depicts a modern representation of DNA. Note that the antiparallel orientation of the polynucleotide chains causes the deoxyribose sugar to point in opposite directions. Figure 12.20(b) provides a space-filling model of the DNA double helix, which twists in a right-handed direction (like a right-handed corkscrew). The sugar-phosphate backbones are located on the outer region of the double helix, while the bases are sequestered on the inside. In conclusion, *the DNA double helix illustrates how the structure of a complex biological molecule is based on chemical principles.*

In 1962, James Watson, Francis Crick, and Maurice Wilkins received the Nobel Prize in Physiology or Medicine for their work on DNA. In 1968, Watson published a bestselling book titled *The Double Helix: A Personal Account of the Discovery of the Structure of DNA.* In this book, Watson captured the thoughts and feelings of his younger self at the University of Cambridge during the exciting time of 1951 to 1953. His account generated considerable controversy because of its frank and sometimes critical descriptions of the major characters,

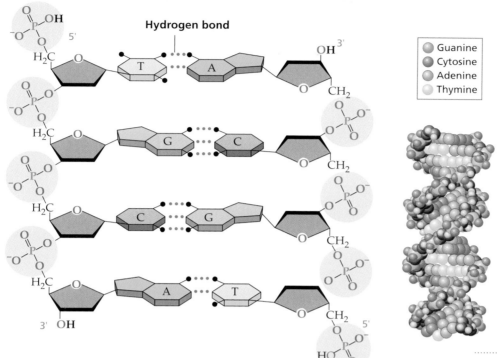

FIGURE 12.20 Molecular structure of DNA. (a) The detailed molecular structure of two antiparallel polynucleotide strands in DNA. Note that the distance between the sugar-phosphate backbones remains constant regardless of the particular base sequence in either strand. (b) The two polynucleotide strands wind together to form the DNA double helix.

especially Rosalind Franklin. Following her work on DNA, Franklin pioneered the application of X-ray diffraction to study the structure of viruses. Sadly, she died of ovarian cancer in 1958, at the young age of 37.

WORKED EXAMPLE 12.2

Question: The sequence of bases on one DNA strand is as follows, along with the 5′- and 3′-ends. Write the sequence of bases on the complementary DNA strand, including the 5′- and 3′-ends.

5′- G C C T A G G T T A T C - 3′

Answer: We can deduce the sequence of bases on the complementary strand by using the base-pairing rules: A pairs with T, and G pairs with C. In addition, the two DNA strands are antiparallel, so the 5′-end of the original strand will be opposite to the 3′-end of the template strand (and vice versa). The complementary sequence is.

3′- C G G A T C C A A T A G - 5′

TRY IT YOURSELF 12.2

Question: The sequence of bases on one DNA strand is shown here, along with the 5′- and 3′-ends. Write the sequence of bases on the complementary DNA strand, including the 5′- and 3′-ends.

5′- T A G G C T T A C G G A - 3′

PRACTICE EXERCISE 12.3

Before Watson and Crick deduced the structure of DNA, a biochemist named Erwin Chargaff (1905–2002) studied the percentage of each base in the DNA of different organisms. The table presents a selection of Chargaff's results.

PERCENTAGES OF A, G, C, AND T BASES IN DIFFERENT ORGANISMS				
ORGANISM	% A	% G	% T	% C
Yeast	31.3	18.7	32.9	17.1
Sea urchin	32.8	17.7	32.1	18.4
Herring	27.8	22.2	27.5	22.6
Human	30.7	19.3	31.2	18.8

Based on these results, Chargaff made some generalizations that became known as *Chargaff's rules*.

(a) Use Chargaff's data to deduce *two rules* that describe the relationships among the proportions of A, G, T, and C bases in DNA.

(b) Discuss how these rules can be explained by the structure of DNA.

Core Concepts

- The four bases in DNA form specific **base pairs**: Adenine pairs with thymine, and guanine pairs with cytosine. The two bases in a pair are joined by hydrogen bonds. Each pair contains one purine and one pyrimidine base, so the total width of the base pair remains constant within the DNA double helix.
- The two polynucleotide strands in DNA are oriented in opposite directions—they are **antiparallel**. This orientation is required to form the base pairs.
- The two polynucleotide strands in DNA form a right-handed double helix. The phosphate groups are located on the outside because they are soluble in water. The bases are located within the interior of the helix, where they form hydrogen-bonded base pairs.

12.4 DNA Replication

Learning Objective:

Analyze the mechanism of DNA replication.

THE KEY IDEA: During DNA replication, each DNA strand serves as a template to synthesize a new strand with a complementary base sequence.

semiconservative The mechanism of DNA replication, such that half of each new double helix is conserved from the parental DNA.

At the conclusion of their famous 1953 publication on the structure of DNA, Watson and Crick included a tantalizing sentence:

It has not escaped our notice that the specific pairing we have postulated immediately suggests a possible copying mechanism for the genetic material.

Watson and Crick realized that the double-helix structure of DNA provided important clues to answering one of the central questions of biology: How does DNA faithfully replicate and pass along its genetic information to the next generation?

In a follow-up publication, Watson and Crick proposed that *each strand of DNA serves as a template to synthesize a matching DNA strand*. This method of *template-directed synthesis* is commonly used for copying biological molecules. The process of DNA replication is illustrated in Figure 12.21. By convention, the original DNA is composed of the "parent strands," whereas the regions of newly synthesized DNA are called the "daughter strands." Because of the base-pairing rules, the identity of a DNA base on the template strand (e.g., adenine) uniquely specifies the complementary base in the new strand (e.g., thymine). This mechanism of DNA replication is called **semiconservative** because half of each new DNA double helix is conserved from the parental DNA.

The hydrogen bonds between the base pairs now become relevant for DNA's biological function. As we saw in Chapter 8, hydrogen bonds are *intermolecular* attractions between two molecules. These bonds are much weaker than covalent chemical bonds within a molecule. Therefore, only a moderate amount of energy is required to "unzip" the DNA molecule by breaking the hydrogen bonds between the base pairs. When the two strands are separated, the bases on each strand are exposed and serve as the template for synthesizing a new DNA strand.

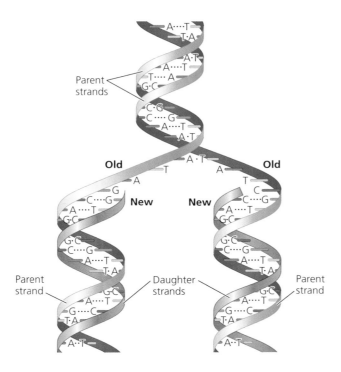

FIGURE 12.21 **The mechanism of DNA replication.** The double helix unwinds, and each parent strand is used as a template to synthesize a new daughter strand. The DNA base-pairing rules ensure that the correct bases are added to make the daughter strands.

Think of the two polynucleotide strands in DNA as two strips of Velcro. The interaction between the Velcro strips is sufficient to hold them together, but a little tug is enough to pull them apart. We now have two sticky unpaired Velcro strips that can be attached to different pieces of Velcro. For DNA molecules, the hydrogen bonding between the bases provides the "stickiness" for the interactions. The base-pairing rules dictate that only particular pairs of bases can "stick" to each other within the DNA double helix.

Watson and Crick's proposal for DNA replication was a hypothesis based on the structure of the DNA double helix. However, the hypothesis initially had no scientific evidence to support it. Five years later, two scientists conducted an experimental test of DNA replication that has been called "the most beautiful experiment in biology." We describe this experiment in *Science in Action*.

WORKED EXAMPLE 12.3

Question: The base-pair sequence within a double-stranded DNA molecule is.

$$5'-AGGCGAAGCTAG-3'$$
$$3'-TCCGCTTCGATC-5'$$

This region of DNA is replicated according to the semiconservative mechanism. Use each parent strand as a template to write the base sequences of the two daughter strands.

Answer: The base sequences of the daughter strands can be deduced using two features of DNA: the base-pairing rules and the antiparallel orientation of complementary DNA strands.

$$\text{Parent strand } 5'-AGGCGAAGCTAG-3'$$
$$\text{Daughter strand } 3'-TCCGCTTCGATC-5'$$

$$\text{Daughter strand } 5'-AGGCGAAGCTAG-3'$$
$$\text{Parent strand } 3'-TCCGCTTCGATC-5'$$

TRY IT YOURSELF 12.3

Question: The base sequence of a daughter DNA strand are given below. What is the base sequence of the original template strand, including the 5′- and 3′-ends?

3′ - T C A A A C C G T G G C - 5′

DNA polymerase The enzyme that creates new DNA polymers by adding new nucleotides to the template strand.

Further investigations into DNA replication revealed new insights. A few years after Watson and Crick's proposal, scientists identified the enzyme responsible for synthesizing the new DNA strand. They named the enzyme **DNA polymerase** to indicate its role in creating new DNA polymers. The operation of this enzyme is illustrated schematically in Figure 12.22. Using the parental DNA strand as a template, DNA polymerase synthesizes the daughter DNA strand by adding new nucleotides to the 3′-end. Initially, the nucleotides that will be added exist in the *triphosphate* form (i.e., a base, deoxyribose, plus three phosphates). The DNA polymerase breaks off two of the phosphates, which provides a source of chemical energy to power the addition of the new nucleotide to the DNA strand. Later research identified several different types of polymerase enzyme that are involved in replicating DNA within cells.

As the DNA is being copied, there is always the possibility of introducing an error. *A change in the sequence of DNA bases is called a mutation.* For example, Figure 12.22 illustrates how DNA polymerase adds a new thymine (T) base to match the adenine (A) base on the template strand. If the enzyme mistakenly attached a nucleotide with a cytosine (C) base, this would constitute a mutation. A mutation can be harmful to an organism if it disrupts the genetic information in DNA, just as a "typo" can change the meaning of a word. To address this problem, DNA polymerase includes a mechanism to ensure the accuracy of DNA replication. After the enzyme adds each nucleotide, it *proofreads* the base to ensure it is correct. If the wrong base has been inserted by mistake, then the enzyme cuts it out and replaces it with the correct one. This proofreading strategy ensures that DNA polymerase has a remarkably low error rate: It produces roughly one mutation in 1 billion copied bases.

In conclusion, DNA illustrates an important principle: *The structure of a biological molecule is directly related to its biological function.* The double-helix structure of DNA, with its two strands of complementary bases, is ideally suited for replication, which is an essential process for any biological organism.

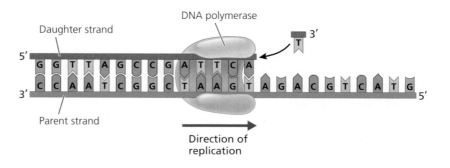

FIGURE 12.22   The role of DNA polymerase in DNA replication. The creation of a new polynucleotide strand during DNA replication is accomplished by DNA polymerase, shown schematically in red. This enzyme uses the parental DNA strand as a template to add new nucleotides to the daughter strand, according to the base-pairing rules.

SCIENCE IN ACTION

How Does DNA Replicate?

How can we test the hypothesis that DNA replicates via a semi-conservative mechanism? According to this mechanism, one of the original strands of DNA is preserved as half of each new double helix. Therefore, we need some way to distinguish between the original and the new DNA strands.

Suppose we can "paint" the nucleotides in different colors. We use blue paint for the nucleotides in the original DNA and red paint for the new nucleotides that are added during DNA replication. This scenario is illustrated in Figure 1, in which each DNA strand is simplified as half of a ladder. During semiconservative DNA replication, the two blue strands unwind, and each one becomes a template for adding new red nucleotides. The result is two daughter DNA helices with one blue strand (original) and one red strand (new).

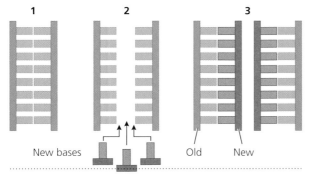

FIGURE 1 **A schematic model of semiconservative DNA replication.** In this model, original DNA nucleotides are painted red and new nucleotides are painted blue.

Of course, in real life we can't actually paint nucleotides with different colors. Therefore, we need a different method to identify the DNA strands. In 1958, Matthew Meselson (b. 1930) and Frank Stahl (b. 1929) used nitrogen isotopes to label ("paint") the DNA nucleotides. Recall that each of the four DNA bases has two or more nitrogen atoms that can be labeled.

The experiment performed by Meselson and Stahl is illustrated in Figure 2. They grew *E. coli* bacteria in a medium containing a *heavy isotope* of nitrogen (^{15}N), which was incorporated into the cell's DNA nucleotides. This procedure labeled the original DNA with the ^{15}N isotope. Next, the scientists transferred the bacterial cells into a different growth medium that contained only the *light isotope* of nitrogen (^{14}N). When the cells replicated their DNA, they incorporated new nucleotides that were labeled with the ^{14}N isotope. Afterward, the bacterial cells were transferred into another growth medium that again contained only the ^{14}N isotope. When the cells replicated their DNA for a second time, they used new nucleotides labeled with ^{14}N.

To detect the presence of the different isotope labels, Meselson and Stahl used a centrifuge to measure the relative density of the DNA double helix. The original DNA, labeled with the ^{15}N

isotope, was dense and floated near the bottom of the centrifuge tube. We can represent the two heavy DNA strands as $^{15}N/^{15}N$. After the first cycle of DNA replication, the newly made DNA floated to a higher level in their tube, indicating that it was less dense than the original DNA. This product, called *hybrid DNA*, was composed of one heavy ^{15}N strand and one light ^{14}N strand ($^{15}N/^{14}N$). The existence of the hybrid DNA meant that half of the original ^{15}N-labeled DNA was conserved during DNA replication.

After the second cycle of DNA replication, Meselson and Stahl detected two types of DNA in the centrifuge. One type was the hybrid DNA ($^{15}N/^{14}N$), and the second type was lighter with two ^{14}N strands ($^{14}N/^{14}N$). This result makes sense if each strand of the hybrid DNA is used as the template to synthesize a new DNA strand. If the ^{15}N strand is the template, DNA replication produces the hybrid DNA ($^{15}N/^{14}N$). If the ^{14}N strand is the template, DNA replication produces the light DNA ($^{14}N/^{14}N$).

In conclusion, the results from the Meselson–Stahl experiment confirmed Watson and Crick's hypothesis—*DNA replicates via a semiconservative mechanism.*

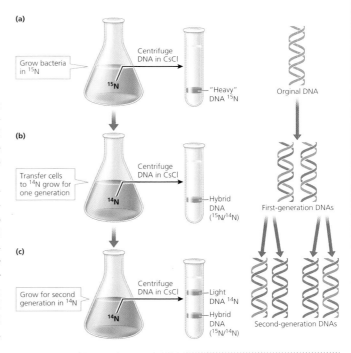

FIGURE 2 **The Meselson-Stahl experiment to study the mechanism of DNA replication.** The scientists identified the DNA strand by labeling the bases with isotopes of nitrogen (^{15}N or ^{14}N).

Core Concepts

- During DNA replication, the two polynucleotide chains in the double helix unwind. Each parental strand serves as a template to synthesize a new daughter strand. This mechanism of DNA replication is called **semiconservative**.
- The base-pairing rules ensure faithful copying of genetic information. Each base on the template strand uniquely specifies the complementary base that is added to the daughter strand.
- DNA replication is performed by an enzyme called **DNA polymerase**. To minimize the occurrence of a mutation, the enzyme has a proofreading function that ensures that the DNA bases are copied accurately. Cells contain several different types of polymerase enzyme.

12.5 DNA Mutations and Cancer

Learning Objective:

Characterize the connections between DNA mutations and cancer.

You likely know a friend or family member who has experienced some type of cancer. In this section, we examine the relationship between DNA mutations and cancer. Today, tens of thousands of scientists and physicians throughout the world are studying the molecular and cellular origins of cancer, with the goal of providing improved diagnosis and treatment. The biology of cancer is a complex topic, and we provide only the most basic foundation in this chapter.

What Is Cancer?

THE KEY IDEA: All types of cancer are caused by the uncontrolled growth of cells.

cancer A group of related diseases that are characterized by the uncontrolled growth of cells.

Cancer is a general term used to encompass a group of related diseases. *All types of cancer are characterized by the uncontrolled growth of cells.* Figure 12.23 shows a magnified image of a breast cancer cell. The protrusions from the cell are used to supply blood and nutrients, which enables the cancer cell to sustain its rapid rate of replication.

Cells grow by dividing into two daughter cells. Normal cell division in the human body is a critical component of maintaining our health. For example, when we cut our skin, the nearby cells begin to divide as a mechanism for healing the wound. Even when we are not injured, cells must divide to replace those that have become old or damaged. Each type of cell in the body has a specific timetable for when to reproduce. Some cells, like the ones in our hair follicles, divide very frequently, whereas others reproduce on a much slower timeframe.

Cell division is a carefully controlled process. Some genes in our body produce proteins that stimulate cells to reproduce. Other genes make proteins that monitor and regulate cell division. For example, if a dividing cell contains damaged DNA, then these regulatory proteins can halt cell division and can even induce the cell to self-destruct. In general, *many cancers arise from DNA mutations in the genes that control cell division.* If the genes that control cell division acquire a serious DNA mutation, then the proteins they produce have a reduced capacity to function. In such cases, cells can be stimulated to divide when they should not. Alternatively, removing the checks on cell division enables cells with damaged DNA to replicate.

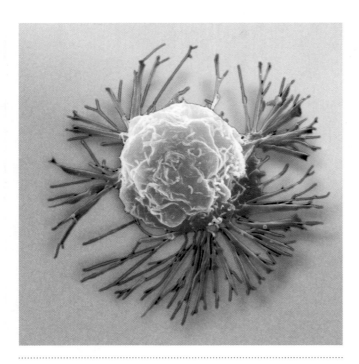

FIGURE 12.23　A magnified image of a breast cancer cell. The cell has been artificially colored to make it more visible.

Core Concepts

- All types of **cancer** are caused by the uncontrolled growth of cells.
- Many types of cancer arises from DNA mutations in the genes that control cell division.

Ultraviolet Light and Skin Cancer

Why should we apply sunscreen before visiting the beach? You have probably heard about the harmful effects of *ultraviolet (UV) light* that is emitted by the sun. Ultraviolet light consists of waves that are shorter than the violet edge of the visible spectrum of light. In general, a shorter wavelength of light corresponds to a higher energy. Because UV light has a very short wavelength, it has sufficient energy to induce mutations in the DNA of our skin cells.

In particular, UV light can initiate a chemical reaction between a pair of neighboring pyrimidine bases within the DNA double helix. The UV light stimulates the formation of new covalent bonds between the adjacent bases, which creates a molecular structure called a **pyrimidine dimer** (*dimer* means "two things"). Figure 12.24 illustrates how UV light induces the formation of a *thymine dimer*. The dimer structure creates a kink in the polynucleotide chain, which distorts the regular shape of the DNA double helix. A similar reaction occurs between cytosine bases to form a *cytosine dimer*. These reactions are very common: A skin cell can experience 50 to 100 pyrimidine dimer formations during every second of sun exposure.

The presence of a pyrimidine dimer makes it difficult for DNA polymerase to accurately copy the base sequence during DNA replication. For example, when the enzyme encounters a cytosine dimer, it often pairs the cytosine with adenine instead of guanine. This replication error introduces a mutation into the daughter DNA strand—instead of GG, the adjacent base sequence now reads AA. Fortunately, our cells contain *DNA repair systems,* which consists of a group of enzymes that detect the kink in the DNA and repair the mutations caused by UV light. However, greater exposure to UV light causes more light-induced mutations, which increases the likelihood that DNA base changes will escape the DNA repair systems. If a DNA mutation occurs within a gene that controls cell division, it can lead to the development of skin cancer.

The connection between UV light and skin cancer is well established. Nevertheless, in the United States, an average of more than 1 million people each day voluntarily expose themselves to high doses of ultraviolet radiation by using an indoor tanning bed. *Chemistry and Your Health* explores the relationship between indoor tanning and skin cancer.

THE KEY IDEA: Ultraviolet light initiates a chemical reaction between two adjacent pyrimidine bases.

pyrimidine dimer   A molecular structure formed by connecting two adjacent pyrimidine bases within DNA.

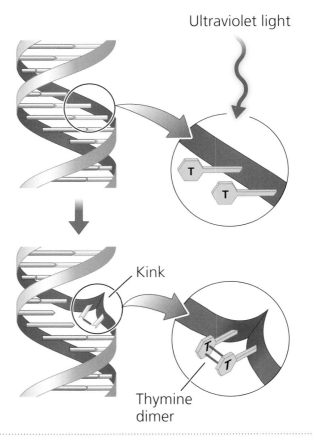

FIGURE 12.24 UV light creates a thymine dimer. High-energy UV light initiates a chemical reaction that forms new chemical bonds between two adjacent thymine bases. The resulting molecular structure is called a thymine dimer. The presence of the dimer creates a distortion (kink) in the double-helix structure of DNA.

Core Concepts

- Ultraviolet (UV) light has sufficient energy to induce mutations in DNA.
- UV light can stimulate the formation of new covalent bonds between adjacent pyrimidine bases. The resulting structures are called **pyrimidine dimers**.
- One example is the formation of a thymine dimer, which induces a kink in the structure of the DNA double helix. This distortion of the DNA structure causes errors in DNA replication.

CHEMISTRY AND YOUR HEALTH

Do Indoor Tanning Beds Increase the Risk of Skin Cancer?

According to a report by the U.S. Surgeon General, skin cancer is the most common type of cancer in the United States. Each year, nearly 5 million people receive medical treatment for this condition. Melanoma, the most dangerous form of skin cancer, is responsible for 9000 deaths each year.

FIGURE 1 An indoor tanning bed. Indoor tanning beds generate high-intensity UV exposure and increase the risk of developing skin cancer.

Given what we know about the connection between UV light and skin cancer, there is growing concern about the use of indoor tanning beds (Figure 1). Despite the mistaken perception that indoor tanning is "safer" than outdoor exposure to the sun, the intensity of UV radiation emitted by a tanning bed is sometimes greater. The U.S. Food and Drug Administration (FDA) warns that "UV radiation in tanning devices poses serious health risks," and the World Health Organization (WHO) has placed indoor tanning beds in its highest cancer risk category.

Despite the known risks, the use of indoor tanning facilities remains very popular. The American Academy of Dermatology provides the following statistics:

- On an average day in the United States, more than 1 million people have a session in a tanning salon.
- Nearly 30 million people use indoor tanning facilities each year. Of this number, 2.3 million are teenagers.
- In a 2014 study, 13% of American adults, 43% of college students, and 10% of teenagers reported using a tanning bed during the past year.
- Nearly 70% of tanning salon users are white women, mostly between the ages of 16 and 29.

Now that you have an understanding of the relationship between UV light and DNA damage, we hope that you will be thoughtful when deciding whether to use an indoor tanning bed.

THE KEY IDEA: Some cancer-causing molecules fit between the bases of the DNA double helix and cause errors in DNA replication.

chemical carcinogens Chemical substances that are capable of causing cancer.

Chemical Carcinogens

The process by which cancer arises is called *carcinogenesis* (*genesis* means "origin"). Some chemical substances are capable of causing cancer; we refer to them as **chemical carcinogens**. In this section, we focus on a potent chemical carcinogen called *benzopyrene*. This molecule is produced when carbon-containing substances are burned. For example, benzopyrene is a component of cigarette smoke, automobile exhaust, and charbroiled food.

As shown in Figure 12.25, benzopyrene is a hydrocarbon molecule that consists of five connected benzene-type rings. Benzopyrene belongs to a class of molecules called *polyaromatic hydrocarbons* (PAH). As we learned in Chapter 4, benzene contains delocalized electrons that produce a planar structure for the molecule. Similarly, benzopyrene is planar because the electrons are delocalized throughout the five rings. After benzopyrene enters the body, either by ingestion or inhalation, it is chemically modified by a type of *cytochrome* enzyme. This enzyme alters one of the carbon rings, changing the double bonds to single bonds and adding three oxygen heteroatoms. The triangular arrangement of atoms with an oxygen at the apex is called an *epoxide* group. The other two oxygen atoms are part of alcohol functional groups.

This chemical modification is important because it enhances the ability of the molecule to react with DNA. The original benzopyrene is a chemically unreactive hydrocarbon that is insoluble in water. However, the addition of oxygen atoms in the modified structure makes the molecule more reactive and water soluble. When the modified benzopyrene interacts with DNA, the epoxide group reacts with a nitrogen atom in a guanine base to form a new covalent bond. The resulting structure is called a *benzoypyrene-DNA adduct*; the term *adduct* refers to the direct addition of two molecules to form a larger one without the loss of

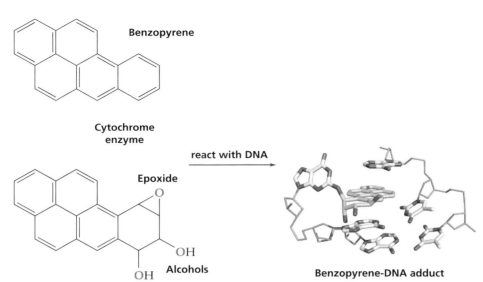

Benzopyrene

Cytochrome enzyme

Epoxide

react with DNA →

Alcohols   OH   OH

Benzopyrene-DNA adduct

FIGURE 12.25 Benzopyrene is a chemical carcinogen. After benzopyrene enters the body, it is chemically modified by a cytochrome enzyme to gain an epoxide group plus two alcohol groups. The epoxide reacts with a guanine base in DNA to form a benzopyrene-DNA adduct, which is a chemical addition of the two molecules. The remaining four aromatic rings slide within the DNA base pairs.

any atoms. The remaining four aromatic rings slide between the DNA base pairs in a process called *intercalation*. The planar structure of these four benzopyrene rings closely resembles the flat structure of the DNA base pairs, a similarity that enables intercalation to occur. The presence of the DNA-benzopyrene adduct produces an error during DNA replication. When DNA polymerase reaches this structure, it misreads the modified guanine as a thymine. This G-to-T mutation is characteristic of benzopyrene's function as a chemical carcinogen.

In 1996, scientists identified benzopyrene in cigarette smoke as a molecular trigger for lung cancer, which is the leading cause of cancer death for both men and women in the United States (Figure 12.26). About 60% of lung cancers contain mutations in a gene called *p53*, which produces a protein that controls cell division. Mutations in the p53 gene can disrupt the protein's ability to safely regulate cellular growth. By surveying all of the p53 mutations that are associated with lung cancer, a team of scientists identified so-called *hotspots* where the DNA bases mutate most frequently. First, they noticed that a large percentage of these hotspots contained a G-to-T mutation, which is the hallmark of benzopyrene. Next, they cultured cells in the laboratory and then added benzopyrene to determine where it binds to the DNA within the p53 gene. After analyzing their results, they observed a match between the location of benzopyrene-DNA adducts and the position of mutation hotspots associated with lung cancer. This scientific discovery was the "smoking gun" that established a causal relationship between smoking and lung cancer.

Core Concepts

- A **chemical carcinogen** is a chemical substance that is capable of causing cancer.
- Benzopyrene is an example of a chemical carcinogen. After benzopyrene enters the body, it is chemically modified by a cytochrome enzyme to become more reactive and water soluble. The modified benzopyrene forms a covalent bond with a guanine base in DNA to create a benzopyrene-DNA adduct.
- By studying the location of DNA mutations, scientists discovered that benzopyrene in cigarette smoke causes lung cancer.

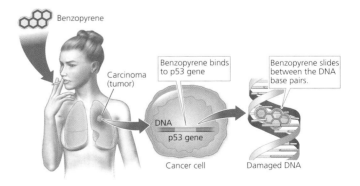

FIGURE 12.26 The benzopyrene in cigarette smoke causes lung cancer. The DNA mutation hotspots in the p53 gene that are associated with lung cancer match the locations where benzopyrene forms adducts with DNA bases.

12.1   How Was the DNA Double Helix Discovered?

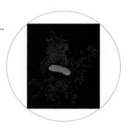

Learning Objective:

Explain why the DNA double helix was an important scientific discovery.

- All cell-based organisms use **DNA** (deoxyribonucleic acid) to store genetic information.

- The double-helix structure of DNA was discovered in 1953 by James Watson and Francis Crick.

12.2   Nucleotides: The Building Blocks of DNA

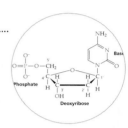

Learning Objective:

Analyze the structures of nucleotides and polynucleotides.

- **Nucleotides** are the molecular building blocks of DNA.

- A nucleotide has three components: a **deoxyribose** sugar, a phosphate, and a base. The carbon atoms on deoxyribose are labeled from 1′ to 5′ (′ means "prime").

- The base is attached to the 1′-carbon of the deoxyribose, and the phosphate is attached to the 5′-carbon.

- There are four varieties of **nucleotide bases** in DNA: **adenine** (A), **thymine** (T), **guanine** (G), and **cytosine** (C).

- Cytosine and thymine are **pyrimidine** bases, whereas guanine and adenine are **purine** bases

- Joining many nucleotides (monomers) creates a **polynucleotide** (polymer).

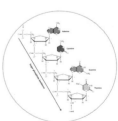

- A polynucleotide chain has two distinct ends: the **5′-end** and the **3′-end**. The **directionality** of a polynucleotide strand is defined in the 5′ → 3′ direction.

- The **sugar-phosphate backbone** of DNA contains an alternating sequence of phosphates and sugars.

- The phosphates are negatively charged, and they give DNA the properties of an acid.

12.3 Discovering the Double Helix

Learning Objective:

Characterize the scientific methods and evidence that led to the discovery of the DNA double helix.

- ⊙ Deducing the structure of the DNA double helix involved using techniques and insights from biology, chemistry, and physics.

- ⊙ In the 1940s, Oswald Avery and his colleagues demonstrated that DNA (and not protein) was responsible for transforming the harmless R strand of *Pneumococcus* bacteria into the infectious S strain.

- ⊙ Linus Pauling pioneered the use of molecular models to determine the structure of biological molecules. This approach greatly influenced Watson and Crick.

- ⊙ X-ray diffraction is an experimental technique that reveals information about the location of atoms within a molecule (prepared as a crystal or a fiber).

- ⊙ Maurice Wilkins obtained the first X-ray diffraction photograph of DNA. Rosalind Franklin demonstrated that DNA can exist as two structures, the A-form and the B-form. She also obtained a high-quality photograph of the B-form (Photo 51), which Watson and Crick used to obtain structural information about DNA.

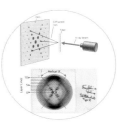

- ⊙ Watson deduced that the DNA bases exist as hydrogen-bonded **base pairs**: adenine pairs with thymine, and guanine pairs with cytosine. Each base pair contains one purine base and one pyrimidine base.

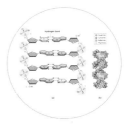

- ⊙ The two polynucleotide strands in DNA are **antiparallel;** that is, they are oriented in opposite directions.

12.4 DNA Replication

Learning Objective:

Analyze the mechanism of DNA replication.

- ⊙ During DNA replication, each strand of the parent DNA serves as a template to synthesize a complementary strand of daughter DNA.

- ⊙ This mechanism of DNA replication is called **semiconservative** because half of the parent DNA is preserved in the double-stranded daughter DNA.

- ⊙ The creation of a new DNA strand during DNA replication is accomplished by an enzyme called **DNA polymerase**.

12.5   DNA Mutations and Cancer

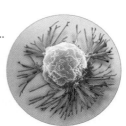

Learning Objective:

Characterize the connections between DNA mutations and cancer.

⊙ **Cancer** is a general term used to describe a group of related diseases. All types of cancer are characterized by the uncontrolled growth of cells.

⊙ Many cancers arise from DNA mutations in the genes that control cell division.

⊙ Ultraviolet (UV) light can stimulate the formation of new covalent bonds between adjacent pyrimidine bases, forming a **pyrimidine dimer**.

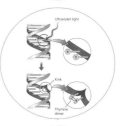

⊙ The pyrimidine dimer creates a kink in the structure of DNA and causes mutations during DNA replication.

⊙ A **chemical carcinogen** is capable of causing cancer. One example is benzopyrene, which is produced by combustion reactions.

⊙ Benzopyrene in cigarette smoke causes lung cancer by inducing mutations in the p53 gene.

LEARNING RESOURCES

Reviewing Knowledge

12.1: How Was the DNA Double Helix Discovered?

1. What is the complete name of DNA?
2. Which two scientists deduced the double-helix structure of DNA?

12.2: Nucleotides: The Building Blocks of DNA

3. What are the three components of a nucleotide?
4. Draw the molecular structure of a deoxyribose sugar, and label the carbon atoms from 1′ to 5′.
5. What are the names of the four DNA bases? Which are pyrimidine bases, and which are purine bases?
6. What name is given to the chemical linkage between two nucleotides?
7. Specify Identify the two ends of a polynucleotide strand. Define the "directionality" of a polynucleotide strand.

12.3: Discovering the Double Helix

8. Describe how experiments on *Pneumococcus* bacteria led to discovery of the "transforming principle."
9. Which scientific method, pioneered by Linus Pauling, did Watson and Crick use to study the structure of DNA?
10. Briefly describe the experimental technique of X-ray diffraction, and explain how it is used to determine the structure of molecules.
11. What did Rosalind Franklin's "Photo 51" reveal about the structure of DNA?

12. What are the base pairs in DNA? What is the importance of having one purine and one pyrimidine base within each pair?
13. What type of bond connects the two bases in a DNA base pair?
14. How are the two polynucleotide strands in DNA oriented with respect to each other?
15. For a DNA double helix in aqueous solution, why are the phosphate groups located on the outside and the bases on the inside of the helix? What is the role of ions in stabilizing the DNA structure?

12.4: DNA Replication

16. Draw a diagram to illustrate how DNA replication occurs. How is the genetic information in the parental DNA preserved within the daughter DNA?
17. Explain why we use the term *semiconservative* to characterize the mechanism of DNA replication.
18. When DNA polymerase creates a new DNA strand, how does the enzyme achieve a very low error rate of one mutation per 1 billion bases?

12.5: DNA Mutations and Cancer

19. What is the characteristic feature of all types of cancer? What is the connection between DNA mutations and cancer?
20. Explain how ultraviolet (UV) light is capable of causing DNA mutations.

21. What is a chemical carcinogen? Describe how benzo-pyrene acts as a chemical carcinogen.

22. Which gene is mutated in 60% of lung cancers? How did scientists establish the connection between genetic mutations and lung cancer?

Developing Skills

23. Based on the information in this chapter, propose a hypothesis for why DNA is double stranded instead of single stranded.

24. Draw the structure of a complete nucleotide by combining these three components: guanine, deoxyribose sugar, and phosphate.

25. *Xanthine* (shown below) is a base that is not part of DNA but is found in human body tissues. Xanthine is produced by an enzyme that removes an amine group from one of the DNA bases. (a) Is xanthine a purine or a pyrimidine base? (b) Which of the DNA bases serves as a source of xanthine?

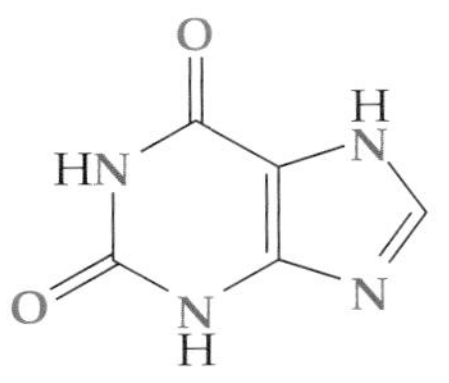

26. Some damage to DNA results from agents that change the bases or sugars of the nucleotides. One chemical "lesion" in DNA is the formation of 8-oxoguanine (shown below). Compare the structure of this mutated base with that of guanine. Identify and circle the site of chemical change in 8-oxoguanine.

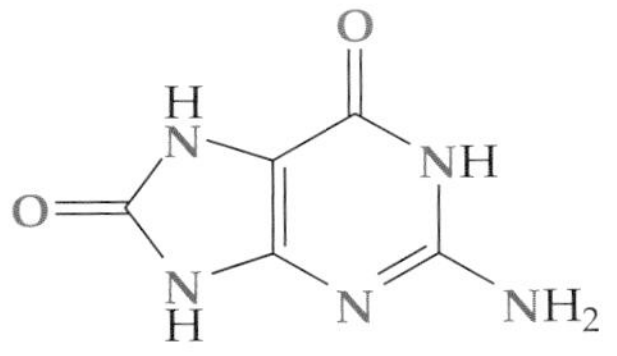

27. Certain viruses have bases that differ from the standard ones. For example, in a particular virus the C is completely replaced by an analog called *hydroxymethyl cytosine* (shown here). Does this base still allow base pairing with G? Explain your answer.

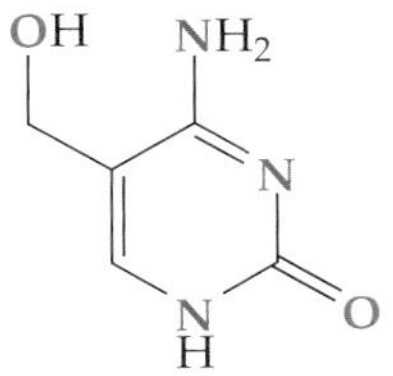

28. The structure of a polynucleotide chain is given in this exercise.
    (a) Identify and label the 5'-end and the 3'-end of the chain.
    (b) Identify the name of each base, and classify the base as either a purine or a pyrimidine.

(c) Write the sequence of DNA bases. Include the 5'-end and 3'-end.

(d) Write the base sequence of the complementary DNA strand. Include the 5'-end and 3'-end.

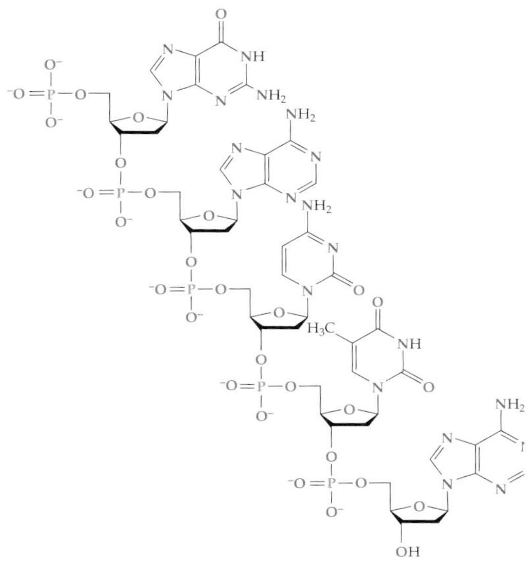

29. For each of the following sequences, write the base sequence on the complementary DNA strand, including the 5'-end and 3'-end.
    (a) 5'-A T T G C C C G A A A T G G C-3'
    (b) 5'-G C A A C G T T T G C A C T G-3'
    (c) 3'-C A G T C C A C G A T T A C A-5'

30. Human DNA is composed of 23% adenine and 27% guanine. What are the percentages of thymine and cytosine?

31. While studying the DNA of a bacterial organism, researchers discovered that 20% of the bases in the gene are thymine. Calculate the percentage composition of the other three bases.

32. Adenine can pair with a base called *uracil*, which is found in a molecule called RNA (ribonucleic acid). The molecular structure of uracil is shown below. Draw the structure of an adenine-uracil base pair, using dashed lines to indicate the hydrogen bonding interactions between the two bases.

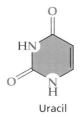
Uracil

33. An *E. coli* bacterium contains roughly 5 million base pairs of DNA. Each time the bacterium divides, all of its DNA is replicated. During this process, the DNA polymerase in *E. coli* has an error rate of approximately one base in 1 billion copied, which creates a mutation in the DNA strand. On average, how many *E. coli* cell divisions are required before the bacterium acquires one DNA mutation?

34. Some laboratory techniques rely on the ability to make random mutations in DNA using "error-prone" polymerases. A well-studied polymerase, called *Taq*, has an estimated error rate of 1 in 3500 base pairs. Suppose you are using the Taq polymerase to copy the DNA from a laboratory strain of *E. coli* bacteria with a total genome size of 4,639,221 base pairs. How many DNA copying errors do you predict?

35. The Polymerase chain reaction (PCR) is an important laboratory technique that is used to amplify a given DNA sequence by multiplying the number of DNA molecules. A PCR experiment doubles the number of DNA molecules in each replication cycle. If we start with one DNA molecule, calculate how many DNA molecules will be present after (a) 5 cycles, (b) 10 cycles, and (c) 30 cycles. (d) What word corresponds to the power of 10 in the answer to part (c) (e.g., thousand, million, etc.).

36. Examine the following DNA sequence. Circle the pairs of nucleotides that are capable of forming a pyrimidine dimer following prolonged exposure to UV light.

    5'-AATCGTAGCCTGAGCAGTA-3'

37. The molecule shown below, called *proflavine*, used as an antiseptic against some types of bacteria. Proflavine can induce mutations in bacterial DNA. Propose a hypothesis for how proflavine causes these DNA mutations.

Exploring Concepts

38. As discussed in the chapter, Watson and Crick used Franklin's data from Photo 51 to construct their model of the DNA double helix. However, they did not acknowledge Franklin's work in their famous 1953 publication of DNA's structure. Do you think that Watson and Crick's actions were unethical? Or do you think that they were justified in omitting Franklin's contribution because of their urgency to make an important scientific discovery? Write a one-paragraph statement to explain and defend your position regarding Watson and Crick's behavior.

39. In humans, DNA is repaired by enzymes that cut out damaged bases and insert new ones. In this process, it is important that the damage take place in only one strand of the DNA and not in both strands at the same site. Explain why damage to a single strand of DNA is readily repaired, whereas damage to both strands is much more severe.

40. *Theobromine* is a molecule that closely resembles the structure of certain DNA bases. Use the Internet to investigate theobromine, and draw its molecular structure. In what type of food is theobromine found? (*Hint:* It's delicious.)

41. Suppose you are investigating replication of the DNA from a newly discovered Martian life form. You suspect that it replicates by breaking down its DNA from one generation and resynthesizing both strands with a mixture of old and new nucleotides. Predict what you would expect to observe if you performed an isotope-labeling experiment like the one used by Meselson and Stahl.

42. *Ethidium* (shown below) is a dye that is used to stain DNA. After the ethidium binds to DNA, it emits fluorescent light when it is illuminated by ultraviolet light. Based on what you have learned in this chapter, predict how ethidium binds to DNA, and identify the region of the molecule that is involved in the binding.

43. You probably know somebody who has experienced cancer. According to the American Cancer Society, the most common types of cancer in the United States include breast cancer, skin cancer, lung cancer, colon cancer, and prostate cancer. Select one type of cancer that has affected a friend, family member, or perhaps you. If you do not know anyone who has experienced cancer, then pick a type of cancer that you find interesting. Use the Internet to investigate the cancer you have chosen, and write a one-page report at a level that is accessible to the general public. In particular, investigate the DNA mutations that are involved in the cancer and how these mutations affect cell division. Cancer is a complicated topic, and your research will reveal scientific terms that you do not recognize. Use your best scientific judgment to understand these terms and translate them into more accessible language.

44. Tobacco smoke contains more than 50 chemical carcinogens that have been evaluated by the International Agency for Research on Cancer. Use the Internet to investigate the chemical carcinogens listed below. For each molecule, draw its two-dimensional structure, and write one to two sentences on your observations concerning its structure. For example, what types of hydrocarbon regions does the molecule contain? Does the molecule contain functional groups that you recognize?
    (a) Hydrazine
    (b) 1,3-Butadiene
    (c) Ethyl carbamate

45. *Doxorubicin* (shown here) is an anticancer drug. Use the Internet to investigate how this drug works, and write a one-paragraph report. In your discussion, connect what you have learned about the drug to the chemical principles discussed in this chapter.

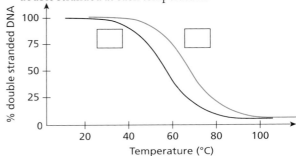

46. (a) Based on what you have learned about UV light and skin cancer, do you think that the use of indoor tanning beds should be banned for minors under 18 years of age? Explain your argument for or against such a ban. (b) Some states have already restricted access to indoor tanning beds. Use the Internet to discover whether there are any restrictions within your home state. Write a one-sentence summary of the current law in your area.

47. Approximately 1 out of every 250,000 children born in the United States has a rare genetic disease called *xeroderma pigmentosum*. Affected individuals experience severe skin damage from exposure to UV light (see below), and they are 1000 times more likely to develop skin cancer. Use the Internet to investigate the cause of *xeroderma pigmentosum*, and write a one- to two-paragraph report summarizing your findings.

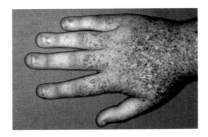

48. DNA "melting" is the process by which double-stranded DNA turns into single strands upon heating. A "melting curve" is a graph that shows the proportion of double-stranded DNA that remains when the temperature is increased. The diagram below shows a melting curve for the following sequence of double-stranded DNA:

5′ - T C C T C C T T T C C T C C T - 3′
3′ - A G G A G G A A A G G A G G A - 5′

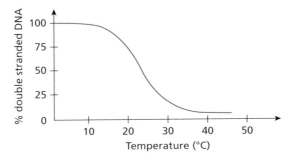

The x-axis of the graph indicates the temperature, and the y-axis shows the percentage of DNA that is double stranded at each temperature.

(a) Draw a dashed line across the graph at the 50% mark. What is the approximate temperature at which 50% of the DNA is double stranded? This is known as the melting temperature of the sequence.
(b) What percentage of the DNA double helix is composed of GC base pairs (%GC content)?
(c) Suppose you were to measure the melting temperature of DNA that is the same length as the one shown above but has a higher %GC content. Do you expect the melting temperature to be higher or lower than your answer in part (a)? Explain.

49. The diagram below shows melting curves for two different sequences of DNA (A and B) that are the same length. The x-axis of the graph indicates the temperature, and the y-axis shows the percent of DNA that is double stranded at each temperature.

A: 5′ - GAT CTT AGA TTA TTA GCT TAA T - 3′
B: 5′ - GGC CGA TTG CGG TAA TGG CTA G - 3′

(a) For each sequence (A and B), write the sequence of the complementary DNA strand. Include the 5′-end and 3′-end.
(b) For the double-helix molecules in part (a), what percentage of each one is composed of GC base pairs (%GC content)?
(c) Based on your calculation in part (b), match the sequence (A or B) to the appropriate melting curve. Write the letter of the sequence in the box, and explain your reasoning.
(d) The melting temperature of a DNA sample is the temperature at which 50% of the DNA is double stranded. Use the graph to determine the melting temperature for each sequence (A and B).

50. DNA melting curves can be used to examine the evolutionary relationships between different biological species. Humans share 98.8% of our DNA (on average) with chimpanzees, and we share 93% of our DNA (on average) with rhesus monkeys. Imagine that we construct a *hybrid DNA molecule*, with one strand composed of human DNA and the other strand composed of DNA from another species. We then perform an experiment to measure the melting temperature of the human–chimp hybrid DNA and the human–rhesus hybrid DNA. (a) Predict which hybrid DNA molecule will have the higher melting temperature and which will melt at the lower temperature. What is the reason for your prediction? (b) What does this experiment tell us about how closely humans are related to chimpanzees and rhesus monkeys? (c) What experiment would you perform to examine how closely humans are related to gorillas?

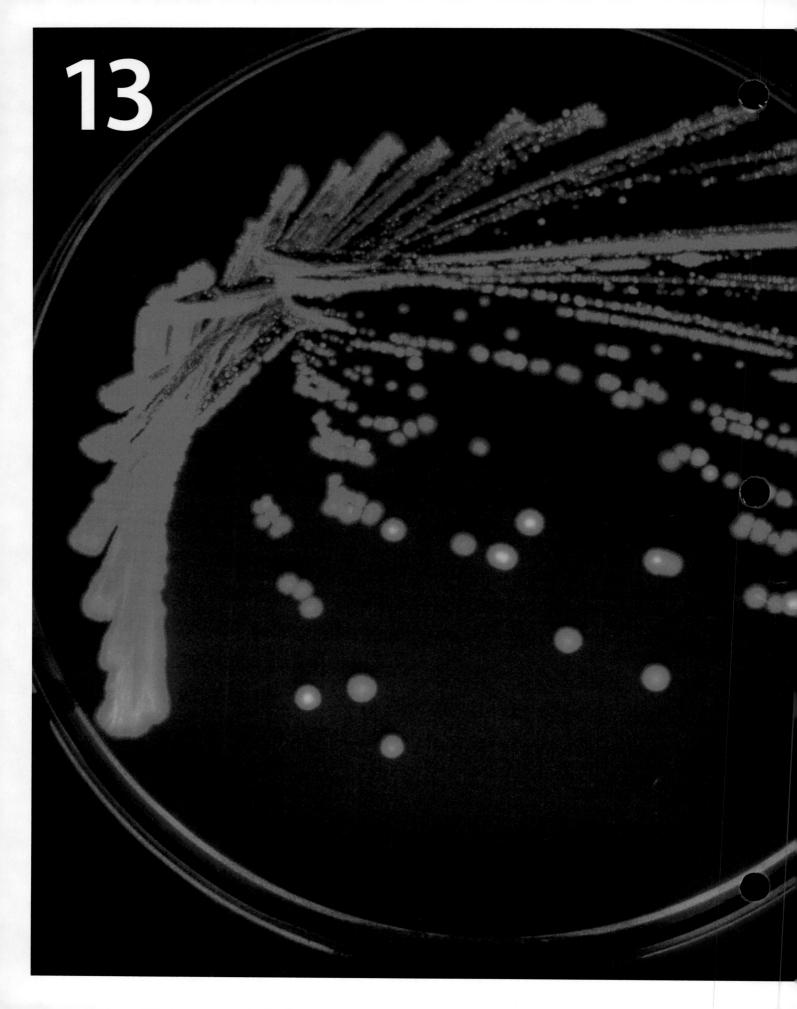

From DNA to Proteins

In 2007, Ryan Clark was playing a football game for the Pittsburgh Steelers in the home stadium of the rival Denver Broncos. During the game, Clark experienced severe pain in his left side and had to be rushed to the hospital. Doctors later performed surgery to remove Clark's spleen and gallbladder, which caused him to miss the remainder of the season. What caused this serious medical emergency?

Learning Objective:

Identify the characteristics of sickle cell disease.

THE KEY IDEA: Sickle cell disease is caused by a DNA mutation in the gene for hemoglobin.

13.1   What Causes Sickle Cell Disease?

At the time of his medical crisis, Ryan Clark had been playing in the National Football League for five years (Figure 13.1). There were no warning signs for his experience on the field in Denver. To understand what happened to Clark, we need to revisit a surprising medical discovery from over a century ago.

In 1910, Dr. James B. Herrick (1861–1954), a professor of medicine in Chicago, published the case of a dental student from the Caribbean. The patient had come to the hospital feeling weak and dizzy. These are typical symptoms of anemia, a disorder caused by a reduced number of red blood cells. When a medical intern examined the patient's blood using a microscope, he reported something unusual to Herrick. Normal red blood cells have a round shape, but some of the patient's cells had a distorted shape that looked like the letter "C" (Figure 13.2). When Herrick published his results, he described these unusual cells as "sickle-shaped," like the curved blade of a sickle that is used to harvest crops. Herrick's journal paper was the first description of **sickle cell disease**, which is also known as *sickle cell anemia*.

In the 1950s, the American chemist Linus Pauling (discussed in Chapter 12) discovered that sickle cell disease is caused by a change in *hemoglobin*, the protein that transports oxygen molecules within red blood cells. Later research pinpointed the origin of this change in a patient's DNA. In general, regions of DNA called *genes* contain the instructions to make proteins, which perform most of the body's important functions. Sickle cell disease is caused by a DNA mutation in the gene for hemoglobin, which affects the structure and function of the protein.

A patient with sickle cell disease inherits two copies of this mutated gene, one from each parent. Ryan Clark did not have sickle cell disease. Instead, he has a related condition called *sickle cell trait*, which is caused by inheriting *one copy* of the mutated hemoglobin gene. This condition was undiagnosed because a person with sickle cell trait does not usually have medical symptoms. However, Denver's home stadium is at an altitude of one mile, the highest in the National Football League

FIGURE 13.1 Ryan Clark was a football player for the Pittsburgh Steelers. During a football game in Denver, Clark experienced a career-threatening medical emergency.

(NFL). At this altitude, there are fewer oxygen molecules in every breath. For somebody with sickle cell trait, performing strenuous exercise at low oxygen levels can induce some red blood cells to turn into a sickle shape. This was the cause of Clark's pain and damaged spleen, which serves as a filter for the body's red blood cells. After this career-threatening emergency, Clark returned to playing football the next season, but the coach sat him on the bench during subsequent games in Denver. Clark retired in 2015, after playing 13 seasons in the NFL and winning a Super Bowl.

Sickle cell disease provides a case study of how the genetic information in DNA is used to make proteins. In addition to its biological importance, this process has particular relevance for medicine. The U.S. Centers for Disease Control and Prevention estimates that sickle cell disease affects approximately 100,000 individuals in the United States. By gaining a deeper understanding of sickle cell disease and of other DNA-based human diseases, we will become better equipped to discover treatments and cures.

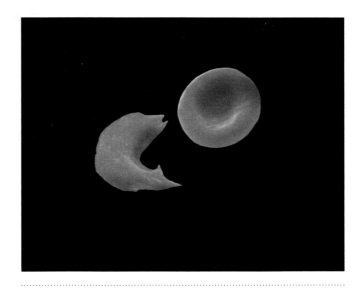

FIGURE 13.2 Normal and sickled red blood cells. In this magnified microscope image, a normal red blood cell (right) has a round shape, whereas a sickled red blood cell (left) is shaped like the letter "C." This distortion of red blood cells is the characteristic symptom of sickle cell disease.

Core Concepts

- The characteristic symptom of **sickle cell disease** is the distortion of red blood cells into a shape that resembles the letter "C."
- Sickle cell disease is caused by a DNA mutation in the gene for hemoglobin.

sickle cell disease Ann inherited disease that is characterized by the distortion of red blood cells into a sickle shape.

13.2 Overview: The Expression of Genetic Information

In Chapter 12, we learned that the genetic information in a cell is stored in the specific sequence of nucleotide bases in its DNA molecules. The biological pathway by which DNA sequences are processed to make proteins is called **gene expression**. This section provides a brief overview of the important steps of gene expression, and further details are provided later in the chapter. We will focus on the processes that occur within human cells, which are summarized in Figure 13.3.

Almost all human cells contain a cell nucleus, which stores the DNA. However, the cell's protein-making factories—called **ribosomes**—are located outside the nucleus in a region of the cell called the *cytoplasm*. Because DNA does not leave the nucleus, another molecule must exist that can be transported outside the nucleus to guide protein synthesis. This molecule is a different type of nucleotide polymer called **ribonucleic acid** (RNA). More specifically, the molecule is called **messenger RNA** or (mRNA) because it carries the genetic message from the DNA. The process of converting DNA's genetic information into mRNA is called **transcription** because the language of DNA and mRNA uses similar "letters" (nucleotides). The mRNA is not an exact copy of the DNA, but it preserves the essential genetic information.

Transcription occurs within the cell nucleus. After the mRNA has been generated, it is trimmed and processed chemically. Next, the mRNA exits the cell nucleus by passing through structures called "pores" that exist in the nuclear envelope. Once outside the nucleus, the mRNA travels to the region of the cell where the ribosomes are located. The ribosomes read the genetic information stored in mRNA to construct a sequence of amino acids in a protein. This process is called **translation** because it involves a conversion between two types of "molecular languages"—nucleotides and amino acids. The protein then folds into a specific three-dimensional shape to carry out its biochemical tasks. The

Learning Objective:
Outline the molecules and processes that are involved in the expression of genetic information.

THE KEY IDEA: The expression of genetic information proceeds from DNA to messenger RNA to protein.

gene expression The biological pathway by which DNA sequences are processed to make proteins.

ribosomes Structures in which proteins are synthesized.

ribonucleic acid (RNA) A polymer composed of RNA nucleotides.

messenger RNA (mRNA) A type of RNA that carries DNA's genetic message.

transcription The process whereby the genetic information contained in DNA is converted into mRNA.

translation The process whereby the information in mRNA is converted into a sequence of amino acids in a protein.

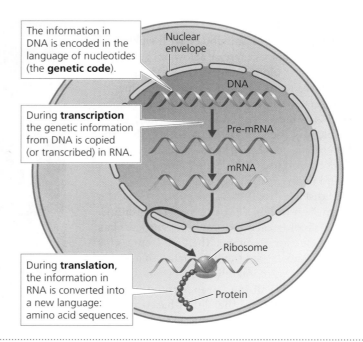

The information in DNA is encoded in the language of nucleotides (the **genetic code**).

During **transcription** the genetic information from DNA is copied (or transcribed) in RNA.

During **translation**, the information in RNA is converted into a new language: amino acid sequences.

Nuclear envelope

DNA

Pre-mRNA

mRNA

Ribosome

Protein

FIGURE 13.3 Overview of gene expression in a human cell. Transcription, which occurs in the cell nucleus, converts DNA's genetic information into a molecule of messenger RNA (mRNA). The mRNA then exits the nucleus and enters the cell's cytoplasm. Translation, which takes place within ribosomes, converts information stored in mRNA into a sequence of amino acids in a protein.

multitude of proteins gives cells their shape and size, their capacity to divide, the ability to generate energy and other attributes of life.

In summary, *gene expression proceeds from DNA to mRNA to protein*. This scientific principle holds for all cells, from bacteria to humans. Understanding how cells process genetic information is essential for comprehending genetic disorders like sickle cell disease. If a mutation occurs in the base sequence of DNA, then this "spelling mistake" is transmitted to the mRNA through transcription. The mRNA mutation can change the identity of one or more of the amino acids that make up the protein, disrupting the protein's normal function within the cell. In the following sections, we examine transcription and translation in greater detail. Later in the chapter, we use this information to explain the molecular origin of sickle cell disease.

Core Concepts

- **Gene expression** is the pathway for utilizing the genetic information in DNA.
- Genetic information is transported outside the nucleus by **messenger RNA** (mRNA), which is a type of **ribonucleic acid** (RNA).
- The conversion of DNA into mRNA is called **transcription**. The conversion of mRNA into a sequence of amino acids in a protein is called **translation**. Proteins are made by **ribosomes**, which are located in the cell's cytoplasm.

13.3 Transcription

Learning Objective:
Characterize the stages of transcription.

This section describes the molecular processes by which genetic information in DNA is converted via transcription into mRNA. However, mRNA is not the only type of RNA that plays a role in gene expression. Later in the chapter, we introduce these other types of RNA and explain their biological roles.

We first examine the structural and chemical differences between DNA and RNA, which apply to *all* types of RNA.

Comparing DNA and RNA

Figure 13.4 illustrates four structural and chemical differences between DNA and RNA. First, the polynucleotide strands of DNA are much longer than those of RNA. The DNA in a human chromosome contains many millions of nucleotides. In contrast, an RNA molecule contains only hundreds or thousands. Second, we learned in Chapter 12 that the structure of DNA is a double helix composed of two polynucleotide strands. By contrast, RNA usually exists as a single strand, although this strand can form internal base pairs that give it a complex three-dimensional shape.

Third, DNA and RNA contain different sugar molecules in their nucleotides: DNA contains deoxyribose, whereas RNA contains **ribose** (which accounts for the "R" in RNA). As illustrated in Figure 13.4(c), ribose has an oxygen atom attached to the 2'-carbon, whereas deoxyribose does not (hence the name "deoxy"). This small difference has a major influence on the molecular structure of RNA. As we learned in Chapter 3, each atom in a molecule occupies a volume based on the number of occupied electron shells. An oxygen atom has electrons in the first and second electron shells, so its atomic volume is much larger than a hydrogen atom, which has only a single electron in the first shell. It is impossible to wind RNA into a tight double helix like DNA because the extra oxygen atom in ribose clashes with other nearby atoms.

Fourth, RNA contains the base **uracil** instead of thymine as found in DNA. (The other three bases are the same in both DNA and RNA.) The difference between the two bases is that thymine has a methyl ($—CH_3$) group attached to the pyrimidine ring, whereas uracil has only a hydrogen atom. This change does not affect the hydrogen-bonding functional groups within the base; therefore, uracil can form hydrogen bonds with adenine in the same way that thymine can.

DNA
Stores information to make RNA and proteins; copies this information to daughter cells.

(a)

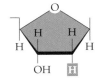
Double stranded and very long

(b)

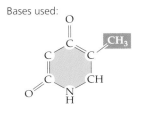
Deoxyribose is the sugar

(c)

Bases used:

Thymine (T)
Cytosine (C)
Adenine (A)
Guanine (G)

RNA
Carries information to make proteins; assists with protein synthesis.

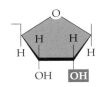
Generally single stranded and shorter

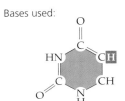
Ribose is the sugar

Bases used:

Uracil (U)
Cytosine (C)
Adenine (A)
Guanine (G)

(d)

FIGURE 13.4 Differences between DNA and RNA. There are four primary differences between DNA and RNA.

Core Concepts

- There are four structural and chemical differences between DNA and RNA:
  1. DNA strands are much longer than RNA strands.
  2. DNA forms a double helix, whereas RNA usually exists as a single strand.
  3. DNA contains deoxyribose as its sugar, whereas RNA contains **ribose**.
  4. DNA contains thymine as a base, which is replaced by **uracil** in RNA.

Synthesizing mRNA

Transcription of a DNA sequence is performed by an enzyme called **RNA polymerase.** This enzyme is similar to DNA polymerase (see Chapter 12), but it synthesizes a polymer of RNA nucleotides instead of DNA nucleotides. As a first step, RNA polymerase binds to a special region of DNA called a *promoter site.* Next, the enzyme unwinds a short region of the DNA double helix, which exposes the sequence of bases in the DNA. One strand of DNA, called the **template strand,** is used by the enzyme as the molecular template for synthesizing a new

THE KEY IDEA: DNA and RNA have structural and chemical differences.

ribose The sugar found in the nucleotides that make up RNA.

uracil A pyrimidine base found in RNA; it replaces the thymine found in DNA.

THE KEY IDEA: During transcription, RNA polymerase produces an mRNA strand that is complementary to the DNA sequence.

FIGURE 13.5 Transcription converts a DNA base sequence into an mRNA base sequence. During transcription, base sequence information within DNA is copied into a new molecule of (mRNA). The RNA polymerase enzyme uses the DNA template strand to create a complementary sequence of mRNA, according to the base-pairing rules. Because of the antiparallel orientation, the growing 3'-end of the mRNA strand is closer to the 5'-end of the template DNA strand.

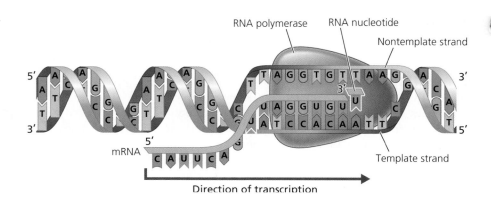

Direction of transcription

RNA polymerase   An enzyme that synthesizes a polymer of RNA nucleotides during transcription.

template strand   The DNA strand that RNA polymerase uses as the molecular template for synthesizing a new strand of mRNA.

strand of mRNA (Figure 13.5). We introduced this principle of *template-directed synthesis* in Chapter 12 when we described DNA replication. The RNA polymerase moves along the DNA, synthesizing the mRNA strand in the 5' → 3' direction by adding new RNA nucleotides to the 3'-end. The newly created mRNA strand has an *antiparallel orientation* with respect to the template DNA.

RNA polymerase synthesizes the mRNA strand by adding nucleotides according to the base-pairing rules. For example, if the enzyme reads a cytosine (C) in the DNA template strand, then it will add an RNA nucleotide with a guanine (G). If the DNA contains adenine (A), then the mRNA will incorporate a nucleotide with uracil (U) instead of thymine (T). As a result of this base pairing, the sequence of bases in the mRNA strand is *complementary* to the sequence of bases in the template strand of DNA. Although the mRNA sequence is not a duplicate of the original DNA sequence, the original genetic information is preserved.

WORKED EXAMPLE 13.1

Question: A template sequence of DNA is provided below. Transcribe this DNA into a sequence of mRNA, and include the 5'- and 3'-ends.

$$3'\text{-}A\,T\,G\,G\,A\,C\,C\,G\,T\,T\,A\,T\,T\,G\,G\,C\,A\,G\text{-}5'$$

Answer: Transcription produces a sequence of mRNA bases that is complementary to the base sequence in the template DNA strand. Recall that adenine (A) in DNA pairs with uracil (U) in mRNA. In addition, the mRNA strand is oriented antiparallel to the template DNA strand, so the 3'-end of DNA is aligned with the 5'-end of mRNA (and vice versa).

DNA   3'- A T G G A C C G T T A T T G G C A G - 5'
mRNA 5'- U A C C U G G C A A U A A C C G U C - 3'

TRY IT YOURSELF 13.1

Question: A template sequence of DNA within a gene is provided below. Transcribe this DNA into a sequence of mRNA, and include the 5'- and 3'-ends.

$$3'\text{-}G\,C\,A\,A\,T\,G\,C\,A\,G\,C\,C\,C\,T\,A\,G\,C\,C\text{-}5'$$

Core Concepts

- To begin transcription, **RNA polymerase** binds to a promoter site and then unwinds the DNA double helix to expose the base pairs. RNA polymerase uses the DNA **template strand** and the base-pairing rules to synthesize a complementary strand of mRNA.
- During transcription, an adenine base in DNA is paired with a uracil base (not thymine) in mRNA.

.4 Amino Acids: The Building Blocks of Proteins

After the mRNA strand has been produced by transcription, the sequence of mRNA bases provides the instructions to generate a sequence of amino acids in a protein. In this section, we examine the molecular structure of amino acids, which are the molecular building blocks for protein structure.

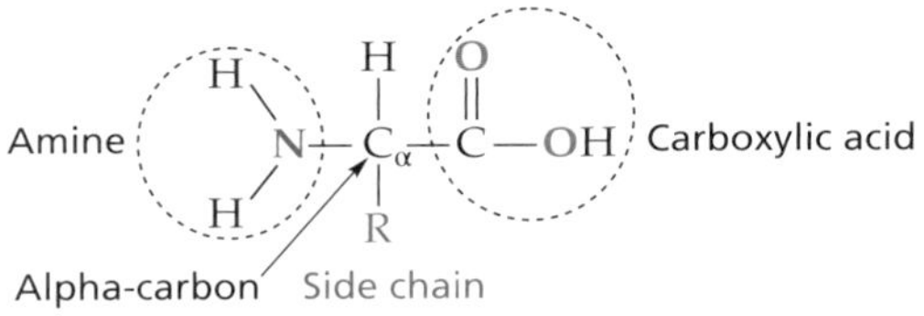

Chemical Diversity of Amino Acids

In Chapter 5, we presented the general structure of an amino acid, which is shown again in Figure 13.6. An amino acid contains a central carbon atom called the *alpha-carbon* (C_α), which forms single covalent bonds with four chemical groups. Two of these groups—the amine and carboxylic acid—are the basis for the term *amino acid*. The third "group" is a hydrogen atom. The fourth group is called the *sidechain* and is represented as R. The sidechain exists as 20 different varieties, which define the 20 standard amino acids commonly found in proteins.

Figure 13.7 provides the molecular structures of all 20 amino acids. In the figure, the carboxylic acid and the amine groups are ionized, which occurs when the pH is at or close to 7. We can classify the amino acids into four categories based on the chemical properties of their sidechains. These categories are called *neutral nonpolar*, *neutral polar*, *acidic*, and *basic*. Table 13.1 describes the key chemical features of each sidechain class. As we shall see later, the specific properties of amino acid sidechains play an important role in explaining the molecular origin of sickle cell disease.

FIGURE 13.6 General structure of an amino acid. An amino acid contains two functional groups, an amine and a carboxylic acid, that are bonded to the alpha-carbon. The sidechain R exists in 20 different varieties that define the identity of the amino acid.

Learning Objective:

Describe the molecular properties of amino acids.

THE KEY IDEA: Amino acids are classified into four groups based on the chemical properties of their sidechains.

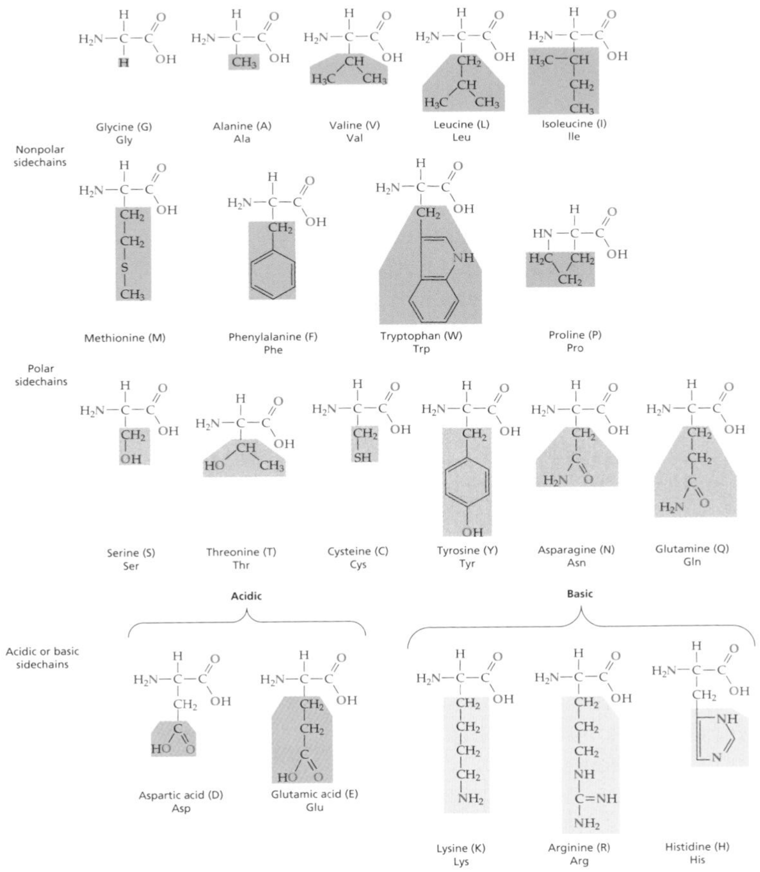

FIGURE 13.7 Molecular structures of the 20 standard amino acids. The amino acids are organized into four categories based on the chemical properties of the sidechains, which are highlighted in various colors. All functional groups are shown in their charge neutral form.

TABLE 13.1   Classification of Amino Acids Based on Their Sidechains

SIDECHAIN TYPE	CHEMICAL PROPERTIES
Neutral nonpolar	Many sidechains in this group are hydrocarbons and nonpolar. The nonpolar sidechains are not soluble in water, a polar solvent ((i.e., they are *hydrophobic*). The NH group in tryptophan is moderately polar, which makes this amino acid a borderline member of the group.
Neutral polar	These amino acid sidechains contain polar functional groups, particularly hydroxyl (—OH) and amide (—CO—NH$_2$). Because these sidechains are polar, they are *hydrophilic* and dissolve easily in water. They also form hydrogen bonds with a wide range of other molecules.
Acidic	There are two amino acids in this group, aspartate and glutamate. Both contain the carboxylic acid (—COOH) functional group in their sidechains, which loses a proton to become the negatively charged carboxylate group (—COO⁻) at pH 7. The acidic sidechains are soluble in water (i.e., they are *hydrophilic*).
Basic	The three sidechains in this group act as bases. Lysine and arginine have a basic nitrogen atom that accepts a proton to become positively charged at pH 7. Histine is a much weaker base, so it is not charged at pH 7. The basic sidechains are soluble in water (they are *hydrophilic*).

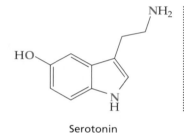

Serotonin

Serotonin is a neurotransmitter, a molecule that transmits chemical signals between nerve cells. Lowered levels of serotonin are associated with depression. The molecular structure of serotonin is shown at the side. Within our body, serotonin is chemically synthesized by using an amino acid as a starting compound. Use Figure 13.7 to identify this amino acid.

Core Concepts

- An amino acid contains a central alpha-carbon, amine and carboxylic acid groups, plus a variable group called a sidechain.
- The 20 standard amino acids can be classified into four categories based on the chemical properties of their sidechains. These categories are neutral nonpolar, neutral polar, acidic, and basic.

⊙➔ THE KEY IDEA: Chiral molecules have isomers that are mirror images of each other and cannot be superimposed.

■ **chiral** An object that cannot be superimposed onto its mirror image.

■ **chiral isomers** Two mirror-image forms of a molecule; they contain the same atoms arranged differently in space.

Amino Acids Are Chiral Molecules

You can purchase amino acids in a nutritional supplement store. If you examine the label of these supplements, you will see the letter "L" before the name of each amino acid. For example, Figure 13.8 shows a container of L-lysine. What does this letter "L" mean?

Amino acids have an important structural property. We say that amino acids are **chiral**, a term derived from the Greek word for "hand." Figure 13.9 illustrates some important principles of chirality using our own hands. For example, *your left and right hands are mirror images of each other.* If you hold your right hand in front of a mirror, then the reflected image matches your left hand. In addition, *mirror images cannot be superimposed.* This statement means that there is no way to orient your right hand in space so that it exactly fits over your left hand (and vice versa).

Because amino acids are chiral, they exist as two varieties: a "left-handed" form and a "right-handed" form. These two forms are called **chiral isomers** because they contain the same atoms that are arranged differently in space (recall our discussion of isomers in Chapter 4). The left-handed isomer is called L, and the right-handed isomer is called D (from the Latin words *laevus* and *dexter*). Technically, the terms *L* and *D* refer to the way the isomers rotate a certain type of light to the left or the right. Figure 13.10 compares the two chiral isomers of alanine, an amino acid that has a methyl group (—CH$_3$) as a sidechain. We can

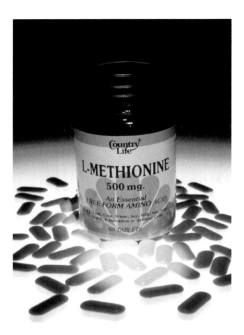

see that the L and D isomers are mirror images of each other, just like our left and right hands. In general, *chiral isomers are mirror images of each other that cannot be superimposed.*

All proteins are constructed from only L-amino acids. This is why nutritional supplement stores sell L-lysine and not D-lysine: Humans cannot use D-amino acids to make proteins. D-amino acids do have some utility in biology; for example, bacteria use them to construct their cell wall, as we shall see in Chapter 15.

Many other biological and nonbiological molecules are chiral. For example, glucose exists as D- and L-isomers, but our bodies use only D-glucose as an energy source. DNA and RNA exclusively use the L-isomer of the sugar in their nucleotides. Chirality is also important in the development of pharmaceuticals. One example is Nexium, an acid reflux medication that contains only one chiral isomer as its active ingredient. This selection was made after laboratory tests and clinical trials revealed that the mirror-image isomer was far less effective as a drug.

How can we tell if a molecule is chiral? For simple molecules, we can apply a general rule: *A molecule is chiral if four different chemical groups are covalently bonded to a carbon atom.* This principle is illustrated in Figure 13.11, using modified versions of methane in which some of the hydrogen atoms have been replaced by fluorine, chlorine, and bromine (all of these atoms are halogens from Group 7 of the periodic table). In molecule A, the carbon atom is bonded to four different atoms (H, F, Cl, and Br). *Molecule A is chiral because it has two mirror-image isomers that cannot be superimposed.* We can demonstrate this property by selecting the isomer to the right of the mirror and rotating it by 180°. The rotated molecule, shown to the far right of the figure, does not overlap with the isomer to the left of the mirror. The carbon atom is highlighted by an asterisk because it serves as a *chiral center* for the molecule.

By contrast, the central carbon atom in Molecule B is bonded to only three types of atoms because the chlorine atom is duplicated. If we select the molecule to the right of the mirror and rotate it by 180°, we generate the

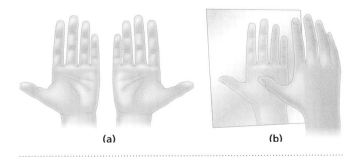

(a)   (b)

FIGURE 13.9 Using hands to illustrate chirality. (a) Your left and right hands are mirror images of each other. (b) The reflection of your right hand in a mirror produces an image of your left hand.

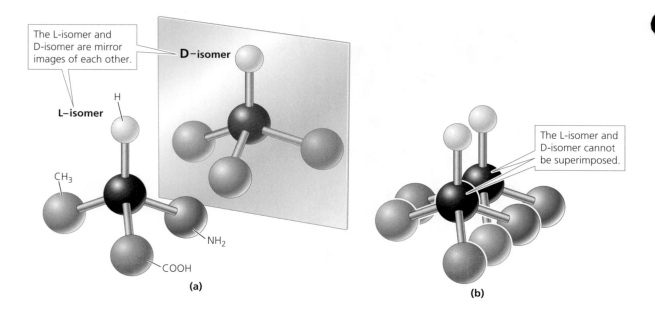

FIGURE 13.10 **The two chiral isomers of an amino acid.** (a) The L- and D-isomers of alanine are mirror images of each other. (b) These two chiral isomers cannot be superimposed by rotating them in space.

same molecule that is on the left of the mirror. *Molecule B is not chiral because the mirror images can be superimposed.* The carbon atom is not marked with an asterisk has because it is *not* a chiral center.

CONCEPT QUESTION 13.1

We said that amino acids are chiral molecules, but there is a caveat to this statement. Of the 20 standard amino acids, 19 of them are chiral, but one is not. Refer to Figure 13.7 to identify the amino acid that is not chiral.

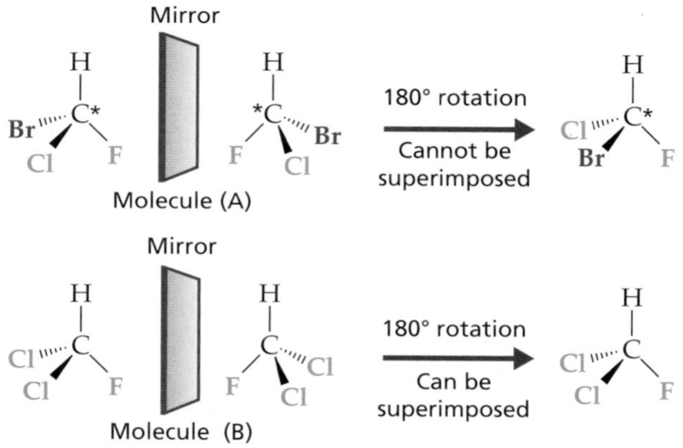

FIGURE 13.11 **Molecule A is chiral because its mirror-image isomers cannot be superimposed, even when one isomer is rotated in space.** The carbon atom is marked with an asterisk because it has a chiral center within the molecule. Molecule B is not chiral because rotating the mirror image produces a structure that can be superimposed on the original.

WORKED EXAMPLE 13.2

Question: Rubbing alcohol, which is sold in drugstores, is an aqueous solution of *isopropanol*. The two-dimensional molecular structure of isopropanol is as follows.

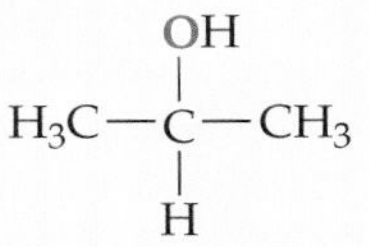

$$H_3C-\underset{\underset{H}{|}}{\overset{\overset{OH}{|}}{C}}-CH_3$$

Isopropanol

(a) Based on this molecular structure, is isopropanol a chiral molecule? Explain your answer.

(b) Using the wedge-and-dash notation, draw a three-dimensional structure of aisopropanol and its mirror image (include a mirror in your drawing). Can the two molecules be superimposed on each other? Use drawings to illustrate your answer. Explain the connection between your answers to parts (a) and (b).

Answer: (a) According to our rule, a molecule is chiral if four different chemical groups are covalently bonded to a carbon atom. The structure of isopropanol has *three* different chemical groups bonded to the central carbon atom. Therefore, isopropanol is *not* a chiral molecule.

(b) Using the principle of valence shell electron pair repulsion (VSEPR), we predict that isopropanol will have a *tetrahedral structure* because it contains four single covalent bonds. Three-dimensional structures of isopropanol and its mirror image are provided below (the two molecules are designated A and B for reference). If we rotate Molecule B by 180°, we obtain a structure that is identical to Molecule A. Consequently, *the mirror-image molecules can be superimposed on each other.* This result is expected because our answer to part (a) identified isopropanol as a molecule that is *not* chiral.

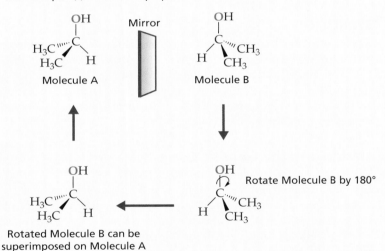

TRY IT YOURSELF 13.2

Question: Lactic acid gives the sour taste to yogurt. Its two-dimensional structure is as follows.

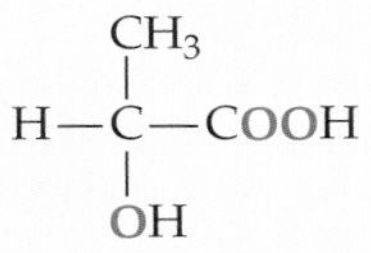

(a) Based on this molecular structure, is lactic acid a chiral molecule? Explain your answer.

(b) Using the wedge-and-dash notation, draw a three-dimensional structure of lactic acid and its mirror image (include a mirror in your drawing). Can the two molecules be superimposed on each other?

(c) Explain the connection between your answers to parts (a) and (b).

$$H-\underset{\underset{OH}{|}}{\overset{\overset{CH_3}{|}}{C}}-COOH$$

Lactic acid

PRACTICE EXERCISE 13.2

Limonene is a hydrocarbon molecule that exists as two chiral isomers. One isomer produces the characteristic smell of lemons, whereas the other produces the distinctive smell of oranges. The line-angle drawing of one chiral isomer of limonene is shown below. In this molecule, the carbon atom that acts as the chiral center (marked with an asterisk) occurs within a hydrocarbon ring. Provide a line-angle drawing of the other isomer.

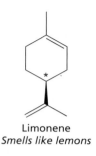

Limonene
Smells like lemons

Core Concepts

- Amino acids are **chiral** and exist as two structural forms: a left-handed form (L) and a right-handed form (D). Proteins use only L-amino acids.
- **Chiral isomers** are mirror images of each other, and they cannot be superimposed.
- A molecule is chiral if it has four different chemical groups bonded to a carbon atom.

13.5 Translation

Learning Objective:

Outline the stages by which mRNA is translated into a sequence of amino acids in a protein.

Amino acids provide 20 different molecular building blocks for making proteins. In this section, we examine the biological processes by which the information stored within an mRNA molecule determines the specific sequence of amino acids in a protein. Translation is much more complex than transcription, just as translating a written text from one language to another is more challenging than simply transcribing another copy.

Translation and the Genetic Code

THE KEY IDEA: Translation uses the genetic code to convert an mRNA base sequence into a sequence of amino acids in a protein.

Translation is the process by which a sequence of bases in mRNA is converted into a sequence of amino acids in a protein. How many mRNA bases are required to specify the 20 standard varieties of amino acid? There are four different mRNA bases (A, U, G, C), so a single base is not enough. Will two bases work? Two bases provide $4 \times 4 = 16$ possibilities, which is still insufficient for 20 amino acids. What about three bases? With three bases, there are $4 \times 4 \times 4 = 64$ different possibilities. This number is sufficient to cover 20 amino acids. In fact, it provides more options than are necessary, which has important implications for how translation occurs.

codon A sequence of three bases in mRNA, which specifies a particular amino acid.

Sequence information in mRNA is read in units of three bases called **codons**. Figure 13.12 illustrates how a sequence of mRNA bases is organized into codons and how each codon specifies a particular amino acid within a protein. How do we know which codon specifies which amino acid? If we wish to translate a written text from one language to another, then we need to use a bilingual dictionary. For biological translation, we have two molecular languages: nucleotides and amino acids. In this case, the role of the dictionary is provided by the **genetic code**. *The genetic code specifies which mRNA codon corresponds to which amino acid*. The original experiments used to discover the genetic code are described in *Science in Action* on page 406.

genetic code The instructions that specify which mRNA codon corresponds to which amino acid.

Figure 13.13 presents the genetic code in the form of a table. You read the table by picking the first letter from the left-hand column, the middle letter from the top row, and the third letter from the right-hand column. For example, the codon GGU, located at the lower right hand of the table, corresponds to the amino acid called *glycine*.

As discussed earlier, four varieties of mRNA bases produce 64 possible codons, which is more than required to specify the 20 different types of amino acid. Consequently, *the genetic code allows most amino acids to be specified by more than one codon*. For example, we learned that the codon GGU corresponds to glycine. However, glycine is also specified by three other codons: GGC, GGA, and GGG. Because of this property, the genetic code is said to be *degenerate*. This scientific use of this term is different from its meaning in everyday speech; a more familiar word is "redundant." Having multiple codons for a single amino acid helps to mitigate the effect of DNA mutations. If a DNA mutation produces another codon that specifies the same amino acid, the mutation has no effect on the construction of the protein. This type of DNA mutation is called a *silent mutation* or a *synonymous mutation*.

Several codons have specialized functions that signal the translation process to start or stop. For example, the AUG codon serves as the **start codon** (it also specifies the amino acid methionine). This codon instructs the ribosome to begin

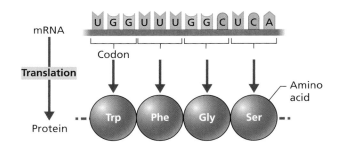

FIGURE 13.12 Translation. During translation, a triplet of mRNA bases—called a codon—specifies which amino acid is included in a protein.

start codon The codon that provides a signal to begin translation.

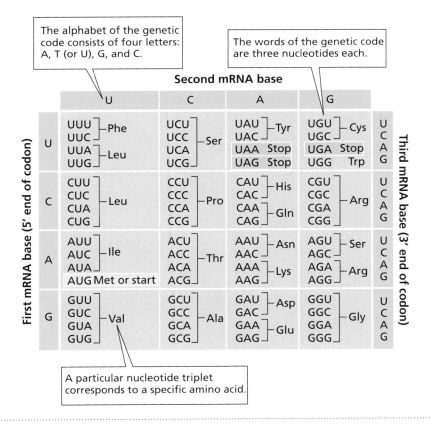

FIGURE 13.13 **The genetic code.** This code provides the correspondence between a specific codon (three mRNA bases) and its amino acid. The genetic code also includes specialized codons that start and stop translation.

stop codon A codon that signals the end of translation.

constructing a protein molecule. In effect, AUG can serve the function of an upper-case letter in the first word of a sentence, which signals that a new sentence is beginning. In addition, three codons—UAA, UAG, and UGA—serve as **stop codons** that signal the end of translation. This function is analogous to a period that indicates the end of a sentence.

CONCEPT QUESTION 13.2

(a) A computer-based analysis of a large number of proteins revealed that tryptophan is the least common amino acid within these proteins. Use the genetic code table to propose a hypothesis for this observation.

(b) Based on your hypothesis for part (a), predict which amino acid(s) will be the most common in proteins.

WORKED EXAMPLE 13.3

Question: Use the genetic code table to answer the following questions:

(a) Which amino acid corresponds to each of the following codons?
CAG
AAG
GCG

(b) Which codons correspond to the amino acid arginine?

Answer: The amino acids are listed next to their corresponding codons.
CAG   glutamine
AAG   lysine
GCG   alanine

(b) The codons that correspond to arginine are CGU, CGC, CGA, CGG, AGA, and AGG.

TRY IT YOURSELF 13.3

Question: Use the genetic code table to answer the following questions:

(a) Which amino acid corresponds to each of the following codons?
CAG
UAC
AAG

(b) Which codons correspond to the amino acid leucine? How many of these codons exist?

WORKED EXAMPLE 13.4

Question: A sequence of template DNA is provided below. (a) Transcribe this DNA into a sequence of mRNA, and include the 5′- and 3′-ends. (b) Translate the mRNA into a sequence of amino acids.

3′-TTCGGTGCCAGGACTGTA-5′

Answer: (a) The mRNA sequence is complementary to the template DNA sequence, and the mRNA strand is oriented antiparallel to the DNA strand.

DNA   3'-T T C G G T G C C A G G A C T G T A -5'
mRNA 5'-A A G C C A C G G U C C U G A C A U -3'

(b) To translate the mRNA sequence, we use the genetic code (Figure 13.13). To highlight the codons, it is helpful to insert spaces between each triplet of mRNA bases. Translation of an mRNA sequence occurs in the 5' → 3' direction.

mRNA 5'-   A A G   C C A   C G G   U C C   U G A   C A U -3'
amino acids   lysine proline arginine serine   STOP

During translation, we encountered a STOP codon (UGA). This codon signifies the end of translation, so we do not assign an amino acid to the triplet of mRNA bases that follows the stop codon.

TRY IT YOURSELF 13.4

Question: A sequence of template DNA within a gene is provided below.
(a) Transcribe this DNA into a sequence of mRNA, and include the 5'- and 3'-ends.
(b) Translate the mRNA into a sequence of amino acids.

3'-A G G C C T C A G T C C T G G C T T -5'

Core Concepts

- Translation is the conversion of a sequence of bases in mRNA into a sequence of amino acids in a protein.
- Sequence information in mRNA is read in units of three bases called **codons**.
- The **genetic code** specifies which codon corresponds to which amino acid.
- The number of codons (64) is greater than the number of amino acids (20) that are used to make proteins. Consequently, the genetic code is degenerate—that is, most amino acids are specified by more than one codon.
- The genetic code has special codons that indicate when translation should begin (**start codon**) and end (**stop codons**).

The Molecular Mechanism of Translation

Translation occurs within the ribosome, a large molecular complex located in the cell's cytoplasm. The two subunits of the ribosome—one large and one small—fit like a clamp around the mRNA strand. This mRNA strand passes through the ribosome, and the genetic information stored within the mRNA codons is used to synthesize a specific sequence of amino acids that make up a protein. To increase the efficiency of transcription, multiple ribosomes are usually attached at different points along a single mRNA strand.

There is no molecular similarity between the three bases in an mRNA codon and its corresponding amino acid. Consequently, a different type of molecule in the cell must act as a "molecular translator," with the capacity to communicate with both the mRNA codon and the amino acid. This "translator" molecule is a type of RNA called **transfer RNA** (tRNA). There is at least one tRNA molecule for each of the 20 amino acids.

The structure of a typical tRNA molecule is provided in Figure 13.14. The tRNA contains a sequence of three bases called the **anticodon**, which binds to the mRNA codon. *The sequence of bases in the tRNA anticodon is complementary to the sequence of bases in the mRNA codon.* This complementary base pairing between the codon and anticodon ensures that the correct tRNA binds to the mRNA codon during translation. The two RNA sequences are antiparallel, which is typical of any pairing between two polynucleotides.

THE KEY IDEA: Translation occurs within the ribosome and is mediated by transfer RNA.

transfer RNA (tRNA) The type of RNA that transfers amino acids to the ribosome.

anticodon A sequence of three bases in tRNA that is complementary to the sequence of bases in the mRNA codon.

SCIENCE IN ACTION

How Did Scientists Crack the Genetic Code?

After Watson and Crick published the structure of DNA in 1953 (see Chapter 12), scientists turned their attention to the next big scientific question: How do cells use DNA's genetic information to make proteins?

An important breakthrough was made in a laboratory run by Marshall Nirenberg (1927–2010) at the U.S. National Institutes of Health (NIH) (Figure 1). These scientists had devised a method of making proteins in a test tube without the need for an intact cell. They created a mix of components that included ribosomes, amino acids, nucleotides, plus adenosine triphosphate (ATP) as a source of chemical energy. The only missing component was RNA. This system allowed them to add artificially synthesized RNA molecules to serve as the genetic message. What amino acids will be produced by specific base sequences of RNA?

When an artificial RNA consisting only of the base uracil (U) was fed into the test tube mixture, it generated a protein made of a single amino acid, phenylalanine. Other scientists had proposed that the information in mRNA is stored in units of three bases—a *triplet code*. Thus, Nirenberg was able to deduce the first word of the genetic code: UUU codes for phenylalanine. Next, he and his team used an mRNA strand consisting only of cytosine (C), which made

a sequence of another amino acid called *proline*. They now had a second word: CCC codes for proline. Using other synthetic RNAs containing mixtures of bases, they could decipher more codons.

The next important step was taken by Har Gobind Khorana (1922–2011) at the University of Wisconsin. Khorana invented a chemical method to synthesize long strands of RNA with exact repeats of specific nucleotides. His first RNA molecule consisted of two alternating nucleotides, UCUCUC, which resulted in a string of alternating amino acids: serine—leucine—serine—leucine, and so on. This discovery provided two codons: UCU codes for serine, and CUC codes for leucine. By synthesizing other combinations of RNA nucleotides, Khorana and his colleagues were able to decipher much of the genetic code. Additional work was needed to identify the stop and start codons, which were deduced using genetic experiments with viruses.

Understanding how the genetic code is used to make proteins has provided insights into fundamental biological processes, revealed the genetic origins of diseases, and enabled the development of genetic engineering. In 1968, Nirenberg and Khorana (together with a third scientist) were awarded the Nobel Prize in Physiology or Medicine.

FIGURE 1   Marshall Nirenberg working in his laboratory.
Nirenberg and his colleagues were the first scientists to deduce the relationship between a specific codon and its corresponding amino acid.

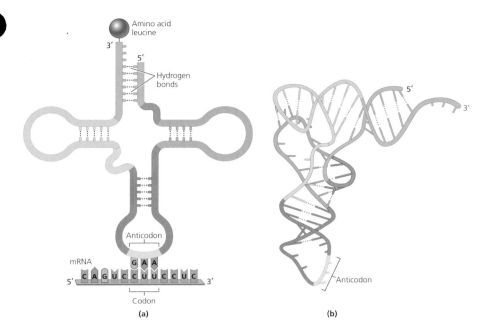

FIGURE 13.14 The structure of transfer RNA (tRNA). (a) A simplified depiction of tRNA, which indicates the anticodon and amino acid binding region. The tRNA anticodon has a base sequence that is complementary to the codon on the mRNA. (b) The three-dimensional structure of tRNA, which resembles an upside down "L." The colors in this structure correspond to the regions shown schematically in part (a).

The other end of the tRNA attaches to the particular amino acid that will be added to the growing protein. For the tRNA in Figure 13.14, the attached amino acid is *leucine*, which is specified by the mRNA codon to which the tRNA binds. Remember that the amino acid attached to the tRNA matches the codon on the mRNA, not the anticodon on tRNA. The relationships among the anticodon, codon, and amino acid are summarized as follows:

anticodon	3′ - G A A - 5′
codon	5′ - C U U - 3′
amino acid	Leucine

Once translation has been initiated by a start codon, the string of amino acids is made from the sequential action of tRNA molecules that bind to the mRNA strand. The specific codon within the mRNA determines which tRNA binds to the strand because only the tRNA with the correct anticodon will match the mRNA codon. Each tRNA carries its appropriate amino acid, which is added to the growing chain. We can think of tRNA molecules as different shuttle buses, each transporting a specific amino acid passenger to the ribosome. After depositing its amino acid, the tRNA is released from the ribosome to create space for the next tRNA molecule. When the ribosome detects the presence of a stop codon in the mRNA, translation is halted and the amino acid chain is released.

WORKED EXAMPLE 13.5

Question: A codon in a sequence of mRNA is 5′-A G U-3′. (a) What are the sequence and directionality of the anticodon on the corresponding tRNA? (b) Which amino acid is attached to this tRNA?

Answer: The anticodon sequence is complementary to the codon sequence, and it has the opposite directionality to the mRNA (i.e., the two mRNA strands are antiparallel).

codon	5′- A G U -3′
anticodon	3′- G C A -5′

(b) The amino acid that is attached to the tRNA is specified by the mRNA codon (not the tRNA anticodon). In the genetic code table, AGU corresponds to the amino acid *serine*. Therefore, serine will be attached to the tRNA molecule.

TRY IT YOURSELF 13.5

Question: A tRNA molecule has the anticodon base sequence 3′-G G G-5′. Which amino acid is attached to the corresponding tRNA?

Core Concepts

- Translation occurs within the ribosome, which synthesizes new proteins based on the genetic information provided by mRNA.
- **Transfer RNA** (tRNA) serves as a "molecular translator" between an mRNA codon and its corresponding amino acid.
- The tRNA molecule contains a sequence of three bases called an **anticodon**. This base sequence is complementary to the mRNA codon. Codon–anticodon binding ensures that the correct amino acid is used.
- A protein is made from the sequential action of tRNA molecules that bind to the mRNA strand, add their amino acid to the growing chain, and are then released.

13.6 Protein Structure

Learning Objective:

Illustrate the formation of a polypeptide chain and its structural properties within proteins.

🔑 **THE KEY IDEA:** Amino acids are joined by peptide bonds to make a polypeptide.

■ **peptide bond** The covalent bond that joins two amino acids.

■ **polypeptide** A polymer composed of amino acids joined by peptide bonds.

Translation at the ribosome produces a linked sequence of amino acids. This section illustrates how amino acids are joined together to make a polymer chain and how this polymer folds into distinctive three-dimensional structures within a protein.

Making a Polypeptide

Amino acids are joined together by a covalent chemical bond called a **peptide bond**. This bond is formed by a *condensation reaction* between the carboxylic acid group (—COOH) of one amino acid and the amine group (—NH$_2$) of a neighboring amino acid. This reaction eliminates an H$_2$O molecule and links the two amino acids. Figure 13.15 illustrates the formation of a peptide bond between two amino acids, alanine and serine. The collection of atoms on either side of the peptide bond—that is, the C, O, N, and H atoms—is called the *peptide group*. This group has the same chemical structure as the amide group presented in Chapter 5. However, the peptide group refers specifically to a linkage between amino acids and not to any other type of molecule.

The molecule shown in Figure 13.15 is called a dipeptide because it is made from two amino acids. Similarly, a molecule that is constructed from three amino acids is a tripeptide. More amino acids can be added by repeating the same condensation reaction until the molecule becomes a polymer containing multiple amino acids. At this point, it is called a **polypeptide.** As with other polymers, we often refer to a *polypeptide chain*. Within the polypeptide chain, each amino acid is linked to two neighbors via peptide bonds. The two exceptions are found at the ends of the chain, which are called the *N-terminus,* where the free —NH$_2$ group is located, and the *C-terminus,* where the free —COOH group is located. During translation, new amino acids are added to the C-terminus of the polypeptide chain, which grows in the N-to-C direction. In this manner, protein synthesis has a specific directionality, just like the synthesis of a polynucleotide strand of DNA or mRNA.

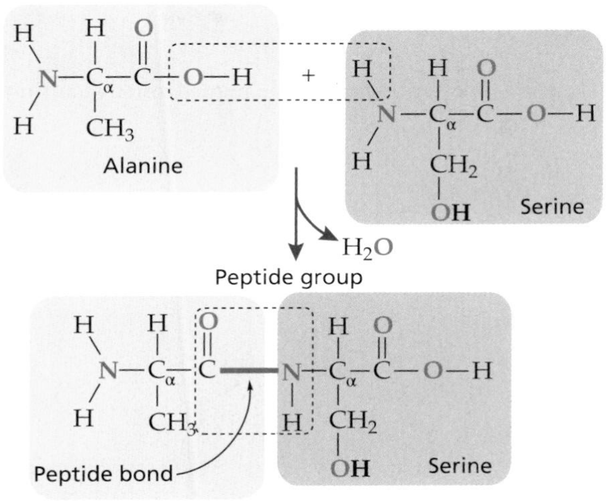

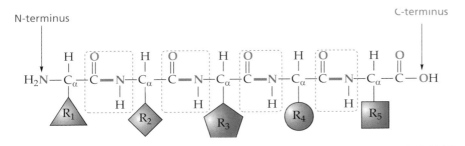

FIGURE 13.15 Formation of a peptide bond. A peptide bond between two amino acids is formed by a condensation reaction between the —COOH group of one amino acid and the —NH₂ group of a neighboring amino acid.

Figure 13.16 illustrates a small polypeptide with five amino acids. The side-chains are represented by different geometric shapes with the labels R_1, R_2, and so on. Note that the *backbone* of the polypeptide chain alternates between a peptide group and a C_a atom that is attached to the amino acid sidechain. Because the backbone remains constant, the uniqueness of each polypeptide chain arises from the specific sequence of amino acid sidechains. This situation is similar to a polynucleotide strand, in which the sugar-phosphate backbone remains constant while the sequence of bases varies (see Chapter 12).

Some chains of amino acids act as chemical messengers that stimulate a biological response in our body. For example, endorphins are polypeptide molecules that are active in your brain. Their primary role is to reduce the sensations of pain, and they can also produce feelings of euphoria. Another example is insulin, a hormone that is discussed in *Chemistry and Your Health* on page 412. Other polypeptide chains are classified as *proteins*; these molecules perform most of the important biological functions within a cell. The next section of the chapter describes the structural properties of proteins.

FIGURE 13.16 A segment of a polypeptide containing five amino acids. The peptide bonds are indicated by thick lines. An amino acid in a polypeptide chain is linked to its neighbors by peptide bonds. Most amino acids have two neighbors with the exception of those amino acids found at the N-terminus and the C-terminus.

PRACTICE EXERCISE 13.3

The structure below shows a peptide molecule made from five amino acids.

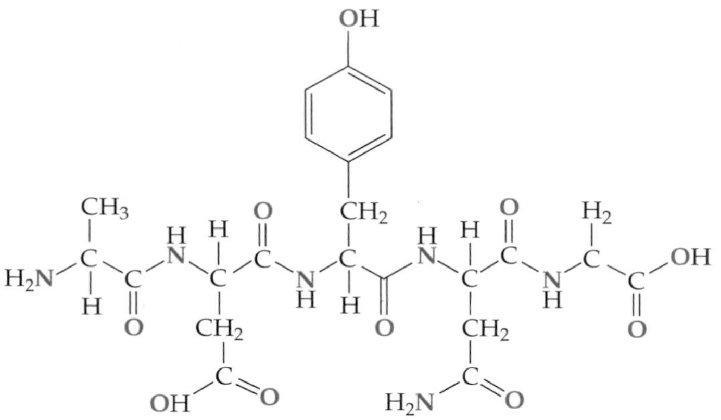

(a) Label the N-terminus and C-terminus of the peptide.
(b) Use Figure 13.7 to identify the amino acids that were combined to make this polypeptide. Write the name of the amino acid next to its sidechain.
(c) Write the amino acid sequence of the polypeptide, beginning at the N-terminus and ending at the C-terminus.

Core Concepts

■ Amino acids are joined together by a **peptide bond**. This bond is formed by a condensation reaction between the carboxylic acid group (—COOH) of one amino acid and the amine group (—NH$_2$) of a neighboring amino acid. The collection of atoms on either side of the peptide bond is called the peptide group.

■ Multiple amino acids can be linked by peptide bonds to form a **polypeptide**. Because the backbone remains constant, the variability within each polypeptide chain arises from the specific sequence of amino acid sidechains.

Four Levels of Protein Structure

THE KEY IDEA: Protein structure is described by a hierarchy of four levels: primary, secondary, tertiary, and quaternary.

primary structure The sequence of amino acids in the polypeptide chain of a protein.

secondary structure A local region within the protein where the polypeptide chain exhibits a regular structure; two examples are alpha-helix and beta-sheet.

Most proteins contain hundreds or thousands of amino acids that are linked together by peptide bonds. The long polypeptide chain folds into a complex three-dimensional shape that enables the protein to perform its biological function. Scientists describe proteins using four levels of structure: primary, secondary, tertiary, and quaternary (Figure 13.17).

The **primary structure** is the sequence of amino acids in the polypeptide chain, beginning with the N-terminus and ending at the C-terminus. For example, a region of primary structure for a protein would be written as alanine-valine-tyrosine, and so on.

The **secondary structure** is a local region within the protein where the polypeptide chain exhibits a regular structure. The two most common types of secondary structure are called the *alpha-helix* and the *beta-sheet*. The alpha-helix is a region of the polypeptide chain that has a right-handed helical twist. The helical structure is stabilized by hydrogen bonds between peptide groups that are oriented above and below one another in the twisted polypeptide backbone. A beta-sheet is formed by hydrogen bonding between adjacent regions of the polypeptide backbone. To visualize a beta-sheet, imagine the polypeptide chain makes a turn and forms a U-shaped structure. The beta-sheet is formed by hydrogen bonding between the two straight regions of the U.

The **tertiary structure** refers to the overall structure of the entire polypeptide chain, including all of its secondary structure components. For proteins with a single polypeptide chain, the structural levels end at the tertiary structure. However, some proteins are constructed from multiple polypeptide chains. These proteins have a **quaternary structure**, which describes the structural relationship between two or more polypeptide chains within a protein.

tertiary structure The overall structure of a single polypeptide chain, including all of its secondary structure components.

quaternary structure The structure formed by two or more polypeptide chains within a protein.

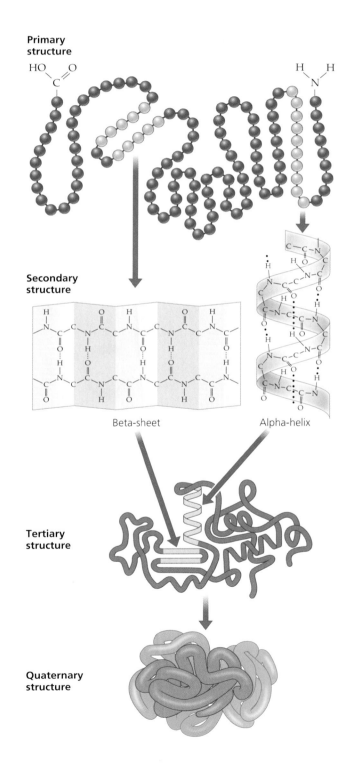

FIGURE 13.17 Four levels of proteins structure. The levels of protein structure are described as primary, secondary, tertiary, and quaternary.

CHEMISTRY AND YOUR HEALTH

Human Insulin for Diabetes Therapy

According to the American Diabetes Association, approximately 1.25 million children and adults in the United States have Type 1 diabetes. Affected individuals are unable to produce sufficient insulin, a polypeptide hormone that regulates the uptake of sugar from the bloodstream into cells (see *Chemistry and Your Health* in Chapter 7). This shortage of insulin arises because the insulin-producing cells in the pancreas are destroyed by the body's immune system.

Type 1 diabetes can be treated by injections of human insulin (Figure 1(a)). The molecular structure of human insulin, shown in Figure 1(b), consists of two polypeptide chains. Chain A contains 21 amino acids, whereas chain B is slightly longer, with 30 amino acids. The two chains are joined by chemical bonds between two cysteine amino acids. The sidechain of cysteine contains a sulfur atom in a *thiol* functional group (—SH). Two neighboring thiols can react to form a covalent bond between sulfur atoms, which is called a *disulfide bond* (S—S). Two of these disulfide bonds link chains A and B, and an additional disulfide bond joins two cysteines within chain A.

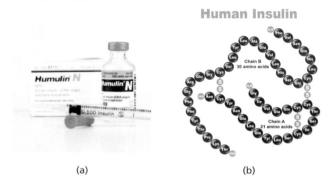

Human Insulin

(a)   (b)

....

FIGURE 1   Human insulin therapy for Type 1 diabetes. (a) Injections of human insulin are used as a treatment for Type 1 diabetes. (b) Human insulin is composed of two polypeptide chains (A and B), each with a specific amino acid sequence. The two chains are joined by covalent bonds between sulfur atoms in cysteine sidechains.

Treating more than 1 million diabetics requires a lot of human insulin. Remarkably, scientists are able to make human insulin using bacteria. This procedure, developed during the 1980s, is called *recombinant DNA technology* because it involves taking DNA from one type of organism and recombining it with DNA from another organism. The method for producing human insulin is illustrated in Figure 2. The first step is to use a DNA-cutting enzyme to snip the insulin gene out of human DNA. The same enzyme is then used to cut a circular region of bacterial DNA, which is called a *plasmid*. Next, the region of DNA containing the insulin gene is inserted into the plasmid. A different enzyme is utilized to glue the two pieces of DNA together by forming new covalent bonds within the sugar-phosphate backbone. We now have a modified bacterial plasmid that contains the DNA for human insulin.

The plasmid is inserted into bacterial cells, which are then placed in a growth medium. When the bacteria reproduce, they use their plasmid DNA to make proteins via transcription and translation. However, these cells now make human insulin in

addition to synthesizing bacterial proteins. The human insulin is extracted from the bacteria, purified, and provided to people with Type 1 diabetes. Insulin that is produced in this manner is called *recombinant insulin*. It contains exactly the same amino acid sequence as natural human insulin because it is made using a human gene. More recently, scientists have developed modified insulins with specific amino acid changes that produce superior therapeutic benefits.

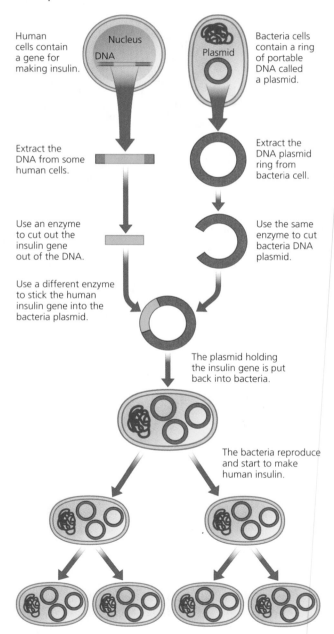

....

FIGURE 2   Producing human insulin from bacteria using recombinant DNA technology. The insulin gene is cut out of human DNA and inserted into *E. coli* bacteria. When the bacteria grow, they express the new gene and produce human insulin.

CONCEPT QUESTION 13.3

When you crack an egg, you see a yellow yolk surrounded by a thick, transparent fluid called the "egg white." This egg white contains many different proteins. The most abundant is called ovalbumin, which constitutes over 50% of the total protein content. When you fry an egg, the egg white becomes opaque and more like a solid (see the photograph). Propose a hypothesis for what happens to the proteins in egg white to produce this change.

Hemoglobin is one protein that has a quaternary structure, as illustrated in Figure 13.18. It contains four polypeptide chains—two alpha-chains and two beta-chains. (To avoid confusion, these terms do *not* refer to alpha-helix and beta-sheet.) The two types of chain are very similar, but the beta-globin chain contains an additional five amino acids. Each of the four polypeptide chains in hemoglobin encloses a heme group, and an oxygen molecule (O_2) binds to the iron ion (Fe^{2+}) at the center of the heme. The quaternary structure of hemoglobin enables the protein to change shape in response to different levels of oxygen. In fact, *hemoglobin changes its quaternary structure when it binds and releases oxygen*. When hemoglobin binds O_2 molecules, it adopts a quaternary structure called *oxyhemoglobin*. When hemoglobin releases its oxygen and no O_2 molecules are bound, the protein switches its shape to a quaternary structure called *deoxyhemoglobin*. This change in quaternary structure involves a rotation of the alpha-chains and beta-chains with respect to each other.

Core Concepts

- Protein architecture is described in a hierarchy of four structural levels: **primary**, **secondary**, **tertiary**, and **quaternary**.
- The two most common examples of secondary structure are the alpha-helix and the beta-sheet.
- Hemoglobin has a quaternary structure consisting of four polypeptide chains: two alpha-chains and two beta-chains. Hemoglobin changes its quaternary structure when it binds and releases oxygen.

Learning Objective:
Outline the molecular origin of sickle cell disease.

THE KEY IDEA: A DNA mutation leads to an amino acid substitution in hemoglobin, which causes the protein to form fibers that distort the shape of red blood cells.

13.7 The Molecular Origin of Sickle Cell Disease

We can now integrate information from throughout this chapter to examine the molecular origin of sickle cell disease. In particular, we want to understand how a DNA mutation in the hemoglobin gene leads to the production of sickle-shaped cells. In this manner, sickle cell disease serves as a paradigm for investigating the molecular causes of other inherited diseases.

As we learned in the previous section, hemoglobin consists of four polypeptide chains: two alpha-chains and two beta-chains. *The DNA mutation that causes sickle cell disease is a single base change in the gene for the beta-chain of hemoglobin.* Figure 13.19 shows the DNA, RNA, and amino sequences for codons 4 to 7 in the beta-chain. In the normal protein, called hemoglobin A (HbA), the DNA template strand corresponding to the sixth codon in the beta-chain has the base sequence CTC. Transcription into mRNA produces the complementary base sequence GAG, which is translated into an amino acid called *glutamic acid*. In the case of sickle

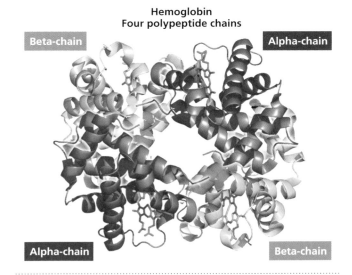

FIGURE 13.18 Hemoglobin has a quaternary structure. Hemoglobin has a quaternary structure composed of four polypeptide chains: two alpha-chains and two beta-chains. Each polypeptide chain encloses a heme group, as discussed in Chapter 2. The quaternary structure of hemoglobin changes when it binds and releases oxygen.

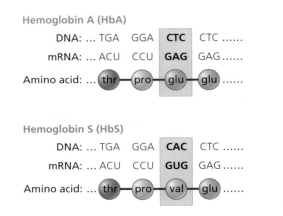

FIGURE 13.19 A DNA mutation produces an amino acid change. Sickle cell disease arises from the mutation of a single base (A → T) in the DNA sequence corresponding to the sixth cordon of the beta-chain in hemoglobin. As a result, the sixth amino acid is changed from glutamic acid to valine.

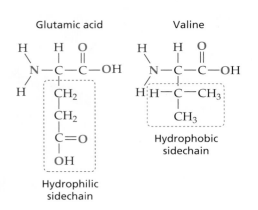

FIGURE 13.20 The molecular structures of glutamic acid (HbA) and valine (HbS). Glutamic acid has a hydrophilic sidechain, whereas valine has a hydrophobic sidechain. This change affects the solubility of the hemoglobin proteins.

cell disease, the middle DNA base of the sixth codon is mutated from T to A, so the DNA template sequence now reads CAC. This mutated DNA sequence is transcribed into a new mRNA codon, GUG, which is then translated into a different amino acid called *valine*. This mutation produces a modified protein, hemoglobin S (HbS), in which valine replaces glutamic acid as the sixth amino acid in the beta-chain.

Why does one amino acid change cause such a major problem? Figure 13.20 compares the structures of glutamic acid and valine. As we learned earlier in the chapter, amino acids are classified according to the chemical structure of their sidechain. Glutamic acid (HbA) is one of the *acidic* amino acids because it contains a carboxylic acid group in its sidechain. By contrast, valine (HbS) is *neutral nonpolar* because its sidechain is composed purely of hydrocarbon. The sidechain of glutamic acid is hydrophilic, whereas the sidechain of valine is hydrophobic.

The difference in solubility of the amino acid sidechains affects the behavior of hemoglobin within the aqueous interior of red blood cells (Figure 13.21). The sidechain of the sixth amino acid is located on the surface of the protein, where it is exposed to water. The sidechain of glutamic acid is hydrophilic, which contributes to the water solubility of HbA. For the HbS protein, valine's sidechain creates a hydrophobic patch on the protein's surface. As we learned in Chapter 9, hydrophobic regions tend to aggregate to shield themselves from water. As a result, the HbS proteins can clump together to form elongated fibers, which bury the valine sidechains within a nonpolar environment. The fibers are created when the hemoglobin protein is in its *deoxy* quaternary structure because this structure is more conducive to aggregation of the proteins. The formation of HbS fibers within a red blood cell distorts it into a sickle shape.

The symptoms of sickle cell disease arise from the behavior of sickled red blood cells. The HbS protein fibers often rupture and destroy these cells, which decreases the number of cells available to transport oxygen in the bloodstream. As a result, individuals with sickle cell disease experience symptoms of low oxygen supply—fatigue, headaches, and shortness of breath. One temporary remedy is to provide patients with pure oxygen to breathe.

In addition, the distorted sickle cells affect the flow of blood throughout the body, as illustrated in Figure 13.22. Normal red blood cells are round and flexible, so they flow easily through narrow blood vessels. By contrast, sickled red blood cells are elongated and rigid. They can become stuck in blood vessels, creating blood clots and reducing blood flow. This restriction in blood circulation often generates severe pain and, over time, can damage vital organs. The spleen is particularly susceptible because it is the organ primarily responsible for filtering and storing red blood cells.

In conclusion, understanding the molecular basis of sickle cell disease has required us to know about DNA, transcription, translation, amino acids, solubility, protein structure, and cellular function. This type of integrative thinking illustrates the power of modern scientific approaches to studying diseases.

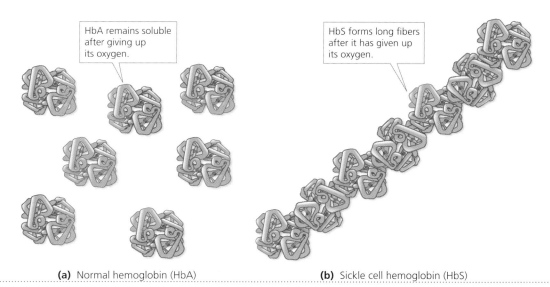

(a) Normal hemoglobin (HbA)   **(b)** Sickle cell hemoglobin (HbS)

FIGURE 13.21 When the oxygen level is low, HbS proteins aggregate to form fibers. (a) Normal hemoglobin (HbA) is soluble within the aqueous environment of the red blood cells. (b) Mutated hemoglobin (HbS) contains a hydrophobic patch that causes the HbS molecules to aggregate and form fibers. This aggregation occurs when the oxygen level is low and the HbS proteins are in their deoxy quaternary structure.

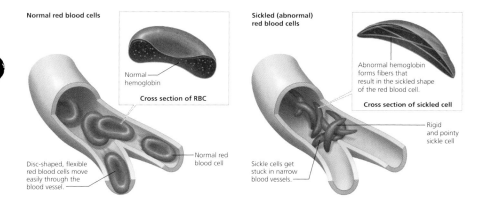

FIGURE 13.22 Sickled red blood cells disrupt blood flow. Normal red blood cells are flexible and flow easily through narrow blood vessels. Sickle cells, in contrast, are rigid and can get stuck in blood vessels, creating blood clots and disrupting blood flow.

CONCEPT QUESTION 13.4

Scientists have identified many mutated forms of hemoglobin. Hemoglobin C (HbC) is produced by a single base mutation in DNA that changes the sixth mRNA codon in the beta-chain from GAG to AAG. (a) Which amino acid is produced by this mutation? (b) Do you predict that HbC will form hemoglobin polymers like HbS? Explain your answer.

Core Concepts

- The DNA mutation that causes sickle cell disease is a single base change (T → A) in the gene for the beta-chain of hemoglobin. This mutation alters the mRNA codon from GAG to GUG, which changes the amino acid from glutamic acid (hydrophilic sidechain) to valine (hydrophobic sidechain).
- When the mutated hemoglobin (HbS) is in its deoxy structure, the proteins aggregate to form long fibers that distort the red blood cell into a sickle shape.
- The HbS fibers can rupture the red blood cells, which decreases their number in the bloodstream. In addition, sickled cells tend to form blood clots in narrow blood vessels, which reduces blood flow and can damage vital organs.

CHAPTER 13
VISUAL SUMMARY

13.1 What Causes Sickle Cell Disease?

Learning Objective:

Identify the characteristics of sickle cell disease.

- The characteristic symptom of **sickle cell disease** is the distortion of red blood cells into a shape that resembles the letter "C."

- Sickle cell disease is caused by a DNA mutation in the gene for hemoglobin.

13.2 Overview: The Expression of Genetic Information

Learning Objective:

Outline the molecules and processes that are involved in the expression of genetic information.

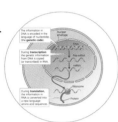

- **Gene expression** is the pathway for utilizing the genetic information in DNA.

- Gene expression proceeds from DNA to **messenger RNA** (mRNA) to protein.

- The conversion of DNA into mRNA is called **transcription**. The conversion of mRNA into a sequence of amino acids in a protein is called **translation.**

13.3 Transcription

Learning Objective:

Characterize the stages of transcription.

- There are four structural and chemical differences between DNA and RNA:

  1. DNA strands are much longer than RNA strands.
  2. DNA forms a double helix, whereas RNA exists as a single strand.
  3. DNA contains deoxyribose as its sugar, whereas RNA contains **ribose**.
  4. DNA contains thymine as a base, which is replaced by **uracil** in RNA.

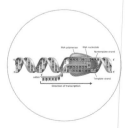

- To begin transcription, **RNA polymerase** binds to a promoter site and then unwinds the DNA double helix to expose the base pairs. RNA polymerase uses the DNA **template strand** and the base-pairing rules to synthesize a complementary strand of mRNA.

- An adenine base in DNA is paired with a uracil base (not thymine) in mRNA.

13.4   Amino Acids: The Building Blocks of Proteins

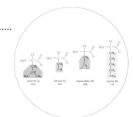

Learning Objective:

Describe the molecular properties of amino acids.

⊙ The 20 standard amino acids can be classified into four groups based on the chemical properties of their sidechain. These groups are neutral nonpolar, neutral polar, acidic, and basic.

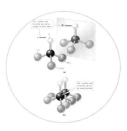

⊙ Amino acids are **chiral** molecules that exist as two **chiral isomers**: a left-handed form (L) and a right-handed form (D). Proteins use only L-amino acids.

⊙ Chiral isomers are mirror images of each other, and they cannot be superimposed.

⊙ A molecule is chiral if it has four different chemical groups bonded to a carbon atom.

13.5   Translation

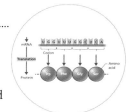

Learning Objective:

Outline the stages by which an mRNA sequence is translated into a sequence of amino acids in a protein.

⊙ Translation is the conversion of a sequence of bases in mRNA into a sequence of amino acids in a protein.

⊙ Three bases in mRNA—called a **codon**—are required to specify each amino acid.

⊙ The **genetic code** specifies which codon corresponds to which amino acid.

⊙ The number of codons (64) is greater than the number of standard amino acids (20). Consequently, most amino acids are specified by more than one codon.

⊙ The genetic code has special codons that indicate when translation should begin (**start codon**) and end (**stop codons**).

⊙ Translation occurs within the ribosome, which synthesizes new proteins based on the genetic information provided by mRNA.

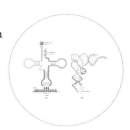

⊙ **Transfer RNA** (tRNA) serves as a "molecular translator" between an mRNA codon and its corresponding amino acid.

⊙ The tRNA molecule contains an **anticodon** with a base sequence that is complementary to the mRNA codon. Codon–anticodon binding ensures that the correct amino acid is inserted into the protein.

⊙ A protein is made from the sequential action of tRNA molecules that bind to the mRNA strand, add their amino acid to the growing chain, and are then discharged.

13.6   Protein Structure

Learning Objective:

Illustrate the formation of a polypeptide chain and its structural properties within proteins.

- ⊙ Amino acids are joined together by a **peptide bond**.
- ⊙ Multiple amino acids can be linked by peptide bonds to form a **polypeptide**. Because the backbone remains constant, the variability within each polypeptide chain arises from the specific sequence of amino acid sidechains.
- ⊙ Protein architecture is described in a hierarchy of four structural levels: **primary**, **secondary**, **tertiary**, and **quaternary**.
- ⊙ The two most common examples of secondary structure are the alpha-helix and the beta-sheet.
- ⊙ Hemoglobin has a quaternary structure consisting of four polypeptide chains: two alpha-chains and two beta-chains. Hemoglobin changes its quaternary structure when it binds and releases oxygen.

13.7   The Molecular Origin of Sickle Cell Disease

Learning Objective:

Outline the molecular origin of sickle cell disease.

- ⊙ The DNA mutation that causes sickle cell disease is a single base change (T → A) in the gene for the beta-chain of hemoglobin. This mutation alters the mRNA codon from GAG to GUG, which changes the amino acid from glutamic acid (hydrophilic sidechain) to valine (hydrophobic sidechain).
- ⊙ When the mutated hemoglobin (HbS) is in its deoxy structure, the proteins aggregate to form long fibers that distort the red blood cell into a sickle shape.

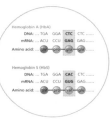

LEARNING RESOURCES

Reviewing Knowledge

13.1: What Causes Sickle Cell Disease?

1. What is the characteristic symptom of sickle cell disease?

13.2: Overview: The Expression of Genetic Information

2. Draw a diagram to illustrate the expression of genetic information in a human cell. Label the key molecules, and name all of the processes involved.

13.3: Transcription

3. Describe the four structural and chemical differences between DNA and RNA.
4. Explain why the oxygen atom on the 2′-carbon of the ribose ring prevents RNA from having a helical structure similar to that of DNA.

5. Describe how RNA polymerase creates an mRNA strand during transcription.
6. In what direction does the mRNA strand grow during transcription?
7. How does transcription preserve the genetic information contained in DNA, even though the mRNA is not identical to the original DNA?
8. In terms of directionality, how is the synthesized mRNA strand oriented with respect to the DNA template strand?

13.4: Amino Acids: The Building Blocks of Proteins

9. Draw the structure of an amino acid, and label its chemical components. Which part of an amino acid changes to produce different varieties?
10. How many different amino acids are commonly found in proteins?

11. What are the four groups of amino acid sidechains?
12. What is the definition of a chiral molecule?
13. Which letters are used to identify the two chiral isomers of amino acids? Which chiral isomer is used to make proteins?

13.5: Translation

14. What is a codon?
15. Explain why the genetic code is called "degenerate."
16. What codons signal the beginning and end of translation?
17. Describe the role of transfer RNA (tRNA) during translation. What is the function of the anticodon?

13.6: Protein Structure

18. Draw a diagram to illustrate how two amino acids are linked by a condensation reaction. Label the peptide bond in your drawing.
19. Which component of the amino acids provides the chemical variation in a polypeptide chain?
20. What are the names of the two ends of a polypeptide chain, and how are they identified?
21. Provide a brief (one- to two-sentence) description of each of the four levels of protein structure.
22. What structural change occurs when hemoglobin binds oxygen?

13.7: The Molecular Origin of Sickle Cell Disease

23. Outline the stages by which a single base mutation in DNA produces distorted red blood cells.

Developing Skills

24. RNA is a single-stranded molecule, whereas DNA needs to be double-stranded to perform its biological function. Propose an explanation for this difference in molecular structure.
25. RNA has a uracil base that replaces the thymine base found in DNA.
    (a) Draw the molecular structures of uracil and thymine. Circle the atoms that differ between the two bases.
    (b) In Chapter 12, we learned that thymine forms hydrogen bonds with adenine. Draw the structure of a uracil–adenine base pair, and use dashed lines to indicate the hydrogen bonds.
    (c) How many hydrogen bonds connect uracil and adenine? Is this the same number of hydrogen bonds that exits in the thymine-adenine base pair?
26. A template sequence of DNA is provided below. Transcribe this DNA into a sequence of mRNA, and include the 5′- and 3′-ends.
    3′-T G C C T C C G A G G C A T T G C C-5′
27. Is the following sequence of bases from DNA or mRNA? Explain your answer.
    5′-A G G C A U C G A C U A C G C U G U-3′
28. A segment of a *nontemplate* DNA strand of a given gene is as follows:
    5′-A T G C T A G C A T G C T T-3′
    (a) Write the sequence of the template strand of DNA. Include the 5′-end and 3′-end.
    (b) Write the sequence of the mRNA strand that is transcribed from the DNA template strand. Include the 5′-end and 3′-end.

(c) Compare the directionality and the base sequence of the mRNA strand and the nontemplate DNA strand. What do you notice about the relationship between these two sequences?
29. (a) Use Figure 13.7 to write all of the codons for the amino acids alanine, valine, and serine. Use a separate column for each amino acid. (b) By studying the codons in each column, what do you notice about the pattern of bases that code for the same amino acid?
30. Use Figure 13.13 to determine the number of silent/synonymous mutations that are possible for the following codons:
    (a) CCU
    (b) GAU
    (c) AUU
31. If the anticodon on a transfer RNA molecule is CCC, then which amino acid does the anticodon carry to the ribosome?
32. The structures of three amino acids are shown here.

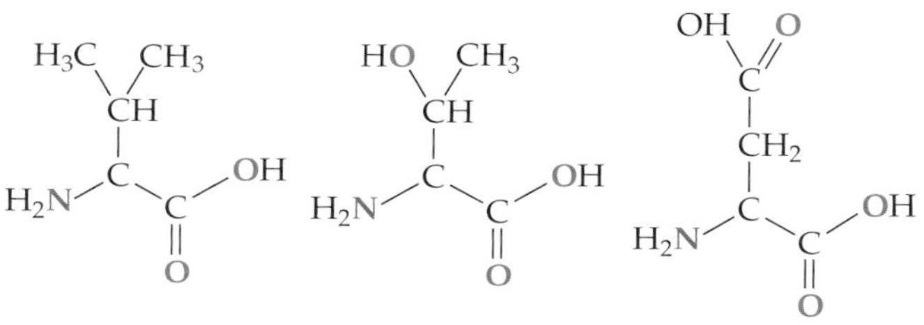

(a) Using Figure 13.7 as a reference, identify each amino acid.
(b) Use an asterisk (*) to label all of the chiral carbon centers in the structures above. Which of these amino acids has a chiral carbon center(s) in the *sidechain*?
33. The following structure shows a peptide molecule made from three amino acids (a tripeptide):

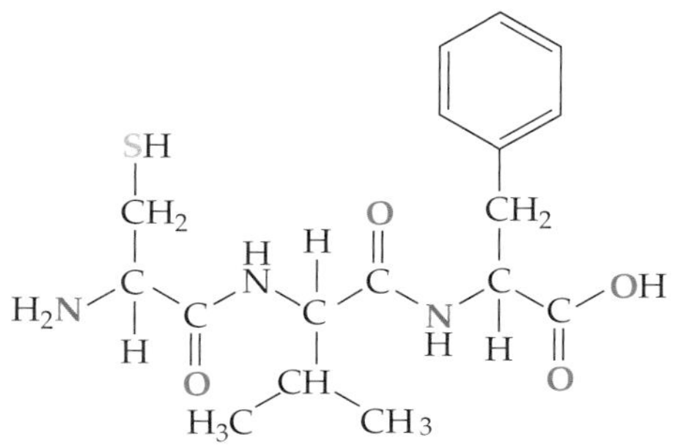

(a) Using Figure 13.7 as a reference, write the amino acid sequence of the tripeptide. Peptide sequences are written from the N-terminus to the C-terminus.
(b) Draw a different tripeptide composed of the following amino acids: alanine – aspartic acid – threonine.
34. The template DNA sequence shown below is part of a gene used to make a particular protein.

    DNA 3′-T A C G T G A T G C A T A G G ...-5′

(a) Write the mRNA sequence that is synthesized from this DNA sequence. Include the 5′-end and 3′-end of the sequence.

(b) Translate the mRNA sequence into amino acids. Translation occurs from the 5′ → 3′ direction in the mRNA strand.

(c) Scientists have identified a DNA mutation in this gene that consists of a single base change from G → C in the final base of this sequence. Surprisingly, this mutation has no observable effect on the protein's function. Explain this observation.

35. Hemoglobin is made up of four polypeptide chains: two α-chains and two beta-chains. Experiments have shown that α-chains and beta-chains prefer to associate with each other, as opposed to association between two alpha-chains or two beta-chains. We can use this principle to predict the organization of alpha-chains and beta-chains in the quaternary structure of hemoglobin. One possible quaternary arrangement of the four chains is shown below. However, this arrangement does not match the actual structure of hemoglobin. Propose another quaternary arrangement of the chains that satisfies the rule for association.

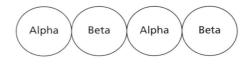

36. The following DNA sequence corresponds to five codons from the middle of the gene called BRCA1. Mutations in this gene produce an increased risk of breast and ovarian cancers.

Codon #   100   101   102   103   104
DNA 3′-...T T A G A G T G T C C A T C G...-5′

(a) Write the mRNA sequence that is synthesized from this DNA sequence. Include the 5′-end and 3′-end of the sequence.

(b) Translate the mRNA sequence into amino acids. Translation occurs from the 5′ → 3′ direction in the mRNA strand.

(c) A particularly serious mutation in the BRCA 1 gene is called 101delAG. This mutation causes the deletion of the second and third bases from codon 101. What effect does this mutation have on the amino acid sequence, and why does it cause a complete loss of protein function?

37. Hemoglobin E is an abnormal form of hemoglobin caused by the mutation of a single DNA base. This mutation changes the 20th amino acid in the beta-chain from glutamic acid to lysine. Use the genetic code table to determine which DNA mutation(s) can produce this amino acid change.

38. As discussed in Chapter 9, cells are surrounded by a membrane that is composed of a phospholipid bilayer. Embedded within this bilayer are membrane proteins, which perform many important biological functions. Membrane proteins often contain alpha-helices that span the phospholipid bilayer, as shown in the figure here. Some of the amino acids in the alpha-helix have sidechains that extend outward, which places then in contact with the hydrocarbon tails of the phospholipids. Use Figure 13.7 to predict *three* amino acids that act in this manner. Explain your reasoning.

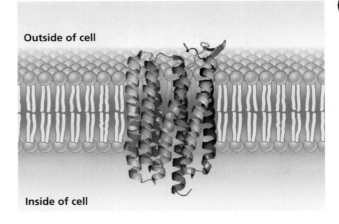

Outside of cell

Inside of cell

Exploring Concepts

39. Our hair is composed of a protein called *keratin*, which forms long fibers. Some hair treatments, such as "perms," involve adding chemicals that modify the properties of keratin. Use the Internet to research this topic, and write a one-paragraph description of the chemistry of a hair perm.

40. Gene therapy is an experimental approach in modern medicine in which patients with inherited disorders are treated by altering their DNA. Use the Internet to research gene therapy, and write a one- to two-paragraph report on this medical technique. You report should include one success and one failure.

41. Gene expression in all cells, from bacteria to humans, proceeds in the following direction: DNA → RNA → protein. However, some types of viruses—called *retroviruses*—do not follow this pathway. One retrovirus is human immunodeficiency virus (HIV), which causes AIDS. Use the Internet to investigate how retroviruses process genetic information, and explain why the viruses are given this name. Write a one- to two-paragraph report on your research.

42. Melittin is the primary component of bee venom. Use the Internet to investigate the structure of this molecule and how it achieves its painful effects. Write a one-paragraph report that summarizes your findings.

43. Prialt is the brand name of a powerful pain medication that doctors use as a last resort when standard drugs such as morphine are not effective. The active ingredient in Prialt is a polypeptide with an unusual origin. Use the Internet to investigate this drug, and write a short account to explain how it was developed.

44. A patient with sickle cell disease has fewer red blood cells than a person with normal hemoglobin. The human body makes a hormone called *erythropoietin* that stimulates the production of red blood cells. Injections of erythropoietin are used to treat individuals with sickle cell disease. However, some endurance athletes have also abused erythropoietin in an attempt to gain a competitive advantage. Use the Internet to investigate erythropoietin, and evaluate whether it actually provides an unfair advantage to athletes who use it. Write a one- to two-paragraph report summarizing your research.

45. The molecule shown here is a peptide hormone called *oxytocin*. In the popular press, oxytocin has sometimes been called the "love molecule." Use the Internet to investigate the experiments that have been performed to study the biological effects of oxytocin. Write a two- to three-paragraph report summarizing your research.

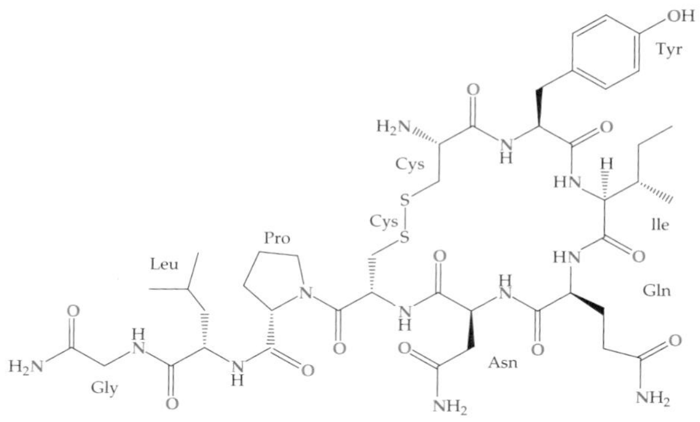

46. Human growth hormone (HGH) is a peptide hormone that is produced by recombinant DNA technology (see *Chemistry and Your Health*). Use the Internet to investigate this topic, and answer the following questions. (a) How many amino acids does HGH contain? Is this hormone larger or smaller than insulin? (b) What are the legitimate medical uses of HGH? (c) HGH is also used by some athletes in an attempt to boost their performance. Select your favorite sport, and investigate the use, regulation, and testing of HGH. Write a two- to three-paragraph report summarizing your findings.

47. Scientists have discovered a connection between sickle cell disease and resistance to infection by malaria. Use the Internet to investigate this relationship, and write a one- to two-paragraph report summarizing your findings.

Chapter 14

Enzymes as Biological Catalysts

Some people enjoy a refreshing glass of cold milk. For many other individuals, however, drinking milk is an unpleasant experience that leads to intestinal cramps, bloating, gas, and sometimes diarrhea. Why are some people affected in this manner, whereas others can drink milk without any ill effects?

14.1 What Causes Lactose Intolerance?

Learning Objective:

Define lactose intolerance.

THE KEY IDEA: Lactose intolerance is an adverse reaction to the lactose sugar in milk.

lactose intolerance The inability to break down the lactose sugar in milk products.

lactase The enzyme that decomposes lactose into two smaller sugars.

enzyme A biological molecule that causes a chemical reaction to occur faster.

Drinking milk provides a good source of nutrients, such as calcium and vitamin D (Figure 14.1). Milk also contains a sugar called *lactose*. As we learned in Chapter 7, lactose is made by joining two smaller sugars called glucose and galactose. It is lactose that causes many people to have an adverse reaction to drinking milk—these individuals have **lactose intolerance**.

To utilize lactose, our body must break the molecule into its constituent sugars. This decomposition is accomplished by a protein called **lactase**. Along with many other proteins, lactase belongs to a class of biological molecule called enzymes. *An **enzyme** is a biological catalyst that accelerates the rate of a chemical reaction.* Figure 14.2 illustrates how lactase facilitates the splitting of lactose into glucose and galactose. What is the connection between lactase and lactose intolerance?

Enzymes are vital to every living organism. Without their assistance, chemical reactions would proceed too slowly to sustain life. This chapter examines the structure and function of enzymes with the goal of understanding the cause of lactose intolerance. Because enzymes affect chemical reactions, we begin with an examination of how reactions occur and how a catalyst increases their speed. We then study enzymes as biological catalysts, including their remarkable effectiveness at accelerating chemical reactions, sometimes more than a trillion times their normal rate. Finally, we apply these principles to the example of lactose intolerance.

FIGURE 14.1 Milk contains lactose sugar. Some people can drink milk without any problems, but others suffer adverse effects because they are lactose intolerant.

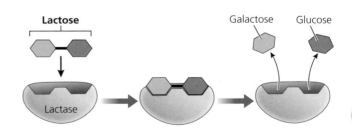

FIGURE 14.2. Lactase and lactose. Lactase is an enzyme that facilitates the breakage of lactose into two smaller sugars, glucose and galactose.

Core Concepts

- **Lactose intolerance** is an adverse reaction to the lactose sugar in milk.
- **Lactase** is an enzyme that decomposes lactose into two smaller sugars, glucose and galactose.
- An enzyme is a biological catalyst that accelerates the rate of a chemical reaction.

14.2 How Do Chemical Reactions Happen?

In Chapter 6, we examined chemical reactions with a focus on the chemical conversion of reactants (the starting materials) into products. In this section, we consider *how* chemical reactions happen, and we discuss the factors that affect their rate.

An Energy Barrier for Chemical Reactions

Chemical reactions often need help to get started. For example, the gasoline in your car does not burn unless you turn on the ignition and provide an energy jolt from an electrical spark. The fact that the gasoline in your tank does not spontaneously combust tells us something interesting about chemical reactions. Most reactions have an *energy barrier* that prevents the reactants from immediately converting into the products. This barrier is called the **activation energy** for the reaction. Overcoming the activation energy requires an input of energy such as a spark. Another familiar example is a match, which does not burn unless we provide an energy input by striking it along a rough surface.

To appreciate the process of overcoming an energy barrier, consider the analogy of how a high jumper gets over a bar. The jumper uses energy from her muscles to run up to the bar and launch herself into the air. Once in the air, she curves her back for a split-second while bringing her feet up over the bar (Figure 14.3). After clearing the bar, the jumper lands on the mat on the far side.

We can draw parallels between the high jumper and the progress of a chemical reaction. We can view the take-off as the "reactant molecule" and the landing as the "product." What happens in between—the jumper's arched back—is a transient state that requires the athlete's body to assume a shape that is different from her take-off and landing shapes. This state has the highest energy, it is unstable, and it exists for only a split second. The high jumper *must* pass through this in-between state to clear the bar.

A chemical reaction also has an in-between state called the **transition state.** *The transition state of a chemical reaction is a transient, high-energy state that is formed during the conversion of reactants into products.* The transition state corresponds to the point of highest energy in a chemical reaction. Like the high jumper in mid-flight, the structure of the transition state has a distinct molecular shape that differs from the structures of both the reactants and the products. The reaction must pass through the transition state in order to convert the reactants into products.

When we write an equation for a chemical reaction, we are often interested only in the identity of the reactants and products.

$$\text{REACTANTS} \implies \text{PRODUCTS}$$

However, this representation does not tell us *how* the chemical reaction happens. In this chapter, we look at chemical reactions as a dynamic process of molecular change. From this perspective, we also need to include the transition state for the reaction.

$$\text{REACTANTS} \implies \text{TRANSITION STATE} \implies \text{PRODUCTS}$$

Chemists commonly portray the stages of a chemical reaction using an *energy diagram*; an example appears in Figure 14.4. The horizontal axis indicates the progress of the reaction, beginning with the reactants and then moving through

Learning Objective:
Characterize the changes that occur during a chemical reaction.

THE KEY IDEA: Most chemical reactions have an energy barrier that must be overcome for the reaction to occur.

activation energy The energy barrier that must be overcome for a chemical reaction to occur.

transition state The transient, high-energy state at that is formed during the conversion of reactants into products in a chemical reaction.

FIGURE 14.3. Getting over the bar. A high jumper arches her back to get over the bar. This fleeting stage of the jump exists for a split second in between the take-off and landing.

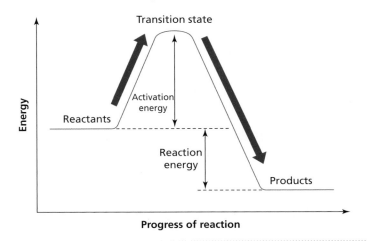

FIGURE 14.4 An energy diagram for a chemical reaction. The diagram shows the energy levels for the products, reactants, and activation energy in a chemical reaction. The thick red arrows indicate the pathway of the chemical reaction. The size of the activation energy determines how rapidly the reaction will proceed. The reaction energy is the energy difference between the reactants and the products.

the transition state to form the products. The vertical axis plots the relative energy of the different reaction components. A reaction component with *lower energy* is represented by a line located nearer the bottom of the energy axis. Conversely, a component with *higher energy* is represented by a line nearer the top of the diagram. Chemists often use the terminology of *stability* in connection with energy changes: Lower-energy components are *more stable*, and those at higher energy are *less stable*. In Figure 14.4, *the activation energy is the energy input that is required to form the unstable transition state for the reaction.*

Returning to our high jumper analogy, we note that the chances of completing a successful jump depend on the height of the bar. If the bar is low, then the jumper can clear it every time. Raising the bar reduces the chances of a successful crossing. Similarly, the completion of a chemical reaction depends on the height of the activation energy. We define the rate of the chemical reaction as the number of times the system can proceed from reactants to products within a given timeframe (usually 1 second). *The rate of a chemical reaction depends on the height of the activation energy.* A small activation energy enables a faster reaction rate; a large activation energy produces a much slower rate.

The height of the activation energy corresponds to the energy of the transition state. A reaction with a very unstable transition state will have a high activation energy. In this case, it is very difficult to convert the reactants into products, so the reaction rate will be low. Conversely, a reaction with a more stable transition state will have a lower activation energy, so the reaction will occur more easily with a faster rate.

Another important feature of a chemical reaction is the amount of energy that is released or absorbed. The *reaction energy* is the energy difference between the reactants and products. As we saw in Chapter 6, an energy input is required to break the chemical bonds in the reactants. Energy is released when bonds are formed in the products. The net difference between the energy input and output determines whether a reaction produces energy (*exothermic reaction*) or absorbs energy (*endothermic reaction*). In many cases, such as combustion of a fuel, an exothermic reaction releases energy in the form of heat.

PRACTICE EXERCISE 14.1

Figure 14.4 shows an energy diagram for an exothermic reaction. Draw a similar energy diagram for an endothermic reaction in which the energy of the products is higher than the energy of the reactants.

Core Concepts

■ Chemical reactions have an **activation energy** that acts as an energy barrier between the reactants and the products.

■ The **transition state** of a chemical reaction is a transient, high-energy state that is formed during the conversion of reactants into products. The activation energy for a reaction is the energy required to form the unstable transition state.

■ The reaction energy is the energy difference between the reactants and the products.

A Chemical Reaction Pathway

We can illustrate the concept of the transition state by examining a specific chemical reaction. Let's consider a molecule of methyl bromide (CH_3Br), which is similar to methane (CH_4) but with one hydrogen atom replaced by a bromine atom. Methyl bromide reacts with a hydroxide ion, OH^-, to yield methanol and a free bromide ion. This type of reaction is called a *substitution reaction* because the OH^- ion substitutes for the Br atom.

$$OH^- \;+\; CH_3Br \longrightarrow CH_3OH \;+\; Br^-$$
hydroxide ion   methyl bromide   methanol   bromide ion

How does this reaction actually occur? The stages by which the reactants are transformed into the products is called the **reaction pathway,** which is shown in Figure 14.5. Although we conventionally write the hydroxide ion as OH^-, the negative charge actually resides on the oxygen atom (see Chapter 11). The reaction pathway includes the molecular structure of the transition state. We now examine each step of this reaction, paying careful attention to the molecular interactions and changes that occur.

In order for the chemical reaction to occur, *the reactant molecules must collide with each other.* But not every collision will be productive. The only collisions that lead to a reaction are ones in which *the molecules interact with the correct spatial orientation.* As shown in Figure 14.5, this occurs when the $H—O^-$ bond in the hydroxide ion is approximately aligned with the C—Br bond in CH_3Br. This type of collision enables the O and C atoms to interact with each other.

The next step of the reaction is the formation of the transition state. After a suitable collision, the O atom in OH^- begins to form a weak covalent bond to the C atom in CH_3Br. We designate this type of bond using a dotted line, that is, O---C. In the quantum mechanical model of bonding, a weak covalent bond is an initial overlap of electron density between the O and C atoms. At this point, however, the strength of this bonding interaction is much less than that in a typical covalent bond. As this new bond begins to form, the existing covalent bond between C and Br begins to weaken.

The central carbon atom now has *five* covalent bonding interactions with other atoms: two weak covalent bonds (O---C and C---Br), plus three existing C—H

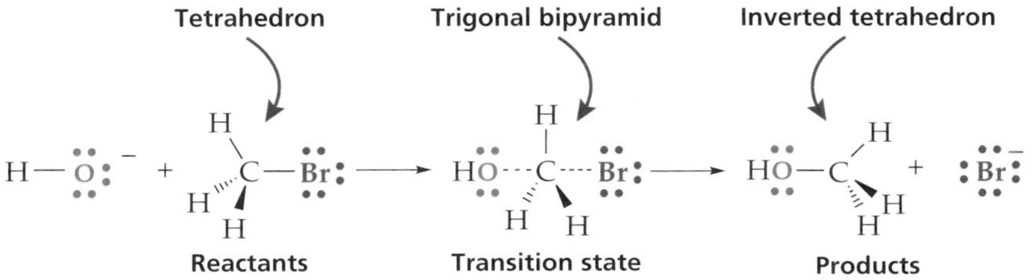

THE KEY IDEA: The molecular structure of the transition state is different from the structures of the reactants and the products.

reaction pathway The stages by which reactants are transformed into products in a chemical reaction.

FIGURE 14.5 The reaction pathway for a substitution reaction. In this reaction, an OH^- ion reacts with CH_3Br to produce CH_3OH and a Br^- ion. The transition state has a molecular structure that is distinct from the structures of the reactants and the products.

bonds. How does this bonding affect the molecular structure? The tetrahedral geometry is no longer optimal because the bonding electron pairs in the C—H bonds are repelled by the new O---C bond. This electron pair repulsion causes the C—H bonds to adjust and form a new shape called *a trigonal bipyramid*. In this shape, the term *trigonal* refers to the position of the three H atoms at the corners of a flat triangle, with the C atom at the center. The term *bipyramid* refers to the fact that each of the O and Br atoms forms a pyramidal structure that extends from the triangular base formed by the C and H atoms. Recall that we used the term *trigonal pyramid* in Chapter 3 to describe the structure of NH_3, in which the N atom forms the apex of the pyramid and the three H atoms form the base. Note that *the molecular structure of the transition state differs from the structures of the reactants and the products.* The transition state is unstable and lasts only a very short time (a few picoseconds, where 1 picosecond = 10^{-12} second).

The final step of the reaction is the formation of the products. The C---Br bond, which has already been weakened in the transition state, now begins to elongate, and it eventually breaks. The Br atom retains both electrons from the covalent bond and forms a bromide ion, Br^-. The departure of bromine leaves a molecule of CH_3OH (methanol), in which the C atom is bonded to four atoms. This is a stable bonding arrangement for carbon, and electron pair repulsion produces a tetrahedral geometry for the molecule. However, there is a subtle but important difference between the tetrahedral structures of CH_3OH and CH_3Br. Note that the orientation of the C—H bonds has flipped during the course of the reaction, analogous to an umbrella turning inside-out in a strong wind. This effect is called *inversion of configuration*; it is a characteristic of this type of reaction.

Figure 14.6 illustrates this reaction pathway as an energy diagram. Although the figure is static, it represents the various stages of a *dynamic* process. This substitution reaction is exothermic, because the energy of the products is lower than the energy of the reactants. The highest energy structure is the transition state because it contains an unfavorable bonding interaction for the central carbon atom. To form the products, the reaction must proceed through this high-energy transition state.

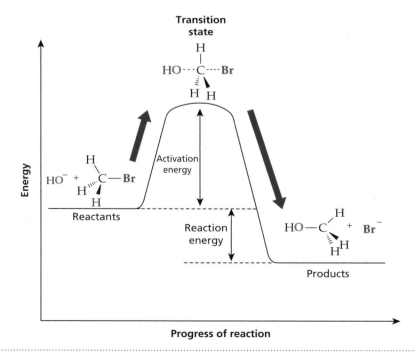

FIGURE 14.6 An energy diagram for a substitution reaction. The diagram shows the relative energies of the reactants, the transition state, and the products. The activation energy for the reaction is the energy required to form the transition state.

PRACTICE EXERCISE 14.2

Figure 14.5 illustrates the substitution reaction between OH^- and CH_3Br to produce CH_3OH. It is possible to chemically modify the CH_3Br and replace all of the H atoms with methyl groups ($—CH_3$). The result is a molecule with the chemical formula $(CH_3)_3CBr$. When we try to react this molecule with OH^-, we observe that no reaction occurs.

(a) Draw the two-dimensional structure of $(CH_3)_3CBr$. Include all the atoms, covalent bonds, and lone pairs.

(b) Based on your drawing, propose a hypothesis for why this molecule does not react with OH^-.

Core Concepts

■ The **reaction pathway** of a chemical reaction describes the stages by which reactants are transformed into products. The substitution reaction between OH^- and CH_3Br provides an example of such a pathway.

■ The reaction is initiated when the OH^- ion collides with CH_3Br in the correct orientation.

■ In this reaction, the transition state has the shape of a trigonal bipyramid. The transition state structure is different from the structures of the reactants and the products.

■ The product of the reaction, methanol CH_3OH, has a tetrahedral geometry, but the orientation of the C—H bonds is inverted as a result of the reaction.

14.3 How Does a Catalyst Work?

In the previous section, we learned that the rate of a reaction is determined by the activation energy. In molecular terms, the reactants must gain sufficient energy to form the high-energy transition state, which is a necessary stage on the way to making the products. How can we make a reaction happen faster? One method, commonly used in the chemistry laboratory, is to heat the reactants. The input of heat energy speeds up the motion of the reactant molecules, enabling more of them to cross the energy barrier and form products. For biological organisms, however, this is not a feasible solution. For example, raising the temperature of the human body by only a few degrees can be dangerous and potentially fatal. Is there another strategy for accelerating chemical reactions?

Learning Objective:
Explain how a catalyst increases the rate of a chemical reaction.

What Is a Catalyst?

A chemical reaction can be accelerated by using a catalyst. *A catalyst increases the rate of a chemical reaction without being permanently altered.* In addition, a catalyst does not change the outcome of the chemical reaction—in other words, the same products are generated with or without the presence of the catalyst. How do catalysts work? In the previous section, we learned that the rate of a chemical reaction is determined by the size of the activation energy. *A catalyst accelerates a chemical reaction by lowering the activation energy.* If the energy barrier is lowered, then the reaction proceeds more quickly because the barrier is easier to overcome.

Figure 14.7 compares the energy diagrams for a chemical reaction in both the absence and the presence of a catalyst. In this case, the catalyst does not affect the energies of the reactants or products, so the reaction energy is the same for both reactions. How is it possible to lower the activation energy? We know that the activation energy is the energy input required to form the transition state. If we can make the transition state more stable, then less energy is needed. Therefore, *a catalyst can lower the activation energy by chemically binding the transition state and reducing its energy.* Atoms in the catalyst form stabilizing chemical interactions with the transition state, which lowers its energy.

THE KEY IDEA: A catalyst increases the rate of a chemical reaction by lowering the activation energy.

Reaction without a catalyst

Reaction with a catalyst

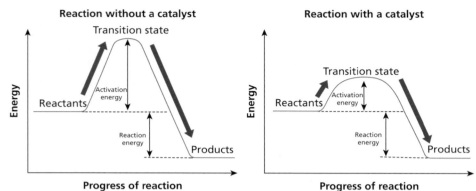

FIGURE 14.7 **A catalyst lowers the activation energy for a reaction.** This figure compares the energy diagrams for a chemical reaction without a catalyst and with a catalyst A catalyst lowers the activation energy, but it does not affect the energies of the reactants or products.

After the activation energy barrier has been overcome, the products are formed, and the catalyst releases them. A catalyst undergoes transient changes during the chemical reaction, but it returns to its original state at the end of the reaction. As a result, the same catalyst can accelerate trillions of chemical reactions without being used up.

Catalysts made in a laboratory can accelerate reactions by factors of hundreds, thousands, or millions, depending on the type of catalyst and the reaction. Worked Example 14.1 illustrates the principle of rate enhancement by considering a thought experiment involving a marathon runner.

WORKED EXAMPLE 14.1

Question: In 2013, Wilson Kipsang of Kenya broke the world record by completing the 26.2-mile Berlin Marathon with a time of 2 hours, 3 minutes, and 23 seconds. How quickly would Wilson Kipsang have completed the marathon if his running rate were increased by (a) a hundredfold, (b) a thousandfold, and (c) a millionfold? Write your numerical answer in the most appropriate time units.

Answer: The first step is to calculate the total number of seconds required to run the marathon.

$$\text{total seconds in 2 hours} = 2 \text{ hours} \times \left(\frac{60 \text{ minutes}}{1 \text{ hour}}\right) \times \left(\frac{60 \text{ seconds}}{1 \text{ minute}}\right)$$
$$= 7200 \text{ seconds}$$

$$\text{total seconds in 3 minutes} = 3 \text{ minutes} \times \left(\frac{60 \text{ seconds}}{1 \text{ minute}}\right) = 180 \text{ seconds}$$

$$\text{total seconds for the marathon race} = 7{,}200 + 180 + 23 = 7403 \text{ seconds}$$

(a) If Wilson Kipsang's running rate were increased by a *hundredfold* (10^2), then the time required for the race would be.

$$\text{race time} = \frac{7403 \text{ seconds}}{10^2} = 74.03 \text{ seconds}$$

This time is slightly under 1¼ minutes (75 seconds).

(b) If his running rate were increased by a *thousandfold* (10^3), then

$$\text{race time} = \frac{7403 \text{ seconds}}{10^3} = 7.403 \text{ seconds}$$

Rounding the answer gives a time of approximately 7.5 seconds, which is faster than the best sprinters can run 100 meters.

(c) If the running rate were increased by a *millionfold* (10^6), then

$$\text{race time} = \frac{7403 \text{ seconds}}{10^6} = 0.007403 \text{ seconds} = 7.403 \times 10^{-3} \text{ seconds}$$

Rounding to two decimal places, we see that the time for the race is now less than 7.5 milliseconds (ms), where 1 ms = 1/1000 of a second. The race would be over immediately after you heard the starting gun!

TRY IT YOURSELF 14.1

Question: A typical nonstop flight from New York to Honolulu, Hawaii, spends 10 hours and 15 minutes in the air. How quickly could you get to Hawaii if the flight speed were accelerated by (a) a hundredfold, (b) a thousandfold, or (c) a millionfold? Write your numerical answers in the most appropriate time units.

Core Concepts

- A catalyst increases the rate of a chemical reaction without being permanently altered.
- A catalyst works by lowering the activation energy for the reaction. The catalyst can chemically bind the transition state, thus reducing its energy
- A catalyst can accelerate multiple reactions without being used up.

The Catalytic Converter

One familiar example of a catalyst is the *catalytic converter* in the exhaust system of automobiles. The energy to power an automobile engine comes from a combustion reaction between gasoline and atmospheric oxygen. The primary component in gasoline is octane (C_8H_{18}), which reacts with O_2 gas to generate carbon dioxide (CO_2) and water (H_2O). However, if an insufficient amount of oxygen reaches the combustion chamber in the engine, then the combustion reaction generates carbon monoxide gas (CO) and water. Carbon monoxide is poisonous because it binds to hemoglobin proteins and reduces the essential transport of oxygen in the bloodstream. Carbon monoxide exposure can quickly lead to sickness and even death (see Chapter 6).

One way to remove the carbon monoxide is to react it with oxygen gas to produce carbon dioxide, which is less harmful. However, if we simply allow CO to mix with atmospheric O_2 in the automobile exhaust, then the reaction will be very slow. The CO and O_2 molecules both exist as gases, and they have to collide with the correct orientation to initiate the reaction. In addition, a high-activation energy needs to be overcome for the reaction to proceed. Owing to the slow speed of the reaction, very little CO would be cleared from the exhaust before being emitted through the tailpipe.

To reduce the emission of carbon monoxide, the exhaust gases from the engine are passed through the catalytic converter. The converter functions as a catalyst by accelerating the chemical reaction between CO and O_2 to produce CO_2. The equation for the chemical reaction that takes place in a catalytic converter can be written as

$$C\equiv O \quad + \quad O=O \quad \rightarrow \quad O=C=O \quad + \quad O$$

carbon monoxide   oxygen gas   carbon dioxide   oxygen atom

As shown in Figure 14.8(a), the inside of a catalytic converter contains a honeycomb structure with a large surface area that is coated with expensive metals

THE KEY IDEA: A catalytic converter uses a metal surface to accelerate the conversion of poisonous carbon monoxide into harmless carbon dioxide.

(a)

FIGURE 14.8 A catalytic converter uses a metal catalyst. (a) A catalytic converter from an automobile tailpipe. The cutaway shows the internal honeycomb structure, which provides a large surface area. (b) The honeycomb region is coated with metals, forming a catalytic surface. The metals that are used—platinum, palladium, and rhodium—belong to the transition metals, which occupy the central regions of the periodic table.

(b)

such as platinum, palladium, and rhodium. These metals belong to a group of elements called the *transition metals*, which are illustrated in Figure 14.8(b). These elements are located in the middle region of the periodic table, in between the main group elements at the left-hand and right-hand sides of the table. The transition metals include well-known elements such as iron (Fe), nickel (Ni), copper (Cu), zinc (Zn), silver (Ag), and gold (Au).

Within the catalytic converter, *the metal surface serves as the catalyst for the conversion of carbon monoxide to carbon dioxide.* A metal surface is a common type of catalyst that is used in numerous applications, especially in the chemical and pharmaceutical industries. Why are the transition metals well suited to serve as a catalyst? The explanation is related to another property of transition metals— they are good conductors of electricity (hence the use of metal wires for electrical circuits). A transition metal is a solid substance composed of a network of bonded atoms. However, some of the valence electrons are not bound tightly to the atomic nucleus, so they are capable of moving through the metal. These *free electrons* are responsible for conducting electrical current. They also enable transition metals to be good catalysts, because they can participate in transient chemical interactions with the reacting molecules on the metal surface.

Core Concepts

- The catalytic converter accelerates the chemical reaction between poisonous carbon monoxide (CO) and atmospheric oxygen (O_2) to produce harmless carbon dioxide (CO_2).
- The inside of a catalytic converter contains a honeycomb structure that is coated with transition metals. The metal surface functions as the catalyst.

14.4 Structure and Function of Enzymes

In this section, we apply our understanding of chemical reactions and catalysts to investigate the structure and function of enzymes. As we shall see, enzymes are superior to metal catalysts in two important ways. First, enzymes can increase the rate of chemical reactions by a factor of 1 billion (10^9), 1 trillion (10^{12}), or even much more. For comparison, a typical metal catalyst used in industrial applications can accelerate a reaction by a factor of roughly a million (10^6). Second, enzymes are very selective for specific types of molecules and chemical reactions. This property enables an enzyme to catalyze a particular reaction in a crowded cellular environment that contains many different types of molecules. By contrast, metal catalysts are far less discriminating because they tend to bind many types of molecules.

The Molecular Structure of Enzymes

As we explained in the chapter introduction, an enzyme is a biological catalyst that accelerates the rate of a chemical reaction. For example, enzymes in our digestive system break down food—carbohydrates, fats, and proteins—into nutrients. Without these enzymes, we could not digest food quickly enough to keep ourselves alive. As another example, *Chemistry in Your Life* discusses an enzyme that protects our cells from damage.

Figure 14.9 illustrates the general structure of an enzyme. Most enzymes are proteins with a specific three-dimensional structure based on the sequence of amino acids in the polypeptide chain (see Chapter 13). The catalytic activity occurs in a relatively small region of the enzyme called the **active site**. The active site binds the **substrate**, which is the term for the reactant molecule in an enzyme-catalyzed reaction. Substrate binding is facilitated by the spatial arrangement of the amino acid sidechains and regions of the polypeptide backbone, which provide chemical groups that interact with the substrate.

Within the active site, the chemical reaction is catalyzed through the participation of a specific set of amino acid sidechains. In some cases, the enzyme requires assistance from an auxiliary molecule or ion, which is called a *cofactor*. For example, the active site shown in Figure 14.9 contains a Zn^{2+} ion that assists with the enzyme's catalytic function. Other cofactors include vitamins, especially the B vitamins. This is why vitamins are an essential component of our diet.

By convention, the names of most enzymes end with the suffix *-ase*. The enzyme name usually includes the substrate that the enzyme binds. For example, *lactase* is the enzyme that binds lactose as a substrate. (The name is shortened to avoid the tongue-twisting name of "lactosease.") In many cases, the name also includes the chemical reaction that is catalyzed by the enzyme. For example, in

THE KEY IDEA: Most enzymes are proteins with a catalytic active site.

active site The region of an enzyme in which the catalytic activity occurs.

substrate The reactant molecule in an enzyme-catalyzed reaction.

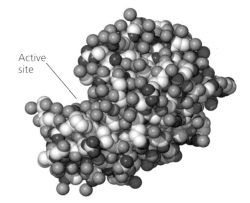

Active site

FIGURE 14.9 The three-dimensional structure of an enzyme. In this depiction of a generic enzyme, the orange "ribbons" represent the folding of the protein's polypeptide chain. The yellow region highlights the enzyme's active site, where the catalytic reaction occurs.

Chapter 5 we learned that *alcohol dehydrogenase* is an enzyme that uses alcohol as a substrate and removes a hydrogen atom from its functional group.

Several decades ago, scientists believed that all enzymes were proteins. However, this generalization was shown to be incorrect by the discovery of some enzymes that are RNA molecules. These RNA enzymes, called *ribozymes*, have the dual capacity of acting as a catalyst and also storing genetic information in their sequence of RNA bases. For this reason, some scientists have proposed that ribozymes were the first enzymes to evolve in the early history of life on Earth. In this chapter, we focus on protein enzymes because they are the more common type.

CHEMISTRY IN YOUR LIFE

Why Does Hydrogen Peroxide Bubble When You Put It on a Cut?

Have you ever used hydrogen peroxide solution to disinfect a cut? If so, then you probably noticed that it formed bubbles at the point of the wound (Figure 1). Why does this happen?

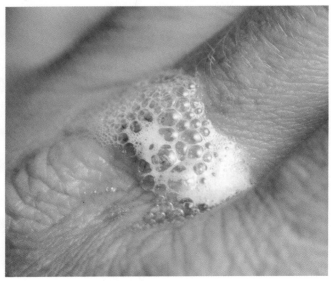

FIGURE 1 Hydrogen peroxide bubbles when you apply it to a wound.

The chemical formula of hydrogen peroxide is H_2O_2. It is an unstable molecule and is therefore highly reactive. If bacteria are present in the wound, then the hydrogen peroxide enters the bacterial cells and reacts indiscriminately with many types of biological molecules. These reactions can cause enough damage to disable the bacterial cells.

Hydrogen peroxide can also harm human cells in the same way. Fortunately, our body contains a highly effective defense mechanism against this chemical damage. All of the cells in our body contain an enzyme called *catalase*, which neutralizes the effects of hydrogen peroxide. The structure of catalase is shown schematically in Figure 2. The protein consists of four subunits, each of which contains a heme group that serves as the site of catalytic activity. Catalase helps break down hydrogen peroxide into two harmless substances: water and oxygen. The bubbles that you see from a treated wound come from oxygen gas. The chemical equation for this reaction

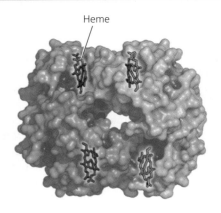

Heme

FIGURE 2 The structure of catalase Catalase is an enzyme that consists of four subunits, indicated by different colors. Each subunit contains a heme group, which is part of the active site.

is shown below; the upward arrow next to O_2 indicates that the oxygen is released as a gas.

$$2\,H_2O_2 \xrightarrow{\text{Catalase}} 2\,H_2O \;+\; O_2 \uparrow$$

Hydrogen   Water   Oxygen
peroxide

Catalase is one of the fastest and most efficient enzymes ever studied. Every second, a single catalase enzyme can convert more than *20 million* hydrogen peroxide molecules into water and oxygen. This very high catalytic rate makes sense because the enzyme's biological function is to disarm hydrogen peroxide before it can damage human cells.

Catalase also has a role in our hair turning gray as we age. Hydrogen peroxide is a common bleaching agent that can remove the color from many substances. As we get older, the amount of catalase produced by our body's cells decreases. As a result, the natural level of hydrogen peroxide in our cells begins to rise because there are fewer enzymes to break it down. When this process occurs in hair follicles, the hydrogen peroxide bleaches the hair from the inside and turns it gray.

Core Concepts

- Most enzymes are proteins with a specific three-dimensional structure.
- The **active site** of the enzyme is a relatively small region that binds the **substrate** and catalyzes the chemical reaction.

Substrate
Straight wire

Transition state
Bent wire

Product
Circular wire

FIGURE 14.10 The stages of a wire-bending reaction. The substrate is a straight wire, the transition state is a bent wire, and the product is a circular wire

The Steps of an Enzyme-Catalyzed Reaction

Biological organisms contain thousands of enzymes, each of which catalyzes a different chemical reaction. Despite this diversity, all enzymes operate by following a series of steps called the **catalytic cycle**. The terms *catalytic* and *catalysis* refer to the acceleration of a chemical reaction by a catalyst. Use of the word "cycle" is important because an enzyme (like all catalysts) is not permanently changed by the chemical reaction. At the end of the catalytic cycle, the enzyme returns to the same state in which it began, ready to catalyze another reaction.

To illustrate the catalytic cycle, we will use a hypothetical enzyme called bendase. Bendase catalyzes a reaction that bends a straight piece of wire into a circle. Figure 14.10 illustrates the stages of the wire-bending reaction. The "substrate" is the straight wire, the "transition state" is a bent wire, and the "product" is a circular wire. Note that *the structure of the transition state is distinct from the structures of the reactants and the products.*

How would bendase catalyze this reaction? The steps of its catalytic cycle are illustrated in Figure 14.11 and summarized below:

Step 1: The substrate binds to the active site of the enzyme.

Step 2: Binding interactions between the enzyme and the substrate form an *enzyme–substrate complex*. These interactions typically involve hydrogen bonds, electrical charge attractions, and nonpolar interactions (dispersion forces) between the substrate and amino acids in the enzyme's active site.

Step 3: The enzyme binds the transition state structure and lowers its energy. To function as an effective catalyst, *the enzyme binds the transition*

THE KEY IDEA: Enzymes operate by following a series of steps called the catalytic cycle.

catalytic cycle A series of steps that all enzymes use to catalyze a chemical reaction.

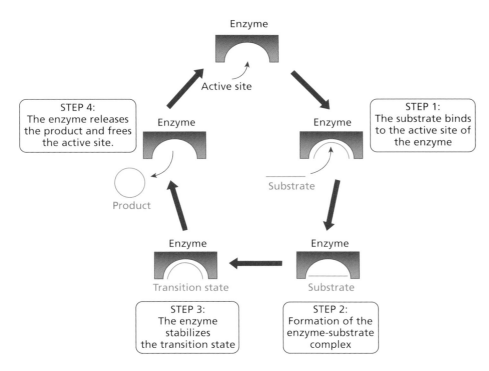

STEP 4:
The enzyme releases the product and frees the active site.

Active site

Enzyme

Enzyme

Enzyme

STEP 1:
The substrate binds to the active site of the enzyme

Substrate

Product

Enzyme

Enzyme

Transition state

Substrate

STEP 3:
The enzyme stabilizes the transition state

STEP 2:
Formation of the enzyme-substrate complex

FIGURE 14.11 The catalytic cycle for the bendase enzyme. The enzyme-catalyzed reaction passes through various stages. After the catalytic cycle is complete, the enzyme is capable of catalyzing another reaction.

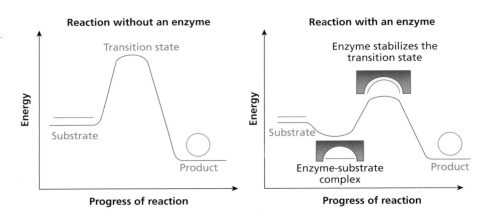

FIGURE 14.12 The effect of an enzyme on the wire-bending reaction. Without an enzyme, the unstable transition state (the bent wire) corresponds to a high-activation energy barrier. In the enzyme-catalyzed reaction, the enzyme first binds the substrate, producing an enzyme-substrate complex. Next, the enzyme binds and stabilizes the transition state, which lowers the activation energy for the reaction. A lower activation energy enables the reaction to proceed more rapidly.

state more tightly than it binds the reactants or the products. This tight binding stabilizes the high-energy transition state, which lowers the activation energy for the reaction and increases its rate.

Step 4: The transition state is converted into products, which are then released by the enzyme. The active site is now empty, and it can catalyze another reaction by binding a new substrate.

Figure 14.12 displays the energy profile of the wire-breaking reaction, which compares the reaction without the enzyme to the reaction that is catalyzed by bendase. The uncatalyzed reaction has a high activation energy, corresponding to the unstable transition state (the bent wire). In the enzyme-catalyzed reaction, the first energy change occurs when the enzyme–substrate complex is formed. The energy of this complex is lowered due to favorable binding interactions between the substrate and the enzyme's active site. However, this energy decrease cannot be too large; otherwise, the substrate would remain stuck in the active site, and the reaction would not proceed. The energy change is greatest when the enzyme binds the transition state, which lowers the energy barrier for the reaction and enables it to proceed more rapidly.

This energy profile for the enzyme-catalyzed reaction is more complex than the simplified impact of a catalyst that was presented in Figure 14.7. To achieve its catalytic function, the enzyme divides the chemical reaction into smaller steps. In effect, the enzyme changes the reaction pathway for wire-bending reaction. However, the central principle remains the same: The enzyme functions as a catalyst by lowering the activation energy for the reaction.

CONCEPT QUESTION 14.1

The Science in Action *feature explains how HIV protease inhibitor drugs are transition state analogs. Suppose that you were asked to design an inhibitor that is a transition state analog for the bendase enzyme that is described in the chapter. Sketch the structure of this inhibitor, and explain your drawing.*

How Does an Enzyme Recognize Its Substrate?

THE KEY IDEA: An enzyme binds to a specific substrate using complementary structural and chemical interactions.

Enzymes are remarkably proficient at binding their specific substrates, picking out the correct molecule from the multitude of different ones within the cell. How do they achieve this feat? As a first approximation, we can view the enzyme as a "lock" and the substrate as a "key." Even though many keys have a similar shape, each one has subtle differences in its ridges and grooves. If you try to put the wrong key in a lock, it likely will not fit. These principles are the basis of the

SCIENCE IN ACTION

Using the Transition State to Design a Drug

Acquired immunodeficiency syndrome (AIDS) is a life-threatening disease caused by human immunodeficiency virus (HIV). Like all viruses, HIV penetrates a host cell in order to make more copies of itself. HIV is particularly damaging because it infects the cells of the body's immune system. Over time, HIV can destroy the immune response and leave a person vulnerable to other infections, which are sometimes fatal.

In the early days of the HIV/AIDS epidemic, infection by the virus was an almost certain death sentence. The prognosis was greatly improved in the 1990s by the introduction of a new class of drug called *HIV protease inhibitors*. These drugs inhibit the function of an enzyme—*HIV protease*—that the virus needs to assemble itself. The development of protease inhibitor drugs was based on a detailed understanding of the molecular structure and reaction mechanism of the HIV protease enzyme.

After HIV infects an immune system cell, it tricks the cell into making more copies of its viral proteins. In order for HIV to replicate, these proteins must assemble to form new viruses. The function of HIV protease is to cut a long polypeptide chain into smaller components. These smaller chains then form proteins that are used to create a protective shell within HIV. After this shell has been formed, the intact virus is capable of infecting another cell.

HIV protease catalyzes a hydrolysis reaction (described in Chapter 7) that uses an H_2O molecule to break a peptide bond. Figure 1 compares the molecular structures of the substrate and the transition state for the enzyme-catalyzed reaction. HIV protease binds a polypeptide chain as its substrate. Within this chain, the structure of the peptide bonds is planar. During the hydrolysis reaction, an extra —OH group (from H_2O) is attached to the carbon atom in the peptide unit to form the transition state. The carbon atom transiently has four covalent bonds, and the principle of VSEPR predicts that the transition state has a *tetrahedral geometry*.

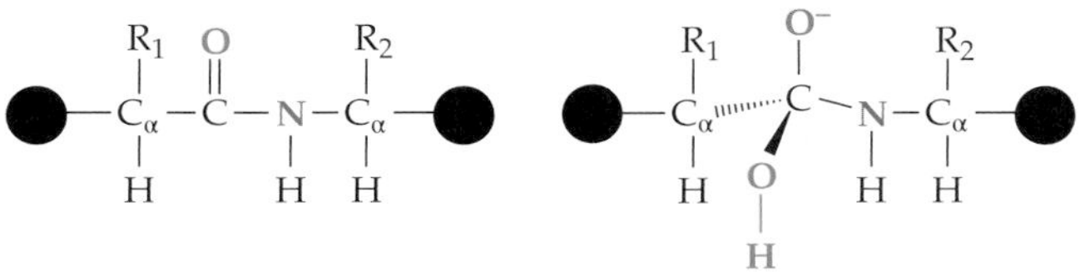

Planar substrate   **Tetrahedral transition state**

FIGURE 1. Structural comparison of the substrate and transition state. HIV protease catalyzes the hydrolysis of a peptide bond in a polypeptide chain. The peptide bonds in the polypeptide substrate are planar. By contrast, the transition state for the enzyme-catalyzed reaction has a tetrahedral structure. The addition of an —OH group in the transition state is highlighted in red.

We have learned in this chapter that enzymes are effective catalysts because they bind the transition state more tightly than the substrate or the products. We can exploit this principle by designing a drug that *mimics the transition state structure*. This type of drug is called a *transition state analog*. The enzyme binds the drug very tightly because of its structural similarity to the transition state.

Figure 2(a) shows the molecular structure of an HIV protease inhibitor (*Crixivan*). This drug is a transition-state analog because it contains a region that mimics the tetrahedral structure of the transition state. Figure 2(b) shows the drug bound tightly within the active site of the HIV protease enzyme.

When the drug is bound, HIV protease cannot perform its usual catalytic function of cutting the polypeptide chain. As a result, all the proteins necessary to fully assemble the virus are not available. In effect, inhibiting the HIV protease enzyme breaks the replication cycle of HIV within the body's immune cells. By applying the principles of enzyme reactions, scientists have been able to design a drug that has benefited millions of people worldwide.

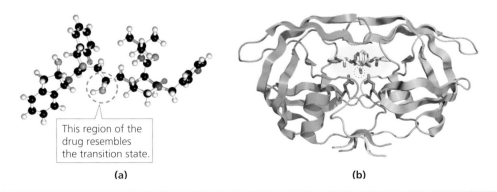

This region of the drug resembles the transition state.

(a)   (b)

FIGURE 2 An HIV protease drug is a transition state analog. (a) The molecular structure of the drug contains a region that mimics the tetrahedral transition state for the enzyme-catalyzed reaction. (b) The drug binds tightly within the active site of HIV protease.

FIGURE 14.13 The lock-and-key model of enzyme function. (a) Only the correct key will fit into a specific lock to form a "lock-key complex." (b) Only the correct substrate will bind to the active site of a specific enzyme to form an enzyme-substrate complex.

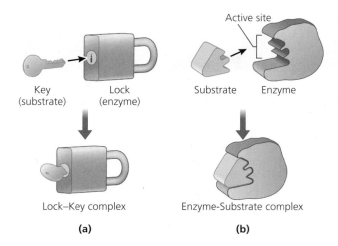

lock-and-key model A model of enzyme function in which the substrate (key) selectively fits into the active site of an enzyme (lock).

lock-and-key model of enzyme functioning, which is illustrated in Figure 14.13. The polypeptide chain of an enzyme folds into a specific three-dimensional shape and creates an active site with specific placements of amino acid sidechains. The size, shape, and chemical properties of the active site constitute the lock. These properties of the active site are *complementary* to the molecular structure of the substrate—that is, the active site of the enzyme is a molecular match for the substrate. This complementary relationship enables the enzyme to selectively bind only one specific substrate to form an enzyme–substrate complex.

Chemical complementarity exists between the enzyme and the substrate. This means that molecular features of the enzyme "lock"— such as polarity, nonpolarity, and charge—make attractive chemical interactions with the corresponding molecular feature on the substrate "key." Table 14.1 lists some types of complementary chemical interactions that occur between an enzyme and its substrate.

The principle of chemical complementarity in lock-and-key binding is illustrated in Figure 14.14 by comparing two very similar substrate molecules. Both substrates are tetrahedral and possess the same chemical groups, but their shapes are mirror images of each other (refer to the discussion of *chirality* in Chapter 13).

TABLE 14.1 Complementary Chemical Interactions Between Enzyme and Substrate

ENZYME	WITH SUBSTRATE	TYPE OF INTERACTION
Polar group	Polar group	Hydrogen bonding
Nonpolar group	Nonpolar group	nonpolar interaction
Positive charge	Negative charge	Charge attraction
Negative charge	Positive charge	Charge attraction

FIGURE 14.14 Chemical complementarity in binding a substrate to an enzyme's active site. (a) Substrate A can achieve tight binding to the enzyme active site through three different types of chemical interactions (polar, nonpolar, and charge). (b) Even though substrate B has the same chemical groups, it can achieve only one favorable chemical interaction with the active site (charge).

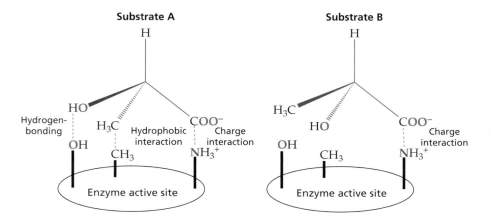

Substrate A binds tightly to the active site because of three favorable interactions: (1) a hydrogen bonding interaction between two polar groups; (2) a nonpolar, hydrophobic interaction between two methyl groups; and (3) a charge attraction between a negative carboxylic acid group and a positive amine group. By contrast, substrate B can achieve only one binding interaction with the active site (a charge attraction). If we rotate the substrate in space, we find that it is impossible to achieve more than one type of favorable interaction. As a result, substrate B binds very weakly to the enzyme, which is insufficient for catalysis. In summary, the asymmetric arrangement of chemical groups in the enzyme's active site makes it complementary to substrate A but not to substrate B. This example illustrates how enzyme active sites can be extremely selective in binding only the correct substrate molecule for catalysis.

PRACTICE EXERCISE 14.3

The drawing below illustrates some binding amino acid sidechains that interact with a substrate in the active site of an enzyme. You are provided with three possible substrates for the enzyme.

(a) Which substrate would bind most tightly to the enzyme active site? Identify the substrate number, and redraw its molecular structure within the active site of the enzyme. [Hint: Rotate the substrates to get the best fit.]

(b) Add labels to specify the type of complementary interactions that occurs between the substrate and each amino acid sidechain in the enzyme active site.

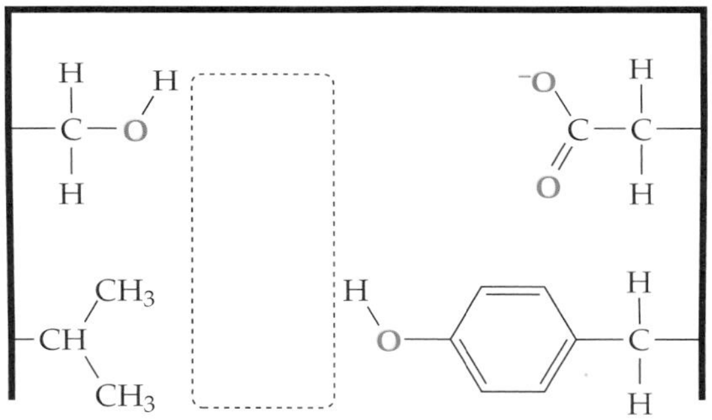

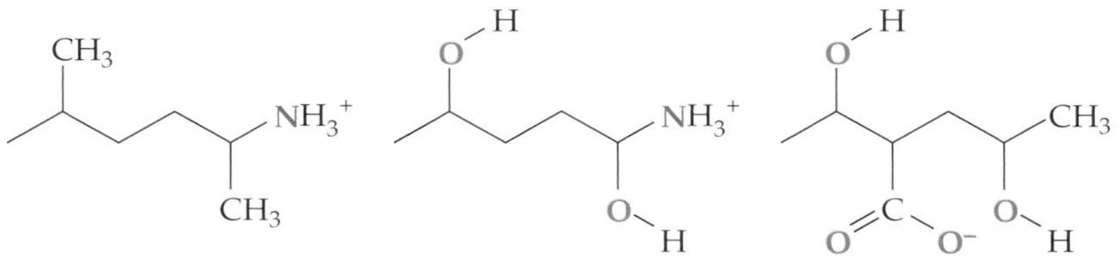

| Substrate 1 | Substrate 2 | Substrate 3 |

The lock-and-key model assumes that both the substrate "lock" and the enzyme "key" are rigid and inflexible. However, as scientists studied enzymes in more detail, a more complex picture emerged. Rather than remaining rigid, *an enzyme changes shape when it binds the substrate*. The interaction with the enzyme also alters the shape of the substrate molecule.

To understand what happens, think of playing baseball or softball. Imagine that the ball is the substrate and that your baseball glove is the active site. Suppose you are standing in the outfield when the batter hits the ball toward you. What will you do to make the catch? If you simply hold your glove in a fixed position, the ball will likely bounce off your palm and drop to the ground. To make the catch, you carefully watch the ball until you feel it touch, then you quickly close your glove over the ball to hold it. During this process, you also squeeze the ball, slightly distorting it from its regular shape.

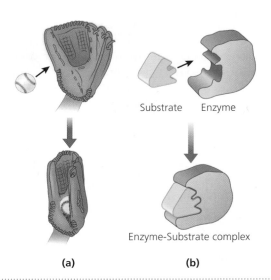

FIGURE 14.15 **The induced fit model of enzyme function.** (a) Catching a baseball requires closing the glove around the ball. (b) The enzyme changes its structure when binding the substrate to create favorable complementary interactions.

induced fit model A model of enzyme functioning in which the enzyme changes its shape when binding the substrate.

The same type of dynamic interaction occurs when an enzyme binds a substrate. This interaction, called the **induced fit model** of enzyme function, is illustrated in Figure 14.15. This figure differs from the lock-and-key-model because the substrate is not a perfect match to the enzyme's active site. *The process of binding the substrate to the enzyme changes the shape of both components to increase the number of complementary interactions.* Like the closing of the baseball glove, the structural change produces tighter enzyme–substrate binding than could be achieved if both components remained rigid.

The induced fit behavior of enzymes is very noticeable in the example of *hexokinase*, an enzyme that adds a phosphate group to a glucose molecule to begin the chemical breakdown of glucose. When hexokinase has no substrate bound, it exists in an "open" form. When it binds glucose, it snaps into a "closed" form to hold the substrate in place (Figure 14.16). The enzyme holds the glucose in the correct position for the addition of the phosphate group, with the chemical energy for the reaction provided by an ATP molecule. The flexibility of the enzyme active site is a major reason why enzymes can achieve rate enhancements that are billions of times faster than chemists' best metal catalysts, which remain rigid throughout the chemical reaction.

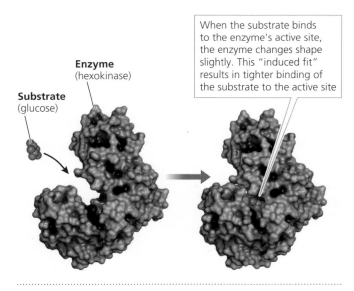

Enzyme (hexokinase)

Substrate (glucose)

When the substrate binds to the enzyme's active site, the enzyme changes shape slightly. This "induced fit" results in tighter binding of the substrate to the active site

FIGURE 14.16 **The hexokinase enzyme provides an example of induced fit.** The enzyme changes its shape upon binding the glucose substrate.

Core Concepts

- The **lock-and-key model** describes the enzyme as the lock and the substrate as the key. An enzyme selects only one substrate among many possible options because of complementary chemical interactions between the active site and the substrate.

- In the **induced fit model,** the enzyme changes its shape upon binding the substrate. This shape change enables the enzyme to bind the substrate more tightly by generating a larger number of complementary chemical interactions.

1.5 Lactose Intolerance: Genes, Enzymes, and Culture

Equipped with our understanding of enzymes, we can now examine the cause of lactose intolerance. If you can digest milk without any ill effects, then you may consider lactose intolerance to be uncommon. In fact, you are in the minority because approximately two-thirds of the Earth's population *cannot* digest lactose. This fact raises a puzzling question. We all drank milk as babies, so why are many adults incapable of digesting lactose? The answer involves a fascinating story that combines elements of science and human culture. We begin with a more detailed examination of lactase and lactose intolerance. We then examine the geographical variation of lactose intolerance within human populations, which provides clues about the origins of this condition.

The Catalytic Activity of Lactase

Recall from Chapter 7 that lactose is composed of two sugar units, glucose and galactose, that are joined by a C—O—C chemical linkage. Lactase catalyzes a hydrolysis reaction that uses an H_2O molecule to break the chemical linkage and release the smaller sugar molecules (Figure 14.17).

The catalytic cycle for lactase, which is illustrated in Figure 14.18, follows the general principles for enzyme catalysis that we discussed in the previous section. Lactose is the substrate, and it binds to the active site of lactase to form an enzyme–substrate complex. Within the active site, an H_2O molecule functions as an additional substrate for the hydrolysis reaction. Functioning as a catalyst, the enzyme stabilizes the transition state for the bond-breaking reaction. Finally, the two products (glucose and galactose) are released from the active site. The active site is now empty, and the enzyme is ready to catalyze another hydrolysis reaction.

Learning Objective:

Illustrate how human culture has influenced the ability to digest lactose.

THE KEY IDEA: Lactase catalyzes the breakdown of lactose into glucose and galactose.

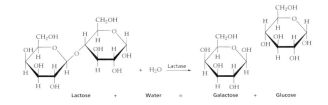

FIGURE 14.17 Lactase catalyzes a hydrolysis reaction. Lactose consists of two sugar units, glucose and galactose, that are joined by a C—O—C linkage. Lactase uses an H_2O molecule to catalyzes the hydrolysis of this linkage, which releases the two smaller sugars.

Core Concepts

- Lactose is composed of two sugar units, glucose and galactose, that are joined by a glycosidic bond.
- Lactase uses an H_2O molecule to catalyze the hydrolysis of the glycosidic bond, which releases the two smaller sugars.
- Lactase breaks down lactose through the standard steps of the catalytic cycle.

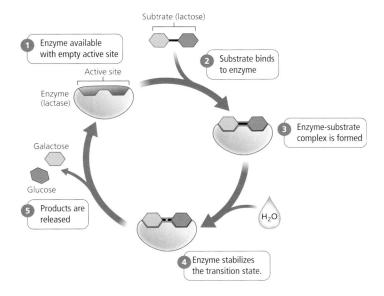

FIGURE 14.18 The catalytic cycle for lactase. The structure of the lactose substrate is shown in schematic form.

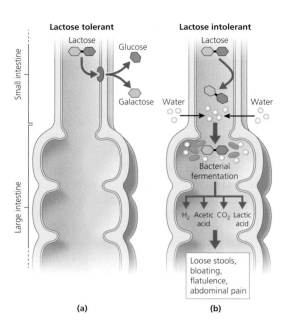

FIGURE 14.19 The physiology of lactose tolerance and intolerance. Lactose tolerance (left) or intolerance (right) depends on the presence or absence of the enzyme lactase in the small intestine. Lactase digests lactose into galactose and glucose. If lactase is not present, then lactose passes into the large intestine, where it is converted by bacteria to lactic acid via a fermentation reaction.

THE KEY IDEA: Individuals with lactose intolerance lack sufficient lactase to break down lactose, so the lactose is used by bacteria within the large intestine.

The Chemistry of Lactose Intolerance

Figure 14.19 compares the chemical processes that occur when milk is consumed by a person who is lactose tolerant versus one who is lactose intolerant. Individuals who can digest milk have an ample supply of lactase, which is attached to the walls of the small intestine. As we just discussed, lactase catalyzes the breakage of lactose into glucose and galactose. After being released by the enzyme, the two smaller sugars are absorbed through the walls of the intestine. They then enter the bloodstream, which circulates them around the body to be used as nutrients.

By contrast, individuals with lactose intolerance either are missing lactase or have an insufficient amount. Because of this enzyme deficiency, lactose is not digested in the small intestine. The intact sugar is not absorbed through the walls of the intestine. Instead, it passes into the large intestine, which contains large amounts of bacteria. The bacteria use lactose as a chemical fuel for a fermentation reaction, which converts lactose into lactic acid. This is the same chemical reaction that occurs when bacteria are added to milk to make yogurt. Molecules produced via fermentation reactions can stimulate the flow of water into the intestine, leading to diarrhea. Bacteria in the large intestine also use the lactose for other chemical reactions. Some of these reactions produce gases—methane and hydrogen—that cause sensations of bloating. In fact, one test for lactose intolerance is to measure the amount of hydrogen gas in a person's breath.

The symptoms of lactose intolerance are not life-threatening, but they are very unpleasant. As a result, individuals with lactose intolerance often stop drinking milk. However, eliminating milk completely from our diet can result in a deficiency of two important nutrients—calcium and vitamin D—that we need for strong bones and teeth. One option is to drink lactose-free milk. Another alternative—although it sounds counterintuitive—is to eat yogurt or cheese. These foods are made by using bacteria to ferment milk, which consumes much of the lactose. Finally, people can obtain calcium from other foods, such as fortified cereals (see Chapter 2) and dark leafy greens (spinach, kale, and collard greens).

CONCEPT QUESTION 14.2

Lactaid® milk does not contain lactose and is therefore suitable for consumption by individuals with lactose intolerance. Using what you have learned in this chapter, propose a method for removing the lactose from milk. Use the Internet to check whether your proposed method is used to make Lactaid® milk.

Core Concepts

- Individuals who are lactose tolerant have the lactase enzyme in the walls of their small intestine. After lactase has decomposed lactose, the glucose and galactose are absorbed through the walls of the intestine and enter the bloodstream.
- Individuals with lactose intolerance do not have sufficient lactase in their small intestine. Consequently, some of the lactose is not digested and it passes into the large intestine, where it is used by bacteria as a fuel for fermentation reactions.

The Evolutionary Origins of Lactase Persistence

The prevalence of lactose intolerance in humans differs significantly throughout the world. Figure 14.20 illustrates the global distribution of lactose intolerance, measured as a percentage of the adult population. Take a moment to study this map and observe the geographical variations. You will notice that most residents of northern Europe are capable of digesting lactose, as are the descendants of Europeans who emigrated to the Americas and Australia. By contrast, most people living in China and Southeast Asia are lactose intolerant, together with populations residing in the southern regions of Africa and South America. What accounts for this geographical variation in lactose intolerance?

To understand the map in Figure 14.20, we need to consider the lifestyle of human populations that lived more than 12,000 years ago. At that time, according to anthropologists, humans lived in hunter-gatherer societies, and their diet consisted of meat and plants. There were no domesticated cows or goats to provide milk. Newborn babies drank their mother's milk, just as they do today, which indicates that all babies have lactase in their stomach to digest the lactose in the milk. After the babies grew and switched to eating solid food, there was no longer any need to digest milk because none was available. After a few years of age, *young children lost their ability to produce lactase*. If this sounds strange, consider the similar example of cats. Young kittens enjoy drinking a saucer of milk, but you should not give milk to an adult cat. The reason is that kittens have lactase in their stomachs but adult cats do not.

What causes this change in the production of lactase? The answer arises from control of *gene expression* for the lactase gene. In Chapter 13, we learned that genetic information in DNA is used to produce a protein with a specific sequence of amino acids. First, the DNA is *transcribed* into a complementary sequence of messenger RNA (mRNA). Next, the mRNA exits the cell nucleus and travels to a ribosome, where it is *translated* into a sequence of amino acids. Humans have a

THE KEY IDEA: The ability to digest lactose into adulthood is an example of gene-culture coevolution.

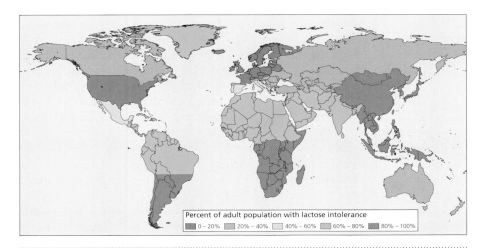

FIGURE 14.20 Global variation of lactose intolerance. This color-coded map shows lactose intolerance in different geographic regions, measured as a percentage of the adult population.

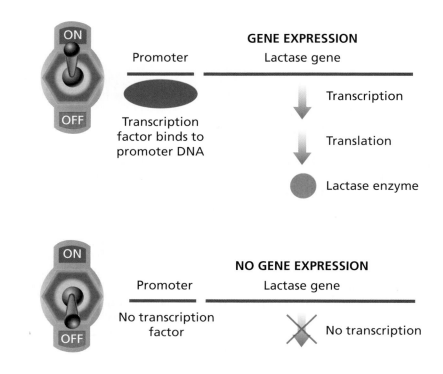

FIGURE 14.21 Expression of the lactase gene is controlled like an ON/OFF switch. The ON condition occurs when a transcription factor binds to a promoter region of DNA, located adjacent to the lactase gene. The promoter activates transcription of the gene, which is followed by translation to generate the lactase enzyme. The OFF condition occurs when the transcription factor is absent, in which case no transcription occurs.

region of DNA containing the lactase gene, which provides instructions to make the specific amino acid sequence for the lactase enzyme. Adjacent to the lactase gene is another region of DNA called a *promoter*, which acts as a control switch (Figure 14.21). The ON position of the switch corresponds to the action of a protein called a *transcription factor*, which binds to the promoter DNA and activates the transcription of the lactase gene. After this step, the mRNA is translated to generate the lactase enzyme. We say that the lactase gene is *expressed* because the corresponding protein is produced. By contrast, when the promoter is switched OFF, the transcription factor does not bind to the promoter, and transcription does not occur. As a result, no protein is produced, and we say that the gene is *not expressed*.

In babies and infants, the control system is switched ON, and sufficient lactase is produced to enable children to digest their mother's milk. When a child reaches the age of eating solid food, the system is switched OFF, and the production of lactase is halted. This biological mechanism informs us that lactose intolerance is the *ancestral state* of all humans. Why do some humans continue to make lactase into adulthood, a trait called *lactase persistence*?

The explanation for lactase persistence has emerged from a combination of research in chemistry, genetics, anthropology, and archaeology (Figure 14.22). Archaeological evidence suggests that human societies began to transition from hunting and gathering to farming around 10,000 to 11,000 years ago in the Middle East. This period of human history is called the *Neolithic*, which means "new stone age." Farmers herded domesticated cattle, which could provide milk. However, these populations could not drink the milk because they were lactose

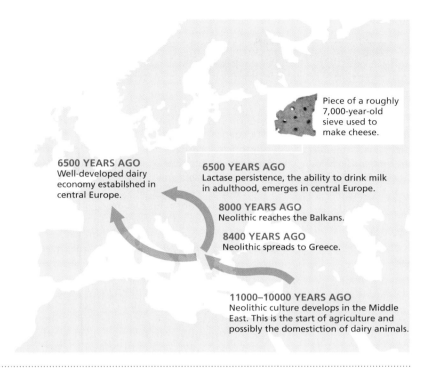

Piece of a roughly 7,000-year-old sieve used to make cheese.

6500 YEARS AGO
Well-developed dairy economy estabilshed in central Europe.

6500 YEARS AGO
Lactase persistence, the ability to drink milk in adulthood, emerges in central Europe.

8000 YEARS AGO
Neolithic reaches the Balkans.

8400 YEARS AGO
Neolithic spreads to Greece.

11000–10000 YEARS AGO
Neolithic culture develops in the Middle East. This is the start of agriculture and possibly the domestiction of dairy animals.

FIGURE 14.22 Lactase persistence developed in central Europe around 7,500 years ago. This condition originally arose as a DNA mutation in the promoter region of the lactase gene. The ability to digest lactose became advantageous in societies that domesticated cattle, so the trait spread through the European population as a result of natural selection.

intolerant. Over time, they discovered an ingenious solution—ferment the milk to produce cheese. As we have seen, during the process of fermentation, bacteria consume most of the milk's lactose as a chemical fuel. As a result, cheese has a low lactose content. Archaeologists have discovered cheese-making pottery dating back 7000 years. Chemical analysis confirmed that the pottery contains residues of fats that are characteristic of milk.

Around 7500 years ago, as farming societies moved into northern Europe, a mutation occurred in the promoter region of DNA located next to the lactase gene. Geneticists can date the origin of the mutation by measuring its distribution in modern humans and extrapolating back in time. This mutation deactivated the switch that shuts off the production of lactase. For individuals with this mutation, the lactase gene is always switched to the ON position, which results in the continued production of lactase into adulthood. Individuals who acquired this mutation were able to consume milk without any of the unpleasant effects of lactose intolerance.

The ability to obtain nutrients from milk was especially valuable during times of famine, when other food supplies were scarce. Individuals with the lactase persistence mutation were able to survive and reproduce more effectively than individuals who lacked the mutation. This scenario provides an example of genetic natural selection driven by changes in cultural practices, which is called *gene-culture coevolution*. Over successive generations, the genetic mutation for lactase persistence spread rapidly, which explains why close to 90% of northern Europeans can drink milk today. In regions of the world where dairy farming was not common, such as East Asia, there was no selective advantage for the lactase persistence mutation. As a result, most modern descendants of these populations are lactose intolerant.

CHEMISTRY AND YOUR HEALTH

What Happens When Enzymes Don't Work?

In addition to lactose intolerance, there are many other examples of enzyme deficiencies. Early in the 20th century, a British doctor named Archibald Garrod (1857–1936) reported on a number of patients whose urine had the strange property of turning black in air. Chemical analysis of the urine revealed one compound that was not present in normal specimens: homogentisic acid, a product of the breakdown of the amino acid tyrosine (Figure 1). Garrod gave the name *alkaptonuria* to this disorder (*alkapton* was the old name for homogentisic acid, and *-nuria* refers to urine).

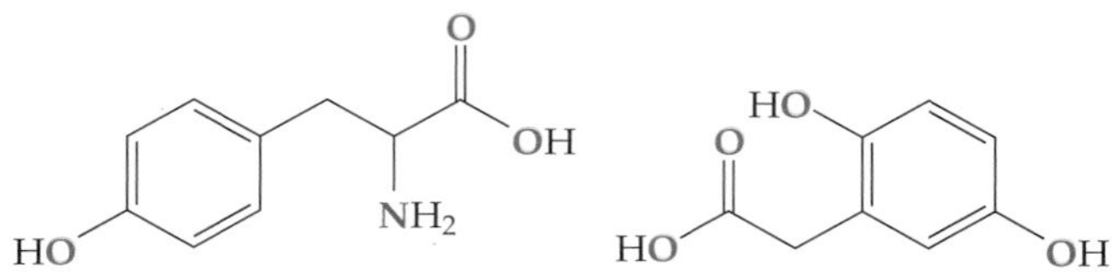

FIGURE 1 Molecular structures of (a) tyrosine and (b) homogentisic acid.

Garrod kept meticulous records of alkaptonuria for several families. At the same time, scientists were rediscovering the earlier work of Gregor Mendel (1822–1884), whose breeding experiments with pea plants explained how certain traits were transmitted to subsequent generations. After learning about Mendel's experiments from a colleague, Garrod realized that alkaptonuria obeyed the same pattern of inheritance. Specifically, alkaptonuria is a trait that appears only when a person inherits a mutated gene from both parents.

Garrod deduced that individuals who produce black urine are missing some biological agent that is present in most people—the phrase he used was an *inborn error of metabolism*. Most people have an enzyme that converts homogentisic acid into a compound that does not turn black in air. Individuals with alkaptonuria are missing this enzyme, so the level of homogentisic acid builds up in the urine to the point where it turns black. In addition, Garrod hypothesized that enzymes must be products of genes, a remarkable prediction that was subsequently confirmed many years later.

While accumulating homogentisic acid has mild health consequences, other mutations in the genes that express human enzymes have severe effects. Genetic disorders involving enzymes are responsible for hemophilia, in which an enzyme in the reactions that leads to formation of a blood clot is missing; Tay-Sachs disease, a lethal neurological disorder; and hundreds of other disorders. An inherited loss of an essential enzyme can be fatal if the product cannot be supplied by diet or if the accumulated substrate is toxic. For example, untreated individuals lacking the enzyme that forms tyrosine from phenylalanine suffer neurological damage if they consume foods containing phenylalanine.

CONCEPT QUESTION 14.3

The map in Figure 14.22 shows that lactose intolerance among Australians is in the range of 20% to 40%. Do you predict that the percentage of lactose intolerance would change if we examined only Australia's Aboriginal populations, whose ancestors populated Australia before the arrival of Europeans? Explain your prediction based on genes and culture.

Core Concepts

- Lactose intolerance varies by global geography. Populations in northern Europe and their descendants have a low percentage of lactose intolerance. By contrast, populations in Southeast Asia have a high percentage.
- Expression of the lactase gene is controlled like an ON/OFF switch. The ON position occurs when a transcription factor binds to a promoter region in DNA, which activates the production of lactase. When the transcription factor is not bound, the switch is in the OFF position, and no lactase is produced.
- Lactose intolerance in adults is the ancestral human condition. Lactase persistence arose as the result of a genetic mutation in the promoter region, which left the gene expression switch in the ON position into adulthood. The ability to digest lactose provided a survival advantage, so the mutation was preserved via natural selection. Lactase persistence is an example of gene-culture coevolution.

14.1 What Causes Lactose Intolerance?

Learning Objective:

Define lactose intolerance.

- ⊙ **Lactose intolerance** is an adverse reaction to the lactose sugar in milk.

- ⊙ **Lactase** is an enzyme that decomposes lactose into two smaller sugars, glucose and galactose.

- ⊙ An **enzyme** is a biological catalyst that accelerates the rate of a chemical reaction.

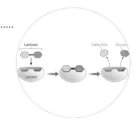

14.2 How Do Chemical Reactions Happen?

Learning Objective:

Characterize the changes that occur during a chemical reaction.

- ⊙ Chemical reactions have an **activation energy** that acts as an energy barrier between the reactants and the products.

- ⊙ The **transition state** of a chemical reaction is a transient, high-energy state that is formed during the conversion of reactants into products The activation energy for a reaction is the energy required to form the transition state.

- ⊙ The reaction energy is the energy difference between the reactants and the products.

- ⊙ The **reaction pathway** describes the stages by which reactants are converted into products. The substitution reaction between OH^- and CH_3Br provides an example of such a pathway.

- ⊙ The structure of the transition state, which has a trigonal bipyramidal geometry, differs from the structures of the reactants and the products.

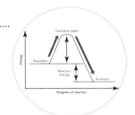

14.3   How Does a Catalyst Work?

Learning Objective:

Explain how a catalyst increases the rate of a chemical reaction.

⊙ A catalyst lowers the activation energy for a reaction by binding and stabilizing the transition state. As a result, the reaction proceeds more rapidly.

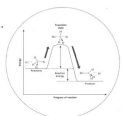

⊙ The reaction energy is not changed by the catalyst.

⊙ The catalytic converter **accelerates** the chemical reaction between poisonous carbon monoxide (CO) and atmospheric oxygen (O_2) to produce harmless carbon dioxide (CO_2).

⊙ The inside of a catalytic converter contains a surface that is coated with transition metals. This metal surface functions as the catalyst.

14.4   Structure and Function of Enzymes

Learning Objective:

Characterize the stages of an enzyme-catalyzed reaction.

⊙ Most enzymes are proteins with a specific three-dimensional structure. Enzymes have an **active site:** a relatively small area that binds the **substrate** molecule and catalyzes the chemical reaction.

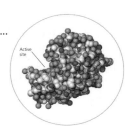

⊙ Enzymes catalyze a reaction by progressing through a **catalytic cycle**.

⊙ The active site of the enzyme binds the substrate to form an enzyme–substrate complex. Next, the enzyme tightly binds and stabilizes the transition state, which lowers the activation energy for the reaction. Finally, the enzyme releases the products of the reaction.

⊙ After the catalytic cycle is complete, the active site is available to bind another substrate and catalyze another reaction.

⊙ The **lock-and-key model** describes the enzyme as the lock and the substrate as the key. This model explains how an enzyme is selective for a specific substrate.

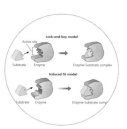

⊙ In the **induced fit model,** the enzyme changes its shape upon binding the substrate. This change of form allows the enzyme to bind the substrate more tightly.

14.5 Lactose Intolerance: Genes, Enzymes, and Culture

Learning Objective:

Illustrate how human culture has influenced the ability to digest lactose.

⊙ Lactase catalyzes the breakage of lactose into glucose and galactose. It facilitates hydrolysis of the bond that joins the two sugar units.

⊙ Lactase proceeds through the steps of the catalytic cycle.

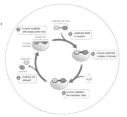

⊙ Individuals who are lactose tolerant have lactase in the walls of their small intestine. After lactase has decomposed lactose, the glucose and galactose enter the bloodstream.

⊙ Individuals who are lactose intolerant lack sufficient lactase in their small intestine to digest lactose. Consequently, the lactose passes into the large intestine, where it is used by bacteria as a fuel for fermentation reactions.

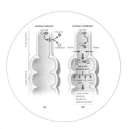

⊙ Populations in northern Europe and their descendants have a low percentage of lactose intolerance. By contrast, populations in Southeast Asia have a high percentage.

⊙ Expression of the lactase gene is controlled like an ON/OFF switch. Lactase persistence arose as the result of a genetic mutation, which left the gene expression switch in the ON position into adulthood. The increase of lactase persistence in some human populations is an example of gene-culture coevolution.

LEARNING RESOURCES

Reviewing Knowledge

14.1: What Causes Lactose Intolerance?

1. What is lactose intolerance?
2. Which enzyme breaks down lactose in our digestive system?

14.2: How Do Chemical Reactions Happen?

3. Why do most chemical reactions (e.g., combustion) not occur spontaneously?
4. Sketch the energy diagram for a chemical reaction. Add labels for the reactants, products, activation energy, and reaction energy.
5. How is the rate of a chemical reaction affected by (a) the activation energy and (b) the reaction energy?

14.3: How Does a Catalyst Work?

6. Use a diagram to explain how a catalyst increases the rate of a chemical reaction.
7. What is the function of a catalytic converter in a vehicle? What elements are used to make the catalytic surface for this device?

14.4: Structure and Function of Enzymes

8. Compare and contrast the ability of a metal surface and an enzyme to catalyze a chemical reaction.
9. What is the role of the active site of an enzyme?
10. Briefly describe the stages in the catalytic cycle of an enzyme.
11. During the catalytic cycle, does the enzyme bind most tightly to the substrate, the transition state, or the products? Explain your answer in terms of how enzymes function as catalysts.
12. Briefly explain two models of enzyme function: lock-and-key and induced fit. In what ways are they similar? In what ways are they different?

14.5: Lactose Intolerance: Genes, Enzymes, and Culture

13. Describe the chemical reaction catalyzed by lactase. Identify all substrates and products, and specify what type of bond is broken.
14. What causes lactose intolerance in humans? What causes the intestinal problems that are characteristic of lactose intolerance?

15. Explain how the expression of the lactase gene is regulated like an ON/OFF switch.

16. Why can some adults continue digesting lactose into adulthood (a trait called lactase persistence)?

17. Explain how human culture affected the spread of lactase persistence in some human populations.

Developing Skills

18. Using the energy diagram below, determine the activation energy of the reaction in kilojoules (kJ). Is the reaction exothermic or endothermic? (See Chapter 6 for a discussion of energy units.)

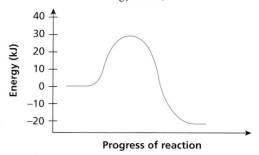

19. A reaction has an activation energy of 30 kJ and a reaction energy of –90 kJ. (a) Using the axes below, sketch the energy diagram for this reaction, and indicate the activation energy and reaction energy using arrows. (b) Is the reaction endothermic or exothermic?

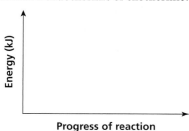

20. It is not usually possible to rotate a C=C double bond, but a sufficiently high input of energy can make this happen. A *cis/trans* isomerization reaction involves twisting a molecule around a C=C double bond to switch between a *cis* and a *trans* isomer (or vice versa). The structure of a molecule containing a *cis* C=C bond (the reactant) is shown here.

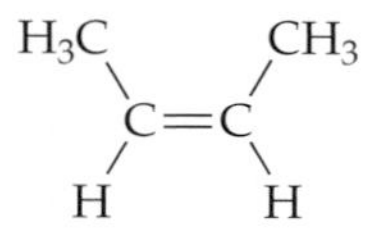

$$H_3C \quad\quad CH_3$$
$$C=C$$
$$H \quad\quad H$$

(a) Draw the structure of the product of the reaction (i.e., the *trans* isomer).

(b) Imagine the molecule changing from the *cis* isomer to the *trans* isomer. What will the transition state for the reaction look like? Draw the 3D structure of the transition state, using the wedge-and-dash notation. Explain the rationale for your drawing.

21. A hypothetical reaction has the following energy diagram (A). Which of the energy diagrams (B–D) represents the catalyzed reaction? Explain your answer in one sentence.

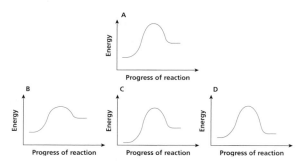

22. Imagine that an outer space probe has discovered a new type of organism that can derive its energy needs from the following chemical reaction:

$$Cl_2 + CH_4 \rightarrow CH_3Cl + HCl$$

Performing this reaction in the laboratory requires high temperatures and a metal catalyst. However, the newly discovered organism can perform the reaction at room temperature. Use the concepts from this chapter to explain how this could be possible.

23. Most simple reactions in a test tube run faster if they are heated. By contrast, heating a solution containing an enzyme actually *slows* the reaction. Explain the difference between these two situations.

24. The *turnover number* of an enzyme is defined as the number of chemical conversions of substrate into product that each active site of an enzyme can accomplish within 1 second. The units of turnover number are per second, written as 1/s or s^{-1}. *Chemistry in Your Life* described the enzyme catalase, which converts hydrogen peroxide (H_2O_2) into water and oxygen gas. The turnover number for catalase has been measured as $2.25 \times 10^7\, s^{-1}$. If catalase is supplied with an unlimited amount of substrate, what amount of time is required for a single catalase enzyme to make 1 billion O_2 molecules?

25. As you learned in Chapter 13, amino acids are joined by peptide bonds to make a protein. A peptide bond can be broken via a hydrolysis reaction. When a molecule made from two amino acids is placed in water at pH 7, the hydrolysis reaction occurs once every 350 years (approximately). However, when an enzyme is added to the solution, it is able to catalyze 1842 hydrolysis reactions every second.

(a) What is the biological importance of having a peptide bond that is resistant to hydrolysis in water?

(b) What *rate enhancement* of the hydrolysis is achieved by the enzyme compared to the uncatalyzed reaction? In other words, how many times faster does the enzyme make the reaction happen?

(c) Is your answer to part (b) closer to one thousand, one million, one billion, or one trillion?

26. Some enzymes have been found to accelerate the uncatalyzed rate by a factor of 10^{18}. If the active enzyme has a turnover rate of 10^8 per second, what would be the

rate of the reaction without catalysis? How many years would be needed to observe one event?

27. Most enzymes are effective catalysts within only a limited pH range, as shown in the graph. In this case, what is the optimal pH for the selected enzyme? What happens to the enzyme activity in more acidic solutions? More basic solutions?

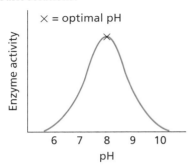

28. Our digestive tract uses different enzymes that catalyze the breakdown of the foods we eat. The stomach becomes highly acidic following a meal, with a pH near 2. Would the enzyme described in question 27 be effective in the stomach? Explain your answer.

29. In pioneering studies of reactions catalyzed by enzymes, scientists noted that as the concentration of substrates increased, the rate of the reaction always reached an upper limit called V_{max}. Without knowing anything about proteins, they concluded that a substrate must bind to the enzyme before catalysis can take place. What is the condition of the enzyme's active site when the maximum rate has been achieved?

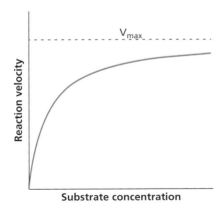

30. The molecule acetylcholine serves as neurotransmitter that conducts chemical signals between nerve cells. After acetylcholine has performed its function, its breakdown is catalyzed by an enzyme called acetylcholinesterase. The enzyme-catalyzed reaction is shown here.

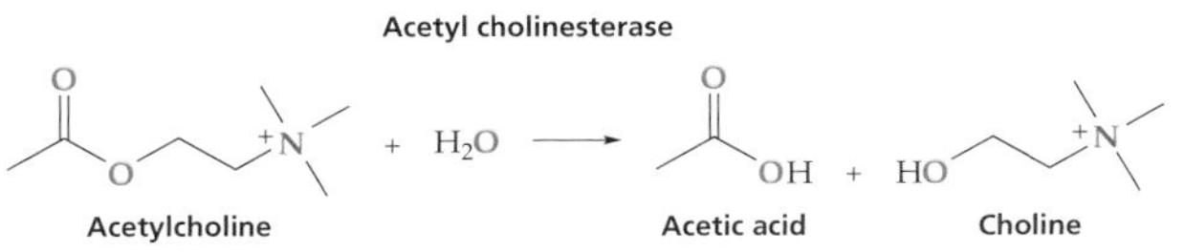

(a) The enzyme has a turnover number of 25,000 s^{-1}, which measures the conversion of substrate to products. How much time is required for the enzyme to break down a single molecule of acetylcholine? How many times could an enzyme carry out its catalytic cycle in 1 millisecond?

(b) VX gas is a very potent nerve toxin that inhibits the function of acetylcholinesterase, producing a lethal buildup of acetylcholine. Enzyme inhibitors frequently resemble the molecular structure of the enzyme substrate. The molecular structure of the VX toxin is shown here. Which chemical groups that are common to VX and acetylcholine? Which are different?

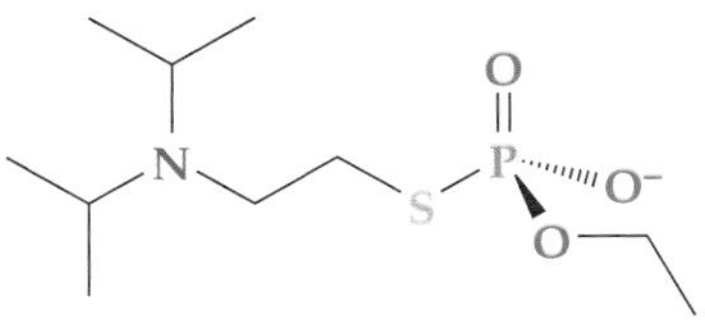

Exploring Concepts

31. Many enzymes contain hundreds or even thousands of amino acids. However, the active site that carries out the catalytic function is much smaller, involving perhaps 10 amino acids. Propose *two possible explanations* for why enzymes are so large.

32. Enzymes are responsible for the replication of DNA and the transcription of RNA from DNA. What might be the fate of a cell in which either of these enzymes becomes inactive through mutation?

33. In this chapter you learned about the origins of lactase persistence. Another example of gene-culture coevolution involves the alcohol dehydrogenase enzyme, which catalyzes the breakdown of alcohol in the liver (see Chapter 5). Use the Internet to investigate this topic, and write a two- to three-paragraph report that explains human variation in the expression of this enzyme.

34. Soybeans are an important source of nutrition. However, they contain a very stable protein molecule that prevents digestion of food proteins in our diet. The inability to digest proteins can lead to gastric distress. Some preparations of tofu, a processed form of soy milk, are treated to block the inhibitors, for example, by heat or acid exposure. How do heat and acid affect the folded structure of proteins?

15

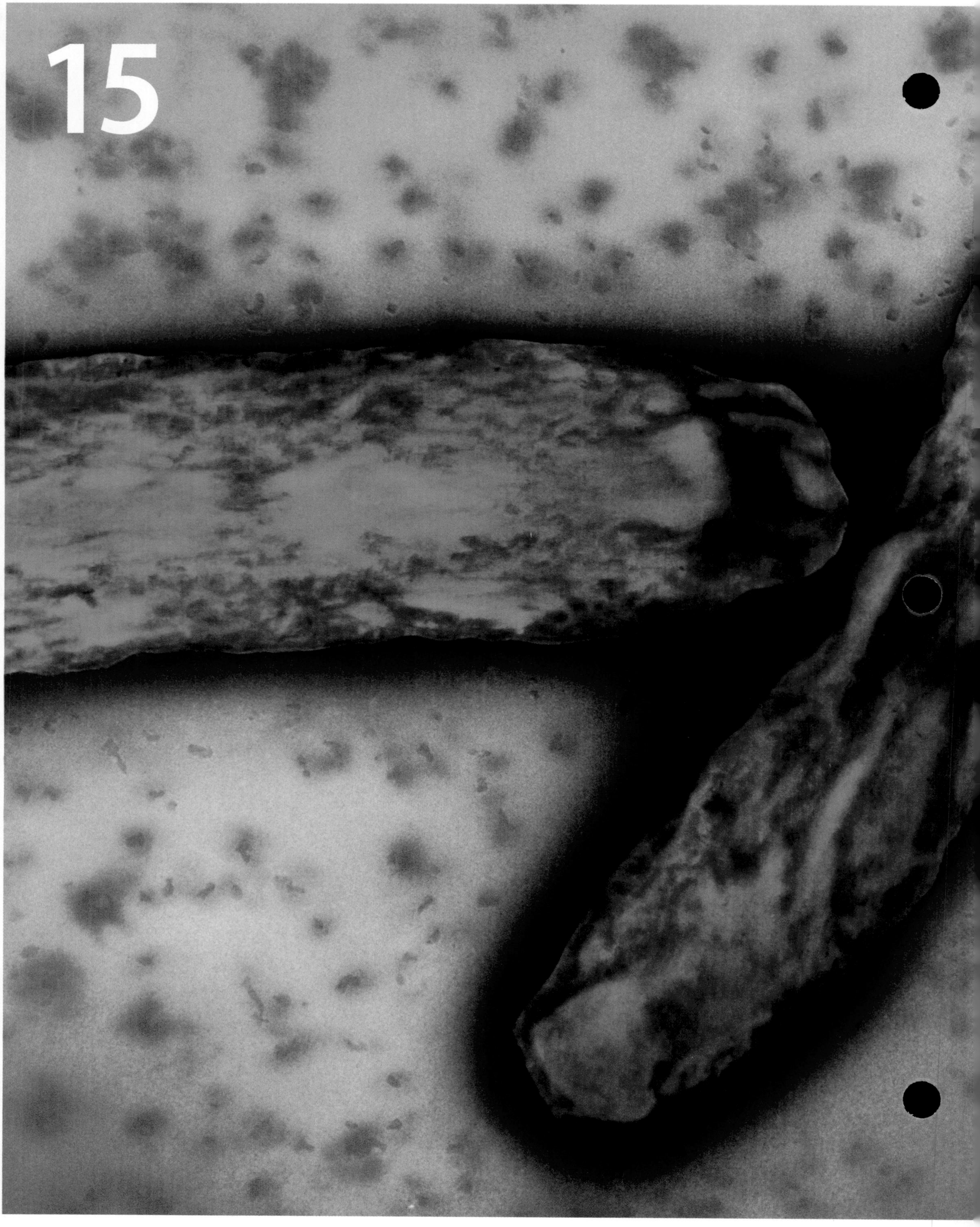

Drug Development

Chapter 15

CONCEPTS AND LEARNING OBJECTIVES

15.1 Why Are Antibiotic-Resistant Infections Increasing? p. 454

Learning Objective: Use tuberculosis as an example of an antibiotic-resistant disease.

15.2 Bacterial Cells, p. 456

Learning Objective: Describe the size and composition of bacterial cells.

15.3 How Do Antibiotics Work? p. 460

Learning Objective: Illustrate how antibiotics use different cellular targets to prevent the growth of bacterial cells.

15.4 How Do Bacteria Become Resistant to Antibiotics? p. 466

Learning Objective: Illustrate the strategies used by bacteria to neutralize the effects of antibiotics.

15.5 Drug-Resistant Tuberculosis, p. 470

Learning Objective: Outline the relationship between the development of new antibiotics and the rise of antibiotic resistance.

15.6 How Are New Drugs Tested and Approved? p. 472

Learning Objective: Outline the stages that are required to develop and evaluate a new pharmaceutical.

FEATURES

- **SCIENCE IN ACTION:** Studying Cells with Microscopes, p. 457

- **CHEMISTRY AND YOUR HEALTH:** How Do Bacteria Keep Us Healthy? p. 468

- **CHEMISTRY IN YOUR LIFE:** Why Can a Popular Painkiller Make You Sick? p. 476

What do the writers John Keats, Edgar Allan Poe, Emily Brontë, D. H. Lawrence, and George Orwell have in common? They all died of tuberculosis, the deadliest disease in human history. The development of antibiotics to treat tuberculosis was a groundbreaking advance in 20th-century medicine. However, tuberculosis is once again becoming a global health threat because antibiotic-resistant infections are increasing.

15.1 Why Are Antibiotic-Resistant Infections Increasing?

Learning Objective:

Use tuberculosis as an example of an antibiotic-resistant disease.

THE KEY IDEA: Antibiotic-resistant diseases are increasing worldwide.

Although the global incidence of tuberculosis has decreased during the past few decades, it still remains a potent killer. In 2014, the World Health Organization estimated that tuberculosis kills 1.5 million people annually, which is over 4000 persons per day. Tuberculosis has now surpassed HIV/AIDS as the world's deadliest infectious disease. The U.S. Centers for Disease Control and Prevention (CDC) recently reported that the number of tuberculosis cases in the United States rose slightly during 2015, after more than two decades of steady declines. In addition, there has been a global increase in tuberculosis infections that are resistant to treatment by the standard repertoire of antibiotics. What causes tuberculosis, how do antibiotics treat it, and why are these antibiotics becoming ineffective?

Tuberculosis is an infectious disease caused by a bacterium called *Mycobacterium tuberculosis*. The bacterium is spread easily through the air by coughing or sneezing. Common symptoms of tuberculosis include lethargy, coughing as the lungs decay, and deformation of the spine and other bones. The last-named effects leave a record of tuberculosis infection that has been observed in prehistoric human remains. Remarkably, the tuberculosis bacterium infects millions of people who do not display any symptoms. This "latent tuberculosis" results from the ability of the immune system to enclose the bacteria in swollen regions called *tubercles*, which is the signature of the disease (Figure 15.1). The thick waxy coat of the bacteria prevents the immune defense from killing tuberculosis cells. Once a patient has a compromised immune system—from HIV, poor diet, or chemotherapy—the latent infection can spread and become lethal.

A tuberculosis infection that is resistant to the two most commonly used antibiotics is classified as *multidrug-resistant tuberculosis* (MDR-TB). Figure 15.2 displays a map of the global distribution of MDR-TB among previously treated TB cases. In total, 20% of these cases show multidrug resistance. Individuals who are infected with MDR-TB must be treated with a long-term therapy that involves multiple drugs. Unfortunately, this treatment leads to unpleasant side effects, which can include nausea, dizziness, and skin rashes. In addition, patients must be quarantined for months.

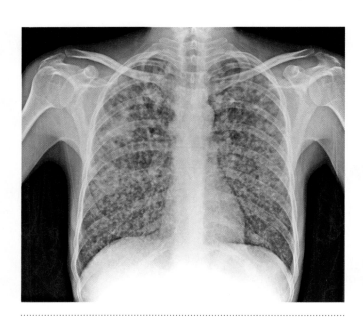

FIGURE 15.1 Tuberculosis bacteria infect the lungs. An artificially colored X-ray photograph of a patient whose lungs show the characteristic signs of tuberculosis. The tubercles (colored pink) consist of tuberculosis bacteria that have been walled off and enclosed by the immune system.

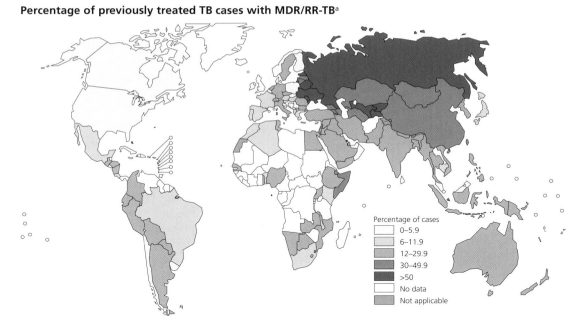

Percentage of previously treated TB cases with MDR/RR-TB[a]

Percentage of cases
- 0–5.9
- 6–11.9
- 12–29.9
- 30–49.9
- >50
- No data
- Not applicable

> [a] Figures are based on the most recent year for which data have been reported, which varies among countries. Data reported before the year 2001 are not shown. The high percentages of previously treated TB cases with MDR-TB in Bahamas, Bahrain, Belize, Bonaire – Saint Eustatius and Saba, French Polynesia and Sao Tomé and Principe refer to only a small number of notified cases (range: 1-8 notified previously treated TB cases).

FIGURE 15.2 Worldwide incidence of MDR-TB. A global map showing the percentage of previously treated TB cases that were discovered to have multidrug resistance (MDR-TB). Data were provided by the World Health Organization in its 2015 *Global Tuberculosis Report*.

In recent years, we have witnessed the emergence of a deadlier form of tuberculosis called *extensively drug-resistant TB* (XDR-TB). This type is resistant to at least four antibiotics, some of which are considered to be the "last resort" for treating the disease.

CONCEPT QUESTION 15.1

(a) According to Figure 15.2, what is the percentage range of MDR-TB cases in the United States? (b) Identify two countries that have particularly high percentages of MDR-TB.

Meanwhile, many other infectious bacteria are becoming resistant to antibiotics that are commonly used as standard medical treatments. In May 2016, researchers at the U.S. Department of Defense reported the case of a woman in Pennsylvania who developed a urinary tract infection. This may sound like a routine medical problem. However, the *E. coli* bacteria that caused the infection were resistant to *colistin*, an antibiotic that has been called a "drug of last resort." Analysis of the bacteria's DNA revealed the presence of 15 genes that conferred resistance to other antibiotics.

In this chapter, we examine the bacterial cell, the ways in which antibiotics act on bacteria, the rise of drug resistance, and the development of future drug therapies to overcome this threat. This chapter serves as a capstone to this textbook by showing how our investigation of antibiotics and resistance draws upon multiple topics discussed in previous chapters.

Core Concepts

- Tuberculosis is an infectious disease caused by a bacterium called *Mycobacterium tuberculosis*. Tuberculosis is the deadliest disease in human history.
- Tuberculosis that is resistant to treatment with antibiotics is classified as multidrug-resistant (MDR-TB) or extensively drug-resistant (XDR-TB). The incidence of both types is increasing worldwide.

15.2 Bacterial Cells

Learning Objective:
Describe the size and composition of bacterial cells.

Mycobacterium tuberculosis is a bacterium. What does this term mean? A bacterium (plural, *bacteria*) is a type of biological organism that can exist as a single cell. One of the great scientific advances in biology was the unifying principle that all life, no matter how complex, is organized into subunits called cells. Bacteria are the most abundant and diverse type of cells on Earth. In the 19th century, some bacteria were identified as the cause of infectious diseases. This section reviews the structure and organization of bacterial cells.

The Cell Is the "Unit of Life"

○┐ THE KEY IDEA: All living organisms are composed of cells.

The biological term *cell* is derived from the Latin *cella*, which means "small room." It was first used in the 17th century by Robert Hooke (1635–1703), a contemporary of Isaac Newton (1643–1727). When Hooke used a rudimentary microscope to study the dried bark of a cork tree, he observed an array of walled cavities that reminded him of the small living spaces occupied by monks in monasteries (Figure 15.3). Hooke used the name "cell" to describe these microscopic structures, and the term has persisted to the present day.

During the 19th century, biologists built better microscopes and developed techniques for staining cells using chemical dyes. These advances led to an important scientific insight called the **cell theory** of life. *According to the cell theory, all living organisms are composed of cells.* In essence, the cell serves as the fundamental unit of life. Bacteria can exist as single cells, as do many other types of organisms. More complex organisms, such as animals and plants, are composed of vast numbers and types of cells. According to one scientific estimate, the human body contains more than 37 trillion cells (1 trillion = 1 million million = 10^{12}).

■ **cell theory** A scientific theory that states all living organisms are composed of basic units called cells.

Core Concepts

■ **Cell theory** states that all living organisms are composed of cells.

How Small Are Cells?

○┐ THE KEY IDEA: A bacterial cell has an approximate size of one-thousandth of a millimeter.

It is possible to see some cells, such as the egg cell of an ostrich, with your naked eye. By contrast, bacterial cells and most human cells are too small to be observed directly. This is why scientists study cells using a *microscope*. One commonly used type of microscope, called a light microscope, uses two glass lenses to create a magnified image of a cell that is large enough to be seen. *Science in Action* describes the operation of a light microscope and a more powerful device called an electron microscope

How small are cells? Imagine looking at the point of a pin. The tip of the point is about 1/10 of a millimeter across, which is the limit of unaided human vision. Cells are even smaller, so we need a unit to describe dimensions smaller than a millimeter. This unit is called the *micrometer* (μm), usually abbreviated as *micron*. One micron is equal to one-thousandth of a millimeter.

$$1 \ \mu m = \frac{1}{10^3} \ mm = 10^{-3} \ mm$$

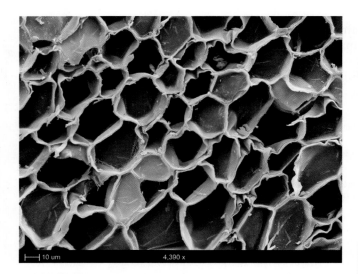

├──┤ 10 um   4,390 x

FIGURE 15.3 A microscopic view of cells. A magnified view of cork bark using a modern microscope, which illustrates the small enclosed spaces that Robert Hooke called "cells."

PRACTICE EXERCISE 15.1

Consider a virus with a diameter of 50 nm. How many viruses, placed size by side, are required to equal the size of a bacterial cell with a length of 2 μm.

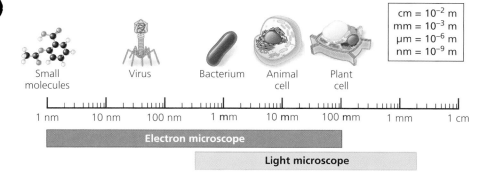

| cm = 10^{-2} m |
| mm = 10^{-3} m |
| μm = 10^{-6} m |
| nm = 10^{-9} m |

Small molecules   Virus   Bacterium   Animal cell   Plant cell

1 nm   10 nm   100 nm   1 mm   10 mm   100 mm   1 mm   1 cm

Electron microscope

Light microscope

FIGURE 15.4 The relative sizes of cells and viruses. The sizes of cells are most conveniently measured in microns (μm), where 1000 μm = 1 mm. Animal and plant cells are larger than a bacterial cell. In turn, bacterial cells are larger than viruses.

Figure 15.4 provides a general comparison of cell sizes. Human cells are in the range of 10 to 100 μm. One exception is red blood cells, which have a diameter of only 8 μm. Bacterial cells are smaller than human cells, with sizes in the range of 1 to 5 μm.

SCIENCE IN ACTION

Studying Cells with Microscopes

A hand lens can magnify small objects to make them more clearly visible to the naked eye. However, a single lens won't suffice to examine something as small as a cell. Instead, we need to use a light microscope that contains two glass lenses (Figure 1(a)). Because it contains two lenses, this device is also called a *compound microscope*. The first lens, called the *objective lens*, is located near the object of study. It creates a magnified image that is magnified further by a second lens close to the viewer's eye, which is called the *ocular lens*.

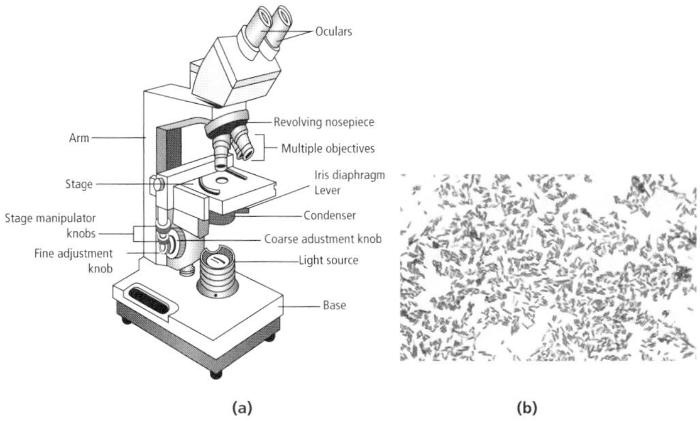

However, bacterial cells are visible only as tiny rods (Figure 1(b)), and viruses cannot be observed at all.

Today, scientists can achieve much higher magnification (50,000×) by using an *electron microscope*. This device exploits the fact that electrons can exhibit the properties of waves. A schematic diagram of an electron microscope is shown in Figure 2(a). A beam of electrons from a high-voltage source is directed down a narrow tube. Instead of glass lenses, the electron microscope uses magnets to focus the beam of electrons and pass it through a sample. The higher magnification of an electron microscope allows us to view the internal structure of bacterial cells (Figure 2(b)). Electron microscopes enable us to directly visualize the magnified images of very small organisms such as viruses.

FIGURE 1 Visualizing cells with a light microscope. (a) A schematic diagram of a compound microscope. A magnified image is produced by two glass lenses called the objective lens and the ocular lens. The light source illuminates the cell, and the focus controls enable the user to produce a sharp image. (b) *E. coli* bacterial cells observed at 1000× using a light microscope. The cells have been stained with a purple dye.

A light microscope can generate high-quality images up to a maximum magnification of roughly a thousand-fold (1000×). This magnification is suitable for viewing large plant and animal cells.

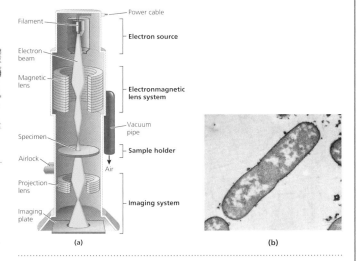

FIGURE 2 Visualizing cells with an electron microscope. (a) A schematic diagram of an electron microscope, which uses magnets to focus electron waves. (b) *E. coli* cells observed at 50,000x magnification using an electron microscope, which reveals the internal structure of the cell.

Viruses are even smaller than bacterial cells, with typical sizes of 10 to 100 nanometers (1 nm = 10^{-9} m). Viruses have a very simple structure that consists of genetic material (DNA or RNA) encased in a protective capsule made of proteins and/or membranes.

Core Concepts

- Cell sizes are measured in units of micrometers or microns (μm). One micrometer is equal to one-thousandth of a millimeter.
- Most human cells have a size range of 10 to 100 μm. Bacterial cells are smaller, with a size range of 1 to 5 μm. Viruses are even smaller, with sizes that typically range from 10 to 100 nm.

THE KEY IDEA: A bacterial cell consists of a cytoplasm interior surrounded by a cell membrane and sometimes a cell wall.

The Composition of a Bacterial Cell

Figure 15.5 presents a cartoon diagram of an *E. coli* bacterial cell. The interior of the cell is a watery medium called the *cytoplasm*, which is densely packed with many different types of molecules and ions. As we discussed in Chapter 12, the bacterium's DNA molecules encode the genetic information that is necessary to make proteins. This is accomplished by first synthesizing molecules of messenger RNA (mRNA), which are then used by the cell's ribosomes to manufacture proteins.

As we learned in Chapter 9, the cell membrane is composed of molecules called *phospholipids*, which self-assemble to create a bilayer structure. The phospholipid bilayer encloses the cytoplasm and provides a barrier to the external environment. In many bacteria (but not all), the cell membrane is surrounded by a rigid, protective cell wall. This cell wall allows the cell to survive in a wide variety of environments, ranging from nearly pure water to salty environments and hot springs.

Bacterial cells often possess finger-like protrusions called *pili* or hairs that enable the cell to stick to surfaces. For example, the *Streptococcus* bacteria that cause strep throat use pili to latch onto the soft tissues inside your throat. Another useful appendage is the *flagellum*, a molecular motor that rotates somewhat like a propeller. The flagellum enables some cells to respond to chemical signals by swimming towards food and away from toxins.

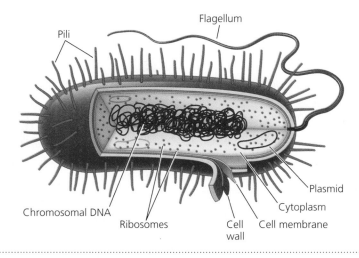

FIGURE 15.5   A cartoon diagram of an *E. coli* cell. The cell contains an interior watery region, the cytoplasm, that is surrounded by a cell wall and a cytoplasmic membrane.

WORKED EXAMPLE 15.1

Question: Consider a bacterium that reproduces once each hour—that is, each cell divides to produce two cells. Imagine that *one bacterial cell* is placed in a growth medium that provides all of the nutrients necessary for cell growth. Plot a graph of the total number of bacterial cells after 1 hour, 2 hours, and so forth, up to 8 hours. Plot time on the *x*-axis and the total number of cells on the *y*-axis. Based on your graph, how many cells are present after 8 hours?

Answer: A graphical plot of the growth of bacterial cells is shown below. Because the cells double every hour, the number begins to increase very rapidly. This is an example of *exponential growth*. After 8 hours, the growth medium contains 256 cells, which all derived from the first cell.

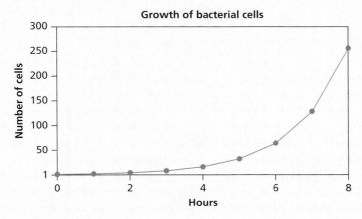

Growth of bacterial cells

TRY IT YOURSELF 15.1

Without using a graph, continue the calculation to determine how many bacteria are present in the growth medium after 24 hours. Assume that the bacteria have enough nutrients and space to continue replicating at the same rate.

PRACTICE EXERCISE 15.2

The graph presented here shows a growth plot of bacteria in a laboratory experiment. The number of bacteria are plotted on the vertical axis using a logarithmic scale, which can represent a large range of numbers. For example, 10^3 bacteria are plotted as "3" on a logarithmic scale (see Appendix C for a review of logarithms).

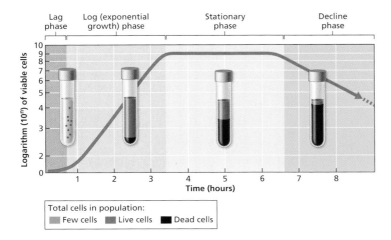

After bacteria are introduced to a fresh growth medium, there is an initial growth delay called the *lag phase* (1). Afterward, the bacteria divide and exhibit an *exponential growth phase* (2) (see Worked Example 15.1). When exponential growth is plotted on a logarithm scale, it produces a straight line. After some time, the exponential growth ends, and the bacteria enter a *stationary phase* (3). Finally, the number of bacteria exhibits a *decline phase* (4).

(a) Propose a hypothesis for why the exponential growth phase ends and the bacteria enter a stationary phase.

(b) Suppose you add fresh growth medium to the bacteria in the middle of the stationary phase. How do you predict the growth curve will change? Sketch a modified growth curve for this new experimental condition.

Core Concepts

- A bacterial cell consists of a watery interior called the cytoplasm, which contains the cell's DNA and ribosomes for making proteins.
- The cytoplasm is surrounded by a cell membrane and a cell wall. Bacterial cells also have pili (which enable the cell to stick to surfaces) and a flagellum (which allows the cell to move).

15.3 How Do Antibiotics Work?

Learning Objective:
Illustrate how antibiotics use different cellular targets to prevent the growth of bacterial cells.

Now that we have studied the composition of bacterial cells, we can examine the structure and function of antibiotics that are used to treat bacterial infections. Many antibiotics were discovered from natural sources such as molds, fungi, and even types of bacteria. One famous example, discussed in Chapter 1, is the discovery of penicillin. These types of antibiotics are called *natural products*. Biological organisms are a good source of antibiotics because they are constantly engaged in "chemical warfare" against one another as part of their struggle to survive and reproduce. In many cases, chemists will take a natural product and modify its molecular structure in the laboratory. For example, they employed this strategy to produce methicillin as a synthetic derivative of penicillin. In other cases, chemists will create a new antibiotic by synthesizing a novel type of molecule that is not found in nature. In this section, we apply our knowledge of chemistry and bacterial cells to investigate how antibiotics work.

Antibiotics Prevent the Growth of Bacterial Cells

Decades of research on infectious diseases by physicians and medicinal chemists have produced an arsenal of antibiotics that have saved millions of lives. The majority of these drugs are small molecules, although some new ones include large proteins or fragments of proteins. In general, an *antibiotic prevents the growth of bacterial cells*. This effect can be accomplished in two ways. A **bactericidal antibiotic** kills the bacterial cells. By contrast, a **bacteriostatic antibiotic** inhibits the replication of bacterial cells without killing them. Both types of drugs have important clinical uses. In addition, we can classify antibiotics in terms of their effect on different types of bacteria. A **narrow-spectrum antibiotic** is effective against a limited number of bacterial types. By contrast, a **broad-spectrum antibiotic** works on a wide range of bacteria.

Figure 15.6 illustrates a *disk diffusion test* for the effectiveness of six different antibiotics against *E. coli* bacteria. The test uses round culture dishes, which are flat containers filled with a solid growth medium that supports the growth of bacterial cells on the surface. To test the effectiveness of an antibiotic, the drug is soaked into a small paper disk and placed on the surface of the growth medium. Over time, the antibiotic spreads into the growth medium around the disk—this process is called *diffusion*. When bacteria are allowed to grow in the culture dish, they produce a cloudy appearance on the surface of the growth medium.

THE KEY IDEA: Antibiotic effectiveness can be measured using various tests.

bactericidal antibiotic A drug that kills bacteria.

bacteriostatic antibiotic A drug that inhibits the replication of bacteria without killing them.

narrow-spectrum antibiotic A drug that is effective against a limited number of bacterial types.

broad-spectrum antibiotic A drug that is effective against a wide range of bacterial types.

However, in some cases, the region around the antibiotic disk remains clear, indicating that no bacteria are growing.

This clear region is called the *inhibition zone* because the presence of the antibiotic in the growth medium inhibits the growth of bacteria. A larger inhibition zone indicates that the antibiotic is more effective. In Figure 15.6, we see that one antibiotic produces the largest inhibition zone. Three other antibiotics produce significant but smaller inhibition zones. However, two of the antibiotics show no inhibition zone, which means that they are ineffective against *E. coli*. These two antibiotics would be a poor selection to treat an *E. coli* infection.

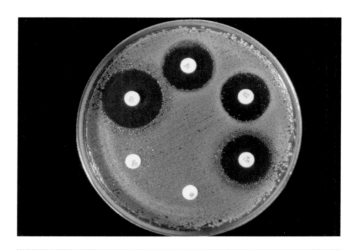

FIGURE 15.6 A disk diffusion test for antibiotic effectiveness. The effectiveness of the antibiotic is measured by the size of the clear inhibition zone around the disc, where the bacteria are not able to grow.

PRACTICE EXERCISE 15.3

A disk diffusion experiment was used to test the effectiveness of 14 different antibiotics against a specific type of bacteria. The results of the test are given in the accompanying figure, with each antibiotic disk labeled by a number.

(a) Which antibiotic is the most effective?

(b) Which antibiotic is the least effective?

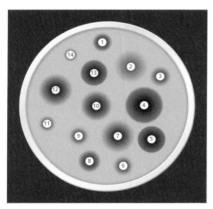

When evaluating antibiotics, another important consideration is *how much* of the drug is required to achieve its antibacterial effect. A drug that is effective only at high doses is more likely to produce harmful side effects. The potency of an antibiotic is measured by its IC_{50}, defined as the *inhibitory concentration* required to prevent the growth of 50% of the bacterial cells. Figure 15.7 shows a bar graph for the IC_{50} of two antibiotics, A and B, that were tested on *E. coli* bacteria. In this example, the units of IC_{50} are micromolar (μM), which is a measurement of solution concentration (see Chapter 10). As a reminder, a 1 molar (1 M) solution contains 1 mole of solute dissolved in 1 liter of solution. Antibiotics function at very small concentrations, and a 1 μM solution is equal to 1 micromole (10^{-6} moles) of antibiotic in 1 liter of solution.

Which of these drugs is more effective? The answer is antibiotic B because it is able to inhibit 50% of bacterial

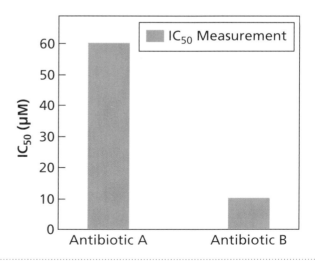

FIGURE 15.7 Measurement of inhibitory concentration. The IC_{50} of an antibiotic measures the inhibitory concentration required to prevent the growth of 50% of the bacterial cells. In this graph, the concentration units for IC_{50} are μM (micromolar).

growth at a *lower concentration* (10 μM). By comparison, antibiotic A requires a higher concentration (60 μM) to achieve the same antibacterial effect. In general, *a more potent antibiotic has a lower IC$_{50}$ value.*

PRACTICE EXERCISE 15.4

This graph shows the IC$_{50}$ values of four antibiotics (C, D, E, and F), as tested against a particular type of bacterium. Rank the four antibiotics from most to least effective.

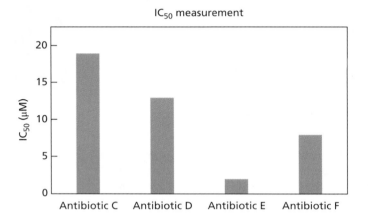

Core Concepts

- Antibiotics prevent the growth of bacterial cells. A **bactericidal antibiotic** kills the bacteria, whereas a **bacteriostatic antibiotic** inhibits the replication of bacteria without killing them.
- A **narrow-spectrum antibiotic** is effective against a limited number of bacterial types, whereas a **broad-spectrum antibiotic** works on a wide range of bacteria.
- The disk diffusion test measures the effectiveness of an antibiotic by the size of the *inhibition zone* in which no bacterial grow.
- The IC$_{50}$ of an antibiotic measures the inhibitory concentration (in **μM**) that is required to prevent the growth of 50% of the bacterial cells. A more potent antibiotic has a lower IC$_{50}$.

Cellular Targets for Antibiotics

The techniques we have described measure the potency of an antibiotic, but they do not tell us *how* the antibiotic works. The next step is to investigate the antibiotic's mechanism of action within a bacterial cell. Does the compound prevent cell wall assembly? Interfere with DNA or RNA synthesis? Halt the construction of proteins? Figure 15.8 illustrates the variety of biological molecules and processes that function as targets for various antibiotics. The following sections provide three case studies of antibiotics, each of which uses a different mechanism to inhibit bacterial growth.

> THE KEY IDEA: A bacterial cell contains multiple targets for antibiotics.

Core Concept

- Bacterial cells contain many different cellular targets for antibiotics.

Antibiotic Target: Constructing the Cell Wall

> THE KEY IDEA: Penicillin blocks the ability of bacterial cells to build their cell wall.

One of the best-known antibiotics is penicillin. To be precise, *penicillin refers to a class of related molecules with similar molecular structures.* Penicillin functions as an antibiotic by disrupting the ability of bacterial cells to build their cell wall. The overall structure of the cell well resembles a mesh wire fence, which is both strong and rigid (Figure 15.9). The cell wall is constructed from long chains of

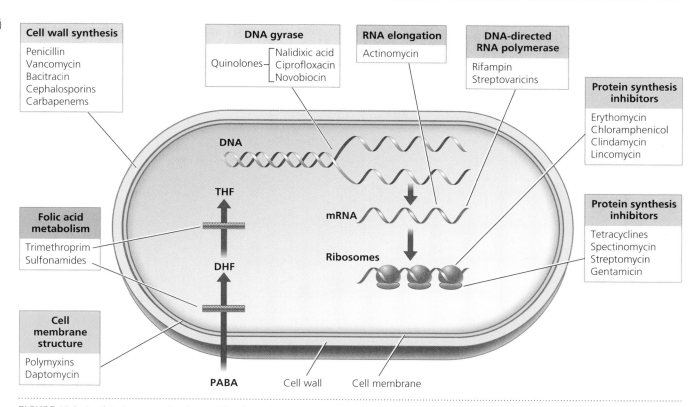

Cell wall synthesis
Penicillin
Vancomycin
Bacitracin
Cephalosporins
Carbapenems

DNA gyrase
Quinolones — Nalidixic acid
Ciprofloxacin
Novobiocin

RNA elongation
Actinomycin

DNA-directed RNA polymerase
Rifampin
Streptovaricins

Protein synthesis inhibitors
Erythomycin
Chloramphenicol
Clindamycin
Lincomycin

Protein synthesis inhibitors
Tetracyclines
Spectinomycin
Streptomycin
Gentamicin

Folic acid metabolism
Trimethroprim
Sulfonamides

Cell membrane structure
Polymyxins
Daptomycin

DNA
THF
DHF
PABA
mRNA
Ribosomes
Cell wall
Cell membrane

FIGURE 15.8 Antibiotic targets in a bacterial cell. A schematic view of some processes in a bacterial cell that are attacked by antibiotics. "Folic acid metabolism" refers to steps the cell takes to synthesize nucleotides to be incorporated into DNA and RNA.

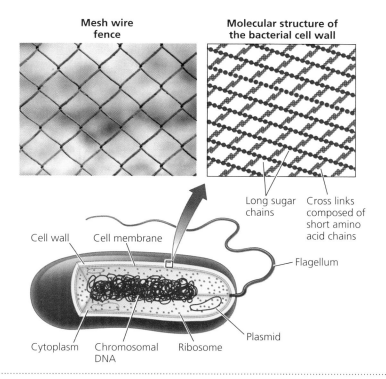

Mesh wire fence

Molecular structure of the bacterial cell wall

Long sugar chains
Cross links composed of short amino acid chains

Cell wall   Cell membrane
Flagellum
Plasmid
Cytoplasm   Chromosomal DNA   Ribosome

FIGURE 15.9 Structure of the bacterial cell wall. Many bacteria are encased by a mesh-like structure called a cell wall. The cell wall is constructed from long chains of sugar molecules joined by shorter crosslinks made from amino acids.

sugar molecules (polysaccharides) that are joined together by shorter chains of amino acids, which serve as crosslinks to toughen its structure.

Like most biological processes, the construction of the cell wall is aided by enzymes. One of these enzymes is called *transpeptidase* because it uses peptides to make linkages between the polysaccharide chains (*trans* means "across"). To make these linkages, the enzyme uses the right-handed form of amino acids called *D-amino acids*. As we learned in Chapter 13, amino acids can exist as chiral isomers (L- and D-) that are mirror images of one another. L-amino acids are used to make proteins, but D-amino acids have specialized roles in the construction of the cell wall and in some natural antibiotics.

Penicillin functions as an enzyme inhibitor that shuts down the catalytic activity of transpeptidase. How does this happen? As we learned in Chapter 14, enzymes contain a region called an *active site* that binds a *substrate* molecule. For transpeptidase, the substrate is a small molecule called D-Ala-D-Ala, which contains two D-alanine amino acids connected by a peptide bond. *Penicillin inhibits transpeptidase because it closely resembles the molecular structure of the enzyme substrate.* Figure 15.10 compares the molecular structures of D-Ala-D-Ala (top left) and penicillin (top right). When the two structures are superimposed, as shown in the lower image, you will notice their close similarity. As a result of this likeness, transpeptidase mistakenly binds penicillin to its active site instead of binding the D-Ala-D-Ala substrate molecule. The penicillin blocks the active site and prevents the enzyme from performing its usual catalytic function.

When the transpeptidase enzyme is shut down by penicillin, the bacterial cell cannot add the strengthening crosslinks when it attempts to construct its cell wall. As a result, the structure of the cell wall is weakened in certain regions. The interior of the cell has a high internal pressure that arises from all of the molecules it contains. This pressure forces the interior contents of the cell to burst through the defective gaps in the cell wall, which kills the bacterium.

Does penicillin damage human cells? Human cells do not have cell walls, so we do not need to make the transpeptidase enzyme that serves as the target for penicillin. As a result, penicillin has no effect on human cells. This is the ideal scenario for an antibiotic—it attacks and kills bacterial cells without harming our own cells. A small percentage of individuals have an allergy to penicillin, which triggers an abnormal reaction by the body's immune system. However, recent medical reports indicate that the frequency of penicillin allergies is overreported.

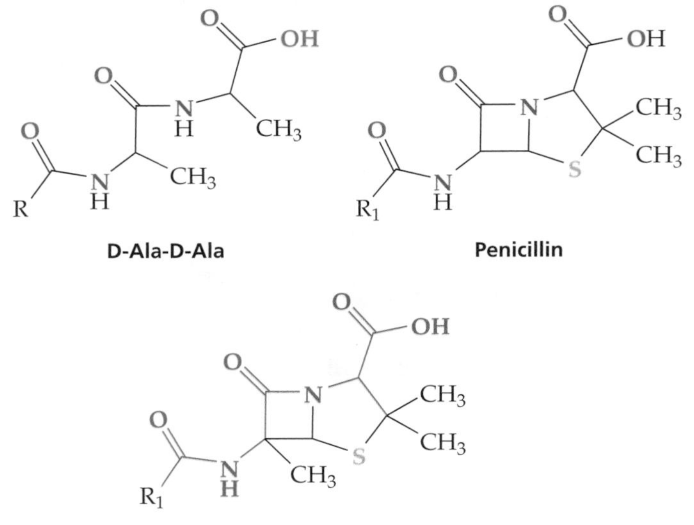

D-Ala-D-Ala   **Penicillin**

Superimposition of D-Ala-D-Ala / Penicillin

FIGURE 15.10 Penicillin closely resembles the substrate for transpeptidase. The molecular structure of penicillin is very similar to the structure of D-Ala-D-Ala, a molecule made from joining two D-alanine amino acids with a peptide bond. The D-Ala-D-Ala molecule serves as the substrate for the transpeptidase enzyme when it adds crosslinks to the bacterial cell wall.

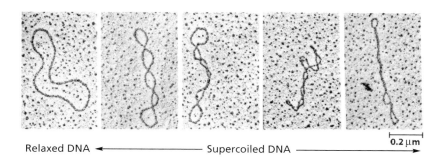

FIGURE 15.11 Magnified images of circular DNA molecules in bacterial cells. The left panel shows a relaxed DNA molecule, whereas the other four panels illustrate different types of supercoiled DNA.

Relaxed DNA ←——————— Supercoiled DNA ———————→

0.2 μm

Core Concepts

- The bacterial cell wall is a strong, mesh-like structure constructed from long sugar chains (polysaccharides) that are crosslinked by shorter chains of D-amino acids.
- The crosslinking is facilitated by an enzyme called transpeptidase, which uses D-Ala-D-Ala as its substrate. The structure of penicillin closely resembles that of D-Ala-D-Ala. Because of this similarity, penicillin binds to the active site of the enzyme and inhibits its function.
- The absence of crosslinks weakens the cell wall, which causes the cell to burst as a result of its internal pressure.

supercoiling A process by which circular DNA gains or loses additional twists.

Antibiotic Target: Replicating DNA

Bacterial DNA consists of circular molecules. When the DNA is relaxed, it forms a circular loop, as shown in the left panel of Figure 15.11. However, it is also possible for the circular DNA to become twisted, as shown in the other four panels. This type of twisting, called **supercoiling**, depends on the extent to which one DNA strand winds around the other. Bacteria make use of supercoiling to package their DNA into the small volume of the cell, because the supercoiled DNA takes up less space than the relaxed DNA. However, supercoiling causes a problem when the bacterium has to replicate its DNA during cell division: DNA polymerase enzymes prefer the relaxed circular structure.

To remedy this problem, bacteria are equipped with *topoisomerase* enzymes (Figure 15.12). This name is based on the word "topology," which refers to the study of shapes. Topoisomerase enzymes bind to DNA, make a cut in the polynucleotide chain, untwist the two strands, and then restore them to their original circular structure. The need for bacterial cells to untwist their supercoiled DNA prior to replication is the basis for several potent antibiotics. These drugs inhibit the function of the topoisomerase enzyme, which leaves the DNA in a tangle that cannot be copied. As a result, the bacterial cells cannot divide. One example of a topoisomerase inhibitor drug is ciprofloxacin, often shortened to Cipro (Figure 15.12). Cipro attacks bacterial cells using both bactericidal and bacteriostatic mechanisms.

THE KEY IDEA: Cipro prevents DNA replication in bacteria.

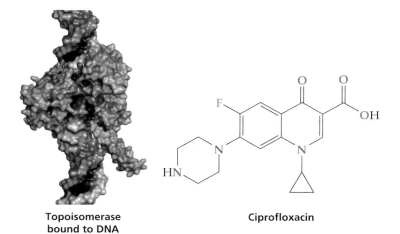

Topoisomerase bound to DNA

Ciprofloxacin

FIGURE 15.12 Topoisomerases are enzymes that bind to DNA and control the twisting of the strands. Ciprofloxacin (Cipro) is a powerful antibiotic drug that inhibits topoisomerase, thereby preventing bacterial cell division.

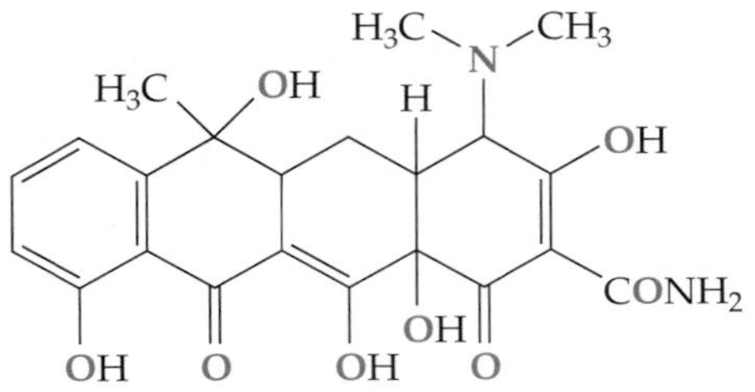

FIGURE 15.13 Molecular structure of tetracycline. Tetracycline is an antibiotic composed of four carbon rings with attached functional groups.

🔑 THE KEY IDEA: Tetracycline inhibits the synthesis of bacterial proteins within the ribosome.

Core Concepts

■ Circular DNA in bacteria can become twisted in a process called **supercoiling**. Before bacterial DNA can be replicated, the supercoiling must be untwisted by a topoisomerase enzyme.

■ Some antibiotics, such as ciprofloxacin, inhibit the function of topoisomerase and thus prevent bacteria from copying their DNA during cell division.

Antibiotic Target: Making Proteins

Many antibiotics block the translation of bacterial mRNA. One example is *tetracycline*, which is commonly used to treat acne (Figure 15.13). The drug's name is based on its structure, which consists of four carbon rings that are joined together (*tetra* is derived from the Greek word for "four"). Tetracycline enters bacterial cells and bonds to the ribosome, the large molecular complex that synthesizes proteins based on instructions provided by mRNA (see Chapter 13). Tetracycline attaches to a specific location in the ribosome and prevents the transfer RNA molecule from binding and participating in translation. This activity halts the synthesis of the bacterial protein. Tetracycline is a bacteriostatic drug because blocking translation prevents the bacterial cells from dividing but does not kill the cells directly.

Human cells also have ribosomes that synthesize proteins. Fortunately, tetracycline binds to bacterial ribosomes and not to human ribosomes because of structural differences between bacterial and human ribosomes. However, there are concerns about some types of drugs that target translation in bacteria. One persistent source of problems in antibiotic therapy involves the mitochondria in animal cells, which are the sites where the cell's ATP is synthesized. These mitochondria have their own DNA, and they make proteins by a mechanism closer to that of bacteria than to that of humans. Consequently, some potent new antibiotics also block mitochondrial protein synthesis, a potentially serious health problem for the person taking the antibiotic.

Core Concept

■ Tetracycline functions as an antibiotic by inhibiting the synthesis of bacterial proteins at the ribosome.

15.4 How Do Bacteria Become Resistant to Antibiotics?

Learning Objective:

Illustrate the strategies used by bacteria to neutralize the effects of antibiotics.

Several decades ago, antibiotics appeared to be so successful that many scientists and physicians believed we had finally conquered infectious diseases. Unfortunately, this has not proven to be the case. Like tuberculosis, other infectious diseases are becoming increasingly common because the bacteria that cause them have developed resistance to our standard arsenal of antibiotics. How do bacteria become resistant to drugs?

🔑 THE KEY IDEA: Bacteria use various strategies to neutralize the effect of antibiotics.

Bacteria Fight Back: Strategies of Antibiotic Resistance

Almost all of the antibiotics we currently use have encountered resistance. Resistant bacteria often first appear in hospitals where patients are situated close to one another and are often susceptible due to compromised immunity. Shortly thereafter, the resistance spreads into the community.

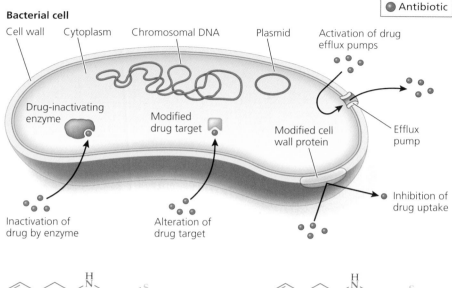

Bacterial cell

Cell wall   Cytoplasm   Chromosomal DNA   Plasmid   Activation of drug efflux pumps

● Antibiotic

Drug-inactivating enzyme

Modified drug target

Modified cell wall protein

Efflux pump

Inactivation of drug by enzyme

Alteration of drug target

Inhibition of drug uptake

FIGURE 15.14 Mechanisms of antibiotic resistance. The figure shows various mechanisms used by bacteria to develop resistance to antibiotics.

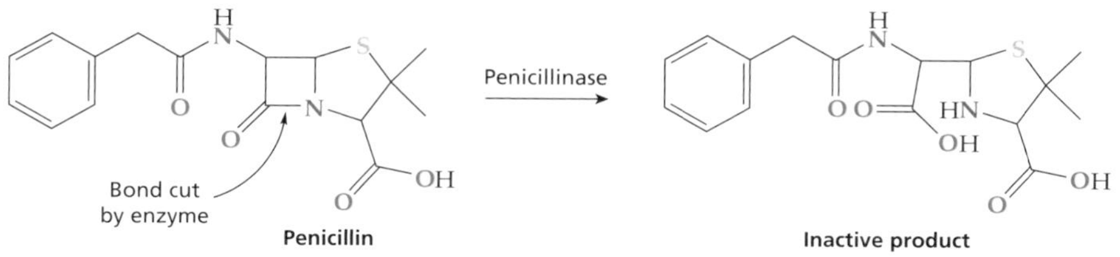

Penicillinase

Bond cut by enzyme

Penicillin

Inactive product

FIGURE 15.15 Inactivation of penicillin. Some antibiotic-resistant strains of bacteria contain penicillinase, an enzyme that cuts a key bond in penicillin and inactivates the drug.

Bacteria have developed diverse strategies to evade antibiotics. The most prevalent method involves developing a pump that transports the drug out of the cell. This action reduces the cell's energy. However, it enables the cell to survive, even if it grows slowly. Other mechanisms involve inactivating the drug. Finally, bacteria can develop resistance by altering the structure of their ribosomes, which reduces the effectiveness of ribosome-binding drugs. These strategies are summarized in Figure 15.14.

As one example, some strains of bacteria have become resistant to penicillin. How can this happen? Bacteria can employ several methods to combat the effects of the drug. One tactic is to quickly eject the drug out of the cell before it can do any damage. This action is accomplished by special proteins that span the cell membrane and pump the drug from the interior of the cell into the outside environment. Another tactic is to attack the drug directly. Some strains of drug-resistant bacteria have an enzyme called *penicillinase* that inactivates penicillin by cutting a key bond within a ring of atoms in the molecule (Figure 15.15). When this bond is broken, the drug loses its unique three-dimensional structure. Significantly, this structure is essential to penicillin's function—if the structure changes, then the molecule no longer binds to the crosslinking enzyme, and it ceases to function as an antibiotic.

Core Concepts

- Over time, bacteria evolve to become resistant to antibiotics. Drug-resistant strains of bacteria have been discovered for almost all current antibiotics.
- The most common defense mechanism used by bacterial cells is to pump the drug out of the cell before it can cause any harm. Other defense strategies include developing an enzyme to inactivate the drug (e.g., penicillinase) and altering the structure of the bacterial ribosome.

Use and Abuse of Antibiotics

Antibiotic treatment of bacterial infections has saved many millions of lives. Because of this success, however, it is tempting to overuse these valuable drugs.

THE KEY IDEA: Antibiotics are used far more than is necessary, which promotes the development of antibiotic resistance.

Some medical investigations have estimated that 50% of antibiotic prescriptions may be unnecessary. For example, one study reported that 60% of patients who visited a doctor's office with a sore throat were prescribed antibiotics. However, bacterial infections (usually *Streptococcus*) cause only 10% of sore throats that require a doctor's visit. This mismatch means that many thousands of individuals are receiving unnecessary antibiotics. In addition to detrimental effects on a person's gut microbiome (see *Chemistry and Your Health*), exposing bacteria to

CHEMISTRY AND YOUR HEALTH

How Do Bacteria Keep Us Healthy?

Thus far, the chapter has focused on the bacteria that cause diseases. This is only one side of the story, however: Bacteria also play an essential role in maintaining our health. For example, our intestines contain roughly 1 kilogram of bacteria that have evolved to perform major nutritional and health functions. Therefore, the condition of this bacterial population has important effects on our own health. To cite one example, the human body on its own cannot break down many kinds of polymers made from sugar fibers. Instead, specialized bacteria in our intestines are adapted to perform this function. The ability to consume these fibers in our diet has definite health benefits.

The large population of bacterial cells in and on our body is called the human microbiome. Advances in the technology of DNA sequencing have enabled scientists to identify the types and numbers of bacteria present in the microbiome. These organisms vary naturally all the time with a person's diet, overall health, age, and even physical activity and emotional state (see Figure 1). In addition, scientists have identified the functions performed by some of these bacteria. For example, specific cells in the microbiome work to suppress inflammation of the bowel. For this reason, your food preferences can also be dictated by

the needs of the bacteria in the gut. In short, humans exist in a *symbiosis*, or mutual dependency, with our microbiomes.

The average U.S. teenager has undergone antibiotic therapy more than a dozen times. Unfortunately, in addition to eliminating the source of the infection, each antibiotic treatment also wipes out many of the beneficial bacteria in our microbiome. The consequences of this systemic elimination of cells that we rely on for our health can be dire: We open up positions in the gut to be recolonized with imperfectly adapted new bacteria, we spread drug resistance throughout the surviving cells, and we upset the balance of activity that enables us to process the foods in our diet.

These harmful effects on the bacteria in our microbiome are one reason why some antibiotics have unpleasant side effects. Any therapy that includes a broad-spectrum antibiotic such as penicillin can severely disrupt the delicate balance between our gut microbiome and our nutritional requirements. Moreover, frequent antibiotic applications increase the likelihood that resistance will develop. One outcome of the overuse of antibiotics is the emergence of multiple strains of drug-resistant bacteria, including MDR-TB, that are difficult to eliminate without severely damaging the microbiome.

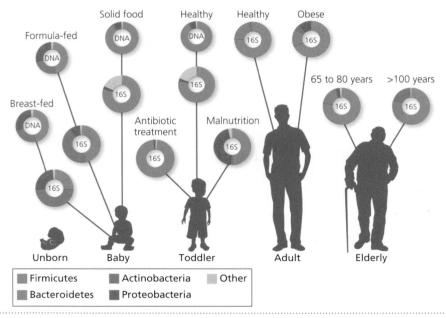

FIGURE 1 The human microbiome. A representation of how the population of bacteria in the human microbiome changes with age. Bacteria were identified using nucleotide bases in DNA or in a type of RNA (16S).

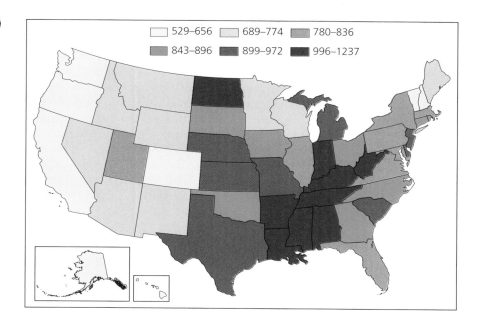

| 529–656 | 689–774 | 780–836 |
| 843–896 | 899–972 | 996–1237 |

FIGURE 15.16 Antibiotic prescriptions according to state in 2010. The shading indicates the number of prescriptions per 1000 persons of all ages.

unwarranted antibiotics facilitates the conditions for developing drug resistance. For this reason, many doctor's offices are now using a *rapid diagnostic test* for *Streptococcus*, which detects characteristic molecules on the surface of the bacterial cell.

A study published in *The New England Journal of Medicine* investigated the prescription of oral antibiotics during outpatient visits. During 2010, health care providers prescribed a total of 258 million courses of antibiotics, which corresponds to 833 prescriptions per 1000 persons. As shown in Figure 15.16, the frequency of antibiotic prescriptions varied significantly among the 50 states. The reasons for this variability are still being investigated.

You may be surprised to learn that *less than 20%* of the antibiotics consumed in the United States are used in human medicine. As shown in Figure 15.17, close to 80% of all antibiotics are consumed by agricultural livestock such as cows, pigs, and chickens. This use of antibiotics accounts for an estimated 13.5 billion kilograms of antibiotics per year. The *therapeutic* use of antibiotics to treat diseased animals accounts for only a small fraction of the total amount. By far the largest agricultural application of antibiotics is *nontherapeutic*. Specifically, they are used to increase the weight of animals grown in industrial-sized feedlots.

Large-scale administration of low doses of antibiotics in agricultural environments, such as the cattle feedlot illustrated in Figure 15.18, has the unfortunate effect of selecting antibiotic-resistant bacteria. Long-term exposure to low levels of different antibiotics is a major cause of multidrug resistance. Today, some strains of bacteria are resistant to as many as a dozen antibiotics.

Scientists have demonstrated that antibiotics stimulate major changes in the population of bacteria in the intestines of animals. As one example, pigs and steers under stress from infection in crowded feedlots can gain weight when fed antibiotics. In addition to stimulating the growth of resistant pathogens, these antibiotics kill off many benign bacteria that the animals need

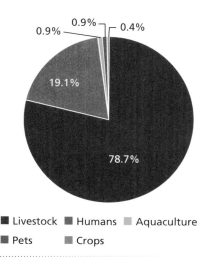

FIGURE 15.17 Annual use of antibiotics in the United States (2013). This graph is based on data from the Food and Drug Administration.

FIGURE 15.18 Use of antibiotics for agricultural livestock. Major beef producers raise cattle in large, crowded feedlots to increase their yield of meat and other products. Antibiotics are often given to cattle (and other farm animals) to prevent disease and stimulate growth.

for survival. After the normal bacteria die out, the animal's intestines become vulnerable to drug-resistant invasive species. The risk of spreading multidrug-resistant bacteria among animals has to be weighed against the increased value of the cattle, pigs, and chickens that are fed these antibiotics. Research confirms that resistant bacteria from animals in feedlots can enter the intestines of the workers who handle the animals and then spread into the human population. Drug-resistant bacteria from farms can also enter the food supply. For example, in 2013 the CDC reported an outbreak of drug-resistant strains of *Salmonella* bacteria from West Coast chicken farms.

CONCEPT QUESTION 15.2

In 2013, the U.S. Food and Drug Administration (FDA) placed restrictions on the nontherapeutic use of antibiotics in healthy farm animals. Use the FDA website to obtain a more detailed description of these regulations, and write a one-paragraph report. Do you think these new FDA regulations will be effective at lowering the agricultural use of antibiotics? Support your position.

Core Concepts

- Less than 20% of antibiotic use in the United States is for medical treatment of humans. Almost 80% of antibiotics are used on agricultural livestock, either to treat sick animals (therapeutic use) or to promote weight gain (nontherapeutic).
- Use of low levels of antibiotics in agriculture leads to the development of drug-resistant bacteria, which can spread into human populations and the food supply.

15.5 Drug-Resistant Tuberculosis: The Coming Plague

Learning Objective:

Outline the relationship between the development of new antibiotics and the rise of antibiotic resistance.

🔑 THE KEY IDEA: Tuberculosis is a case study of a bacterial disease that was once controlled by antibiotics but has now become drug-resistant.

We are exposed to vast numbers of bacteria every day, regardless of how clean our environment is. If our immune system is functioning normally, as it does in healthy individuals, then this exposure poses little threat. For example, scientists estimate that roughly one-third of the human population has been exposed to tuberculosis bacteria, yet only a small fraction suffers symptoms. However, when people are malnourished or imprisoned, or their immunity is otherwise weakened, the incidence of the disease increases. Children and elderly people are particularly vulnerable.

How did scientists discover that tuberculosis is a bacterial disease spread by inhaling infected air, drinking infected fluids, and eating infected foods? For centuries, tuberculosis was considered a wasting disease with no obvious cause. Then in the 1880s, the German doctor and microbiologist Robert Koch (1843–1910) succeeded in identifying the tuberculosis bacterium. This was a remarkable achievement because *Mycobacterium tuberculosis* has an unusual waxy coat that shields the bacterium from attack by the immune system. This coat also resists staining by any of the chemical agents that scientists typically use to view bacteria with the microscope. Working in his living room, constantly risking infection, Koch found a new way to stain tuberculosis cells and prove that these cells are a type of bacteria. Figure 15.19 shows a highly magnified image of tuberculosis bacteria.

A major advance in the treatment of tuberculosis was made in the 1940s by Selman Waksman (1888–1973), a microbiologist working at Rutgers University in New Jersey. Waksman and his colleagues initiated a program of testing common soil samples for their ability to prevent tuberculosis cells from growing.

Using this method, they discovered a natural compound produced by fungi that live in certain soils. They purified the active compound and injected it into infected animals to test its ability to work in a realistic environment. This compound was called *streptomycin*, and it was the first antibiotic found to be effective against tuberculosis bacteria. Later studies revealed that streptomycin binds to the ribosome of bacterial cells and blocks their ability to synthesize proteins. Although streptomycin does not kill tuberculosis bacteria, it prevents them from reproducing. Streptomycin cured tuberculosis in many patients, and it offered the first real hope that the disease could be overcome. In 1952, Waksman was awarded the Nobel Prize in Physiology or Medicine for his research (Figure 15.20). Within a few years, however, it was observed that some patients stopped responding to streptomycin after several months of therapy. This effect was due to the presence of tuberculosis bacteria that had become resistant to the drug.

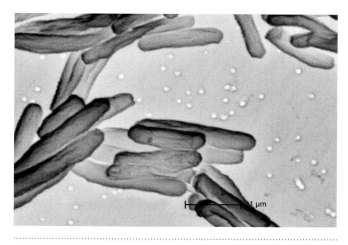

FIGURE 15.19 An image of tuberculosis bacteria, magnified with an electron microscope. Note the scale bar indicating a size of 1 μm.

Soon after the discovery of streptomycin, the search for new anti-tuberculosis drugs led to a molecule called *isoniazid*, which blocked the ability of tuberculosis bacteria to produce their protective cell coating (Figure 15.21). Isoniazid rapidly became a "magic bullet" against tuberculosis. From the 1970s onward, scientists came to believe that tuberculosis was a cured disease, with its total eradication only a matter of time.

Unfortunately, following several years of effective use, tuberculosis bacteria developed resistance to isoniazid. Scientists then turned to mixtures of antibiotics rather than to single drugs in the effort to suppress resistance. "Cocktails" containing four drugs, each acting on a different cell target, became the standard treatment. The drugs are classified into two categories. "First-line" drugs are the most effective, and they encounter less resistance. "Second-line drugs" provide additional antibiotics for combination treatments. Table 15.1 summarizes some of these drugs and the cellular targets responsible for their effect. As you can see from the table, antibiotics attack a variety of cellular processes that are essential for bacterial growth and survival. When taken in combination, modern "cocktails" are extremely potent against most forms of tuberculosis. Unfortunately, multiple strains of tuberculosis have developed that are able to resist essentially all of the available combinations of first- and second-line drugs.

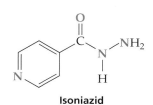

Streptomycin

(a)   **(b)**   **Isoniazid**

FIGURE 15.20 A treatment for tuberculosis. (a) Selman Waksman discovered streptomycin, the first drug that was effective against tuberculosis. (b) The molecular structure of streptomycin, which was isolated from a soil bacterium.

FIGURE 15.21 Molecular structure of isoniazid. Isoniazid is an antibiotic that selectively blocks the ability of tuberculosis cells to make their waxy coating.

TABLE 15.1 Antibiotic Treatments for Tuberculosis	
DRUG AND YEAR OF INTRODUCTION	**CELLULAR TARGET**
FIRST-LINE DRUGS	
Isoniazid, 1952	Synthesis of cell wall coating
Rifampicin, 1963	RNA transcription
Pyrazinamide, 1954	Protein synthesis
Ethambutol, 1961	Cell wall synthesis
SOME SECOND-LINE DRUGS	
Para-amino salicylic acid, 1948	DNA synthesis
Streptomycin, 1944	Protein synthesis
Ofloxacin, 1980	DNA synthesis

Although the number of MDR-TB cases in the United States is currently low, increased global travel makes it likely that the disease will become more widespread in the future. Treatment of patients infected by MDR and XDR strains requires many months of therapy, and the drugs have unpleasant side effects. As a consequence, patients tend to avoid completing their full treatment cycles, a situation that facilitates resistance development. Exposure of populations of bacteria to low concentrations of drugs that are insufficient to kill the cells favors growth of resistant populations.

Core Concepts

- Selman Waksman and his research team discovered streptomycin, the first antibiotic that was effective against the tuberculosis bacterium. Streptomycin binds to the ribosome and inhibits the ability of the bacterium to synthesize new proteins.
- A later antibiotic, isoniazid, inhibits the ability of tuberculosis cells to make their protective coating.
- Cocktail drug treatments contain a mixture of four antibiotics, each of which acts on a different bacterial target.

15.6 How Are New Drugs Tested and Approved?

Learning Objective:
Outline the stages that are required to develop and evaluate a new pharmaceutical.

THE KEY IDEA: The Food and Drug Administration is responsible for ensuring that pharmaceuticals are safe and effective.

The global increase of drug-resistant bacteria has created an urgent need for new drugs. However, developing a new pharmaceutical for public use is a long, complicated, and expensive process. In this section, we examine the stages that are required to bring a new pharmaceutical to the public.

How Do We Know that the Drugs We Take Are Safe and Effective?

When we take a medication, either prescription or over the counter, we assume that the drug has been evaluated for safety and effectiveness. This was not always the case. In the 19th century, there were no restrictions on what could be sold as a "cure" or on the claims that could be made about how well it worked. In some cases, the treatment was more deadly than the disease. The phrase "snake oil salesman" is a legacy from this time.

By the beginning of the 20th century, the U.S. government had decided that it had to protect the public from unscrupulous salesmen and their unproven elixirs. In 1906, the Federal Food and Drugs Act was passed by the U.S. Congress and signed into law by President Theodore Roosevelt. This law led to the formation of the federal agency that is now called the Food and Drug Administration (FDA). Over the next 50 years, Congress imposed increasingly stricter regulations on drug manufacturers.

Despite this progress, the level of scrutiny imposed by the FDA in the mid-20th century still was much less than we expect today. A major turning point occurred in response to use of a drug called *thalidomide*. From 1957 to 1962, thalidomide was prescribed in Europe and Japan as a sleep-inducing sedative. Thousands of pregnant women took the drug to combat morning sickness. Soon after its introduction, women who had used the medication began to give birth to babies with partially formed limbs (Figure 15.22). In total, an estimated 10,000 babies were born with this type of deformity. Why did this tragedy happen?

Many drug molecules with complex structures contain one or more carbon atoms that are *chiral centers*. In these cases, the molecule can exist as two *chiral isomers* that are mirror images of each other (see Chapter 13). The molecular structure of thalidomide also has a chiral center; its two chiral isomers are illustrated in Figure 15.23. When thalidomide was prescribed in the 1950s and 1960s, it contained a mixture of the two isomers. Unfortunately, scientists did not realize that the two isomers had very different effects in the human body. One of the isomers, designated R-thalidomide, produces the desired sedative effect. However, the other isomer, designated S-thalidomide, had the unanticipated outcome of disrupting the development of fetal limbs within the womb.

Unlike Europe and Japan, the United States did not suffer from the devastating effects of thalidomide. The credit can be given to an FDA regulator named Frances Kelsey (1914–2015). Born in Canada, Kelsey obtained a Ph.D. in toxicology before joining the FDA as a junior evaluator. When she reviewed the application to approve thalidomide in the United States, the harmful effects of the drug were not yet known. Based on her scientific knowledge, however, Kelsey had concerns about thalidomide's possible toxicity. She therefore rejected the application and demanded more evidence of the drug's safety. Kelsey was harshly criticized for her decision, but she held her ground. By the time the drug company resubmitted the application, evidence was emerging from overseas about the suspected effects of thalidomide. As a result, the drug was never sold in the United States.

Kelsey's intervention prompted Congress to pass a new set of laws that required drug companies to meet much stricter requirements to demonstrate that their products are safe and effective. This legislation launched the modern concept of a controlled clinical trial, which is now the accepted standard for evaluating any new drug.

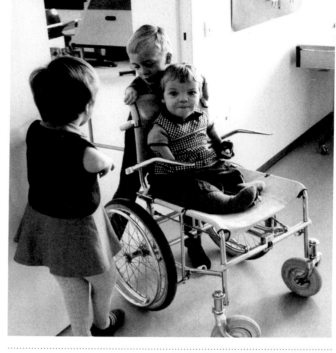

FIGURE 15.22 The harmful effects of thalidomide. Some women who took thalidomide during pregnancy gave birth to children without fully developed arms.

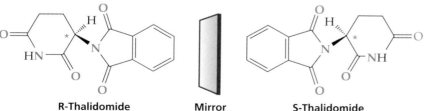

R-Thalidomide   **Mirror**   **S-Thalidomide**
(induces sleep)   **(disrupts fetal development)**

FIGURE 15.23 Thalidomide exists as two chiral isomers. The molecular structure of thalidomide contains a carbon atom, marked with an asterisk (*), that is a chiral center. One of the chiral isomers, R-thalidomide, functions as a sedative. The other chiral isomer, S-thalidomide, disrupts the formation of limbs in a developing fetus.

Core Concepts

- The Federal Food and Drugs Act, passed in 1906, was designed to protect the public from disreputable "cures" and claims about their effectiveness. This law led to the agency that is now called the Food and Drug Administration (FDA).
- In the 1950s and 1960s, thalidomide was prescribed in Europe and Japan as a sedative. Unexpectedly, this drug caused thousands of children to be born without limbs. Thalidomide exists as two chiral isomers that are mirror images of each other. One isomer induces sleep, but the other disrupts the formation of limbs in a developing fetus.
- Dr. Francis Kelsey, an FDA regulator, prevented thalidomide from being sold to the public in the United States. As a result of this example, the FDA demanded more rigorous testing of new pharmaceuticals through the use of clinical trials.

The Stages of Drug Development and Approval

⚷ THE KEY IDEA: The development and approval of a new pharmaceutical is a lengthy and expensive process with a high failure rate.

clinical trials Rigorous tests of new drugs performed on human subjects.

Obtaining FDA approval for a drug involves exhaustive evaluation of its effects on laboratory animals, followed by rigorous tests called **clinical trials** that involve human volunteers. Only 1 out of every 10,000 candidate drugs ultimately receive FDA approval. On average, guiding a new drug from discovery through approval takes 12 years and costs approximately $1 billion. Figure 15.24 illustrates the multiple stages of drug testing. For consistency we consider the case of a drug that is being examined as an enzyme inhibitor. The stages of testing would be similar for other types of pharmaceuticals.

The first tests of a potential drug are usually performed in the laboratory, a stage called *preclinical research*. These tests are performed *in vitro*, meaning, literally, "in glass," or a test tube. One common experiment is to assess how effectively the drug inhibits the function of its target enzyme. Scientists measure the rate of the enzymatic reaction—that is, the conversion of substrate to product—in the presence of the drug, focusing specifically on whether the rate decreases. These tests indicate how tightly the drug binds to the enzyme, which determines how much of the drug will be needed to achieve a therapeutic effect. Another factor that is examined at this stage is the solubility of the molecule, which is especially important if the drug is to be taken orally and transported by the bloodstream.

Promising candidate drugs proceed to *in vivo* (literally "in something alive") tests using laboratory animals. Significantly, a drug that effectively inhibits the target enzyme may have unwanted side effects that will render it unsuitable as a medicine. By exposing laboratory animals to the drug, scientists obtain data on how long the drug is retained in the body, the pathway for its chemical breakdown, and its effects on various organs such as the liver, kidneys, and brain. Scientists also need to determine the dose at which the drug becomes toxic. They accomplish this task by determining the LD50 dose, which is the minimum amount of the compound that kills half of the test animals. ("LD" means "lethal dose," and the "50" refers to the 50% of animals that are killed.) Even commonly used drugs can be dangerous at high doses. *Chemistry in Your Life* provides one example.

FIGURE 15.24 The stages of development and testing for a new drug. Drug development begins with preclinical research, followed by animal testing. Promising drug candidates are then subjected to three phases of clinical studies. Approval of a new drug is granted only after a thorough review of all relevant data by the FDA.

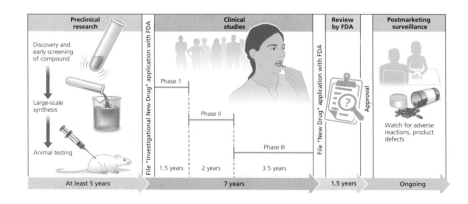

The results of the *in vivo* tests go to the FDA for review. If the animal studies reveal low levels of toxicity and side effects, then the FDA may grant approval to test the candidate drug in human subjects. These clinical trials involve three phases:

> **Phase 1:** The first objective is to ensure that the candidate drug is *safe.* Various doses are administered to healthy volunteers who are carefully monitored for health problems. This phase typically involves around 50 participants.

> **Phase 2:** After the drug's safety has been established, the next step is to examine its *efficacy*: How effective is the drug in treating patients with a specific condition? The answer to this question is provided by *double-blind trials.* In these trials, patients are divided into two groups that are matched as closely as possible in terms of basic characteristics such as sex, age, and overall health. One group receives the drug, while the other group receives an inactive imitation known as a *placebo.* The double-blind format ensures that neither the patients nor the researchers know who is receiving which treatment. In some trials, the new drug is tested against currently available medications to assess whether there is an improvement in treatment. This phase typically involves between 100 and 300 participants.

> **Phase 3:** If the drug is effective in treating patients during Phase 2, then the clinical trials proceed to Phase 3. These large-scale trials are designed to include different groups of subjects—men, women, populations of various ethnicities, people with preexisting conditions, and so forth—because each group may react differently to the drug. A Phase 3 trial can include between 1000 and 3000 participants. It is the largest, costliest, and most complicated phase of the clinical trial.

The FDA approves drugs only after it has rigorously scrutinized the results of all the clinical trials. Even after the FDA approves a drug for sale, the agency continues to monitor physician reports. Some problems become apparent only after the drug is used by a large number of people. For example, an adverse effect that occurs for only 1 in 10,000 people might not appear during the clinical trials. As one example, in February 1992 the FDA approved a new antibiotic named temafloxacin after clinical trials involving more than 4000 patients. Less than four months later, the drug was withdrawn from the market after 50 patients developed serious health problems from the medication, including three fatalities. Only after the drug was used by a much larger population did such rare but serious side effects become apparent.

CONCEPT QUESTION 15.3

The process of developing, testing, and approving a new pharmaceutical can take 10 to 15 years. During this time, patients who might benefit from these drugs cannot use them because they have not been approved by the FDA. To address this situation, several states have introduced "Right to Try" laws that allow terminally ill patients to request unapproved drugs after early-stage testing.

Consider the scenario of a new anticancer drug whose Phase 2 clinical trial has shown some promise for shrinking brain tumors and prolonging the life of these patients. The pharmaceutical company that makes the drug is now planning a Phase 3 clinical trial that will take 2 years to complete, plus another year for the FDA to review. Should a terminally ill patient with a brain tumor be allowed to use this experimental drug while it is being tested and evaluated?

Write two arguments in favor of allowing the patient to use the drug and two arguments against it. Your arguments can be used as the basis for a class discussion of this difficult scientific and ethical decision.

CHEMISTRY IN YOUR LIFE

Why Can a Popular Painkiller Make You Sick?

We think of medicines as beneficial and poisons as harmful. However, many substances function as both a medicine *and* a poison. The difference depends on *how much* of the substance you take, which is called the *dose.*

Consider the example of *acetaminophen* (Figure 1), a common ingredient in many popular over-the-counter (OTC) medications for headache and cold symptoms. For example, acetaminophen is an active ingredient in Tylenol, Theraflu, Benadryl, Robitussin, and DayQuil. At low doses, acetaminophen has the beneficial effect of reducing inflammation and fever. At high doses, however, it can damage the liver, the organ that removes toxins from our body. A medical study revealed that approximately 25% of patients treated in hospitals for severe liver failure had unintentionally overloaded their bodies with excess amounts of acetaminophen. In an effort to alleviate their colds, these patients had taken multiple medications containing acetaminophen. Without realizing it, they exceeded the maximum recommended dose of 4 grams (4000 milligrams) per day. Elevated acetaminophen levels can trigger acute liver failure, especially in individuals whose liver is already stressed from drinking alcohol.

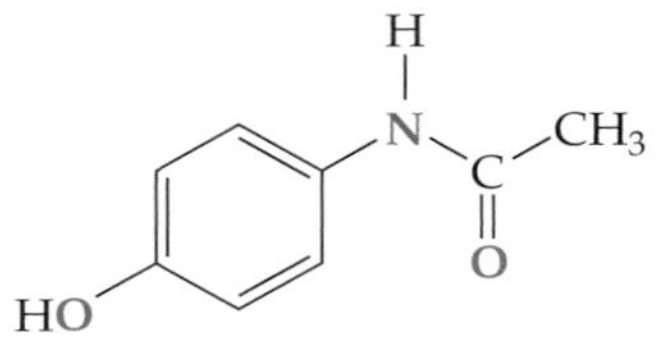

FIGURE 1 The structure of acetaminophen. Many over-the-counter medicines for headache, cold, and flu contain acetaminophen as an active ingredient.

Let's consider another example. If somebody gave you rat poison, would you swallow it? The obvious answer is no! Yet thousands of people ingest rat poison every day, and it improves their health. The active ingredient in rat poison is *warfarin,* a molecule that causes severe internal bleeding at high doses. At much lower doses, however, this same substance can help prevent the formation of lethal blood clots in the lungs, heart, and brain. Warfarin can act either as a poison or as a medicine, depending on the dose. Because rats are smaller than humans, they are also more sensitive to large doses of the drug. Consequently, a dose that is therapeutic for humans can be fatal for rats. Figure 2 illustrates these two uses of warfarin, along with the drug's molecular structure.

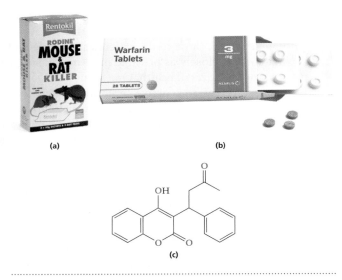

(a)   (b)

(c)

FIGURE 2 Warfarin is both a poison and a medicine. (a) Rat poison and (b) medication for patients at risk of stroke both contain (c) warfarin as their active ingredient. The difference between a poison and a medicine is the *dose* of warfarin.

These two examples—acetaminophen and warfarin—illustrate why understanding chemical quantities is so important. Every time you receive a prescription, your doctor specifies not just the *type* of medication but also the *dose* you should take. The dose of any drug must be specified precisely because too little will be ineffective and too much can be harmful. The size of the dose is based on clinical trials—it is chosen to produce benefits for the maximum number of patients while minimizing the risks. An *overdose* of a medication can cause serious harm, either from the direct effects of excess drug molecules or from the chemical conversion of the drug into other poisonous molecules within the body.

Core Concepts

- The first stage of drug development is to conduct preclinical research in which potential new compounds are discovered and tested. The next stage is to evaluate the toxicity of the drug using animal studies to measure the LD_{50} dose.
- Clinical trials of a drug are performed on human subjects and typically occur in three phases. Phase 1 evaluates the safety of the drug using healthy volunteers. Phase 2 tests the efficacy of the drug in a small population by comparing it to a placebo or an existing medication. Phase 3 gives the drug is given to a larger and more diverse population to further test its efficacy and to identify side effects that did not appear in the smaller study.

CHAPTER 15
VISUAL SUMMARY

15.1 Why Are Antibiotic-Resistant Infections Increasing?

Learning Objective:

Use tuberculosis as an example of an antibiotic-resistant disease.

⊙ Tuberculosis has affected humans for thousands of years, and it is the deadliest disease in human history.

⊙ Tuberculosis that is resistant to treatment with antibiotics is classified as multidrug-resistant (MDR-TB) or extensively drug-resistant (XDR-TB). Both types of tuberculosis are increasing worldwide.

15.2 Bacterial Cells

Learning Objective:

Describe the size and composition of bacterial cells.

⊙ Cell sizes are measured in units of micrometers, or microns (µm). One micrometer is equal to one-thousandth of a millimeter.

⊙ Most human cells have a size range of 10 to 100 µm. Bacterial cells are smaller, with a size range of 1 to 5 µm. Viruses are 10 times smaller than bacterial cells.

⊙ A bacterial cell consists of cytoplasm (a watery interior) that is surrounded by a cell membrane and a cell wall.

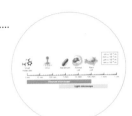

15.3 How Do Antibiotics Work?

Learning Objective:

Illustrate how antibiotics use different cellular targets to prevent the growth of bacterial cells.

⊙ Antibiotics are substances that prevent the growth of bacterial cells.

⊙ A **bactericidal antibiotic** kills bacterial cells, whereas a **bacteriostatic antibiotic** prevents the replication of bacterial cells without killing them.

⊙ A **narrow-spectrum antibiotic** is effective against a limited number of bacterial types, whereas a **broad-spectrum antibiotic** works on a wide range of bacteria.

⊙ The effectiveness of an antibiotic can be tested by the disk diffusion test, or by measuring its IC_{50}.

⊙ Bacterial cells contain many different cellular targets for antibiotics.

⊙ The bacterial cell wall is a strong, mesh-like structure constructed from long sugar chains (polysaccharides) that are crosslinked by shorter chains of amino acids.

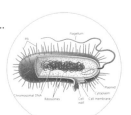

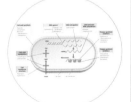

- The crosslinking is facilitated by an enzyme called transpeptidase. Penicillin binds to this enzyme and inhibits its function—it acts as an enzyme inhibitor.

- The absence of crosslinks weakens the cell wall, which causes the cell to burst as a result of its internal pressure.

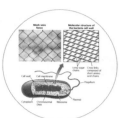

- Circular DNA in bacteria can become twisted in a process called **supercoiling.** Before bacterial DNA can be replicated, the supercoiling must be untwisted by a topoisomerase enzyme.

- Some antibiotics, such as ciprofloxacin, inhibit the function of topoisomerase and thus prevent bacteria from copying their DNA during cell division.

- Tetracycline functions as an antibiotic by inhibiting the synthesis of bacterial proteins at the ribosome.

15.4   How Do Bacteria Become Resistant to Antibiotics?

Learning Objective:

Illustrate the strategies used by bacteria to neutralize the effects of antibiotics.

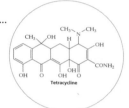

- Over time, bacteria evolve to become resistant to antibiotics. Drug-resistant strains of bacteria have been discovered for almost all current antibiotics.

- The most common defense mechanism used by bacterial cells is to pump the drug out of the cell before it can cause any harm. Other defense strategies include developing an enzyme to inactivate the drug (e.g., penicillinase) and altering the structure of the bacterial ribosome.

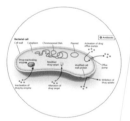

- Less than 20% of antibiotic use in the United States is for medical treatment of humans. Almost 80% of antibiotics are used on agricultural livestock.

- Use of low levels of antibiotics in agriculture promotes the development of drug-resistant bacteria, which can spread into human populations and the food supply.

15.5   Drug-Resistant Tuberculosis

Learning Objective:

Outline the relationship between the development of new antibiotics and the rise of antibiotic resistance.

- The tuberculosis bacterium is covered by a thick waxy coat that protects it against attacks by the immune system.

- Selman Waksman and his research team discovered streptomycin, the first antibiotic that was effective against the tuberculosis bacterium. Streptomycin inhibits the ability of the bacterium to synthesize new proteins.

- A later antibiotic, isoniazid, inhibits the ability of tuberculosis cells to make their protective bacterial coat.

15.6 How Are New Drugs Tested and Approved?

Learning Objective:

Outline the stages that are required to develop and evaluate a new pharmaceutical.

- ⊙ The Federal Food and Drugs Act, passed in 1906, led to the formation of the agency that is now called the Food and Drug Administration (FDA).

- ⊙ Thalidomide was a drug marketed as a sedative in Europe and Japan. Unexpectedly, pregnant women who took thalidomide gave birth to babies with deformed limbs. Thalidomide exists as two chiral isomers—one isomer induces sleep, but the other disrupts fetal limb development.

- ⊙ Dr. Frances Kelsey, an FDA regulator, prevented thalidomide from being sold to the public in the United States.

- ⊙ The first stage of drug development is preclinical research in which potential new compounds are discovered and tested. The next stage is to evaluate the toxicity of the drug using animal studies to measure the LD_{50} dose.

- ⊙ **Clinical trials** of a drug are performed on human subjects. They typically occur in three phases. Phase 1 tests the drug on healthy volunteers. Phase 2 evaluates the efficacy of the drug compared to a placebo or a currently available medication. Phase 3 extends the efficacy trials to a larger and more diverse population.

LEARNING RESOURCES

Reviewing Knowledge

15.1: Why Are Antibiotic-Resistant Diseases Increasing?

1. Which infectious agent causes tuberculosis?
2. Tuberculosis is usually diagnosed using a chest X-ray. What is the characteristic feature of the disease that is observed in an X-ray photograph?
3. What do the abbreviations MDR-TB and XDR-TB stand for?

15.2: Bacterial Cells

4. Which length unit is used to measure the size of cells?
5. Compare the sizes of a typical human cell, bacterial cell, and virus.
6. Draw the structure of a bacterial cell, and label its components.

15.3: How Do Antibiotics Work?

7. Penicillin is an example of a "natural product." Why do biological organisms provide a good source for antibiotics?

8. What is an antibiotic?
9. What is the difference between a bactericidal and a bacteriostatic antibiotic?
10. What is the difference between a broad-spectrum and a narrow-spectrum antibiotic?
11. Describe the disk diffusion test to evaluate the effectiveness of an antibiotic. Illustrate your answer with a sketch that shows the experimental results for three antibiotics with (a) no effect, (b) a moderate effect, and (c) a large effect.
12. What is the IC_{50} for an antibiotic? Does a potent antibiotic have a low or high IC_{50}?
13. List the principal antibiotic targets in a bacterial cell.
14. Sketch a diagram of the bacterial cell wall, and label its molecular components.
15. Describe how penicillin disrupts the construction of the bacterial cell wall.
16. What is DNA supercoiling? What happens to supercoiled bacterial DNA during DNA replication?
17. How does ciprofloxacin (Cipro) function as an antibiotic?
18. How does tetracycline function as an antibiotic?

15.4: How Do Bacteria Become Resistant to Antibiotics?

19. List three strategies that bacteria use to neutralize the effects of antibiotics.
20. How do some bacteria inactivate penicillin before it can harm the cell?
21. What proportion of total antibiotic consumption in the United States is used for human medicine? For which purposes are the majority of antibiotics used?
22. Explain how the use of antibiotics in agriculture facilitates the development of antibiotic-resistant bacteria.

15.5: Drug-Resistant Tuberculosis

23. Which structural feature of tuberculosis bacteria enables them to evade attack by the immune system?
24. What was the first antibiotic shown to be effective against tuberculosis bacteria? What is the source of this antibiotic?
25. How does isoniazid function as an antibiotic for tuberculosis bacteria?
26. What is a "cocktail" antibiotic therapy? Why is this strategy used?

15.6: How Are New Drugs Tested and Approved?

27. Why did Congress pass the Federal Food and Drugs Act of 1906?
28. What condition was the drug thalidomide used to treat, and what were the unintended side effects?
29. How does the molecular structure of thalidomide explain how the drug can have both beneficial and harmful effects?
30. A clinical trial of a new drug typically occurs in three phases. Outline the goals of each phase and the approximate number of patients who are involved.

Developing Skills

31. *E. coli* cells can exhibit a doubling time of 20 minutes at 37°C. If you start with 100 cells, how many minutes will it take for your bacterial culture to grow to 3200 cells? Assume that the bacteria have enough nutrients and space to continue replicating at the same rate.
32. Scientists have discovered certain bacteria that grow without a cell wall. One example is called *Mycoplasma*. Predict whether these cells will be harmed by each of three antibiotics: (a) penicillin G, (b) ciprofloxacin, and (c) tetracycline. Explain your answer for each antibiotic.
33. You are testing the effectiveness of two antibiotics, A and B. When each antibiotic is used at its optimal dose, it inhibits the growth of bacteria. You now test the combination of both antibiotics, A + B, with each at its optimal dose. You observe that the combination is twice as effective as either A or B alone. Hypothesize whether the two antibiotics act on the same cellular target or on different targets. Explain your reasoning.
34. Imagine that you are supervising an investigation of new antibiotics, labeled A through D. You ask four scientists in your lab to each measure the IC_{50} for one of

the drugs. Unfortunately, the scientists report their IC_{50} results using different concentration units. (See the data table of results.) Rank the antibiotics in order of potency, from highest to lowest. (M = molar, mM = millimolar, μM = micromolar, nM = nanomolar).

ANTIBIOTIC	IC_{50}
A	0.05 mM
B	3.0×10^{-5} M
C	10,000 nM
D	40 μM

35. Penicillin drugs are effective against replicating (growing) bacteria, but they do not harm bacteria that are not replicating. Propose a hypothesis for this observation.
36. Bacteria are generally classified into two groups, gram-negative and gram-positive, based on their ability to be stained by a dye called Gram stain. Gram-negative bacteria typically have an outer membrane that surrounds their cell wall, whereas gram-positive bacteria lack this second membrane. Based on this classification, propose a hypothesis to explain why existing antibiotics are typically *less* effective against gram-negative bacteria than they are against gram-positive bacteria.
37. Chemists have synthesized new antibiotic, which is shown here. Predict whether this new drug will be active against bacteria that are sensitive to penicillin, and explain your answer. Refer to the structure of penicillin in Figure 15.10.

38. As discussed in the chapter, antibiotics can be classified as narrow spectrum or broad spectrum. (a) In what circumstances would a broad-spectrum antibiotic be appropriate to treat a patient? In what circumstances would a narrow-spectrum antibiotic be appropriate to treat a patient? (b) What do you predict will be the effect of each type of antibiotic on the "good" bacteria that are discussed in the *Chemistry and Your Health* feature?
39. Development of resistance can be revealed by experiments in which the IC_{50} values of a drug against a bacterial strain are measured at several generations after exposing the bacteria to low levels of the drug that do not kill most cells. If the IC_{50} values increase, then the cells are acquiring resistance. If they do not, and the IC_{50} remains constant, then resistance does not develop. The figure below shows the results of this experiment for three antibiotics. One is a new trial drug,

and the other two are currently in use. The *x*-axis of the plot shows the number of generations of *Staphylococcus aureus* that have grown in the presence of sublethal doses of the antibiotics. The *y*-axis shows the increase in IC_{50} of samples of the bacteria from different generations against the three drugs.

(a) Following 400 generations of growth, which antibiotic shows the greatest change in stimulating resistant cells? By how much has its IC_{50} increased?

(b) One drug shows essentially no change in IC_{50}. How do you interpret this behavior?

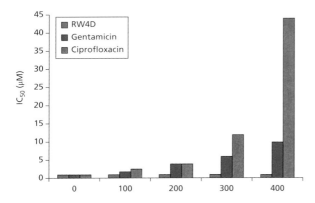

40. The *therapeutic index* of a drug is a measure of the harmful toxicity of a drug relative to the dose needed for therapeutic activity. The therapeutic index is defined as a ratio, TD_{50}/ED_{50}, in which:

TD_{50} = toxic dose in 50% of patients

ED_{50} = effective dose for 50% of patients.

Suppose you are evaluating two drugs. Drug A has a therapeutic index of 1000, and drug B has an index of 10. Which drug is preferable? Explain your answer.

Exploring Concepts

41. *Yersinia pestis* is the bacterium responsible for the Black Plague, which killed at least one-third of Europeans in the Middle Ages. It is carried by fleas that infect animals such as rats and ferrets. Evidence that *Y. pestis* was the cause of the plague comes from sequencing the DNA in corpses from plague cemeteries in Europe. Search the Internet to discover if this plague is still active and, if so, where.

42. Recent research on the gut microbiome has suggested that the types of bacteria in our gut can affect our mood. Use the Internet to write a one-paragraph description of possible mechanisms for how the make-up of bacteria in our intestines can affect the way we feel.

43. Drug costs have escalated dramatically in recent years. For example, one drug for treating hepatitis C viral infections costs $30,000 for a course of therapy. Insurance companies and the federal government have expressed concerns over this problem. Use the Internet to prepare a position statement on the following debate topic: Drug companies overcharge patients for new medications. Do you agree or disagree with this statement?

44. Warfarin is a drug that is used to treat the risk of blood clot formation. Its therapeutic index is low (see question 42 for a discussion of the therapeutic index). Warfarin is responsible for a large number of emergency hospital visits because patients using warfarin risk experiencing severe bleeding if they accidentally exceed their dose. A substitute drug has been developed that has both a better therapeutic index and an available treatment to stop its action in an emergency. However, the drug is more expensive than warfarin. Which drug would you request from your health provider? Explain.

45. The diversity of bacteria in our intestines is reflected in our fecal matter. Physicians are now experimenting with "fecal transplants," which involve using fecal bacteria from one person to treat another person. Use the Internet to investigate the use of fecal transplants to treat patients who are infected with antibiotic-resistant bacteria. Select one case study, and write a one- to two-paragraph report about the type of bacterial infection being treated and the effectiveness of the fecal transplant.

46. The chapter described the devastating effects of thalidomide on thousands of children. You may be surprised to learn, then, that the FDA has approved thalidomide for specific therapies. Search the Internet to discover how thalidomide is still being used as a drug, and write a one- to two-paragraph report on your findings.

47. Scientists have recently discovered new strains of *M. tuberculosis* that are even more resistant to antibiotics than those classified as MDR-TB. These new strains have provisionally been called "totally drug-resistant tuberculosis," or TRD-TB. Use the Internet to investigate the global locations where TDR-TB has been observed. How does TDR-TB differ from MDR-TB? Write one- to two-paragraph report on your findings.

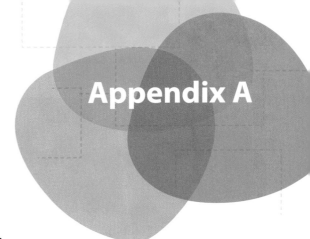

Scientific Notation and Units

In Chapter 1, you learned about antibiotic-resistant bacteria. Bacteria are one type of microorganism—a tiny, single-celled organism that cannot be seen with the naked eye ("micro" means *small*). Our body and our environment are teeming with billions of microorganisms, but we usually give them little thought because we cannot see them.

This book will discuss many things that you cannot see directly—not just bacteria, but also atoms and molecules. In order to describe these entities, we need to use mathematical tools called scientific notation and scientific units. This Appendix introduces these tools and explains their usefulness. For those who have studied this material before, these pages will serve as a refresher. If you are encountering these topics for the first time, we encourage you to become familiar with these important topics that we will use throughout the book. To begin our discussion, we ask you to answer Concept Question A.1.

CONCEPT QUESTION A.1

When people talk about something very small, they may use the phrase "as thin as a human hair." In fact, the width of a human hair is roughly the size limit of what we can see with our naked eye. A magnified microscope image of a single hair is shown.

Estimate the width of a human hair and write this dimension using the most size-appropriate unit. It will be helpful to examine one of your own hairs for this exercise!

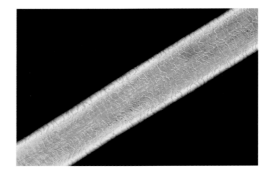

Scientific Notation

In everyday life, we tend to use numbers that are easy to imagine and communicate. In science, however, we frequently encounter numbers that are very large or very small. For example, Figure A1.1 shows a magnified image of red blood cells, which enable the transport of oxygen through our blood stream. Each red blood cell is shaped like a donut with a filled center. What is the size of a red blood cell?

The *meter* is the standard unit of length in science. Using this unit, we can write the approximate diameter of a red blood cell as:

0.000008 m

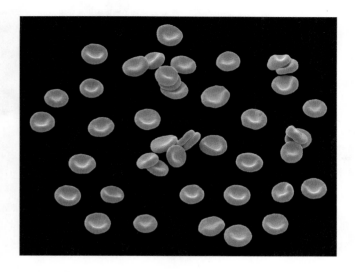

FIGURE A.1   A magnified image of human red blood cells. Each red blood cell has an approximate diameter of 0.000008 m.

scientific notation The representation of a number in terms of a coefficient multiplied by a power of 10 (the exponent); an example is 2.5×10^6.

In words, this size is eight millionths of a meter. At the other extreme, here is the approximate number of carbon atoms in a 12.0 g sample of carbon:

$$602,200,000,000,000,000,000,000$$

This number amounts to a little more than six hundred thousand billion trillion atoms. It is called *Avogadro's number*, and it is the basis of a chemical unit called the *mole*, which we examine in Chapter 6.

Using numbers in this form is inconvenient and cumbersome. It is also easy to make mistakes when trying to keep track of all the zeroes. For these reasons, scientists have developed a system of representing numbers that is called **scientific notation**. *Scientific notation expresses any quantity as a number between 1.00 and 9.99 that is multiplied by a power of ten.* The number between 1.00 and 9.99 is called the *coefficient*, and the power of 10 is called the *exponent*.

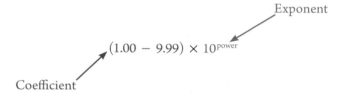

Let's look at the two examples we mentioned in this section and convert them into scientific notation. For the red blood cell,

$$0.000008 \text{ m} = 8 \times 10^{-6} \text{ m}$$

In this case, the *negative exponent* 10^{-6} indicates that *the decimal point should be moved six places to the left* to write the full number. For Avogadro's number,

$$602,200,000,000,000,000,000,000 = 6.022 \times 10^{23}$$

The *positive exponent* 10^{23} indicates that *the decimal point should be moved 23 places to the right* to write the full number.

Calculations Using Scientific Notation

Scientific notation makes calculations much easier. Three examples in this section illustrate multiplication, division, and addition/subtraction. The numbers have been chosen to give simple results because we want to focus on the method of the calculations. In most cases, the numbers will not work out so neatly, and you will need to consider the rules for *significant figures* that presented in Appendix B. In the following examples, the coefficients and exponents are highlighted in different colors.

Multiplication

Question 1: Multiply 2.5×10^3 by 3.0×10^5.

Answer: When multiplying two numbers in scientific notation, *multiply the coefficients* (2.5 and 3.0) and *add the exponents* (10^3 and 10^5).

$$(2.5 \times 10^3) \times (3.0 \times 10^5) = (2.5 \times 3.0) \times (10^3 \times 10^5)$$
$$= (7.5) \times 10^{3+5}$$
$$= 7.5 \times 10^8$$

Division

Question 2: Divide 8.8×10^5 by 1.6×10^3.

Answer: When dividing two numbers in scientific notation, *divide the coefficients* (8.8 and 1.6), and *subtract the exponents* (10^5 and 10^3).

$$\frac{(8.8 \times 10^5)}{(1.6 \times 10^3)} = \frac{8.8}{1.6} \times \frac{10^5}{10^3}$$
$$= 5.5 \times 10^{5-3}$$
$$= 5.5 \times 10^2$$

Addition/Subtraction

Question 3: Subtract 3.0×10^4 from 4.8×10^5.

Answer: To add or subtract numbers in scientific notation, first *convert the numbers so that they all have the same power of ten*. After this step, you can add or subtract the coefficients.

$$(4.8 \times 10^5) - (3.0 \times 10^4) = (4.8 \times 10^5) - (0.30 \times 10^5)$$
$$= (4.8 - 0.30) \times 10^5$$
$$= 4.5 \times 10^5$$

Scientific Units

We can simplify our representation of numbers even further by using **scientific units**. For example, a meter (which is roughly 3 feet and 3 inches) is the appropriate length unit for measuring a person or the size of a building. However, it is not convenient for measuring sizes that are very small (e.g., the diameter of a red blood cell) or distances that are very large (e.g., the distance to the sun).

For numbers that are much larger or much smaller than our standard unit of measurement, *we can define a new scientific unit that incorporates a power of 10 into its definition*. For example, the straight-line distance between London and Paris is approximately 343,000 meters (m). We can express this in a different way introducing a new unit—the *kilometer*—that is equal to one thousand meters.

$$\text{distance between London and Paris} = 343 \times \underbrace{(10^3\,\text{m})}_{1\,\text{km}} = 343 \text{ km}$$

We learned that the diameter of a red blood cell is 8×10^{-6} m. We can rewrite this size using a new unit called a *micrometer* (μm), which is equal to one millionth of a meter.

$$\text{diameter of red blood cell} = 8 \times \underbrace{(10^{-6}\,\text{m})}_{1\,\text{μm}} = 8 \text{ μm}$$

Table A.1, lists some of the common prefixes for scientific units that we will use throughout the text. Note that the reference point of 1 is listed in the middle, with larger numbers written above and smaller numbers below.

scientific unit A unit in which the power of 10 is included in the definition of the unit; an example is 1 mm = 10^{-3} m.

TABLE A.1	Powers of 10, prefixes, and symbols commonly used for scientific units.	
POWER OF 10	**PREFIX**	**SYMBOL**
10^{12}	tera-	T
10^{9}	giga-	G
10^{6}	mega-	M
10^{3}	kilo-	k
1		
10^{-2}	centi-	c
10^{-3}	milli-	m
10^{-6}	micro-	μ
10^{-9}	nano-	n
10^{-12}	pico-	p

With the exception of the centimeter, the power of 10 changes by three between each scientific unit. What should we do when a measured quantity falls between two possible units? For example, a length of 1×10^{-7} m could be written using the micrometer unit.

$$1 \times 10^{-7} \text{ m} = 10^{-1} \times (10^{-6} \text{ m}) = 0.1 \text{ μm}$$

Alternatively, the same length could be written using the nanometer unit:

$$1 \times 10^{-7} \text{ m} = 10^{2} \times (10^{-9} \text{ m}) = 100 \text{ nm}$$

As a rule of thumb, *we select the scientific unit that gives a number larger than one*. In this case, we write the length as 100 nm instead of 0.1 μm.

Unit Conversion

unit conversion The conversion of a quantity from one unit to another unit: the quantity remains the same, but the unit changes.

In many cases, it is necessary to change from one scientific unit to another in order to describe a scientific object or quantity; this procedure is called **unit conversion**. It is analogous to converting your money from one currency to another when you travel abroad. As we write this appendix, 100 dollars in U.S. currency converts to 87 euros in France or 1800 pesos in Mexico (we'll ignore fees on currency conversions). Note how the number preceding the currency unit increases or decreases depending on which currency we are using. We will perform conversions using currency to show how they follow the same rules as conversions using scientific units.

conversion factor A ratio of units used to perform a unit conversion; the conversion factor is always equal to 1.

Converting money from one currency to another requires a **conversion factor**. This is a ratio of numbers and currency that is always equal to 1. Our exchange rate from dollars to euros can be used to create two conversion factors, which are written below. To distinguish between the currencies, the dollars and euros are highlighted in different colors.

$$\frac{1 \text{ dollar}}{0.87 \text{ euro}} = 1 \quad \text{and} \quad \frac{0.87 \text{ euro}}{1 \text{ dollar}} = 1$$

Which conversion factor should we use? The answer depends on the type of conversion that you want to perform. For example, suppose that you have $400 to spend on a European vacation and you want to convert this amount to euros. In general, you want a conversion factor with the *currency you have* (dollars) on the *bottom of the ratio* (the denominator), and the *currency you want* (euros) on the *top of the ratio* (the numerator). This conversion factor enables you to convert from dollars to euros:

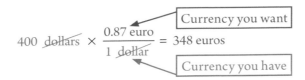

$$400 \; \text{dollars} \times \frac{0.87 \; \text{euro}}{1 \; \text{dollar}} = 348 \; \text{euros}$$

Currency you want

Currency you have

Note how the dollars in the denominator of the conversion factor cancel with the dollars in the starting amount. The outcome of the calculation is in euros because this currency is in the numerator of the conversion factor and is not canceled out. The numerical answer makes sense based on the conversion factor: We end up with fewer euros than our starting number of dollars.

PRACTICE EXERCISE A.1

You are traveling back to the United States after a vacation in Mexico, and you have 450 pesos left. You want to know how many dollars you will get from the currency exchange.

(a) Write the appropriate *conversion factor* for converting pesos to dollars.

(b) Calculate the number of dollars after the currency conversion.

While currency conversion rates vary over time, conversions between scientific units remain constant. Unit conversion in science proceeds through the same method: *Use a conversion factor whose ratio of numbers and units is always equal to 1.* A sample question is provided below:

Question 4: Use a conversion factor to convert 0.00005 m to units of micrometers.

Answer: First, we express the distance in scientific notation.

$$5 \times 10^{-5} \; \text{m}$$

Next, we need to convert from meters to micrometers. There are two possible conversion factors that contain these two units, which are written below. To distinguish between the currencies, meter and micrometer are highlighted in different colors.

$$\frac{10^{-6} \; \text{m}}{1 \; \mu\text{m}} = 1 \quad \text{and} \quad \frac{1 \; \mu\text{m}}{10^{-6} \; \text{m}} = 1$$

Which conversion factor should we choose? By analogy with the currency example, the meter is the unit we have in the given quantity, and the micrometer is the unit we want in the new answer. Therefore, we pick the conversion factor with meters in the denominator and micrometers in the numerator.

$$\frac{1 \; \mu\text{m}}{10^{-6} \; \text{m}} = 1$$

We can now use this conversion factor to make the unit conversion:

$$5 \times 10^{-5} \; \text{m} \times \frac{1 \; \mu\text{m}}{10^{-6} \; \text{m}} = 5 \times 10^{-5} \times 10^{6} \; \mu\text{m} = 5 \times 10^{1} \; \mu\text{m} = 50 \; \mu\text{m}$$

New units in answer

Given units in question

Note that the meter unit cancels because it is present in the original quantity and in the denominator of the conversion factor. The answer 5×10^1 µm is technically correct, but it is more convenient to write the number as 50 µm.

WORKED EXAMPLE A.1

Question: Scientists have measured the size of the virus—called a *rhinovirus*—that causes the common cold. The approximate diameter of a rhinovirus is 3.0×10^{-8} m. Use a conversion factor to rewrite this number using a more convenient scientific unit.

Answer: The most convenient unit to use for the diameter of a virus is the *nanometer* (nm). According to Table A.1, 1 nm = 10^{-9} m.

$$\text{diameter of virus} = 3.0 \times 10^{-8} \ \cancel{m} \times \frac{1 \ nm}{10^{-9} \ \cancel{m}} = 3.0 \times 10^{-8} \times 10^9 \ nm$$

$$= 3.0 \times 10^1 \ nm = 30 \ nm$$

TRY IT YOURSELF A.1

Question: The approximate diameter of a carbon atom is 0.000000000170 m. Use a conversion factor to rewrite this number using a more convenient scientific unit.

PRACTICE EXERCISE A.2

In chemistry, the volume of liquid samples is measured in liters (L). Suppose that you have 2.5 L of a liquid, and you need to divide it into smaller samples of 5 milliliters (mL) each.

(a) What is the total volume of the liquid in milliliters?

(b) How many 5 mL samples can you obtain from this volume?

Developing Skills

1. Express the following quantities in scientific notation:
   (a) 1500 m
      the length of a middle-distance race in the Olympics
   (b) 0.000000000074 m
      the length of the chemical bond in a hydrogen molecule.
   (c) 41300000000000000 m
      the distance to Alpha Centuari, one of the sky's brightest stars.
2. A 10 mL blood sample is found to contain 5×10^{10} red blood cells. Calculate the number of red blood cells per liter (L).
3. The speed of sound in air at sea level is 343 meters per second (m/s). Only one land vehicle has ever reached a high enough speed to break the sound barrier—a speed called *Mach 1*. What speed did this vehicle need to reach in *miles per hour? 1 mile = 1.61 km*
4. The distance between the Earth and the sun is approximately 93 million miles. Light travels at a very fast speed of 3.00×10^8 (m/s). How many minutes does it take for light emitted by the sun to reach the Earth? *1 mile = 1.61 km*

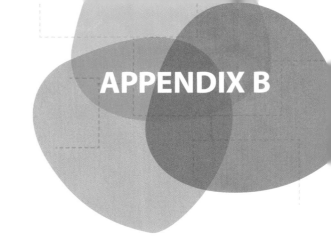
Significant Figures

Consider the following conversation:

"The dinosaurs were killed off by a meteor impact that happened 65 million and four years ago."

"How do you know?"

"I read in a newspaper article that 'the dinosaurs were killed off by a meteor impact 65 million years ago.' I read that article four years ago."

This dialogue illustrates the mistake of thinking that numbers are more accurate than they really are. Obviously, the author of the newspaper article did not mean that the dinosaurs were killed *exactly* 65 million years ago, Instead, they were killed off *approximately* 65 million years ago (give or take a million years).

Significant Figures

Whenever we read or use a number, we must be clear about its level of certainty. This certainty can be communicated by the way we write the number using **significant figures**. For a particular numerical result, *the significant figures are all of the precisely known digits plus a final digit that is estimated*. There is always some uncertainty in a measurement, which is embedded within the *last significant digit of the result*.

Suppose that we are measuring the dimensions of a room in order to buy a new carpet. We have a measuring tape from a hardware store that is divided into meters (m), centimeters (cm), and millimeters (mm). We begin by measuring the length of the room, and we confidently specify the length as just over 4 m and 47 cm. But measuring the length to the smallest unit—the mm—is more difficult, so we cannot specify this value with the same degree of confidence. We write the length of the room as

$$\text{length} = 4.473 \text{ m}$$

which contains *four significant figures*. We are certain about the first three digits but acknowledge that some inaccuracy exists in the fourth digit. In fact, the correct value may be 1 millimeter higher or lower. We could write this

significant figures The number of digits included in a quantity that reflects the level of certainty in its measurement.

inaccuracy explicitly by stating that the measurement is certain to ± 0.001 m (i.e., ± 1 millimeter).

What would you think if someone else measured the room and reported that the length is 4.47319468 m? Including all these figures is obviously unreasonable in terms of a person's ability to measure the room with a measuring tape. The extra digits following the 3 therefore contain no meaningful information—that is, they are *not significant*.

Let's repeat the procedure for the width of the room, which we measure as:

$$width = 2.756 \text{ m}$$

This number is also written with four significant figures using the same logic as above. What is the area of the room that we need to cover with carpet? We multiply the width and length using a calculator, which displays the following numerical result:

$$area = 12.327588 \text{ m}^2$$

If you asked for this amount of carpet in a store, you would get some puzzled looks. The output from the calculator gives eight digits, but not all these digits contain meaningful information. We saw earlier that we can measure the length and width to the nearest 1 mm, so *the calculated area can be only as certain as these initial measurements*. Since the length and width contained four significant figures, the calculated area must be rounded to four significant figures; that is,

$$area = 12.33 \text{ m}^2$$

The relationship among these quantities is illustrated in Figure B.1.

Whenever you perform any numerical calculation, it is important to consider the appropriate number of significant figures in your answer. It is tempting to write every digit given by your calculator because that seems to be the "correct" result. But not all these digits are significant, and so writing them does not convey any useful information.

Calculations in chemistry involve adding or multiplying numbers that represent masses, moles, numbers of molecules, and other quantities. In some cases, it may not be obvious whether or not a calculated answer has the appropriate degree of certainly. We therefore need rules to help us determine how many significant figures to include in a measurement or a calculated result.

Which Digits Are Significant?

Rule 1: For a measured quantity, the number of significant figures should reflect all certain digits plus the first uncertain digit. In a properly reported measurement, *all nonzero digits are significant*.

> **Example**: The mass of sodium chloride measured on a tabletop balance can be written as 3.54 g (three significant figures). The same amount of sodium chloride measured on a more precise analytical balance can be written as 3.5437 g (five significant figures). In the second example, the extra digits are included because the measuring instrument has a higher precision.

Rule 2: Zeroes are significant if they are part of the measurement. However, zeroes are *not* significant if they are not measured and are used only to position the decimal point.

Here are some examples to illustrate whether or not a zero should be counted as significant in a number that contains a decimal point.

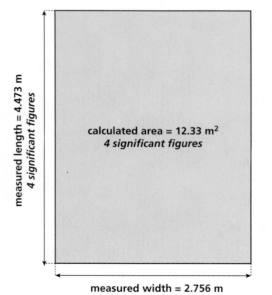

measured length = 4.473 m
4 significant figures

calculated area = 12.33 m²
4 significant figures

measured width = 2.756 m
4 significant figures

FIGURE B.1 Measuring the dimensions of a room. The length and width of the room can be measured with a certainty of four significant figures. Therefore, the calculated area of the room must also be specified with a certainty of four significant figures.

- Zeroes that are used before the decimal point are *not significant*.
    - **Example:** For 0.276, the first zero is *not significant* (three significant figures)
- Zeroes that precede the first nonzero digit are *not significant*.
    - **Example:** For 0.000079, the zeroes are *not significant* (two significant figures)
- Zeroes that occur between other nonzero digits *are significant*.
    - **Example:** For 52.007 the two zeroes *are significant* (five significant figures)
- Zeroes at the end of a number *are significant* if they are to the right of a decimal point.
    - **Example:** For 0.300, the first zero before the decimal point is *not significant*, but the two zeroes after the decimal point *are significant* (three significant figures).

Rule 3: Zeroes at the end of the number without a decimal point are *ambiguous*—they may be significant or they may not be. The number should be restated in *scientific notation* to remove the ambiguity.

Example: Suppose that we write a length of "500 m." It is not clear whether we mean *exactly* 500 m (500.0 m) or *roughly* 500 m (i.e., 500 m ± 10 m). To avoid this ambiguity, we can write the number in scientific notation and use the decimal point to indicate the number of significant digits. The length "500 m" can be written in several ways, depending on the certainty of the measurement. The error of the measurement is contained within the last significant digit.

5.000×10^2 m   *4 significant figures with an error of ±0.1 m*

5.00×10^2 m   *3 significant figures with an error of ±1 m*

5.0×10^2 m   *2 significant figures with an error of ±10 m*

Determine the number of significant figures for each of the following quantities:

(a) 0.0176 mg

(b) 55.05 cm^3

(c) 100 s

(d) 100.0 m

(e) 0.00037 g

Multiplication and Division

For the result of a multiplication or division, the answer is given as *the number of significant figures corresponding to the least certain input value*. In other words, the answer should contain the *lowest number of significant figures* used for any of the given input values.

For example, suppose we want to calculate the density of an object from its mass and volume. We have a good balance, so we can measure the mass with high certainty as 14.659 g. By contrast, our graduated cylinder for measuring volume has a limited number of markings, so we can measure the volume only to a certainty of 3.8 cm^3. What is the density in g/cm^3? We can enter the numbers in a calculator to obtain:

$$\text{density} = \frac{\text{mass}\,(\text{g})}{\text{volume}\,(\text{cm}^3)} = \frac{14.659\,\text{g}}{3.8\,\text{cm}^3} = 3.857631579\,\text{g}\,/\,\text{cm}^3 \qquad \textit{Is this correct?}$$

Not all of these digits can be significant given the quality of the input measurements. Because we know the certainty of the volume to only two significant figures, we must round off the calculated density to two significant figures—that is, 3.9 g/cm^3.

Addition and Subtraction

For the result of an addition or subtraction, *the answer has the same number of decimal places as the measurement with the fewest decimal places.* As before, the least certain measurement determines the overall certainly of the calculation.

For example, let's add 20.789 g of sodium chloride to 84.3 g of water and calculate the total mass. We can easily put the numbers into a calculator, and it may be tempting to write:

$$
\begin{aligned}
\text{total mass} = \quad & 20.437 \text{ g NaCl} \\
+ \quad & 84.3 \quad \text{ g H}_2\text{O} \\
\hline
= \quad & 104.737 \text{ g}
\end{aligned}
\qquad \textit{Is this correct}
$$

In fact, this answer is *incorrect.* The mass of H$_2$O has been measured to one decimal place, which limits the certainty of the total measurement. In the addition shown above, the last two digits in the mass of NaCl have no corresponding digits in the mass of H$_2$O. Following the rule, we round off the result to 104.7 g, which contains one decimal place.

Suppose that we were able to measure the mass of the water with the same certainty as the mass of the sodium chloride. We now have two mass measurements with three decimal places, so they can be added in the following way:

$$
\begin{aligned}
\text{total mass} = \quad & 20.437 \text{ g NaCl} \\
+ \quad & 84.300 \text{ g H}_2\text{O} \\
\hline
= \quad & 104.737 \text{ g}
\end{aligned}
$$

The result of the second addition is 104.737 g, which has 6 significant figures. We are more certain of this addition compared to our earlier result of 104.7 g.

Rounding Off Numbers

The procedure for rounding off numbers involves discarding the digits that are not significant and adjusting the last remaining digit.

The rules for adjusting the final digit are:

- If the first discarded digit is *less than 5,* the remaining digits are not changed.

- If the first discarded digit is *5 or greater,* the last significant digit is increased by 1.

Although more sophisticated rounding rules exist, these practical versions will suffice for our purposes. In general, *retain* all the digits when performing the intermediate steps of your calculation, and then *round off* the final answer in the last step to the correct number of significant figures. This avoids the "rounding" error that can accumulate if the intermediate steps in the calculation are rounded.

In conclusion, the correct use of significant figures is an essential part of any numerical calculation. The number of significant figures reflects the degree of certainty of any numerical statement, which is just as important as the number itself.

WORKED EXAMPLE B.1

Question: Perform the following calculations. First write the numerical result given by your calculator, Next, identify the least certain measurement in the calculation. Finally, provide the answer of the calculation using the correct unit and number of significant figures. Report the number of significant figures in your final answer.

(a) 2.14 g + 0.376 g + 1.7 g

(b) 12.73 m × 1.82 m

(c) $\dfrac{34.86 \text{ g}}{3.7 \text{ cm}^3}$

Answer:

(a) The numerical answer given by a calculator is 4.216. The least certain measurement is 1.7 g, which has only one decimal place. Therefore, we round the answer to 4.2 g, which contains one decimal place and two significant figures.

(b) The numerical answer given by a calculator is 23.1686. The least certain number in the calculation is 1.82 m, which contains 3 significant figures. Therefore, we round the answer to 23.2 m^2, which also contains three significant figures.

(c) The number given by a calculator is 9.421621622. This number clearly has too many digits! The least certain number in the calculation is 3.7 cm^3, which contains two significant figures. Therefore, we round the answer to 9.4 g/cm^3, which also contains two significant figures.

TRY IT YOURSELF B.1

Question: Perform the following calculations. Provide the answer with the correct unit and number of significant figures.

(a) 1.362 cm × 8.487 cm × 6.89 cm

(b) $\dfrac{25.325 \text{ g} + 7.8 \text{ g}}{2.445 \text{ mL}}$

Developing Skills

1. How many significant figures are contained in each of the following quantities?
    (a) 73.04 kg
    (b) 0.00623 km
    (c) 20.0 L
    (d) 0.03 g

2. Round each of the following numbers to three significant figures:
    (a) 52.143
    (b) 1.007
    (c) 0.00004698
    (d) 3.14159265358979323384626433832.....   the numerical value of π

3. For the following calculations, provide the answer using the correct unit and number of significant figures:
    (a) 108.25 m + 25.3 m + 3.475 m
    (b) 11.23 cm × 3.45 cm × 5.372 cm
    (c) 8.235 g − 3.69 g
    (d) $\dfrac{12.36 \text{ cm}}{3.4 \text{ cm}}$

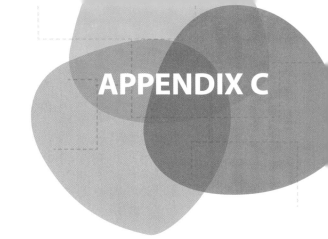

Logarithms and the pH Scale

What Are Logarithms?

The use of pH to measure acidity is an example of a **logarithmic scale**. *A logarithm is a mathematical method that represents a number in terms of its power of 10.* Logarithms are particularly useful for expressing numbers that vary over a wide range of values.

As we learned in Appendix A, any number can be expressed in terms of a power of 10 by using an exponent. For example,

$$1{,}000{,}000 = 10^6$$

$$0.0001 = 10^{-4}$$

where the exponent is 6 in the first example and −4 in the second example. Consider this question: What power of 10 did we need to generate the number 10^6? The obvious answer is 6. This simple observation is the essence of logarithms. If we use 10 as the "base" of the logarithm, we can write the following equation:

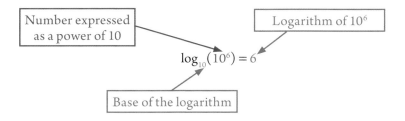

Number expressed as a power of 10

Logarithm of 10^6

$$\log_{10}(10^6) = 6$$

Base of the logarithm

Similarly, if we calculate the logarithm of 10^{-4} we obtain:

$$\log_{10}(10^{-4}) = -4$$

We can generalize these relationships using the fundamental equations for logarithms:

$$\text{if} \quad x = 10^y \quad \text{then} \quad \log_{10} x = y$$

logarithmic scale A numerical scale based on powers of 10; an example is the pH scale.

Table C.1 shows a list of base-10 logarithm values for numbers ranging from 10^6 to 10^{-6}. We can make some general observations based on this table.

- Any number greater than 1 has a positive logarithm, while any number smaller than one has a negative logarithm.

- The logarithm of 1 is zero; in other words, $10^0 = 1$.

- If a number increases by a *factor of 10* (e.g., from 10^4 to 10^5), *the logarithm increases by 1* (e.g., from 4 to 5). If a number *decreases by a factor of 10* (e.g., from 10^{-3} to 10^{-4}), *the logarithm decreases by 1* (e.g., from −3 to −4).

TABLE C.1   Powers of 10 and Their Base-10 Logarithms

NUMBER	POWER OF 10	LOGARITHM
1,000,000	10^6	6
100,000	10^5	5
10,000	10^4	4
1,000	10^3	3
100	10^2	2
10	10^1	1
1	10^0	0
0.1	10^{-1}	−1
0.01	10^{-2}	−2
0.001	10^{-3}	−3
0.0001	10^{-4}	−4
0.00001	10^{-5}	−5
0.000001	10^{-6}	−6

Not all numbers can be written as simply as the ones in Table C.1. For example, what is the logarithm of 3.5×10^3? This number is larger than 10^3 and smaller than 10^4, so we expect the logarithm to be larger than 3 and less than 4. We can calculate the exact value using the logarithm button on our calculator (usually labeled with the word "log"). Calculators vary in how they implement the logarithm function, so we suggest that you become familiar with your particular calculator.

$$\log_{10}(3.5 \times 10^3) = 3.54$$

This number occurs within the range that we predicted based on our understanding of logarithms. We recommend that you perform these checks for your calculations since it will enable you to catch mistakes that arise from pressing the wrong buttons on your calculator.

PRACTICE EXERCISE C.1

Calculate the base-10 logarithm of the following numbers:

(a) 6.2×10^{-2}

(b) 5.9×10^5

(c) 8.3×10^{-9}

(d) 7.1×10^4

It is also possible to perform the reverse calculation, which is called an *anti-logarithm*. For example, if the logarithm is 5, what is its corresponding number expressed as a power of 10? Based on the fundamental equations given above, the answer is 10^5. However, not all examples will be this easy. Suppose that the logarithm is 5.78—what is the corresponding antilogarithm?

We can write an equation with x as the unknown:

$$\log_{10} x = 5.78 \quad \text{so} \quad x = 10^{5.78}$$

By convention, scientific notation only uses powers of 10 that are integers such as 5, 6, and so on. We can convert this number into scientific notation using the calculator button marked "10^x". Again, you should become familiar with how this function works on your specific calculator. Using the calculator button, we obtain

$$10^{5.78} = 6.0 \times 10^5$$

Does this answer make sense? We expect the value of $10^{5.78}$ to be greater than 10^5 and less than 10^6, and the calculated number falls within this range.

PRACTICE EXERCISE C.2

Calculate the base-10 logarithm of the following numbers:

(a) 2.54

(b) −1.86

(c) −6.22

(d) −3.26

The Logarithmic pH Scale

We can use logarithms to create a scale to measure scientific quantities. Before discussing pH, let's consider another logarithmic scale—the *Richter scale* that is used to measure the intensity of earthquakes. Each increase of one unit in the Richter scale corresponds to a *10-fold increase* in earthquake intensity. This means that an earthquake of magnitude 5 is 10 times more powerful than one of magnitude 4. A two-unit change in the Richter scale (e.g., from 4 to 6) corresponds to a *100-fold increase* in earthquake intensity

These same principles apply to the pH scale in chemistry, which is used to measure the acidity of a solution. Acidity arises from the concentration of aqueous hydronium ions, written as $[H_3O^+]$ and measured in moles per liter (molarity, M). Typically, the solution values of $[H_3O^+]$ are small and written as *negative powers of 10*. For pure water and neutral solutions, $[H_3O^+] = 10^{-7}$ M (see Chapter 11). A dilute solution of acid may contain $[H_3O^+] = 10^{-3}$ M, while a dilute solution of base may contain $[H_3O^+] = 10^{-10}$ M.

The pH scale is used to replace negative powers of 10 by a positive number. The relationship between $[H_3O^+]$ and pH is given by two fundamental equations.

$$[H_3O^+] = 10^{-pH} \qquad \text{or} \qquad pH = -\log_{10}[H_3O^+]$$

Note the similarities between these equations for pH and the fundamental equations for logarithms given in the previous section. In these equations, x has been replaced by $[H_3O^+]$, and y has been replaced by −pH. The first equation enables us to calculate $[H_3O^+]$ based on the pH value, which is present in the exponent. The second equation allows us to use logarithms to calculate the pH of a solution based on the numerical value of $[H_3O^+]$. The pH equations contain a negative sign because a typical value for $[H_3O^+]$ is written as a negative power of 10. Consequently, the presence of a negative sign enables the pH value to be a positive number.

WORKED EXAMPLE C.1

Question: Calculate the pH of the following solutions:

(a) Coffee, $[H_3O^+] = 3.0 \times 10^{-6}$ M

(b) Household bleach, $[H_3O^+] = 2.5 \times 10^{-13}$ M

Answer: To answer these questions, we use the equation

$$pH = -\log_{10}[H_3O^+]$$

(a) $pH = -\log_{10}(3.0 \times 10^{-6}) = -(-5.52) = 5.52$

(b) $pH = -\log_{10}(2.5 \times 10^{-13}) = -(-12.60) = 12.60$

TRY IT YOURSELF C.1

Question: Calculate the pH of the following solutions:

(a) Battery acid, $[H_3O^+] = 0.10$ M

(b) Liquid drain cleaner, $[H_3O^+] = 2.0 \times 10^{-14}$ M

WORKED EXAMPLE C.2

Question: Calculate the concentration of $H_3O^+(aq)$ ions in the following solutions:

(a) Household ammonia cleaner, pH = 11.2

(b) Orange juice, pH = 3.5

Answer: To answer these questions, we use the equation

$$[H_3O^+] = 10^{-pH}$$

(a) $[H_3O^+] = 10^{-11.2} = 6 \times 10^{-12}$ M

(b) $[H_3O^+] = 10^{-3.5} = 3 \times 10^{-4}$ M

TRY IT YOURSELF C.2

Question: Calculate the pH of the following solutions:

(a) Tomato juice, pH = 4.2

(b) Milk, pH = 6.7

PRACTICE EXERCISE C.3

Calculate the concentration of $H_3O^+(aq)$ ions in the following solutions:

(a) Milk of Magnesia, pH = 10.5

(b) Vinegar, pH = 2.4

PRACTICE EXERCISE C.4

Calculate the pH of the following solutions:

(a) Coca-Cola, $[H_3O^+] = 2.5 \times 10^{-3}$ M

(b) Eggs, $[H_3O^+] = 1.3 \times 10^{-8}$ M

Logarithms and Significant Figures

Appendix B presented rules for using significant figures with calculations that use addition, subtraction, multiplication, and division. There are also rules for the appropriate number of significant figures when using logarithms.

Rule 1: When taking the logarithm of a number, the *number of significant figures* in the number should equal *the number of digits after the decimal place in the logarithm.*

For example, suppose that $[H_3O^+] = 4.5 \times 10^{-5}$ M. If we put this number into our calculator, we obtain the answer

$$pH = \log_{10}(4.5 \times 10^{-5}) = 4.346787486$$

However, not all of the digits after the decimal place are significant. Because the original measurement of $[H_3O^+]$ has two significant figures, we round the pH measurement to two decimal places.

$$pH = 4.35$$

Rule 2: When taking the power of 10 of a number, the *number of digits after the decimal place* in the number should equal *the number of significant figures in the calculated answer.*

For example, suppose that we measure the acidity of a solution and discover that pH = 3.8. We can calculate the value of $[H_3O^+]$ in the solution by using the pH value as the negative exponent. Using our calculator gives

$$[H_3O^+] = 10^{-3.8} = 1.584839192 \times 10^{-4}$$

However, the pH measurement contains only *one digit after the decimal point.* Consequently, we must report the hydronium ion concentration using *one significant figure.*

$$[H_3O^+] = 2 \times 10^{-4}\ M$$

PRACTICE EXERCISE C.5

For the following calculations, provide the answer using the correct number of significant figures:

(a) $10^{-7.439}$

(b) $\log_{10}(2.1 \times 10^{-5})$

(c) $10^{2.1359}$

(d) $\log_{10}(9.43 \times 10^{-8})$

Developing Skills

1. Calculate the concentration of $H_3O^+(aq)$ ions in the following solutions:
   (a) Sea water, pH = 8.3
   (b) Acid rain, pH = 4.3

2. Calculate the pH of the following solutions:
   (a) Toothpaste, $[H_3O^+] = 1.3 \times 10^{-10}$ M
   (b) Tomato juice, $[H_3O^+] = 7.8 \times 10^{-5}$ M

3. For the following calculations, provide the numerical answer using the correct number of significant figures:
   (a) $10^{8.88}$
   (b) $\log_{10}(7.524 \times 10^{-2})$
   (c) $10^{-4.1}$
   (d) $\log_{10}(8.95 \times 10^6)$

Answers to Try It Yourself and Practice Exercises

CHAPTER 2

Try It Yourself 2.1

(a) $\text{Density } (g/cm^3) = \dfrac{36.0 \text{ g}}{3.00 \text{ cm}^3} = 12.0 \text{ g/cm}^3.$

Sample A will sink in water because its density is greater than 1.00 g/cm^3.

(b) $\text{Density } (g/cm^3) = \dfrac{125.0 \text{ g}}{500.0 \text{ cm}^3} = 0.2500 \text{ g/cm}^3.$

Sample B will float on the surface of water because its density is less than 1.00 g/cm^3.

(c) $\text{Density } (g/cm^3) = \dfrac{55.0 \text{ g}}{20.0 \text{ cm}^3} = 2.75 \text{ g/cm}^3.$

Sample C will sink in water because its density is greater than 1.00 g/cm^3.

Try It Yourself 2.2
(a)   Pizza is a heterogeneous mixture.
(b)   Coffee is a homogeneous mixture.
(c)   Air is a homogeneous mixture.
(d)   A bowl of mixed nuts is a heterogeneous mixture.

Try It Yourself 2.3
We predict that the chemical formula is PH_3 because both N and P belong to Group 5A of the periodic table.

Try It Yourself 2.4
The oxygen-16 isotope contains 6 protons and 8 neutrons $(16 - 8 = 8)$. The oxygen-18 isotope also has 6 protons, but it contains 10 neutrons $(18 - 8 = 10)$.

Try It Yourself 2.5

(a) The ^{35}Cl isotope has a higher natural abundance (75.78%) than the ^{37}Cl isotope (24.22%). Therefore, we predict that the relative atomic mass of naturally-occurring chlorine will be closer to the mass of the ^{35}Cl isotope.

(b)
$$\begin{aligned}\text{relative atomic mass} &= (34.97 \times 0.7578) + (36.97 \times 0.2422) \\ &= 26.50 + 8.954 \\ &= 35.45\end{aligned}$$

The calculation matches the prediction from part (a). The higher natural abundance of the ^{35}Cl isotope makes a greater contribution to the calculation of the weighted average.

Try It Yourself 2.6

$$Mg \rightarrow Mg^{2+} + 2\ e^-$$

The net charge of the magnesium ion is +2.

Practice Exercise 2.1

(a) $\text{Density}\ (g/cm^3) = \dfrac{0.895\ g}{5000\ cm^3} = 0.000179\ g/cm^3$. Sample A is helium.

(b) $\text{Density}\ (g/cm^3) = \dfrac{448.0\ g}{50.0\ cm^3} = 8.96\ g/cm^3$. Sample B is copper.

(c) $\text{Density}\ (g/cm^3) = \dfrac{675.0\ g}{250.0\ cm^3} = 2.700\ g/cm^3$. Sample C is aluminum.

Practice Exercise 2.2

Group 6A

Practice Exercise 2.3

(a) $\text{Diameter of red blood cell}\ (\mu m) = 8 \times \cancel{10^{-6}}\ \cancel{m} \times \dfrac{1\ \mu m}{\cancel{10^{-6}}\ \cancel{m}} = 8\ \mu m$

(b)
$$\begin{aligned}\text{Diameter of cold virus (nm)} &= 3 \times 10^{-8}\ \cancel{m} \times \frac{1\ nm}{10^{-9}\ \cancel{m}} \\ &= 3 \times 10^{-8} \times 10^{9}\ nm = 3 \times 10^{1}\ nm\ \text{ or }\ 30\ nm\end{aligned}$$

Practice Exercise 2.4

ISOTOPE	ATOMIC SYMBOL	NUMBER OF PROTONS	NUMBER OF NEUTRONS
Uranium-235	$^{235}_{92}U$	92	143
Iron-56	$^{56}_{26}Fe$	26	30
Sulfur-32	$^{32}_{16}S$	16	16
Krypton-86	$^{86}_{36}Kr$	36	50

Practice Exercise 2.5

(a) From the periodic table, the relative atomic mass of boron (B) is 10.81.

(b) Use m to denote the unknown mass of the boron isotope. The equation for calculating the relative atomic mass of boron using a weighted average is given below. Convert the percentages to decimals for the calculation.

$$\left(11.01 \times 0.8022\right) + \left(m \times 0.1978\right) = 10.81$$

Rearrange the equation to isolate the unknown mass on the left-hand side.

$$\left(m \times 0.1978\right) = 10.81 - \left(11.01 \times 0.8022\right) = 1.98$$

Perform a division to calculate m:

$$m = \frac{1.98}{0.1978} = 10.0$$

Strictly speaking, we can only report this result using three digits, which is the allowable number of *significant figures*. The topic of significant figures is presented in Appendix B.

Practice Exercise 2.6

(a)

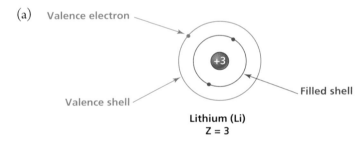

Lithium (Li)
Z = 3

(b) The electron configuration shows that a potassium atom has one valence electron in the fourth electron shell (the first, second, and third electron shells are filled). Lithium and sodium also have one valence electron. Consequently, potassium has similar chemical properties to lithium and sodium.

Practice Exercise 2.7

ATOM	ION	NUMBER OF PROTONS	NUMBER OF ELECTRONS
Sulfur (S)	S^{2-}	16	18
Aluminum (Al)	Al^{3+}	13	10
Iodine (I)	I^{-}	53	54
Hydrogen (H)	H^{+}	1	0

CHAPTER 3

Try It Yourself 3.1

STEP 1: *Write the Lewis dot structures for each of the atoms in the molecule.*

Nitrogen   Iodine

STEP 2: *Identify which atom should be placed at the center of the molecule.*
Iodine needs only one more electron to form a stable octet, so it will form only one covalent bond. Nitrogen has five valence electrons, so it can form three covalent bonds. Therefore, we place N at the center of the molecule.

Nitrogen

STEP 3: *Determine how many additional electrons are required for nitrogen to achieve a stable electron configuration.*
Nitrogen can achieve a stable octet by sharing electrons with three atoms. Consequently, the N atom will combine with three I atoms.

STEP 4: *Draw the complete Lewis dot structure for the molecule, which includes stable valence electron configurations for all the atoms.*
The Lewis dot structure for NI_3 is shown below. Note the lone pair on the N atom.

Try It Yourself 3.2

(a) Lewis dot structure

(b) Two-dimensional structure

H—S—H

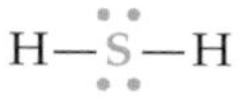

(c) Three-dimensional structure

Both bonds
in the plane

H   S   H

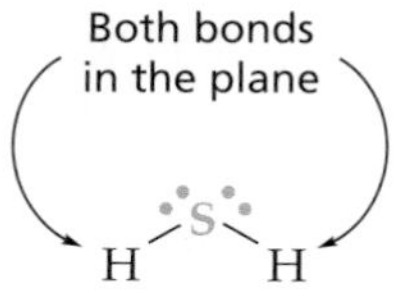

(d) Space-filling model

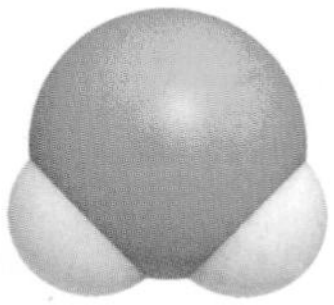

Try It Yourself 3.3

(a) Lewis dot structure

$$H \! : \! C \! : : : \! N \! :$$

(b) Two-dimensional structure

$$H—C≡N \! :$$

The triple covalent bond is represented by three straight lines joining the C and N atoms.

(c) Three-dimensional structure

$$H—C≡N \! :$$

The three-dimensional structure of HCN is *linear*. The central C atom does not have any lone pairs, so there is no electron repulsion that would produce a bent molecule.

Try It Yourself 3.4

(a) $MgSO_4$

(b) $NaHCO_3$

(c) K_3PO_4

(d) $Ca(OH)_2$

(e) NH_4NO_3

Practice Exercise 3.1

Lewis dot structure   Two-dimensional structure

The chemical formula is H_2O, which is a molecule of *water*. The H_2O molecule has two nonbonding electron pairs on the O atom.

Practice Exercise 3.2

(a) The Lewis dot structure for SF_6 is shown below. There are *12 electrons* in the valence shell of the sulfur atom.

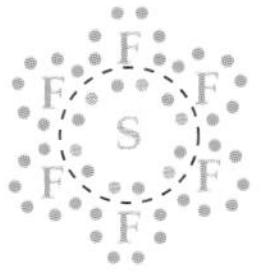

(b) To apply VSEPR to predict the three-dimensional structure, we note that there are six bonding electrons pairs and no lone pairs on the central S atom. VSEPR predicts the three-dimensional structure shown below. This molecular shape is called *octahedral* because the spatial arrangement of S—F bonds creates a geometrical shape called an octahedron (with eight faces).

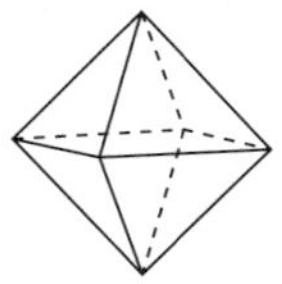

Practice Exercise 3.3

The two polyatomic ions that show expanded valence are the sulfate ion, SO_4^{2-}, and the phosphate ion, PO_4^{3-}. In the sulfate ion, the S atom violates the octet rule because it is surrounded by 12 valence electrons (we count 2 valence electrons for each covalent bond). In the phosphate ion, the P atom violates the octet rule because it is surrounded by 10 electrons.

CHAPTER 4

Try It Yourself 4.1

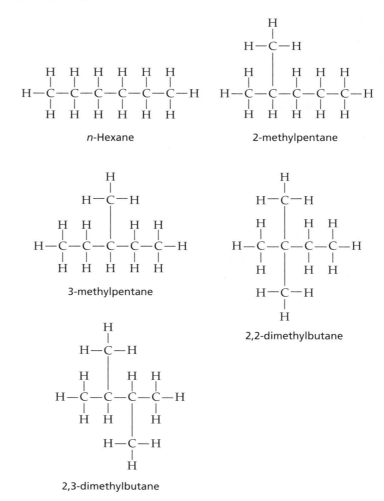

n-Hexane

2-methylpentane

3-methylpentane

2,2-dimethylbutane

2,3-dimethylbutane

Try It Yourself 4.2

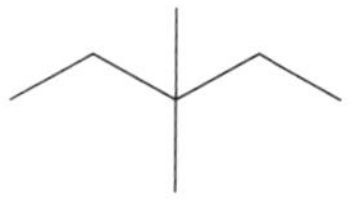

Try It Yourself 4.3

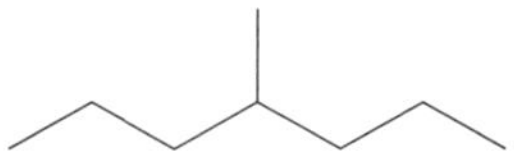

Try It Yourself 4.4

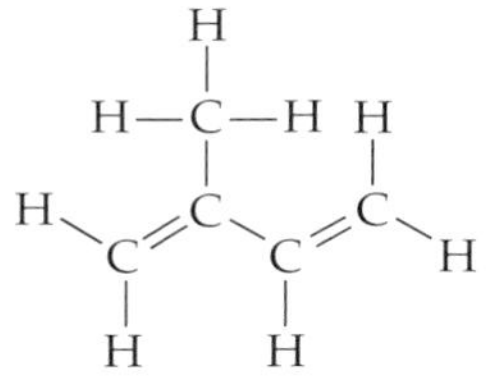

The chemical formula of isoprene is C_5H_8.

Practice Exercise 4.1

The Lewis dot structure depicts the valence electrons in the molecule. The two-dimensional structure illustrates the covalent bonds between the atoms but does not convey the accurate three-dimensional geometry. The three-dimensional structure represents the spatial geometry of the molecule using the wedge-and-dash notation. The ball-and-stick model depicts atoms as balls and covalent bonds as sticks. Finally, the space-filling model displays the atomic radius for each atom within the molecule and therefore shows the total volume of the entire molecule.

Practice Exercise 4.2

$CH_3CH_2CH_2CH_2CH_3$ *n*-pentane

$CH_3CH(CH_3)CH_2CH_3$ 2-methylbutane

$CH_3C(CH_3)_2CH_3$ 2,2-dimethypropane

Practice Exercise 4.3

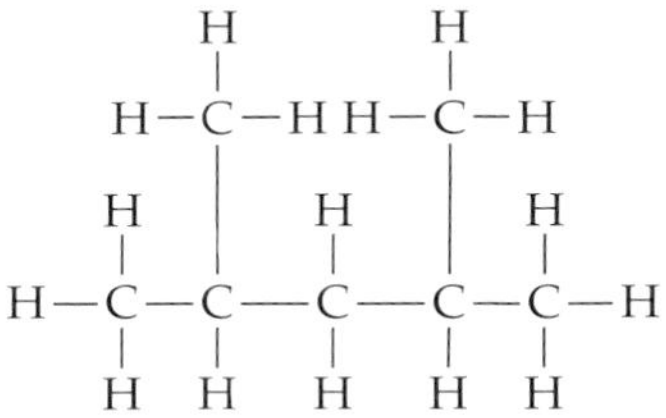

C_7H_{16}

Practice Exercise 4.4

(a) $C_{10}H_8$

(b)

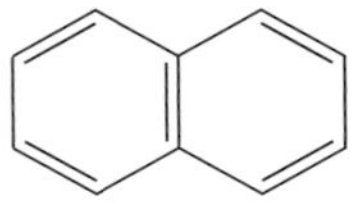

(c) Naphthalene is planar like benzene because it has delocalized electrons throughout the two connected rings. As with benzene, the delocalized electrons constrain the molecule to be planar.

Practice Exercise 4.5

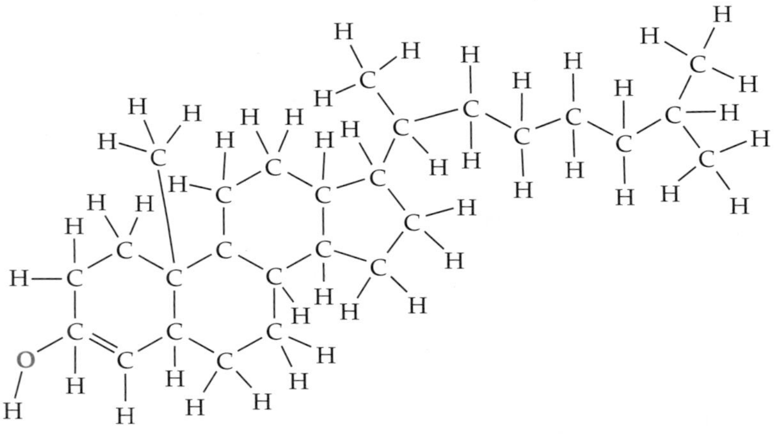

The chemical formula of cholesterol is $C_{27}H_{46}O$.

CHAPTER 5

Try It Yourself 5.1

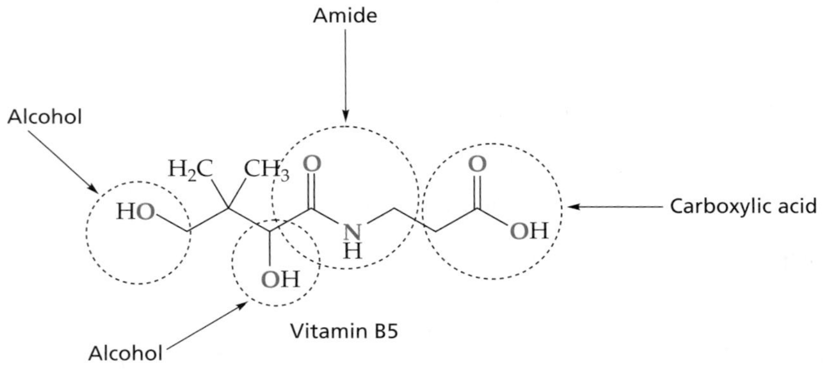

Try It Yourself 5.2

(a) Vitamin B3 contains a pyridine heterocycle.

(b) A carboxylic acid group is attached to the heterocycle.

Practice Exercise 5.1

(a) The ester in the strawberry has a different composition because it contains a sulfur (S) atom that replaces an oxygen atom within the functional group.

(b)

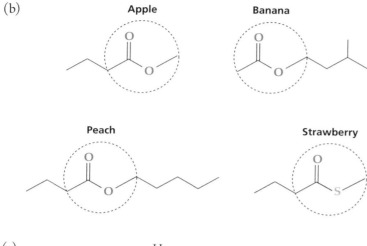

(c)

The chemical formula is $C_7H_{14}O_2$.

Practice Exercise 5.2

(a) The nitrogen heterocycle in coenzyme A is derived from a purine.

(b)

CHAPTER 6

Try It Yourself 6.1

(a) $N_2 + O_2 \rightarrow 2 NO$

ATOM	REACTANTS	PRODUCTS
N	2	2
O	2	2

(b) $2 NO + O_2 \rightarrow 2 NO_2$

ATOM	REACTANTS	PRODUCTS
N	2	2
O	2	2

Try It Yourself 6.2

(a) $2 C_8H_{18} + 17 O_2 \rightarrow 16 CO + 18 H_2O$

(b)

ATOM	REACTANTS	PRODUCTS
C	16	16
H	36	36
O	34	34

(c) When oxygen is plentiful, 2 octane molecules react with 25 O_2 molecules (a combining ratio of 1:25/2). When oxygen is limited, 2 octane molecules react with only 17 O_2 molecules (a combining ratio of 1:17/2). Consequently, less oxygen is consumed when CO is the product.

Try It Yourself 6.3

(a) chemical amount of Fe (mol) = $0.018 \text{ g Fe} \times \dfrac{1 \text{ mol Fe}}{55.8 \text{ g Fe}} = 3.2 \times 10^{-4}$ mol Fe

(b) number of Fe atoms = $3.2 \times 10^{-4} \text{ mol Fe} \times \dfrac{6.022 \times 10^{23} \text{ atoms Fe}}{1 \text{ mol Fe}}$

$= 1.9 \times 10^{20}$ atoms Fe

Try It Yourself 6.4

$$\text{molar mass of } C_2H_5OH = 46.1 \text{ g/mol}$$

$$\text{volume of ethanol} = 1 \text{ mol} \times \frac{46.1 \text{ g}}{1 \text{ mol}} \times \frac{1 \text{cm}^3}{0.789 \text{ g}} = 58.4 \text{ cm}^3$$

Try It Yourself 6.5

As explained in Appendix B, a mass of 1200 mg has an ambiguous number of significant figures. This is the quantity given on the supplement tablet bottle, so we have to make a judgment call about how many significant figures we will use. If we assume that quality control is good, there may be some error in the last digit of the mass. Consequently, we assume that the mass is accurate to four significant figures. In other words, the mass is 1.200×10^3 mg or 1.200 g.

$$\text{molar mass of } CaCO_3 \text{ (g/mol)} = 100.1 \text{ g/mol}$$

$$\text{number of formula units} = 1.200 \text{ g } CaCO_3 \times \frac{1 \text{ mol } CaCO_3}{100.1 \text{ g } CaCO_3} \times \frac{6.022 \times 10^{23} \text{ } CaCO_3}{1 \text{ mol NaCl}} = 7.219 \times 10^{21} \text{ } CaCO_3$$

Practice Exercise 6.1

$$S_8 + 8 \, O_2 \rightarrow 8 \, SO_2$$

Practice Exercise 6.2

(a) $2 \, C_2H_2 \ + \ 5 \, O_2 \ \rightarrow \ 4 \, CO_2 \ + \ 2 \, H_2O$

(b)

ATOM	REACTANTS	PRODUCTS
C	4	4
H	4	4
O	10	10

Practice Exercise 6.3

(a) 1.008 g/mol

(b) 14.00 g/mol

(c) 39.10 g /mol

(d) 55.85 g/mol

(e) 238.0 g/mol

Practice Exercise 6.4

$$\text{mass of propane (g)} = 20 \ \cancel{lb} \times \frac{1 \ \cancel{kg}}{2.2 \ \cancel{lb}} \times \frac{1000 \ g}{1 \ \cancel{kg}} = 9091 \ g$$

$$\text{molar mass of propane} = 44.1 \ g/mol$$

The balanced chemical equation for the combustion of propane is:

$$C_3H_8 + 5 \ O_2 \rightarrow 3 \ CO_2 + 4 \ H_2O$$

We can perform the calculation in a single step.

$$\text{mass of } CO_2 \text{ (g)} = 9091 \ \cancel{g \ C_3H_8} \times \frac{1 \ \cancel{mol \ C_3H_8}}{44.1 \ \cancel{g \ C_3H_8}} \times \frac{3 \ \cancel{mol \ CO_2}}{1 \ \cancel{mol \ C_3H_8}} \times \frac{44.0 \ g \ CO_2}{1 \ \cancel{mol \ CO_2}} = 27{,}211 \ g \ CO_2$$

Convert the mass back to pounds.

$$\text{mass of propane (lb)} = 27{,}211 \ \cancel{g} \times \frac{1 \ \cancel{kg}}{1000 \ \cancel{g}} \times \frac{2.2 \ lb}{1 \ \cancel{kg}} = 60 \ lb$$

Practice Exercise 6.5

(a) $2H_2 + O_2 \rightarrow 2H_2O$

(b) $2 \ H{-}H + O{=}O \longrightarrow 2 \ \overset{O}{\underset{H \quad H}{\diagup \ \diagdown}}$

(c) energy input to break bonds in H_2 (kJ) $= 2 \ \cancel{mol} \times \left(\dfrac{+436 \ kJ}{1 \ \cancel{mol}} \right) = +872 \ kJ$

   energy input to break bonds in O_2 (kJ) $= 1 \ \cancel{mol} \times \left(\dfrac{+498 \ kJ}{1 \ \cancel{mol}} \right) = +498 \ kJ$

   total energy input $= 872 \ kJ + 498 \ kJ = +1370 \ kJ$

(d) energy output when forming bonds in H_2O (kJ) $= 4 \ \cancel{mol} \times \left(\dfrac{-467 \ kJ}{1 \ \cancel{mol}} \right)$

   $= -1868 \ kJ$

(e) energy change for reaction $= 1370 \ kJ - 1868 \ kJ = -498 \ kJ$

The overall energy change for the chemical reaction is negative ($-498 \ kJ$), which indicates that the reaction releases energy: It is exothermic.

Practice Exercise 6.6

$$6\,CO_2 + 6\,H_2O \xrightarrow{\text{light energy}} C_6H_{12}O_6 + 6\,O_2$$

The gas in the photograph is oxygen (O_2).

CHAPTER 7

Try It Yourself 7.1

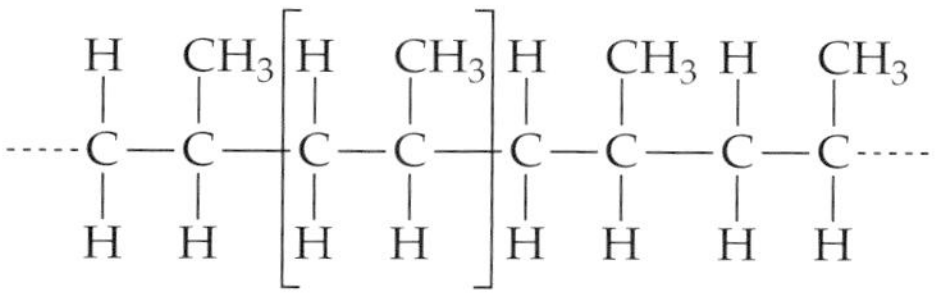

Try It Yourself 7.2

(a)

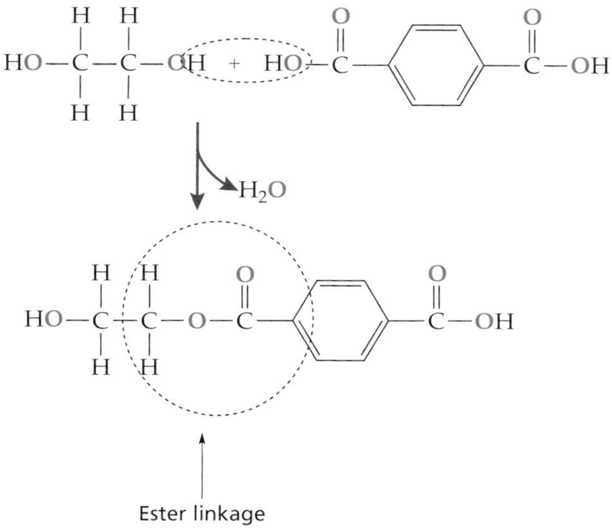

Ester linkage

(b) An ester linkage is produced by joining the monomers. This linkage is circled in part (a).

(c)

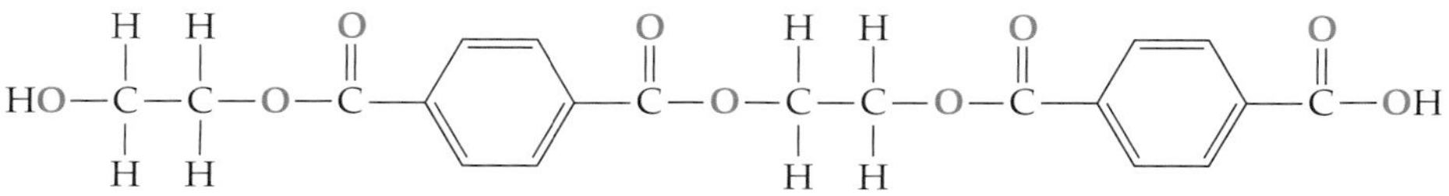

Try It Yourself 7.3

It is not possible to use acetic acid as a monomer to make a condensation polymer called polyacetic acid. To make a polymer, the monomer must have two reactive functional groups that can participate in a condensation reaction. By contrast, acetic acid has only one such group, a single carboxylic acid. The methyl group ($-CH_3$) cannot participate in a condensation reaction.

Practice Exercise 7.1

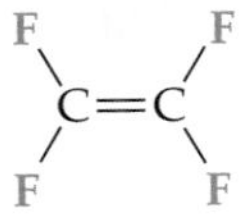

$$
\begin{array}{c}
\underset{F}{\overset{F}{\diagdown}} C = C \underset{F}{\overset{F}{\diagup}}
\end{array}
$$

Practice Exercise 7.2

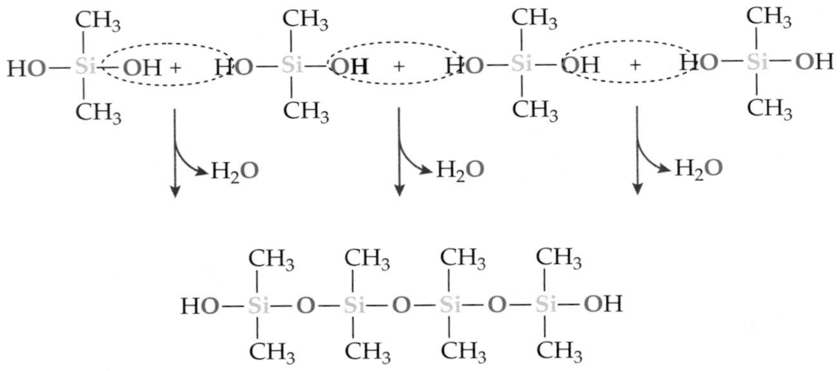

Practice Exercise 7.3

$$
\text{glycemic load} = \frac{(\text{glycemic index}) \times (\text{grams of carbohydrates})}{100}
$$

$$
= \frac{93 \times 26}{100}
$$

$$
= 24 \ (\text{rounded to the nearest whole number})
$$

CHAPTER 8

Try It Yourself 8.1

(a) $\Delta EN = 4.0 - 2.1 = 1.9$ (polar)

(b) $\Delta EN = 3.0 - 2.1 = 0.9$ (polar)

(c) $\Delta EN = 2.8 - 2.1 = 0.7$ (polar)

As the atomic number of the halogen increases, the polarity of the bond decreases.

Try It Yourself 8.2

(a) For hydrogen and iodine, $\Delta EN = 2.5 - 2.1 = 0.4$. The bond is nonpolar covalent.

(b) For lithium and fluorine, $\Delta EN = 4.0 - 1.0 = 3.0$. The bond is ionic.

(c) For lithium and iodine, $\Delta EN = 4.0 - 2.1 = 1.9$. The bond is ionic because lithium is a metal.

Try It Yourself 8.3

(a) Use the equation

$$Q = m \cdot c \cdot \Delta T$$

For the copper pan,

$Q = 10,000 \, \text{J}$ $m = 250 \, \text{g}$ $c = 0.386 \, \text{J/g}^\circ\text{C}$ (from Table 8.2) $\Delta T = ?$

Because ΔT is unknown, we rearrange the equation to isolate ΔT on the left side.

$$\Delta T = \frac{Q}{mc} = \frac{10,000 \, \cancel{J}}{250 \, \cancel{g} \times 0.386 \, \cancel{J}/\cancel{g}^\circ\text{C}} = 104^\circ\text{C}$$

(b) For water, we perform a similar calculation with $c = 4.184 \, \text{J/g}^\circ\text{C}$ (from Table 8.2).

$$\Delta T = \frac{Q}{mc} = \frac{10,000 \, \cancel{J}}{250 \, \cancel{g} \times 4.184 \, \cancel{J}/\cancel{g}^\circ\text{C}} = 9.56^\circ\text{C}$$

(c) For the same input of heat energy (10,000 J), the copper pan increases in temperature by 104°C, whereas the water increases in temperature by only 9.56°C. This difference is a consequence of their respective specific heat capacities. As a result, the pan will become hotter to the touch much faster than the water it contains.

Practice Exercise 8.1

(a) The C—O bond is polar ($\Delta EN = 1.0$), and the O—H bond is polar ($\Delta EN = 1.4$).

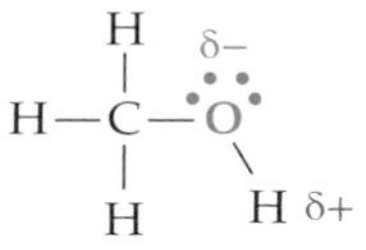

(b) The C—H bonds in methanol do not form hydrogen bonds because these bonds are nonpolar. Consequently, the H atoms do not acquire the significant partial charges required to form a hydrogen bond.

(c)

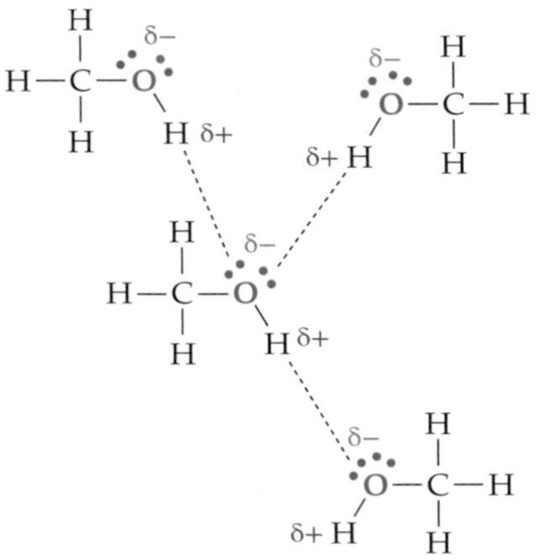

Each methanol molecule can form a maximum of three hydrogen bonds. The molecule has one H atom in a polar covalent bond that can serve as a hydrogen bond donor, plus an O atom that can serve as a hydrogen bond acceptor for two such bonds.

(d) We predict that the boiling point of methanol will be lower than that of liquid water. Methanol can form a maximum of three hydrogen bonds, whereas H_2O can form four hydrogen bonds. As a result, the hydrogen bonding network in methanol is less than that of water. Less heat energy is required to break all the hydrogen bonds in liquid methanol, so the boiling point will be lower.

Practice Exercise 8.2

Use the equation

$$Q = m \cdot c \cdot \Delta T$$

$$Q = 480\,J \quad m = 20.0\,g \quad c = ? \quad \Delta T = 40.0°C$$

Because c is unknown, we rearrange the equation to isolate c on the left side:

$$c = \frac{Q}{m\,\Delta T} = \frac{480\,J}{20.0\,g \times 40.0\,°C} = 0.600\,J/g°C$$

CHAPTER 9

Try It Yourself 9.1

Vitamin D3 is fat soluble because it is composed mostly of nonpolar hydrocarbon regions. Vitamin B6 is water soluble because it contains numerous polar functional groups.

Try It Yourself 9.2

First, write the chemical equations for the ionization of each compound when dissolved in water.

$$KCl(s) \xrightarrow{H_2O} K^+(aq) + Cl^-(aq)$$

$$AgNO_3(s) \xrightarrow{H_2O} Ag^+(aq) + NO_3^-(aq)$$

According to the solubility rules, chlorides are soluble except compounds formed with mercury, silver, or lead. Therefore, we predict that mixing the solutions will form a precipitate of silver chloride, AgCl.

Practice Exercise 9.1

(a)

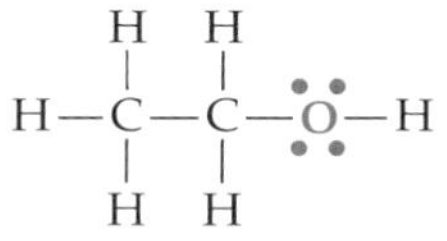

Ethanol is soluble in water because it contains a polar —OH group, which can form hydrogen bonds with H_2O molecules.

(b)

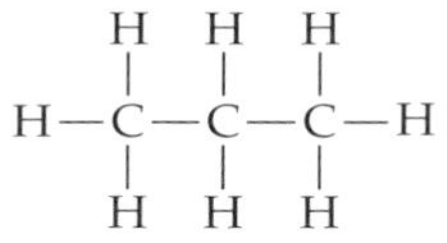

Propane is not soluble in water because it is a hydrocarbon molecule with nonpolar C—C and C—H bonds. None of the atoms in propane can form a hydrogen bond with an H_2O molecule.

(c)

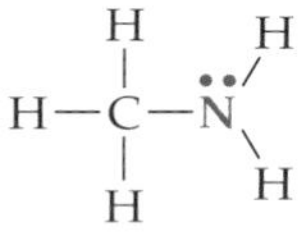

Methylamine is soluble in water because it contains a polar —NH$_2$ group, which can form hydrogen bonds with H_2O molecules.

Practice Exercise 9.2

(a) The molecules are ordered from most to least soluble in water (left to right).

   Most soluble   C   A   D   B   *Least soluble*

(b) The molecules are ordered from most to least soluble in hexane (left to right).

   Most soluble   B   D   A   C   *Least soluble*

Note that the sequence in part (b) is the reverse of the sequence in part (a).

Practice Exercise 9.3

For vitamins C and B6, the relative solubility in water is determined by the number of polar regions within the molecule. For example, vitamin C has a larger number of polar —OH groups compared to vitamin B6. These —OH groups can form hydrogen bonds with H_2O molecules, which contributes to the greater water solubility of vitamin C. For vitamins A and E, the relative solubility in fats is determined by the regions of the molecule that are nonpolar. Vitamin E has a greater number of carbon atoms in nonpolar hydrocarbon regions of the molecule, including a long hydrocarbon tail containing 16 carbon atoms. As a result, vitamin E is more soluble in fat compared to vitamin A.

Practice Exercise 9.4

Like detergents, stearic acid has a polar head group and a long hydrocarbon chain. A collection of these molecules will self-assemble to form micelles. The center of the micelle provides a hydrophobic region for dissolving nonpolar grease, which is solubilized and washed away.

Practice Exercise 9.5

The majority of the cholesterol molecule is composed of nonpolar hydrogen regions, so this molecule is soluble within the hydrophobic interior of the lipid bilayer. Cholesterol contains a single polar —OH group, which will interact with the polar head groups of the phospholipids.

Practice Exercise 9.6

First, write the chemical equations for the ionization of each compound when dissolved in water.

$$BaCl_2(s) \xrightarrow{H_2O} Ba^{2+}(aq) + 2\,Cl^-(aq)$$

$$(NH_4)_2SO_4(s) \xrightarrow{H_2O} 2\,NH_4^+(aq) + SO_4^{2-}(aq)$$

According to the solubility rules, sulfates are soluble except for compounds that are formed with certain cations, which include barium. We predict that mixing the solutions will form a precipitate of barium sulfate, $BaSO_4$.

CHAPTER 10

Try It Yourself 10.1

$$3.0\% = \frac{\text{volume of hydrogen peroxide (mL)}}{946 \text{ mL}} \times 100\%$$

$$\text{volume of hydrogen peroxide (mL)} = \frac{3.0\% \times 946 \text{ mL}}{100\%} = 28 \text{ mL}$$

Try It Yourself 10.2

First, convert the volume to milliliters.

$$0.500 \text{ L} \times \frac{1000 \text{ mL}}{1 \text{ L}} = 500 \text{ mL}$$

We can now use the equation for % (m/v).

$$5.0\% \text{ (m/v)} = \frac{\text{mass of glucose (g)}}{500 \text{ mL}} \times 100\%$$

$$\text{mass of glucose (g)} = \frac{5.0\% \times 500 \text{ mL}}{100\%} = 25 \text{ g}$$

Try It Yourself 10.3

(a) From the definitions of ppm and ppb, 1 ppm = 1000 ppb. We can now convert the arsenic measurement from ppb to ppm.

$$100 \text{ ppb} \times \frac{1 \text{ ppm}}{1000 \text{ ppb}} = 0.1 \text{ ppm}$$

(b) According to Table 10.1, the MCL for arsenic is 0.01 ppm. Therefore, the concentration of arsenic in the water sample is 10 times higher than the MCL.

Try It Yourself 10.4

First, calculate the molar mass of glucose $(C_6H_{12}O_6)$.

$$
\begin{aligned}
\text{molar mass of } C_6H_{12}O_6 \text{ (g/mol)} = {} & (6 \times \text{molar mass of C}) \\
& + (12 \times \text{molar mass of H}) \\
& + (6 \times \text{molar mass of O}) \\
= {} & (6 \times 12.0 \text{ g/mol}) + (12 \times 1.0 \text{ g/mol}) \\
& + (6 \times 16.0 \text{ g/mol}) \\
= {} & 180.0 \text{ g/mol}
\end{aligned}
$$

Next, calculate the chemical amount of glucose in units of moles.

$$\text{chemical amount of } C_6H_{12}O_6 \text{ (mol)} = \frac{9.0 \text{ g}}{180.0 \text{ g/mol}} = 0.050 \text{ mol}$$

Before calculating the concentration, we first convert the volume from milliliters to liters.

$$\text{volume (L)} = 200 \text{ mL} \times \frac{1 \text{ L}}{1000 \text{ mL}} = 0.200 \text{ L}$$

Finally, we calculate the molar concentration of the solution.

$$\text{molarity (M)} = \frac{0.050 \text{ mol}}{0.200 \text{ L}} = 0.25 \text{ M}$$

Try It Yourself 10.5

(a) First, convert the volume from milliliters to liters.

$$\text{volume (L)} = 50.0 \text{ mL} \times \frac{1 \text{ L}}{1000 \text{ mL}} = 0.0500 \text{ L}$$

We now calculate the mass of F^- ions in the solution.

$$\text{mass of F}^- \text{ (g)} = 0.0500 \text{ L} \times \frac{1.00 \text{ mg F}^-}{1 \text{ L}} \times \frac{10^{-3} \text{ g}}{1 \text{ mg}} = 5.00 \times 10^{-5} \text{ g F}^-$$

According to the periodic table, one F atom has a mass of 19.0 amu. To a good approximation, this is the same as the mass of an F^- ion. We can now calculate the number of F^- ions in the sample.

$$\text{number of F}^- \text{ ions} = 5.00 \times 10^{-5} \text{ g} \times \frac{1 \text{ amu}}{1.66 \times 10^{-24} \text{ g}} \times \frac{1 \text{ F}^- \text{ ion}}{19.0 \text{ amu}}$$

$$= 1.59 \times 10^{18} \text{ F}^- \text{ ions}$$

We can perform a similar calculation for the Pb^{2+} ions. From the periodic table, one Pb atom has a mass of 207.2 amu, and we can use this mass for the Pb^{2+} ion. The other quantities in the calculation remain the same.

$$\text{number of Pb}^{2+} \text{ ions} = 5.00 \times 10^{-5} \text{ g} \times \frac{1 \text{ amu}}{1.66 \times 10^{-24} \text{ g}} \times \frac{1 \text{ Pb}^{2+} \text{ ion}}{207.2 \text{ amu}}$$

$$= 1.45 \times 10^{17} \text{ Pb}^{2+} \text{ ions}$$

The numbers of F^- and Pb^{2+} ions are not the same. We conclude that two solutions with the same volume and the same ppm concentration contain different numbers of F^- and Pb^{2+} ions. There are approximately 10 times more F^- ions than there are Pb^{2+} ions, which arises from the fact that the mass of an F^- ion is approximately 10 times smaller than the mass of a Pb^{2+} ion.

(b) We begin with the F^- ions. By definition, 1.00 mM = 1.00 mmol/1 L = 10^{-3} mol/1 L.

$$\text{number of F}^- \text{ ions} = 0.0500 \text{ L} \times \frac{10^{-3} \text{ mol F}^-}{1 \text{ L}} \times \frac{6.02 \times 10^{23} \text{ F}^- \text{ ions}}{1 \text{ mol F}^-}$$

$$= 3.01 \times 10^{19} \text{ F}^- \text{ ions}$$

We can perform a similar calculation for the Pb^{2+} ions.

$$\text{number of } Pb^{2+} \text{ ions} = 0.0500 \text{ } L \times \frac{10^{-3} \text{ mol } Pb^{2+}}{1 \text{ } L} \times \frac{6.02 \times 10^{23} \text{ } Pb^{2+} \text{ ions}}{1 \text{ mol } Pb^{2+}}$$

$$= 3.01 \times 10^{19} \text{ } Pb^{2+} \text{ ions}$$

The numbers of F^- and Pb^{2+} ions are the same, even though their masses are different. In other words, two solutions with the same volume and the same 1.00 mM concentration contain the same number of F^- and Pb^{2+} ions.

Try It Yourself 10.6

Use the equation $C_1V_1 = C_2V_2$. We can use volume in milliliters, provided that we are consistent on both sides of the equation.

$$3.00 \text{ M} \times 100 \text{ mL} = C_2 \times 500 \text{ mL}$$

$$C_2 = \frac{3.00 \text{ M} \times 100 \text{ mL}}{500 \text{ mL}} = 0.600 \text{ M}$$

Try It Yourself 10.7

The concentration of Na^+ ions in blood serum is 135 to 145 mM. For our calculation, we will take the midpoint concentration of 140 mM in a drink volume of 240 mL.

$$\text{chemical amount of } Na^+ \text{ (mol)} = \frac{140 \times 10^{-3} \text{ mol } Na^+}{1 \text{ } L} \times 0.240 \text{ } L$$

$$= 3.36 \times 10^{-2} \text{ mol } Na^+$$

$$\text{mass of } Na^+ \text{ (g)} = 3.36 \times 10^{-2} \text{ mol } Na^+ \times \frac{23.0 \text{ g } Na^+}{1 \text{ mol } Na^+} = 0.773 \text{ g}$$

This mass of sodium required for an isotonic solution is 773 mg, which is approximately 7 times the amount that is in a typical sports drink. Companies do not put this much sodium in a sports drink because it would taste too salty and therefore be unpleasant to drink.

Practice Exercise 10.1

$$1 \text{ mm} = 0.1 \text{ cm, so } 1 \text{ mm}^3 = 0.1 \text{ cm} \times 0.1 \text{ cm} \times 0.1 \text{ cm} = 10^{-3} \text{ cm}^3$$

By definition, $1 \text{ L} = 10^3 \text{ cm}^3$. Therefore,

$$1 \text{ mm}^3 = 10^{-3} \text{ cm}^3 \times \frac{1 \text{ L}}{10^3 \text{ cm}^3} = 10^{-3} \times 10^{-3} \text{ L} = 10^{-6} \text{ L}$$

$10^{-6} = 1/10^6$ is the same as one millionth.

Practice Exercise 10.2

$$1 \text{ ppb} = \frac{1 \text{ g solute}}{10^9 \text{ g solution}}$$

By definition, $1 \text{ g} = 10^6 \text{ mg}$. Therefore, we can replace 1 g in the equation.

$$1 \text{ ppb} = \frac{\cancel{10^6} \text{ μg solute}}{\cancel{10^6} \times 10^3 \text{ g solution}} = \frac{1 \text{ μg solute}}{10^3 \text{ g solution}} = \frac{1 \text{ μg solute}}{1 \text{ L solution}}$$

This equation applies to dilute solutions for which the density is very close to 1.00 g/cm^3.

Practice Exercise 10.3

The salinity of seawater can be expressed as $35 \text{ ppt} = \dfrac{35 \text{ g salt}}{1000 \text{ g seawater}}$. We introduce kilograms by noting that $1000 \text{ g} = 1 \text{ kg}$. Therefore,

$$\text{mass of salt} = \frac{35 \text{ g salt}}{1 \cancel{\text{ kg seawater}}} \times 10^3 \cancel{\text{ kg seawater}} = 35 \times 10^3 \text{ g salt} = 35 \text{ kg salt}$$

Practice Exercise 10.4

The estimated half-life of the drug is $T_{1/2} = 1\frac{1}{2}$ hours.

CHAPTER 11

Try It Yourself 11.1

Like HNO_3, $HClO_4$ has a single hydrogen atom that can function as an acidic hydrogen. Therefore, we predict that $HClO_4$ is an acid that can donate a proton to H_2O.

$$HClO_4(aq) + H_2O(l) \longrightarrow H_3O^+(aq) + ClO_4^-(aq)$$

Try It Yourself 11.2

(a) Unlike ammonia (NH_3), the ammonium ion (NH_4^+) does not have any lone pairs that enable it to accept an H^+ ion. However, the ammonium ion can donate an H^+ ion to an H_2O molecule, so it functions as a Brønsted-Lowry acid.

(b)

(c) The H_2O molecule acts as a proton acceptor, so it is a Brønsted-Lowry base.

Try It Yourself 11.3

$$2\,H_3PO_4(aq) + 3\,Mg(OH)_2(aq) \rightarrow Mg_3(PO_4)_2(aq) + 6\,H_2O(l)$$

Try It Yourself 11.4

(a) The neutralization reaction is

$$\underset{\text{Acid}}{HBr(aq)} + \underset{\text{Base}}{LiOH(aq)} \xrightarrow{\text{Neutralization}} \underset{\text{Salt}}{LiBr(aq)} + \underset{\text{Water}}{H_2O(l)}$$

(b) To identify the spectator ions, we write the chemical equation showing all the dissolved ions.

$$\underbrace{H^+(aq) + Br^-(aq)}_{HBr(aq)} + \underbrace{Li^+(aq) + OH^-(aq)}_{LiOH(aq)} \xrightarrow{\text{Neutralization}} \underbrace{Li^+(aq) + Br^-(aq)}_{LiBr(aq)} + H_2O(l)$$

The spectator ions are $Li^+(aq)$ and $Br^-(aq)$.

Try It Yourself 11.5

$$HCO_3^-(aq) + H_2O(l) \rightleftharpoons H_3O^+(aq) + CO_3^{2-}(aq)$$

The conjugate base is the carbonate ion (CO_3^{2-}).

Try It Yourself 11.6

We use the equation $K_w = [H_3O^+][OH^-] = 10^{-14}$. Because we are given $[OH^-] = 10^{-8}$ M, we need to solve for $[H_3O^+]$.

$$[H_3O^+] = \frac{K_w}{[OH^-]} = \frac{10^{-14}}{10^{-8}} = 10^{-14+8} = 10^{-6}\ M$$

Try It Yourself 11.7

(a) If $[H_3O^+] = 10^{-12}$ M, the solution is basic because $[H_3O^+] < 10^{-7}$ M.

(b) If $[H_3O^+] = 10^{-7}$ M, the solution is neutral because $[H_3O^+] = [OH^-] = 10^{-7}$ M.

(c) If $[OH^-] = 10^{-10}$ M, we first need to determine the value of $[H_3O^+]$. Using the equation $K_w = [H_3O^+][OH^-] = 10^{-14}$,

$$[H_3O^+] = \frac{K_w}{[OH^-]} = \frac{10^{-14}}{10^{-10}} = 10^{-14+10} = 10^{-4}\ M$$

This solution is acidic because $[H_3O^+] > 10^{-7}$ M.

Try It Yourself 11.8

(a) First, express the concentration in scientific notation: $0.010M = 1.0 \times 10^{-2}$ M. We then use the equation $pH = -\log_{10}[H_3O^+] = -\log_{10}(1.0 \times 10^{-2}) = 2.00$.

(b) The new concentration of H_3O^+ ions is $[H_3O^+] = 5.0 \times 10^{-2}$ M, which is 5 times the original amount. The new pH is $pH = -\log_{10}[H_3O^+] = -\log_{10}(5.0 \times 10^{-2}) = 1.30$. The addition of extra acid has decreased the pH value.

Try It Yourself 11.9

(a) A pH of 7.40 is a little above 7.00, which means that blood is slightly basic.

(b) If the pH of blood were exactly 7.00, then $[H_3O^+] = 10^{-7}$ M. This is our prediction for the approximate concentration of H_3O^+ ions in blood. However, the actual concentration of H_3O^+ ions will be slightly lower than our prediction because the pH is slightly higher than 7.0

(c) Calculate the concentration of H_3O^+ ions in blood.

$$[H_3O^+] = 10^{-pH} = 10^{-7.40} = 4.0 \times 10^{-8} \text{ M}$$

(d) Our prediction of 10^{-7} M is close to the calculation concentration of H_3O^+ ions. As expected, the calculated value is slightly less than the prediction.

Practice Exercise 11.1

$$HBr(aq) + H_2O(l) \longrightarrow H_3O^+(aq) + Br^-(aq)$$

Practice Exercise 11.2

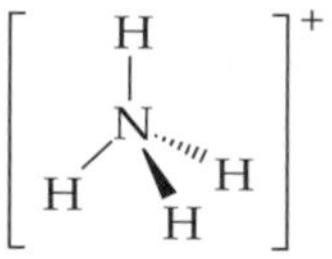

Practice Exercise 11.3

(a) $NaHCO_3(aq) + HCl(aq) \xrightarrow{\text{Neutralization}} NaCl(aq) + H_2O(l) + CO_2(g)$

(b) The gas produced is carbon dioxide.

Practice Exercise 11.4

(a) $HCl(aq) + NH_3(aq) \xrightarrow{\text{Neutralization}} NH_4Cl(g)$

(b) The lone pair of electrons on the nitrogen atom in ammonia accepts an H^+ from HCl to form an ammonium ion (NH_4^+)

Practice Exercise 11.5

$$H_3PO_4(aq) + H_2O(l) \longrightarrow H_3O^+(aq) + H_2PO_4^-(aq)$$

Practice Exercise 11.6

Photograph A depicts zinc in acetic acid, and Photograph B depicts zinc in hydrochloric acid. The extent of the reaction between zinc and the acid, revealed by the bubbles of hydrogen gas, depends the number of H_3O^+ ions in the acidic solution. Hydrochloric acid is a strong acid, which means that it is 100% ionized in aqueous solution (to a good approximation). The concentration of H_3O^+ ions in solution is very close to 1.0 M, which makes the reaction proceed more vigorously. Acetic acid is a weak acid that ionizes only a small amount. Consequently, the concentration of H_3O^+ ions in the acetic acid solution is much less than 1.0 M, which reduces the extent of the reaction.

Practice Exercise 11.7

$$[HCl] = [H_3O^+] = 10.0\, M$$
$$pH = -log[H_3O^+] = -log[10.0] = -1.000$$

This pH is unusual because it is a negative number.

Practice Exercise 11.8

$$[H_3O^+] = 10^{-pH} = 10^{-2.0} = 0.01\, M$$

Practice Exercise 11.9

Histamine acts as a base because the nitrogen atoms each contain a lone pair, which can accept an H^+ ion.

Practice Exercise 11.10

Solution A can form a buffer because the mixture contains both a weak acid, HF, and its conjugate base, F^- (which is available from the presence of NaF). Solution B cannot form a buffer because the solution does not contain the conjugate base of HNO_2.

CHAPTER 12

Try It Yourself 12.1

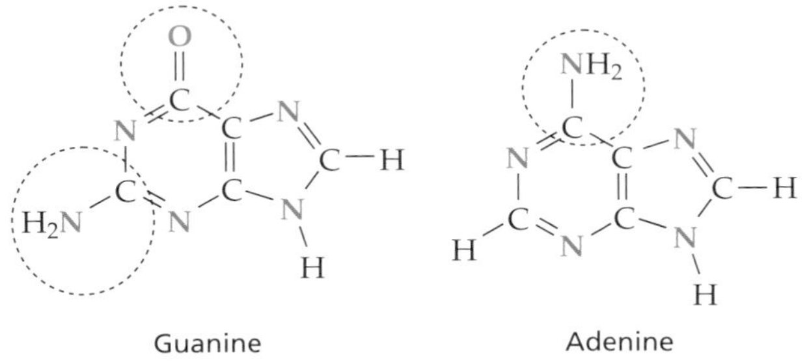

Guanine   Adenine

Try It Yourself 12.2

The sequence of bases on the complementary strand are provided below, including the 5'- and 3'-ends.

$$3'-A\,T\,C\,C\,G\,A\,A\,T\,G\,C\,C\,T-5'$$

Try It Yourself 12.3

The base sequence of the original template strand is given below, including the 5'- and 3'-ends.

$$5' - A\,G\,T\,T\,T\,G\,G\,C\,A\,C\,C\,G - 3'$$

Practice Exercise 12.1

Uric acid contains two fused rings and resembles a purine heterocycle. Therefore, we deduce that purine bases are the source for making uric acid.

Practice Exercise 12.2

The answers to parts (a), (b) and (c) are provided in the following figure.

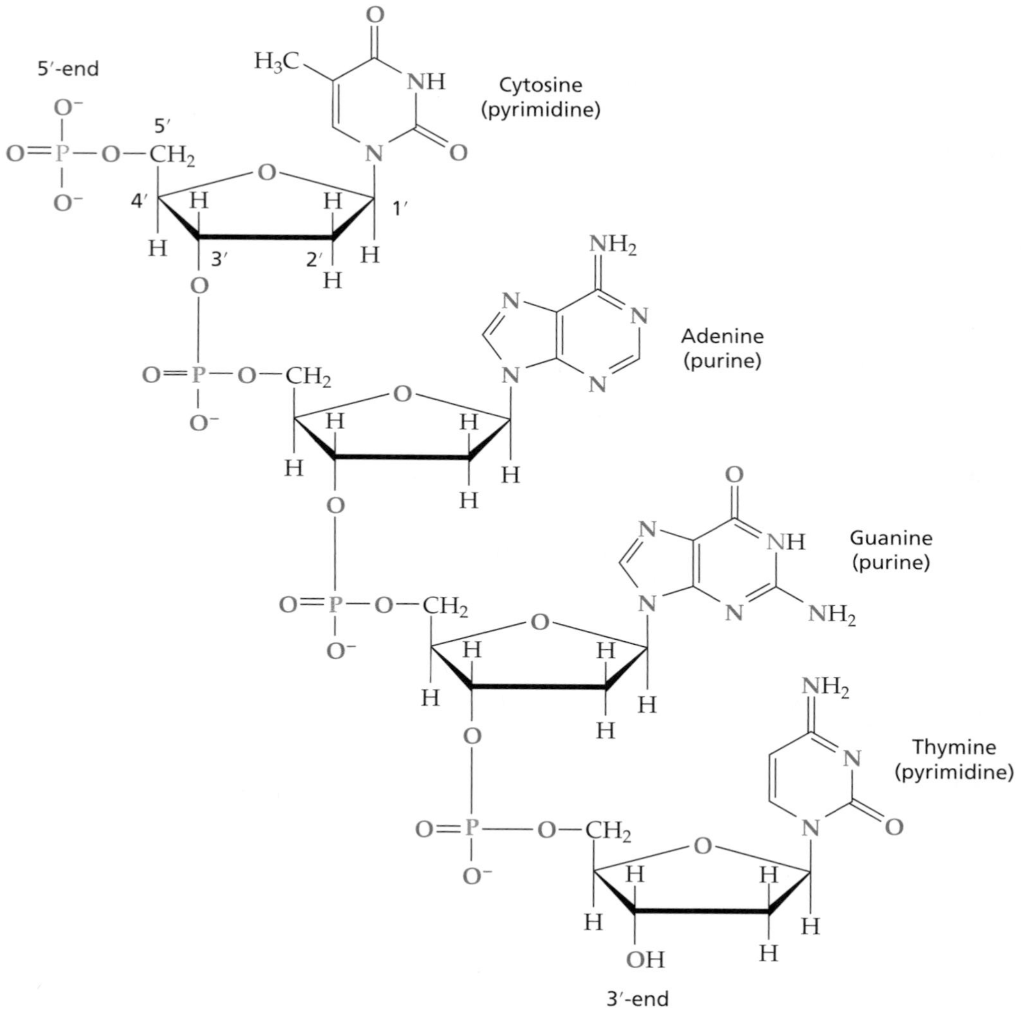

(d) The base sequence is 5' - C A G T - 3'.

Practice Exercise 12.3

(a) Based on these percentages of bases, we can deduce the following two rules:

- The percentage of A bases is (approximately) equal to the percentage of T bases.
- The percentage of G bases is (approximately) equal to the percentage of C bases.

(b) In the structure of DNA, the four bases form specific base pairs: A pairs with T, and G pairs with C. This base pairing within DNA explains Chargaff's rules. Every time an A base appears on one DNA strand, a T base must be present on the complementary DNA strand. Consequently, the percentages of A and T bases within the DNA must be equal. The same argument also applies to the G and C bases.

CHAPTER 13

Try It Yourself 13.1

The mRNA sequence is given below, including the 5′- and 3′-ends.

$$5′\text{-}C\,G\,U\,U\,U\,A\,C\,G\,U\,C\,G\,G\,G\,A\,U\,C\,G\,G\text{-}3′$$

Try It Yourself 13.2

(a) Based on the molecular structure, lactic acid is a chiral molecule because four different chemical groups are attached to the central carbon atom.

(b) A three-dimensional structure of lactic acid and its mirror image are shown below.

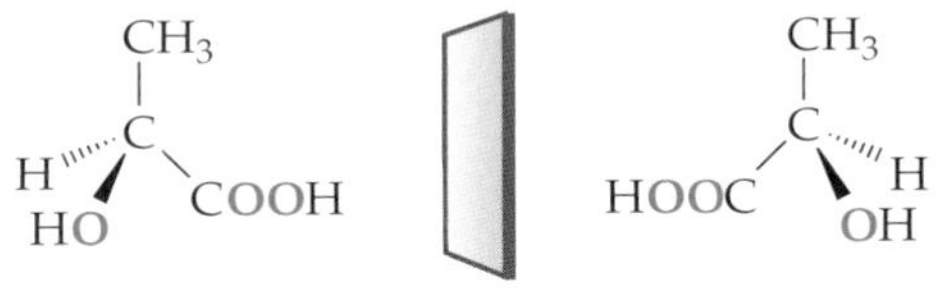

These two mirror-image molecules cannot be superimposed on each other. For example, if we overlap the —CH_3 and —COOH groups for both molecules, then the —OH and —H groups do not align.

(c) In part (a), we deduced that lactic acid is a chiral molecule. This observation is consistent with the answer to part (b), because the two mirror-image molecules cannot be superimposed.

Try It Yourself 13.3

(a) CAG Glutamine   UAC Tyrosine   AAG Lysine

(b) UUA, UUG, GUU, CUC, CUA, CUG
There are six codons that correspond to leucine.

Try It Yourself 13.4

(a) DNA   3′-AGG   CCT   CAG   TCC   TGG   CTT-5′

(b) mRNA   5′-UCC   GGA   GUC   AGG   ACC   GAA-3′

amino   serine   glycine   valine   arginine threonine   glutamic
acids   acid

Try It Yourself 13.5

anticodon sequence 3′ - G G G - 5′

codon sequence 5′ - C C C - 3′

amino acid   proline

Practice Exercise 13.1

The amino acid is tryptophan.

Practice Exercise 13.2

The other chiral isomer of limonene is shown to the right. The C—C bond attached to the chiral carbon is now oriented downward with respect to the plane of the paper (as indicated by the dashed wedge).

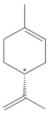

Practice Exercise 13.3

Parts (a) and (b) are answered in the following figure.

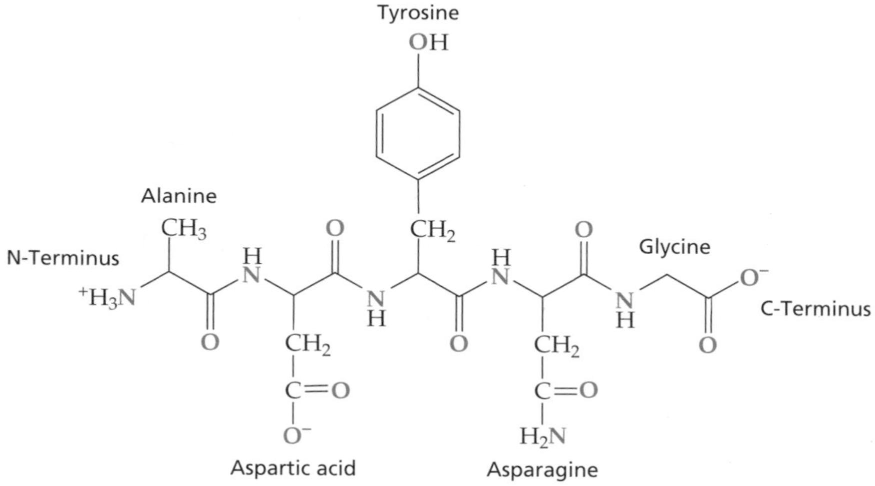

Tyrosine
OH

Alanine
CH₃   CH₂   Glycine
N-Terminus   O⁻
⁺H₃N   C-Terminus

CH₂   CH₂
C=O   C=O
O⁻   H₂N
Aspartic acid   Asparagine

(c) The amino acid sequence is
N—alanine—aspartic acid—tyrosine—asparagine-glycine—C

CHAPTER 14

Try It Yourself 14.1

First, calculate the flight time in minutes.

$$\text{flight time (minutes)} = \left(10 \ \cancel{\text{hours}} \times \frac{60 \text{ minutes}}{1 \ \cancel{\text{hour}}} \right) + 15 \text{ minutes}$$

$$= 615 \text{ minutes}$$

(a) If the flight speed were accelerated by a hundredfold,

$$\text{flight time (minutes)} = \frac{615 \text{ minutes}}{10^2} = 6.15 \text{ minutes}$$

(b) If the flight speed were accelerated by a thousandfold,

$$\text{flight time (minutes)} = \frac{615 \text{ minutes}}{10^3} = 0.615 \text{ minutes}$$

This time is equal to 36.9 seconds.

(c) If the flight speed were accelerated by a millionfold,

$$\text{flight time (minutes)} = \frac{615 \text{ minutes}}{10^6} = 6.15 \times 10^{-4} \text{ minutes}$$

This time is equal to 0.0369 seconds, or 36.9 milliseconds (ms).

Practice Exercise 14.1

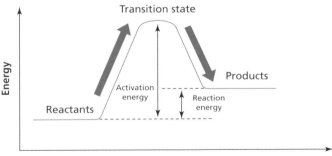

Practice Exercise 14.2

(a) The two-dimensional structure of $C(CH_3)_3Br$ is shown below.

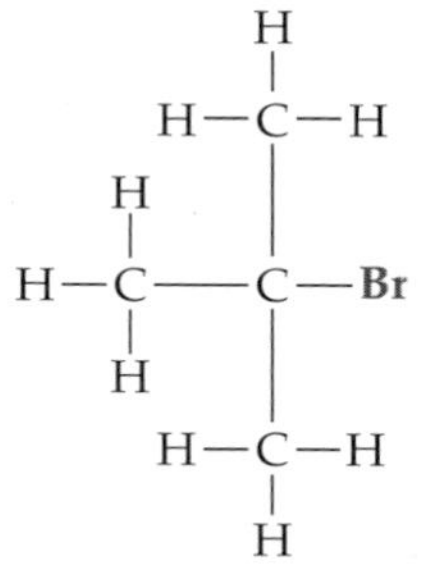

$$
\begin{array}{c}
H \\
| \\
H\!-\!C\!-\!H \\
\end{array}
$$

(structure)

$$
\begin{array}{ccc}
 & H & \\
 & | & H \\
 & H\!-\!C\!-\!H & \\
H & | & \\
| & & \\
H\!-\!C\!-\!\!-\!\!-\!C\!-\!Br \\
| & | \\
H & \\
 & H\!-\!C\!-\!H \\
 & | \\
 & H
\end{array}
$$

(b) The presence of the three methyl groups ($-CH_3$) means that the region around the central C atoms is very crowded with atoms. For the reaction to occur, the OH^- ion must collide with the $C(CH_3)_3Br$ molecule with the correct orientation, which facilitates an interaction between the O and C atoms. The hydrogen atoms in the methyl groups prevent the OH^- ion from reaching the C atom. Without the appropriate collision between the reactants, the reaction will not proceed.

Practice Exercise 14.3

(a) Substrate 2 would bind most tightly to the active site because it can form the largest number of complementary chemical interactions with the amino acid sidechains. A drawing of the substrate in the active site is provided below. Note that the substrate has been flipped and rotated relative to the orientation given in the question.

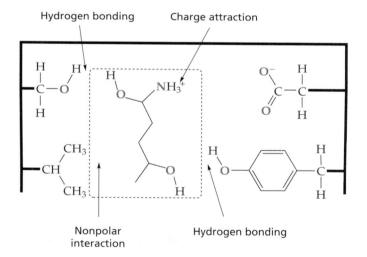

(b) Labels have been added to the drawing in part (a) to indicate the type of complementary chemical interaction.

CHAPTER 15

Try It Yourself 15.1

If the bacteria double every hour, there are 2 bacteria after 1 hour, 4 (2^2) bacteria after 2 hours, 8 (2^3) bacteria after 3 hours, and so on. In general, the number of bacterial cells after n hours is equal to 2^n. After 24 hours, the number of bacteria is $2^{24} = 16,777,216$ cells. Remarkably, a single original cell can generate over 16 million cells after 24 hours of bacterial growth. This calculation assumes that the bacteria have enough nutrients and space to continue replicating at the same rate.

Practice Exercise 15.1

Divide the length of the bacterial cell by the diameter of the virus.

$$\text{number of viruses} = \frac{2\ \mu m}{50\ nm} = \frac{2 \times 10^{-6}\ m}{50 \times 10^{-9}\ m} = 40\ \text{viruses}$$

Practice Exercise 15.2

(a) We hypothesize that exponential growth ends because the bacteria have depleted one or more of the important nutrients (e.g., sugars) that are required for cell division. The lack of these nutrients means that the bacteria are not able to grow at the same rate, so they enter a stationary phase in which their replication has slowed.

(b) Adding fresh growth medium in the middle of the stationary phase will supply the cells with more nutrients. Following an initial lag phase, the bacterial cells will resume their exponential growth. This growth will continue until the fresh nutrients have been depleted, at which point the cells will again enter a stationary phase.

Practice Exercise 15.3

Antibiotics that are more effective produce a larger inhibition zone around each disk. Based on this principle, (a) antibiotic 4 is the most effective, and (b) antibiotic 14 is the least effective.

Practice Exercise 15.4

Antibiotics that are more effective against bacteria have a lower IC_{50}. Based on this principle, the ranking of the antibiotics is E, F, D, C (from most effective to least effective).

APPENDIX A

Try It Yourself A.1

First, we convert the diameter of the carbon atom to scientific notation.

$$0.000000000170\ m = 1.70 \times 10^{-10}\ m$$

The most convenient unit to use is the *picometer* (nm). According to Table A.1, 1 pm = 10^{-12} m. Therefore

$$\text{diameter of carbon atom} = 1.70 \times 10^{-10}\ \cancel{m} \times \frac{1\ \text{pm}}{10^{-12}\ \cancel{m}}$$

$$= 1.70 \times 10^{-10} \times 10^{12}\ \text{pm}$$

$$= 1.70 \times 10^{2}\ \text{pm} = 170\ \text{pm}$$

Practice Exercise A.1

(a) $\dfrac{1\ \text{dollar}}{18\ \text{pesos}} = 1$

(b) $450\ \cancel{\text{pesos}} \times \dfrac{1\ \text{dollar}}{18\ \cancel{\text{pesos}}} = 25\ \text{dollars}$

Practice Exercise A.2

(a) $\text{volume of liquid} = 2.5\ \cancel{L} \times \dfrac{1\ \text{mL}}{10^{-3}\ \cancel{L}} = 2.5 \times 10^{3}\ \text{mL}$

$\text{number of samples} = 2.5 \times 10^{3}\ \cancel{\text{mL}} \times \dfrac{1\ \text{sample}}{5\ \cancel{\text{mL}}} = 500\ \text{samples}$

APPENDIX B

Try It Yourself B.1

(a)

$$1.362\ \text{cm} \times 8.487\ \text{cm} \times 6.89\ \text{cm} = 79.6\ \text{cm}^{3}$$

The answer contains three significant figures because the least certain input value contains this number of significant figures.

(b)

$$\frac{25.325\ \text{g} + 7.8\ \text{g}}{2.445\ \text{mL}} = \frac{33.1\ \text{g}}{2.445\ \text{mL}} = 13.5\ \text{g/mL}.$$

The result of the addition has one decimal place, because the second number in the addition contains one decimal place. The result of the division has three significant figures, because the least certain input has three significant figures.

Practice Exercise B.1

(a) 0.0176 mg contains three significant figures.

(b) 55.05 cm^3 contains four significant figures.

(c) 100 s is ambiguous; the number of significant figures is not obvious.

(d) 100.0 m contains four significant figures.

(e) 0.00037 g contains two significant figures.

APPENDIX C

Try It Yourself C.1

(a) $pH = -\log_{10}(0.10) = -(-1.00) = 1.00$

(b) $pH = -\log_{10}(2.0 \times 10^{-14}) = -(-13.70) = 13.70$

Try It Yourself C.2

(a) $\left[H_3O^+\right] = 10^{-4.2} = 6 \times 10^{-5} \ M$

(b) $\left[H_3O^+\right] = 10^{-6.7} = 2 \times 10^{-7} \ M$

Practice Exercise C.1

(a) $\log_{10}(6.2 \times 10^{-2}) = -1.21$

(b) $\log_{10}(5.9 \times 10^5) = 5.77$

(c) $\log_{10}(8.3 \times 10^{-9}) = -8.08$

(d) $\log_{10}(7.1 \times 10^4) = 4.85$

Practice Exercise C.2

(a) $10^{2.54} = 3.5 \times 10^2$

(b) $10^{-1.86} = 1.4 \times 10^{-2}$

(c) $10^{6.22} = 1.7 \times 10^6$

(d) $10^{-3.26} = 5.5 \times 10^{-4}$

Practice Exercise C.3

(a) $\left[H_3O^+\right] = 10^{-10.5} = 3 \times 10^{-11} \ M$

(b) $\left[H_3O^+\right] = 10^{-2.4} = 4 \times 10^{-3} \ M$

Practice Exercise C.4

(a) $\text{pH} = -\log_{10}\left(2.5 \times 10^{-3}\right) = -(-2.60) = 2.60$

(b) $\text{pH} = -\log_{10}\left(1.3 \times 10^{-8}\right) = -(-7.89) = 7.89$

Practice Exercise C.5

(a) According to a calculator, $10^{-7.439} = 3.63915036 \times 10^{-8}$. Because the exponent contains three decimal places, we must round the answer to three significant figures. The correct answer is 3.64×10^{-8}.

(b) According to a calculator, $\log_{10}\left(2.1 \times 10^{-5}\right) = -4.67778070527$. Because the original number contains two significant figures, we must round the answer to two decimal places. The correct answer is -4.68.

(c) According to a calculator, $10^{2.11} = 1.288249552 \times 10^{2}$. Because the exponent contains two decimal places, we must round the answer to two significant figures. The correct answer is 1.3×10^{2}.

(d) According to a calculator, $\log_{10}\left(9.43 \times 10^{-8}\right) = -7.02548830726$. Because the original number contains three significant figures, we must round the answer to three decimal places. The correct answer is -7.025.

Answers to Selected Learning Resources (End-of-Chapter Questions)

CHAPTER 1

Reviewing Knowledge

1. MRSA is the acronym for methicillin-resistant *Staphylococcus aureus*. MRSA infections can be dangerous because these bacteria are resistant to the therapeutic effect of methicillin and related antibiotics.

3. Ethanol is used as an antiseptic in hand sanitizers. Ethanol kills bacterial cells by (a) dissolving the lipids that constitute the cell membrane, thereby destroying the membrane's structure; and (b) causing proteins to unfold and lose their structure, producing a denatured protein that no longer functions.

5. Howard Florey and Ernst Chain developed new laboratory equipment and procedures to isolate the active penicillin from a complex mixture. This breakthrough enabled penicillin to be tested as a drug

7. Penicillin disrupts the formation of the cell wall that surrounds many bacterial cells. Penicillin interferes with the function of an enzyme that forms molecular linkages within the cell wall. As a result, the linkages are not formed, the cell wall is weakened, and the interior contents of the cell explode through a hole in the cell wall.

9. Developing new antibiotic drugs is challenging for several reasons. First, the process of developing a new drug and obtaining FDA approval is lengthy, complex, and expensive, with no guarantee of success. Second, bacteria that are resistant to existing antibiotics are often resistant to related variants of these drugs. Third, bacteria may quickly become resistant to the new antibiotic, sometimes within only a few years, which will render the drug ineffective. Fourth, antibiotics are only prescribed for a few days and are a less profitable financial investment than developing a new drug that a patient will take for many years.

Developing Skills

11. An antibiotic that is more effective at preventing bacterial growth has a larger clear zone surrounding the antibiotic disc. Using this criterion, we can rank the antibiotics in order of increasing effectiveness, from lowest to highest: B, D, A, C.

CHAPTER 2

Reviewing Knowledge

1. The RDA of a nutrient is the recommended dietary allowance, which is the daily amount that is required to maintain good health.

3. Iron deficiency inhibits hemoglobin's ability to transport oxygen and lowers the body's production of red blood cells. Symptoms include pale skin, fatigue, headaches, and shortness of breath. These conditions can be remedied by eating iron-rich foods or by taking an oral iron supplement.

5. An extensive property changes as the amount of material changes (e.g., mass). An intensive property remains the same regardless of how much material is present (e.g., temperature).

7. A physical change describes a change in physical properties without any modification in the composition of the substance. A chemical change involves the transformation of one substance into another.

9. The components of a mixture can be separated by physical methods; one example is boiling.

11. The elements in the periodic table are organized into groups based on their similar chemical properties.

13. (a) A proton and neutron each has a mass of approximately 1.0 amu, whereas the mass of an electron is only 0.00055 amu. (b) A proton has a positive charge of +1, an electron has a negative charge of −1, and a neutron has no charge (it is electrically neutral).

15. Isotopes are variants of an atom that have the same atomic number but different mass numbers, which arise from different numbers of neutrons in the nucleus. Examples include carbon-12 (6 protons, 6 neutrons) and carbon-14 (6 protons, 8 neutrons).

17. For the first electron shell, the maximum capacity is 2 electrons. For the second electron shell, the maximum capacity is 8 electrons. For the third electron shell, the maximum capacity is 8 electrons. However, the rule for the third shell is an approximation, and it is sometimes possible to fit 18 electrons into this shell.

19. Helium and neon are both unreactive noble gases because the atoms of each element have filled electrons shells.

21. In the quantum mechanical model of the atom, the electrons are described as waves. In the electron shell model, the electrons are described as particles.

23. Calcium ions have an essential role in forming the solid structures of bones and teeth.

Developing Skills

25. (a) On the Earth's surface, a mass of 1.0 kg has a weight of 2.2 pounds (lb). The weight of the astronaut on Earth can be calculated using a conversion factor:

$$\text{weight on Earth (lb)} = 70 \; \cancel{\text{kg}} \times \frac{2.2 \, \text{lb}}{1 \; \cancel{\text{kg}}} = 154 \, \text{lb}$$

(b) On the surface of Mars, the astronaut's mass will remain as 70 kg. However, the astronaut's weight will be different on the surface of Mars because this planet has a lower gravitational attraction compared to that of the Earth. To calculate the weight on Mars, multiply the Earth weight by the fraction of the Earth's gravity that exists on Mars (38%).

$$\text{weight on Mars (lb)} = 154 \, \text{lb} \times \frac{38}{100} = 58.5 \, \text{lb}$$

27. (a) 2.1×10^{-5} m

(b) cell size $(\mu m) = 2.1 \times 10^{-5} \; \cancel{m} \times \dfrac{1 \, \mu m}{10^{-6} \; \cancel{m}} = 2.1 \times 10^{-5} \times 10^{6}$

$= 2.1 \times 10^{1} \, \mu m$ or $21 \, \mu m$

29. (a) volume of water $(cm^3) = 1000 \; \cancel{g} \times \dfrac{1 \, cm^3}{1.00 \; \cancel{g}} = 1000 \, cm^3$

(b) volume of ethanol $(cm^3) = 1000 \; \cancel{g} \times \dfrac{1 \, cm^3}{0.789 \; \cancel{g}} = 1267 \, cm^3$

Because the density of ethanol is lower than that of water, a larger volume is required to reach the same mass.

31. (a) H_2 is an element.

(b) N_2O is a compound.

(c) Li is an element.

(d) CO is a compound.

(e) S_8 is an element.

33. Silicon belongs to Group 4A of the periodic table. In the third period, germanium (Ge) also belongs to Group 4A. Therefore, germanium will have chemical properties that are most similar to those of silicon.

35. Number of neutrons $= Z - A = 266 - 109 = 157$

37. The element is phosphorus, and the atomic symbol is $^{31}_{15}P$.

39.

MASS NUMBER	ATOMIC SYMBOL	NUMBER OF PROTONS	NUMBER OF NEUTRONS
32	$^{32}_{16}S$	16	16
33	$^{33}_{16}S$	16	17
34	$^{34}_{16}S$	16	18
36	$^{36}_{16}S$	16	20

41. The ^{28}Si isotope has a very high natural abundance (92.23%), which is much larger than the natural abundances of the other two isotopes. Consequently, the ^{28}Si isotope will make the largest contribution to the relative atomic mass of naturally occurring silicon. Because the two other isotopes have a larger mass, we predict that the relative atomic mass of natural silicon will be slightly higher than the mass of the ^{28}Si isotope. Our numerical estimate is 28.10.

We calculate the relative atomic mass as a weighted average of all three isotopes

relative atomic mass $= (27.98 \times 0.9223) + (28.98 \times 0.04686) + (29.97 \times 0.03092)$

$= 25.81 + 1.358 + 0.9267$

$= 28.09$

This result is very close to our estimate of 28.10.

43. The halogens are found within Group 7A of the periodic table. Therefore, we predict that a bromine atom has seven valence electrons in its valence shell.

45. (a) K^+

    (b) Cl^-

    (c) S^{2-}

    (d) Argon does not form an ion.

47. We predict that the chemical formula of arsine is AsH_3, which is similar to the chemical formula for phosphine (PH_3).

CHAPTER 3

Reviewing Knowledge

1. The two most abundant gases are nitrogen (78%) and oxygen (21%).

3. Inhaled air, containing oxygen gas, passes through the trachea, which delivers the oxygen to our lungs. Next, the oxygen passes through the thin lining of the lungs and enters the bloodstream.

5. Two H atoms each contribute one electron to produce a shared electron pair, which generates an attraction between the H atoms. The shared electron pair constitutes a covalent bond, which connects the two H atoms to form an H_2 molecule.

7. In the quantum mechanical model, a covalent bond corresponds to increased electron density between two atomic nuclei.

9. The octet rule states that many atoms tend to acquire eight electrons in their valence shell, which then resembles the stable electron configuration of a noble gas.

11. The general principle is valence shell electron pair repulsion (VSEPR), which states that electron pairs within the valence shell repel each other to become as far apart as possible. VSEPR predicts that CH_4 has a tetrahedral geometry, NH_3 has a trigonal pyramidal geometry, and H_2O has a bent geometry.

13. The N atom has a larger atomic radius because it has electrons in both the first and second electron shells. Because the second electron shell is farther from the nucleus, the N atom occupies a larger volume. By contrast, an H atom has only one electron in the first electron shell, which is closer to the nucleus, so its atomic radius is smaller.

15. N_2 has larger bond energy because the two N atoms are joined by a triple covalent bond, consisting of three shared electron pairs. By contrast, the two O atoms in O_2 are joined by a double covalent bond, consisting of two shared electron pairs. More energy is required to break a triple bond compared to a double bond.

17. Electrons are more stable in pairs, so a radical with an unpaired electron readily reacts with another atom or molecule to obtain another electron.

19. An ionic bond is the electrical attraction between two oppositely charged ions.

21. A salt is composed of ions that are held together by ionic bonds. By contrast, a molecule contains atoms that are held together by covalent bonds.

Developing Skills

23.     H:O:O:H   H—O—O—H

     Lewis dot structure   Two-dimensional structure

25. (a)  :F:N:F:
             :F:

     (b) NF_3

     (c) 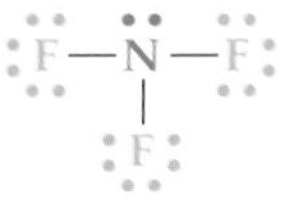 :F—N—F:
               |
             :F:

     (d)    ..
               N
              / \\""F
            F   F

27. (a) 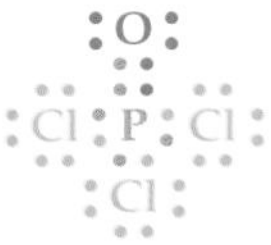   :O:

          :Cl:P:Cl:
             :Cl:

     (b) 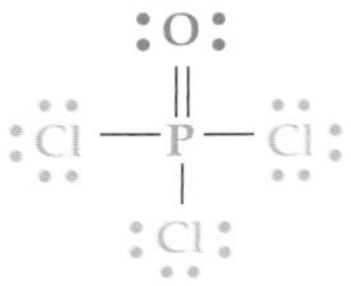   :O:
                 ||
          :Cl—P—Cl:
                |
              :Cl:

     (c) 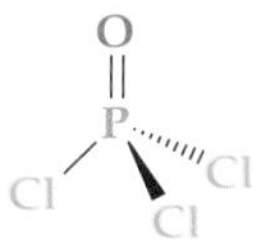   O
                 ||
                 P""""Cl
              / \\
            Cl   Cl

29.  H:C:::C:H

     A triple covalent bond is used to join the carbon atoms

31. (a)  H:B:H
             H

     (b) BH_3 violates the octet rule because the central boron atom is sur-
          rounded by only six valence electrons, which is two less than the
          number required by the octet rule.

     (c) 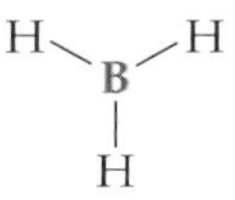 H   H
            \\ /
             B
             |
             H

33. $Al(NO_3)_3$
     $Mg(OH)_2$
     Na_2CO_3
     NH_4Cl

Chapter 4

Reviewing Knowledge

1. The RDA of saturated fat is lower than that of unsaturated fat.

3. Foods that contain saturated fat include red meats and dairy products. Foods that contain unsaturated fat include avocados and fatty fish (salmon).

5. At room temperature, a saturated fat is more likely to be solid than an unsaturated fat.

7. Hydrocarbons provide the structural framework for many biological molecules and pharmaceuticals.

9. (a)

(b)

(c)

11. The presence of C=C bond causes the geometry of ethene to be planar because it requires a large input of energy to twist a double bond. This property arises from how the electrons are distributed within the double bond.

13. A biological role of fat in the body is to store energy.

15.

17. The chemical process used to convert unsaturated hydrocarbons into saturated ones is called hydrogenation.

19. Trans fats contain unsaturated hydrocarbon tails in which the carbon atoms attached to the C=C bond are in the *trans* configuration as opposed to the usual *cis* configuration. Trans fats are typically formed during the partial hydrogenation of unsaturated vegetable oils. Hydrogenation changes the shape of the fatty acid from bent to linear, which now resembles the molecular structure of saturated fats. Trans fats are associated with elevated cholesterol and an increased risk of heart diseases.

21. Benzene, C_6H_6, can be drawn in two equivalent resonance structures with different placement of the electron in the covalent bonds. The resulting resonance hybrid molecule contains six identical carbon—carbon bonds, with a length that is intermediate between the typical values for single and double bonds.

Developing Skills

23.

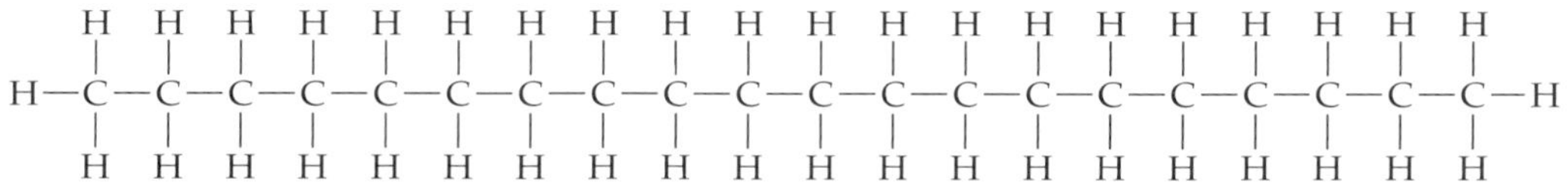

$C_{20}H_{42}$

25. (a)

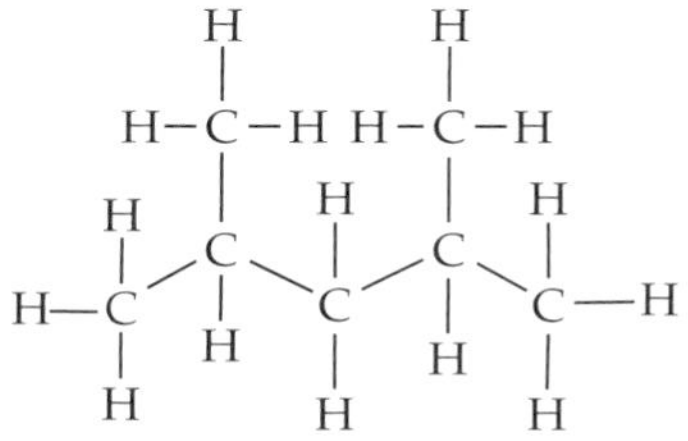

C_7H_{16}

(b)

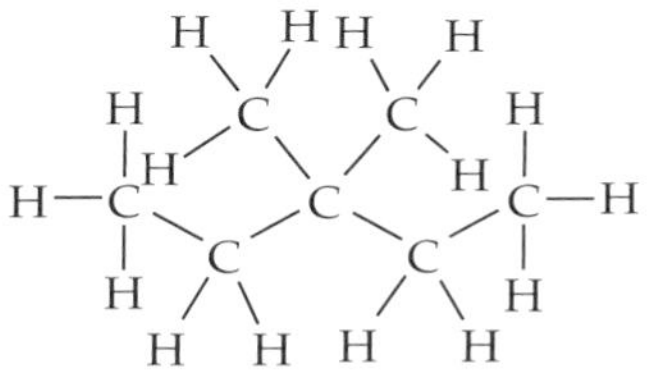

C_7H_{16}

(c)

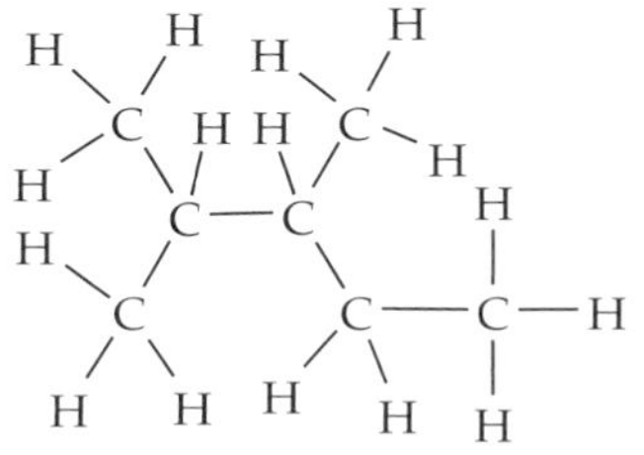

C_7H_{16}

(d)

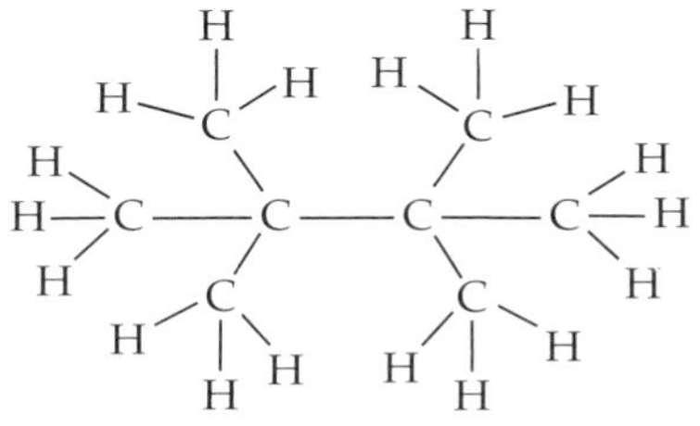

C_8H_{18}

27. Examples of possible answers are provided below.
    (a) Linear hydrocarbon

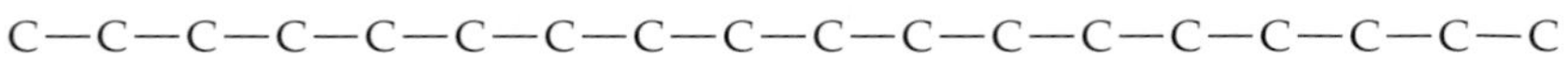

    C—C—C—C—C—C—C—C—C—C—C—C—C—C—C—C—C—C

    (b) Hydrocarbon with a single branch

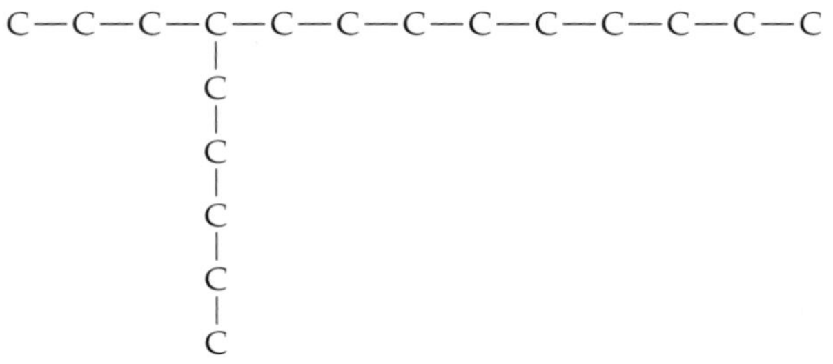

    C—C—C—C—C—C—C—C—C—C—C—C—C
           |
           C
           |
           C
           |
           C
           |
           C
           |
           C

    (c) Hydrocarbon with both a linear segment and a ring sigment

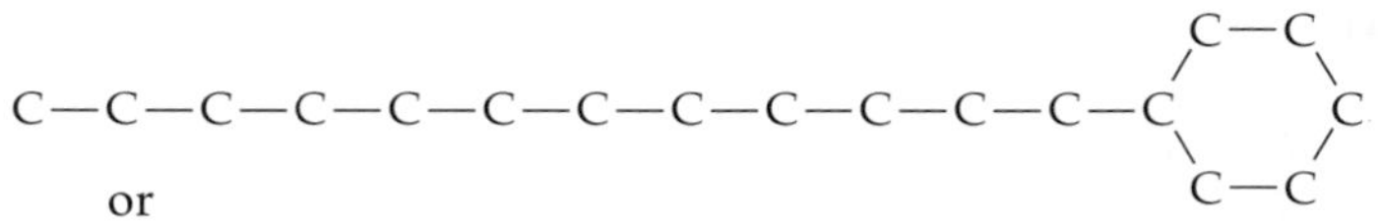

    or

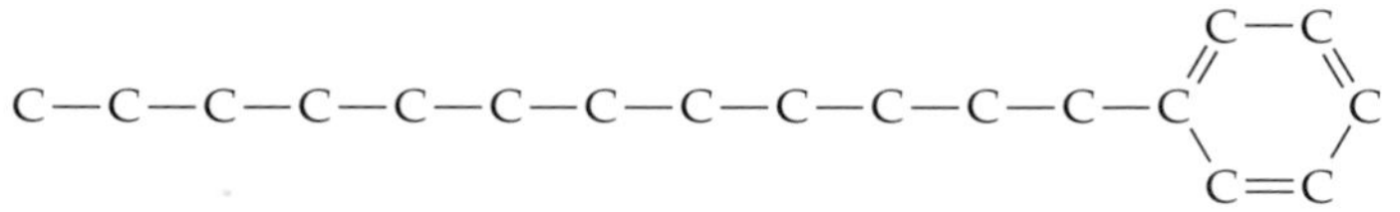

29.

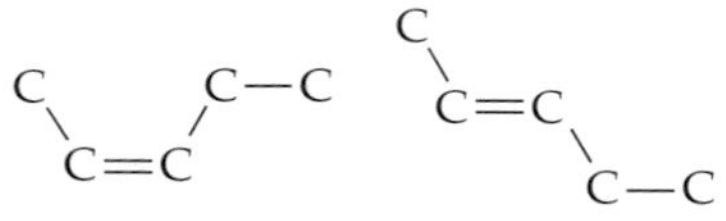

     Cis isomer   *Trans* isomer

31.

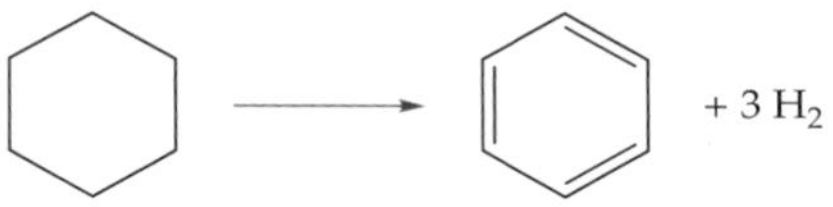

+ 3 H$_2$

33. (a)

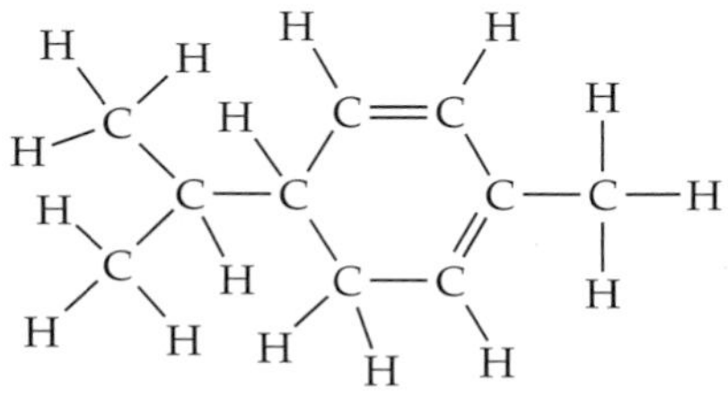

    (b) C$_{10}$H$_{16}$

35.

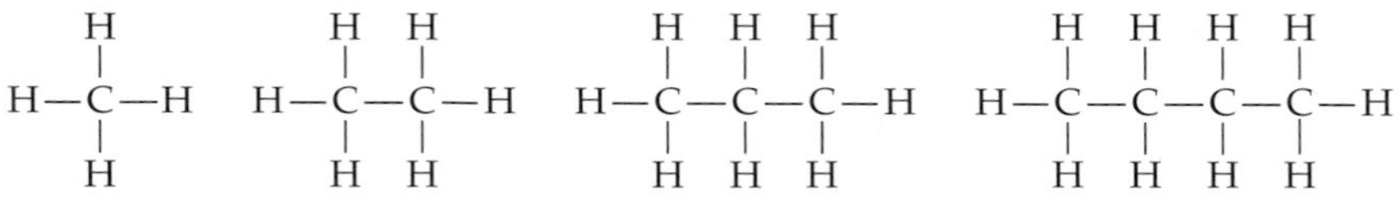

    The formula that predicts the number of hydrogens for a linear alkane molecule is C$_n$H$_{2n+2}$.

Chapter 5

Reviewing Knowledge

1. A heteroatom is an atom that is different from carbon and hydrogen. A functional group is a collection of atoms within a molecule that has a characteristic structure and chemical behavior.

3. Functional groups are important in pharmaceuticals and biological molecules because they have a strong influence on their chemical and biological properties.

5. The first molecule used as a medical anesthetic was diethyl ether.

7. An ester molecule contributes to the fragrance of pineapples.

9.

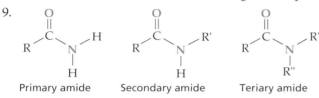

Primary amide   Secondary amide   Teriary amide

11.

R—S—H   R—S—R'

Thiol   Thioether   Thioester

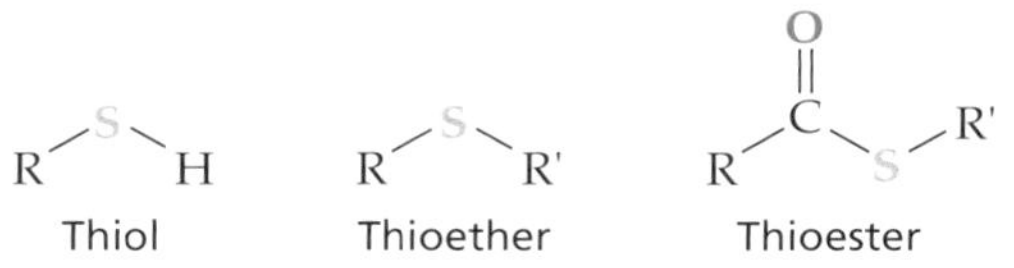

In each of these molecules, a sulfur (S) atom replaces an oxygen (O) atom. This is possible because S and O are in the same group of the periodic table, so the two atoms have similar chemical properties.

13. A heterocycle is a cyclic (ring-shaped) molecule in which one or more carbon atoms have been replaced by a heteroatom.

Developing Skills

15. (a) Alcohol
    (b) Thiol
    (c) Amine (primary)
    (d) Carboxylic acid
    (e) Amide (secondary)
    (f) Aldehyde

17.

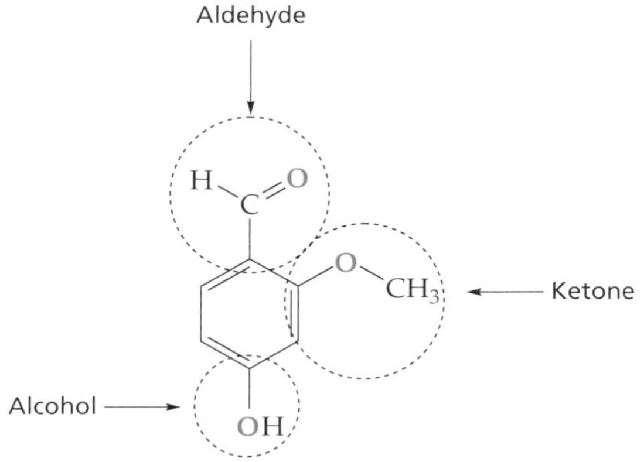

Aldehyde

Alcohol

Ketone

Vanillin

19.

Naproxen
A painkiller used in Aleve

Novocaine
A dental anesthetic

21.

23. The structure of uracil is based on the pyrimidine heterocycle.

25. (a) The nitrogen-containing heterocycle is derived from a purine.
    (b)

Chapter 6

Reviewing Knowledge

1. "Carb" is short for "carbohydrate." This term is used for sugar molecules because they are formed from multiples of C, H, and O atoms, with the general formula $C_x(H_2O)_y$.

3. A chemical reaction produces changes in matter and energy.

5. The stoichiometry of a chemical reaction describes the numerical ratios of reactants and products.

7. $2\,C_4H_{10} + 13\,O_2 \rightarrow 8\,CO_2 + 10\,H_2O$

9. The molar mass of an atom is defined as the mass of 1 mole (Avogadro's number) of atoms. The molar mass, in units of grams per mole (g/mol), is numerically equal to the relative atomic mass provided for each element in the periodic table.

11.

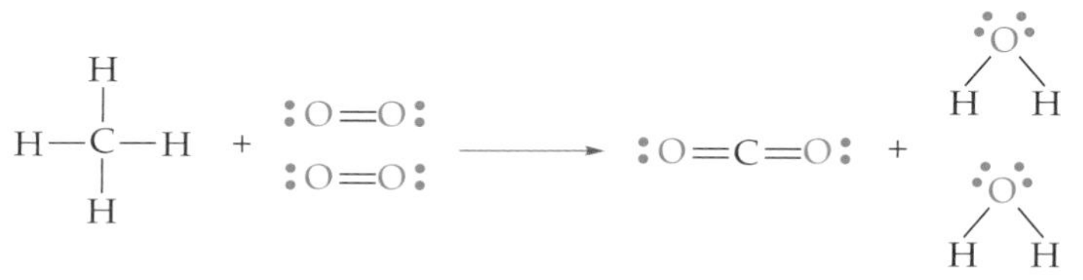

$$H-\underset{\underset{H}{|}}{\overset{\overset{H}{|}}{C}}-H \quad + \quad \begin{matrix} :O=O: \\ :O=O: \end{matrix} \quad \longrightarrow \quad :O=C=O: \quad + \quad \begin{matrix} \overset{..}{\underset{H \quad H}{O}} \\ \overset{..}{\underset{H \quad H}{O}} \end{matrix}$$

The products contain C=O bonds (in CO_2) and O—H bonds (in H_2O). These bonds are not present in the reactants.

13. The bond energy of a covalent bond is defined as the energy required to break 1 mole of these bonds.

15. $C_6H_{12}O_6 + 6\,O_2 \longrightarrow 6\,CO_2 + 6\,H_2O + \textbf{CHEMICAL ENERGY}$

The change in matter that occurs during cellular respiration is the same as the change that happens during glucose combustion. Specifically, glucose and oxygen (reactants) are chemically transformed into carbon dioxide and water (products). However, the energy output is different for the two reactants. Cellular respiration generates chemical energy that is stored within ATP molecules, whereas glucose combustion releases energy as heat.

17. ATP releases chemical energy through a reaction that breaks off one of the phosphate groups. This reaction forms a molecule of ADP. ATP can be re-generated by using energy from cellular respiration to convert ADP to ATP.

19. Sprinters have large muscles because they need to generate a quick burst of energy that is sustained for only a short time. Their large muscle mass enables sprinters to store a large reservoir of ATP, and to generate ATP anaerobically via lactic acid metabolism.

Developing Skills

21. $C_6H_{12}O_6 \rightarrow 2\,C_2H_5OH + 2\,CO_2$

ATOM	REACTANTS	PRODUCTS
C	6	6
H	12	12
O	6	6

23. Helium 4.003 g Sulfur 32.06 g Copper 63.54 g Mercury 200.6 g.

The volume of 1 mol of each element decreases as the atomic mass increases.

25. The atomic mass of gold (197.0 amu) is nearly twice that of silver (107.9 amu). Thus, 20 g of silver contains more atoms than 20 g of gold.

27. (a) 26.0 g/mol (b) 48.0 g/mol, and (c) 100.1 g/mol

29. 8.1×10^{21} formula units of KCl

31. 14.9 cm^3

33. (a) $C_6H_{12} + 9\,O_2 \rightarrow 6\,CO_2 + 6\,H_2O$ (b) 77.0 g CO_2

35. An equation that is not correctly balanced will prevent us from estimating accurately the energy change during the reaction. We need the correct numbers of each reactant and product molecules from the balanced equation in order to calculate the total numbers of bonds that are broken in the reactants and the total number of bonds that are formed in the products. If the numbers of bonds are incorrect, the calculated energy change for the reaction will also be incorrect.

Chapter 7

Reviewing Knowledge

1. Simple carbohydrates are small and consist of one or two sugar molecules. Complex carbohydrates are large and contain hundreds or thousands of sugar molecules.

3. Starch is a polymer made from linking multiple copies of a glucose monomer.

5. The repeat unit in the polymer has a different chemical formula than the monomer because new covalent bonds are formed when joining the monomers to make the polymer.

7. High-density polyethylene (HDPE) is made from a linear polymer. These polymer chains pack tightly together to form a tough, rigid plastic. In contrast, low-density polyethylene is made from a branched polymer. These polymers cannot pack closely together, and so they produce a soft, flexible plastic.

9. Polyesters and polyamides get their names from the type of chemical linkage that exists within the polymer.

11.

13. Lactose (milk sugar) is made from two monosaccharides, β-galactose and β-glucose. These monosaccharides are joined by a condensation reaction.

15.

17. Amylose is a linear polymer of α-glucose that forms a helical structure. The center of this helix provides a cavity for the triodide ion (I_3^-). This ion generates the blue-black color that is characteristic of a positive I_2/KI test for starch.

19. The glycemic index of a food measures the effect of the food on a person's blood glucose level.

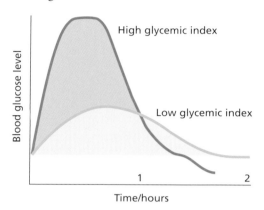

21. The glycemic index is a number that indicates the effect of a food on a person's blood glucose level. The glycemic index is measured using a standard amount of consumed carbohydrate, usually 50 g. However, the foods that we regularly consume contain different amounts of carbohydrate. The glycemic load includes the amount of carbohydrate in a serving of food, and it provides a more accurate indicator of how blood sugar is affected by our everyday diet.

23. Humans are unable to digest cellulose because the cellulose polymers pack together tightly to form tough cellulose fibers. These fibers are resistant to digestion by enzymes in the human digestive system.

Developing Skills

25. The number of ethylene monomers in the polymer chain is 3571 (rounded to the closest whole number)

27. (a)

The polymer is nonplanar because each carbon has four single covalent bonds, which will produce a tetrahedral geometry according to the predictions of VSEPR.

29. (a)

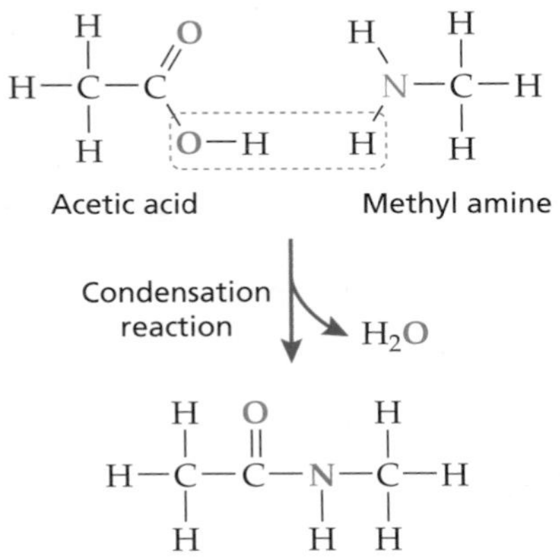

Acetic acid   Methyl amine

Condensation reaction → H_2O

This is a **condensation** reaction.

(b) In order to form a long polymer chain, the monomers must be capable of making a new chemical linkage at both ends. But after the reaction shown in part (a), the two ends of the molecule ($—CH_3$) cannot participate in another condensation reaction to form an amide linkage.

(c) The monomers could be modified so they each have two functional groups as shown below.

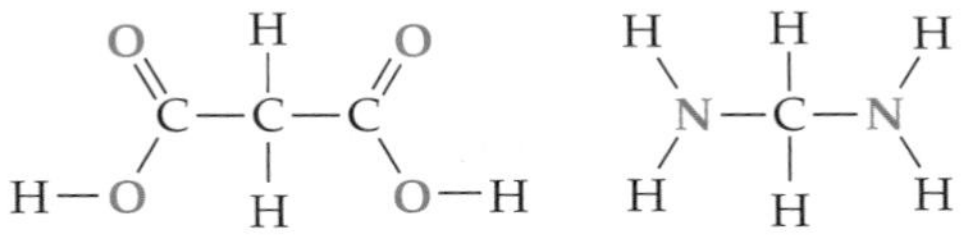

After two of these monomers are joined using a condensation reaction, the resulting molecule still has reactive functional groups at either end that can be used to make new chemical bonds to other monomers to form a polyamide chain.

31. As we learned in Chapter 4, large amounts of energy are required to twist a double bond. The presence of the double covalent bond in the repeat unit means that polybutadiene chains cannot rotate easily. As a result, the chains cannot pack together closely, which causes polybutadiene to have elastic properties.

Chapter 8

Reviewing Knowledge

1. Water is a liquid at room temperature, which provides an environment in which molecules are mobile and can interact. In addition, water is capable of dissolving a wide variety of molecules and ions, and all organisms require a diversity of dissolved substances in their cells.

3. The bent geometry of the H_2O molecule is caused by electron pair repulsion (VSEPR) among the two bonding pairs and the two lone pairs.

5. In an H—H covalent bond, the electron density is distributed equally between the two atoms. In an H—F bond, the electron density is not distributed equally because the F atom exerts a stronger attraction for the shared electron pair. Because of the unequal distribution of electron density, the F atom becomes slightly more negative and the H atom becomes slightly more positive.

7. Electronegativity measures the ability of an atom in a molecule to attract a shared pair of electrons. The Pauling electronegativity scale is used to assign electronegativity values to the elements.

9. A dipole is formed when positive and negative charges are separated by a short distance. An H—F bond has a dipole because the H atom has a partial positive charge (δ+), whereas the F atom has a partial negative charge (δ-). An H—H bond does not have a dipole because the H atoms do not have a partial charge.

11. The electronegativity of elements increases from left to right across a period, and the electronegativity of elements within a group decreases from top to bottom.

13.

15. A hydrogen bond is an electrical attraction between a hydrogen atom in a polar bond of one molecule and a nonbonded electron pair in another highly electronegative atom (e.g., N, O, F). A hydrogen bond is often an intermolecular bond that connects two different molecules, whereas a covalent bond is an intramolecular bond within a single molecule. In addition, a hydrogen bond is much weaker than a covalent bond.

17. Methane (CH_4) is a gas at room temperature because its nonpolar C—H bonds do not form hydrogen bonds. By contrast, methanol (CH_3OH) is a liquid at the same temperature because the polar —OH group in each molecule is capable of forming hydrogen bonds with other molecules.

19. When heat energy is added to ice, the kinetic energy of the water molecules increases and water molecules begin to distort the rigid structure of ice's hydrogen-bonding network. After some time, the kinetic energy of the molecules is sufficient to disrupt the hydrogen-bonding network in ice; at this point, the ice melts to form liquid water. Hydrogen bonds among H_2O molecules are still present in liquid water, but it is a dynamic hydrogen-bonded network in which bonds are continually being made and broken.

21. H_2O molecules form a liquid at room temperature because of the combined effect of many hydrogen-bonding interactions that cause the molecules to stick together. CH_4 and NH_3 are gases at room temperature because these molecules do not form the same type of hydrogen-bonded network.

23. Periodic trends for hydride compounds predict that water should boil at −100°C, but it actually boils at +100°C. This large discrepancy arises from extensive intermolecular hydrogen bonding within liquid water.

25. Adding extra heat energy does not increase the temperature because the heat input is used to produce a phase change between liquid water and gaseous water vapor. The heat energy is used to break the hydrogen-bonding network in liquid water, which liberates the H_2O molecules to become a vapor.

Developing Skills

27. (a) $O^{\delta-} \longleftarrow N^{\delta+}$

    (b) $N^{\delta-} \longleftarrow H^{\delta+}$

    (c) $Cl^{\delta+} \longrightarrow O^{\delta-}$

    (d) $S^{\delta+} \longrightarrow N^{\delta-}$

29. Na and O: $\Delta EN = 3.5 - 0.9 = 2.6$ (ionic bond)

    N and Cl: $\Delta EN = 3.0 - 3.0 = 0$ (nonpolar covalent bond)

    Ca and F: $\Delta EN = 4.0 - 1.0 = 3.0$ (ionic bond)

    S and O: $\Delta EN = 3.5 - 2.5 = 1.0$ (polar covalent bond)

31. (a) The N-H bond is polar because the $\Delta EN = 3.0 - 2.1 = 0.9$

    (b)

    (c) NH_3 does have a molecular dipole. Because of the trigonal pyramidal geometry, all three of the polar N—H bonds contribute to the molecule dipole of NH_3.

33. If the temperature drops below zero degrees Celsius, the water in the pipes can freeze. Because ice is less dense than water, the volume of ice is greater than the volume of water from which it is formed. This increase in volume can cause the water pipes to burst.

35. The concrete sidewalk will warm up more quickly than the water. Because water has a very high heat capacity, it requires a large amount of energy to raise its temperature.

37. HF has a higher boiling point than HCl because of the electronegativity difference between the atoms in each molecule. The H—F bond has a ΔEN of 1.9, whereas the H—Cl bond has a ΔEN of 0.9. As a result, HF molecules will exhibit stronger intermolecular hydrogen bonding compared to HCl molecules. Hydrogen bonding among HF molecules is responsible for its unusually high boiling point.

39. (a) For water, $\Delta T = 65.0°C - 25.0°C = 40.0°C$

$$Q = 10.0 \text{ g} \times 4.184 \frac{J}{g \cdot °C} \times 40.0 °C = 1673 \text{ J}$$

    (b) For ethanol, $\Delta T = 65.0°C - 25.0°C = 40.0°C$

$$Q = 10.0 \text{ g} \times 2.460 \frac{J}{g \cdot °C} \times 40.0 °C = 984 \text{ J}$$

    (c) A larger amount of heat is required to raise the temperature of water by the same amount.

Chapter 9

Reviewing Knowledge

1. Many vitamins are "helper molecules" that help an enzyme perform its role as a biological catalyst. Other vitamins enable cells to synthesize important molecules that are necessary for cells to function.

3. A solution is a homogeneous mixture in which the solute particles are distributed uniformly throughout the solution. In a heterogeneous mixture, the components are not evenly mixed, and the different components are distinct from each other.

5. Forcing nonpolar methane (CH_4) molecules into liquid water causes the polar H_2O molecules to form a cage-like clathrate that surrounds the solute.

7. Hydrophobic molecules tend to aggregate in water because they are nonpolar and cannot form hydrogen bonds with H_2O molecules.

9. Dispersion forces generate an intermolecular attraction among hexane molecules. A momentary fluctuation in the electron density of a nonpolar molecule produces an instantaneous dipole. This unequal distribution of charge generates an induced dipole in a nearby molecule. The dispersion forces arise from intermolecular attractions among the induced dipoles.

11. Water-soluble vitamins contain several polar functional groups, whereas fat-soluble vitamins primarily consist of nonpolar hydrocarbon regions.

13. The two regions of a detergent molecule are a polar head group and a nonpolar tail composed of a hydrocarbon chain. The head group is hydrophilic, and the tail is hydrophobic.

15. A phospholipid contains a polar head group and two nonpolar tails composed of hydrocarbon chains.

17. A cell membrane consists of a phospholipid bilayer that contains embedded membrane proteins.

19. A solution of table salt in water conducts electricity because the solid crystals of NaCl dissolve in water to produce aqueous ions, $Na^+(aq)$ and $Cl^-(aq)$. These ions are mobile in water, and they conduct an electrical current. Purified water contains very few aqueous ions, so it does not conduct an electrical current.

21. $NaCl(s) \xrightarrow{H_2O} Na^+(aq) + Cl^-(aq)$

23. Sucrose is not an electrolyte because it does not dissociate in water to produce aqueous ions.

25. The electrical charge on ions makes them insoluble within the hydrophobic interior of a cell membrane. Ion transport proteins provide a pathway for ions to cross the cell membrane.

Developing Skills

27. Hexane is a nonpolar solute that cannot form hydrogen bonds with H_2O molecules in the solvent. As a result, the presence of a hexane molecule in water will induce the surrounding H_2O molecules to form a more ordered structure that maximizes their hydrogen-bonding interactions. Because hexane is larger than methane, the H_2O molecules will not be able to form a clathrate that surrounds the solute molecule.

29. Helium gas is nonpolar, so we predict that it will be more soluble within the nonpolar environment of liquid hexane.

31. Vitamin K is a fat-soluble vitamin because it is composed mostly of nonpolar hydrocarbon regions. Vitamin B6 is a water-soluble vitamin because it has a polar amide functional group that can form hydrogen bonds with H_2O molecules.

33. Sodium chloride is more soluble in liquid water because H_2O molecules are capable of forming a larger number of favorable charge interactions with aqueous Na^+ and Cl^- ions. By contrast, a methanol molecule has

only one polar —OH group, which limits the number of charge interactions with dissolved ions in liquid methanol. In addition, methanol contains a —CH_3 group that does not interact with the dissolved ions.

35. (a) Salt changes in solubility only slightly as the temperature increases. By contrast, the solubility of sugar increases significantly.

    (b) The solubility of sugar is much greater in hot water than in cold water.

    (c) % increase $= \dfrac{(480 - 200)}{200} \times 100\% = 140\%$

37. The salt is more likely to be Na_2SO_4.

CHAPTER 10

Reviewing Knowledge

1. The runner's death in the 2002 Boston Marathon was caused by hyponatremia, which produced swelling of cells in her brain.

3. The standard unit for measuring the volume of solution is the liter. One liter is the volume of a cube that measures 10 centimeters along each side.

5. $1 \text{ ppm} = \dfrac{1 \text{ mg solute}}{1 \text{ L solution}}$

   $1 \text{ ppb} = \dfrac{1 \text{ µg solute}}{1 \text{ L solution}}$

7. The molarity (M) is the molar concentration of a solution, which is equal to the chemical amount of solute (measured in moles) present in 1 liter of solution.

$$\text{molarity (M)} = \frac{\text{chemical amount of solute (mol)}}{\text{volume of solution (L)}}$$

9. Osmosis is the spontaneous flow (diffusion) of H_2O molecules from a region of higher solvent concentration to a region of lower solvent concentration. This flow occurs through a selectively permeable membrane, which allows H_2O molecules to cross but blocks the passage of solutes.

11. Hyponatremia is defined as a sodium ion concentration of 135 mM or less. Critical hyponatremia is diagnosed when the sodium ion concentration falls to 120 mM or less.

Developing Skills

13. (a) Concentration of cadmium = 0.0080 ppm. This concentration exceeds the MCL.

    (b) Concentration of lead = 0.012 ppm. This concentration does not exceed the MCL.

15. (a) 1.00 M

    (b) 0.100 M

    (c) 1.00 M

    (d) 0.100 M

17. (a) 0.080%

    (b) 80 ppm

    (c) 4.5×10^{-4} M or 0.45 mM

19. 6.14 M

21. (a) 300 mL

    (b) 18 mL

23. Dispense 2.50 mL of the 12.0 M stock solution into a container and add water to obtain a volume of 100 mL.

25. 50 μg of arsenic in 1.0 L of water corresponds to a concentration of 0.050 ppm. This concentration exceeds the MCL for arsenic.

27. 2.24 g

29. 0.25 g

Chapter 11

Reviewing Knowledge

1. Acid reflux is caused by the backflow of stomach acid into the esophagus, where it can cause a painful burning sensation.

3.
$$HCI(aq) \quad + \quad H_2O(l) \quad \longrightarrow \quad H_3O^+(aq) \quad + \quad CI^-(aq)$$

   Hydrochloric acid   Water   Hydronium ion   Chloride ion

5. When sodium hydroxide ($NaOH$) is dissolved in water, it ionizes to produce $Na^+(aq)$ and $OH^-(aq)$ ions. The aqueous hydroxide ion acts as a base by accepting an H^+ ion from a H_2O molecule.

7. $HCI(aq) + NaOH(aq) \xrightarrow{\text{neutralization}} NaCI(aq) + H_2O(l)$

   The spectator ions are $Na^+(aq)$ and $Cl^-(aq)$. These ions appear on both sides of the chemical equation and do not participate in the neutralization reaction.

9. $$CH_3COOH(aq) \quad + \quad H_2O(l) \quad \rightleftharpoons \quad H_3O^+(aq) \quad + \quad CH_3COO^-(aq)$$

   Acetic acid   Water   Hydronium ion   Acetate ion

   The ionization of acetic acid is a reversible reaction because the products can react together to generate the reactants.

11. A 0.1 M solution of HCl exhibits stronger acidic properties than a 0.1 M solution of CH_3COOH because HCl generates a larger concentration of $H_3O^+(aq)$ ions. HCl is a strong acid and dissociates almost 100% in aqueous solution, producing a concentration of $H_3O^+(aq)$ ions that is very close to 0.1 M. This high concentration of $H_3O^+(aq)$ ions is responsible for the strong acidic properties of the HCl solution. By contrast, only 1% of CH_3COOH dissociates, which produces a much lower concentration of $H_3O^+(aq)$ ions.

13. $K_w = [H_3O^+][OH^-] = 10^{-14}$. In pure water, $[H_3O^+] = [OH^-]$, which means that $[H_3O^+] = 10^{-7}$ M and $[OH^-] = 10^{-7}$ M

15. $[H_3O^+] = 10^{-pH}$ and $pH = -\log_{10}[H_3O^+]$

17. The biological role of stomach acid is to aid the digestion of food.

19. Histamine receptor blockers attach to the H_2 receptor and prevent histamine from binding. This binding disrupts the sequence of chemical signals that activates the proton pump. Proton pump inhibitors bind directly to the proton pump and prevent it from producing H^+ ions.

21. (a) $$CO_2(g) + H_2O(l) \rightleftharpoons H_2CO_3(aq)$$

    Carbon dioxide   Carbonic acid

(b) $H_2CO_3(aq)$ + $H_2O(l)$ ⇌ $H_3O^+(aq)$ + $HCO_3^-(aq)$

Carbonic acid   Bicarbonate ion

Blood is a buffer because it contains a mixture of a weak acid (carbonic acid, H_2CO_3) and its conjugate base (bicarbonate ion, HCO_3^-).

Developing Skills

23. $HNO_3(aq), + NaOH(aq) \rightarrow NaNO_3(aq) + H_2O(l)$

25. (a) $H_2CO_3(aq) + 2\,NaOH(aq) \rightarrow Na_2CO_3(aq) + 2\,H_2O(l)$
    (b) Na^+ and CO_3^{2-}

27. (a) $[H_3O^+] = 10^{-10}$ M
    (b) $[H_3O^+] = 10^{-8}$ M
    (c) $[H_3O^+] = 10^{-6}$ M
    (d) $[H_3O^+] = 10^{-3}$ M
    (e) $[H_3O^+] = 1$ M

29. pH = 9

31. Adding NaOH to this buffer solution will generate $OH^-(aq)$ ions, which will neutralize some of the acetic acid (CH_3COOH) and lower its concentration. This reaction increases the concentration of acetate ions (CH_3COO^-). The concentration of $H_3O^+(aq)$ will decrease slightly, but not significantly, due to the buffering effect.

33. The solution mixture with a 1 M concentration of the weak acid and conjugate base will have a higher buffer capacity. In general, the buffer capacity increases with increasing concentration of the weak acid and conjugate base. Higher concentrations of the buffer components are able to react with larger amounts of H_3O^+ and OH^- ions that are added to the buffer.

35. 1.0 M NaOH has the highest pH. 1.0 M HCl has the lowest pH.

37. (a) $S(s) + O_2(g) \rightarrow SO_2(g)$
    (b) $2\,SO_2(g) + O_2(g) \rightarrow 2\,SO_3(g)$
    (c) $SO_3(g) + H_2O(l) \rightarrow H_2SO_4(aq)$.
    (d) Sulfuric acid.

Chapter 12

Reviewing Knowledge

1. The complete name of DNA is deoxyribonucleic acid.

3. The three components of a nucleotide are: a deoxyribose sugar, a phosphate, and a nucleotide base.

5. The four DNA bases are adenine (A), thymine (T), guanine (G), and cytosine (C). Thymine and cytosine are pyrimidine bases, whereas guanine and adenine are purine bases.

7. The two ends of a polynucleotide strand are the 5′-end and the 3′-end. The directionality of a polynucleotide is defined in the 5′ → 3′ direction.

9. Watson and Crick used molecular models to represent the molecular structure of DNA.

11. The X-pattern in Rosalind Franklin's "Photo 51" revealed that B-form DNA has a helical structure. In addition, specific features of the

photograph provided structural parameters of the helix, such as the pitch and the number of nucleotides per helical turn.

13. The two bases in a DNA base pair are joined by hydrogen bonds.

15. The phosphate groups are located on the outside of the helix because they are soluble in water. Positively charged ions (cations) such as Na^+ and Mg^{2+} stabilize DNA's structure by neutralizing the negative charges on the phosphates.

17. During DNA replication, each parent DNA strand serves as a template to synthesize a new daughter strand. The replicated DNA molecule contains half of the original DNA, so the mechanism is called semiconservative.

19. All types of cancer are characterized by the uncontrolled growth of cells. Cancer arises from DNA mutations in the genes that control cell division.

21. A chemical carcinogen is a chemical substance that is capable of causing cancer. One example is benzopyrene, which is a type of polyaromatic hydrocarbon. After benzopyrene enters the body, it is chemically modified by a cytochrome enzyme to become more reactive and water soluble. The modified benzopyrene forms a covalent bond with a guanine base in DNA to create a benzopyrene-DNA adduct. The presence of this adduct causes errors during DNA replication, especially G-to-T mutations.

Developing Skills

23. The double-stranded structure of DNA is important for DNA replication. When double-stranded DNA unwinds, two polynucleotide chains are available to serve as templates to synthesize a complementary strand. In this manner, one double-stranded DNA replicates to create two double-stranded DNA molecules. This duplication strategy would not work if DNA were a single-stranded molecule.

25. (a) Xanthine is a purine base.

    (b) Guanine serves as the source of xanthine.

27. The hydroxymethyl cytosine base can still form a base pair with guanine (G). The addition of the extra chemical group to the ring does not affect the atoms that form hydrogen bonds with guanine.

29. (a) 3′ – T A A C G G G C T T T A C C G – 5′

    (b) 3′ – C G T T G C A A A C G T G A C – 5′

    (c) 5′ – G T C A G G T G C T A A T G T – 3′

31. Adenine = 20%, guanine = 30%, cytosine = 30%.

33. 200 cell divisions.

35. (a) 32 DNA molecules

    (b) 1204 DNA molecules

    (c) 1.07×10^9 DNA molecules

    (d) 10^9 corresponds to 1 billion.

37. We hypothesize that the planar structure of proflavine enables it to intercalate within the planar base pairs of the bacterial DNA. The presence of proflavine within the base pairs can cause mutations in bacterial DNA during its replication.

Chapter 13

Reviewing Knowledge

1. The characteristic symptom of sickle cell disease is the distortion of red blood cells into a sickle shape (like the letter "C").

3. Three structural and chemical differences between DNA and RNA are as follows.

   • DNA forms a double helix, whereas RNA exists as a single strand.

   • DNA contains deoxyribose as its sugar, whereas RNA contains ribose.

   • DNA contains thymine as a base, which is replaced by uracil in RNA.

5. To begin transcription, RNA polymerase binds to a promoter site and then unwinds the DNA double helix to expose the base pairs. RNA polymerase uses the DNA template strand and the base-pairing rules to synthesize a complementary strand of mRNA.

7. The genetic information contained in DNA is preserved during transcription because the sequence of bases in the mRNA strand is complementary to the sequence of bases in the template strand of DNA.

9.

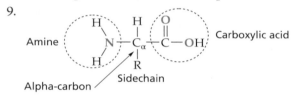

The sidechain (R) changes to produce different varieties of amino acid.

11. The four groups of amino acid sidechains are neutral nonpolar, neutral polar, acidic, and basic.

13. The letters used to identify the two chiral isomers of amino acids are L (left-handed form) and D (right-handed form). Proteins use only L-amino acids.

15. The genetic code is called "degenerate" because most amino acids can be specified by more than one codon.

17. During translation, the transfer RNA (tRNA) serves as a "molecular translator" between an mRNA codon and its corresponding amino acid. The tRNA molecule contains a sequence of three bases called an anticodon. The anticodon base sequence is complementary to the mRNA codon, and codon–anticodon binding ensures that the correct amino acid is added to the growing amino acid chain.

19. The chemical variation in a polypeptide chain arises from the specific sequence of amino acid sidechains.

21. The primary structure is the sequence of amino acids in the polypeptide chain, beginning with the N-terminus and ending at the C-terminus. The secondary structure is a local region within the protein where the polypeptide chain exhibits a regular structure. The two most common types of secondary structure are called alpha-helix and beta-sheet. The tertiary structure refers to the overall structure of the entire protein, including all of its secondary structure components. The quaternary structure describes the structural relationship among several polypeptide chains.

23. A single base mutation in DNA (T→A) is passed along to mRNA via transcription. The mRNA mutation (A→U) changes the sixth codon in the beta-globin chain from GAG to GUG. When the mRNA is translated, valine is inserted into the polypeptide chain in place of glutamic acid. The sidechain of valine is hydrophobic, which induces the deoxy form of the

Hb S protein to aggregate. This process creates protein fibers, which distort red blood cells into a sickle shape.

Developing Skills

25. (a)

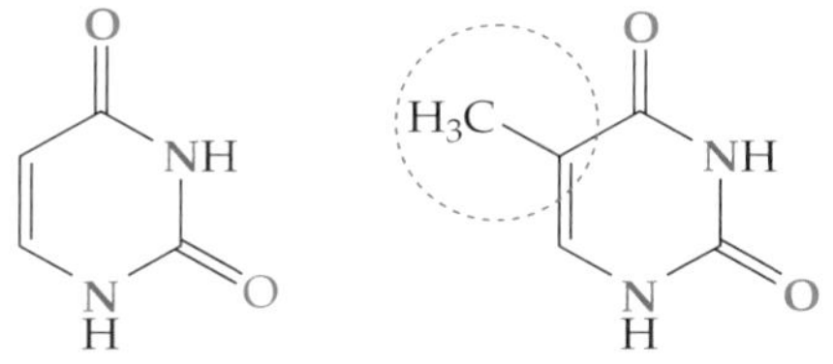

Uracil

Thymine

(b)

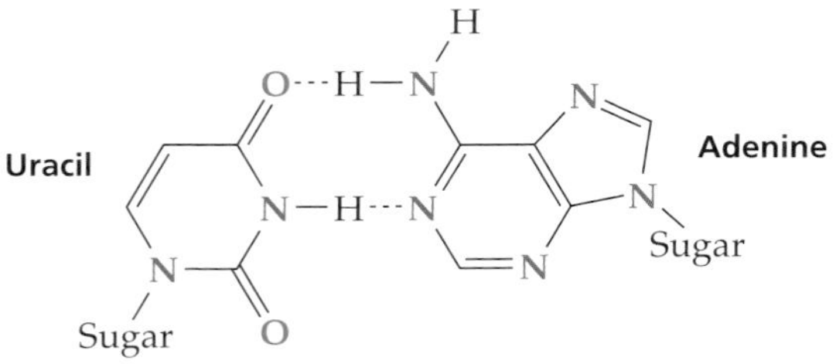

Uracil

Adenine

(c) The uracil–adenine base pair is connected by two hydrogen bonds. This is the same number of hydrogen bonds in the thymine–adenine base pair.

27. The base sequence is from RNA because it contains U (uracil), which does not exist in DNA.

29. (a)

ALANINE	VALINE	SERINE
GCU	GUU	UCU
GCC	GUC	UCC
GCA	GUA	UCA
GCG	GUG	UCG

(b) In each group of codons, the first two bases in the codon remain the same whereas the third base varies. For example, the codons for alanine can be written as GCX, where X = U, C, A, or G.

31. Glycine

33. (a) N—cysteine—valine—phenylalanine—C

(b)

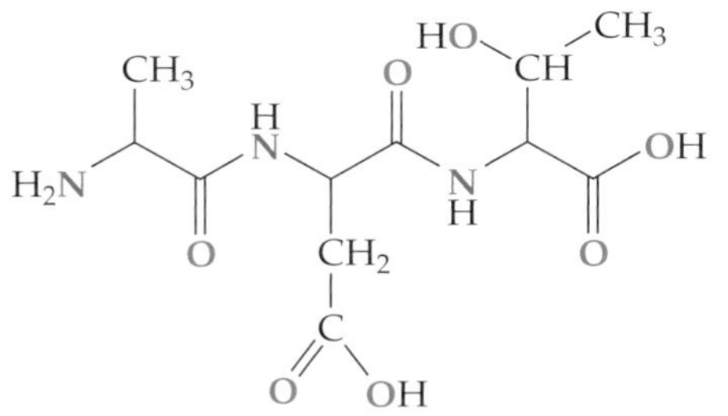

35.

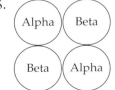

37. The DNA mutation is a G-to-T base change in the first position of the 20th codon.

Chapter 14

Reviewing Knowledge

1. Lactose intolerance is an adverse reaction to the lactose sugar in milk.

3. Most chemical reactions do not occur spontaneously because there is an energy barrier that prevents the reactants from immediately converting into the products.

5. (a) A larger activation energy corresponds to a slower reaction rate, and a smaller activation energy corresponds to a faster reaction rate.

   (b) The reaction energy does not affect the rate of a chemical reaction.

7. The catalytic converter in a vehicle accelerates the chemical reaction between poisonous carbon monoxide (CO) and atmospheric oxygen (O_2) to produce harmless carbon dioxide (CO_2). The surface of a catalytic converter is coated with transition metals such as platinum, palladium, and rhodium.

9. The active site of an enzyme is the region that binds the substrate and catalyzes the chemical reaction.

11. During the catalytic cycle, the enzyme binds the transition state more tightly than it binds the reactants or the products. Because this tight binding stabilizes the high-energy transition state, the enzyme lowers the activation energy for the reaction and increases its rate.

13. Lactase uses an H_2O molecule to catalyze the hydrolysis of a C—O—C linkage in lactose. This reaction releases two smaller sugars, glucose and galactose, as products.

15. The ON position for the lactase gene occurs when a transcription factor binds to a promoter region in DNA, which activates the expression of the lactase gene to produce the enzyme. The OFF position occurs when the transcription factor is not bound and no lactase enzyme is produced.

17. When farmers began to keep dairy cows, milk became an important nutrient. Lactase persistence enabled adult humans to consume milk and milk-based products without experiencing any ill effects. The ability to tolerate milk provided an evolutionary advantage during times of famine, when lactase-persistent individuals were more likely to survive and reproduce. Because of this advantage, the mutation that causes lactase persistence increased within certain human populations. This is an example of gene-culture coevolution, which occurs when a genetic mutation (the one causing lactase persistence) undergoes evolutionary selection because of a cultural behavior (dairy farming).

Developing Skills

19. (a) The energy diagram for the reaction is given below.

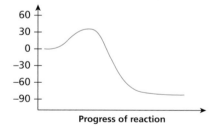

   (b) The reaction is exothermic.

21. Energy diagram B represents the enzyme-catalyzed reaction. Compared to the original energy diagram, the activation energy is smaller, but the reaction energy remains the same.

23. Simple reactions in a test tube occur when the reactants collide with each other. When the temperature is increased, the reactants acquire more kinetic energy (energy of motion), and it is more likely that the reaction will overcome its activation energy. By contrast, an enzyme-catalyzed reaction requires the substrate molecule(s) to bind the active site of the enzyme. Raising the temperature will perturb the three-dimensional structure of the protein that acts as the enzyme, including the structure of the active site. If an enzyme cannot adopt its optimal structure, its capacity for catalyzing the reaction is reduced. Consequently, the rate of the enzyme-catalyzed reaction is slowed at higher temperature.

25. (a) As we learned in Chapter 13, peptide bonds serve as the linkages between amino acids in proteins. Most proteins exist within an aqueous environment. Consequently, a peptide bond must be resistant to hydrolysis; otherwise, the links between amino acids in the protein would be easily broken in water. If this happens, the protein would no longer exist as a biological polymer and would be incapable of performing its biological function.

    (b) To calculate the rate enhancement of the enzyme, we need to compare the time required to break one peptide bond with and without the enzyme. The enzyme is able to catalyze 1842 hydrolysis reactions per second. On average, the time to catalyze one hydrolysis reaction is $1/1842$ seconds $= 5.429 \times 10^{-4}$ s. In the absence of the enzyme, the time required for one hydrolysis reaction is 350 years. The number of seconds in 350 years is approximately equal to $350 \times 365 \times 24 \times 60 \times 60 = 1.10 \times 10^{10}$ s. The rate enhancement is obtained by dividing the times of the uncatalyzed reaction and the catalyzed reactions, which gives 1.10×10^{10} s $/ 5.429 \times 10^{-4}$ s $= 2.03 \times 10^{13}$.

    (c) The answer is closer to 1 trillion, which is defined as 10^{12}. In fact, the enzyme-catalyzed reaction is approximately 20 trillion times faster than the uncatalyzed reaction.

27. The optimal pH for the enzyme is pH 8, which corresponds to the maximum enzyme activity. The enzyme activity decreases in more acidic solutions (pH < 8) and in more basic solutions (pH > 8).

29. In order for the enzyme to catalyze the reaction, the enzyme must bind to the active site. This is the first step of the catalytic cycle. At low-substrate concentration, there are many enzymes available relative to the number of substrate molecules. Any substrate can gain access to the active site of an enzyme, so the reaction rate increases with an increase in substrate concentration. As the substrate concentration begins to increase, more of the enzymes will have a substrate bound to its active site. If the active site is already filled, another substrate molecule cannot bind to the enzyme. At a certain concentration of substrate, all of the enzyme active sites are saturated with substrate molecules. This is the maximum rate that the enzyme can achieve for the catalyzed reaction.

Chapter 15

Reviewing Knowledge

1. Tuberculosis is caused by the bacterium *Mycoplasma tuberculosis*.

3. MDR-TB is multiple drug-resistant tuberculosis, which is resistant to two of the standard antibiotics. XDR-TB is extensively drug-resistant tuberculosis, which is resistant to all four of the standard antibiotics.

5. Typical sizes are 50 μm for a human cell, 1 μm for a bacterial cell, and 50 nm for a virus.

7. Biological organisms are a good source of antibiotics because they are constantly engaged in "chemical warfare" against one another as part of their struggle to survive and reproduce.

9. A bactericidal antibiotic kills bacteria, whereas a bacteriostatic antibiotic prevent the growth of bacterial cells without killing them.

11. The disk diffusion test compares the size of the inhibition zone around each antibiotic disk. The inhibition zone is an area in which the growth of bacteria has been prevented by the presence of the antibiotic. A larger inhibition zone indicates that the antibiotic is more potent against bacterial growth. In the diagram below, antibiotic A has no effect on the bacteria, antibiotic B has a moderate effect, and antibiotic C has a large effect.

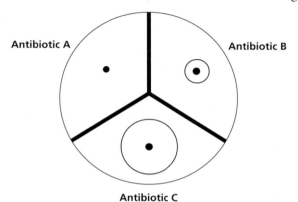

13. The antibiotic targets in a bacterial cell are the key biochemical processes that the cell needs to function and divide. These four targets are (1) synthesis of the bacterial cell wall, (2) synthesis of proteins at the ribosome, (3) DNA replication, and (4) synthesis of folic acid, which is necessary for the cell to make the nucleic acid bases in DNA.

15. Penicillin inhibits an enzyme called transpeptidase, which catalyzes a chemical reaction that forms crosslinks between polymers in the bacterial cell well. Without these crosslinks, the cell wall is weakened.

17. Ciprofloxacin functions as an antibiotic by inhibiting the DNA topoisomerase enzyme that unwinds supercoiled DNA within the bacterial cell. Without the functional enzyme, the bacterial cell cannot replicate its DNA correctly.

19. The three strategies are: (1) pump the drug out of the target cell; (2) inactivate the drug as by synthesizing an enzyme that destroys it (for example, penicillinase); and (3) alter the structure of the antibiotic target, such as modifying the ribosomes that are normally sensitive to antibiotics.

21. Less than 20% of antibiotics are used for human medicine. Most antibiotics are used in agriculture and given to animals such as cows, pigs, and chickens.

23. Tuberculosis bacteria are encased by a waxy cell coating that prevents our immune system from destroying the cells.

25. Isoniazid blocks the ability of tuberculosis bacteria to produce their protective cell coating.

27. In the 19th century, inactive or even harmful medications were often sold to unsuspecting customers. To address this problem, Congress passed the Federal Food and Drugs Act of 1906 to regulate what could be sold as a "medicine."

29. Thalidomide is an example of a molecule with two chiral isomers, which are mirror images of each other. One chiral isomer achieved the desired sedative effect for pregnant mothers. However, the other isomer was harmful to the developing fetus and caused developmental deformities.

Developing Skills

31. 100 minutes

33. We hypothesize that the drugs work on different targets. The two drugs, A and B, are each being used at their optimal doses. If they worked on the same target, it would be unlikely to see a stronger effect from their combination.

35. Bacterial cells that are not replicating already have an intact cell wall, so they do not need to synthesize a new one. Because penicillin targets the process of building the cell wall, it has no effect on nonreplicating cells that already have a cell wall.

37. The new drug has the same core molecular structure as penicillin. Therefore, we predict that this drug will be effective against bacteria that are sensitive to penicillin.

39. (a) Cipro shows the greatest change. The IC_{50} of Cipro has increased by 45 times the original value.

    (b) Development of resistance to this drug might require changes in cells that are not possible or that would kill the resistant cells.

APPENDIX A

Developing Skills

1. (a) 1.5×10^3 m

   (b) 7.4×10^{-11} m

   (c) 4.13×10^{16} m

3. 772 miles per hour

Appendix B

Developing Skills

1. (a) Four significant figures

   (b) Three significant figures

   (c) Three significant figures

   (d) One significant figure

3. (a) 137.0 m

   (b) 2.08 cm³

   (c) 4.55 g

   (d) 3.6

Appendix C

Developing Skills

1. (a) $[H_3O^+] = 5 \times 10^{-9}$ M

   (b) $[H_3O^+] = 5 \times 10^{-5}$ M

3. (a) 7.6×10^8

   (b) -1.1236

   (c) 8×10^{-5}

   (d) 6.952

Glossary

3′-end The end of the polynucleotide at which an —OH group is attached to the 3′-carbon of deoxyribose.

5′-end The end of the polynucleotide at which a phosphate group is attached to the 5′-carbon of deoxyribose.

A

acid A substance that can donate a hydrogen ion (H^+) to another substance.

acid reflux The backflow of stomach acid into the esophagus.

acidic solution A solution in which the concentration of H_3O^+ ions is greater than that of OH^- ions.

activation energy The energy barrier that must be overcome for a chemical reaction to occur.

active site The region of an enzyme in which the catalytic activity occurs.

addition polymerization A chemical process in which monomers are "added together" to form a polymer with no addition or removal of atoms.

Adenine One of the four nucleotide bases in DNA.

adenosine diphosphate (ADP) A molecule formed by the removal of one phosphate group from ATP.

adenosine triphosphate (ATP) A molecule containing three phosphate groups that stores chemical energy for use in cellular processes.

aerobic A chemical reaction that uses oxygen.

alcohol A functional group with the general formula R—OH.

aldehyde A functional group with the general formula R—CO—H

alkanes Hydrocarbons that contain only single bonds.

alkenes Hydrocarbons that contain at least one C=C double bond.

amide A functional group that contains both nitrogen and oxygen atoms.

amine A functional group that contains a nitrogen atom.

amino acid A group of molecules that are the building blocks of proteins.

anaerobic A chemical reaction that takes place in the absence of oxygen.

anion A negatively charged ion; for example, Cl^-.

antibiotic A drug that kills bacteria or inhibits their growth.

anticodon A sequence of three bases in tRNA that is complementary to the sequence of bases in the mRNA codon.

antiparallel The spatial arrangement of two polynucleotide chains that are oriented in opposite directions.

antiseptic A chemical substance that prevents infection.

aqueous ion An ion dissolved in water.

aqueous solution A solution in which water is the solvent.

aromatic Compounds that contain a benzene ring, or related structures.

atom The fundamental unit of matter.

atomic mass The weighted average of all the naturally occurring isotopes of an element, based on their mass and relative abundance.

atomic mass unit (amu) The unit used to measure the mass of subatomic particles; 1 amu is defined as 1/12 the mass of a ^{12}C atom.

atomic number (Z) The number of protons in the nucleus of an atom.

atomic symbol A notation for an atom that includes the element, the atomic number, and the mass number.

Avogadro's number The number of atoms in exactly 12 g of carbon-12; it is equal to 6.022×10^{23}.

B

bactericidal antibiotic A drug that kills bacteria.

bacteriostatic antibiotic A drug that inhibits the replication of bacteria without killing them.

ball-and-stick model A molecular model that represents the atoms as balls and the bonds between them as sticks.

base A substance that can accept a hydrogen ion (H^+) from another substance.

base pairs Specific pairings of the DNA bases within the double helix; adenine pairs with thymine, and guanine pairs with cytosine.

basic solution A solution in which the concentration of OH^- ions is greater than that of H_3O^+ ions.

bent geometry The shape of an upside-down "V."

boiling point The temperature at which a liquid changes into a vapor.

bond dipole The dipole of an individual covalent bond within a molecule.

bond energy The energy required to break a chemical bond.

bond polarity The degree to which a covalent bond is polar or nonpolar; it is determined by the electronegativity difference between the two bonded atoms.

branched polymer A polymer in which some of the monomers branch sideways from the primary polymer chain.

broad-spectrum antibiotic A drug that is effective against a wide range of bacterial types.

buffer A mixture of chemical substances in solution that resists changes in pH; in most cases, it consists of a weak acid and its conjugate base.

C

cancer A group of related diseases that are characterized by the uncontrolled growth of cells.

carbohydrate A sugar molecule composed of C, H, and O atoms, with the general formula $C_x(H_2O)_y$.

carbon-only drawing A simplified representation of molecules that includes only the carbon atoms and their bonds.

carboxylic acid A functional group with the general formula R—COOH; it has the chemical properties of an acid.

catalyst A substance that increases the rate of a chemical reaction.

catalytic cycle A series of steps that all enzymes use to catalyze a chemical reaction.

cation A positively charged ion; for example, Na^+.

cell The basic unit of life.

cell membrane A protective barrier surrounding a cell.

cell theory A scientific theory that states that all living organisms are composed of basic units called cells.

cell wall A tough, mesh-like structure that surrounds many bacterial cells.

cellular respiration A chemical reaction in cells between glucose and oxygen that generates chemical energy stored in ATP molecules.

cellulose An indigestible polysaccharide composed of linear chains of β-glucose molecules; it provides fiber in our diet.

chemical amount The amount of a chemical substance, measured in units of moles.

chemical bond A bond formed between the atoms in a molecule.

chemical carcinogens Chemical substances that are capable of causing cancer.

chemical change The transformation of one substance to another via a chemical reaction.

chemical equation A representation of a chemical reaction that uses chemical formulas to depict the reactants and products.

chemical equilibrium The balance point of a reversible reaction that describes the relative amounts of reactants and products.

chemical formula A representation of a chemical substance that identifies the types and number of atoms contained in that substance.

chemical property A characteristic of a substance that can be observed during a chemical reaction.

chemical reaction A process that produces changes in matter and energy.

chemical symbol An abbreviation for an element that uses one or two letters.

chiral An object that cannot be superimposed onto its mirror image.

chiral isomers Two mirror-image forms of a molecule; they contain the same atoms arranged differently in space.

***cis* isomer** A structural isomer in which two chemical groups are on the same side of a C=C bond.

clathrate A cage-like structure of H_2O molecules that completely surrounds a nonpolar solute molecule.

clinical trials Rigorous tests of new drugs performed on human subjects.

codon A sequence of three bases in mRNA, which specifies a particular amino acid.

coefficients Numbers placed in front of a chemical formula to indicate how many times that molecule must be counted in a balanced chemical equation.

combustion A chemical reaction that occurs when a fuel combines with oxygen, which releases heat energy.

complex carbohydrate A carbohydrate that contains hundreds or thousands of sugar units.

compound A pure substance formed from two or more different elements that are chemically joined in a fixed ratio; such as water (H_2O).

concentration A measure of the amount of substance dissolved in a specific volume of solution.

condensation polymer A polymer formed via condensation reactions.

condensation reaction A chemical reaction that joins two molecules to form a larger product, while releasing a smaller molecule such as H_2O.

conjugate acid-base pair An acid-base pair in which the two substances differ only by an H^+ ion.

conservation of mass A law stating that the total mass remains unchanged during a chemical reaction

conversion factor A ratio of two different units that is equal to one.

covalent bond A bond that is formed when two atoms share an electron pair.

cyclic hydrocarbons Hydrocarbons in which the carbon atoms form a closed ring.

cytosine One of the four nucleotide bases in DNA.

D

delocalized electrons Electrons that are not associated with a specific atom or covalent bond.

density The amount of mass contained in a unit of volume.

deoxyribose The sugar contained in a DNA nucleotide.

detergent A molecule used for cleaning; it has a hydrophilic head group plus a long hydrocarbon tail.

dilution The addition of solvent to a solution to lower the concentration of the solution.

dipole A pair of positive and negative charges separated by a short distance.

directionality The orientation of a polynucleotide strand; it is defined in the 5'-to-3' direction.

disaccharide A sugar that consists of two linked monosaccharides.

dispersion forces Forces that produce mutual attractions among nonpolar molecules.

DNA Deoxyribonucleic acid

DNA polymerase An enzyme that creates new DNA polymers by adding new nucleotides to the template strand.

double covalent bond A covalent bond formed when atoms share two pairs of electrons.

E

eclipsed conformation The conformation in which two methyl groups line up with each other.

electrolyte A dissolved solute that conducts electricity.

electron A negatively charged subatomic particle that surrounds the nucleus.

electron configuration The location of electrons within specific shells of an atom.

electron density The probability of finding an electron at a specific location within an orbital.

electron orbital A quantum mechanical description of an electron's probable location.

electron shell A quantized energy level where electrons are located.

electronegativity The ability of an atom to attract the shared pair of electrons in a chemical bond.

element A substance composed of only one type of atom, which cannot be broken down into simpler substances using chemical methods.

endothermic A chemical reaction that absorbs energy from its surroundings.

energy The capacity to do work.

entropy A measure of the degree of disorder or randomness within a system.

enzyme A biological molecule that causes a chemical reaction to occur faster.

ester A functional group with the general formula R—COO—R'

ether A functional group with the general formula R—O—R'.

evolution by natural selection A process by which certain individuals in a population preferentially survive and reproduce within a specific environment.

exothermic A chemical reaction that releases energy to its surroundings.

expanded valence A molecule containing a central atom that has more than eight electrons in its valence shell.

experiment A procedure designed to test a hypothesis.

extensive property A property that depends on the amount of material; examples are length, mass, and volume.

F

fat A class of molecules that the body uses for energy storage.

fat-soluble vitamin A vitamin that dissolves in regions of the body that store fat molecules.

fatty acid An acid that consists of an extended hydrocarbon chain plus a carboxylic acid group.

formula unit The simplest ratio of the components in a compound.

full-atom drawing A representation of molecules that includes every atom in the molecule.

functional group A collection of atoms within a molecule that has a characteristic structure and chemical behavior.

G

gene expression The biological pathway by which DNA sequences are processed to make proteins.

genetic code The instructions that specify which mRNA codon corresponds to which amino acid.

glucose A sugar used by living organisms to obtain energy; its chemical formula is $C_6H_{12}O_6$.

glycemic index A number that indicates the effect of a food on a person's blood glucose level.

glycemic load A measure of blood glucose that includes the glycemic index and the amount of carbohydrates consumed.

group A column in the periodic table that contains elements with similar chemical properties.

H

heat energy A form of energy that can be transferred between objects.

hemoglobin A protein found in red blood cells that transports oxygen to the body's tissues.

heteroatom An atom that is different from carbon and hydrogen; for example, O, N, P, and S.

heterocycle A cyclic molecule in which one or more of the carbon atoms has been replaced with a heteroatom.

heterogeneous mixture A mixture in which the composition is not uniform, enabling the individual components to be identified.

homogeneous mixture A mixture in which the composition is the same throughout.

hydrocarbons Molecules that contain only hydrogen and carbon atoms.

hydrogen bond An electrical attraction between a hydrogen atom in a polar bond and another highly electronegative atom (e.g., N, O, F).

hydrogen bond acceptor A highly electronegative atom with a pair of nonbonding electrons that attracts the hydrogen bond donor to form a hydrogen bond.

hydrogen bond donor The H atom that forms a hydrogen bond; this H atom is attached to a highly electronegative atom.

hydrogenation A chemical reaction for adding hydrogen atoms to an unsaturated hydrocarbon.

hydrolysis reaction A chemical reaction in which water is used to break the bond between two molecules.

hydronium ion An ion that is formed when a water molecule accepts a proton; its chemical formula is H_3O^+.

hydrophilic The ability to mix with water.

hydrophobic The inability to mix with water.

hydrophobic effect A principle that describes the association of nonpolar solutes in water to minimize their interaction with the solvent.

hyponatremia A health-threatening condition caused by a lowered amount of sodium ions in the blood.

hypothesis An initial attempt to explain an observation.

I

induced fit model A model of enzyme functioning in which the enzyme changes its shape when binding the substrate.

insoluble The relative difficulty of dissolving a solute in a solvent.

intensive property A property that remains the same regardless of the amount of material (e.g., temperature).

intermolecular bond A chemical bond between two molecules.

intramolecular bond A chemical bond within a single molecule.

ion An atom (or group of atoms) that has lost or gained one or more electrons and acquired an electrical charge.

ionic bond The electrical attraction between two oppositely charged ions.

ionic compound A compound composed of oppositely-charged ions; for example, NaCl (table salt).

ionization The decomposition of a neutral compound into charged ions; it is sometimes called *dissociation*.

isotope Variants of an atom that have the same atomic number but different mass numbers (i.e., different numbers of neutrons).

K

ketone A functional group with the general formula R—CO—R'.

kinetic energy A form of energy that arises from the motion of an object.

L

lactase The enzyme that decomposes lactose into two smaller sugars.

lactose intolerance The inability to break down the lactose sugar in milk products.

lactose The disaccharide found in milk; it is made by joining α-glucose and β-galactose.

latent heat The amount of heat energy required to produce a phase change without an increase in temperature.

Lewis dot structure A representation of an atom or molecule that depicts valence electrons as dots.

linear polymer A polymer in which the monomers are attached end-to-end.

line-angle drawing A simplified representation of molecules that uses lines to indicate bonds between carbon atoms and the angles of those bonds.

lipids Fat-like substances that do not dissolve easily in water.

liter The standard unit for measuring the volume of a solution; $1 \text{ L} = 10^3 \text{ cm}^3$.

lock-and-key model A model of enzyme function in which the substrate (key) selectively fits into the active site of an enzyme (lock).

logarithmic scale A numerical scale based on powers of 10; an example is the pH scale.

lone pair See nonbonding (electron) pair.

M

mass The amount of matter that an object contains.

mass number (A) The sum of the protons and neutrons in the nucleus of an atom.

mass/volume percent A measure of concentration that uses a percentage ratio of the mass of solute compared to the volume of solution.

matter Any substance that occupies space and has mass.

messenger RNA (mRNA) A type of RNA that carries DNA's genetic message.

methicillin-resistant _Staphylococcus aureus_ (MRSA) A type of infectious bacteria that is not harmed by the antibiotic methicillin and related drugs.

methyl group An group of atoms with the formula $-CH_3$.

micelle A structure formed by the self-assembly of molecules with both hydrophilic and hydrophobic regions.

mixture A physical combination of multiple substances that are not chemically bonded to one another.

molar mass The mass of 1 mole (Avogadro's number) of atoms, molecules, ions, or other particles.

molarity The molar concentration of a solution; it is equal to the chemical amount of solute, measured in moles, that is present in 1 liter of solution.

mole ratio The ratio of moles of product to moles of reactants in a balanced chemical equation.

mole The chemical amount of substance that contains Avogadro's number of atoms, molecules, ions, or other particles.

molecular dipole The directional sum of all the bond dipoles in a molecule.

molecular self-assembly The spontaneous assembly of a collection of molecules to form an organized structure.

molecule A group of two of more atoms joined by chemical bonds to form a specific structure.

molecule A group of two or more atoms bonded together in a specific arrangement.

monomer A single molecule that can be linked together many times to form a large polymer.

monosaccharide A sugar that consists of a single molecule.

monounsaturated An unsaturated fatty acid that contains only a single $C=C$ double bond.

MRSA See _methicillin-resistant Staphylococcus aureus (MRSA)_.

mutation A change in the DNA of an organism.

N

narrow-spectrum antibiotic A drug that is effective against a limited number of bacterial types.

neutral solution A solution in which the concentrations of H_3O^+ and OH^- ions are equal.

neutralization A chemical reaction between an acid and a base.

neutron An electrically neutral subatomic particle located in the nucleus of atom.

noble gas An unreactive (or mostly unreactive) element in Group 8A of the periodic table.

nonbonding (electron) pair An unshared pair of valence electrons from a single atom.

nonpolar covalent bond A covalent bond that does not form a dipole.

nonpolar solvent A solvent in which the molecules are nonpolar, such as liquid hexane.

nucleotide The molecular building block of DNA; it consists of a deoxyribose sugar, a phosphate group, and a nitrogen-containing base.

nucleotide base The nitrogen-containing base in a nucleotide; it exists in four varieties: adenine, thymine, guanine, and cytosine.

nucleus A small region at the center of the atom that contains almost all of the atom's mass.

O

observation The identification of a natural phenomenon that requires an explanation.

octet rule The tendency of atoms to acquire eight electrons—an octet—in their valence shell.

organic chemistry The chemistry of molecules containing carbon.

osmosis The net flow of water through a selectively permeable membrane, from a region of higher solvent concentration to one of lower solvent concentration.

P

partial charge A small electrical charge acquired by an atom as a result of unequal sharing of the electron density in a covalent bond.

parts per billion A measure of concentration equal to 1 part of solute per billion parts of solution; 1 ppb = 1 μg/L.

parts per million A measure of concentration equal to 1 part of solute per million parts of solution; 1 ppm = 1 mg/L.

penicillin An antibiotic that is produced by the *Penicillium* mold.

peptide bond The covalent bond that joins two amino acids.

period A row in the periodic table.

periodic table An organized arrangement of the elements based on their physical and chemical properties.

pH scale A logarithmic scale for measuring the acidity of a solution that is based on the concentration of $H_3O^+(aq)$ ions.

pharmacokinetics The analysis of how drugs move through the body.

phase change A transition between two phases of matter; for example, solid ice melts to form liquid water.

phosphate group A functional group in which a phosphorus atom is bonded to four oxygen atoms.

phosphate ion An ion with the chemical formula, PO_4^{3-}.

phosphodiester A chemical linkage composed of a phosphorus atom and two modified ester bonds.

phospholipid A molecule composed of two fatty acids, a glycerol unit, a phosphate group, and a head group with polar components.

phospholipid bilayer A double layer of phospholipids that forms the structural foundation of the cell membrane.

physical change A change in the physical properties of a substance that does not modify its composition.

physical property A characteristic that can be measured without changing the composition of the substance.

planar A molecular geometry in which all of the atoms lie in a two-dimensional plane.

plastic A solid substance that can be molded into different shapes.

polar covalent bond A covalent bond that forms a dipole.

polar molecule A molecule that contains a molecular dipole (e.g., H_2O).

polyamide A polymer in which the monomers are joined by amide linkages.

polyatomic ion An ion composed of two or more atoms that are held together by covalent bonds.

polyester A polymer in which the monomers are joined by ester linkages.

polyethylene A polymer constructed from multiple copies of the monomer ethylene.

polymer A large molecule that is constructed from many copies of a smaller monomer.

polymerization The chemical process whereby multiple monomers are linked to form a polymer.

polynucleotide A polymer containing many linked nucleotides.

polypeptide A polymer composed of amino acids joined by peptide bonds.

polysaccharide A sugar made from many monosaccharides.

polyunsaturated An unsaturated fatty acid that contains two or more C=C double bonds.

potential energy A form of energy that derives from an object's position.

precipitate An insoluble ionic solid formed by the combination of an aqueous cation and anion.

precipitation reaction A chemical reaction that forms an insoluble precipitate.

primary amine An amine in which the nitrogen atom is bonded to one carbon atom.

primary structure The sequence of amino acids in the polypeptide chain of a protein.

products The substances formed by a chemical reaction.

protein A biological molecule constructed from a long chain of amino acids.

proton A positively charged subatomic particle located in the nucleus of an atom.

pure substance A substance with a constant composition and characteristic chemical properties; it cannot be separated into components by physical methods.

purine A nitrogen-containing heterocycle comprised of two fused rings; it serve as the foundation for adenine and guanine.

pyrimidine A nitrogen-containing heterocycle comprised of a single ring; it serves as the foundation for cytosine and thymine.

pyrimidine dimer A molecular structure formed by connecting two adjacent pyrimidine bases within DNA.

Q

quantized A characteristic of electron shells in which they can have only integer values.

quantum mechanics A theory that describes electrons as waves.

quaternary structure The the structure formed by two or more polypeptide chains within a protein.

R

radical A reactive molecule that contains an unpaired electron.

reactants The starting components of a chemical reaction

reaction energy The net change in energy for a chemical reaction; it is the difference between the energy input to break bonds in the reactants, and the energy output from making bonds in the products.

reaction pathway The stages by which reactants are transformed into products in a chemical reaction.

relative atomic mass A unitless number that compares the mass of one atom of the element to 1 amu.

repeat unit The collection of atoms, derived from the monomer, that is replicated many times within the molecular structure of the polymer.

resonance A theory that includes more than one bonding arrangement in a molecule.

resonance hybrid A hybrid structure that incorporates the different resonance structures for the molecule; this structure is the most accurate representation of the chemical bonding.

resonance structures Two or more equivalent bonding arrangements for a molecule.

reversible reaction A two-way chemical reaction in which some of the products are transformed back into the reactants.

ribonucleic acid (RNA) A polymer composed of RNA nucleotides.

ribose The sugar found in the nucleotides that make up RNA.

ribosomes Structures in which proteins are synthesized.

RNA polymerase An enzyme that synthesizes a polymer of RNA nucleotides during transcription.

S

saccharide A chemical term for a sugar.

salt An ionic compound.

saturated fat A fat that contains only single covalent bonds between its carbon atoms.

scanning tunneling microscope (STM) A microscope that uses a moving metal probe to monitor the landscape of atoms arranged on a surface.

scientific law A statement that summarizes a collection of scientific observations but does not provide an explanation for them.

scientific method A systematic process for investigating the natural world.

scientific notation The expression of a quantity as a number that is multiplied by 10 raised to a power.

scientific unit A unit of measurement that incorporates a power of 10.

secondary amine An amine in which the nitrogen atom is bonded to two carbon atoms.

secondary structure A local region within the protein where the polypeptide chain exhibits a regular structure; two examples are alpha-helix and beta-sheet.

semiconservative The mechanism of DNA replication, such that half of each new double helix is conserved from the parental DNA.

sickle cell disease An inherited disease that is characterized by the distortion of red blood cells into a sickle shape.

significant figures The number of digits in a quantity that reflects the level of certainty in its measurement.

simple carbohydrate A carbohydrate that consists of only one or two sugar molecules.

simple ion An ion that is formed when a single atom gains or loses electrons.

solubility The ability of a solute to dissolve in a solvent.

soluble The relative ease of dissolving a solute in a solvent.

solute The component of a solution that is dissolved in the solvent.

solution A homogeneous mixture that consists of a solute dissolved in a solvent.

solvent The component of a solution that dissolves the solute.

space-filling model A molecular model that depicts the volume of each atom.

specific heat capacity The amount of heat energy required to raise the temperature of 1 gram of a substance by 1 degree Celsius.

spectator ion An ion that appears on both sides of a chemical equation and does not participate in the chemical reaction.

staggered conformation The conformation in which two methyl groups in are twisted relative to each other.

starch A polysaccharide composed of many α-glucose molecules, linked together as linear or branched chains.

start codon The codon that provides a signal to begin translation.

stoichiometry The numerical ratios of reactants and products in a chemical reaction.

stop codon The codon that signals the end of translation.

strains Variations of an organism that arise from differences in DNA.

strong acid An acid that ionizes almost completely in aqueous solution.

structural isomers Various molecules that have the same chemical formula but different molecular structures; they are also called constitutional isomers.

substrate The reactant molecule in an enzyme-catalyzed reaction.

sucrose The disaccharide commonly known as table sugar; it is made by joining α-glucose and β-fructose.

sugar-phosphate backbone An alternating sequence of sugars and phosphates; it links the nucleotides in a polynucleotide chain and supports the DNA bases.

supercoiling A process by which circular DNA gains or loses additional twists.

T

temperature A measurement of how hot or cold a substance is, relative to a standard scale.

template strand The DNA strand that RNA polymerase uses as the molecular template for synthesizing a new strand of mRNA.

tertiary amine An amine in which the nitrogen atom is bonded to three carbon atoms.

tertiary structure The overall structure of a single polypeptide chain, including all of its secondary structure components.

tetrahedral geometry The shape of a tetrahedron, in which all bond angles are 109.5°.

theory An explanation of the natural world that has been confirmed by multiple experiments.

thioester A functional group in which a sulfur atom replaces one of the oxygens in an ester; it has the general formula R—CO—S—R'.

thioether A functional group in which sulfur replaces oxygen in an ether; it has general formula R—S—R'.

thiol A functional group in which sulfur replaces oxygen in an alcohol; it has the general formula R—SH.

three-dimensional structure The spatial arrangement of atoms in a molecule.

thymine One of the four nucleotide bases in DNA.

trans fat An unsaturated fatty acid containing a trans C=C bond; this bond causes the hydrocarbon chain to be linear.

***trans* isomer** A structural isomer in which two chemical groups are on opposite sides of a C=C bond.

transcription The process whereby the genetic information contained in DNA is converted into mRNA.

transfer RNA (tRNA) The type of RNA that transfers amino acids to the ribosome.

transition state The transient, high-energy state that is formed during the conversion of reactants into products in a chemical reaction.

translation The process whereby the information in mRNA is converted into a sequence of amino acids in a protein.

triglyceride A dietary fat made from three fatty acids attached to a glycerol molecule.

trigonal pyramidal geometry The shape of a three-sided pyramid.

triple covalent bond A covalent bond formed when atoms share three pairs of electrons.

two-dimensional structure A flat drawing of a molecule that uses straight lines to represent covalent bonds.

U

unit conversion The conversion of a quantity from one unit to another unit; the quantity remains the same, but the unit changes.

unsaturated fat A fat that contains at least one double covalent bond between its carbon atoms.

uracil A pyrimidine base found in RNA; it replaces thymine found in DNA.

V

valence electrons The electrons located in the outer shell; they determine the chemical properties of the atom.

valence shell The outer electron shell of an atom.

valence shell electron pair repulsion (VSEPR) A principle to predict the three-dimensional geometry of molecules; it states that all valence electron pairs in a molecule repel one another so as to maximize their separation in space.

vapor A gas that arises from a liquid or a solid.

vitamin An organic molecule that plays an essential role in many biological processes.

volume The amount of space occupied by an amount of matter.

volume/volume percent A measure of concentration that uses a percentage ratio of the volume of solute compared to the volume of solution.

W

water-soluble vitamin A vitamin that dissolves in any watery environment within the body.

weak acid An acid that ionizes to a limited extent in aqueous solution.

wedge-and-dash drawing A drawing that depicts a three-dimensional perspective for molecular structures.

weight The force exerted on an object by gravity.

X

X-ray crystallography A technique in which scientists bombard crystals of a substance with X-rays to determine the structure of the substance.

Credits

Chapter 1

p. 2 (top) AndrewMcClenaghan/Science Source; **p. 2 (bottom)** Molekuul/Science Source; **1.1** Used with permission of Mayo Foundation for Medical Education and Research. All rights reserved.; **1.2** DAVID MCCARTHY/ SCIENCE PHOTO LIBRARY; **1.3 (b)** Molekuul/Science Source; **1.5** anyaivanova/Shutterstock.com; **1.6** tab62/ Shutterstock.com; **Concept Question 1.2** Marina Lohrbach/ Shutterstock.com; **1.8** Alfred Eisenstaedt/Pix Inc./The LIFE Picture Collection/Getty Images; **1.9 (a)** Wellcome Images/Science Source; **1.9 (b)** Dr. Christine Case/Visuals Unlimited, Inc; **1.10** YANGCHAO/Shutterstock.com; **1.11 (b)** Molekuul/Science Photo Library/Getty; **Learning Resources, Question 11** CDC/Dr. JJ Farmer (PHIL #3031), 1978.

Chapter 2

p. 24 © 2003 Richard Megna - Fundamental Photographs; **2.1** © 2012 General Mills; **Worked Example 2.1** DJM-photo/Shutterstock.com; **Concept Question 2.1** © 2008 Richard Megna - Fundamental Photographs; **2.5** © manx_in_the_world/iStock.com; **2.6** © Richard Menga, Fundamental Photographs, NYC; **2.8 (a)** Heritage Images/ Alamy; **2.8 (b)** Emilio Segrè Visual Archives/American Institute of Physics/Science Source; **2.11** Andrew Dunn/ Alamy; **Concept Question 2.3** Kenneth Sponsler/ Shutterstock.com; **2.26** © Peterjunaidy/Fotolia; **2.28** Majority World/UIG via Getty Images; **Chemistry and Your Health, Figure 1** © 1992 Richard Megna-Fundamental Photographs; **Visual Summary 2.5 Osteoporosis** © Peterjunaidy/Fotolia.

Chapter 3

p. 60 Wolfgang Amri/iStock Photo; **3.3 (a)** Charles Winters/Science Source; **3.3 (b)** Science Source; **3.9** Science Source; **Chemistry in Your Life Figure 1 (a)** INTERFOTO/Alamy Stock Photo; **Chemistry in Your Life Figure 1 (b)** Photos.com/ThinkStock; **3.23** Laguna Design/Science Photo Library; **Science in Action Figure 1** Hank Morgan/Science Source; **Science in Action Figure 2** Encyclopaedia Britannica/UIG/Getty Images; **3.27** © 1993 Paul Silverman, Fundamental Photographs, NYC; **3.29** Magnetix/Shutterstock.com; **Visual Summary 3.6** Magnetix/ Shutterstock.com.

Chapter 4

p. 92 JPC-PROD/Shutterstock.com; **4.1** A.D.A.M. Images; **4.2 (b)** alexpro9500/Shutterstock.com; Craevschii Family/ Shutterstock.com; **4.3** imagedb.com/Shutterstock.com; **4.5** molekuul_be/Shutterstock.com; **4.19 (b)** Leonid Andronov/Shutterstock; **Chemistry and Your Health, Figure 1** Shevs/Shutterstock.com; **4.21** SOMMAI/ Shutterstock.com; nexus 7/Shutterstock.com; **4.22 (a)** Cordelia Molloy/Science Source; FDA; **4.24.** Dr. Mark J. Winter/Science Source **4.26** Science Photo Library; **4.30** Science Photo Library; **Learning Resources, Question 36** Marcel Clemens/Shutterstock.com; Tyler Boyes/ Shutterstock.com.

Chapter 5

p. 126 Yasonya/Shutterstock.com; **5.1** Warren Price Photography/Shutterstock.com, Warren Price Photography/ Shutterstock.com, George W. Bailey/Shutterstock.com; **5.8** Boston Medical Library in the Francis A. Countway Library of Medicine; **5.12** Dionisvera/Shutterstock.com; Nata-Lia/ Shutterstock.com; **5.14 (a)** Shelia Fitzgerald/Shutterstock. com; **5.17 (a)** Juanmonino/iStockPhoto.com; **5.18 (a)** Diana Taliun/Shutterstock.com; **5.20 (a)** Murina Natalia/ Shutterstock.com; **Practice Exercise 5.1** Dionisvera/ Shutterstock.com; Maks Narodenko/Shutterstock.com; Anna Kucherova/Shutterstock.com; Anna Kucherova/ Shutterstock.com; **Chemistry in Your Life, Figure 1** National Institute of Health; **5.31 (a)** Eric Isselee/Shutterstock. com; **5.39 (a)** Keith Homan/Shutterstock.com; **Learning Resources Question 19** jiangdi/Shutterstock.com; **Learning Resources, Question 20** joy_stockphoto/Shutterstock.com.

Chapter 6

p. 154 Science Photo Library/Science Source; **6.1** bikeriderlondon/Shutterstock.com; **6.2** Science Photo Library/Science Source; **6.3** pixelsnap/Shutterstock.com; **Science In Action, Figure 1** © FineArt/Alamy Stock Photo; **Science in Action, Figure 2** MarcelClemens/ Shutterstock.com ; Andrew Lambert Photographer/Science Photo Library; **Practice Exercise 6.2** KAMONRAT/ Shutterstock.com; **Chemistry and Your Health, Figure 2** Danny E. Hooks/Shutterstock.com; **6.6** © 2001 Richard Megna, Fundamental Photographs, NYC; **6.7** ArturNyk/ Shutterstock.com; **6.9** © 2001 Richard Megna, Fundamental Photographs, NYC; **Worked Example 6.4** © 2001 Richard Megna, Fundamental Photographs, NYC; **Practice Exercise 6.4** Joe_Potato/iStock Photo; **Practice Exercise 6.5** NASA; **Practice Exercise 6.6** Nigel Cattlin/Science Photo Library; **6.27** © PCN Photography/Alamy Stock Photo; **Visual Summary 6.1** bikeriderlondon/Shutterstock. com; **Visual Summary 6.3** © 2001 Richard Megna, Fundamental Photographs, NYC; **Learning Resources Question 20** Wang Song/Shutterstock.com; **Learning Resources, Question 37** © istock.com/Gary Sludden.

Chapter 7

p. 196 Maximilian Stock Ltd/Science Photo Library; **7.1** windu/Shutterstock.com; **7.6** Photo Melon/Shutterstock. com; Carlos Gawonski/iStock Photo; **Worked Example**

7.1 Devonyu/iStock Photo; **Science in Action Figure 7.1 (a)** Popperfoto/Getty Images; **Worked Example 7.2** Dusan Kostic/iStock Photo; **Try It Yourself 7.2** Daniya Melnikova/iStock Photo; **7.13** George Manga/iStock Photo; **7.15** Mizina/iStock Photo; **Figure 7.16** Aleksandr Ryzhov/Alamy Stock Photo; **7.20 (a)** GIPhotoStock/Science Source; **Figure 7.25** travellight/Shutterstock.com; **Concept Question 7.3** Oleh_Slobodeniuk/iStock Photo; **Figure 7.28** David McCarthy/Science Source.

Chapter 8

p. 228 Everett Historical/Shutterstock.com; **8.1** Aleksandr Markin/Shutterstock.com; **Science in Action, Figure 1** NASA/JPL; **Science in Action, Figure 2** NASA/JPL-Caltech/Univ. of Arizona; **8.11** Sciencephotos/Alamy; **8.14** ASHLEY COOPER/SCIENCE PHOTO LIBRARY; **8.16** Kenneth Libbercht/Shutterstock.com; **8.18** © Ryerson Clark/iStock.com; **Chemistry and Your Health, Figure 1** Getty Images/Sasha Radosavljevic.

Chapter 9

p. 256 duangnapa_b/Shutterstock.com; **9.1** © Pamela Moore/Istock.com; **9.3 (a)** GIPhotoStock/Science Photo Library; **9.3 (b)** Alexandre Dotta/Science Photo Library; **9.4** Mikkel Juul Jensen/Science Photo Library; **9.5 (a)** Michael D Brown/Shutterstock.com; **9.5 (b)** Serg_dibrova/Shutterstock.com; **9.6** © asbe/iStock.com; **Concept Question 9.3** Copyright © 1996 Richard Megna, Fundamental Photographs, NYC; **Concept Question 9.4** windu/Shutterstock.com; **9.13** Biophoto Associates/Science Photo Library; **Chemistry in Your Life, Figure 1** Zerbor/Shutterstock.com; molekuul_be/Shutterstock.com; **Chemistry in Your Life, Figure 2** JGA/Shutterstock.com; **9.15** Nevodka/Shutterstock.com; **9.21 (a)** Copyright © 1995 Richard Megna, Fundamental Photographs, NYC; **9.21 (b)** Copyright © 1995 Richard Megna, Fundamental Photographs, NYC; **Chemistry and Your Health, Figure 1** Layland Masuda/Shutterstock.com; **9.23 (a)** Copyright © 1996 Richard Megna, Fundamental Photographs, NYC; **9.23 (b)** Copyright © 1996 Richard Megna, Fundamental Photographs, NYC; **Worked Example 9.2** © 2000 Richard Megna - Fundamental Photographs; **9.24 (a)** nikkytok/Shutterstock.com; **9.24 (b)** picsfive/Shutterstock.com; **9.24 (c)** antpkr/Shutterstock.com; **Learning Resources, Question 39** US Geological Survey/Science Photo Library.

Chapter 10

10.1 Cultura RM Exclusive/Frank and Helena/Getty Images; **10.6** Charles D. Winters/Science Photo Library; **10.8 (a)** dny3d/Shutterstock.com; **10.8 (b)** Copyright © 2009 Richard Megna, Fundamental Photographs, NYC; **10.9** Copyright © 1989 Chip Clark, Fundamental Photographs, NYC; **Science in Action, Figure 1** Blausen.com staff. "Blausen gallery 2014". Wikiversity Journal of Medicine.; KTSDESIGN/Getty Images; **Science in Action, Figure 2** Phil Degginger/Science Source.

Chapter 11

p. 322 © 2006 Richard Megna, Fundamental Photographs, NYC; **11.2** Sheila Fitzgerald/Shutterstock.com; dcwcreations/Shutterstock.com; Jeffrey B. Banke/Shutterstock.com; **11.3** futureimage/iStock Photo; Warren Price Photography/Shutterstock.com; Magone/iStock Photo; victoriya89/iStock Photo; **11.4** Jeffrey B. Banke/Shutterstock.com; traveler1116/iStock Photo; jfmdesign/iStockPhoto; Keith Homan/Shutterstock; **11.5** © 1994 Richard Megna - Fundamental Photographs, NYC; **Chemistry and Your Health, Figure 1** Susan Leavines/Science Source; Science Source; **11.11 (a)** © 1993 Richard Megna, Fundamental Photographs, NYC; **11.11 (b)** © 1993 Richard Megna, Fundamental Photographs, NYC; **11.11 (c)** © 1993 Richard Megna, Fundamental Photographs, NYC; **Practice Exercise 11.3** traveler1116/iStock Photo; **Practice Exercise 11.4** Charles Winters/Science Source; **11.13** © 1995 Richard Megna, Fundamental Photographs, NYC; **Science in Action, Figure 1** Barry Marshall; **Learning Resources, Question 46** Science Source.

Chapter 12

p. 358 ktsdesign/Shutterstock.com; **12.1** DR GOPAL MURTI/SCIENCE PHOTO LIBRARY; **12.2** Courtesy of The James D. Watson Collection, Cold Springs Harbor Laboratory Archives; **12.3** Ted Kinsman/Science Source; **12.10** Creative Commons Avery, O. T. et al. Studies on the chemical nature of the substance inducing transformation of pneumococcal types. Journal of Experimental Medicine. 79, 137–157 (1944).; **12.11** World History Archive/Alamy Stock Photo; **12.12** National Library of Medicine/Science Photo Library; **12.13 (b)** KING'S COLLEGE LONDON ARCHIVES/SCIENCE PHOTO LIBRARY; **12.14 (a)** Science Source; **12.14 (b)** World History Archive/Alamy Stock Photo; **12.15** Science Source/Science Photo Library; **12.16** Crystallographic photo of Sodium Thymonucleate, Type B. "Photo 51." May 1952.; **12.19** A. Barrington Brown/Science Source; **12.23** Science Source/Science Photo Library; **Chemistry and Your Health, Figure 1** Ridofranz/iStock Photo; **Learning Resources, Question 47** © Dr. Ken Greer/Visuals Unlimited.

Chapter 13

p. 390 Jim Occi, Fundamental Photographs, NYC; **13.1** REX/Shutterstock; **13.2** Dr. Stanley Flegler/Getty Images; **13.8** Charlie Winters/Science Source; **Science in Action, Figure 1** Science Photo Library; **Concept Question 13.3** Valentina Razumova/Shutterstock.com.

Chapter 14

p. 422 Darren Pullman/Shutterstock.com; **14.1** Butsaya/iStock Photo; **14.3** OJO_Images/iStock Photo; **Worked Example 14.1** 360b/Shutterstock.com; **14.8 (a)** Dorling Kindersley/UIG/Science Photo Library; **Chemistry in Your Life, Figure 1** schankz/Shutterstock.com; **Concept Question 14.2** Helen Sessions/Alamy Stock Photo.

Chapter 15

p. 452 A. DOWSETT, HEALTH PROTECTION AGENCY/SCIENCE PHOTO LIBRARY; **15.1** Du Cane Medical Imaging LTD/Science Photo Library; **15.3** Ted Kinsman/Science Source; **15.4** © 1997–2004 Department of Biochemistry and Molecular Biophysics, University of Arizona; **Science in Action, Figure 1 (b)** Dr. Gladden Willis/Visuals Unlimited, Inc.; **Science in Action, Figure 2 (b)** Biology Pics/Science Source; **15.6** John Durham/Science Photo Library; **15.9** Reddogs/Shutterstock.com; **15.16** Friom The New England Journal of Medicine, Lauri A. Hicks, Thomas H. Taylor, Robert J. Hunkler, "U.S. Outpatient Antibiotic Prescribing, 2010," 368;15: 1461–1462. Copyright © 2013 Massachusetts Medical Society. Reprinted with permission from Massachusetts Medical Society.; **15.18** ewais/Shutterstock.com; **15.19** Scott Camazine/Science Source; **15.20 (a)** Science Source; **15.22** Otfried Schmidt/ullstein bild via Getty Images; **Chemistry in Your Life, Figure 2 (a)** Public Health England/Science Source; **Chemistry in Your Life, Figure 2 (b)** © Christine Whitehead/Alamy Stock Photo; **Visual Summary, 15.1** Reprinted from Global Tuberculosis report 2015. World Health Organization, © Copyright (2015).; **Learning Resources, Question 36** Youst/iStock Photo.

Appendix A

Concept Question A.1 FRANK FOX/SCIENCE PHOTO LIBRARY; **p. 484** STEVE GSCHMEISSNER/SCIENCE PHOTO LIBRARY.

Index

Page numbers followed by *f* and *t* refer to figures and tables, respectively.

A

A (*See* Mass number)
A (adenine), 362, 372
Absorbance, 309, 309*f*
Absorption, of heat energy, 245–48
Absorption phase (pharmacokinetic), 315
Acetaldehyde dehydrogenase, 317
Acetaminophen:
    amide group in, 140, 140*f*
    dose and health effects of, 476, 476*f*
    molecular structure of, 128*f*
    oral bioavailability of, 274, 274*f*
Acetate ion, 336
Acetic acid:
    as carboxylic acid, 134, 135*f*
    strength of, 335–37, 335*f*
    use and structure of, 325, 326*f*
Acetone, 133, 133*f*
Acid–base chemistry, 323–54
    biological applications of, 346–52
    and causes of acid reflux disease, 324, 325*f*
    classification of acids and bases, 325–35
    and pH scale, 340–46
    relative strengths of acids and bases, 335–40
Acidic hydrogen atom, 327
Acidic sidechains, 397, 398*t*, 414
Acidic solutions, 341–44, 342*f*
Acidity, measuring (*See* pH scale)
Acidosis, 351
Acid reflux, 324
Acid reflux disease:
    defined, 324, 325*f*
    origin of, 347, 347*f*
    treatment of, 348–51
Acids, 325–35
    behavior of, 133, 133*f*
    Brønsted-Lowry theory of, 327–28
    in conjugate acid–base pairs, 337–39, 339*f*
    examples of, 325, 326*f*
    in neutralization reactions, 332–35
    relative strengths of, 335–40
    See also specific acids and specific types
Activation energy:
    effect of catalyst on, 429–30, 430*f*
    as reaction energy barrier, 425–26
Active site, 433

Addition:
    and scientific notation, 485
    significant figures in, 492
Addition polymerization, 200
Adducts, 382–83
Adenine (A), 362, 372
Adenosine, 147, 147*f*
Adenosine diphosphate (ADP), 188, 189*f*
Adenosine triphosphate (ATP), 188–90, 188*f*–190*f*
Adipose tissue, 266–67
ADP (adenosine diphosphate), 188, 189*f*
Aerobic reactions, 190
A-form of DNA, 371, 371*f*
Agre, Peter, 321
Agriculture, antibiotic use in, 469–70, 470*f*
Air:
    breathing, 62–63, 63*f*
    composition of, 62, 62*f*
Alcohol:
    concentration of, in beverages, 295, 295*f*
    metabolism of, 134
    pharmacokinetic profile of, 316–17, 316*f*
    See also Ethanol (ethyl alcohol)
Alcohol dehydrogenase:
    in alcohol metabolism, 134, 134*f*
    and blood alcohol concentration, 317
    substrate for, 434
Alcohols (functional group), 130, 130*f*
Aldehyde dehydrogenase, 134, 134*f*
Aldehydes, 132, 132*f*
Alkaloids, 330
Alkalosis, 351
Alkanes, 97–107
    branched, 102
    butane, 101–3
    and definition of hydrocarbons, 97
    drawing, 105–7
    ethane, 98–99
    linear, 102
    methane, 97–98
    naming of, 99–100, 100*t*
    propane, 100–101
    structural isomers of, 101–5
Alkaptonuria, 446